T0135317

Advances in Intelligent Systems and Computing

Volume 758

Series editor

Janusz Kacprzyk, Systems Research Institute, Polish Academy of Sciences,
Warsaw, Poland
e-mail: kacprzyk@ibspan.waw.pl

The series "Advances in Intelligent Systems and Computing" contains publications on theory, applications, and design methods of Intelligent Systems and Intelligent Computing. Virtually all disciplines such as engineering, natural sciences, computer and information science, ICT, economics, business, e-commerce, environment, healthcare, life science are covered. The list of topics spans all the areas of modern intelligent systems and computing such as: computational intelligence, soft computing including neural networks, fuzzy systems, evolutionary computing and the fusion of these paradigms, social intelligence, ambient intelligence, computational neuroscience, artificial life, virtual worlds and society, cognitive science and systems, Perception and Vision, DNA and immune based systems, self-organizing and adaptive systems, e-Learning and teaching, human-centered and human-centric computing, recommender systems, intelligent control, robotics and mechatronics including human-machine teaming, knowledge-based paradigms, learning paradigms, machine ethics, intelligent data analysis, knowledge management, intelligent agents, intelligent decision making and support, intelligent network security, trust management, interactive entertainment, Web intelligence and multimedia.

The publications within "Advances in Intelligent Systems and Computing" are primarily proceedings of important conferences, symposia and congresses. They cover significant recent developments in the field, both of a foundational and applicable character. An important characteristic feature of the series is the short publication time and world-wide distribution. This permits a rapid and broad dissemination of research results.

More information about this series at http://www.springer.com/series/11156

Janmenjoy Nayak · Ajith Abraham
B. Murali Krishna · G. T. Chandra Sekhar
Asit Kumar Das
Editors

Soft Computing in Data Analytics

Proceedings of International Conference on SCDA 2018

Editors
Janmenjoy Nayak
Department of Computer Science and Engineering
Sri Sivani College of Engineering
Srikakulam, Andhra Pradesh
India

G. T. Chandra Sekhar
Department of Electrical and Electronics Engineering
Sri Sivani College of Engineering
Srikakulam, Andhra Pradesh
India

Ajith Abraham
Machine Intelligence Research Labs (MIR Labs)
Scientific Network for Innovation and Research Excellence
Washington
USA

Asit Kumar Das
Department of Computer Science and Technology
Indian Institute of Engineering Science and Technology (IIEST), Shibpur
Howrah, West Bengal
India

B. Murali Krishna
Department of Mechanical Engineering
Sri Sivani College of Engineering
Srikakulam, Andhra Pradesh
India

ISSN 2194-5357 ISSN 2194-5365 (electronic)
Advances in Intelligent Systems and Computing
ISBN 978-981-13-0513-9 ISBN 978-981-13-0514-6 (eBook)
https://doi.org/10.1007/978-981-13-0514-6

Library of Congress Control Number: 2018941982

This Springer imprint is published by the registered company Springer Nature Singapore Pte Ltd.
The registered company address is: 152 Beach Road, #21-01/04 Gateway East, Singapore 189721, Singapore

Preface

The modern era of advanced computing is still striving to develop consciousness-based machines. The used methods are quite close to reasoning process of a human. The insidious use of soft computing (SC) in diversified engineering applications makes it an essential tool in the development of products that have implications for the human society. The term "soft computing" refers to many of the useful techniques like fuzzy logic, neural computing, probabilistic reasoning, evolutionary computation. In recent years, soft computing has been extended by adding many other subdisciplines and it became quite obvious that, this has been frequently used in various engineering domains. Starting from computer science engineering to other engineering domains like electrical, civil, mechanical, soft computing has become one of the premier choices for all researchers to solve complex as well as uncertain problems.

With the substantial growth of large volumes of industries, the present scenario of storing the data is quite complex in the modern database due to the availability and popularity of the Internet. For an efficient decision making, the information needs to be summarized and structured in order to maintain the consistency. When there are problems in handling the quantity of data, dimension and complexity in between the relationships at human level, there will be an immediate need of intelligent data analysis technique, which may discover useful knowledge from data. Soft computing is such an efficient technique to handle all the irregularities and complexities of data in any of the residing form.

The first International Conference on "Soft Computing in Data Analytics (SCDA 2018)" is organized by Sri Sivani College of Engineering, Andhra Pradesh, India, on March 10 and 11, 2018. SCDA is an international forum for representation of research and developments in the fields of soft computing, data mining, computational intelligence, machine learning, etc. Many of the prospective authors had contributed their valuable research work to the conference. After a thorough double-blind peer-review process, editors have selected 80 papers which are being reviewed by highly experienced subject experts chosen from the country and abroad. The Proceedings of SCDA is a volume comprising papers from latest findings in diversified areas, where soft computing has been elegantly applied.

The selected papers are in the areas of the applications of soft computing techniques in cloud computing, image analysis, classification, clustering, intrusion detection, pattern recognition, power electrical and electronics problems, circuit design and analysis. It is a great honor for us to edit the proceedings with different dimensions and features of this book. We have enjoyed considerably working in cooperation with the International Advisory, Program and Technical Committees during a call for papers, review of papers and finalization of papers to be included in the proceedings.

We extend our deep sense of gratitude to all for their warm encouragement, inspiration and continuous support for making it possible.

Hope all of us will appreciate the good contributions made and justify our efforts.

Srikakulam, India — Dr. Janmenjoy Nayak
Washington, USA — Dr. Ajith Abraham
Srikakulam, India — Dr. B. Murali Krishna
Srikakulam, India — Dr. G. T. Chandra Sekhar
Howrah, India — Dr. Asit Kumar Das

Committees

Honorary General Chair

Prof. (Dr.) Ajith Abraham, Director, Machine Intelligence Research Lab, USA and Professor, Technical University of Ostrava, Czech Republic
Prof. (Dr.) V. Ganesan, Indian Institute of Technology Madras, India

General Chair

Prof. (Dr.) V. E. Balas, *Senior Member IEEE*, Aurel Vlaicu University of Arad, Romania
Prof. (Dr.) J. M. Mallikarjuna, Indian Institute of Technology, Madras, India

Honorary Advisory Chair

Prof. (Dr.) Sankar Kumar Pal, *Padma Shri awardee*, in Science and Engineering, Distinguished Scientist and Former Director, ISI, Kolkata
J. C. Bose National Fellow, Raja Ramanna Fellow

Chief Patron

Sri. V. M. M. Sainadh Reddy, President, Sri Sivani Group of Colleges

Patrons

Sri. J. Suryachandra Rao, Vice President, Sri Sivani Group of Colleges
Sri. K. Durga Srinivas, Secretary, Sri Sivani Group of Colleges
Sri. D. Venkata Rao, Treasurer, Sri Sivani Group of Colleges
Sri. P. Durgaprasad Raju, Joint Secretary, Sri Sivani Group of Colleges
Dr. S. Srinivasa Rao, Correspondent, Sri Sivani Group of Colleges

Program Chairs

Dr. B. Murali Krishna, Professor and Principal, Sri Sivani College of Engineering, Srikakulam, Andhra Pradesh, India
Prof. (Dr.) Adel M. Alimi, Professor, REGIM-Lab., ENIS, University of Sfax, Tunisia

Dr. Asit Kumar Das, Indian Institute of Engineering Science and Technology (IIEST), Shibpur, Kolkata, India
Dr. H. S. Behera, Veer Surendra Sai University of Technology, Burla, Odisha, India

Convenor(s)

Dr. Janmenjoy Nayak, Associate Professor, Sri Sivani College of Engineering, Srikakulam, Andhra Pradesh, India
Dr. G. T. Chandra Sekhar, Associate Professor, Sri Sivani College of Engineering, Srikakulam, Andhra Pradesh, India

International Advisory Committee

Dr. Ramesh C. Bansal, University of Pretoria, South Africa
Dr. Ramazan Bayindir, Gazi University, Turkey
Dr. Madan M. Gupta, University of Saskatchewan, Canada
Dr. Mohan Lal Kolhe, University of Agder, Norway
Dr. Zhao Xu, Hong Kong Polytechnic University, Hong Kong
Dr. Ahmad Taher Azar, Faculty of Computers and Information, Benha University, Egypt
Dr. B. L. Deekshatulu, IISc, Bangalore, India
Dr. A. Sreedharan, IISc, Bangalore, India
Dr. Chi-Chang Chang, Chung Shan Medical University
Dr. N. P. Padhy, IIT Roorkee, India
Dr. S. K. Panada, National University of Singapore, Singapore
Dr. Kenji Suzuki, University of Chicago
Dr. P. Mohapatra, University of California
Dr. M. S. Obaidat, Monmouth University, USA
Dr. R. Govinda Rajulu, IIIT Hyderabad
Dr. Sukumar Mishra, IIT Delhi
Dr. A. K. Pradhan, IIT Kharagpur
Dr. B. K. Panigrahi, IIT Delhi
Dr. Dinesh Thakur Singh, DIET, Pune
Dr. M. S. Prasad Babu, AU, Visakhapatnam
Er. Rosa Bolanos, IEE Quito
Dr. Ruven, EPN, Venezuela
Dr. C. Chandra Sekhar, IIT Madras
Dr. M. Narasimha Murty, IISc, Bangalore
Dr. M. Manzoor Hussain, JNTU Hyderabad
Dr. G. Ranga Janardhana, JNTU Hyderabad
Dr. V. V. S. S. Srikanth, HCU, Hyderabad
Dr. T. Narasinga Rao, ARCI, Hyderabad
Dr. G. Appa Rao, IIT Madras
Dr. K. V. L. Subramaniam, IIT Hyderabad
Dr. M. Viswanadham, JNTU Hyderabad
Dr. G. P. Raja Sekhar, IIT Kharagpur

Dr. S. K. Nimbhorkar, Dr. Babasaheb Ambedkar Marathwada University, Aurangabad
Dr. Chen Yuqun, South China Normal University, China
Dr. V. Uma Maheswara Rao, AU, Visakhapatnam
Dr. Aqeel ur-Rehman, Hamdard University
Dr. Israr Qureshi, IE Business School, Hong Kong
Dr. Bindu Chawla, Graduate School of Education, Greater New York
Dr. P. Bheema Rao, San Diego, California

International Technical Committee Members

Dr. Alfanso Zozaya, EPN, Quito
Dr. Li Wenzheng, Beijing Institute of Technology
Er. Javier, IEE Quito
Dr. D. V. L. N. Somayajulu, NIT Warangal
Dr. Yepez, IEE Quito
Dr. Gwendolyn Boyd, Alcorn State University, Lorman, Mississippi
Dr. Lina Girdauskiene, COMMEDIA, Lithuania
Dr. R. B. V. Subramanyan, NIT Warangal
Dr. B. Prabhakar Rao, JNTU Kakinada
Dr. P. S. Avadhani, AU, Visakhapatnam
Dr. N. V. S. Narasimhasarma, NIT Warangal
Dr. Sidhartha Panda, VSSUT, Burla
Dr. Rabindra Kumar Sahu, VSSUT, Burla
Dr. N. P. Patidar, NIT Bhopal
Dr. Girish Parmar, Rajasthan Technical University, Kota, Rajasthan
Dr. Nidul Sinha, NIT Silchar
Dr. S. Siva Nagaraju, JNTU Kakinada
Dr. R. Srinivasa Rao, JNTU Kakinada
Dr. P. Malli karjuna Rao, AU, Visakhapatnam
Dr. K. Rama Sudha, AU, Visakhapatnam
Dr. N. Prema Kumar, AU, Visakhapatnam
Dr. Alok Kanti Deb, IIT Kharagpur
Dr. R. Rajeswara Rao, JNTUK-UCE Vizianagaram
Dr. R. Vijayasanthi, AU, Visakhapatnam
Mr. Baidyanath Bag, NIT Kurukshetra
Dr. A. S. N. Chakravarthy, JNTUK-UCE Vizianagaram
Dr. S. S. Dash, SRM University, Chennai
Dr. Swathi Sharma, Jodhpur National University, Rajasthan
Dr. G. Sreenivasan, SRIT, Anantapur
Dr. K Chandra Sekaran, NITK Surathkal
Dr. B. Ramadoss, NIT Tiruchirappalli
Dr. M. Indira Rani, JNTU Hyderabad
Dr. P. Prasanna, JNTU Hyderabad

Dr. K. V. Krishna, IIT Guwahati
Dr. D. Linga Raju, JNTU Hyderabad
Dr. Y. Vasudeva Rao, IIT Bhubaneswar
Dr. G. Swami Naidu, JNTUK-UCE Vizianagaram
Dr. N. V. Uma Mahesh, NIT Warangal
Dr. Ram Seshu, NIT Warangal
Dr. G. C. Rao, AU, Visakhapatnam
Dr. Bighnaraj Naik, VSSUT, Burla, Odisha, India
Dr. Vijayasree, Director (Research), QIS Engineering College, Ongole
Dr. K. S. Kusuma, MCRC, New Delhi
Dr. P. Subba Rao, JNTU Kakinada
Dr. S. Srinivas Kumar, JNTU Kakinada
Dr. K. Chandra Bhushana Rao, JNTUK-UCE Vizianagaram

Publicity Chair(s)

Dr. T. Kondala Rao, Professor, Department of H & BS, Sri Sivani College of Engineering
Sri. P. Pradeep, Assistant Professor, Department of CSE, Sri Sivani College of Engineering
Sri. S. Papa Rao, Assistant Professor, Department of CSE, Sri Sivani College of Engineering

Logistic Chair(s)

Dr. B. Srinivasa Rao, Professor, Department of H & BS, Sri Sivani College of Engineering
Sri. P. Rama Rao, Head of the Department, MCA, Sri Sivani College of Engineering
Sri. P. Ramana Murthy, Head of the Department, H & BS, Sri Sivani College of Engineering
Sri. K. Prem Kumar, Head of the Mechanical Department, Sri Sivani College of Engineering
Sri. Y. Jagadeesh Kumar, Assistant Professor, Department of CSE, Sri Sivani College of Engineering
Sri. T. Balaji, P. D., Sri Sivani College of Engineering
Ms. J. Divya Jyothi, Assistant Professor, Department of MBA, Sri Sivani College of Engineering

Finance Chair(s)

Dr. G. T. Chandra Sekhar, Head of the Department, EEE, Sri Sivani College of Engineering
Dr. B. Prasada Rao, Associate Professor, Department of MBA, Sri Sivani College of Engineering
Smt. H. Swapna Rekha, Associate Professor, Department of CSE, Sri Sivani College of Engineering

Registration Chair(s)

Dr. M. P. Suri Ganesh, Associate Professor, Department of MBA, Sri Sivani College of Engineering
Smt. G. Himala Kumari, Head of the Department, Department of Civil, Sri Sivani College of Engineering
Smt. P. Lalitha Kumari, Associate Professor, Department of CSE, Sri Sivani College of Engineering
Smt. S. Halini, Assistant Professor, Department of EEE, Sri Sivani College of Engineering
Smt. S. Kiranmayi, Assistant Professor, Department of Mechanical Engineering., Sri Sivani College of Engineering
Sri. K. Koti Babu, Assistant Professor, Department of H & BS, Sri Sivani College of Engineering

Sponsorship Chair(s)

Dr. G. Rajendra Kumar, TPO and Professor, Department of CSE, Sri Sivani College of Engineering
Dr. P. G. Prasad, Professor, Department of H & BS, Sri Sivani College of Engineering
Sri. K. Ravi Kumar, Head of the Department, Department of MBA, Sri Sivani College of Engineering
Sri. T. Mankyala Rao, Head of the Department, Department of ECE, Sri Sivani College of Engineering

Publication Chair(s)

Dr. V. Chenchu Raju, Professor, Department of H & BS, Sri Sivani College of Engineering
Smt. P. Indira, Librarian, Sri Sivani College of Engineering
Smt. B. G. Lakshmi, Head of the Department, Department of CSE, Sri Sivani College of Engineering
Sri. Ch. Jagan Mohana Rao, Assistant Professor, Department of EEE, Sri Sivani College of Engineering
Sri. N. Naga Srinivas, Assistant Professor, Department of EEE, Sri Sivani College of Engineering
Sri. G. V. L. Narayana, Assistant Professor, Department of CSE, Sri Sivani College of Engineering

Event Chair(s)

Dr. A. Lakshmana, Professor, Department of H & BS, Sri Sivani College of Engineering
Dr. T. Panduranga Vital, Associate Professor, Department of CSE, Sri Sivani College of Engineering
Smt. U. Suvarna, Associate Professor, Department of CSE, Sri Sivani College of Engineering

Sri. B. V. G. D. Krishna Rao, Head of the Department, CRT, Sri Sivani College of Engineering
Sri. K. Madhu Kumar, Associate Professor, Department of H & BS, Sri Sivani College of Engineering

Local Organizing Committee

Sri. D. Chiranjeevulu, Associate Professor, Sri Sivani College of Engineering
Sri. G. Parameswara Rao, Associate Professor, Sri Sivani College of Engineering
Sri. P. Mohana Krishna, Assistant Professor, Sri Sivani College of Engineering
Sri. Ch. Mohana Rao, Assistant Professor, Sri Sivani College of Engineering
Smt. A. Santhi, Assistant Professor, Sri Sivani College of Engineering
Sri. B. Sesha Sai, Assistant Professor, Sri Sivani College of Engineering
Ms. M. Sai Sudha, Assistant Professor, Sri Sivani College of Engineering
Sri. Asish Kumar Panigrahi, Assistant Professor, Sri Sivani College of Engineering
Sri. Panigrahi Jagannadh, Assistant Professor, Sri Sivani College of Engineering
Sri. B. Srirammurthy, Associate Professor, Sri Sivani College of Engineering
Smt. J. Kalpana, Assistant Professor, Sri Sivani College of Engineering
Smt. A. Suneetha, Assistant Professor, Sri Sivani College of Engineering
Sri. K. Chandan Rao, Assistant Professor, Sri Sivani College of Engineering
Sri. S. V. N. Sunil Kumar, Assistant Professor, Sri Sivani College of Engineering
Sri. P. Revanth Kumar, Assistant Professor, Sri Sivani College of Engineering
Sri. K. Mani Kantha, Assistant Professor, Sri Sivani College of Engineering
Ms. G. Navya, Assistant Professor, Sri Sivani College of Engineering
Sri. K. Venkatesh, Assistant Professor, Sri Sivani College of Engineering
Smt. M. Dhana Lakshmi, Assistant Professor, Sri Sivani College of Engineering
Sri. N. Seethayya, Assistant Professor, Sri Sivani College of Engineering
Sri. K. Dileep Kumar, Assistant Professor, Sri Sivani College of Engineering
Smt. Dabbeeru Priyanka, Assistant Professor, Sri Sivani College of Engineering
Sri S. Papa Rao, Assistant Professor, Sri Sivani College of Engineering
Ms. H. V. Bhagyasri, Assistant Professor, Sri Sivani College of Engineering
Ms. J. V. Priyanka, Assistant Professor, Sri Sivani College of Engineering
Ms. M. Mownika Rayudu, Assistant Professor, Sri Sivani College of Engineering
Smt. B. Padmavathi, Associate Professor, Sri Sivani College of Engineering
Smt. D. Padmini Priyadarsini, Associate Professor, Sri Sivani College of Engineering

Technical Reviewer's Committee

Prof. Arun Agarwal, Siksha 'O' Anusandhan University, BBSR
Dr. M. Marimuthu, Coimbatore Institute of Technology, Coimbatore
Dr. H. S. Behera, VSSUT, Burla
Dr. Asit Kumar Das, IIEST Shibpur, Howrah
Prof. Ranjan Kumar Barik, VSSUT, Burla, India
Dr. Harihar Kalia, Seemanta Engineering College
Dr. Ajanta Das, University of Engineering and Management, Kolkata
Prof. Andrews Samraj, Department of Information Technology, Mahendra Engineering College, Namakkal
Dr. RanjitRajak, Dr. Hari Singh Gour Central University, Sagar (MP)
Dr. R. K. Sahu, VSSUT, Burla
Dr. Prasanna Ramakrisnan, Institute of Neo Education (iNED)
Prof. Chandan Jyoti Kumar, Cotton College State University
Prof. Ripal Patel, BVM Engineering College, Vallabh Vidyanagar, Gujarat, India
Dr. P. Sivakumar, SKP Engineering College, Thiruvannamalai, Tamil Nadu, 60611
Prof. Partha Garai, Kalyani Government Engineering College, Kalyani
Dr. Jyotismita Chaki, Jadavpur University
Prof. Sawon Pratiher, IIT Kharagpur
Dr. R. V. S. Lalitha, Aditya College of Engineering and Technology, Surampalem
Dr. Narayan Joshi, Parul University, Vadodara
Dr. A. Rawal, Bhilai Institute of Technology, Durg
Prof. S. Vaithyasubramanian, Sathyabama University, Chennai, Tamil Nadu
Prof. Abhishek Kumar, Lovely Professional University, Phagwara, Jalandhar, Punjab, India
Dr. K. M. Azharul Hasan, Khulna University of Engineering and Technology (KUET), Bangladesh
Prof. Sudan Jha, IT Entrepreneur, Nepal
Dr. Narender Singh, Chhaju Ram Memorial Jat College, Hisar
Dr. P. M. K. Prasad, GVP College of Engineering for Women, Visakhapatnam
Dr. A. K. Verma, Thapar University, Patiala

Prof. Mohd Arif Siddique, Maya Academy of Advanced Cinematics, Kamla Nagar, Delhi

Dr. Srinivas Sethi, Indira Gandhi Institute of Technology, Sarang

Dr. Mahmood Ali Mirza, DMS SVH College of Engineering, Machilipatnam, Andhra Pradesh, India

Dr. Sachin D. Ruikar, Walchand College of Engineering, Vishrambag, Sangli, 416415, Maharashtra

Prof. Prashant Kumar, Zeal College of Engineering and Research, Pune-411033

Dr. Sunanda Das, Neotia Institute of Technology, Management and Science, Diamond Harbour, West Bengal, India

Dr. R. Manjula Devi, Kongu Engineering College, Perundurai-638 052

Prof. Rahul Paul, University of South Florida

Dr. Imon Mukherjee, IIIT Kalyani

Prof. Rama Krushna Das, National Informatics Centre, Berhampur-760 004, Odisha, India

Dr. Chinta Someswara Rao, S R K R Engineering College, Bhimavaram

Dr. Jyothi Pillai, Bhilai Institute of Technology, Durg, Chhattisgarh

Prof. Devendra Potnuru, GMR Institute of Technology, Rajam

Dr. A. Meenakabilan, Sri Sairam Engineering College

Prof. Pilla Ramana, GMR Institute of Technology

Prof. L. V. Suresh Kumar, GMR Institute of Technology

Prof. Bighnaraj Naik, VSSUT, Burla

Dr. Sripada Rama Sree, Aditya Engineering College (AEC), Surampalem, East Godavari District, Andhra Pradesh

Dr. Santosh Majhi, VSSUT, Burla

Acknowledgements

The International Conference on Soft Computing in Data Analytics (SCDA 2018) is organized to help researchers to gain insight in various methods of soft computing in the analysis of data at various research domains. This conference has resulted in this volume, and this work would not have been possible without the prolific contributions of our dedicated authors. The theme and relevance of SCDA have attracted more than 300 researchers/academicians around the globe for submitting their valuable research works, which enabled us to select good quality papers and serve to demonstrate the popularity of the SCDA conference for sharing ideas and research findings with truly national and international communities. Thanks to all those researchers/academicians who have contributed in producing such a comprehensive conference proceedings of SCDA.

The organizing committee believes that we have been true to the spirit of collegiality that members of SCDA value even as also maintaining an elevated standard as we have reviewed papers, provided feedback and present a strong body of published work in this collection of proceedings. Thanks to all the members of the organizing committee for their heartfelt support and cooperation.

We have been fortunate enough to work in cooperation with a brilliant international as well as national advisory, reviewer, program and technical committees' members consisting of eminent academicians.

All submitted papers have been double-blind peer-reviewed before the final submission to be accepted as a paper for oral presentation at the conference venue. We would like to express our heartfelt gratitude and obligations to the benign reviewers for sparing their valuable time and putting in an effort to review the papers in a stipulated time and providing their valuable suggestions and appreciation in improvising the presentation, quality and content of the papers. The eminence of these papers is an accolade not only to the authors but also to the reviewers who have guided toward perfection.

Also, many thanks to the management members, teaching and non-teaching staff members of Sri Sivani College of Engineering, Srikakulam, for helping and extending cooperation and support in every step of the conference.

Last but not least, the editorial members of Springer Publishing deserve a special mention and our sincere thanks to them not only for making our dream come true in the shape of this proceedings, but also for its hassle-free and in-time publication in the reputed Advances in Intelligent Systems and Computing, Springer.

The SCDA conference and proceedings are a credit to a large group of people, and everyone should be proud of the outcome.

Contents

About the Editors

Dr. Janmenjoy Nayak is working as Associate Professor, Sri Sivani College of Engineering, Srikakulam, Andhra Pradesh, India. He has been awarded INSPIRE Research Fellowship from the Department of Science and Technology, Government of India (both as JRF and SRF levels), for doing his doctoral research in the Department of Computer Science Engineering and Information Technology, Veer Surendra Sai University of Technology, Burla, Odisha, India. He completed his M. Tech. (gold medalist and topper of the batch) in computer science from Fakir Mohan University, Balasore, Odisha, India, and M.Sc. (gold medalist and topper of the batch) in computer science from Ravenshaw University, Cuttack, Odisha, India. He has published more than 50 research papers in various reputed peer-reviewed international conferences, referred journals and chapters. He has also published one textbook on "Formal Languages and Automata Theory" in Vikash Publishing House Pvt. Ltd., which has been widely acclaimed throughout the country and abroad by the students of all levels. He has been serving as Active Member of Reviewer Committee of various reputed peer-reviewed journals such as *IET Intelligent Transport Systems*, *Journal of Classification*, *Springer*, *International Journal of Computational System Engineering*, *International Journal of Swarm Intelligence*, *International Journal of Computational Science and Engineering*, *International Journal of Data Science*. He is Life Member of some of the reputed societies like Computer Society of India (CSI), India, Orissa Information Technology Society (OITS), Orissa Mathematical Society (OMS), IAENG (Hong Kong). His research interests include data mining, nature-inspired algorithms and soft computing.

Prof. Ajith Abraham received his M.S. from Nanyang Technological University, Singapore, and Ph.D. in computer science from Monash University, Melbourne, Australia. He is currently Research Professor at the Technical University of Ostrava, Czech Republic. He is also Director of Machine Intelligence Research Labs (MIR Labs), Scientific Network for Innovation and Research Excellence, which has members from more than 75 countries. He serves/has served the Editorial Board of over 50 international journals and has also guest-edited 40 special issues on various topics. He is author/co-author of more than 700 publications, and some of the works have also won best paper awards at international conferences and also received several citations. Some articles are available in the ScienceDirect top 25 hottest articles. He serves IEEE Computer Society's Technical Committee on Scalable Computing and was General Chair of the 8th International Conference on Pervasive Intelligence and Computing (PICOM 2009) and the 2nd International Conference on Multimedia Information Networking and Security (MINES 2010). He is also General Co-chair of MINES 2011. Since 2008, he is Chair of IEEE Systems Man and Cybernetics Society Technical Committee on Soft Computing. He is Senior Member of IEEE, IEEE Computer Society, the Institution of Engineering and Technology (UK), the Institution of Engineers Australia (Australia), etc. He is Founder of several IEEE-sponsored annual conferences, which are now annual events—Hybrid Intelligent Systems (HIS—11 years); Intelligent Systems Design and Applications (ISDA—11 years); Information Assurance and Security (IAS—7 years); Next Generation Web Services Practices (NWeSP—7 years), Computational Aspects of Social Networks (CASoN—3 years), Soft Computing and Pattern Recognition (SoCPaR—3 years), Nature and Biologically Inspired Computing (NaBIC—3 years) are some examples.

Dr. B. Murali Krishna received his Ph.D. from Department of Mechanical Engineering, *Indian Institute of Technology Madras*, Chennai, in 2010; he did his post-graduation from *NIT Warangal* and graduation in mechanical engineering from *Institution of Engineers* (India), Kolkata. He was Fellow and Life Member of FIE, ISTE, IAENG, ISAET, IACSIT, etc. His total research and teaching experience was 18 years at various positions: Engineer R&D, Assistant Professor, Associate Professor, Professor, HoD, Professor and Principal; at present, he is working for *Sri Sivani College of Engineering, Srikakulam, AP, India*, as Professor and Principal. He has published more than 75 research papers, articles, book chapters in reputed publications at national and international levels like *ASME, SAE, Springer, IEEE, ELSEVIER, AIP, JAFM, EBSCO International, Tata McGraw-Hill, Pearson, etc.* He has visited *Paris* and *London* for presenting research work. He has been awarded/selected/nominated for *Best Paper Award* for World Congress on Engineering, London; *Best Indian Educationist* and *Bharat Excellence* awards with *gold medals* received from honorable Governor of Sikkim, *Indira Gandhi Gold Medal* Award, *National Award of Excellence, Indian Global Golden, Sparkling Indian Award*, Green ThinkerZ-*Outstanding Researcher Award, Academic Leadership Award and Best Researcher Award* for his contribution in the relevant fields of engineering and service. He is the one among the 50 shortlisted at national level for AICTE-UKIERI *Further Education Leadership Development Programme* (FELDP). He was Chairman and Advisory Board Member for several reputed conferences and technical events. Also, he received I2OR *Academic Leadership Award* from NITTTR Chandigarh and IIT Bombay, MHRD, Government of India, in association with Green ThinkerZ Society, India. His research interests include thermal engineering, internal combustion engines, automobile pollution, CFD, in-cylinder flow studies, image processing, alternate fuels, renewable energies, energy from waste, automobile engineering.

Dr. G. T. Chandra Sekhar is working as Associate Professor and Head of the Department in Electrical and Electronics Engineering, Sri Sivani College of Engineering, Chilakapalem, Srikakulam, Andhra Pradesh. He has received his bachelor's and master's degrees from JNTU Hyderabad, and JNTU, Kakinada, Andhra Pradesh, respectively. He received his doctoral degree in electrical engineering from Veer Surendra Sai University of Technology, Burla, Odisha, India. He has published more than 30 research papers in various international journals and international/national conferences. He is the recipient of Best Young Researcher in electrical engineering by ITSR Foundation Award 2017 and Young Leader in Engineering during the Higher Education Leadership Meet-HELM 2018 organized by Venus International Foundation. His research interests include soft computing application in power system engineering. He has acted as Member of Editorial/Reviewer Board of various international journals/conferences. Moreover, he has been serving as Active Member of Reviewer Committee of various reputed peer-reviewed journals such as *IET Journals* (GTD), *Springer*, *International Journal of Computational System Engineering*, *International Journal of Swarm Intelligence*, *Elsevier*, *Taylor & Francis*. His research interests include soft computing application in power system engineering, power quality and FACTS devices. He is Life Member of ISTE and AMIE (India), IAENG (Hong Kong), ISAET, and Associate Member of IRED.

Dr. Asit Kumar Das is currently Associate Professor at the Department of Computer Science and Technology, Indian Institute of Engineering Science and Technology, Shibpur, Howrah, West Bengal, India. He received his Ph.D. in engineering from the Department of Computer Science and Technology, Bengal Engineering and Science University, Shibpur, Howrah. He teaches discrete structures, database management systems, design and analysis of algorithm, introduction to data mining, etc. His primary research interests include data mining, bioinformatics, evolutionary algorithms, audio and video data analysis, text mining and social networks.

Optimal Control Method for Interleaved High-Power Flyback Inverter Using PV

Prashant Kumar

Abstract The interleaved flyback converter used to increase the power level, in which current ripples can get reduced and reduces size of components as well as reduces total cost of system. The main motive is to design flyback converter for high-power application. The design of flyback inverter is rated as 12 Kw by interleaving of two-stage flyback converter. The main advantages of flyback converter are less size of filtering component. This converter system works at discontinuous current mode (DCM) operation as it provides fast dynamic response, easy control, and no turn-on losses. The designed system converter is simulated in MATLAB software.

Keywords Discontinuous current mode (DCM) · Flyback converter topology Interleaved · PV inverter

1 Introduction

PV source is most significant energy source in the market of power generation system because it gives light from the sun and it is available everywhere freely [1]. The low cost is important for industrialization as well as residential applications [2–6]. Therefore, the main motive of this paper is the advancement in the photovoltaic (PV) inverter technology with the help of flyback inverter. The simplest structure of the flyback inverter and simple power flow control with improved power quality output at grid are the main requirements of this system. The flyback converter is having low cost, and it requires less number of elements. This is because the inductor is combined along with the transformer so that the voltage ratios are multiplied with an advantage of isolation. So that, inductor is required for storage of energy and the transformer is responsible for transfer of energy [7]. Here, combination of inductor and transformer can eliminate the bulky system and cost

P. Kumar (✉)
Zeal College of Engineering and Research, Pune 411041, India
e-mail: prashant2685@gmail.com

J. Nayak et al. (eds.), *Soft Computing in Data Analytics*,
Advances in Intelligent Systems and Computing 758,
https://doi.org/10.1007/978-981-13-0514-6_1

reduction and reduction in size of the converter take place. A flyback converter has constructed by using a transformer which has very high leakage flux and has less transfer efficiency. So, the flyback converter is not suitable for very high-power applications. Flyback converters are interleaved to make it suitable for high-power applications. Another advantage of interleaving is that the frequency of ripple components is increased in proportion to number of cells which allows easy filtering.

The solar power must be used very properly through this method so maximum power point tracker can be used for the PV system. Another important factor for system is selection of mode of operation, and it is discontinuous current mode (DCM) which has several advantages mentioned as follows:

- It gives instant dynamic response.
- It has the absence of turn-on losses.
- The transformer size is small, so cost is low.
- It is easy to control.

The disadvantage of discontinuous current mode (DCM) is higher form factor which is compared to continuous current conduction mode (CCM) which leads to large amount of power losses. The solution for this is paralleling the device. In this system, we provide interleaving of converters which decreases the peak value of current with high discontinuity. This discontinuity is removed by connecting all cells together at one common point. In the existing system, three interleaved transformers were implemented. It makes circuit more bulky and size of circuit is also increased circuit complexity and cost increase due to use of more components are drawbacks of those system. In the proposed system, size of circuit and complexity is reduced by removing switch and transformer. The capability of existing system can be developed in proposed system by increasing the turns of the transformers [3, 8].

2 System Descriptions

The block diagram of the proposed system contains main five blocks. The photovoltaic source acts as an input source and is fed to the flyback converter integrating decoupling capacitor. DC output of the converter is applied to the full-bridge inverter; the inverter converts the DC output of the flyback converter to AC output. The low-pass filter is provided to reduce the harmonic content of the inverter output current. The two control techniques are provided

1. To regulate DC input photovoltaic current and voltage;
2. It should also provide control strategy to convert direct current into alternating current for power injection at the grid interface (Fig. 1).

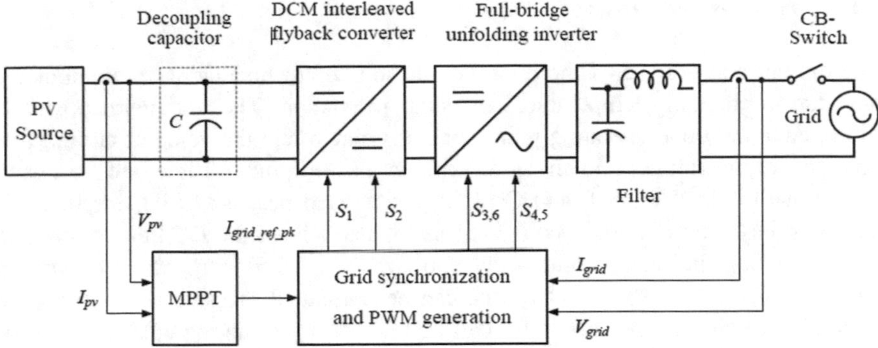

Fig. 1 Block diagram of interleaved flyback converter system

The maximum power point tracking algorithm is developed to extract the maximum power from the photovoltaic array and is fed to the unfolding full-bridge inverter by using interleaved flyback converter.

2.1 Maximum Power Point Tracking [MPPT]

The relationship between operating environment of photovoltaic cells in solar panel and maximum power produced by it is very complex. The V–I characteristics in PV panel produce a nonlinear output efficiency which can be also observed on the I–V curve. Because of this complex relationship, it is a very rare occasion when a PV system is operating at peak power. The operating point of PV panel can be moved by varying the impedance of panel. DC/DC converter is used to transform the impedance of one circuit to another. A change in the duty ratio of DC/DC converter results in the change of impedance in panel. The operating point will produce peak power at particular impedance. This operating point is also known as maximum power point, i.e., MPP. The MPP of panel varies with a variation in operating environment such as temperature and irradiation. It is not feasible to fix the duty ratio of DC/DC converter with such varying conditions. Therefore, MPPT, i.e., maximum power point tracking, algorithm is used to extract maximum available power from PV panel in certain conditions. It is also used to track changing maximum power point. In PV power generation system, an electronic device is set to track the maximum power point which utilizes algorithms to frequently sample voltage and current of panel. The MPPT adjusts the duty ratio of panel as needed after applying algorithm logic to these voltage and current samples. MPPT is a very useful technique to find the maximum power point and keep the load characteristic of panel at that point.

2.1.1 Perturb and Observe

The current and voltage of the photovoltaic source are fluctuating throughout the day due to changing temperature and solar irradiation. The maximum power is extracted at only one operating point such as point where the value of current and voltage are maximum known as maximum power point. There are so many algorithms for MPPT but P and O algorithm is used because of its simplicity. It unsettles duty ratio of flyback converter, and it also adjusts the DC link current and voltage among the photovoltaic array and flyback converter to extract maximum power [1–3, 7]. The PV panel power can be computed measure of photovoltaic current and voltage. The duty ratio can be reduced by changing small amount of voltage and power, but the change should be positive. The duty ratio can be increased by keeping the power positive, and change in voltage is negative. The appropriate duty ratio can be calculated by using MPPT algorithm, and firing pulses of flyback converter are generated.

2.2 *Flyback Converter*

In a flyback topology, the main switch and flyback transformer are connected in series. Energy fed by PV module is stored in an inductor which is connected in parallel with flyback transformer which boosts up the voltage. The flyback transformer not only boosts up the voltage but also provides isolation between PV module and load or grid. The inductor configuration used in flyback converter is divided into two parts to form a transformer. It is also a buck–boost converter, this topology provides high voltage conversion ratio by multiplying turn ratio of the transformer, and it is also capable of providing isolation. This configuration consists of magnetizing inductance 'Lm' with transformer of turn ration N1/N2. The losses in the transformer are negligible so losses are neglected. The mode of operation of this converter depends on magnitude of magnetizing inductance (Fig. 2).

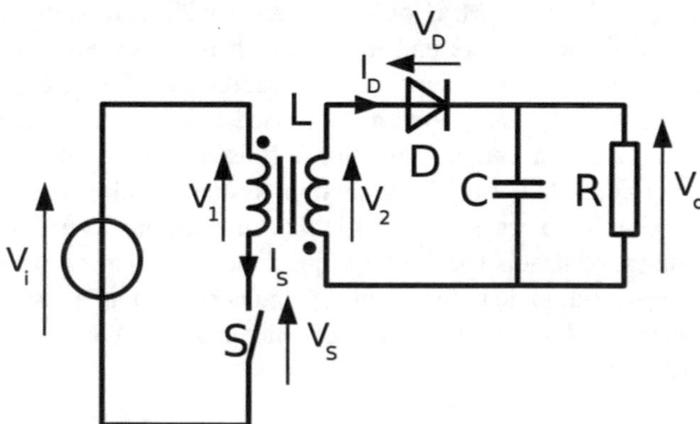

Fig. 2 Flyback converter

The MOSFET is used as a switch. The primary side of the transformer is connected to the input supply. The magnetic flux is increased, and potential is stored in it. Reversed biased condition of the output diode results in negative voltage induced in it. Output filter capacitor is required to supply the energy to the load. The topology in which MOSFET is turned off automatically, current and flux in the primary side are reduced, and the diode on the secondary side is forward biased results in positive voltage induced in secondary side of the transformer. The transformer containing stored energy is successfully supplied to the load. When MOSFET is turned on, the output capacitor discharges energy to the load by delivering load current, where its value should be high in this converter [1–3, 7].

3 Operation of Proposed System

The flyback converter is used in both AC/DC and DC/DC conversion with galvanic isolation between input and many outputs. The flyback converter is a buck–boost converter with the inductor split to form a transformer, so that the voltage ratios are multiplied with an additional advantage of isolation. The flyback converter is an isolated power converter [7–11]. As shown in Fig. 3 the circuit configuration of proposed converter. The photovoltaic array as an input source, the photovoltaic current (Ipv) is applied to the two stage interleaved flyback converter through the decoupling capacitor (C). The decoupling capacitor is provided to eliminate harmonics present in the photovoltaic current and also provide balance to the system. The low on-state resistance power electronic switch (MOSFET) is used in the primary side of the flyback converter as a flyback switch. During turn-on period of MOSFET switches S1, S2, and S3, a current from photovoltaic array is passed through the magnetizing inductance (L_m) and primary winding of the flyback converter. The main function of magnetizing inductance is to store the magnetic field energy. During turn-on period of switches, they cannot flow through the secondary side due to the current blocked by the reversed biased diode. During this period, the energy is supplied to the grid by C_f and L_f. During turn-off period

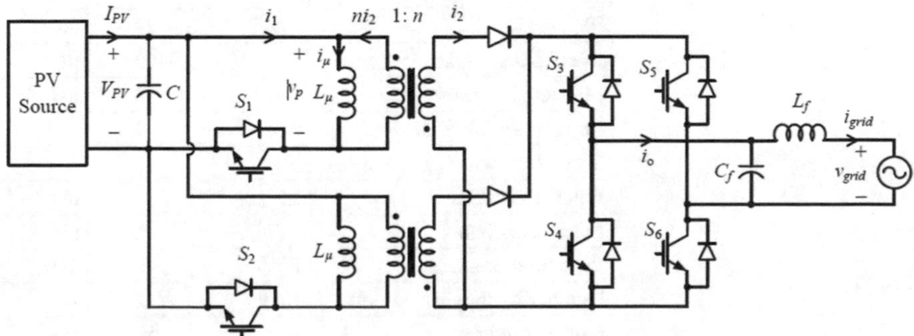

Fig. 3 Circuit diagram of the PV inverter system based on two-cell interleaved flyback converter topology

of switches, the magnetic field store energy of the magnetizing inductance is delivered to the grid in the form of current. So, the operation of the above configuration is based on the flyback inverter and behaves as a voltage-controlled current source. The proposed converter is operated in DCM and provides easy as well as steady-state generation at AC current at the grid interface. While operating of proposed converter in the DCM mode with open loop control generates triangular current pulses for each switching period. If SPWM method is used for control, the proposed system can provide sinusoidal current in phase with grid voltage. The unfolding full-bridge inverter is only responsible for unfolding the sinusoidally modulated DC current packs into AC at the right moment of the grid voltage. The grid frequency is used to operate the inverter switches. The grid frequency is low. So, the turn-on and turn-off losses of IGBT are negligible. The conduction losses are considered due to the high on-state resistance power electronic switches (IGBT) of the inverter. The control signal for flyback converter's MOSFET switch, primary side voltage of flyback transformer (Vp), magnetizing current (i_m).

4 Simulation Result and Discussion

To verify the design as well as to determine hardware requirements, simulations should be done. Like, current ratings, power required etc. can be easily getting from the simulation results. Figure 3 shows simulation model used for the simulation studies.

4.1 Simulation Parameters

Following parameters are used for simulation of proposed system (Table 1).

Table 1 Simulation parameters [1]

Parameter	Symbol	Value	Unit
PV voltage	Vpv	88	V
PV current	Ipv	22	A
Decoupling capacitor	C	9400	μF
Magnetizing inductance	Lm	8	μH
Primary winding turns	Np	4	–
Secondary winding turns	Ns	18	–
Filter inductance	Lf	250	μH
Filter capacitance	Cf	1	μF
Switching frequency	Fs	40	kHz
Number of interleaved cell	Ncell	2	–
Grid frequency	F	50	Hz
Grid voltage	V	240	V
Decoupling capacitor	Vpv	88	V

4.2 Model of Proposed System

In order to test the performance of the proposed converter, the system shown in Fig. 4 is simulated (Figs. 5, 6 and 7).

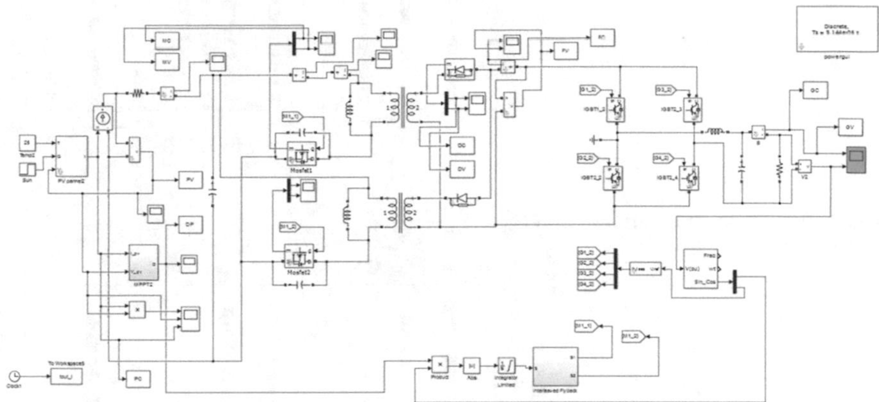

Fig. 4 MATLAB model for the proposed interleaved flyback inverter

Fig. 5 Grid output current and voltage

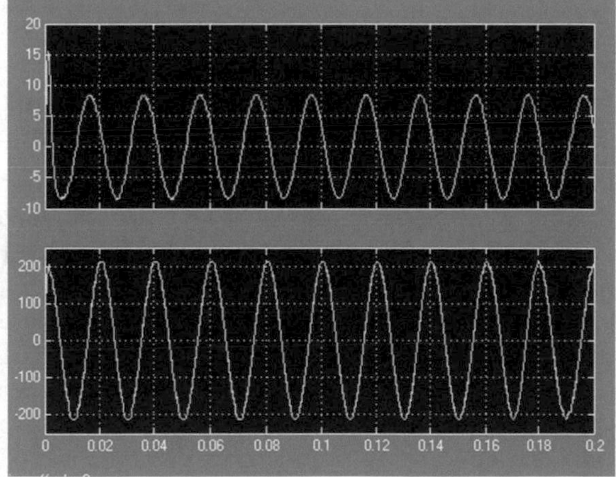

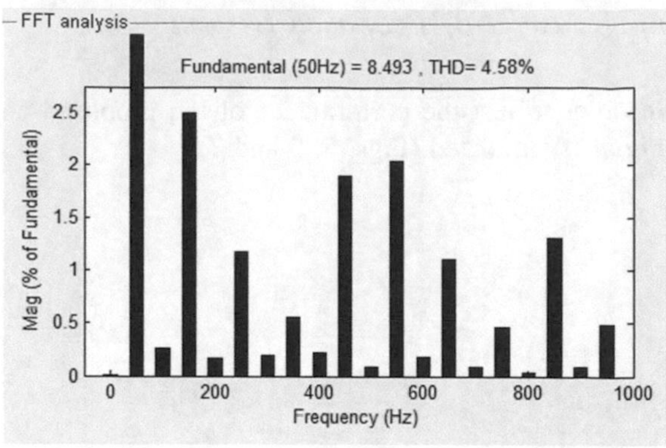

Fig. 6 THD analysis for grid current

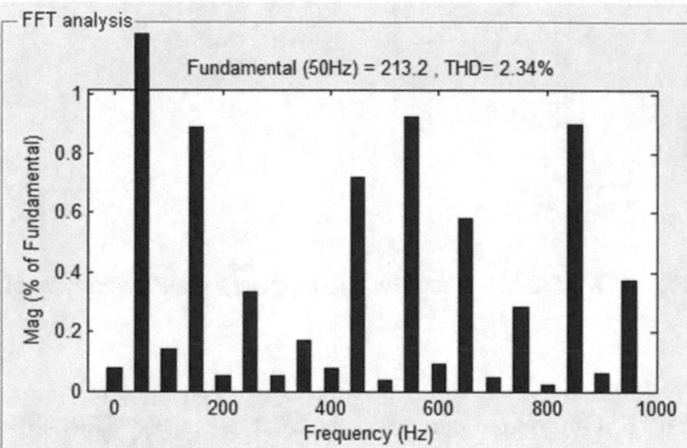

Fig. 7 THD analysis for grid voltage

5 Hardware Implementation and Result Validation

In this paper, we describe details about the prototype model of interleaved flyback inverter using the driving circuit of MOSFET and controlled by PIC [PIC16F877A]. And also, it includes the hardware results. Following Fig. 8 shows the experimental setup of proposed interleaved flyback inverter for PV application. This experimental setup consists of solar panel, driver circuit, transformer, rectifier diode, capacitor filter, single inverter circuit, PIC16F877 controller board, relay, and load.

Figure 8 shows the experimental setup of interleaved flyback inverter for PV applications. The solar panel is used as input source, to use a 12 W solar panel these converts light energy to electrical energy. And this electrical energy is used. Depending upon solar radiations intensity, harmonics will generate at input side, so switching stresses may occur. To eliminate we are using flyback converter technology. The switch, inductor, diode, and transformer are the combination of flyback converter. The flyback switch creates more stresses, and this can be overcome by flyback inductor. All harmonics are dropped at inductor, so at the secondary side of transformer, there are no losses or harmonics, so we have getting pure sinusoidal (Figs. 9 and 10).

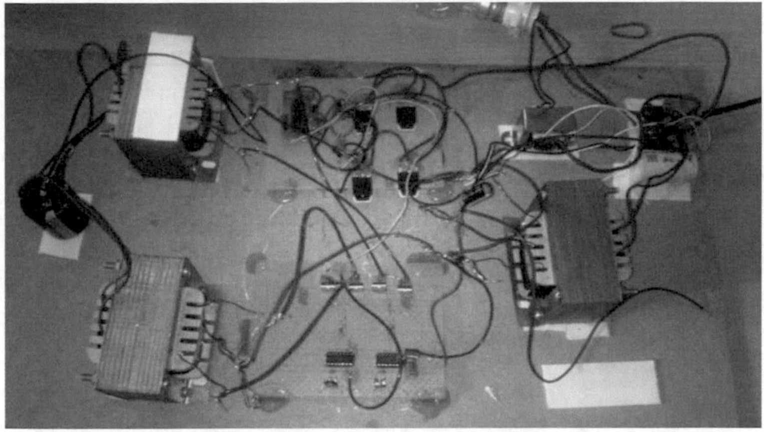

Fig. 8 Experimental setup

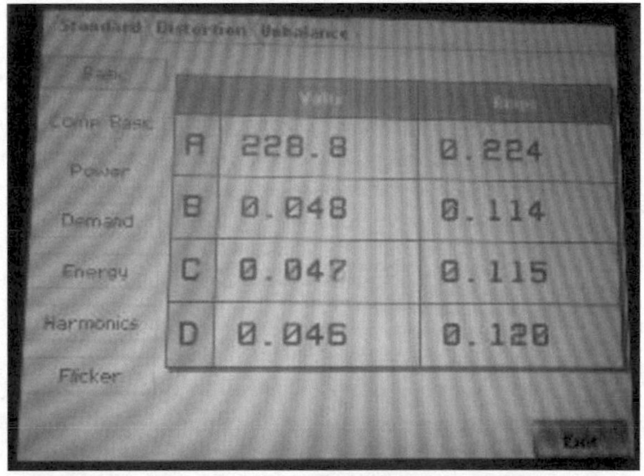

Fig. 9 Output of DSO showing inverter output and grid

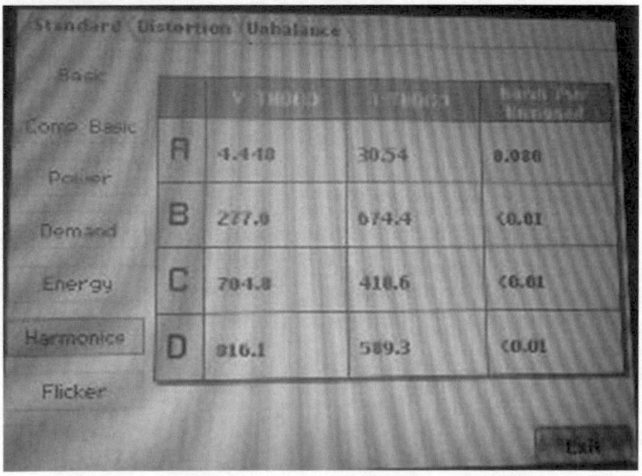

Fig. 10 THD value is 4.5%

Standard Distortion Unbalance
Basic
Comp Basic
Power
Demand
Energy
Harmonics
Flicker
A 4.448 30.54 0.080
B 277.0 674.4 <0.01
C 704.8 418.6 <0.01
D 816.1 589.3 <0.01
Exit

6 Conclusion

An interleaved flyback inverter has become an important component because it can operate light loads. The conduction losses can be reduced in switches by using this concept. Ultimately, an improved interleaved high-power flyback is best solution for all the photovoltaic applications. It having efficiency of 90.16%, THD of 4.5%, and power factor of 0.998 is very advantageous when compared to other inverters. The effectiveness of this proposal is confirmed by the simulation result.

Acknowledgements I wish to thank my loving and supportive wife, Shipra Kumari, and my parents, who provide unending inspiration.

References

1. Xue, Y., Chang, L., Kjaer, S.B., Bordonau, J., Shimizu, T.: Topologies of single-phase inverters for small distributed power generators: an overview. IEEE Trans. Power Electron. **19**(5), 1305–1314 (2004)
2. Li, Y., Oruganti, R.: A low cost flyback CCM inverter for AC module application. IEEE Trans. Power Electron. **27**(3), 1295–1303 (2012)
3. Kasa, N., Iida, T., Chen, L.: Flyback inverter controlled by sensor less current MPPT for photovoltaic power system. IEEE Trans. Ind. Electron. **52**(4), 1145–1152 (2005)
4. Michalewicz, Z.: Genetic Algorithms + Data Structures = Evolution Programs, 3rd edn. Springer-Verlag, Berlin Heidelberg New York (1996)
5. Olalla, C., Clement, D., Rodriguez, M., Maksimovic, D.: Architectures and control of submodule integrated DC–DC converters for photovoltaic applications. IEEE Trans. Ind. Appl. **28**(6), 2980–2997 (2013)
6. Tan, G.H., Wang, J.Z., Ji, Y.C.: Soft-switching flyback inverter with enhanced power decoupling for photovoltaic applications. Electr. Power Appl. **1**(2), 264–274 (2007)
7. Chen, Y.M., Liao, C.Y.: Three-port flyback-type single-phase micro inverter with active power decoupling circuit. In: Proceedings of the IEEE Energy Conversion Congress and Exposition, pp. 501–506 (2011)
8. Nanakos, A.C., Tatakis, E.C., Papanikolaou, N.P.: A weighted efficiency-oriented design methodology of flyback inverter for AC photovoltaic modules. IEEE Trans. Power Electron. **27**(7), 3221–3233 (2012)
9. Tamyurek, B., Torrey, D.A.: A three-phase unity power factor single stage AC–DC converter based on an interleaved flyback topology. IEEE Trans. Power Electron. **26**(1), 308–318 (2011)
10. Tamyurek, B., Kirimer, B.: An interleaved flyback inverter for residential photovoltaic applications. In: Proceedings of the 15th European Conference on Power Electronics and Applications, pp. 1–10 (2013); Michalewicz, Z.: Genetic Algorithms + Data Structures = Evolution Programs, 3rd edn. Springer, Berlin Heidelberg New York (1996)
11. Kumar, P., Kamble, S.: An improved interleaved high power flyback inverter for photovoltaic application. In: International Conference on Computation of Power, Energy, Information and Communication, pp. 853–864, 22nd–23rd Mar 2017

A Data Mining Approach for Sentiment Analysis of Indian Currency Demonetization

Maddukuri Sree Vani, Rajasree Sutrawe and O. Yaswanth Babu

Abstract Demonetization is big step initiated by Government of India that is withdrawal of the particular currency from circulation and replacing it with a new currency. In the period of demonetization, people used debit cards, credit cards, mobile wallet, and net banking for their day-to-day need. Many people across the world posted their opinions in all social media. With this, large amount of data is gathered about demonetization. In this paper, sentiment analysis is performed to discover how people feel about a demonetization. Twitter tweets of 11843 people were collected, and data mining was conducted to understand the real sentiment of the people of India during the demonetization period using naive Bayes classifier algorithm. The results of analysis were very convincing and strong that the people had a positive sentiment toward the Indian currency demonetization.

Keywords Demonetization · Sentiment analysis · Data mining
Twitter data

1 Introduction

Social network Web sites have evolved to become a source of different types of information, where people post real-time messages about their opinions on a mixture of topics, discuss current issues, complain, and express their sentiment for

M. Sree Vani (✉)
Department of CSE, Mahatma Gandhi Institute of Technology, Gandipet, Hyderabad 500075, India
e-mail: osreevani@gmail.com

R. Sutrawe
Department of CSE, Gurunanak Institutions Technical Campus Autonomous, Hyderabad 501506, India
e-mail: raj.sutrawe@gmail.com

O. Yaswanth Babu
Software Engineer, TCS, Gachibowli, Hyderabad 500032, India
e-mail: oyaswanth@gmail.com

J. Nayak et al. (eds.), *Soft Computing in Data Analytics*,
Advances in Intelligent Systems and Computing 758,
https://doi.org/10.1007/978-981-13-0514-6_2

products they use in general. In fact, companies manufacturing such products have started to poll these social networks to get a sense of general sentiment for their product. One challenge is to build technology to detect and summarize an overall sentiment. Demonetization is withdrawal of a particular form of currency from circulation. In India, the demonetization that took place in November 2016 was mainly to control the fake currency, tackle black money, corruption, curb financing of terrorism, and form the cashless economy. The people across India may post their opinions on demonetization in social media. In this paper, Twitter data will be considered and build models for classifying tweets into positive, negative, and neutral sentiment [1]. Then, analyze the Twitter data to find the effect of demonetization on the people [2, 3]. One advantage of this data is that the tweets are collected in a streaming fashion and therefore represent a true sample of actual tweets in terms of language use and content. The sentiment of the Twitter tweets can be analyzed using naive Bayes algorithm and Bayes theorem [4]. Every tweet is taken, and each word is matched with the database of positive and negative words [5]. Section 2 discusses related work, Sect. 3 describes sentiment classification of demonetization data, and Sect. 4 discusses sentiment analysis of demonetization data followed by conclusion.

2 Related Work

In [6], the present and future challenges of demonetization have been discussed. The darker side of the economy and the black money is discussed on [7], while demonetization and their impacts on the society are discussed on paper [8]. Sentiment analysis can be used to find the customer response to a product and news, and it measures like, dislike sentiments of the people. The paper [5] takes Twitter data as training data set and trains the sentiment classifier. This is a very challenging task because there is no explicit rating system present in previous work. Classifying the sentiment of Twitter messages is most similar to sentence-level sentiment analysis [9] for the limited sized tweets and the paragraph-level sentiment analysis for more than one sentence. However, the Twitter follows micro-blogging nature and small tweets and also supports different natural languages, so Twitter is the best source for the sentiment analysis [10]. In paper [4], word presence and word frequency are calculated, and then identified word presence is more effective than word frequency for sentiment analysis. Location of a word in the given sentence is more effective to identify strength of a particular word. The paper [11] discuses the method for collecting corpus tweet data and segregates as positive and negative sentiments. This is similar to [5] by using emotions for positive and negative tweets. Classification of tweets with polarity method provides good measurement of positive (+1 weak positive to +5 extreme positive) and negative (−1 weak negative to −5 extreme negative) sentiments [12].

3 Sentiment Analysis of Tweets

Sentiment analysis can be carried out by two methods such as lexicon analysis and naive Bayes classifier to get higher accuracy. Lexicon analysis contains a words list which is positive and negative. Each tweet is parsed and matched with the words from lexicon list.

To identify sentiment behind the tweet, count positive and negative words and assign a score for each tweet. Based on the score, the tweet will be classified as positive, negative, and neutral. Polarity scores are also assigned to each tweet based on polarity and emotions of tweets. Different polarities are positive, negative, and neutral. Different emotions are anger, disgust, fear, joy, sadness, surprise, and other unknown emotions. Figure 1 describes the overview of lexicon approach.

3.1 Data Preprocessing

Demonetization data is collected from Twitter from November 8, 2016, to January 27, 2017. Though there were limitations of the access to the Twitter API, 6000 tweets were collected successfully. The collected data was further processed. Data such as numbers, symbols, non-English characters, and extra spaces was removed so that the data is perfect to perform data mining. The collected data is stored in the database of R, and various functions are utilized to conduct sentiment analysis [5].

The tweets gathered from Twitter are a mixture of URLs and other non-sentimental data like hashtags "#", annotation "@", and retweets "RT". To obtain n-gram features, we first have to tokenize the text input. Tweets pose a problem for standard tokenizers designed for formal and regular text. Figure 2

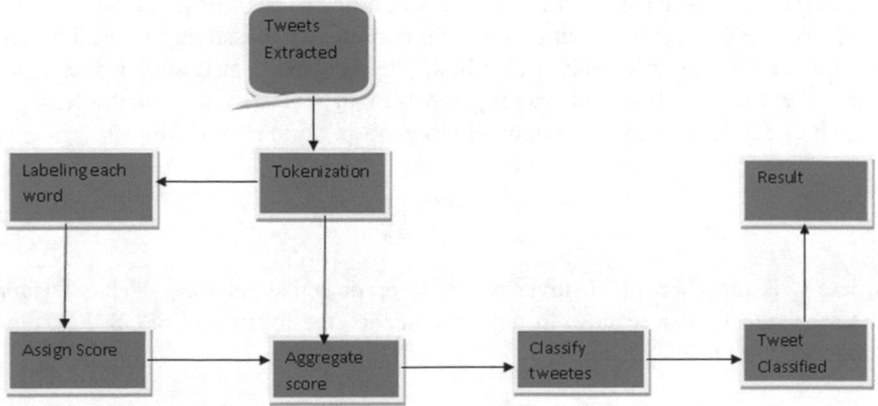

Fig. 1 Overview of general approach

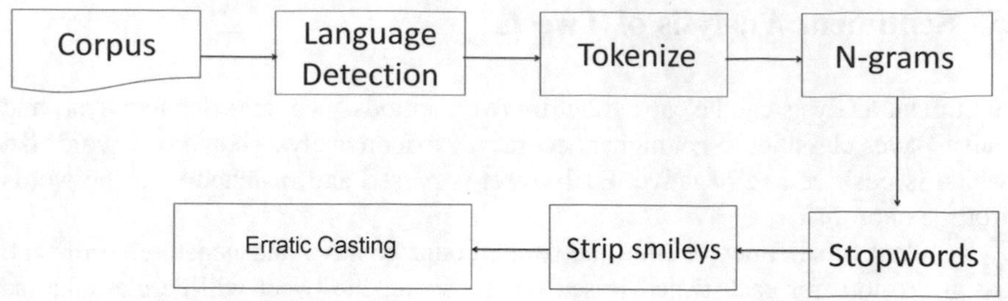

Fig. 2 Data preprocessing steps

displays the various intermediate processing feature steps. The intermediate steps are the list of features to be taken into account of by the classifier.

3.2 AFINN Dictionary

The AFINN is a dictionary which consists of 2500 words which are rated from +5 to −5 depending on their meaning. The Twitter words are analyzed using AFINN dictionary, and rate the Twitter words as per their meaning from +5 to −5. Initially load the dictionary into pig by using the below statement:

$$\text{dictionary} = \text{load}' /\text{AFINN.txt}' \text{ using PigStorage}('/\text{t}') \text{ As}(\text{word: char array, rating: int});$$

3.3 Naive Bayes Classifier

Naive Bayes classifier is used to predict the particular class of given words. It is used because of its easiness in both during training and classifying steps. To train the input set using naïve Bayes classifier, preprocessed data is supplied as input data. After training phase, test data is given as input to classify sentiment words. As shown in Eq. 1, the Bayes theorem shows good performance as follows.

$$P(C|m) = P(C) \prod_{i=1}^{n} P(f_i|C) \tag{1}$$

where C is the class of positive, negative, or objective sets, *mψ* is the Twitter message, and *fψ* is a feature. In our experiments, the features are POS tags, unigrams, or bigrams.

4 Results and Discussion

The word scores of the features are tested based on chi-square method. Frequency distribution of positive sentiment and negative sentiment words is generated. Finally, the number of positive and negative words as well as the total number of words and the dictionary of word score based on chi-square test is found. The performance of naïve Bayes classifier will be evaluated using accuracy. Number of correctly selected positive and negative words divided by total number of words will produce accuracy as given in Eq. 2.

$$Accuracy = \frac{\sum True\ Positive + \sum True\ Negative}{\sum Total\ number\ of\ words} \quad (2)$$

where TP is number of tweets identified as positive and TN is number of tweets identified as negative, respectively. Figure 3 describes sentimental score for keywords based on chi-square method. Table 1 shows experimental results.

Sentiment analysis of tweets based on polarity was performed, and the result was plotted as shown in Fig. 4.

Sentiment analysis of tweets based on emotions was performed, and the result was plotted as shown in Fig. 5. The different emotions such as anger, disgust, fear, joy, sadness, the surprise were identified. Emotions such as joy, sadness, and anger were the emotion of the most of the people during demonetization of currency in India.

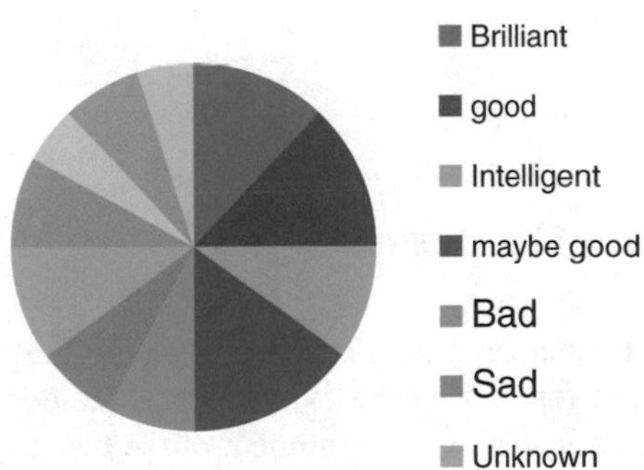

Fig. 3 Wordscore tweets

Table 1 Experimental results

Sentiment	No. of samples	TP	TN
Positive	3400	2336	1064
Negative	2476	1739	737
Neutral	4233	3919	314

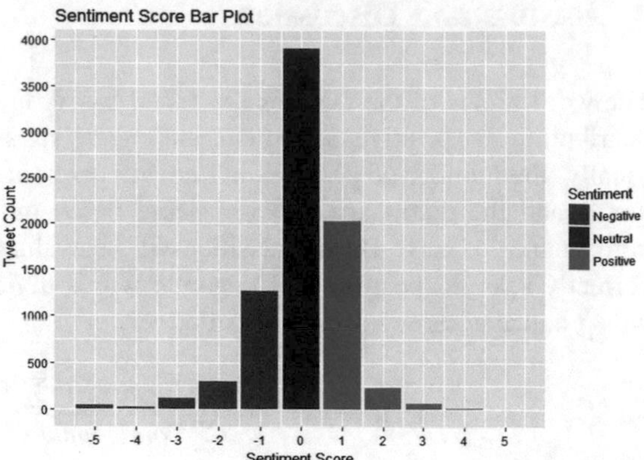

Fig. 4 Sentiment analysis of tweets on polarity

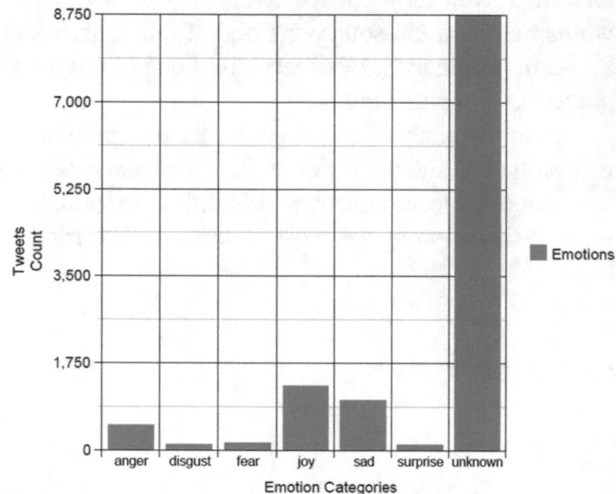

Fig. 5 Sentiment analysis of tweets on emotions

5 Conclusion

Twitter sentiment analysis to understand the people's sentiment on demonetization put forward by the government. Methods such as naïve Bayes and lexicon are used for analyzing the sentiment of the tweets. The analysis results gave strong evidence that the people of India are having positive sentiment toward the Indian currency demonetization. Finally, it is important to recall that opinion is multidimensional object. Tweets may have mixed positive and negative opinions. As in future work, sentiment analysis needs to incorporate negation handling and emphasis handling in order to improve its classification accuracy.

References

1. Pandey, A.: Sentiment analysis on demonetization, 12 December 2016. https://bigishere.wordpress.com/2016/12/12/sentiment-analysis-on-demonetization-pig-use-case/
2. Mining Data from Twitter from AbhishangaUpadhyay, Luis Mao, Malavika Goda Krishna (PDF)
3. Shukla, R.K.: Sentiment analysis on demonetization using R on twitter data, 17 November 2016. https://www.linkedin.com/pulse/sentiment-analysis-demonetization-using-r-twitter-data-shukla
4. Liu, B., Blasch, E., Chen, Y.: Dan Shen and summarized report about the opinion from Twitter. It is Genshe Chen, 2013. In: Big Data, 2013 IEEE International Conference on Scalable Sentiment Classification Using Naive Bayes Classifie, pp. 99–104. IEEE, 2013
5. Go, A., Bhayani, R., Huang, L.: Twitter sentiment classification using distant supervision. Processing **150**(12), 1–6 (2009)
6. What is Demonetization. http://www.investopedia.com/terms/d/demonetization.asp
7. Jose, T.: What is the impact of demonetization 16 November 2016. http://www.indianeconomy.net/splclassroom/309/what-are-the-impacts-of-demonetisation-on-indian-economy/
8. Ananthram, P.: Sentiment analysis of tweets, 2 month ago. https://www.kaggle.com/priyaananthram/d/arathee2/demonetization-in-india-twitter-data/sentiment-analysis-of-tweets
9. Mane, S.B., Assar, K., Sawant, P., Shinde, M.: Product rating using opinion mining. Int. J. Comput. Eng. Res. Trends **4**(5), 161–168 (2017)
10. Vinodhini, G., Chandrasekaran, R.M.: Sentiment analysis and opinion mining: a survey. Int. J. Adv. Res. Comput. Sci. Softw. Eng. **2**(6) (2012). ISSN: 2277 128X
11. Pak, A., Paroubek, P.: Twitter as a corpus for sentiment analysis and opinion mining. In: Chair, C.C., Choukri, K., Maegaard, B., Mariani, J., Odijk, J., Piperidis, S., Rosner, M., Tapias, D. (eds.) Proceedings of the Seventh International Conference on Language Resources and Evaluation (LREC'10)
12. Thelwall, M., Buckley, K., Paltoglou, G., Cai, D., Kappas, A.: Sentiment strength detection in short text. J. Am. Soc. Inf. Sci. Technol. **61**(12), 2544–2558 (2010)

An Efficient Solution of an Optimization Problem in Financial Engineering

Nawdha Thakoor, Dhiren Kumar Behera, Désiré Yannick Tangman and Muddun Bhuruth

Abstract A popular model for the pricing of financial derivatives is the stochastic alpha beta rho model. The model has the capabilities for fitting various volatility structures observed in options markets. An optimization problem needs to be solved for estimating parameters in the model. This work considers a computational partial differential equation approach for this calibration process. It is shown that the partial differential equation method outperforms methods based on analytical price approximations.

Keywords Nonlinear optimization · Calibration · SABR · Arbitrage-free Implied volatility

1 Introduction

The valuation of financial derivatives consistent with market volatility structures observed in option prices is of much importance for hedging risks associated with movement of the underlying asset prices. The inability of the Black–Scholes model [1] to empirically fit the market volatility has led to the development of alternative option pricing models. One such model is the constant elasticity of variance (CEV) model [2] which has the capability of fitting a skew-shaped volatility structure.

N. Thakoor · D. Y. Tangman · M. Bhuruth
Department of Mathematics, University of Mauritius, Reduit, Mauritius
e-mail: n.thakoor@uom.ac.mu

D. Y. Tangman
e-mail: y.tangman@uom.ac.mu

M. Bhuruth
e-mail: mbhuruth@uom.ac.mu

D. K. Behera (✉)
Mechanical Engineering Department, Indira Gandhi Institute of Technology, Sarang 759146, India
e-mail: dkb_igit@rediffmail.com

J. Nayak et al. (eds.), *Soft Computing in Data Analytics*,
Advances in Intelligent Systems and Computing 758,
https://doi.org/10.1007/978-981-13-0514-6_3

The stochastic alpha beta rho (SABR) model [3] extends the CEV process by incorporating a stochastic volatility process. SABR is capable of fitting Black-Scholes implied volatilities in the market. Pricing of derivatives under the CEV and SABR models for equity derivatives is considered in recent works [4–7].

The availability of an analytical approximation for the implied volatility for the SABR model [3] made the SABR a popular model to manage smile risk. However, the analytical asymptotic approximation formula for the implied volatility are only first-order accurate in maturity. Also, there is a positive probability for the forward price to hit zero, resulting in arbitrage opportunities. To avoid arbitrage opportunities, an absorbing boundary condition must be imposed at zero [8]. The probability density of SABR can be approximated by solving a one-dimensional arbitrage-free partial differential equation (pde) [9]. A closed-form approximation for European calls and puts in terms of the non-central chi-square distribution is derived in [10].

For addressing the shortcomings of the constant volatility Black–Scholes model, Dupire [11] developed a local volatility model which takes as input the entire volatility smile and then constructs an implied local volatility function which is consistent with the volatility smile and skews observed in the market. A forward pde was also derived in terms of the strike prices to calibrate the model to market data in order to obtain the deterministic implied volatility.

In this work, we use the explicit forward volatility function in the Dupire's forward equation to calibrate the SABR model to a set of strike and maturity-dependent market options data which is then solved recursively by an implicit finite difference scheme. Since only one time step is involved, the update in the optimization process is fast. This pde calibration is then compared with the calibration obtained using two analytical formulas.

The remainder of this paper is organized as follows. In Sect. 2, the dynamics of the SABR model is described and the analytical approximations for option pricing under this model are presented. In Sect. 3, the problem resulting from the pde discretization process is described, and in Sect. 4, numerical results related to the different approaches are described. Finally, Sect. 5 concludes our work.

2 The SABR Model

Consider the SABR model given by

$$\begin{aligned} dF_t &= \alpha_t F_t^{\beta} dW_t^1, \quad F_0 = f, \\ d\alpha_t &= \nu \alpha_t dW_t^2, \quad \alpha_0 = \alpha, \end{aligned} \tag{1}$$

where $0 \leq \beta \leq 1$ is the exponent parameter, α_t is the volatility of the forward price F_t, ν is the constant volatility of the volatility parameter, dW_t^1 and dW_t^2 are two correlated Wiener processes with correlation of ρ such that $dW_t^1 dW_t^2 = \rho\, dt$. The SABR implied volatility is the volatility parameter in the Black–Scholes option formula that yields a market price. An analytical approximation for the implied volatility $\sigma_B(E, f)$ is given by [3]

$$\sigma_B(E, f) = \frac{\alpha\left[1 + \left(\frac{(1-\beta)^2\alpha^2}{24(fE)^{1-\beta}} + \frac{\rho\beta\nu\alpha}{4(fE)^{(1-\beta)/2}} + \frac{(2-3\rho^2)\nu^2}{24}\right)T\right]}{(fE)^{(1-\beta)/2}\left(1 + \frac{(1-\beta)^2}{24}\left(\ln\frac{f}{E}\right)^2 + \frac{(1-\beta)^4}{1920}\left(\ln\frac{f}{E}\right)^4\right)} \times \frac{z}{w(z)}, \quad (2)$$

where $z = \frac{\nu}{\alpha}(fE)^{(1-\beta)/2}\ln(f/E)$, $w(z) = \ln\left((\sqrt{1-2\rho z+z^2}+z-\rho)/(1-\rho)\right)$. The time zero price $V(f, 0)$ of a European call option with strike E and maturity date T on a forward contract with dynamics (1) can then be easily computed using Black's formula given by

$$V(f, 0) = e^{-rT}\left[f\Phi(d_+) - E\Phi(d_-)\right],$$

where Φ is the cumulative distribution function of the standard normal distribution and $d_\pm = \left(\ln\left(\frac{f}{E}\right) \pm \frac{1}{2}\sigma_B^2 T\right)/(\sigma_B\sqrt{T})$.

2.1 Arbitrage-Free One-Dimensional SABR PDE

The marginal density $Q(F, T)$ defined by $Q(F, T)dF = \mathbb{P}\{F < F(T) < F + dF\}$ can be obtained by solving the one-dimensional equation [9]

$$\frac{\partial Q}{\partial T} = \frac{1}{2}\frac{\partial^2}{\partial F^2}\left[D^2(F)G(F, T)Q\right], \quad (3)$$

where

$$D(F) = \sqrt{\alpha^2 + 2\alpha\rho\nu y(F) + \nu^2 y^2(F)}F^\beta, \quad G(F, T) = e^{\rho\nu\alpha\Gamma(F)(T-t)},$$

where $y(F) = \left(F^{1-\beta} - f^{1-\beta}\right)/(1-\beta)$, $\Gamma(F) = \left(F^\beta - f^\beta\right)/(F-f)$ and $Q \to \delta(F-f)$ as $t \to 0^+$ with δ denoting the Dirac delta function. Since the SABR model has a barrier at the lower boundary $F = 0$, the absorbing boundary condition is imposed by setting $D(0) = 0$. Several numerical schemes for the solution of (3) are considered in [12].

2.2 Closed-Form Arbitrage-Free Approximation

Using transformations of the two-dimensional SABR pde, it is shown in [10] that the option price $V(f, 0)$ is given by

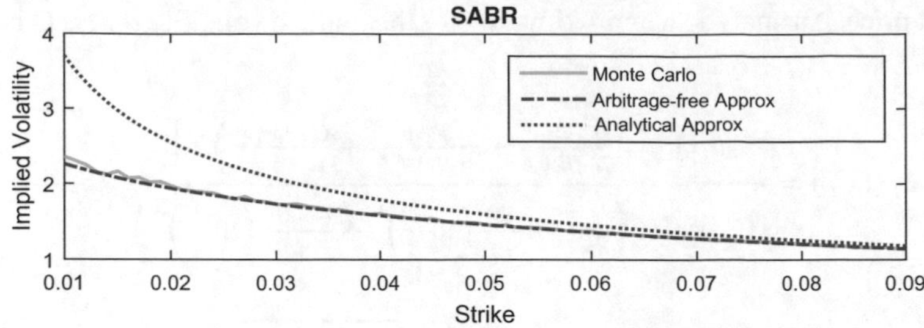

Fig. 1 Implied volatilities for $f = 0.05$, $T = 1$, $\alpha = 0.1$, $\beta = 0.1$, $\rho = 0$, $\nu = 0.1$

$$V(f,0) = \left(1 - \chi^2\left(\frac{E^{2(1-\beta)}}{\alpha^2 T(1-\beta)^2}, \frac{3-2\beta}{1-\beta}, \frac{f^{2(1-\beta)}}{\alpha^2 T(1-\beta)^2}\right)\right) f - \chi^2\left(\frac{f^{2(1-\beta)}}{\alpha^2 T(1-\beta)^2}, \frac{1}{1-\beta}, \frac{E^{2(1-\beta)}}{\alpha^2 T(1-\beta)^2}\right) E, \quad (4)$$

where $\chi^2(y, k, l)$ is the cumulative distribution function of the non-central chi-squared random variable with non-centrality parameter l and k degrees of freedom.

Figure 1 compares the implied volatility computed using the formula (2) and those implied by formula (4). Comparisons are also done using Monte Carlo simulations with 100,000 time steps. It is observed that the arbitrage-free price of [10] is in close agreement to the Monte Carlo simulation results, while the Hagan et al. [3] approximation deviates from the latter for low strikes.

3 A Computational PDE Optimization

The approximation of the volatility by a piecewise time-dependent and piecewise constant function is considered in [13]. A different pde-based approach for constructing the volatility is considered here. Let $\tau = T - t$, and consider the arbitrage-free volatility expansion

$$\vartheta(E,\tau) = D(E)\,(G(E,\tau))^{\frac{1}{2}}\,. \quad (5)$$

Given a set of arbitrage-free European option prices $V(E, T)$, the arbitrage-free volatility (5) is used to generate option prices as the solution to the Dupire forward pde [11] given by

$$\frac{\partial V}{\partial \tau}(E,\tau) = -\frac{1}{2}\vartheta^2(E,\tau)\frac{\partial^2 V}{\partial E^2}(E,\tau), \quad (6)$$

$$V(E,0) = (f-E)^+. \quad (7)$$

For a finite difference solution, the problem is localized to the finite domain $\Omega = [E_{\min}, E_{\max}] \times [0, T]$. Thus, for a call option, Eq. (6) must be solved using the initial condition (7). Considering a set of grid points (E_m, τ_n) where $E_m = mh$ with $h = \left(E_{\max} - E_{\min}\right)/M$ and $\tau_n = nk$ with $k = T/N$, let $V_m^n = V(E_m, \tau_n)$ and consider central-difference approximations to the second-order derivative given by $\delta_E^2 V_m = (V_{m+1} - 2V_m + V_{m-1})/h^2$. This leads to a semi-discrete scheme of the form

$$\left(V_\tau\right)_m = -\frac{1}{2}\vartheta^2(E_m, \tau)\delta_E^2 V_m.$$

Using an implicit time stepping and letting $\vartheta(E_m, \tau_n) = (\vartheta(E_m))^n$ gives the scheme

$$\left(1 + \frac{\Delta\tau}{2}(\vartheta^2(E_m))^n \delta_E^2\right) V_m^{n+1} = V_m^n, \quad 0 \le n \le N-1.$$

At the left and right boundaries, the boundary conditions $V_{EE}(E, \tau) = 0$ as $E \to \infty$ and $V_{EE}(E, \tau) = 0$ as $E \to 0$ are used and the absorbing boundary condition is implemented as $D(E_{\min}) = 0$. This leads to the system of equations

$$\mathbf{A}^n V^{n+1} = V^n, \quad 0 \le n \le N-1, \tag{8}$$

where

$$\mathbf{A}^n = \begin{pmatrix} 1 & 0 & & & \\ \gamma_1 & 1-2\gamma_1 & \gamma_1 & & \\ & \gamma_2 & 1-2\gamma_2 & \gamma_2 & \\ & \ddots & \ddots & \ddots & \\ & & \gamma_{M-1} & 1-2\gamma_{M-1} & \gamma_{M-1} \\ & & & 0 & 1 \end{pmatrix},$$

with

$$\gamma_m = \frac{\Delta\tau}{2h^2}(\vartheta^2(E_m))^n, \quad m = 0, 1, \ldots, M.$$

Using initial guesses for α, β, ρ, ν, we solve the linear system given by (8). For a set of observed option quotes, the system (8) is solved recursively and the parameters are estimated by minimizing the squared difference between the observed market option prices for different maturities and strikes and the computed option prices. The minimization problem is casted as

$$\left(\hat{\alpha}, \hat{\beta}, \hat{\rho}, \hat{\nu}\right) = \underset{\alpha,\beta,\rho,\nu}{\operatorname{argmin}} \sum_i \left\{V_i^{\text{mkt}} - V_i\right\}^2. \tag{9}$$

Equation (9) is then solved using the least squares nonlinear function "`lsqnonlin`" in MATLAB.

4 Numerical Results

In this section, a numerical example illustrating the calibration of the SABR model to a given set of strikes and maturity-dependent option prices is considered. All numerical experiments have been performed using MATLAB R2015a on a Core i5 laptop with 4 GB RAM and speed 4.60 GHz.

Consider the 75 call option prices on the S&P 500 Index for different maturities and strikes obtained from the market as given in [14]. The initial stock price is $S_0 = 1124.47$, and we set $E_{\min} = 100$ and $E_{\max} = 2500$. Estimates of thc SABR model parameters, α, β, ρ, ν, for different maturities are obtained by solving (8) with initial guesses of the model parameters as $\alpha = 0.2$, $\beta = 0.4$, $\rho = -0.2$ and $\nu = 0.2$ and calibrating to the market data. The calibration procedure (9) is used to estimate the SABR parameters by different approaches. The first approach is based on using the scheme (8) which is denoted by "PDE approach" to compute the options prices consistent to market option prices. Secondly, the same minimization procedure is used to calibrate the analytical approximation (2) to market data, and thirdly, the data is calibrated to the closed-form expression (4).

Figure 2 shows how the calibrated model fits the observed market prices. The circles represent the market prices, and the plus signs represent the calibrated model prices. Figure 2a shows the results obtained by calibrating the CEV model [2] which is a special case of the SABR model obtained by setting $\nu = 0$ in (1). This model admits a closed-form formula for European option prices [15], and this formula is used to calibrate the model to the market data. Secondly, we show the models fit for the analytical formula (2) and the no-arbitrage analytical approximation (4) in Fig. 2b, c. Finally, Fig. 2d shows the calibrated results for the pde approach (8). It is observed that the pde method provides a more accurate fit of the market data than the two analytical approximations.

The estimates of the parameters α, β, ρ, and ν obtained from the calibration procedure for each method with the corresponding cpu timings are given in Table 1. For comparison purposes, an estimate of the goodness fit, the root mean square error (RMSE) between the market prices, and fitted prices for each method are reported using the formula

$$\text{RMSE} = \sqrt{\frac{\sum \left(V^{\text{Mkt}} - V^{\text{Model}}\right)^2}{\text{number of options}}}. \tag{10}$$

Figure 2 and Table 1 show that the pde method provides a better fit to the market data with an RMSE of 2.2176. The RMSEs computed using (10) in Table 1 also show that the SABR model provides a better fit to market data than the Black–Scholes and CEV models.

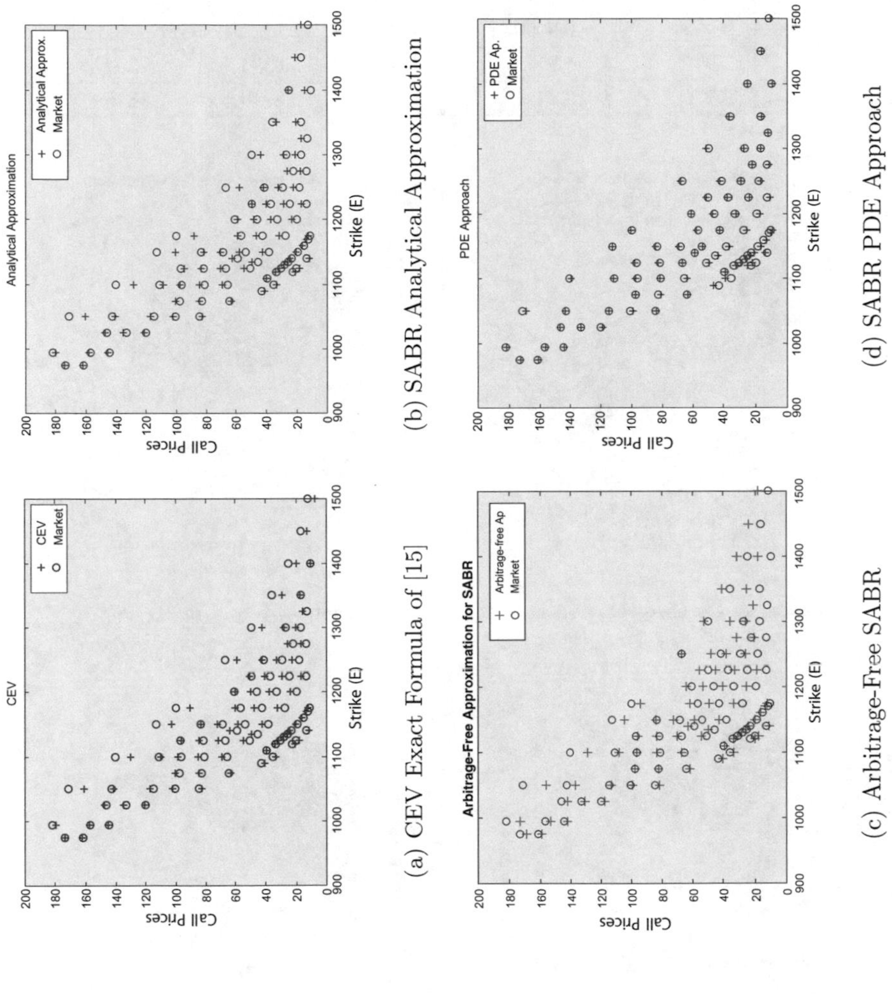

(a) CEV Exact Formula of [15]

(b) SABR Analytical Approximation

(c) Arbitrage-Free SABR

(d) SABR PDE Approach

Fig. 2 Fitting to market data for different strikes and maturities

Table 1 Calibrated parameters for the different methods

Estimates	SABR			CEV	BS
	Hagan et al. [3]	Yang et al. [10]	PDE approach		
α	0.9000	0.9908	0.9000	$\alpha = 0.1959$ $\beta = -1.9087$	$\alpha = 0.1891$
β	0.7854	0.7649	0.7443		
ρ	−0.5158	−0.5000	−0.9625		
ν	0.9000	0.5000	0.4883		
RMSE	3.9455	5.3445	2.2176	3.8394	6.2697
cpu(s)	0.2868	410.24	0.1943	0.8710	0.8468

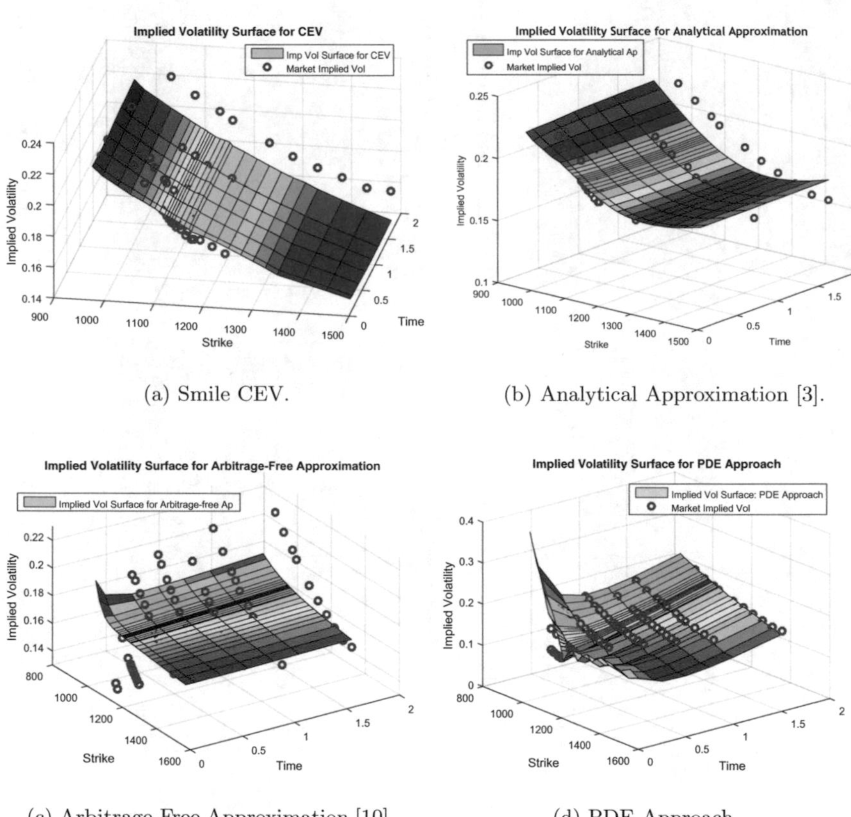

(a) Smile CEV.

(b) Analytical Approximation [3].

(c) Arbitrage-Free Approximation [10].

(d) PDE Approach.

Fig. 3 Implied volatility surfaces

4.1 Implied Volatilities

Figure 3 shows the resulting implied volatility surfaces for the different methods. The implied volatilities for the CEV model, arbitrage-free approximation, and the pde approach are obtained by inverting the calibrated option prices.

For the CEV model (Fig. 3a), a skew-shaped implied volatility curve fitting only some of the market implied volatilities is obtained. Although the analytical approximation (2) produces a smoother smile-shaped volatility surface (Fig. 3b), as the maturity increases and the strike price decreases, the surface does not fit the market implied volatilities. The arbitrage-free approximation (4) is also not able to provide an accurate fit to market implied volatilities. On the other hand, the calibration process via the pde method (Fig. 3d) yields a volatility surface which more accurately fits the observed implied volatilities.

5 Conclusion

This work considered calibrating the SABR model to market data using different approaches. Numerical results showed that the solution of the Dupire pde combined with a minimization problem is fast and has lower RMSE than two analytical approaches.

References

1. Black, F., Scholes, M.: The pricing of options and other corporate liabilities. J. Polit. Econ. **81**, 637–654 (1973)
2. Cox, J.: The constant elasticity of variance option pricing model. J. Portfolio Manage. **22**, 16–17 (1996)
3. Hagan, P.S., Kumar, D., Lesniewski, A.S., Woodward, D.E.: Managing smile risk. Wilmott Mag. **September**, 84–108 (2002)
4. Thakoor, N., Tangman, D.Y., Bhuruth, M.: A new fourth-order scheme for option pricing under the CEV model. Appl. Math. Lett. **26**, 160–164 (2013)
5. Thakoor, N., Tangman, D.Y., Bhuruth, M.: Efficient and high accuracy pricing of barrier options under the CEV diffusion. J. Comput. Appl. Math. **259**, 182–193 (2014)
6. Thakoor, N., Tangman, D.Y., Bhuruth, M.: Fast valuation of CEV American options. Wilmott Mag. **2015**, 54–61 (2015)
7. Thakoor, N.: A non-oscillatory scheme for the one-dimensional SABR model. Pertanika J. Sci. Technol. **25**, 1291–1306 (2017)
8. Rebonato, R., McKay, K., White, R.: The SABR/LIBOR Market Model Pricing, Calibration and Hedging for Complex Interest-Rate Derivatives. Wiley, West Sussex, United Kingdom (2009)
9. Hagan, P.S., Kumar, D., Lesniewski, A.S., Woodward, D.E.: Arbitrage-free SABR. Wilmott Mag. **January**, 60–75 (2014)
10. Yang, N., Chen, N., Liu, Y., Wan, X.: Approximate arbitrage-free option pricing under the SABR model. J. Econ. Dyn. Control **83**, 198–214 (2017)

11. Dupire, B.: Pricing with a smile. Risk **January**, 18–20 (1994)
12. Le Floc'h, F., Kennedy, G.: Finite difference techniques for arbitrage-free SABR. J. Comput. Financ. **20**, 51–79 (2017)
13. Andreasen, J., Huge, B.: Volatility interpolation. Risk **March**, 86–89 (2011a)
14. Schoutens, W.: Lévy Processes in Finance: Pricing Financial Derivatives. John Wiley and Sons Ltd, West Sussex, England (2003)
15. Schroder, M.: Computing the constant elasticity of variance option pricing formula. J. Financ. **44**, 211–219 (1989)

Open-Loop-Data-Based Integer- and Non-integer-Order Model Identification Using Genetic Algorithm (GA)

Abdul Wahid Nasir and Arun Kumar Singh

Abstract Modeling of the process is very important aspect of engineering which helps us to understand the process behavior under different conditions. Also from control point of view, a good process model always proves to be vital in designing a good controller. Based on the order of the model, the process can be modeled into two categories, i.e., integer- and non-integer-order models. As non-integer modeling provides improved precision of the process model by offering more flexibility in model identification, a good number of researchers are utilizing this concept to obtain better results. Therefore, in the present work an attempt has been made to identify models for some processes based on its open-loop data. Therefore, for open-loop-data-based model identification, both integer- and non-integer-order models are estimated by minimizing the integral error criteria using genetic algorithm (GA). Comparative analysis ratifies that the non-integer model is able to capture process dynamics more accurately as compared to integer-order model.

Keywords Fractional calculus · Fractional-order modeling · Model identification · Genetic algorithm (GA)

1 Introduction

Model of the process serves as the basic building block for the research [1], in different domains of engineering. Model is needed for the development of solutions to problems associated with some phenomenon, generally carried out in simulation environment. Therefore, nowadays modeling of the process has become an integral part of any technologically driven program and also needed for its evolution. Hence, the modeling of the process is a very interesting topic and important as well.

A. W. Nasir (✉) · A. K. Singh
Electrical and Electronics Engineering Department, NIT Jamshedpur, Jamshedpur, India
e-mail: 2014rsee003@nitjsr.ac.in

A. K. Singh
e-mail: aksingh.ee@nitjsr.ac.in

J. Nayak et al. (eds.), *Soft Computing in Data Analytics*,
Advances in Intelligent Systems and Computing 758,
https://doi.org/10.1007/978-981-13-0514-6_4

With the increase in application of fractional calculus in past few decades in engineering and technology [2] due to the advent of different solutions pertaining to the problems involving such fractional operators, the modeling aspect has attained new dimension. Actually, the idea of fractional calculus is not a new concept and is as old as conventional integer calculus. It is just the generalization of order of differential operators from integer to non-integer, thus making it more flexible. After going through various literatures, it is observed that the term fractional order and non-integer conveys the same entity and used interchangeably, and so is the present case. It is also observed that non-integer operators are comparatively more efficient in modeling the dynamics of real-world systems [3–6]. At present, model is estimated from open-loop experimental data. To find out the optimal model parameters, genetic algorithm (GA) is employed in order to minimize integral error criterion. Equation (1) gives the definition of such non-integer differential, where α is the operator, 'a' is the lower limit of the operator, and 't' is the upper limit of the operator.

$$
{}_aD_t^\alpha = \begin{cases} d^\alpha/dt^\alpha & ,\alpha > 0 \\ 1 & ,\alpha = 0 \\ \int_a^t (d\tau)^{-\alpha} & ,\alpha < 0 \end{cases}. \tag{1}
$$

The rest of the paper is organized as follows. The brief introduction of fractional calculus is given in Sect. 2. Methodology adopted for modeling is discussed in Sect. 3. Section 4 gives the fundamentals of GA and its usage in model identification. The obtained results are explored in Sect. 5. Based on the results, conclusion is drawn in Sect. 6.

2 Brief Description of Fractional Calculus

During seventeenth century, Isaac Newton and Gottfried Wihelm Leibniz developed a very important tool in mathematics known as Calculus, which revolutionized the approach of dealing with engineering problems. During the same period, a thought triggered in the mind of L'Hospital for the order of differential operator that why it cannot be non-integer? At this very moment, fractional calculus came into existence. Different experts have given the definition of this fractional operator, among which three popular definitions are given below [7].

The Riemann–Liouville definition for non-integer derivative of function f(t) in the form of integral is given by (2),

$$
D^\alpha = \frac{d^m}{dt^m}\left[\frac{1}{\gamma(m-\alpha)}\int_0^t \frac{f(\tau)}{(t-\tau)^{\alpha-m+1}}d\tau\right] \tag{2}
$$

where $m - 1 < \alpha < m$, and m belongs to integer number.

Another common definition proposed by Caputo, which captures the initial state conditions but required integration of mth-order derivatives of function f(t), is given below,

$$D^{\alpha} = \frac{1}{\gamma(m-\alpha)} \int_{0}^{t} \frac{f^{(m)}(\tau)}{(t-\tau)^{\alpha-m+1}} d\tau \tag{3}$$

where $m - 1 < \alpha < m$, and m belongs to integer number.

The Grünwald–Letnikov definition is given below.

$$D^{\alpha} = \sum_{k=0}^{m} \frac{f^{(k)}(0^{+})t^{k-\alpha}}{\gamma(m+1-\alpha)} + \frac{1}{\gamma(m+1-\alpha)} \int_{0}^{t} (t-\tau)^{m-\alpha} f^{(m+1)}(\tau) d\tau \tag{4}$$

where $m > \alpha - 1$, and m belongs to integer number.

And $\gamma(m)$ is the gamma function involved in all three definitions of fractional derivative, i.e., (2), (3), and (4) for positive real m which is defined as given in (5).

$$\gamma(m) = \int_{0}^{\infty} t^{m-1} e^{-m} dt \tag{5}$$

For a single-input single-output (SISO) linear system, the differential equation relating the input x(t) and output y(t) can be defined as,

$$\begin{aligned} & a_n \frac{d^{\alpha_n} y(t)}{dt^{\alpha_n}} + a_{n-1} \frac{d^{\alpha_{n-1}} y(t)}{dt^{\alpha_{n-1}}} + \cdots + a_0 \frac{d^{\alpha_0} y(t)}{dt^{\alpha_0}} \\ & = b_m \frac{d^{\beta_m} x(t)}{dt^{\beta_m}} + b_{m-1} \frac{d^{\beta_{m-1}} x(t)}{dt^{\beta_{m-1}}} + \cdots + b_{m-1} \frac{d^{\beta_{m-1}} x(t)}{dt^{\beta_m}} \end{aligned} \tag{6}$$

The transfer function of the system represented by (6) is given in (7).

$$G(s) = \frac{Y(s)}{X(s)} = \frac{b_m s^{\beta_m} + b_{m-1} s^{\beta_{m-1}} + \cdots + b_0 s^{\beta_0}}{a_n s^{\alpha_n} + a_{n-1} s^{\alpha_{n-1}} + \cdots + a_0 s^{\alpha_0}} \tag{7}$$

where $(\alpha_n > \alpha_{n-1} > \cdots > \alpha_0 \geq 0) \in R$, $(\beta_m > \beta_{m-1} > \cdots > \beta_0 \geq 0) \in R$, and $a_p (p=0, 1, 2, \ldots, n), b_q (q=0, 1, 2, \ldots m)$ are constants.

3 Integer- and Non-integer-Order Modeling

The open-loop step response data are collected for two different linear processes from experimental setup which are given in Table 1 and Table 2, respectively, considering their input and output. For first process, PROCESS I, a step change of

Table 1 Open-loop data for PROCESS I

Time (in s)	Y (output)
0	0.00
0.1	1.07
0.2	2.30
0.3	3.39
0.4	3.67
0.5	106.92
0.6	279.41
0.7	381.79
0.8	438.52
0.9	478.50
1.0	505.40
1.1	524.81
1.2	537.43
1.3	547.26
1.4	553.91
1.5	557.58
1.6	561.89
1.7	564.00
1.8	567.24
1.9	569.14
2	569.14

Table 2 Open-loop data for PROCESS II

Time (in s)	Y (output)
0	32.00
76	32.51
152	33.02
228	34.53
304	35.53
380	36.55
456	37.05
532	38.57
608	39.58
684	40.09
760	41.60
836	42.11
912	43.13
988	44.13
1064	44.63
1140	45.14
1216	46.15
1292	46.65
1368	47.66
1444	48.17
1520	48.17

25 units is made, while for second process, PROCESS II, a step change of 20 units is made. Note that both the tables do not show all the data, but present the data trend throughout the input–output range due to space constraints. It is noticed that step response of these processes in open-loop configuration is very much similar to step response of first-order plus delay (FOPD) system. Hence from input–output data, the model parameters are optimally estimated with the help of genetic algorithm (GA). Model parameters associated with integer-order model are gain (K), time constant (τ), and delay (L) as given by Eq. (8). Also model parameters associated with non-integer-order model are gain (K), time constant (τ), delay (L), and order (α) as well given by Eq. (9). Estimation of model parameters based on available input–output data yields less modeling error while using non-integer model as compared to integer one, as the former is more flexible, thus fitting the data curve more accurately.

$$G_{integer} = \frac{K}{\tau s + 1} e^{-Ls} \tag{8}$$

$$G_{non-integer} = \frac{K}{\tau s^{\alpha} + 1} e^{-Ls} \tag{9}$$

4 Model Parameter Estimation Using GA

Genetic algorithm, a very effective and popular optimization algorithm, has been employed to find out the model parameters, whose fundamentals are nature-inspired. In 1975, Holland proposed this search technique which basically involves three processes similar to selection, crossover, and mutation [8]. After some time, in year 1985, David Goldberg further elaborated this concept [9]. GA can be used very efficiently for solving different types of problems, whether it be continuous or discrete, linear or nonlinear, or time-varying or time-invariant [10, 11]. Through the process of selection, the fittest individuals are identified to reproduce best offsprings. After that, the process of reproduction of new individuals takes place from the parents, known as crossover. And the process of mutation is required in maintaining diversity among populations which provides strength to algorithm in dealing with specific problems. Flowchart for the discussed GA is depicted in Fig. 1. Before initializing the search for parameters optimally, firstly a set of possible solution is considered, and also a fitness function or objective function is constructed in such manner that its fitness value depends on these chosen parameters. Here, integral of squared error (ISE) is taken objective function as given by Eq. (10), is being minimized to obtain the solution. Comparing the step response data with open-loop data gives the error signal needed for determining ISE that is required by GA at every simulation run as shown in Fig. 2. This process

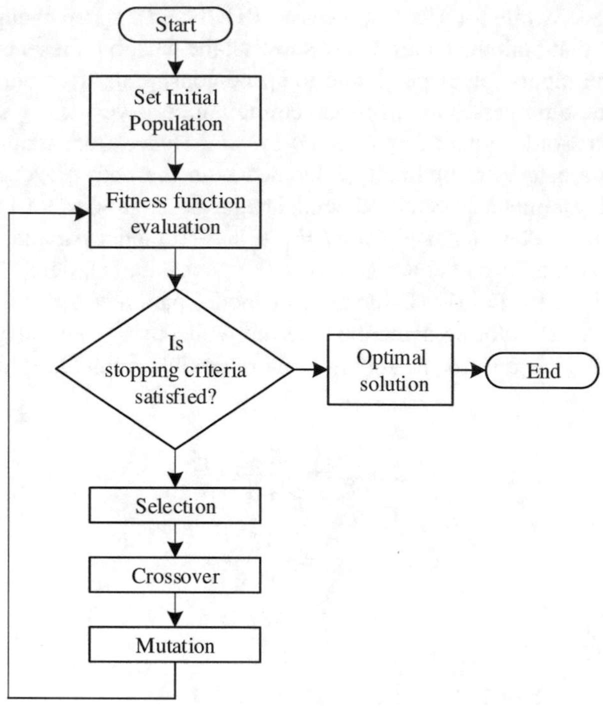

Fig. 1 GA flowchart

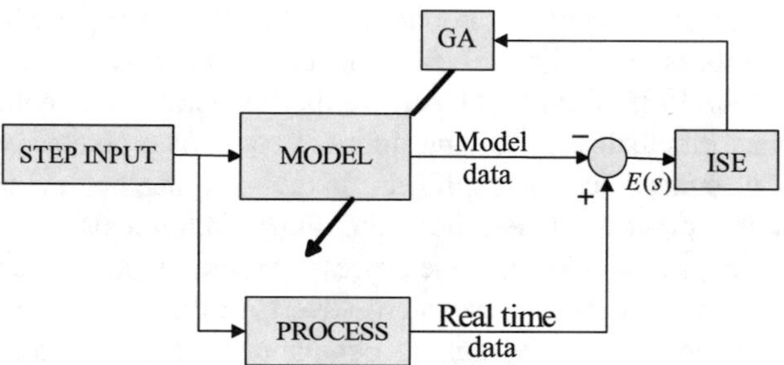

Fig. 2 Implementation of GA for model identification

keeps on running in loop until optimum minimum is attained. For the model parameter identification through GA, MATLAB optimization toolbox has been used. Also, FOMCON toolbox [12] has been used whenever simulation work regarding non-integer-order system is required in MATLAB.

Table 3 GA settings

Population size	50
Fitness scaling function	Rank
Crossover function	0.8
Crossover fraction	Scattered
Migration fraction	0.2
Ending criterion	Function tolerance of 1e−5

The settings for GA for the work carried out here are tabulated in Table 3.

$$ISE = \int_0^{\infty} \{y_{sp}(t) - y(t)\}^2 dt = \int_0^{\infty} \{e(t)\}^2 dt \tag{10}$$

5 Results and Discussions

As discussed in previous section both lower-order integer- and non-integer-order models are identified. Table 4 gives the parameters for integer and non-integer transfer function model along with their corresponding ISE values for PROCESS I. Table 5 gives the same for PROCESS II.

The open-loop step response of PROCESS I with its non-integer and integer FOPD model step response is presented in Fig. 3. Figure 4 gives the same for PROCESS II.

Tables 4 and 5 clearly indicate that the ISE value is less for non-integer-order models rather than integer order, validating the supremacy of non-integer-order modeling over integer order. It can be also inferred from Fig. 3 and Fig. 4 that the non-integer-order model response is closer to real system response for both the processes, respectively.

Table 4 Transfer function models for PROCESS I

Nature of model	Transfer function	ISE
Integer order	$\frac{22.58}{0.235s+1}e^{-0.347s}$	440
Non-integer order	$\frac{23.358}{0.25s^{0.91}+1}e^{0.364s}$	154

Table 5 Transfer function models for PROCESS II

Nature of model	Transfer function	ISE
Integer order	$\frac{1.114}{1170s+1}e^{-122s}$	690
Non-integer order	$\frac{1.122}{1313s^{1.021}+1}e^{-112s}$	230

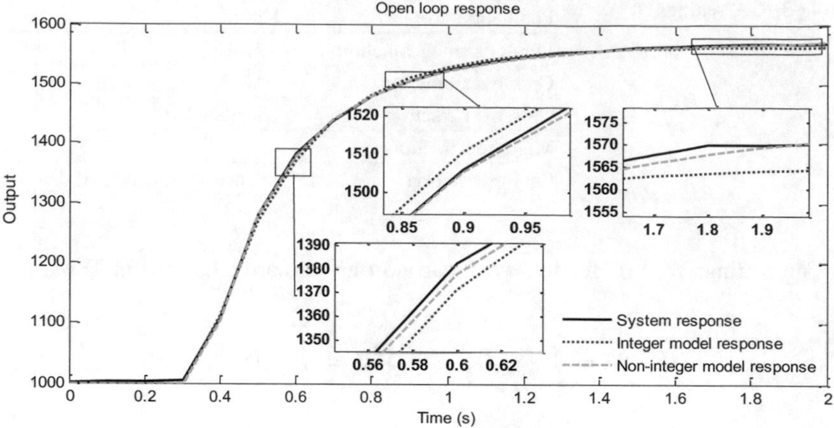

Fig. 3 System and model open-loop step responses for PROCESS I

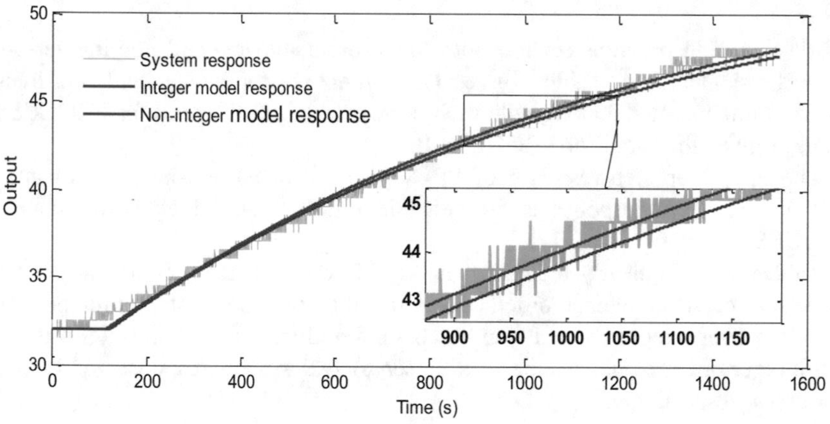

Fig. 4 System and model open-loop step responses for PROCESS II

6 Conclusion

In this paper, open-loop-data-based modeling for two different processes was carried out. The objective of present work is to establish the fact that non-integer-order model is more efficient than integer-order model in capturing the dynamics of real-world systems, which is well validated by the obtained results. The performance index, i.e., ISE in this case, is found to be lesser for non-integer-order model as compared to integer-order model.

References

1. Duka, V., Zeidmane, A.: Importance of mathematical modelling skills in engineering education for master and doctoral students of Latvia University of Agriculture. In: 2012 15th International Conference on Interactive Collaborative Learning (ICL), Villach, pp. 1–6 (2012)
2. Valério, D., Machado, J.T., Kiryakova, V.: Some pioneers of the application of fractional calculus. Fract. Calc. Appl. Anal. **17**(2), 552–578 (2014)
3. Tepljakov, A.: Fractional-order modeling and control of dynamic systems, Ph.D. Thesis, Springer International Publishing AG (2017)
4. Xu, B., Chen, D., Zhang, H., et al.: Dynamic analysis and modeling of a novel fractional-order hydro-turbine-generator unit. Nonlinear Dyn. **81**(3), 1263–1274 (2015)
5. Jalloul, A., Trigeassou, J.C., Jelassi, K., et al.: Fractional order modeling of rotor skin effect in induction machines. Nonlinear Dyn. **73**, 801–813 (2013)
6. Nasir, A.W., Kasireddy, I., Singh, A.K.: IMC based fractional order controller for three interactimg tank process. TELKOMNIKA **15**(4), 1723–1732 (2017)
7. Monje, C.A., Chen, Y., Vinagre, B.M., Xue, D., Feliu, V.: Fractional Order Systems and Controls: Fundamental and Applications, pp. 09–34. Springer, London (2010)
8. Holland, J.H.: Adaption in Natural & Artificial Systems. MIT Press, Cambridge MA (1975)
9. Goldberg, D.E.: Genetic Algorithms in search Optimization and Machine Learning. Addison-Wesley, Boston, MA (1989)
10. Oduguwa, V., Tiwari, A., Roy, R.: Genetic algorithm in process optimisation problems. In: Hoffmann, F., Köppen, M., Klawonn, F., Roy, R. (eds.) Soft Computing: Methodologies and Applications. Advances in Soft Computing, vol. 32. Springer, Berlin, Heidelberg (2005)
11. Langdon, W.B., Poli, R., McPhee, N.F., Koza, J.R.: Genetic programming: an introduction and tutorial, with a survey of techniques and applications. In: Fulcher, J., Jain, L.C. (eds.) Computational Intelligence: A Compendium. Studies in Computational Intelligence, vol. 115. Springer, Berlin, Heidelberg (2008)
12. Tepljakov, A., Petlenkov E., Belikov, J.: FOMCON: fractional-order modeling and control toolbox for MATLAB. In: Proceedings of the 18th International Conference Mixed Design of Integrated Circuits and Systems—MIXDES 2011, Gliwice, pp. 684–689 (2011)

An Intelligent Fault Locator for 400-kV Double-Circuit Line of Chhattisgarh State: A Comparative Study

V. Ashok, A. Yadav, C. C. Antony, K. K. Yadav, U. K. Yadav and Sandeep Kumar Sahu

Abstract The designing of fault location algorithms for double-circuit transmission lines is more complex due to the presence of mutual coupling. In this paper, a comparative study has been exemplified based on an intelligent fault location algorithm using illustrious feature extraction techniques such as discrete Fourier transform (DFT), discrete cosine transform (DCT) and discrete wavelet transform (DWT) for a 400-kV double-circuit transmission line of Chhattisgarh State in MATLAB/Simulink software. The proposed intelligent algorithm uses standard deviations of discrete coefficients of three-cycle data of three-phase current and voltage signals from relaying point to edify the hidden connotations of neural structures. Extensive simulation studies have been performed for all types of common shunt faults at widespread circumstances by varying different fault parameters in the presence of mutual coupling. However, the mean error in fault location estimation is within $\pm 1\%$; henceforth, the outcomes are worthy to get a research insight while designing relaying algorithms.

V. Ashok (✉) · A. Yadav · S. K. Sahu
Department of Electrical Engineering, NIT Raipur, Raipur 492010, CG, India
e-mail: ashokjntuk@gmail.com

A. Yadav
e-mail: anamikajugnu4@gmail.com

S. K. Sahu
e-mail: sndpsahu01@gmail.com

C. C. Antony
Technical Advisor Group, CSPTCL, Raipur 492099, CG, India
e-mail: antonycseb@gmail.com

K. K. Yadav
MRT Division, CSPTCL, 400 kV Sub-Station Bhilai-3, Bhilai 492013, CG, India
e-mail: yadavkarunesh@gmail.com

U. K. Yadav
Extra High Tension, C&M Division, CSPTCL, Raipur 492001, CG, India
e-mail: kantuma006@gmail.com

J. Nayak et al. (eds.), *Soft Computing in Data Analytics*,
Advances in Intelligent Systems and Computing 758,
https://doi.org/10.1007/978-981-13-0514-6_5

Keywords Intelligent fault locator · DFT-ANN · DWT-ANN DCT-ANN · Double-circuit line · Distance relay

1 Introduction

In recent days, the new-fangled intelligent techniques have come up to tackle various complex problems in power system protection. One of those renowned intelligent techniques is ANN, which is most dominant in dealing with pattern detection, cataloguing and clustering. These ANNs are more expedient to use in power system since they can be edified using recorded/collected data and adaptable at typical circumstances as it is having adequate learning/memorizing capability. By virtue of excellent attributes such as noise imperviousness, sturdiness and fault lenience, ANN will exhibit decisiveness at indeterminate conditions. Therefore, the decisiveness of ANN-based protective scheme will not be pretentious to inconsistencies of system constraints. Accordingly, different ANN-based algorithms were designed and employed for fault location estimation in power transmission system concurrently [1, 2]. An ANN-based faulty phase identification and fault location were described [3]. The generalization of fault resistance using neural network for guesstimate of fault location was reported in [4]. A wavelet fuzzy neural network-based single line–ground fault location method was employed in the distribution lines of an industrial system in [5]. An adaptive distance relaying scheme for a double-circuit transmission line expending zero-sequence impedance (Thevenin's equivalent) with appropriate restitution factor in the presence of mutual coupling is reported [6]. Back-propagation (BP) scheme centred on Levenberg–Marquardt learning algorithm has been employed to estimate fault location [7]. In [8], fault locator for SLG faults in double-circuit lines with ANN was described. An ANN-based SLG faults' classification in a double-circuit transmission line is employed in [9]. Fault classification and location using RBF neural network were exemplified [10]. A non-iterative fault location scheme using two-terminal unsynchronized data has been described in [11]. The distance protection scheme in a parallel transmission lines in case of ground fault has been investigated and compared with traditional double-circuit transmission line protection schemes in [12]. Innovative decision tree regression-based fault location estimation in case of shunt faults in double-circuit line has been narrated and tested with standard IEEE network model in [13]. Though, above specified all distance relaying schemes possessing certain merits and demerits convincingly while adapting to practical power system.

This paper presents a comparative study based on an intelligent fault algorithm using illustrative feature extraction techniques such as DFT, DCT and DWT, and further simulation studies have been performed on existing 400-kV double-circuit transmission line of Chhattisgarh State in MATLAB/Simulink software.

The proposed fault location scheme employed on three-phase voltage and current signals of double-circuit line of one-terminal data and comprising the influence of mutual coupling. The performance of different intelligent fault locators has been assessed at wide range of fault scenarios. The simulation outcome shows that all the fault locator modules can be correctly estimated fault location within one cycle from the inception of fault. As much as concerning about protection of practical power system, this comparative study has not been narrated previously for fault location which gives research insight to design and adapt efficient distance relaying schemes in double-circuit line (Tables 1, 2 and 3).

Table 1 Parameters used for generation of data set to edify fault locator modules

Parameter	Training	Testing
Fault type	LG: A1G, B1G, C1G, LLG: A1B1G, B1C1G, A1C1G, LL: A1B1, B1C1, A1C1, LLLG: A1B1C1G	LG: A1G LLG: A1B1G LL: A1B1 LLLG: A1B1C1G
Fault location (L_f)	(2–196)-km line in steps of 2 km	(1–197)-km line in steps of 2 km
Fault inception angle (Φ_i)	0°, 90° and 270°	0°, 90° and 270°
Fault resistance (R_f)	0, 50 and 100 Ω	0, 50 and 100 Ω
No. of fault cases	LG: 2646 LLG: 2646 LL: 2646 LLLG: 0882	LG: 891 LLG: 891 LL: 891 LLLG: 891

Table 2 Training architecture of different intelligent fault locator modules

Type	Module	Size of architecture	No. of epochs	Performance goal	Training time (min)
DFT	1	10-10-20-20-1	2591	9.995e-8	2:56
	2	10-10-10-10-10-1	10630	7.2407e-07	45:09
	3	10-10-10-10-1	4556	2.9210e-09	12:07
	4	10-10-10-1	4219	9.9955e-10	2:56
DWT	1	16-16-16-16-1	17727	1e-9	25:55
	2	18-18-18-18-1	7363	1e-8	20:04
	3	16-16-20-20-1	8458	1e-9	22:35
	4	12-12-12-1	3459	1e-8	00:44
DCT	1	16-16-16-16-1	4283	1e-8	06:34
	2	16-16-16-16-1	2697	1e-8	04:08
	3	16-16-18-18-1	4153	1e-8	07:45
	4	16-16-16-1	1279	9.98e-8	00:53

Table 3 Test results of different intelligent fault locator modules

Module	Fault type	R_f (Ω)	Φ_i (°)	Mean error (%)		
				DFT-ANN	DWT-ANN	DCT-ANN
1	A1G	0	0	0.0839	−0.9646	0.0552
		0	90	0.0319	−1.9760	-0.1584
		0	270	0.0320	−1.975	-0.1584
		50	0	0.1601	−0.2720	0.0359
		50	90	−0.0451	−0.8958	0.036494
		50	270	−0.0457	−0.9025	0.0364
		100	0	0.0388	−0.0708	0.0459
		100	90	−0.1098	−0.1183	0.0390
		100	270	−0.1110	−0.1148	0.03906
2	A1B1G	0	0	−0.0027	0.02050	−0.0195
		0	90	−0.0014	−0.34219	−0.1371
		0	270	−0.0013	−0.3433	−0.1370
		50	0	0.0022	0.1381	1.3923e-4
		50	90	0.0019	−0.0538	0.0013
		50	270	0.0016	−0.0517	0.0014
		100	0	−2.7250e-04	0.03077	-0.0044
		100	90	4.3588e-4	−0.1149	−3.9331e-4
		100	270	−1.3825e-04	−0.1158	−1.3145e-4
3	A1B1	0	0	−3.8791e-4	−0.3531	0.0091
		0	90	−9.9260e-4	0.0919	0.0090
		0	270	−0.0011	0.09216	0.0090
		50	0	−7.9797e-4	0.0595	−0.0219
		50	90	−0.0013	−0.0212	0.0147
		50	270	−0.0014	−0.0155	0.0146
		100	0	0.0029	0.005	0.0061
		100	90	1.4826e-4	0.0034	0.0035
		100	270	6.0940e-5	0.0051	0.0034
4	A1B1C1	0	0	−3.4474e-4	0.0969	−0.0177
		0	90	−6.002-4	0.0719	−0.0094
		0	270	−5.9889-4	0.0719	−0.0093
		50	0	0.0029	0.0043	0.0059
		50	90	0.0015	−0.0031	0.0063
		50	270	0.0014	−0.0023	0.0063
		100	0	−0.0011	3.6551e-4	0.0087
		100	90	−5.6927e-4	0.0021	0.0096
		100	270	−4.9600e-4	−0.0019	0.0097

2 Power System Network Under Study

A 400-kV, 50-Hz double-circuit transmission lines has been chosen from Chhattisgarh State transmission network which is shown in Fig. 1 with corresponding network parameters given in Appendix: Tables 4 and 5, and at bus-1 (KSTPS/NTPC) with two power plants (Unit-I with 4 × 500 MW and Unit-II with 3 × 210 MW), four double-circuit transmission lines (towards Vindyachal with 215 km, Raipur/PGCIL-B with 220 km, Bhilai/Khedamara with 198 km Birsinghpur with 231 km) and three single-circuit transmission lines (towards Korba-West with 17 km, Bhatpara with 100 km, Sipat with 60 km) and at bus-2 (Bhilai/Khedamara) with two triple circuit lines (from Raipur/Raita with 65.68 km and another towards Bhilai 220-kV grid bus), one double-circuit line (from KSTPS/NTPC 198 km) and seven single-circuit lines (from Korba-West with 212 km, Bhatpara with 90 km, towards Raipur/PGCIL-A with 20 km, Seoni with 250 km, Koradi with 272 km, Marwa with 170 km and Bhadrawati with 322 km). A 198-km double-circuit transmission lined between bus-1 and bus-2 (KSTPS/NTPC and Bhilai/Khedamara) which has been modelled in MATLAB/Simulink software. The proposed intelligent algorithm-based fault locators are employed at bus-1 (KSTPS/NTPC) and simulated different shunt faults at widespread circumstances.

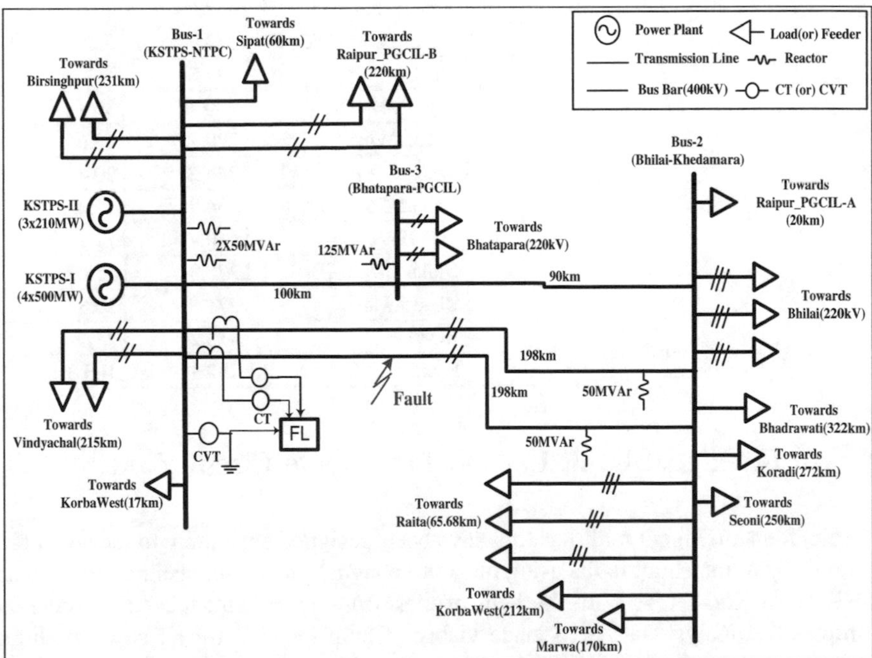

Fig. 1 Single-line diagram of 400-kV double-circuit line of Chhattisgarh State

Table 4 Power system network parameters

Generator			Transmission line	
Parameter	KSTPS-I	KSTPS-II	Parameter	(KSTPS to Khedamara)
Rating (MVA)	600	250	Length (km)	198
Xl (pu)	0.15	0.116	R^1 (pu)	0.0297
Ra (pu)	0.00192	0.0016	X^1 (pu)	0.332
Xd (pu)	2.31	1.7	$B^1/2$ (pu)	1.73e-6
Xq (pu)	2.19	1.62	R^0 (pu)	0.3237
Xd1 (pu)	0.267	0.256	X^0 (pu)	1.089
Xq1 (pu)	0.7	0.245	$B^0/2$ (pu)	1.26e-6
Xd11 (pu)	0.212	0.185	–	–
Xq11 (pu)	0.233	0.147	–	–
Td01 (s)	6.69	4.8	–	–
Tq01 (s)	2.5	0.5	–	–
Td011 (s)	0.038	0.0437	–	–
Tq011 (s)	0.05	0.141	–	–
H (s)	3	4.129	–	–

Table 5 Loads/feeders connected at bus-1 (KSTPS/NTPC) and bus-2 (Bhilai/Khedamara)

Bus. no	Load/Feeder	P (MW)	Q (MVAr)
1	Korba-West	166.7	5.1
1	Vindyachal (D/C)	467.5	27.6
1	Birsinghpur (D/C)	250.7	68.8
1	Sipat	134.8	4.2
1	PGCIL-B (D/C)	200.1	39.7
2	Korba-West	229.2	78.6
2	Marwa	244.2	20.4
2	Raipur/Raita (T/C)	134.7	15.2
2	PGCIL-A	420.7	72.1
2	Bhilai (T/C)	226.3	119.7
2	Bhadrawati	371.8	19.8
2	Koradi	473.2	27.7
2	Seoni	283.1	100.8

3 An Intelligent Fault Locator for Double-Circuit Line

Herein, four intelligent fault locators have been designed separately to locate all ten types of common shunt faults using illustrative signal processing techniques such as DFT, DWT and DCT. With the help of these intelligent fault locator modules, a comparative analysis has been made viably of fault location for all types of shunt faults by employing one-terminal data only and flowchart of the proposed fault location algorithm has been depicted in Fig. 2. The simulation studies have been

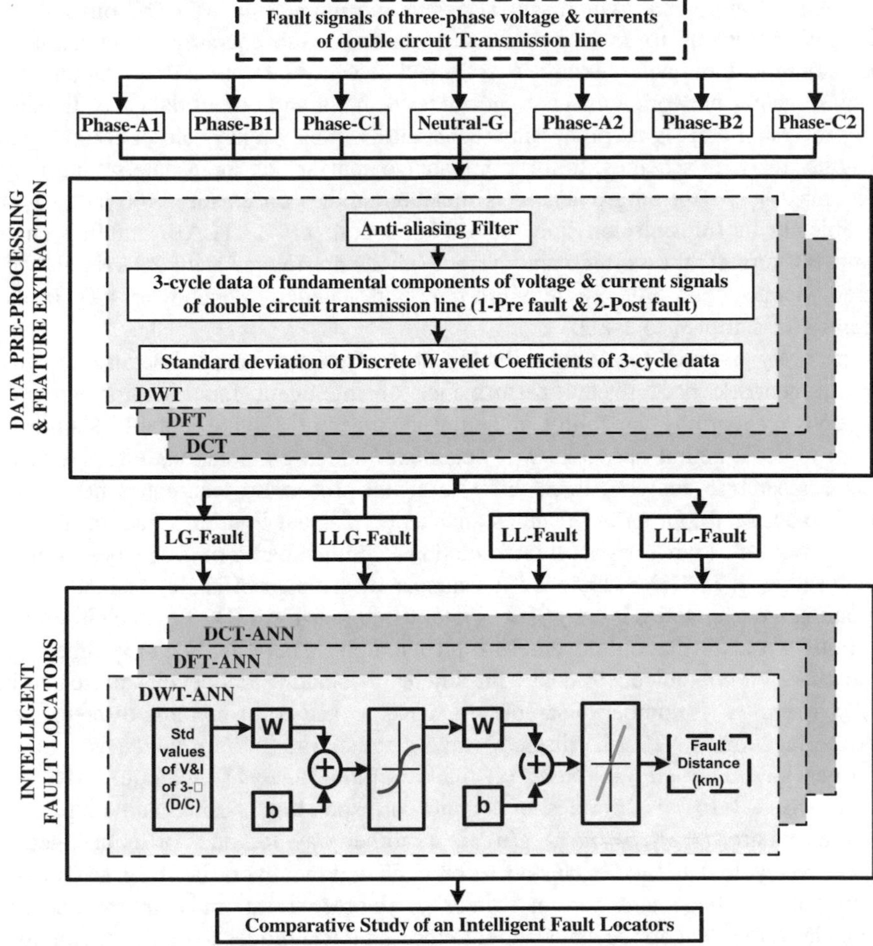

Fig. 2 Flowchart of the proposed fault location algorithm

conducted in MATLAB/Simulink environment at widespread circumstance to assess %error in fault location estimation.

3.1 Data Preprocessing and Design of Input/Output Data Set for Training/Testing of Intelligent Fault Locator Modules

Designing of an intelligent fault locator using different signal processing techniques is more concerned to preprocessing of faulty signal which was recorded at relaying point with different fault conditions. These faulty signals consist of miscellaneous frequency components which will also give the impression of a DC offset component conferring to the advancement of time. Subsequently, a numeral of non-sinusoidal frequency elements reformed at various faulty points. As the efficacy of neural network subjects to attributes of input and output data sets, the data preprocessing and signal processing techniques will play a key role in designing of distance relaying schemes. In this comparative analysis, an input data set has been generated by performing extensive simulation studies on existing 400-kV, 50-Hz double-circuit transmission line of 198 km length in MATLAB/Simulink environment; further, the voltage and current signals are logged with 20 kHz and filtered with a second-order Butterworth filter with an edge frequency of 400 Hz and then down sampled to 1 kHz.

It is very essential to extract suitable features from the recorded data set to edify neural network because the performance of intelligent fault locator modules depends on learning capability of neural architecture. Giving suitable feature as input to edify neural networks will determine hidden connotations effectively. In this comparative analysis, three different signal processing techniques have been used to abstract appropriate features such as (i) discrete Fourier transform (DFT), which is a well-known signal transformation technique based on frequency-domain analysis given in '(1),' where 'N' is number of samples of signal 'x' and corresponding Fourier series is 'X'; (ii) discrete cosine transform (DCT) which is similar to Fourier transformation; nonetheless, it contemplates periodicity and symmetry of real and even coefficients to deal with different oscillating frequencies presented in '(2),' where 'N' is number of samples of signal 'x' and corresponding discrete even cosine function is 'X'; (iii) discrete wavelet transform (DWT), which is a time–frequency-domain analysis based on multiresolution analysis techniques given in '(3),' where 'x(n)' is samples of signal, corresponding scaling and translation parameters are $a = a_0^m, b = na_0^m$, 'g(n)' is a mother wavelet, and 'm' is an integer. The three-cycle data (one-cycle data of pre-fault and two-cycle data of post-fault) of three-phase voltage and current signals of double-circuit transmission line are considered to adapt above-stated signal processing techniques to eradicate different oscillating components, and further standard deviations of appropriate coefficients are used to design an input data set to edify an intelligent fault locator module. Various parameters have been used to generate input/output data set for training and testing of fault locator modules which are reported in Table 1. Therefore, totally ten numbers of inputs are given to each fault locator module as presented in '(4).' There is one output corresponding to the input data set of fault locator module. Thus, the faulty phase location can be estimated as shown in '(5).'

$$DFT: X_k = \sum_{n=0}^{N-1} x_n \cdot e^{-i2\pi/N} \tag{1}$$

$$DCT: X_k = (1/2)(x_n + (-1)^k x_{N-1}) + \sum_{n=0}^{N-2} x_n \cdot Cos[\pi nk/(N-1)] \tag{2}$$

$$DWT: dwt[m,k] = (1/\sqrt{a_0^m}) \sum_n x(n) \cdot g\left[(k - na_0^m)/a_0^m\right] \quad (3)$$

$$X = [Ia1, Ib1, Ic1, Va1, Vb1, Vc1, Ig1, Ia2, Ib2, Ic2] \quad (4)$$

$$Y = [L] \quad (5)$$

3.2 Design of Training Architecture to Edify Fault Locator Modules

After design of input/output data sets with pragmatic features for training of different fault location modules, primary thing is designing of appropriate neural network structure by choosing the size of learning architecture arbitrarily and should be ensured for minimum training error. Since there is no empirical/standard notation available for choosing the size of learning architecture predominantly, the designing of training architecture is a heuristic method. So, the size of learning architecture was chosen by random experimentation of 1, 5, …, 50 neurons. Subsequently, the activation functions were applied such as tansig and purelin with the Levenberg–Marquardt (LM) learning algorithm to edify intelligent fault location modules separately. The hidden architecture and corresponding training parameters have been reported in Table 2. Thus, the total four intelligent fault location modules have been designed/trained to locate all common shunt faults using three illustrative signal processing techniques such as DFT-ANN, DCT-ANN, DWT-ANN and corresponding intelligent fault locator modules are module_1 for LG (AG, BG, CG) faults, module_2 for LLG (ABG, ACG, BCG) faults, module_3 for LL (AB, AC, BC) faults and module_4 for LLL (ABC/ABCG) faults. All ten types of shunt faults have been simulated in the three-phase double-circuit transmission line (between 0 and 100% of line length) at diverse fault locations by changing fault resistance (0, 50, 100) Ω and fault inception angles (0°, 90°, 270°) to ensure accuracy of intelligent fault locators.

4 Test Results of Different Intelligent Fault Locator Modules

Once the intelligent fault locators have been edified properly, it should be tested substantially with separate data sets comprising of widespread fault situations at different fault locations which have not been used earlier in edification. The input/test data set was designed as such by changing corresponding fault type, fault location, fault resistance and fault inception angle to examine the influence of fault parameters on efficacy of the proposed intelligent scheme. An edified neural network of different fault locator modules has been tested concurrently, and further comparative assessment has been done at wide range of fault scenarios, i.e., fault location (Lf = 1–197 km in steps of 2 km), fault resistances (R_f = 0, 50, 100 Ω) and fault inception angles (Φ_i = 0°, 90°, 270°). As exemplified in Table 3, outcomes of fault locator modules for different types of fault cases are properly located and mean error in percentage for 99 different locations throughout the line was reported. Then, estimated error for fault location 'E' is calculated in percentage with respect to the total line length L_t (6).

$$\%\mathrm{Error} = \left[\left(L_{f(\mathrm{Actual})} - L_{f(\mathrm{Estimated})}\right)/L_{t(\mathrm{Line\ Length})}\right] \times 100 \tag{6}$$

5 Conclusion

In this paper, a comparative study has been done using different intelligent fault locators which are employed with illustrative signal processing techniques such as DFT, DWT and DCT; standard deviation of discrete coefficients of three-cycle data of three-phase current and voltage signals of double-circuit transmission line is given as input to the fault locator modules to edify fault location of SLG faults, LLG faults, LL faults and LLL faults separately. The performance of the proposed fault location algorithm has been investigated at widespread fault scenarios in the presence of mutual coupling. The mean error in fault location estimation of different intelligent fault locators is within ±1% which is acceptable. This comprehensive analysis is noteworthy to get research insight for an engineers and researchers while designing distance relaying schemes for practical power system network.

Acknowledgements The authors acknowledge the financial support of Central Power Research Institute, Bangalore, for funding the project. no. RSOP/2016/TR/1/22032016, dated July 19, 2016. The authors are thankful to the Head of the Institution as well as Head of the Department of Electrical Engineering, National Institute of Technology Raipur, for providing the research facilities to carry this research project. The authors are grateful to the local power utility (Chhattisgarh State Power Transmission Company Limited) for their cooperation in providing valuable data to execute this research work.

Appendix

See Tables 4 and 5.

References

1. Mazon, A.J., et al.: A new method of fault location on double-circuit two-terminal transmission lines. Electr. Power Syst. Res. J. **35**(3), 213–219 (1995)
2. Aggarwal, R., Song, Y.: Artificial neural networks in power systems. III. Examples of applications in power systems. Power Eng. J. **12**(6), 279–287 (1998)
3. Jain, A, Kale, V.S., Thoke, A.S.: Application of artificial neural networks to transmission line faulty phase selection and fault distance location. In: Proceedings of the IASTED International Conference on Energy and Power System, vol. 262–267, pp. 526–803, Chiang Mai, Thailand, 29–31 March 2006
4. Khaparde, S.A., Warke, N., Agarwal, S.H.: An adaptive approach in distance protection using an artificial neural network. Electr. Power Syst. Res. **37**(1), 39–46 (1996)
5. Fan, C., Li, K.K., Chan, W.L., Yu, W., Zhang Z.: Application of wavelet fuzzy neural network in locating single line to ground fault (SLG) in distribution lines. Int. J. Electr. Power Energy Syst. **29**(6), 497–500 (2007)
6. Jongepier, A.G., Van Der Sluis, L.: Adaptive distance protection of double circuit lines using artificial neural networks. IEEE Trans. Power Deliv. **12**(1), 97–105 (1997)
7. Mazon, A.J., et al.: A new approach to fault location in two-terminal transmission lines using artificial neural networks. Electr. Power Syst. Res. J. **56**, 261–266 (2000)
8. Jain, A., Thoke, A.S., Patel, R.N.: Double circuit transmission line fault location using artificial neural network. CSVTU J. **2**(1), 40–45 (2009). (Bhillai, C.G. India)
9. Jain, A., Thoke, A.S., Patel, R.N.: Classification of single line to ground faults on double circuit transmission line using ANN. Int. J.Comput. Electr. Eng. (IJCEE) **1**(2), 199–205 (2009)
10. Mahanty, R.N., Gupta, P.B.D.: Application of RBF neural network to fault classification and location in transmission lines. IEE Proc. Gen. Trans. Distr. **151**(2), 201–212 (2004)
11. Izykowski, J., Rosolowski, E., Balcerek, P., Fulczyk, M., Saha, M.M.: Accurate noniterative fault-location algorithm utilizing two-end unsynchronized measurements. IEEE Trans. Power Deliv. **26**(2), 547–555 (2011)
12. Jia, K., Tianshu, B., Li, W., Yang, Q.: Ground fault distance protection for paralleled transmission lines. IEEE Trans. Ind. Appl. **51**(6), 110–119 (2015)
13. Swetapadma, A., Yadav, A.: A novel decision tree regression-based fault distance estimation scheme for transmission lines. IEEE Trans. Power Deliv. **32**(1), 234–245 (2017)

Comparison of SCA-Optimized PID and P&O-Based MPPT for an Off-grid Fuel Cell System

Shashikant and Binod Shaw

Abstract This paper presents a simulation and modelling of a fuel cell system to extract maximum power from an array for an off-grid stand-alone system. A stand-alone system consists of fuel cell stack/array, DC–DC boost converter and a load. A DC–DC boost converter is needed to boost the voltage as per the requirement or application. Sine Cosine Algorithm is adopted to enhance the output of the system and to find out the controller gain parameters. SCA-optimized PID-based MPPT controller is validated over conventional P&O-based MPPT scheme, to regulate the pulse width of the DC–DC boost converter to enhance the output power, voltage and current. The gain parameters of the PID controller are tuned/selected in such that it gives best result, since gain parameters highly influence the system performance. The model is simulated in MATLAB 2015a, and the obtained result is validated that the SCA-optimized PID controller is better than the conventional P&O based MPPT.

Keywords Perturbation and optimization (P&O) · Sine Cosine Algorithm (SCA) Maximum power point tracking (MPPT) · Proportional integral and derivative controller (PID) · Proton exchange membrane fuel cell (PEMFC)

1 Introduction

Recent years show a great advancement in fuel cell such as development in vehicles, fast battery charging, electricity generation to meet peak load demands, unmanned aerial vehicles (UAVs). Fuel cell is used as a distributed generation in power system. There are mainly three categories in which fuel cell is used; they are

Shashikant (✉) · B. Shaw
Department of Electrical Engineering, National Institute of Technology Raipur, Raipur 492010, India
e-mail: shashikantkaushaley20@gmail.com

B. Shaw
e-mail: binodshaw2000@gmail.com

J. Nayak et al. (eds.), *Soft Computing in Data Analytics*, Advances in Intelligent Systems and Computing 758,
https://doi.org/10.1007/978-981-13-0514-6_6

transportation, generation and stationary. This paper concerns about generation only. The main objective is to develop an isolated system so that we can save conventional source of energy and its resources. The basic concepts of generation of electricity in fuel cell and its types are illustrated in [1]. The development of fuel cell is needed in off-grid connection. The main area to focus for development is MPPT. The different types of algorithms such as voltage- and current-based MPPT, resistance matching algorithm-based MPPT are illustrated in [2, 3]. Comparison of different types of optimization techniques is illustrated in [4]. Tuning of PID controller to extract maximum power point tracking is illustrated in [5]. Further development in fuel cell energy is illustrated in [6]. History of fuel cell and its development are illustrated in [7]. To enhance the capability of fuel cell, a boost converter is applied to the output of fuel cell. The collaboration of fuel cell and boost converter is illustrated in [8]. To extract MPPT with optimization technique is one of the powerful tools. Researchers show the development of algorithm for fuel cell in [9]. The researchers did not stop in isolated system only, they further went to grid-connected system which is illustrated in [10, 11].

In this present work, Fig. 1 illustrates the system description of 1.3 KW capacity in which MPPT block is mainly concerned. A recently developed optimization technique, Sine Cosine Algorithm (SCA) is applied to PID controller to find its gain parameters to extract maximum power from the fuel cell. PWM technique is used to give proper pulses to the gate of chopper. The output of conventional method such as P&O is compared with SCA-optimized PID controller. The result shows that there is an enhancement in the output power and output voltage.

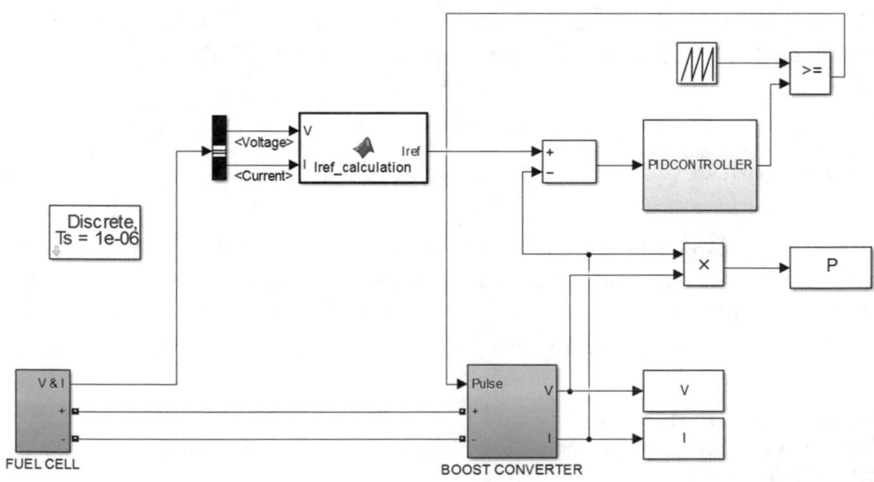

Fig. 1 Simulink model of PID controller-based fuel cell

2 System Description

The Simulink model of fuel cell module is shown in Fig. 1, and the proposed fuel cell model consisting of a fuel cell, DC–DC boost converter, MPPT controller are shown in Fig. 2. Boost converter boosts the voltage up to a certain level which depends on switching, i.e. ON and OFF period. The time ON/OFF can be controlled through gate signal. This gate control is achieved by MPPT controller. The advancement of MPPT controller is that it enhances the system efficiency. The calculation of boost converter parameter is illustrated in [4], and its parameters are given in Table 1.

3 PID-Based MPPT Algorithm

The term PID controller refers to proportional, integrator and derivative controller which is a feedback control loop mechanism and is widely used in industries. It generally calculates an error value *e(t)* continuously. The difference of desired value and processed value is known as error value. PID controller gives desired result in an optimized way, without overshoot and time delay; i.e. its response is fast. Hence, it is applicable for automatic control. The block diagram of PID controller is shown in Fig. 3. The proportional gain Kp deals with present values of error; if error is large, then its control output will be proportionately large. The integral gain Ki with integrator integrates its past values of error over time. Due to this integral term, it eliminates error by control output, which is based on its present cumulative error

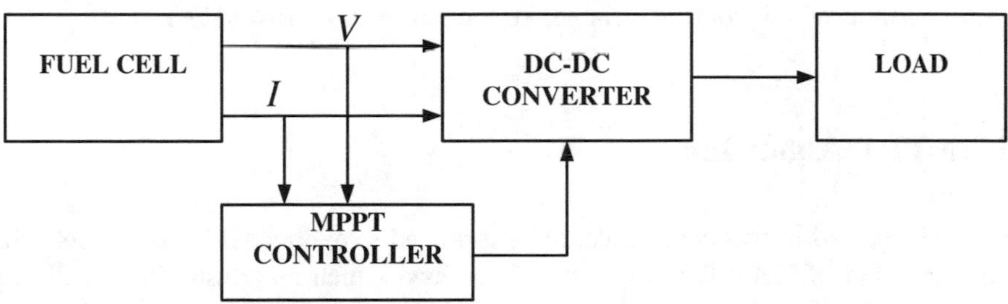

Fig. 2 Block diagram of fuel cell

Table 1 Boost converter parameters

Model components	Parameters
Inductance, L	100 mH
Capacitance, C	3300 μF
Load, R	25 Ω
DC voltage, V_{dc}	24 V
Switching frequency	10,000 Hz

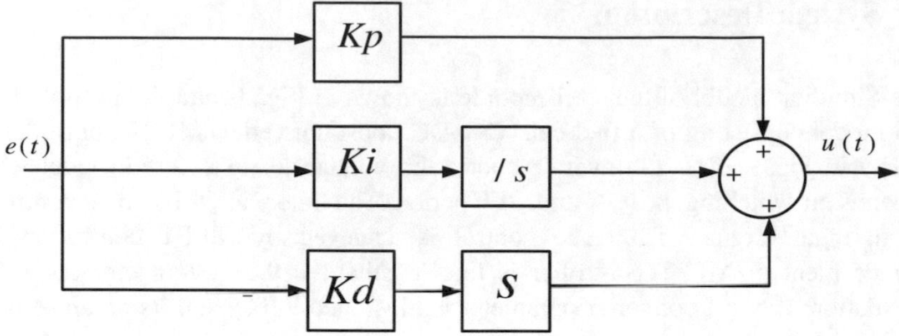

Fig. 3 Block diagram of PID controller

value. The derivative gain with differentiator eliminates error by generating control signal proportional to the rapid change of current. It can be expressed mathematically as Eq. (1).

$$u(t) = Kp \cdot e(t) + Ki \int_0^t e(t) \cdot dt + Kd \frac{d}{dt} e(t) \tag{1}$$

Its equivalent Laplace transform in Laplace domain can be written as in Eq. (2).

$$U(s) = KpE(s) + \frac{Ki}{S} \cdot E(s) + Kd \cdot S \cdot E(s) \tag{2}$$

where

Kp = proportional gain constant, Ki = integral gain constant, Kd = differential gain constant, u(t) = control variable, i.e. output, e(t) = error value.

4 MPPT Algorithm

Power is limited in fuel cell; hence, there is a need to extract MPPT from fuel cell. The operation of fuel cell mainly depends on load which may result in a nonlinear operation of fuel cell. Due to this dynamic behaviour of load, fuel cell operation changes dynamically which might damage the fuel cell stack or equipment connected to the load, in near future. Hence, a proper control algorithm such as an optimization technique is used to solve this problem.

In this present work, SCA algorithm is used to properly optimize the fuel cell operating parameter. The concept of MPPT arise from solar energy, in which MPPT is used to control the flow of charge to charge the batteries. In the same way, MPPT is used in fuel cell, to control the output current. If there is a variation in load,

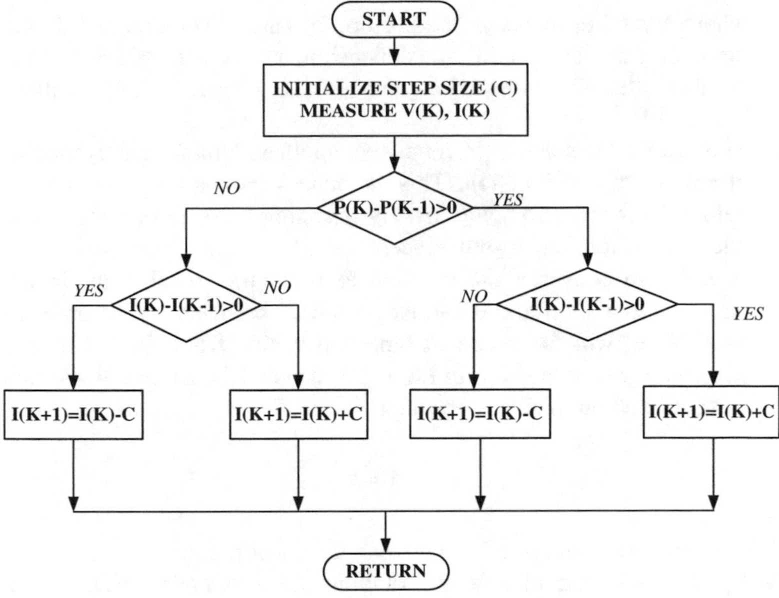

Fig. 4 Flow chart of MPPT algorithm

then the output voltage changes. Figure 4 represents the flow chart of MPPT algorithm. V(K), I(K) and P(k) are the present values of voltage, current and power. V(K − 1), I(K − 1) and P(K − 1) are the past values of voltage, current and power. C is the step size which depends on the user. In this present work, C is taken as 0.1.

5 Sine Cosine Algorithm

A single objective function is used to optimize the problem. The objective is to minimize the error by using ITAE as an objective function. One of the recent algorithms is Sine Cosine Algorithm (SCA), which is a mathematical model of sine and cosine functions. SCA involves several steps which are illustrated as:

Step 1 Initialize number of population and number of variables which are your design parameters. In this paper, number of population is taken as 50 and number of variables as 3.

Step 2 It involves finding the promising region for search space which involves exploration and exploitation which is formulated as an equation:

$$X_i^{t+1} = \begin{Bmatrix} x_i^t + r_1 \times \sin(r_2) \times \left| r_3 P_i^t - X_i^t \right| \\ x_i^t + r_1 \times \cos(r_2) \times \left| r_3 P_i^t - X_{ii}^t \right| \end{Bmatrix} \quad (3)$$

where X_i^t and P_i represent the position of current solution and destination point of j-th iteration in i-th dimension; r1, r2 and r3 are the random numbers; the effects of random number r_i for (i = 1, 2, 3, 4) are illustrated in [11, 12].

This allows the solution to reposition another solution in a cyclic position of sine or cosine function. This guarantees the exploitation of the space defined between two solutions. By changing the sine or cosine function, one can explore the search space.

Step 3 In order to converge the solution to optimum global point in feasible region, exploration and exploitation should be balanced. In order to balance these two, the range of sine and cosine from Eq. (3) is changed adaptively which is given in Eq. (4) in which T is maximum iteration, t is current iteration, and a is constant.

$$r1 = a - t\frac{a}{T} \tag{4}$$

Step 4 Save the best values and assign as a destination point.

Step 5 Update the result of another solution and also update sine and cosine ranges so that the search space increases for exploitation as per the iteration increases.

Application of SCA is also introduced in soft computing neural networks which is illustrated in [13], and some other applications are illustrated in [14, 15].

6 Result and Discussion

The simulation model of fuel cell as shown in Fig. 1 is simulated in MATLAB 2015a. A conventional P&O method is compared with SCA-optimized PID controller, and the corresponding voltage, current and power waveforms are obtained which are shown in Figs. 5a, b, c.

It is seen that SCA-optimized PID controller shows better result as compared with conventional PID controller. Conventional PID controller gives output power around 343 W, whereas SCA-optimized PID controller gives 1.09 KW for a system rating if 1.3 KW at 24 V DC. The comparison of conventional P&O and SCA-optimized PID controller is given in Table 2.

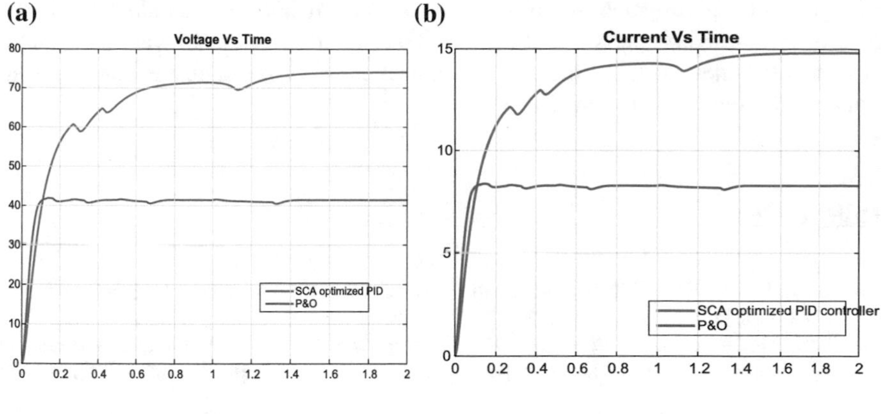

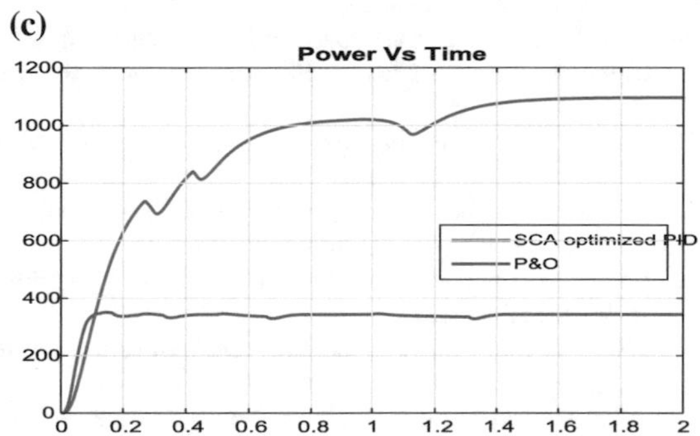

Fig. 5 Comparison of output waveform of P&O with SCA-optimized PID controller for **a** voltage and current **b** current versus time and **c** power versus time

Table 2 Comparison of P&O and SCA-optimized PID controller

Comparison	Voltage (V)	Current (A)	Power (W)
Conventional P&O	41.41	8.25	342.9
SCA-optimized PID controller	74.03	14.76	1096

7 Conclusion

The PEMFC 1.3 KW, 24 V DC is simulated in MATLAB 2015a system to extract maximum power from an array for an off-grid stand-alone system. A stand-alone system consists of fuel cell stack/array, DC–DC boost converter and a load. A DC–DC boost converter is needed to boost the voltage as per the requirement or application. Sine Cosine Algorithm is adopted to enhance the output of the system. As from the output waveform and result discussion, it is noted that the conventional P&O algorithm gives an output power of 342.9 W, whereas the SCA-optimized

PID controller gives 1096 W output. To get better result, gain parameters of PID controller is tuned properly to get enhanced output. The results are validated, and it is concluded that SCA-optimized PID controller gives better result than conventional P&O algorithm for MPPT.

References

1. Huria, T.: Dy. Chief Mechanical Engineer, DMU Care Base, Northern Railway, Jalandhar City Introduction. Fuel Cells (1958)
2. Sarvi, M., Soltani, I.: Voltage and current based MPPT of fuel cells for fuel consumption optimization and mismatching compensation. Tech. J. Eng. Appl. Sci. ©2013 TJEAS J **3** (2013)
3. Anbarasan, M., Latha, K.: Resistance matching algorithm for MPPT of fuel cell system. Int. J. Adv. Res. Electr. Electron. Instrum. Energy, 32–40 (2014)
4. Mussetta, M., Grimaccia, F., Zich, R.E.: Comparison of different optimization techniques in the design of electromagnetic devices. In: 2012 IEEE Congress on Evolutionary Computation, pp. 1–6 (2012)
5. Derbeli, M., et al.: Control of proton exchange membrane fuel cell (PEMFC) power system using PI controller
6. Derbeli, M., Sbita, L., Farhat, M., Unis, E.A., Barambones, O., De Ingenieria, E.U.: PEM fuel cell green energy generation—SMC efficiency optimization
7. Perry, M.L., Fuller, T.F.: A historical perspective of fuel cell technology in the 20th century. J. Electrochem. Soc. **149**(7), S59 (2002)
8. Tourkia, L.: Interaction between fuel cell and DC/ DC converter for renewable energy management (2017)
9. Derbeli, M., Sbita, L., Farhat, M., Unis, E.A., Barambones, O., De Ingenieria, E.U.: Proton exchange membrane fuel cell–a smart drive algorithm
10. Giustiniani, A., Petrone, G., Spagnuolo, G., Vitelli, M.: Low-frequency current oscillations and maximum power point tracking in grid-connected fuel-cell-based systems. IEEE Trans. Ind. Electron. **57**(6), 2042–2053 (2010)
11. Meshkat, M., Parhizgar, M.: A novel weighted update position mechanism to improve the performance of sine cosine algorithm. In: 2017 5th Iranian Joint Congress on Fuzzy and Intelligent Systems, pp. 166–171 (2017)
12. Mirjalili, S.: SCA: a sine cosine algorithm for solving optimization problems. Knowl. Based Syst. **96**, 120–133 (2016)
13. Sig, I.: Training feedforward neural networks using sine-cosine algorithm to improve the prediction of liver enzymes on fish farmed on nano-selenite, Techncal Report, vol. 2009, no. 5, pp. 5–8 (2009)
14. Banerjee, A., Nabi, M.: Re-entry trajectory optimization for space shuttle using sine-cosine algorithm. In: 2017 8th International Conference Recent Advance Space Technology, pp. 73–77 (2017)
15. Hafez, A.I., Zawbaa, H.M., Emary, E., Hassanien, A.E.: Sine cosine optimization algorithm for feature selection. In: International Symposium on Innovations in Intelligent Systems and Applications, pp. 1–5 (2016)

Implementation of Quasi-Oppositional-Based GHS Optimized Fractional Order PID Controller in Deregulated Power System

Jyoti Ranjan Nayak and Binod Shaw

Abstract This work is a persuasive exertion to contemplate the demeanor of AGC in an interconnected thermal power system in deregulated scenario by conceding the constraints of generation. The constraints considered in the system are GRC and boiler dynamics. ITAE is adopted as objective function of this system by conceding the frequency and tie-line power deviation. The basic aim is to lessen the area of the deviations of the system. Fractional order PID (FOPID) controller is validated to enhance the performance or to attain the lesser functional value. The performance is hugely influenced by gain parameters of the controller. group hunt search (GHS) and quasi-oppositional-based GHS (QOGHS) algorithms are adopted for the purpose to extract relevant pair of gain parameters of FOPID controller. The analysis to validate the controller and optimization algorithm is executed by implementing a load divergence 0.01 p.u. in area-1. Finally, QOGHS optimized FOPID controller is concluded as a better controller for this system in deregulated scenario over GHS optimized FOPID controller.

Keywords Automatic generation control (AGC) · Proportional–integral–derivative controller (PID) · Fractional order PID (FOPID) · Group hunting search (GHS) · Quasi-oppositional-based GHS (QOGHS)

1 Introduction

In power system, the basic objective is to counterbalance the generated power and demand power comprising power loss. In interconnected power system, diversity of load demand in one area may cause frequency and tie-line power deviations in that

J. R. Nayak (✉) · B. Shaw
Department of Electrical Engineering, National Institute of Technology Raipur, Raipur 492010, India
e-mail: bapi.jyoti.2@gmail.com

B. Shaw
e-mail: binodshaw2000@gmail.com

J. Nayak et al. (eds.), *Soft Computing in Data Analytics*,
Advances in Intelligent Systems and Computing 758,
https://doi.org/10.1007/978-981-13-0514-6_7

area and may spread over other areas. This may cause the power system unstable. Secondary control scheme (AGC) achieves a fast and stable control over huge and long duration deviations in frequency and tie-line power. The fast response of AGC enhances the capability of the system to handle continuous deviation of load [1, 2].

In deregulated power system, ISO regulates the open electricity market. DISCOs, GENCOs, and TRANSCOs are three individual imperative players in the electricity market. DISCOs have freedom to get contracts with GENCOs in any areas. AGC control scheme is an imperative approach in deregulated power system scenario. The basic objectives of the competitive electricity markct provided by deregulation power system are as follows

i. Economically affordable.
ii. Efficient to provide emergency support from other areas at critical situation.
iii. Freedom of DISCOs to make contract with GENCOs.

Many concepts to enhance the ability of AGC have been proposed by many authors from last few decades. Conventional PID controller is validated over I and PI controllers optimized by ICA algorithm as illustrated in [3]. The cascade combination of PI and PD controllers is adopted as inner and outer controller loop in multiarea power system. The cascade PI–PD controller is validated as better controller over conventional PID controller and the parameters of the controller are tuned by FPA to enhance the performance of the controller in [4]. The superiority of Fuzzy-PID controller optimized by various algorithms and hybrid algorithms over PID as AGC is validated in [5–9]. Fractional order PID controller (FOPID), tilted integral derivative controller (TID), and fuzzy-FOPID controller optimized by different algorithms are adopted as AGC in [10–12].

The basic purpose of this paper is to design AGC for two area thermal power system in restructured environment with generation rate constraint (GRC) with saturation limit of ±0.05. FOPID controller optimized by GHS and QOGHS algorithms is adopted as controller in the system to minimize the objective function by concerning frequency and power deviations.

2 System Investigated

The proposed system is a two area interconnected thermal power system with 5% (±0.05) generation rate constraint (GRC) with two GENCOs (GENCO1 and GENCO2) and two DISCOs (DISCO1 and DISCO2) as portrayed in Fig. 1. This paper is validated by concerning three case studies. The elements of DISCO

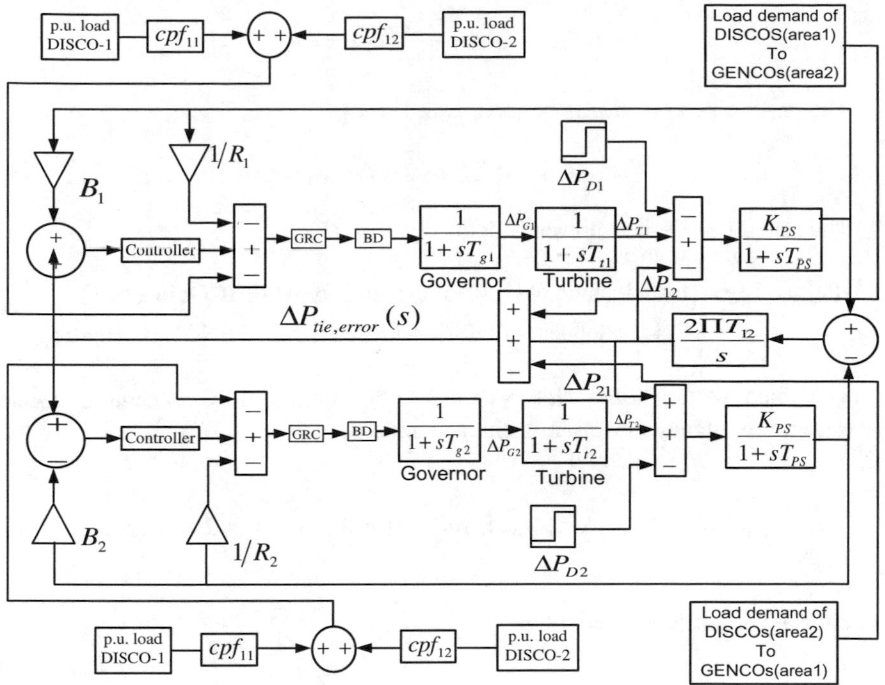

Fig. 1 Power system model in deregulated environment [7]

participation matrix (DPM) are the contract participation factors (cpf_{ij}) between ith GENCO and jth DISCO. DPM may be characterized as

$$DPM = \begin{bmatrix} cpf_{11} & cpf_{12} \\ cpf_{21} & cpf_{22} \end{bmatrix}$$

cpf_{ij} correlates to the factor of which $GENCO_i$ is contracted with $DISCO_j$ to supply power. Total power demand from a particular GENCO is 1 p.u., i.e., $\sum_i cpf_{ij} = 1$. ACEs in both the areas are defined in Eqs. (1) and (2).

$$ACE_1 = B_1 \Delta f_1 + \Delta P_{tie,\,error} \tag{1}$$

$$ACE_2 = B_2 \Delta f_2 + \Delta P_{tie,\,error} \tag{2}$$

Where B_1 and B_2 are the bias factors of frequency. The deviations of frequency with respect to nominal value in area-1 and area-2 are Δf_1 and Δf_2, respectively. The deviation of power in tie-line is entitled as ΔP_{tie}, actual and is characterized as in Eq. (3)

$$\Delta P_{tie,\,actual} = \frac{2\pi T_{12}}{s}(\Delta f_1 - \Delta f_2) \tag{3}$$

The tie-line power error is defined as in Eq. (4).

$$\Delta P_{tie,\,error} = \Delta P_{tie,\,actual} - \Delta P_{tie,\,scheduled} \tag{4}$$

The scheduled tie-line power is defined in Eq. (5)

$$\begin{aligned}\Delta \mathrm{P}_{\mathrm{tie,\,scheduled}} = &\ (\text{load demand of DISCOs in area 2 from GENCOs in area 1}) \\ &- (\text{load demand of DISCOs in area 1 from GENCOs in area 2})\end{aligned} \tag{5}$$

The objective function of this system is to be minimized by concerning tie-line power and frequency deviation is characterized as

$$ITAE = \int_0^T t(\Delta f_1 + \Delta f_2 + \Delta P_{tie,\,error}),\ \text{Subject to: } 0.001 \le K_i \le 2\ i = 1, 2, \ldots, n \tag{6}$$

where n is the designed variables.

3 Controller Structure

The performance of the system mostly relies upon the controller design. Picking up the appropriate pair of gain parameters of controllers is also very significant factor.

FOPID Controller

Fractional order PID controller is a novel approach recommended from the fractional calculus. The orders of the integration and differentiation (λ and μ) are fractional values. λ and μ values may not be integer. The transfer function of the FOPID controller is described in Eq. (7).

$$G_C(s) = K_P + \frac{K_I}{s^\lambda} + s^\mu K_D \tag{7}$$

Due to fractional order, it has supremacy control over PID controller to maintain stability of the system.

4 Quasi-Oppositional-Based Group Hunting Search Algorithm (QOGHS)

In this paper, GHS and QOGHS algorithms are adopted to tune the parameters of FOPID controller individually to validate the performance of FOPID controller. The basic purpose of optimization is to lessen ITAE by hunting the parameters of controllers within the specified limit as defined in Eq. (6).

The relation between predator (group hunters, i.e., lions, wolves.) and prey is beautifully expressed as optimization technique by Oftadeh et al. [13]. GHS algorithm is derived from the strategy of hunting a prey by concerning the group hunting technique. Unity of group members adopts an approach to trap the prey by circumscribing it. The member of the group near to the prey is adopted as leader and all other member follows leader to move toward the prey (optimum solution). If any of the group member amends by a better position compared to the recent leader then it becomes leader in the next generation. The opposition of the current solution is established by the mirror position of current population from the center of the search space. The stride of the QOGHS is as

1. Initialize the group of hunters of size $X_{[NP \times D]}$ within the limit 0.001–2.
2. Evaluate the opposite number of the initial vector as follows

$$M_j = \frac{para^{\min} + para^{\max}}{2}$$

M_j is the center of the search space. The opposite number of initial vector (X) is defined as

$$OX = para^{\min} + para^{\max} - X$$

Where $para^{min}$ and $para^{max}$ are the minimum and maximum of the search space. The quasi-oppositional number may be characterized in Eq. (8).

$$QOX = M + rand \times (M - X) \tag{8}$$

3. The best NP hunters among QOX are adopted as hunters and the worst hunters are wiped out. The best fitted hunter among the group is adopted as leader.
4. The hunter's positions are refurbished toward the leader. The mathematical expression is defined in Eq. (9).

$$\begin{aligned} X_i^{k+1} &= X_i^k + rand \times MML \times (X_i^L - X_i^k) \\ MML &= 0.6 - \left(it \times \left(\frac{0.6}{iter\max} \right) \right) \end{aligned} \tag{9}$$

Where 'it' is the current iteration, 'itermax' is the maximum iterations and X_i^L is the position of leader.

5. The position of hunters is corrected by concerning hunter's group consideration rate (HGCR = 0.3) as described in Eq. (10) and distance radius (0.0001 ≤ R_a ≤ 1 × 10^{-6}) are represented in Eq. (11).

$$X_i^{k+1} = \begin{cases} X_i^{k+1} \in \{X_i^1, X_i^2, \ldots, X_i^{HGS}\} \; with \quad probability \quad HGCR \\ X_i^{k+1} \pm R_a \, with \quad probability \quad (1 - HGCR) \end{cases} \tag{10}$$

$$Ra(it) = Ra_{\min}(\max(X_i) - \min(X_i)) \exp\left(\frac{\ln\left(\frac{Ra_{\min}}{Ra_{\max}}\right) \times it}{iter\max}\right) \tag{11}$$

Ra is an exponential decay function and may be defined as in Eq. (11).

6. Identify the group to avoid the algorithm to be trapped into local optima. It may be characterized in Eq. (12).

$$X_i^{k+1} = X_i^L \pm rand(\max(X_i) - \min(X_i)) \times \alpha \exp(-\beta \times EN) \tag{12}$$

Where EN is the numbers of epochs. EN is estimated by matching the difference of leader and worst hunter with a small value.

7. Repeat steps 3–6 up to termination criteria satisfied. In this problem, maximum iteration (100) is treated as termination criteria.

5 Results and Discussion

FOPID controller is implemented in both areas individually. QOGHS and GHS algorithms are executed with 60 numbers of hunters for 100 iterations to tune the controller parameters by concerning ITAE as an objective function. The numbers of parameters to be tuned by QOGHS and GHS algorithm of FOPID controllers are 10. The above parameters are within a specified perimeter of 0.001–2. The simulation study is executed with three case studies base case, bilateral transaction case, and contract violation case. The optimal gain parameters of the controllers for both the case studies are tabulated in Table 1.

Case 1 Base case:

In this scenario, the load demand of DISCOs in area-1 is fulfilled by the GENCOs in area-1. The DISCOs in one area is contracted with the GENCOs of same area. In this system, DISCO-1 is in contract with GENCO-1. The DPM matrix for this scenario is $DPM = \begin{bmatrix} 1 & 0 \\ 0 & 0 \end{bmatrix}$

Table 1 Gain parameters of the FOPID controller optimized by GHS and QOGHS algorithm

Gains of FOPID controller												
Base case					Bilateral transaction case				Contract violation			
	Area-1		Area-2		Area-1		Area-2		Area-1		Area-2	
	GHS	QOGHS	GHS	QOGHS	GHS	QOGHS	GHS	QOGHS	GHS	QOGHS	GHS	QOGHS
K_P	1.6520	1.7291	0.9312	1.3099	0.5857	0.0255	2.0000	0.0100	0.9349	0.1515	1.1664	2.0000
K_D	0.7346	1.3816	0.8627	1.0262	0.5713	0.3061	0.3990	0.9729	1.4579	2.0000	0.4912	1.7976
K_I	1.0050	0.8074	0.4748	1.4069	2.0000	2.0000	0.9253	1.9259	1.3376	2.0000	0.9585	0.0476
μ	1.0075	0.3137	0.9413	0.7281	0.7548	0.8318	0.3272	0.3955	1.0037	0.2498	1.0342	0.5659
λ	0.3767	0.5308	0.9855	0.5861	0.1856	0.5049	1.2232	0.6376	0.3115	0.6562	1.1193	0.7186

Table 2 Case 1 transient parameters

Controllers	Transient responses	Δf_1 (Hz)	Δf_2 (Hz)	ΔP_{tie} (pu)
GHS FOPID	U_{sh} ($\times$ 10^{-3})	−6.8135	−2.1754	−0.0833
	O_{sh} ($\times$ 10^{-3})	7.1293	1.2392	0.4680
	T_s	3.1524	7.2214	5.7254
QOGHS FOPID	U_{sh} ($\times$ 10^{-3})	−6.8206	−2.0312	−0.0823
	O_{sh} ($\times$ 10^{-3})	2.6285	0.0669	0.0266
	T_s	1.2589	3.1325	1.2802

GHS and QOGHS optimized gain parameters of the controller are illustrated in Table 2. The system responses in case 1 in terms of tie-line power deviation and frequency deviations of both areas by employing FOPID controller are portrayed in Figs. 2, 3, and 4.

Case 2 (Bilateral based transaction)

In this scenario the load demand of DISCOs in any area is free to make contract with GENCOs in any area. The DISCOs in one area is contracted with the GENCOs of either same or different area. The DPM matrix for this scenario is $DPM = \begin{bmatrix} 0.6 & 0.25 \\ 0.4 & 0.75 \end{bmatrix}$

GHS and QOGHS optimized gain parameters of FOPID controllers implemented in case 2 are illustrated in Table 3. The scheduled tie-line power by concerning ΔP_{L1} and ΔP_{L2} as 0.05 is characterized as $\Delta P_{tie} = 0.25 * \Delta P_{L1} - 0.4 * \Delta P_{L2} = -0.0075$. The system responses of case 2 are portrayed in Figs. 5, 6 and 7.

Case 3 (Contract violation)

In this scenario, the load demand of DISCOs is higher than the contracted load demand. This additional load demand is contributed by the local area GENCOs.

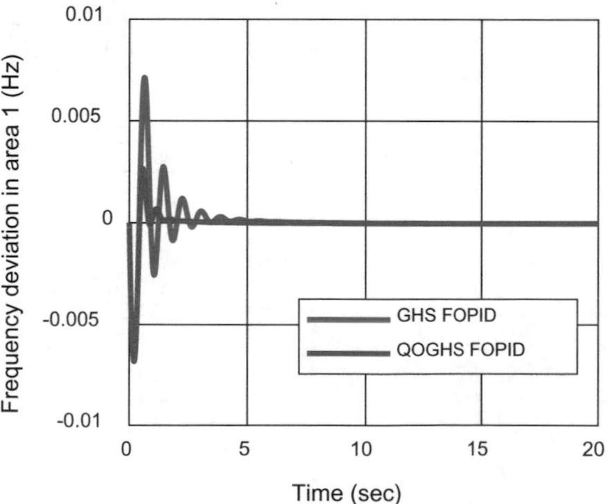

Fig. 2 Frequency deviation in area-1 for case 1

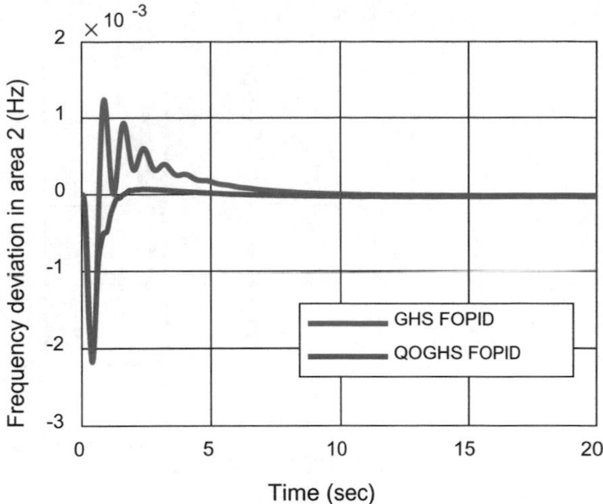

Fig. 3 frequency deviation in area-2 for case 1

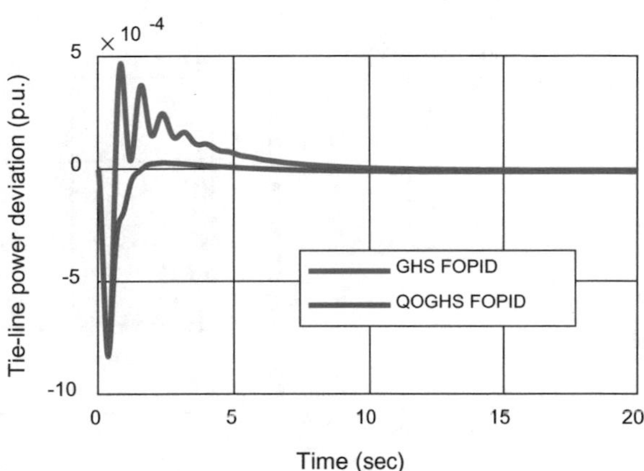

Fig. 4 Tie-line power deviation for case 1

Table 3 Case 2 transient parameters

Controllers	Transient responses	Δf_1 (Hz)	Δf_2 (Hz)	ΔP_{tie} (pu)
GHS FOPID	U_{sh} ($\times$ 10^{-3})	−18.9564	−3.2198	−0.351
	O_{sh} ($\times$ 10^{-3})	1.3985	10.6324	0
	T_s	4.0785	4.1521	4.3000
QOGHS FOPID	U_{sh} ($\times$ 10^{-3})	−17.0425	−3.4062	0
	O_{sh} ($\times$ 10^{-3})	0	6.3032	0
	T_s	3.5821	1.3120	4.0701

The DPM matrix is identical as case 1. The system responses of case 3 are portrayed in Figs. 8, 9, and 10. The numerical values of system responses of case 3 are tabulated in Table 4.

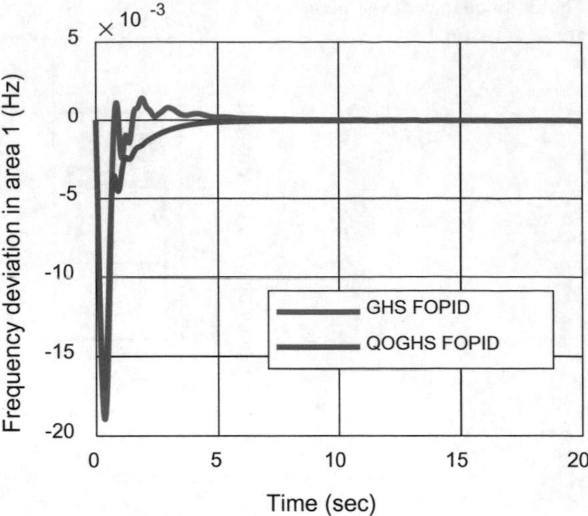

Fig. 5 Frequency deviation in area-1 for case 2

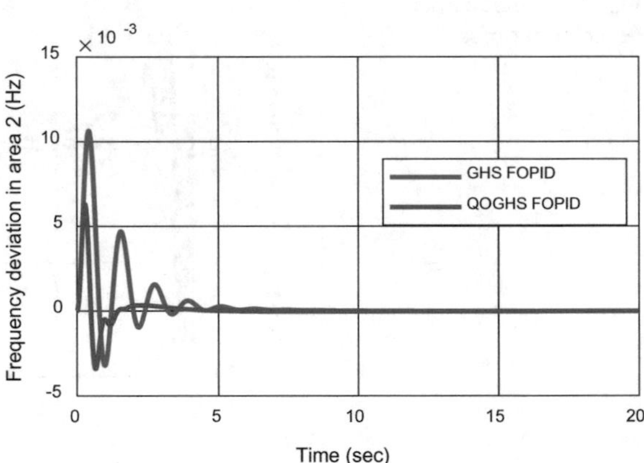

Fig. 6 Frequency deviation in area-2 for case 2

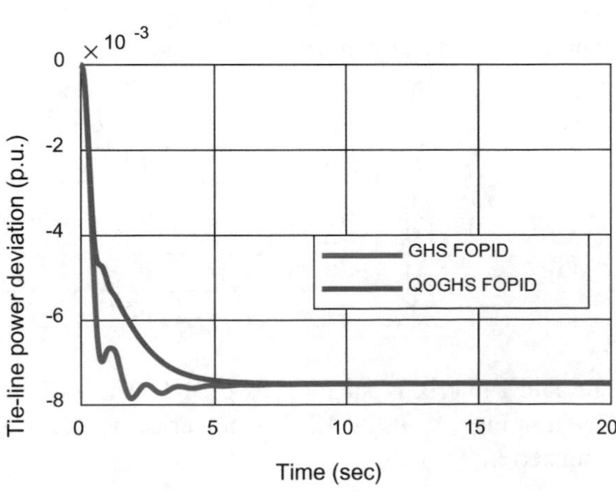

Fig. 7 Tie-line power deviation for case 2

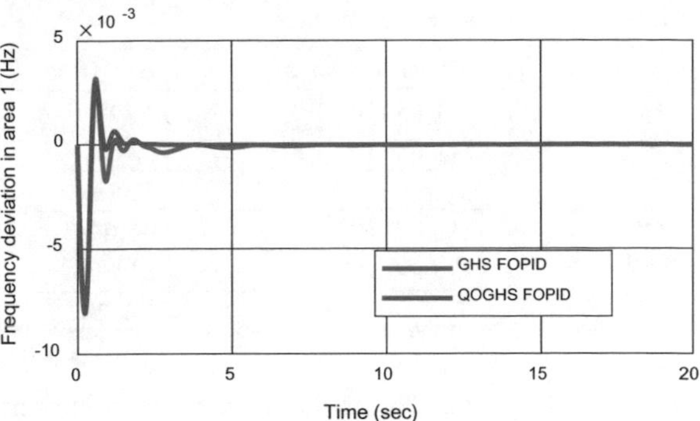

Fig. 8 Frequency deviation in area-1 for case 3

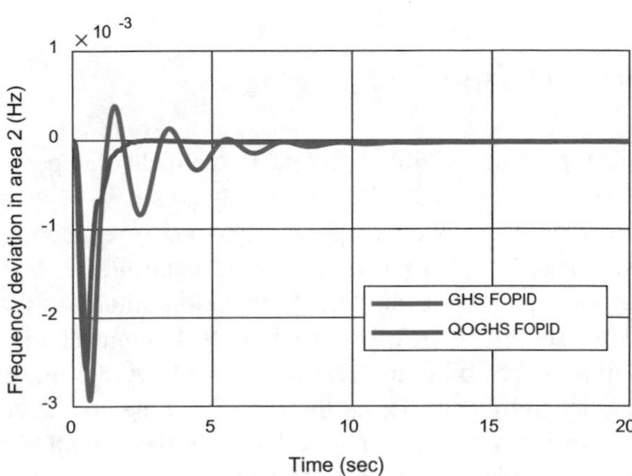

Fig. 9 Frequency deviation in area-2 for case 3

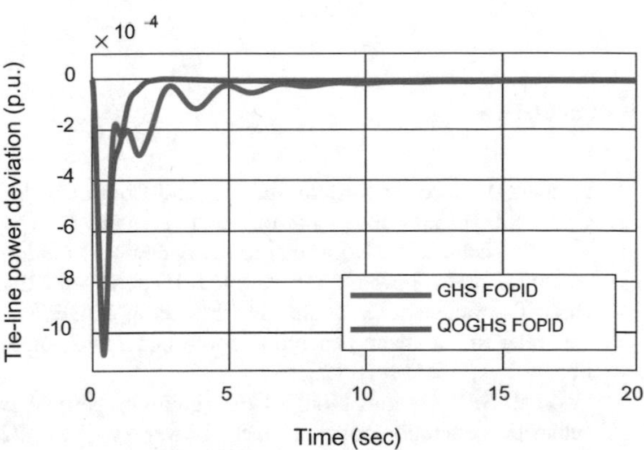

Fig. 10 Tie line power deviation for

The settling time (T_s) peak overshoot (O_{sh}), and peak undershoot (U_{sh}) are the ingredients which are used to discriminate the performances of the controllers. Settling time is evaluated by considering a dimension of ±0.05% (0.0005) of final

Table 4 Case 3 transient parameters

Controllers	Transient responses	Δf_1 (Hz)	Δf_2 (Hz)	ΔP_{tie} (pu)
GHS FOPID	U_{sh} ($\times$ 10^{-3})	−7.8041	−2.9325	−1.0861
	O_{sh} ($\times$ 10^{-3})	3.1858	0.3834	0
	T_s	5.5816	8.7845	1.6245
QOGHS FOPID	U_{sh} ($\times$ 10^{-3})	−8.0732	−2.5821	−1.0361
	O_{sh} ($\times$ 10^{-3})	2.9325	0	0
	T_s	1.5420	1.8842	6.0914

value. T_s, U_{sh}, and O_{sh} of the system are minimum with FOPID controller optimized by QOGHS algorithm as reported in Tables 2 and 3.

6 Conclusion

The purpose of this paper is to validate the performance of FOPID controller optimized by QOGHS algorithm as an improved secondary controller of the interconnected thermal power system by concerning GRC in deregulated environment. For this purpose, FOPID controller is optimized by QOGHS and GHS algorithm by conceding the termination criteria as maximum iterations (100). With 1% load disturbance in area-1, FOPID controller is validated better controller to enhance the ability to get better control over tie-line power deviation and frequency deviation by considering their settling time, undershoot, and overshoot. The above validations of FOPID controller optimized by QOGHS algorithm over GHS algorithm are executed by conceding two case studies entitled as base case and bilateral transaction case.

References

1. Kundur, P.S.: Power System Stability and Control, vol. 46 (2011)
2. Cohn, N.: Tie-Line Bias Control, pp. 1415–1436 (1957)
3. Khodabakhshian, A., Hooshmand, R.: A new PID controller design for automatic generation control of hydro power systems. Int. J. Electr. Power Energy Syst. **32**(5), 375–382 (2010)
4. Dash, P., Saikia, L.C., Sinha, N.: Flower pollination algorithm optimized PI-PD cascade controller in automatic generation control of a multi-area power system. Int. J. Electr. Power Energy Syst. **82**, 19–28 (2016)
5. Ghoshal, S.P.: Optimizations of PID gains by particle swarm optimizations in fuzzy based automatic generation control. Electr. Power Syst. Res. **72**(3), 203–212 (2004)
6. Sahu, B.K., Pati, T.K., Nayak, J.R., Panda, S., Kar, S.K.: A novel hybrid LUS-TLBO optimized fuzzy-PID controller for load frequency control of multi-source power system. Int. J. Electr. Power Energy Syst. **74**, 58–69 (2016)
7. Sahu, B.K., Pati, S., Mohanty, P.K., Panda, S.: Teaching-learning based optimization algorithm based fuzzy-PID controller for automatic generation control of multi-area power system. Appl. Soft Comput. J. **27**, 240–249 (2015)

8. Nayak, J.R., Shaw, B.: Load frequency control of hydro-thermal power system using fuzzy PID controller optimized by hybrid DECPSO algorithm. IJPAM **114**(I), 3–6 (2017)
9. Sahu, R.K., Panda, S., Chandra Sekhar, G.T.: A novel hybrid PSO-PS optimized fuzzy PI controller for AGC in multi area interconnected power systems. Int. J. Electr. Power Energy Syst. **64**, 880–893 (2015)
10. Pan, I., Das, S.: Fractional order AGC for distributed energy resources using robust optimization. **7**(5), 2175–2186 (2016)
11. Morsali, J., Zare, K., Hagh, M.T.: Comparative performance evaluation of fractional order controllers in LFC of two-area diverse-unit power system with considering GDB and GRC effects. J. Electr. Syst. Inf. Technol. (2017)
12. Kumar Sahu, R., Panda, S., Biswal, A., Chandra Sekhar, G.T.: Design and analysis of tilt integral derivative controller with filter for load frequency control of multi-area interconnected power systems. ISA Trans. **61**, 251–264 (2016)
13. Oftadeh, R., Mahjoob, M.J., Shariatpanahi, M.: A novel meta-heuristic optimization algorithm inspired by group hunting of animals: hunting search. Comput. Math. Appl. **60**(7), 2087–2098 (2010)

Optimization of Pole Placement for Adaptive Pitch Control

Shrabani Sahu and Sasmita Behera

Abstract This paper represents an optimization procedure used in adaptive control method for wind energy conversion system (WECS) control purpose. Pole placement (PP) method is used to design the control of pitch angle of the wind turbine to get the desired output. The setting of the parameters of the PP controller design method used for pitch regulation is very difficult. So a particle swarm optimization (PSO) technique is exploited to choose the parameter to achieve desired power level. MATLAB–SIMULINK is used to assess the efficiency of the proposed PSO technique. In order to verify the validity and the performance of the proposed method, the simulation results for a WECS using a PP method optimized by the PSO technique is compared with that obtained using adaptive controller without PSO.

Keywords WECS · Adaptive control · PP · PSO · Pitch control

1 Introduction

In the course of recent years, wind energy has acquired huge consideration as a standout among the most promising sustainable power sources. Because of the clean and endless source of energy, wind energy has turned out to be one of the most important and fastest developing sustainable power sources on the planet. Depending upon the nature of the wind speed, the wind energy conversion system (WECS) is classified into two types, i.e., fixed speed wind turbine system and variable speed wind turbine system. Variable speed wind turbines can extract

S. Sahu (✉)
Department of Electrical Engineering, Veer Surendra Sai University of Technology, Burla 768018, India
e-mail: shrabanisahu@gmail.com

S. Behera
Department of Electrical & Electronics Engineering, Veer Surendra Sai University of Technology, Burla 768018, India
e-mail: sasmitabehera2000m@gmail.com

J. Nayak et al. (eds.), *Soft Computing in Data Analytics*, Advances in Intelligent Systems and Computing 758,
https://doi.org/10.1007/978-981-13-0514-6_8

additional energy and offer better output power quality, but also brings cost and reliability concerns. So some advanced control strategies are introduced to compensate the cost and reliability issues. A wind turbine benchmark model utilized here is detailed in [1]. The throughout analysis is executed in MATLAB/SIMULINK environment. To achieve maximum power in full load region some control techniques are utilized. Two examples of control design method, i.e., fuzzy modeling and control and adaptive control with application to a complex nonlinear wind turbine model are represented in [2]. In [3], a control scheme is developed by using nonlinear adaptive control algorithm to achieve smooth rotor speed tracking. A MATLAB toolbox for design and verification of industrial controllers is represented in [4, 5].

To make the WECS highly efficient and reliable, it is necessary to choose an efficient controller with suitable controller parameters. Kinetic energy reserve (KE) control by using PSO is discussed in [6] to provide frequency regulation for permanent magnet synchronous generator (PMSG) wind turbine. The performance of the WECS by implementing an optimized RST controller using PSO meta-heuristic technique is analyzed in [7]. It is also shown that the proposed PSO technique causes reduction of ripples in torque. PSO-based wind turbine rotor speed setpoint algorithm is used in [8] for capturing maximum power from wind and to reduce turbine speed error. A numerical optimization method is also described which can be used to improve wind turbine performance leading to a lower cost of energy. In [9], optimal parameters of PI controller are found by using the particle swarm optimization algorithm to improve the transient performance of wind turbine generator system.

Previously, PSO has been effectively implemented in many research areas and it is observed that PSO obtains improved outcomes compared to other techniques [10]. So in this paper, an adaptive controller is used to control the pitch angle of a wind turbine system. The parameters of the adaptive controller are updated by using an optimization technique. Section 2 explains the details about the wind turbine benchmark model used here. Pole placement method to design the adaptive controller is described in Sect. 3. Recommended optimization procedure is presented in Sect. 4. Simulation results are shown and described in Sect. 5. Lastly, some conclusion is given in Sect. 6.

2 Wind Turbine Benchmark Model

The wind turbine benchmark model considered in this paper is layout in block diagram in Fig. 1 [1]. It has been implemented in MATLAB/SIMULINK environment and proposes a realistic simulator for a wind turbine system with some common fault scenarios. It represents a three-bladed horizontal axis variable speed pitch controlled turbine with a full converter generator. According to the general principle of wind turbine, the kinetic energy of wind is converted to mechanical energy in respect of a rotating shaft. This energy conversion can be controlled by pitching the blades or by controlling the rotational speed of the turbine relative to

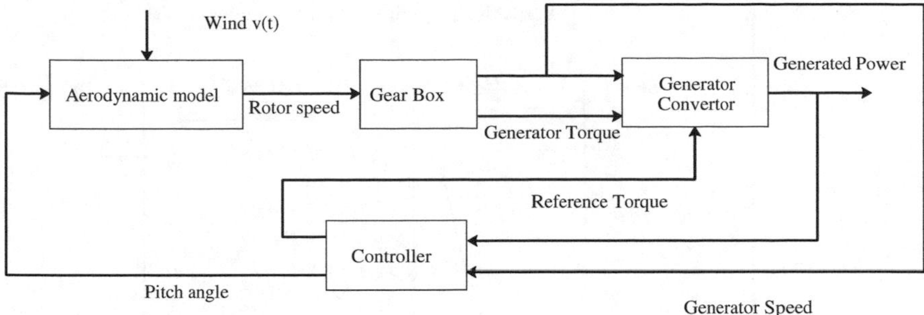

Fig. 1 Block diagram of WECS

the wind speed. In order to control the pitch angle and generator torque of the wind turbine, the generated power and the generator speed act as input to the controller. Here, the purpose is to follow the reference power level, i.e., 4.8 MW in constant power region and the reference value of generator speed is taken as 162 rad/s.

The power graph of a wind turbine can be divided into four regions which are shown in Fig. 2. The transition between the partial load region and constant power region is regulated by a bumpless transfer mechanism based on generator speed.

A self-tuning controller (STC) used in the constant power region is designed by using PP approach which is also adaptive in nature. This controller controls the pitch angles of the blades which in turn forces the generator speed to its nominal value ω_{nom} that implies the generation of reference power. STC controller utilizes three signals of the process: the reference signal, the output of the plant, and the present input to the plant. Here, the controller is controlling the blade pitch angle so at *j*th time period ω_{nom} is reference signal $w(j)$, ω_g is the output signal $y(j)$, and β_r is taken as the controller input $u(j)$. The tracking error is given as $e(j) = \omega_{nom} - \omega_g(j)$. Output from the controller is the control signal which is input to the controlled plant.

3 Pole Placement Method

A pole placement approach is utilized to achieve the preset poles of the characteristic polynomial. To make a process to be stable, the poles have to lie in the left half of s plane and the value of damping factor (ξ) and natural frequency (ω_n) must be greater than zero. The controller output for the system in (1) is:

$$u(j) = q_0 e(j) + q_1 e(j-1) + q_2 e(j-2) + (1-\gamma)u(j-1) + \gamma u(j-2) \quad (1)$$

The controller coefficient q_0, q_1, q_2 can be calculated as (2–9)

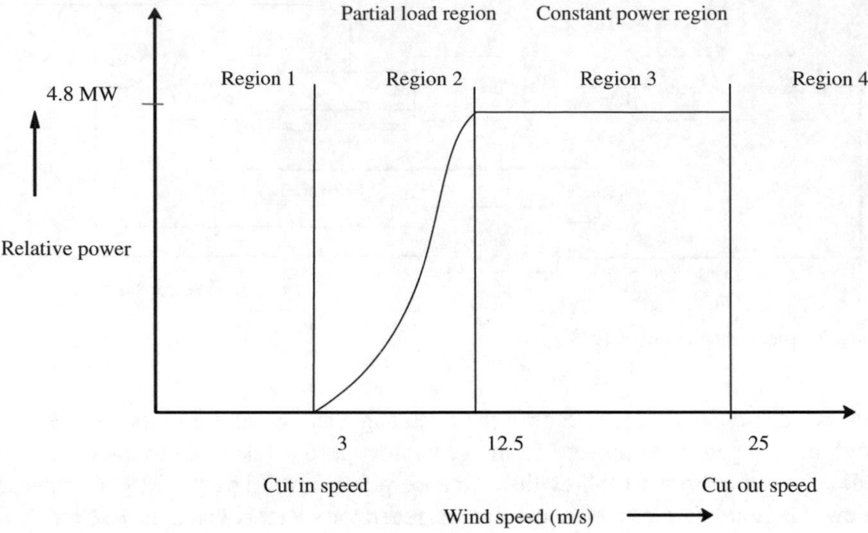

Fig. 2 Power graph of wind turbine benchmark model

$$q_0 = \frac{1}{b_1}(d_1 + 1 - a_1 - \gamma) \tag{2}$$

$$q_1 = \frac{a_2}{b_2} - q_2\left(\frac{b_1}{b_2} - \frac{a_1}{a_2} + 1\right) \tag{3}$$

$$q_2 = \frac{-s_1}{r_1} \tag{4}$$

$$d_1 = \begin{cases} -2e^{-\xi\omega_n T_0}\cos(\omega_n T_0\sqrt{1-\xi^2}), & \text{if } \xi \le 1, \\ -2e^{-\xi\omega_n T_0}\cosh(\omega_n T_0\sqrt{\xi^2 - 1}), & \text{if } \xi > 1, \end{cases} \tag{5}$$

$$d_2 = e^{-2\xi\omega_n T_0} \tag{6}$$

$$s_1 = a_2[(b_1 + b_2)(a_1 b_2 - a_2 b_1) + b_2(b_1 d_2 - b_2 d_1 - b_2)] \tag{7}$$

$$r_1 = (b_1 + b_2)\left(a_1 b_1 b_2 + a_2 b_1^2 + b_2^2\right) \tag{8}$$

$$\gamma = q_2 \frac{b_2}{a_2} \tag{9}$$

Here, T_0 is sampling time period and parameter vectors a_1, a_2, b_1, b_2 are estimated using online identification approach.

The user will have to adjust the value of ξ and ω_n to get better response. Unfortunately, selection of these parameters manually is quite tedious. So the values of ξ and ω_n are found out by utilizing an optimization technique. Here particle swarm optimization (PSO) algorithm, using objective function ITAE (Integral Time Absolute Error), is used to select the controller parameters ξ and ω_n so as to improve the performance of wind turbine system.

4 Particle Swarm Optimization

This technique was motivated by swarming tendency of birds or fish and evolutionary computation which is used for resolving nonlinear optimization problem. Here, it is used to calculate the parameters value of the pole placement controller which often has to be found by trial and error method. In PSO algorithm, fitness of each particle is updated by using (10) and (11).

$$v_k(j+1) = wv_k(j) + m_1 r_1[\hat{x}_k(j) - x_k(j)] + m_2 r_2[g(j) - x_k(j)] \tag{10}$$

$$x_k(j+1) = x_k(j) + v_k(j+1) \tag{11}$$

Here, $v_k(j)$ is the particle's velocity, $x_k(j)$ is the particle's position, $\hat{x}_k(j)$ is the particle's individual solution and $g(j)$ is swarm's best solution as of time j. Also k is defined as particle's index, w is inertial coefficient, m_1 and m_2 are acceleration coefficients, r_1 and r_2 are random values [0, 1].

In PSO algorithm, first the particles are initialized and for each iterations the fitness of each particle is evaluated based on previous best and global best positions to achieve the best performance. Flowchart of PSO algorithm is given in Fig. 3.

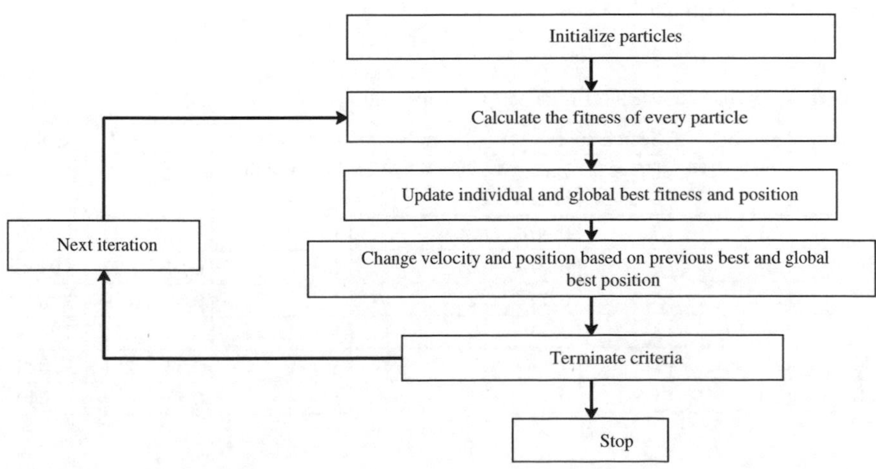

Fig. 3 Flow chart of PSO algorithm

Here, in pole placement design approach pole locations are chosen by utilizing PSO technique.

5 Simulation and Results

The simulation studies are carried out by taking wind velocity as input to the wind turbine benchmark model which is shown in Fig. 4.

In order to get maximum power in full load region the blade pitch angle is controlled by using pole placement design approach. The controller's potential is found out by percent normalized sum of squared tracking error (NSSE) (12) [2].

$$NSSE\% = 100\sqrt{\frac{\sum_{i=1}^{N}(\mathrm{w(j)} - \mathrm{y(j)})^2}{\sum_{i=1}^{N} w^2(\mathrm{j})}} \tag{12}$$

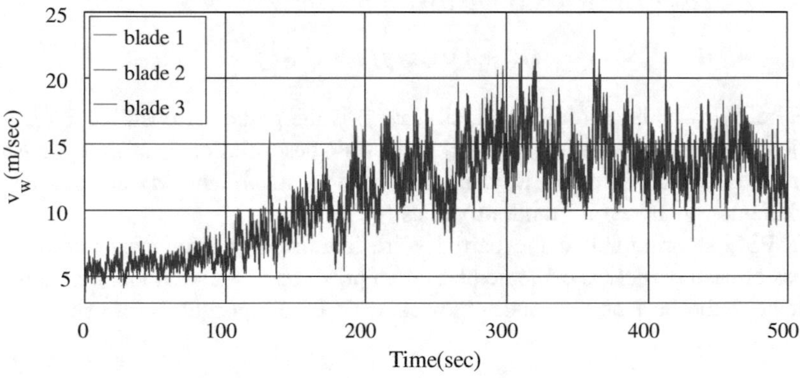

Fig. 4 Wind velocity profile

Table 1 PSO characteristics for different parameter settings

Swarm size	No. of evaluation	Search space		Swarm's best weight	Particle's best weight	ξ	ω_n	NSSE %
		Upper bound	Lower bound					
2	2	5	0.001	2	1	0.91	1.31	3.9778
10	10	3	0.001	2	1	0.54	2.77	3.9328
20	20	3	0.001	2	1	0.56	0.79	3.7878
30	30	3	0.001	2	1	0.55	2.21	3.8988
30	30	1.5	0.5	2	1	0.52	0.81	3.7820
20	20	1.5	0.5	2	1	0.53	1.32	3.8640

Table 2 Optimized pole location parameter

Name of parameters	Optimized value
ξ	0.52
ω_n	0.81

Table 3 Comparison of results

Model type	NSSE% value
Adaptive pole placement controller without optimization technique	6.6324
Adaptive pole placement controller with using optimization technique	3.7820

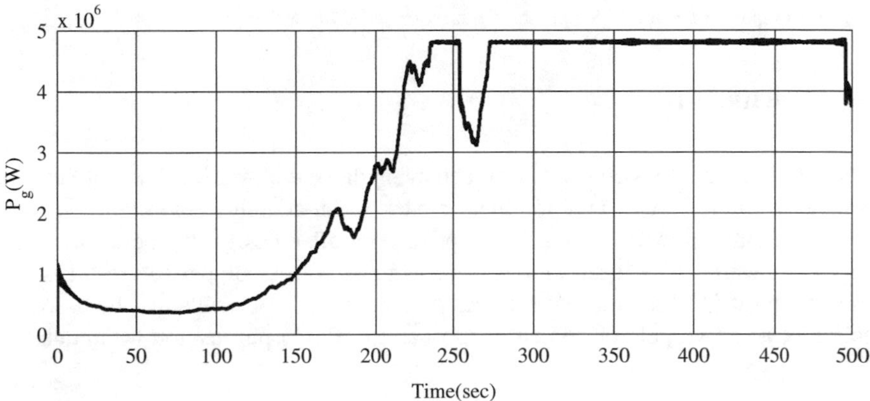

Fig. 5 Variation of generated power P_g with time

Different parameters settings were taken and each setting was run for ten times to obtain an appropriate one. The best NSSE% value of each parameter setting is given in Table 1.

By utilizing the optimization approach the best values of ξ and ω_n are selected according to the NSSE% value as shown in Table 2. The search space was reduced and the number of evaluation was increased to get the suitable result. All the results were assessed by using Monte Carlo analysis. A comparison of NSSE% value of wind turbine benchmark model with and without using optimization technique is shown in Table 3.

The tracking of reference value of generated power is shown in Fig. 5 and a comparative result of generator speed of a wind turbine with and without using PSO are represented in Fig. 6.

From Fig. 6, it is realized that an adaptive controller designed by pole placement method can give better results while the controller parameters are obtained from PSO optimization technique rather than selecting the parameters randomly.

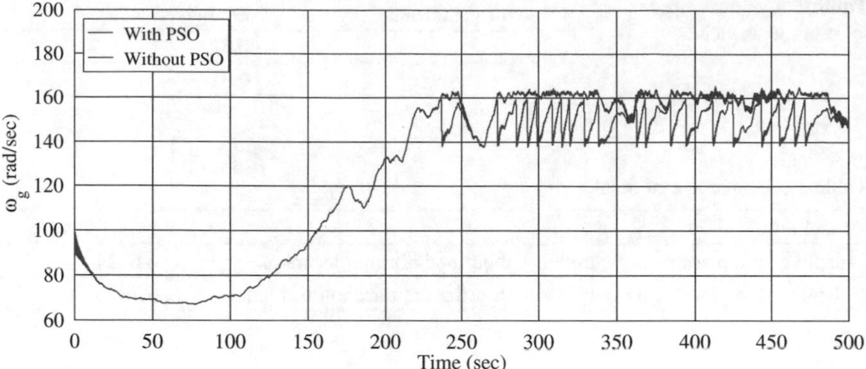

Fig. 6 Comparative result of variation of generator speed ω_g with time

6 Conclusion

Here, the pitch angle is controlled for achieving desired power level in wind turbine benchmark model. As it is a nonlinear model, so an adaptive controller is used to get the desired power level. For adaptive controller design by pole placement approach desired pole location was set utilizing PSO. A comparison result is also shown, and it is observed that a better performance can be obtained by a combination of adaptive pole placement controller and PSO optimization technique.

References

1. Odgaard, P.F., Stoustrup, J., Kinnaert, M.: Fault tolerant control of wind turbines–a benchmark model. In: IFAC Proceedings Volumes, vol. 42, no. 8, pp. 155–160 (2009)
2. Simani, S., Castaldi, P.: Data-driven and adaptive control applications to a wind turbine benchmark model. Control Eng. Pract. **21**(12), 1678–1693 (2013)
3. Song, Y.D., Dhinakaran, B., Bao, X.Y.: Variable speed control of wind turbines using nonlinear and adaptive algorithms. J. Wind Eng. Ind. Aerodyn. **85**(3), 293–308 (2000)
4. Bobál, V., Chalupa, P.: New modification of MATLAB-toolbox for CAD of simple adaptive controllers
5. Bobál, V., Böhm, J., Fessl, J., Machácek, J.: Digital Self-tuning Controllers: Algorithms, Implementation and Applications. Springer Science & Business Media (2006)
6. Kim, M.K.: Optimal control and operation strategy for wind turbines contributing to grid primary frequency regulation. Appl. Sci. **7**(9), 927 (2017)
7. Tahir, K., Belfedal, C., Allaoui, T., Gerard, C., Doumi, M.: Optimal tuning of RST controller using PSO optimization for synchronous generator based wind turbine under three-phase voltage dips
8. Tibaldi, C., Hansen, M.H., Henriksen, L.C.: Optimal tuning for a classical wind turbine controller. J. Phys. Conf. Ser. **555**(1), 012099 (2014)

9. Qiao, W., Venayagamoorthy, G.K., Harley, R.G.: Design of optimal PI controllers for doubly fed induction generators driven by wind turbines using particle swarm optimization. In: International Joint Conference on Neural Networks, IJCNN'06, pp. 1982–1987 (2006)
10. Belghazi, O., Douiri, M.R., Cherkaoui, M., Ainane, T., Sobh, M.: The application of particle swarm optimization control in pitch wind turbine. J. Eng. Sci. Technol. Rev. **7**(3) (2014)

Design of Off-grid Fuel Cell by Implementing ALO Optimized PID-Based MPPT Controller

Sonu Kumar and Binod Shaw

Abstract The main objective of proposed paper is to implement an off-grid fuel cell array in combination with DC-to-DC boost chopper and MPPT. To extract the maximum power from the fuel cell array under different conditions, tuned PID-based MPPT technique is adopted. The design parameters of PID controller have critical role to improve the performance of the system. Ant lion optimizer (ALO) algorithm is taken up to get the optimum values of PID parameters to supply suitable duty ratio for DC-to-DC boost chopper to optimize the output power and voltage of the proposed system. P&O-based MPPT technique is carried out to validate the supremacy of PID-based MPPT to upgrade the response of the system. In this paper, the proposed ALO optimized PID controller-based MPPT technique is performed better over conventional P&O technique in terms of the output voltage, current, and power of the fuel cell array.

Keywords Maximum power point tracking (MPPT) · Perturb and observe (P&O) · Proportional–integral–derivative (PID) controller · Ant lion optimizer (ALO) algorithm

1 Introduction

In present scenario, renewable energy source plays a significant approach to meet the fast-growing load demand. Fuel cells are considered as one of the most assuring renewable power generator. Fuel cell is imperative concern among renewable energy due to its noise-free, eco-friendly, and high-energy density with impressive lifespan. Fuel cell system needs to minimize cost and increase the efficiency.

S. Kumar (✉) · B. Shaw
Department of Electrical Engineering, National Institute of Technology, Raipur, Raipur 492010, India
e-mail: singh.sonu1076@gmail.com

B. Shaw
e-mail: binodshaw2000@gmail.com

J. Nayak et al. (eds.), *Soft Computing in Data Analytics*,
Advances in Intelligent Systems and Computing 758,
https://doi.org/10.1007/978-981-13-0514-6_9

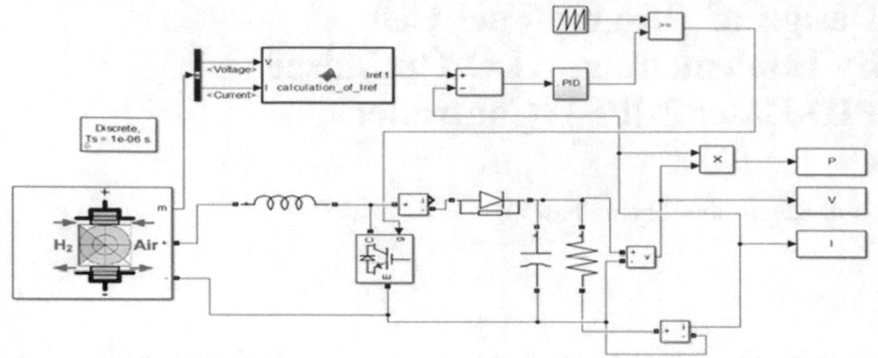

Fig. 1 Simulink model of proposed off-grid fuel cell array system

The maximum power point (MPP) is the extraction of power from fuel cell under specific circumstances. Maximum power point tracking (MPPT) is the process to track the maximum power by optimizing the load resistance properly in any environmental condition.

In this proposed paper, Fig. 1 points up about the off-grid fuel cell array of 6 kW capacity in which MPPT block is mainly concerned. A recently developed optimization technique, ant lion optimizer (ALO), is applied to PID controller to get its gain parameters to extract maximum power from fuel cell. PWM technique is used to give appropriate pulses to the gate of DC-to-DC converter. The output of traditional method such as P&O is compared with ALO optimized PID controller. The result represents the enhancement in the output power and output voltage.

2 Literature Survey

Many researchers have implemented various techniques to enhance the efficiency of the fuel cell by enhancing the MPPT techniques from last few decades. Esram and Chapman [1] have contributed 19 different MPPT techniques and provided a fair interpretation for researchers to adopt relevant techniques. The P&O of variable step size is proposed by Al-Diab and Sourkounis [2], and the step size is tuned automatically and compared with the conventional method. Ishaque and Salam [3] have contributed a brief literature to design MPPT by adopting soft computing methods during partial shading. The variable CS MPPT algorithm is validated by comparing with conventional P&O and PSO MPPT algorithm in three distinct case studies and is described in [4]. The fuzzy logic controller-based MPPT algorithm with 8-bit microcontroller is compared with conventional P&O MPPT algorithm in [5 and 6]. Neural network-based MPPT is implemented in 230 W PV system in [7]. A brief literature survey on MPPT design is described beautifully in [8, 9]. The P&O algorithm is optimized to enhance the efficiency of the MPPT technique in

[10, 11]. The brief study of fuel cell and the power electronics converter is completed by studying the paper from [12–17].

In this paper, ALO algorithm [18] optimized PID controller-based MPPT technique is strived to validate over P&O technique to enhance the power and voltage of the system by contributing gate pulse of DC-DC boost converter. The proposed experiment is executed in MATLAB/Simulink environment.

3 System Investigated

The MATLAB/Simulink model of fuel cell array with PID-based MPPT controller is portrayed in Fig. 1. The proposed isolated off-grid fuel cell array system is illustrated in Fig. 2 which basically subsists of fuel cell array, DC-to-DC boost chopper, and MPPT controller.

MPPT controller regulates gate pulse of boost converter by conceding the voltage and current of the fuel cell array. The output of the MPPT is compared to sawtooth waveform, which results in a regulated gate signal for the DC-to-DC boost chopper. The regulated pulse of the converter enhances the efficiency of the off-grid fuel cell array system.

3.1 Fuel Cell Array

Recent years show a great advancement in fuel cell such as development in vehicles, fast battery charging, electricity generation to meet peak load demands, unmanned aerial vehicles (UAVs). Fuel cell is used as a distributed generation in power system. The basic structure of fuel cell is represented by the following equation which takes place during the energy generation in the fuel cell. Hydrogen and oxygen combine to form water as stated in Eq. (1).

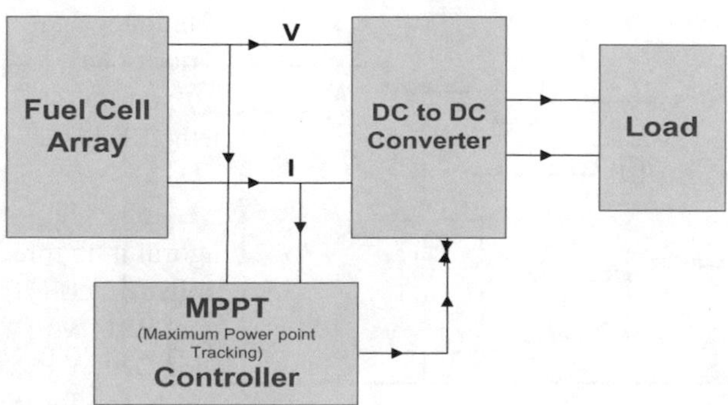

Fig. 2 Block diagram of fuel cell system

$$2H_2 + O_2 = 2H_2O \tag{1}$$

Anode reaction of an acid electrolyte fuel cell—the hydrogen gas ionizes, releasing electrons and creating H^+ ions (or protons) which are represented in Eq. (2).

$$2H_2 = 4H^+ + 4e^- \tag{2}$$

Cathode reaction of the fuel cell—oxygen reacts with electrons taken from the electrode, and H⇓← ions from the electrolyte, to form water as represented in Eq. (3).

$$O_2 + 4e^- + 4H^+ = 2H_2O \tag{3}$$

3.2 DC-to-DC Boost Chopper

The basic objective of design of boost converter is to raise the output voltage of the dc system. The output of the converter is remarkably influenced by the switching frequency (gate pulse). The parameters of boost converter are tabulated in Table 1.

Table 1 Model parameters of boost converter

Model components	Parameters
Inductance, L	1 μH
Capacitance, C	3000 μF
Load, R	24 Ω
DC voltage, Vdc	45 V
Switching frequency	1 kHz

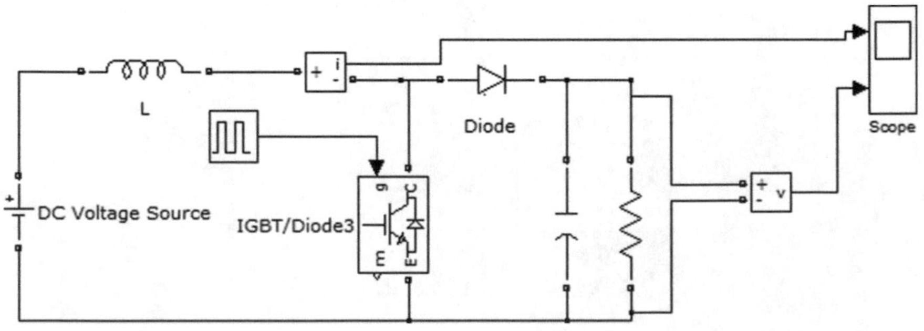

Fig. 3 Boost converter

Figure 3 represents the boost chopper, and output of the chopper is characterized in Eq. (4)

$$V_{out} = \frac{1}{1-D} V_{in} \tag{4}$$

where, D represents duty ratio of the boost chopper and is characterized in Eq. (5).

$$D = \frac{t_{on}}{t_{on} + t_{off}} \tag{5}$$

where, t_{on} and t_{off} represent ON time and OFF time of IGBT switch of the chopper and are characterized in Eqs. (6) and (7).

$$t_{on} = D * T_s \tag{6}$$

$$t_{off} = (1 - D) * T_s \tag{7}$$

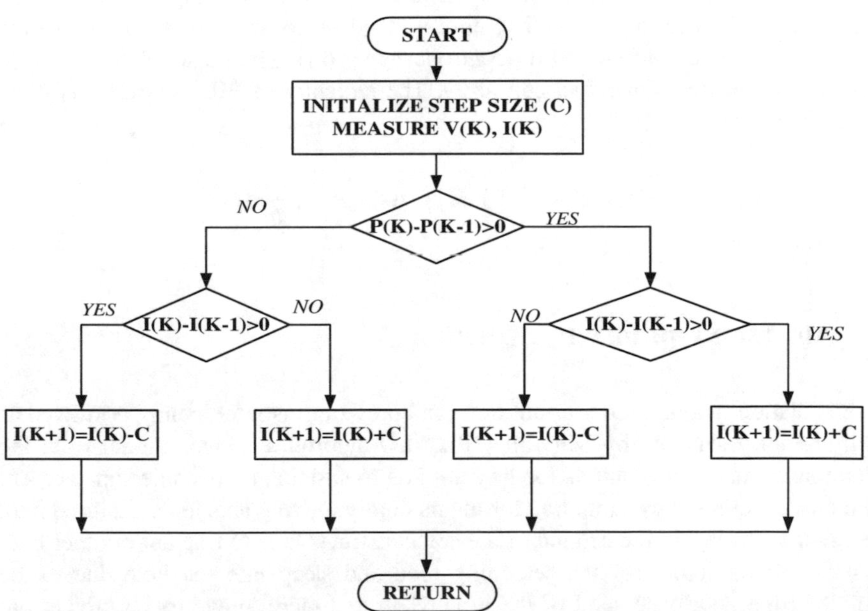

Fig. 4 Algorithm for V_{ref} calculation

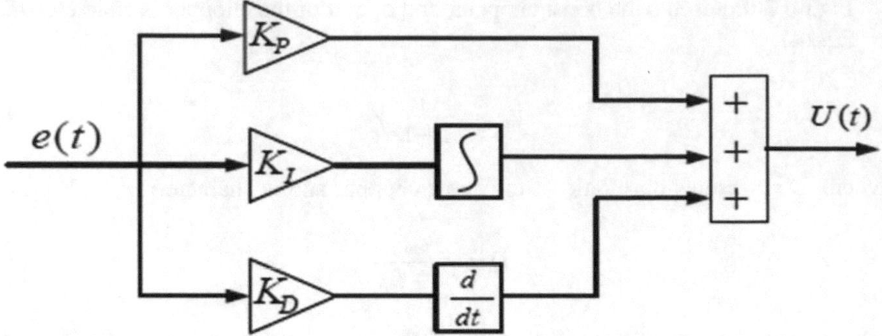

Fig. 5 Structure of PID controller

3.3 PID-Based MPPT Controller

The primary objective of MPPT technique is to track the maximum power from the fuel cell array by concerning the fuel cell array voltage and power. In proposed paper, reference voltage (V_{ref}) is developed by comparing the present power (P_k) and previous power (P_{k-1}) as portrayed in Fig. 4. The error signal obtained by comparing reference voltage with output voltage of boost converter is fed to the PID controller. The output of PID controller is used as gate pulse of the switch to enhance the power of the fuel cell array. The structure of PID controller is illustrated in Fig. 5.

$$u(t) = K_p * e(t) + K_i \int e(t)dt + K_d \frac{d}{dt} e(t)$$

4 Ant Lion Optimizer Algorithm

The affiliation among predator (ant lion) and prey (ant) is intelligently portrayed as optimization technique by Mirjalili [18]. ALO algorithm is derivational from the planning of hawking of ant as food by ant lion to sustain and become capable. Ant lion creates reversed pyramid trap for the randomly moving ants to be captured into. Ant lion downtime in the ground of the soil constructs hole to trap ant or other bug. Ants move randomly for the searching food and sleep into the hole due to the pointed edge and loose sand of the hole. Here and there, preys try to protect out from the opening; however, ant lion impels sands to the edge of the gap to make the prey slip into its jaw. The extent of opening specifically relies on the starvation of ant lion. The upgrade of the span of opening improves the likelihood to get nourishment. The steps followed for ALO algorithm is described as:

1. The components of the framework which holds the places of preys is introduced arbitrarily with estimate in (8)

$$[M_{pray}]_{NP\times D} \tag{8}$$

So also, introduction of ant lion position grid is resolved arbitrarily with same size represented in (9)

$$[M_{Antlion}]_{NP\times D} \tag{9}$$

where NP and D are the population and measurement of plan factors separately. For this issue irregular in statement is in the middle of 0–2. Useful estimations of the every ant lion and prey are dictated by (10) and (11).

$$F_{pray} = f(M_{pray}) \tag{10}$$

$$F_{Antlion} = f(M_{Antlion}) \tag{11}$$

where F_{prey} is a variety of wellness estimations of arbitrarily introduced M_{prey} and $F_{Antlion}$ is the variety of wellness estimations of $M_{Antlion}$.

2. The ant lion with best fitness is allocated as best.
3. Roulette wheel is utilized to choose ant lions which gives higher likelihood of fit ant lions to chase preys.
4. The base and greatest vector of *i*th factors C_i^g and d_i^g individually are modified as in Eqs. (12) and (13), respectively.

$$C_j^i = AL^g + C^g \tag{12}$$

$$d_i^g = AL_i^g + d^g \tag{13}$$

where AL^g is the position of the *i*th ant lion at *g*th generation. C^g and d^g might be described as: $c^g = \frac{c^g}{I}$, $d^g = \frac{d^g}{I}$, and $I = 10^w \frac{g}{n}$

where w is a round number chosen between 2 and 6 based on new generation.

5. The activity of preys is random in nature and may be shown in Eq. (14).

$$X(g) = [0, cumsum(2r(g_1) - 1), cumsum(2r(g_2) - 1), \ldots, cumsum(2r(g_n) - 1) \tag{14}$$

where cumulative sum is found by cumsum. g and n are the generation and peak generation number, respectively. r is a random probability distribution function described in Eq. (15).

$$r(g) = \begin{cases} 1 & if \quad rand > 0.5 \\ 0 & if \quad rand \leq 0.5 \end{cases} \tag{15}$$

6. The status of preys is modified by Eq. (16).

$$P_i^g = \frac{R_P^g + R_E^g}{2} \tag{16}$$

where R_p^g and R_E^g are the strange changes around ant lion and best, respectively.

7. The functional values of preys are determined.
8. The ant lion is updated by its analogous fitter prey as described in Eq. (17).

$$AL_i^g = P_i^g \quad if\, f(P_i^g) > \quad f(AL_i^g) \tag{17}$$

9. The practical value of preys is determined as explain previously.
10. Elite value is updated by the ant lion with fitness value better than elite.
11. Steps from 4 to 10 are repeated until maximum generation reached.

In the present work, the gains of PID controllers are tuned by ALO algorithm to enhance the power of the fuel cell system.

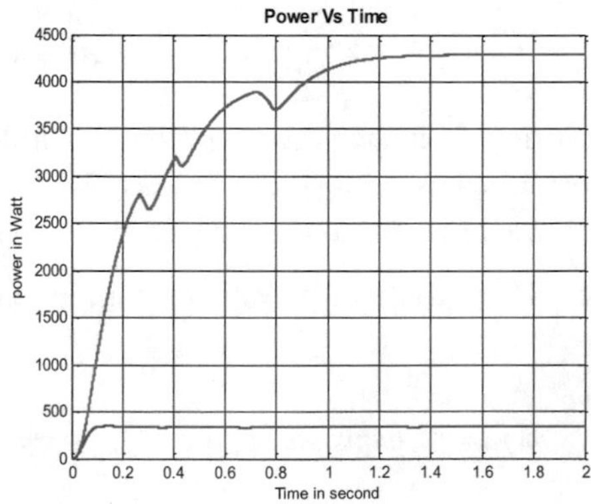

Fig. 6 Power versus time graph

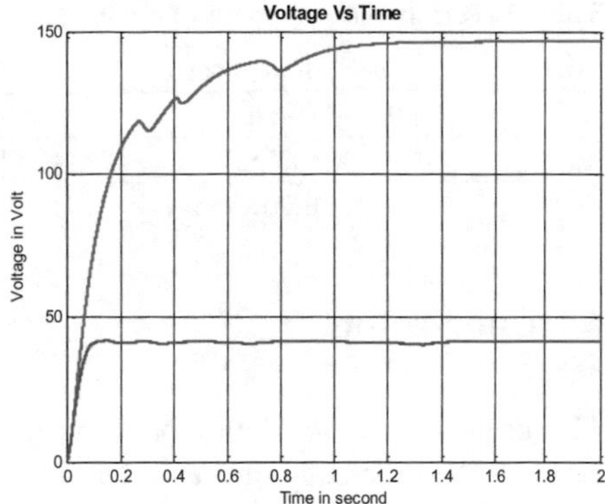

Fig. 7 Voltage versus time graph

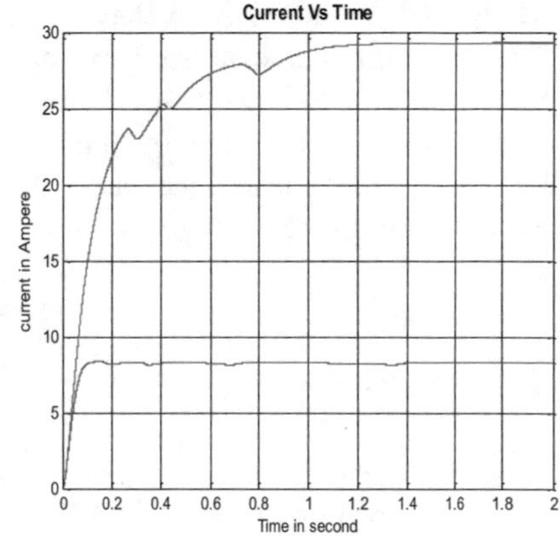

Fig. 8 Current versus time graph

5 Results and Discussion

ALO algorithm is executed for 100 iterations with 50 populations to resolve the steps to discover the optimal gain parameters of PID controllers. The objective of the algorithm is to hunt the parameters within a specified limit as described in Eq. (18).

$$0.001 \leq K_p, K_i \text{ and } K_d \leq 2 \tag{18}$$

The optimal values of Kp, K_i, and K_d are 0.0726, 0.9706, and 1.1160, respectively. The performance of fuel cell by conceding power, voltage, and current are portrayed in Fig. 6, Fig. 7, and Fig. 8, respectively. Performance response of the output is represented in Table 2.

Table 2 Performance response of output

Sl.no.	Fuel cell parameter	P&O-based MPPT	PID-based MPPT
1.	Power (Watt)	389.495	4232.5150
2.	Voltage (Volt)	47.9893	146.1524
3.	Current (Ampere)	8.1163	28.9596

6 Conclusion

The prime objective of this paper is to design a off-grid fuel cell system and to enhance the efficiency of the system by implementing PID-based MPPT technique. The PID-based MPPT controller optimized by ALO algorithm is validated as an improved controller over P&O-based MPPT controller of the fuel cell array system. This PID- and P&O-based MPPT controller are executed on a fuel cell array having rating of 45 V and 6 kW. A pure resistive load of 5 Ω is connected in the load side. The error signal is evaluated by the contrast of reference voltage and measured process voltage to achieve relevant gate pulse of the converter. The output voltage is enormously influenced by duty ratio of the gate pulse. The ALO optimized PID-based MPPT techniques are validated over P&O technique to achieve higher power, current, and voltage.

References

1. Esram, T., Chapman, P.L.: Comparison of photovoltaic array maximum power point tracking techniques. IEEE Trans. Energy Convers. **22**(2), 439–449 (2007)
2. Al-Diab, A., Sourkounis, C.: Variable step size P&O MPPT algorithm for PV systems. In: Proceedings of International Conference on Optimization of Electrical and Electronic Equipment (OPTIM), pp. 1097–1102 (2010)
3. Ishaque, K., Salam, Z.: A review of maximum power point tracking techniques of PV system for uniform insolation and partial shading condition. Renew. Sustain. Energy Rev. **19**, 475–488 (2013)
4. Ahmed, J., Salam, Z.: A maximum power point tracking (MPPT) for PV system using cuckoo search with partial shading capability. Appl. Energy **119**, 118–130 (2014)
5. Giustiniani, A., Petrone, G., Spagnuolo, G.: Low-frequency current oscillations and maximum power point tracking in grid-connected fuel-cell-based systems. IEEE Trans. Ind. Electron. **57**(6), 2010
6. Yang, Y.R.: A fuzzy logic controller for maximum power point tracking with 8-Bit Microcontroller, vol. 5, pp. 1078–1086 (2011)
7. Liu, Y.H., Liu, C.L., Huang, J.W., Chen, J.H.: Neural-network-based maximum power point tracking methods for photovoltaic systems operating under fast changing environments. Sol. Energy **89**, 42–53 (2013)
8. Kamarzaman, N.A., Tan, C.W.: A comprehensive review of maximum power point tracking algorithms for photovoltaic systems. Renew. Sustain. Energy Rev. **37**, 585–598 (2014)
9. Reza, A., Hassan, M., Jamasb, S.: Classification and comparison of maximum power point tracking techniques for photovoltaic system: a review. Renew. Sustain. Energy Rev. **19**, 433–443 (2013)

10. Femia, N., Petrone, G., Spagnuolo, G., Vitelli, M.: Optimization of Perturb and Observe Maximum Power Point Tracking Method, vol. 20, no. 4, pp. 963–973 (2005)
11. Hoffmann, S., Koehl, M.: Investigation on the Impact of Macro- and Micro-Climate on the Potential Induced Degradation, pp. 1822–1825 (2013)
12. Ali, E.S., AbdElazim, S.M., Abdelaziz, A.Y.: Ant Lion Optimization algorithm for renewable distributed generations. Energy **116**, 445–458 (2016)
13. Todorovic, M.H., Palma, L., Enjeti, P.N.: Design of a wide input range DC–DC converter with a robust power control scheme suitable for fuel cell power conversion. IEEE Trans. Ind. Electron. **55**(3), 1247–1255 (2008)
14. Derbeli, M., Farhat, M., Barambones, O., Sbita, L.: Control of Proton Exchange Membrane Fuel Cell (PEMFC) Power System Using PI Controller. IEEE (2017)
15. Derbeli, M., Farhat, M., Barambones, O., Sbita, L.: PEM Fuel Cell Green Energy Generation —SMC Efficiency Optimization. IEEE (2017)
16. Tourkia, L.: Interaction Between Fuel Cell and DC/DC converter for Renewable Energy Management. IEEE (2017)
17. Lajnef, T., Adid, S., Ammous, A.: Modeling, Control, and Simulation of a Solar Hydrogen/Fuel Cell Hybrid Energy System for Grid-Connected Applications, Hindawi Publishing Corporation: Advances in Power Electronic, vol. 2013 (2013)
18. Mirjalili, S.: The Ant Lion Optimizer. Adv. Eng. Softw. **83**, 80–98 (2015)

Power System Security Improvement by Means of Fuzzy Adaptive Gravitational Search Algorithm-Based FACTS Devices Under Fault Condition

S. Venkata Padmavathi, Sarat Kumar Sahu and A. Jayalaxmi

Abstract In deregulated power market, the parameters of the power system are to be operated nearer to the rated values to meet up the load. To improve system security with existing power lines, the flexible AC transmission system (FACTS) devices are the best alternatives. In this work, the line power flow factor and bus voltage deviations are considered as power system security indices for the system security problem. The proposed algorithm to improve the security is fuzzy adaptive gravitational search algorithm (FAGSA), as gravitational search algorithm (GSA) is wearisome to resolve the best operating setting of FACTS due to the selection of gravitational constant (G). To alleviate this difficulty, FAGSA is selected to tune the G with the help of fuzzy rules. This method is validated on standard IEEE 14 bus system, and the effectiveness is also tested under three-phase symmetrical fault condition.

Keywords GSA · FAGSA · FACTS · SVC · TCSC · Symmetrical fault

S. V. Padmavathi (✉)
EEE Department, GITAM (Deemed to be University), Hyderabad campus, Hyderabad, TS, India
e-mail: sv.padmavathi@gmail.com

S. K. Sahu
EEE Department, M.V.G.R Engineering College, Vizianagaram, AP, India
e-mail: sahu.sarat@gmail.com

A. Jayalaxmi
EEE Department, JNT University, Kukatpally, Hyderabad, TS, India
e-mail: ajl1994@yahoo.co.in

J. Nayak et al. (eds.), *Soft Computing in Data Analytics*, Advances in Intelligent Systems and Computing 758,
https://doi.org/10.1007/978-981-13-0514-6_10

1 Introduction

Economic, effective and secured operation of electrical power system is a vital problem to meet the system demand. With the increased everyday demand and competition in deregulated power industry, the power system will become complex and less secure. If the present operating state is found insecure, action should be taken [1]. In this situation, to meet the demand, either the existing power lines must be used more effectively or new transmission lines are to be added in the system. The latter is regularly not practical because of the reason of time-consuming process and environmental factors [2, 3]. To lessen this problem, FACTS devices are designed.

Generally, these devices are power electronics-based converters that have the capability to govern various transmission network parameters [4]. The different types of FACTS devices are: thyristor-controlled series compensator (TCSC) [5], static VAR compensator (SVC) [6], static synchronous series compensator (SSSC) [7], static compensator (STATCOM), unified power flow controller (UPFC) [7], etc.

The increased concern in these FACTS is mainly for two reasons: effective in cost and increased demand of power networks [8]. Finding the best placement of FACTS devices to enhance the system security is a problem. The heuristic methods [9–12] and intelligence techniques [13, 14] can be used to alleviate this problem. The FACTS devices are used effectively with proper planning of reactive power [15, 16]. Some of the literature papers have been followed the approaches to obtain the system security and stability, which is as follows in [17], and an optimal placement method for FACTS devices for market-based power systems is considered congestion ease and voltage stability. In [18], SCOPF for an economical operation of FACTS devices in liberalized energy markets is presented. In [19, 20], VAR planning and fault analysis are presented.

In this work, the proposed algorithm is FAGSA, which helps to solve the optimal setting of FACTS devices and also tested this device under three phase to ground (LLLG) fault. Here the gravitational constant (G) is tuned by fuzzy "IF/THEN" rules to lessen the problems in GSA. The GSA is explicated in Sect. 4, and the FAGSA method is explicated in Sect. 5. Fault analysis is discussed in Sect. 6. Before that, power system security assessment is explicated in Sect. 3 and the FACTS devices are modelled in Sect. 2.

2 Modelling of FACTS Devices

In this segment, the modelling of two types of FACTS devices is presented. The circuit diagram models of TCSC [5] and SVC [6] are shown in Figs. 1 and 2.

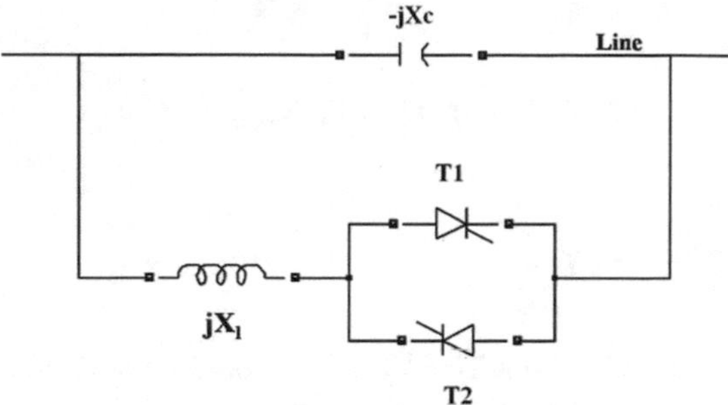

Fig. 1 TCSC model

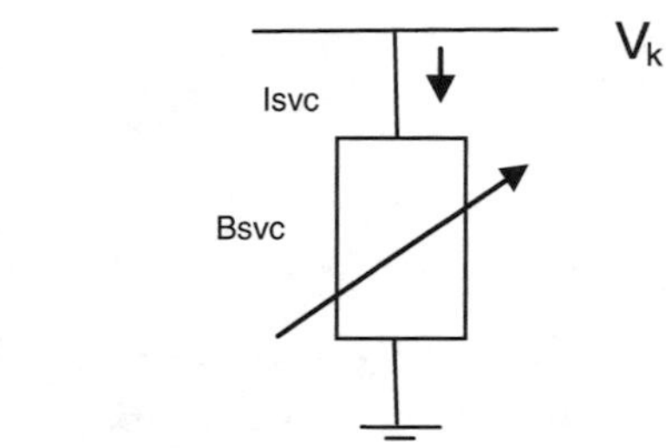

Fig. 2 SVC equivalent variable susceptance model

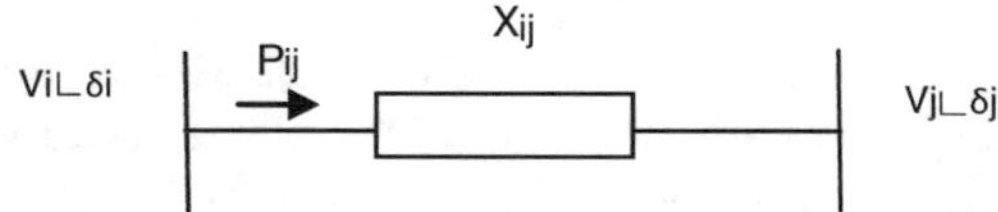

Fig. 3 A simple transmission line

From Fig. 3, the power flow (P_{ij}) through the transmission line is inversely proportional to the reactance (X_{ij}) of line and directly proportional to bus voltages (V_i, V_j) and angles (δ_i, δ_j).

It is given as [8] in Eq. (1).

$$P_{ij} = \frac{V_i V_j \sin(\delta_i - \delta_j)}{X_{ij}}. \tag{1}$$

2.1 *TCSC*

TCSC controls the line reactance. This alters the power flow in the line due to change in line series reactance. From Fig. 1, TCSC is modelled [5] as following Eq. (2).

$$X_{ij} = X_{line} + X_{TCSC}. \tag{2}$$

Where X_{line}: Line reactance, X_{TCSC}: TCSC reactance.

2.2 SVC

The FACTS device SVC [6, 9] could be utilized for inductive as well as capacitive compensation. From Fig. 2, SVC is modelled as a reactive power generation device at ith bus. It is given [21] as in Eq. (3).

$$\Delta Q_i = \frac{V_j\left(V_j - V_i\cos\left(\delta_i - \delta_j\right)\right)}{X_{ij}}. \tag{3}$$

3 Power System Security Index and Formulation of Objective Function

The major aim of the system security is to preserve the voltage profile [2, 12, 13, 15, 16] and not to violate transmission line power flow. These two security objectives are modelled using voltage security index (VSI) and line security index (LSI), respectively. Both the indices [5, 22] are given as in Eqs. (4) and (5), and constraints are shown in Eq. (6).

$$VSI = \sum_i w_i \left|V_i - V_{i,ref}\right|^2. \tag{4}$$

$$LSI = \sum_j w_j \left(\frac{P_j}{P_{j,\max}}\right)^2. \tag{5}$$

$$\left.\begin{array}{l} V_i^{\min} \le V_i \le V_i^{\max} \\ P_j \le P_j^{\max} \end{array}\right\}. \tag{6}$$

Where V_i, w_i and $V_{i,ref}$ are voltage amplitude, weighting factor and nominal voltage magnitude for ith bus, respectively. P_j, w_j and $P_{j,\max}$ are real power, associated weighting factor and the nominal real power of jth line, respectively. The weighting factor is mainly used to reflect the significance of lines, and here it is selected as 1. The cost functions of TCSC and SVC are shown [9] in Eq. (7).

$$\left.\begin{aligned} C_{TCSC} &= 0.0015s^2 - 0.7130s + 153.75 \\ C_{SVC} &= 0.0003s^2 - 0.3051s + 127.38 \end{aligned}\right\}. US\$/KVAR. \tag{7}$$

Where s: FACTS device operating size in MVAR. Therefore, the total installation cost index (ICI), for two FACTS devices, is given as in Eq. (8).

$$ICI = C_{TCSC} + C_{SVC}. \tag{8}$$

The total power loss is given in Eq. (9).

$$PLI = \sum_{k=1}^{NL} g_k \left[V_i^2 + V_j^2 - 2V_iV_j\cos\left(\delta_i - \delta_j\right)\right]. \tag{9}$$

Where NL: total number of lines and g_k: conductance of the kth line. The main object of the proposed work is to allocate these devices optimally in the power system to improve the system security (VSI and LSI) and economic operation of FACTS devices (ICI) with the reduced total power losses (PLI). Therefore, the single objective function is as given in Eq. (10).

$$SOF = \{(LSI) + (VSI) + (ICI) + (PLI)\}. \tag{10}$$

The objective or fitness function is subjected to some constraints.

Line power limits: $P_{ij} \leq P_{ij}^{max}$
where Pij: power flow between the nodes i and j.
Voltage constraint: $0.95 \leq V_b \leq 1.05$
where V_b: system bus voltage.

The ranges of FACTS devices are given in Eqs. (11) and (12).

$$-0.8X_L \leq X_{TCSC} \leq 0.2X_L. \tag{11}$$

$$-1\,p.u \leq Q_{svc} \leq 1\,p.u. \tag{12}$$

where X_{TCSC} is the TCSC reactance. X_L is the transmission line reactance, and Q_{SVC} is the SVC injected reactive power.

4 Gravitational Search Algorithm (GSA)

GSA is an evolutionary algorithm introduced recently which is stimulated by Newton gravitational law by Rashedi et al. in 2009 [10, 11]. It is best known for its searcher particles or agents termed as group of masses. In physics, the word gravitation is explained by objects with the mass accelerate towards each other and particle attracts the other particles with a force named as the gravitational force and is directly proportional to the product of the masses and is inversely proportional to distance square between them [23].

5 Fuzzy Adaptive Gravitational Search Algorithm (FAGSA)

It is an innovative stochastic optimization search algorithm [14], which is developed to apply the engineering problems. FAGSA method is used to adjust the gravitational constant (G) with the use of fuzzy [13] "IF/THEN" rules. Appropriate selection of constant G gives a balance among the global exploration, exploitation and local exploration, which provides lesser iterations to get the best solution. The G describes the performance of the particles, and in addition, the experience says that the search failure or success.

Therefore to overcome these problems, FAGSA is proposed to tune the value of G using the fuzzy rules to solve the optimization problem. The input variables considered for fuzzy inference system (FIS) are present best performance evaluation called as normalized fitness value (NFV) and the present constant G. For output, it is taken as the variation or change in the gravitational constant (ΔG) as shown in Fig. 4.

The fuzzy logic system consists of four main components; those are explained in the subsequent subsections.

5.1 *Fuzzification*

Two input fuzzy sets are considered to get the better G value. The inputs are (i) NFV (ii) G, and the output variable is considered as the change in gravitational constant (ΔG). Here triangular membership functions are considered for fuzzification of input and output with three linguistic parameters: Small (S), Medium (M) and Big (L), whereas for the output, the three linguistic values are: Negative (NG), Zero (Z) and Positive (PV).

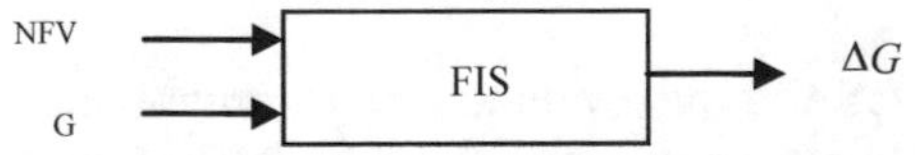

Fig. 4 Fuzzy inference system

Table 1 Fuzzy rules for ΔG

Rule no.	NFV	G	ΔG
1	S	S	Z
2	S	M	NG
3	S	L	NG
4	M	S	PV
5	M	M	Z
6	M	L	NG
7	L	S	PV
8	L	M	Z
9	L	L	NG

5.2 *Fuzzy Rules*

These are utilized to prepare conditional statements, and Mamdani-type rule base implication is used.

Here two inputs having three linguistic values are considered. Therefore, nine (3 * 3 = 9) fuzzy rules can be constructed as shown in Table 1.

5.3 *Fuzzy Reasoning*

This methodology is utilized to map the input variables to the output variables. Usually, the operator AND is employed to join the fuzzy membership values for every rule to produce the fuzzy membership value for the output variable.

The NFV [13] is defined as shown in Eq. (13):

$$NFV = \frac{SOF - SOF_{\min}}{SOF_{\max} - SOF_{\min}}. \tag{13}$$

Here the lower value of the NFV indicates the better solution. Single objective function (SOF) is a fitness function calculated from Eq. (10). The range of G has been taken between 0.4 and 1.0, the range of *NFV* has been taken between 0 and 1.0, and change in the gravitational constant range is taken between −0.1 and +0.1. And present value of gravitational constant is updated as shown in Eq. (14).

$$G^{t+1} = G^{t} + \Delta G. \tag{14}$$

5.4 Defuzzification

For defuzzification, the centroid method is used for the given membership functions.

6 Symmetrical Fault Analysis

In power system, 5% of transmission line faults are symmetric. It is the most severe fault occurred infrequently. This problem is solved using per-phase basis.

So far, we studied short circuit calculations for very simple systems whose network can easily be reduced. Now this study extended to n bus system. Initial step is to get the pre-fault voltages and currents in the network using load flow studies. At the time of fault, current entering at each bus except the fault bus is taken as zero and fault current equation is shown in Eq. (15).

$$\begin{aligned} & I_{Bus}(F) = Y_{Bus}\Delta V_{Bus}. \\ & \Delta V_{Bus} : \textit{Voltage change} \end{aligned} \tag{15}$$

$$V_{Bus}(F) = V^0_{Bus} + \Delta V_{Bus}. \tag{16}$$

Equation (16) is the fault bus voltage and is sum of before-fault bus voltage and change in bus voltage during fault. The short circuit current is given in Eq. (17).

$$\mathrm{I_{ij}(F) = V_i(F) - V_j(F)/(Z_{ij})}. \tag{17}$$

7 Simulation Results and Discussion

Initially, the population is produced randomly within the limits. The variables are FACTS location and size. The fitness or objective function (SOF) is given in Eq. (10). It consists of four terms. In the SOF, the four terms are security indices (Jp and Jv), FACTS investment cost and network loss. Here for every vector of population, the power line data is upgraded for TCSC device and bus data is upgraded for SVC. Next run the Newton–Raphson power flow and find the bus or node voltages, power flows. By utilizing these results, objective or fitness is calculated. This process is continued till maximum iterations are arrived. The proposed FAGSA method for best location and setting of FACTS is developed in the MATLAB software and tested on IEEE 14 bus system. The GSA and FAGSA performance is analysed as follows. Table 1 shows fuzzy rules used in fuzzy logic toolbox rule editor. Table 2 shows GSA parameters. Tables 3, 4 and 5 show the feasible location of FACTS devices and security indices for various cases.

Table 2 GSA parameters

NP	G_o	Iterations
50	100	100

Table 3 Performance of GSA in terms of VSI and LSI

Scenario	Location	LSI	VSI	Cost ($)	PLI (MW)
1 (base case)	–	27.162	0.0179	–	13.58
2 (with TCSC)	1–5	26.993	0.01630	129790	12.40
3 (with SVC)	14	27.150	0.0174	120521	12.25

Gravitational constant G is a function of value (G_o), and G_o is assumed initially with high value as shown in Table 2, and to control the search accuracy, the G_0 should be reduced with the time. NP is the population size generated randomly. If the population is more, the search accuracy gets improved.

By observing Table 3, the security indices are improved in case of TCSC. The values improved from 27.162, 0.0179 to 26.993 and 0.01630, and in case of SVC, the values improved from 27.162, 0.0179 to 27.150, 0.0174. By observing these values, the security indices are reducing with reference to base case security index. The lesser security index means improved power system security and reduction in losses with reference to base case.

By observing Table 4, the security indices are improved in case of TCSC. The values improved from 27.162, 0.0179 to 26.05, 0.01620, and in case of SVC, the values improved from 27.162, 0.0179 to 26.140, 0.0144. By observing these values, the security indices are reducing with reference to base case security index. The lesser security index means improved power system security and reduction in losses with reference to base case. When fault occurs at bus 5, the bus voltages are heavily reduced compared to the pre-fault voltages and the loss is increased as shown in Tables 5 and 6.

Now to check how effectively the FACTS devices operating under fault condition is as follows: Under fault, the power flows through lines will be reduced due to reduced bus voltages. So the LSI index automatically reduced, and so it should be increased by maximizing it in the objective function by changing it as 1/(1 + LSI).

By observing Table 6, under balanced three-phase fault, the security indices are slightly improved. In case of TCSC, the values improved from 1.254, 5.145 to 1.667, 5.140, and in case of SVC, the values improved from 1.254, 5.145 to 1.521,

Table 4 Performance of FAGSA in terms of VSI and LSI

Scenario	Location	LSI	VSI	Cost ($)	PLI (MW)
1 (base case)	–	27.162	0.0179	–	13.58
2 (with TCSC)	2–5	26.05	0.0162	141402	12.20
3 (with SVC)	14	26.14	0.0144	122870	11.98

Table 5 Bus voltages during fault

S.no	Pre-fault voltage	Voltage during fault
1	1.06	0.4997
2	1.035	0.4776
3	1.010	0.5225
4	1.008	0.2041
5	1.015	0
6	1.020	0.4417
7	1.017	0.4295
8	1.060	0.6652
9	0.995	0.4019
10	0.992	0.4087
11	1.002	0.4247
12	1.004	0.4389
13	0.998	0.4359
14	0.997	0.4166

Table 6 Performance of FAGSA in terms of VSI and LSI under fault

Scenario	Location	LSI	VSI	Cost ($)	PLI (MW)
1 (base case)	–	1.254	5.145	–	67.8
2 (with TCSC)	2–3	1.667	5.140	927100	66.45
3 (with SVC)	4	1.521	5.130	950880	66.56

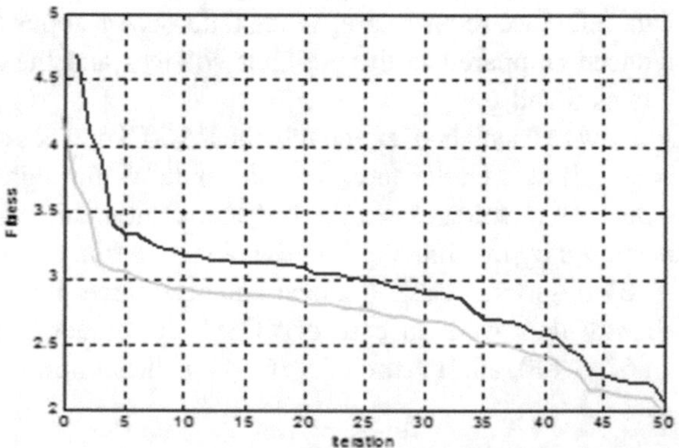

Fig. 5 Fitness versus iteration with SVC

5.130. So under fault condition, losses are more and FACTS devices improved the security by injecting reactive power. Here LSI should increase because at the time of fault-less power will flow through the line so index is reduced. So to increase the power nearer to the line limit, the LSI will be increased to improve security.

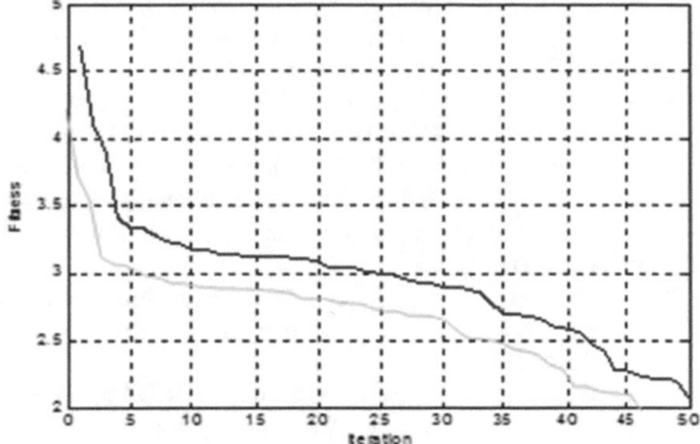

Fig. 6 Fitness versus iteration with TCSC

The results are shown that the installation of FACTS in the system could improve security and reduce system losses simultaneously. The performance characteristics of proposed FAGSA algorithm are compared with GSA. It is observed from Tables 3 and 4 that the proposed FAGSA is giving lesser power loss and security indices. And also the fitness variation graphs are improved by observing the graphs in Figs. 5 and 6 using FAGSA. Hence, the parameters attained from FAGSA are best compared to GSA technique. In graphs as shown in Figs. 5 and 6, the upper line indicates GSA algorithm and lower line for FAGSA technique.

8 Conclusion

In this work, a FAGSA technique is used for the best allocation and setting of devices SVC and TCSC. The result of GSA mainly depends on G, and the method has a problem of struck in local optima. To overcome this problem, the value of G has been adjusted by using fuzzy rules to reach global optima. From the results, it has been shown that the FAGSA proposed method has placed FACTS devices in optimal locations which improved the security indices for with and without fault condition, and hence, the power system security is enhanced.

References

1. Ranganathan, S., Kalavathi, M.S., Christober Asir Rajan, C.: Self-adaptive firefly algorithm based multi-objectives for multi-type FACTS placement. IET Gener. Trans. Distrib. **10**, 2576–2584 (2016)
2. Momoh, J.A., Zhu, J.Z., Boswell, G.D., Hoffman, S.: Power system security enhancement by OPF with phase shifter. IEEE Trans. Power Syst. **16**, 287–293 (2001)

3. Shakib, A.D., Balzer, G.: Optimal location and control of shunt FACTS for transmission of renewable energy in large power systems. In: IEEE Proceedings of Mediterranean Electro technical Conference, pp. 890–895 (2010)
4. Hingorani, N.G., Yugyi, L.G.: Understanding FACTS, 1st edn. IEEE Power Engineering Society, Standard Publishers Distributors, Delhi (2001)
5. Baghaee, H.R., Vahidi, B., Jazebi, S., Gharehpetian, G.B.: Power system security improvement by using differential evolution algorithm based FACTS allocation. In: Proceedings of International Conference on Power System Technology and IEEE Power India conference, pp. 1–6 (2008)
6. Ambriz-perez, H., Acha, E., Fuerte-Esquivel, C.R.: Advanced SVC models for Newton Raphson load flow and Newton optimal power flow studies. IEEE Trans. Power Syst. **26**, 129–136 (2000)
7. Anitha, C., Arul, P.: New modelling of SSSC and UPFC for power flow study and reduce power Losses. Int. J. Sci. Mod. Eng. **1**, 7–11 (2013)
8. Gerbex, S., Cherkaoui, R., Germond, A.J.: Optimal location of multi-type FACTS devices in a power system by means of genetic algorithms. IEEE Trans. Power Syst. **16**, 537–544 (2001)
9. Saravanan, M., Slochanal, S.M.R., Venkatesh, P., Prince Stephen Abraham, J.: Application of particle swarm optimization technique for optimal location of FACTS devices considering cost of installation and system loadability. Electric Power Syst. Res. **77**, 276–283 (2007)
10. Duman, S., Sonmez, Y., Guvenc, U., Yorukeren, N.: Optimal reactive power dispatch using a gravitational search algorithm. IET Gener. Trans. Distrib. **6**, 563–576 (2012)
11. Bhattacharya, A., Roy, P.K.: Solution of multi-objective optimal power flow using gravitational search algorithm. IET Gener. Trans. Distrib. **6**, 751–763 (2012)
12. Yorino, N., El-Araby, E.E., Sasaki, H., Harada, S.: A new formulation for FACTS allocation for security enhancement against voltage collapse. IEEE Trans. Power Syst. **18**, 3–10 (2003)
13. Su, C.-T., Lin, C.-T.: Fuzzy-based voltage/reactive power scheduling for voltage security improvement and loss reduction. IEEE Trans. Power Deliv. **16**, 319–323 (2001)
14. Vijaya Kumar, J., Vinod Kumar, D.M., Edukondalu, K.: Strategic bidding using fuzzy adaptive gravitational search algorithm in a pool based electricity market. Appl. Soft Comput. **13**, 2445–2455 (2013)
15. Za´rate-Minano, R., Conejo, A.J., Milano, F.: OPF-based security redispatching including FACTS devices. IET Gener. Trans. Distrib. **2**, 821–833 (2008)
16. Berizzi, A., Delfanti, M., Marannino, P., Pasquadibisceglie, M.S., Silvestri, A.: Enhanced security-constrained OPF with FACTS devices. IEEE Trans. Power Syst. **20**, 1597–1605 (2005)
17. Wibowo, R.S., Yorino, N., Eghbal, M., Zoka, Y., Sasaki, Y.: FACTS devices allocation with control coordination considering congestion relief and voltage stability. IEEE Trans. Power Syst. **26**, 2302–2310 (2011)
18. Lehmköster, C.: Security constrained optimal power flow for an economical operation of FACTS-devices in liberalized energy markets. IEEE Trans. Power Deliv. **17**, 603–608 (2002)
19. Kumar, S.K.N., Renuga, P.: Optimal VAR planning using FACTS. Int. J. Power Energy Syst. **32** (2012)
20. Seetharamayya, K., Venkateswararao, M.V.: Evaluation of fault voltage and current in a symmetric power network. Int. J. Innov. Res. Electr. Electron. Instrum. Control Eng. **3**, 26–35 (2015)
21. Acha, E., Fuerte-Esquivel, C.R., Ambriz-perez, H., Angeles-Camacho, C.: FACTS modelling and simulation in power networks, pp. 1–403. Wiley (2004)
22. Venkata Padmavathi, S., Sahu, S.K., Jayalaxmi, A.: Hybrid differential evolution algorithm based power system security analysis using FACTS. J. Electr. Eng. (JEE), **15**, 1–10 (2015)
23. Venkata Padmavathi, S., Sahu, S.K., Jayalaxmi, A.: Application of gravitational search algorithm to improve power system security by optimal placement of FACTS devices. J. Electr. Syst. **11**, 326–342 (2015)
24. Saada, H.: Power System Analysis, edition 2002. WCB/McGraw-Hill (1999) 1-690

Early Estimation of Protest Time Spans: A Novel Approach Using Topic Modeling and Decision Trees

Satyakama Paul, Madhur Hasija, Ravi Vishwanath Mangipudi and Tshilidzi Marwala

Abstract Protests and agitations have long been used as means for showing dissident toward social, political, and economic issues in civil societies. In recent years, we have witnessed a large number of protests across various geographies. Not to be left behind by similar trends in the rest of the world, South Africa in recent years has witnessed a large number of protests. This paper uses the English text description of the protests to predict their time spans/durations. The descriptions consist of multiple causes of the protests, courses of actions, etc. Next, we use unsupervised (topic modeling) and supervised learning (decision trees) to predict the duration of protests. The results are very promising and close to 90% of accuracy in early prediction of the duration of protests. We expect the work to help public services departments to better plan and manage their resources while handling protests in future.

Keywords Protests and agitations · Early prediction · Duration · South Africa Topic modeling · Latent Dirichlet allocation · Perplexity · Decision trees

---

S. Paul (✉)
Oracle India Pvt. Ltd., Prestige Technology Park, Bengaluru 560087, India
e-mail: satyakama.paul@gmail.com

M. Hasija
Abzooba India Infotech Pvt. Ltd, Office 201 Amar Neptune, Pune 411045, India
e-mail: madhur.hasija@gmail.com

R. V. Mangipudi
Synechron Technologies Pvt. Ltd, Adarsh Palm Retreat, Bengaluru 560103, India
e-mail: ravi.vishwanath.m@gmail.com

T. Marwala
University of Johannesburg, Auckland Park, Johannesburg 2006, South Africa
e-mail: tmarwala@uj.ac.za

J. Nayak et al. (eds.), *Soft Computing in Data Analytics*,
Advances in Intelligent Systems and Computing 758,
https://doi.org/10.1007/978-981-13-0514-6_11

1 Introduction

Protests and public dissents are a critical part of democratic civil societies. Not to be left behind when compared with the rest of the world, South Africa in recent years has also seen a massive increase in public protests. The causes of these protests were varied and have ranged from service delivery, labor-related issues, crime, education, to environmental issues.

While in the past multiple studies and news articles have analyzed the nature and cause of such protests, this research uses a combination of unsupervised (topic modeling using probabilistic latent Dirichlet allocation (pLDA)) and supervised (single and ensemble decision trees) learning to predict the duration of future protests. We develop an approach in which a user inputs a description of a protest in free-flowing English text, and the system predicts the duration of the protest to a high (close to 90%) degree of accuracy. We expect that an early correct prediction of the duration of the protest by the system will allow police and other security services to better plan and allocate resources to manage the protests.

2 Problem Statement

The objective of this research is to provide an early estimation of the time span/ duration of a protest based on South African protest data. The master dataset is obtained from the Web site—Code for South Africa [1]. It consisted of 20 features (columns) describing 876 instances (rows) of protests over the period of February 1, 2013, to March 3, 2014. Among the 20 features, the statistically important and hence selected ones are shown in Table 1. The rest are repeated codification of the important features that convey the same statistical information as the important features. Hence, they are ignored. Also, detailed addresses of the location (town or city name, first street, cross street, suburb area place name, etc.) of the protests are not considered. Instead, the more accurate measures—coordinates (latitudes and longitudes)—are used.

Tables 2, 3, 4, and 5 show the overall descriptive statistics of the protests during the above-mentioned period. From the 876 rows, three are removed for which the

Table 1 Important statistical features

Whether metro or not	Police station
Coordinates	Start date
End date	Cause of protest
Status of protest (violent or not)	Reason for protest (text data)

Table 2 Percentage of protests vis-a-vis provinces

Provinces	% of protests	Provinces	% of protests
Gauteng	37	Limpopo	6
Western Cape	18	Mpumalanga	5
KwaZulu-Natal	14	Free State	3
Eastern Cape	9	Northern Cape	2
North West	6		

Table 3 Percentage of protests vis-a-vis issues

Issue	% of protests	Issue	% of protests
Service delivery	31	Political	4
Labor	30	Transport	3
Crime	12	Xenophobia	2
Election	6	Individual causes	1
Vigilantism	5	Environment	1
Education	5		

Table 4 Percentage of protests vis-a-vis state

State	% of protests
Peaceful	55
Violent	45

one or more columns are missing.[1] Thus, our modeling exercise is based on 873 instances of protests. Table 2 shows that Gauteng[2] as the seat of commerce, and President and Cabinet, and Western Cape[3] as the legislative capital have the largest concentration of protests, followed by the others. Table 3 shows that the three largest issues of protests are from service delivery, labor-related issues, and crime related at, respectively, 31%, 30%, and 12% of the total. From Table 4, it can be seen that the difference between peaceful and violent protests is low at 55% and 45%, respectively. However, the most interesting insight comes from the duration of protests.[4] From Table 5, it can be observed that the majority of protests (at 74.34% of the total) last for less than 24 h. Thus, this feature is highly skewed.

[1]Since an insignificant percentage (0.34%) of our data is missing, we conveniently remove them without doing missing value imputation.

[2]Gauteng has in it Johannesburg and Pretoria. Johannesburg is the commercial hub of South Africa, and the office of President and Cabinet is in Pretoria.

[3]Cape Town as the seat of South African parliament is situated in Western Cape.

[4]Duration of protests is the difference between the End Day and the Start Day of the protest.

Table 5 Percentage of protests vis-a-vis duration in days

Duration	% of protests
0 (less than 24 h)	74.34
1	11.34
2	4.58
3	2.06
4	2.17
5–13,19, 21–23, 31, 34, 37, 39, 57, 65	Less than 1%

The idea behind this work is to use only the text description of the protests (predictor variable) to predict the duration of future protest(s) (response variable). Some typical examples of the descriptions of protests are as follows: flagged as service delivery and violent—"residents of both towns Butterworth and Centane blockaded the R-47 between the two towns, accusing the Mnquma Municipality of ignoring their request for repairs to the road." Eyeballing the text does not indicate any violence. A second example flagged as service delivery and peaceful protest is "ANGRY community leaders in four North West villages under the Royal Bafokeng Nations jurisdiction protested this week against poor services and widespread unemployment among the youth. Now, they not only demand their land back, but want a 30% stake in the mines which are said to employ labor from outside the villages." While the first line of the text referred to the cause of the protest as service delivery, the second line referred to political (demand for return of land) and labor issues (3% stake in mines and corresponding increase in employment). In this sense, we believe that strict flagging of protests into one category or another restricts the knowledge of the protests. We also believe that it is normal for human social concerns to spill from one area to another during protests that are not well captured by only one restrictive flag attached to one protest. Thus, we drop the categorical features (with strict class labels) and consider only the text descriptions of the protests as a predictor of its duration. Another important advantage of using text descriptions is that they give flow/progress of events that occurred during a protest and other relevant details. Figure 1 shows the word cloud of the entire protest corpus. Prior building the word cloud, the usual preprocessing on the text corpus such as removal of punctuation, numbers, common English stopwords, white spaces has been carried out. The cloud consists of 75 words,[5] and words that occur with a minimum of at least 25 times are included. The more the frequency of the words occurring in the text corpus, the bigger is its font size. Lastly due to word stemming, resid is created from words like residence and residing.

[5]The number of words is kept low for better visibility.

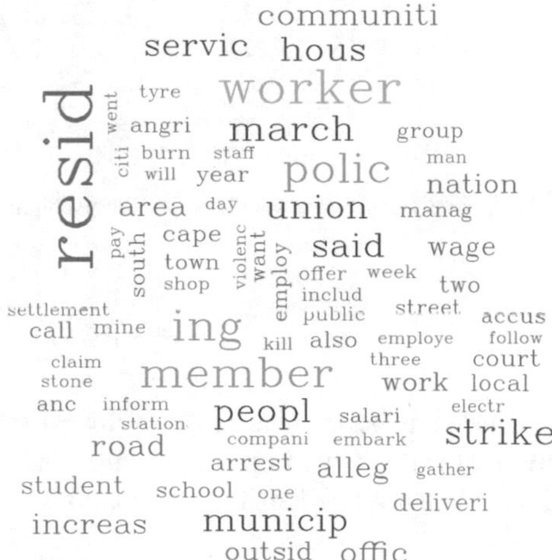

Fig. 1 Word cloud of the entire text corpus of the protests

3 Theoretical Concept—Probabilistic Latent Dirichlet Allocation

In this subsection, we provide a short introduction to pLDA. LDA is an unsupervised generative probabilistic model primarily used for topic discovery. It states that for a collection of words in a large number of documents, each document is a mixture of a number of "latent topics"[6] and each word in a document is a result of a latent topic. Following [2, 3], a document is a random mixture of latent topics and each topic[7] in turn is created by the distribution of words. Mathematically the LDA model can be stated as follows. For each document **w** in a corpus D

1. Choose $N \sim Poisson(\xi)$.
2. Choose $\theta \sim Dir(\alpha)$.
3. For each of the N words w_n:
    (a) Choose a topic $w_n \sim Multinomial(\theta)$.
    (b) Choose a word w_n from $p(w_n|z_n, \beta)$, a multinomial probability conditioned on the topic z_n.

where a document **w** is a combination of N words, for example $\mathbf{w} = (w_1, w_2, \ldots, w_N)$. A corpus D is a collection of M documents such that $D = (\mathbf{w}_1, \mathbf{w}_2, \ldots, \mathbf{w}_M)$. α and β are parameters of the Dirichlet prior on per document topic and per topic word distribution, respectively. z is a vector of topics. The central idea of LDA is to find the posterior distribution (θ) of the topics (z) when the document (**w**) is given, i.e.,

[6]Note that the topics are latent because they are neither defined semantically nor epistemologically.

[7]Topic and latent topic are the same and used interchangeably.

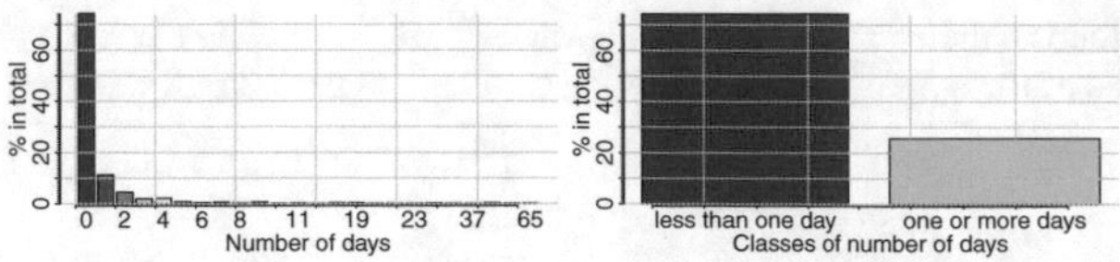

Fig. 2 Setting of two-class classification problem

$$p(\theta, z|\mathbf{w}, \alpha, \beta) = \frac{p(\theta, z, \mathbf{w}|\alpha, \beta)}{p(\mathbf{w}|\alpha, \beta)} \tag{1}$$

Since it is beyond the scope of this paper to derive the detailed formula, we summarize the two other important results that will subsequently be required in our analysis. The marginal distribution of a document is:

$$p(\mathbf{w}|\alpha, \beta) = \int p(\theta|\alpha)(\prod_{n=1}^{N}) \sum p(z_n|\theta)p(w_n|z_n, \beta)d\theta \tag{2}$$

The probability of a corpus is:

$$p(D|\alpha, \beta) = \prod_{d=1}^{m} \sum p(\theta_d|\alpha)(\prod_{n=1}^{N_d} p(w_{dn}|z_{dn}, \beta)d\theta) \tag{3}$$

4 Experimental Setup

As seen in Table 5, our response variable duration of protests is highly skewed. Protests lasting for less than 24 h are 74.34% of the total number of protests. Protests above one day are less than 5% of the total number. In effect, it means that in reality, South Africa rarely experiences protests that stretches beyond one day. So for practical purposes, we couple protests lasting above 24 h into one class and compare it against protests lasting below 24 h. The right panel of Fig. 2 shows the percentage of protests falling in less than one day (74.34%) and one or more days (25.66%). Thus, the binary classification problem now is to correctly predict the response variable (less than one day against one or more days) for the text corpus of the protest descriptions.

However before getting into the classification exercise, we need to perform two tasks. One, we need to find the optimal number of latent topics from the text corpus. Two, since classification algorithms per se cannot take text documents, we need to extract a set of latent topics for each text description of a protest. In the next section, we discuss the above tasks and their results.

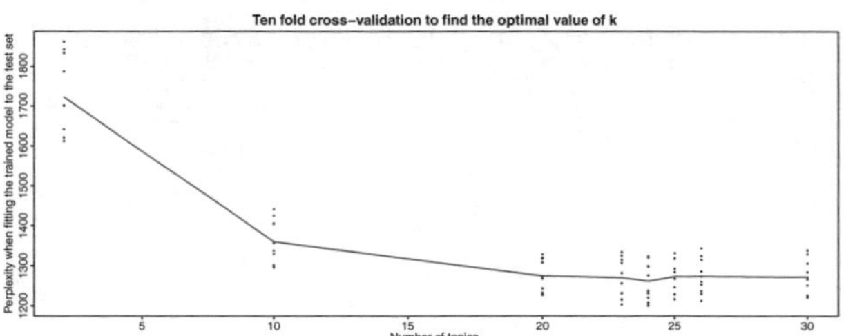

Fig. 3 Finding the optimal number of topics

5 Results and Discussions

Finding the optimal number of hidden topics from the text corpus is an important task. If the chosen number of topics is too low, then the LDA model is unable to identify the accurate classifiers. However, if the number is too high, the model becomes increasingly complex and thus less interpretable [4, 5]. In contrast to the often used procedure of intelligently guessing the optimal number of topics in a text corpus, following [2] we use the perplexity approach to find the optimal number of topics. Often used in language/text-based models, perplexity is a measure of "on average how many different equally most probable words can follow any given word" [6]. In other words, it is a measure of how well a statistical model (in our case LDA) describes a dataset [4] where a lower perplexity denotes a better model. It is mathematically denoted by:

$$perplexity(D_{test}) = exp\{-\frac{\sum_{d=1}^{M} logp(\mathbf{w}_d)}{\sum_{d=1}^{M} N_d}\} \tag{4}$$

where the symbols have the same meaning as in Sect. 3. Using a trial number of topics that range between 2 and 30,[8] we use perplexity to find the optimal number of topics from our text corpus. The text corpus is broken into a training and test set. A tenfold cross-validation is carried out using 1000 iterations. The black dots show the perplexity score for each fold of cross-validation on the test set for various numbers of topics. The average perplexity score is shown by the blue line in Fig. 3, and the score is least for 24 latent topics. Thus, the optimal number of topics for our text corpus is 24.

Next, we perform a LDA with 24 topics on our text corpus. Table 6 shows an example of a text description and the probabilities associated with the various topics.

The topic names are given in the bottom of the table. The topics court and resid have the highest probabilities for the text. It might also be noted that since the topics are multinomial distributed to the entire text corpus and the words in corpus, hence there is no direct or visible relationship that connects the topics with the text

[8]X-axis of Fig. 3 shows the trial number of topics.

Table 6 An example of a text description and its associated topic probabilities

Text	$P(T_1)$..	..$P(T_8)$	...	..$P(T_{20})$	..$P(T_{24})$
The residents wanted Sterkspruit to be moved from the Senqu Municipality and be a municipality on its own	0.0027	0.15	...	0.79	0.0028

$P(T_n)$ refers to probability of the *n*th topic where n varies from 1 to 24. Here, T_1 = shop, T_8 = March, T_{20} = municip, and T_{24} = anc

Table 7 An example of text description and its associated highest probability topics

Text	LP	2nd **LP**	3rd **LP**	4th **LP**
The residents wanted Sterkspruit to be moved from the Senqu Municipality and be a municipality on its own	Municip	March	Resid	Hospit

$P(T_n)$ refers to probability of the *n*th topic where n varies from 1 to 24. Here, T_1 = shop, T_8 = March, T_{20} = municip, and T_{24} = anc

descriptions. We can only assume that the topics are complexly related to the text descriptions.

Table 7 shows the topics with four largest probabilities (largest, second largest, ...fourth largest) associated with a text description. With computational concerns in mind, we restrict the largest probabilities to only four.

It might be recollected from the last paragraph of Sect. 4 where we stated that classification algorithms per se cannot take text documents as a predictor variable. Table 7 shows a way in which a single text description can be represented as a set of most relevant[9] topics. Thus for an entire corpus of 873 text documents, the predictor side of the classification model would consist of a topic matrix of 873×4 dimensions. Including the response variable, our modeling data would have 873×5 dimensions.

In addition, it might also be recalled that our response variable duration of protest is unbalanced with the percentage of values falling in the class—less than one day at 74.34% and one or more days at 25.66%. Since unbalanced classes are not learned well by decision trees, so we use both way balanced sampling strategies to balance

[9]The most relevant topics are the ones with highest probabilities.

Table 8 Performance metrics of various algorithms on the test set

	C5.0	Treebag	Random forest
Balanced accuracy (%)	79.38	88.40	89.69
Kappa	0.590	0.769	0.795
Sensitivity (%)	87.03	94.59	95.68
Specificity (%)	72.41	82.76	84.24

P(T_n) refers to probability of the *n*th topic where n varies from 1 to 24. Here, T_1 = shop, T_8 = March, T_{20} = municip, and T_{24} = anc

the modeling data. Next, the modeling data is split into a training and test set in the 7:3 ratio. The dimensions of the training and test sets are 910×5 and 388×5, respectively.

In the next step, we use decision tree-based classification algorithms to model the relationship between the topic matrix and the duration of the protests. Specifically, we use three algorithms—C5.0, treebag, and random forest—and do a tenfold cross-validation with five repeated iterations. The prediction results for the class one or more days on the test dataset are shown in Table 8.

Treebag with a number of decision trees is able to better predict the response variable than a single tree in C5.0 because multiple trees help in reducing variance without increasing bias. Again, random forest performs better than treebag because in addition to multiple trees it also randomly selects a subset of the total number of features at each node. Further best split feature from the subset is used to split each node of the tree. This salient feature of random forest is absent in treebag. The combination of a number of trees and random selection of a subset of features at each node of the tree helps in further reducing the variance of the model. Thus in effect, random forest performs best in predicting the duration of the protests on the test dataset.

6 Limitations

The first limitation of this research is that LDA does not allow for evolution of topics over time. This means that if the nature and scope of the protests remain pretty stagnant over time then our model is expected to perform fairly well. Second, since it is a bag of words model, so sentence structures are not considered. And third, topics are not independent of one another. Thus, it might happen that the same word represents two topics creating a problem of interpretability.

7 Conclusion

This paper is a combination of unsupervised and supervised learning to predict the duration of protests in South African context. Protests and agitations being social issues have multiple nuances in terms of causes, courses of actions, etc., that cannot be very well captured by restrictive tags. Thus, we discard the approach of restrictive characterization of protests and use free-flowing English texts to understand its nature, cause(s), course of action(s), etc. Topic discovery (unsupervised learning using pLDA) and subsequent classification (supervised learning using various decision tree algorithms) provide promising results in predicting the duration of protests. We expect that the implementation of the framework can help police and other security services in better allocating resources to manage the protests in future.

References

1. Code for South Africa, *Protest Data*. https://data.code4sa.org/dataset/Protest-Data/7y3u-atvk
2. Blei, D.M., Ng, A.Y., Jordan, M.I.: Latent Dirichlet Allocation. J. Mach. Learn. Res. **3**, 993–1022 (2003)
3. Reed, C.: Latent Dirichlet Allocation: Towards a Deeper Understanding. http://obphio.us/pdfs/ldatutorial.pdf
4. Zhao, W., Chen, J.J., Perkins, R., Liu, Z., Ge, W., Ding, Y., Zou, W.: A Heuristic Approach to Determine an Appropriate Number of Topics in Topic Modeling, 12th Annual MCBIOS Conference, March 2015
5. Zhao, W., Zou, W., Chen, J.J.: Topic modeling for cluster analysis of large biological and medical datasets, BMC Bioinform. **15**(Suppl 11) (2014)
6. NCLab, *Perplexity*. http://nclab.kaist.ac.kr/twpark/htkbook/node218_ct.html

Mathematical Model for Characterization of Lung Tissues Using Multiple Regression Analysis

D. Lakshmi and R. Niruban

Abstract Tuberculosis (TB) and lung cancer are major problems of the lung, and their occurrence as co-morbidities is dealt in many studies. World Health Organization (WHO) states that over 95% of deaths occur due to TB in low and middle income countries. This work presents a retrospective cohort study to derive the mathematical model characterizing the tissues like fibrosis, TB, and carcinoma. The cohort includes 113 normal cases, 103 fibrosis cases, 185 carcinoma cases, and 39 suspicious of tuberculosis cases. Multiple Regression Analysis (MRA) is performed on Gray-Level Co-occurrence Matrix (GLCM)- and Gray-Level Run Length Matrix (GLRLM)-based features extracted from CT images for the characterization of lung tissues. MRA of these 18 numbers of GLCM and 44 numbers of GLRLM-based features gives R^2 value of 0.8827 and 0.9456 with mean square error (MSE) of 0.01979 and 0.009852, respectively.

Keywords Lung abnormalities · Mathematical modeling · Multiple Regression Analysis

D. Lakshmi (✉)
Department of Electronics and Instrumentation Engineering, St. Joseph's College of Engineering, Old Mamallapuram Road, Semmencherry, Chennai 600119, Tamil Nadu, India
e-mail: lakhramdevan@gmail.com

R. Niruban
Department of Electronics and Communication Engineering, St. Joseph's College of Engineering, Old Mamallapuram Road, Semmencherry, Chennai 600119, Tamil Nadu, India
e-mail: nirubanme@gmail.com

J. Nayak et al. (eds.), *Soft Computing in Data Analytics*,
Advances in Intelligent Systems and Computing 758,
https://doi.org/10.1007/978-981-13-0514-6_12

1 Introduction

According to WHO, lung cancer and tuberculosis are among the top 10 causes of deaths globally in 2015 [1]. Tuberculosis (TB) is an infectious disease caused by the bacterium Mycobacterium tuberculosis. It is the second leading disease due to a single infectious agent [2]. It affects the lungs and other parts of the body. One-third of the world's population is thought to be infected with TB. In developing countries, effective therapies are used to reduce the infectious pulmonary tuberculosis. It prevails as public health problem [3], due to lack of clarity in the complex overlapping structures of TB in chest CT images. It is also imperative to study the differentiation of lung cancer with TB.

Currently, lung abnormalities are diagnosed and confirmed only by histopathological test. Assessment by histopathological test is indispensable at an early stage of lung abnormalities. An inexpensive tool called chest radiograph is used to identify the signs or symptoms in the clinical practice [4]. However, the CT utilization rate increases in the emergency department of every hospital [5].

Clinical diagnostics of these diseases becomes difficult due to vague symptoms or inexperience and insufficient human resources. A similarity in the appearance of these lung tissues in the CT images envisages higher competence algorithm to differentiate the lung cancerous tissue from TB tissues. Timely diagnosis of these tissues concedes therapeutic procedures. Hence, there is a need for a consistent non-invasive tool for characterization of these lung abnormalities.

The statistical textural feature-based model is proposed in this paper. The textural pattern features are extracted from CT image showing statistical distribution of pixel intensity. Two set of features are used in this model, namely GLCM and the GLRLM [6–9]. GLCM is a tabulation of the frequencies. GLRLM is a matrix based on length of pixel co-occurrences. Thus, regression analysis of these features results in extracting valuable data, deriving model rules and regulations, patterns and making precise decision for diagnosis and treatment of lung diseases from the database.

The main contribution of this paper is to obtain mathematical equations to characterize the lung tissue extracted from CT image without prior knowledge of Region of Interest (ROI). The proposed system involves preprocessing and morphological processing in CT images to get lung tissues, and the textures are extracted using GLCM- and GLRLM-based features. Finally, the extracted textures are used as input to MRA to obtain the mathematical model.

2 Materials and Methods

The system comprises three main steps as shown in Fig. 1. In step 1, the lung CT undergoes preprocessing and morphological processing to obtain the image containing the tissues inside the lung. In step 2, the textural features, namely GLCM

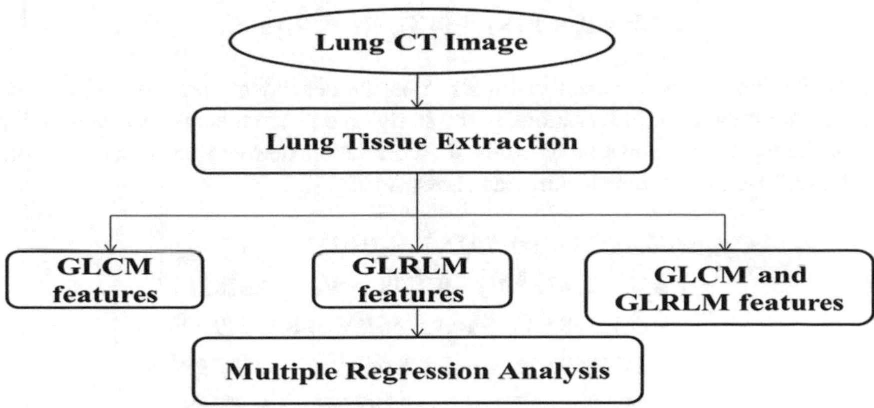

Fig. 1 Proposed system

and GLRLM, are extracted in order to enable the characterization of the tissues using CT images. In step 3, the extracted features are given to Multiple Regression Analysis. Here, steps 1 and 2 are carried by MATLAB tool and step 3 uses National Cruncher Statistical Software (NCSS).

2.1 The Proposed System

The original CT image is processed to extract the tissue gray levels present inside the lung [10]. Subsequently, mathematical model is derived from the GLCM- and GLRLM-based feature set for these tissues. A study by Chan et al. [11] shows the effectiveness of dependence matrix for the classification of masses and normal breast tissues from mammograms. The Gray-Level Co-occurrence Matrix approach (known as Spatial Dependence Matrix) introduced by Haralick et al. [12] and Run Length Matrix by Tang [13], a well-known statistical method for extracting texture-based features, are used for this study.

2.2 Mathematical Model

Multiple Regression Analysis (MRA) is a statistical technique used to derive the mathematical model stating the relationship between the dependent variable and two or more independent variables. Apart from regression analysis, studies reveal that quantifying tumor heterogeneity by standardized uptake value (SUV) is limited [14]. The goal of our research is to search regular patterns among the features by the MRA and to study the straight line relationship among two or more variables as shown in Eq. (1).

$$\mathbf{Y = \beta_0 + \beta_1 X_{1j} + \beta_2 X_{2j} + \cdots \beta_p X_{pj} + \varepsilon_j} \tag{1}$$

where Xs are the independent variables, Y is the dependent variable, suffix j represents the number of observations in the study, and βs are the unknown regression coefficients. MRA allows us to study the correlation between the tissue category and GLCM-based texture features as shown in Eq. (2).

$$\begin{aligned} \mathbf{C_{glcm}} = {} & \mathbf{56.62753 - 2.91797 * V_1 + 0.21517 * V_2} \\ & \mathbf{+ 75.427871 * V_3 + 0.93793 * V_4 - 74.82851 * V_5} \\ & \mathbf{- 2.33708 * V_6 + 14.93986 * V_7 + 8.75949 * V_8} \\ & \mathbf{- 12.66692 * V_{10} - 5.57816 * V_{11} - 30.81198 * V_{12}} \\ & \mathbf{+ 25.69261 * V_{14} - 64.02298 * V_{15} + 0.97608 * V_{16}} \\ & \mathbf{+ 7.47759 * V_{18}} \end{aligned} \tag{2}$$

where V_1, $V_2 \ldots V_{18}$ are the 18 variables representing GLCM-based features. It is to be noted that the variables V_9, $V_{13,}$ and V_{17} are ruled out in the estimated model, since the regression coefficients of these variables are zero. As per MM-1, these variables represent correlation, energy, and homogeneity in 90 directions and play no role for the characterization of the lung tissues taken for study. MRA on GLRLM-based features results as stated in Eq. (3). The model shows high correlation between the tissues category with the GLRLM-based texture features.

$$\begin{aligned} \mathbf{C_{glrlm}} = {} & \mathbf{16.16032 + 31.94339 * W_1 + 18.27784 * W_2} \\ & \mathbf{- 0.93134 * W_3 + 4.56757 * W_4 - 4.09438 * W_5} \\ & \mathbf{- 1.64760 * W_6 - 7.94566 * W_7 + 0.96557 * W_8} \\ & \mathbf{+ 8.93718 * W_9 + 0.65292 * W_{10} + 0.34035 * W_{11}} \\ & \mathbf{- 28.80494 * W_{12} - 5.15651 * W_{13} + 0.16630 * W_{14}} \\ & \mathbf{+ 28.38590 * W_{15} - 26.60958 * W_{16} - 0.47404 * W_{17}} \\ & \mathbf{- 23.62012 * W_{18} + 0.23924 * W_{19} + 24.47087 * W_{20}} \\ & \mathbf{+ 0.38098 * W_{21} + 1.00814 * W_{22} - 10.75567 * W_{23}} \\ & \mathbf{- 13.47880 * W_{24} - 0.02258 * W_{25} - 41.45759 * W_{26}} \\ & \mathbf{+ 40.03802 * W_{27} - 1.05187 * W_{28} + 38.89070 * W_{29}} \\ & \mathbf{+ 0.69702 * W_{30} - 38.31160 * W_{31} + 0.26327 * W_{32}} \\ & \mathbf{- 0.79732 * W_{33} - 3.54282 * W_{34} - 8.70673 * W_{35}} \\ & \mathbf{+ 0.33144 * W_{36} + 8.46983 * W_{37} - 7.76350 * W_{38}} \\ & \mathbf{- 1.54844 * W_{39} - 35.47847 * W_{40} + 0.95102 * W_{41}} \\ & \mathbf{+ 31.78052 * W_{42} + 1.32912 * W_{43} + 0.10245 * W_{44}} \end{aligned} \tag{3}$$

where W_1, W_2, W_3, $\ldots W_{44}$ are the variables representing GLRLM-based features.

Table 1 R^2 value and MSE

Model	Feature	R^2 value	MSE
MM-1	GLCM-based features	0.8827	0.01979
MM-2	GLRLM-based features	0.9456	0.009852

3 Results and Discussions

A total of 440 subjects including healthy, fibrosis, suspicious of TB, and carcinoma are used for the development of mathematical model. Two mathematical models, namely MM-1 and MM-2, are estimated for the characterization for lung tissues using GLCM- and GLRLM-based features. Both MM-1 and MM-2 obtain results demonstrating a strong correlation between the tissue category and GLRLM-based features or GLCM-based features as stated in Table 1.

Our mathematical model MM-1 provides a reasonable R^2 value of 0.8827 with MSE of 0.01979. But, our MM-2 competes with MM-1 giving significant R^2 value of 0.9456 with MSE of 0.009852.

4 Conclusion

It is interesting to note that GLCM- and GLRLM-based features are able to characterize the lung tissues, namely fibrosis, carcinoma, and suspicious of TB from healthy cases [9]. Figures 2 and 3 depict that the residual values of these two model outputs are distributed normally with less deviation giving closeness of fit with 95% confidence level [15]. However, MRA of these variables gives the mathematical model with their regression coefficient. The value of this coefficient contributes considerably toward the predicted output.

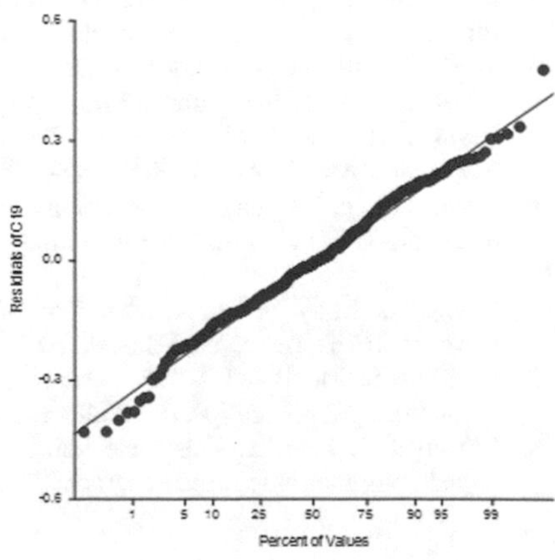

Fig. 2 Normal probability plot of residual of MM-1 output

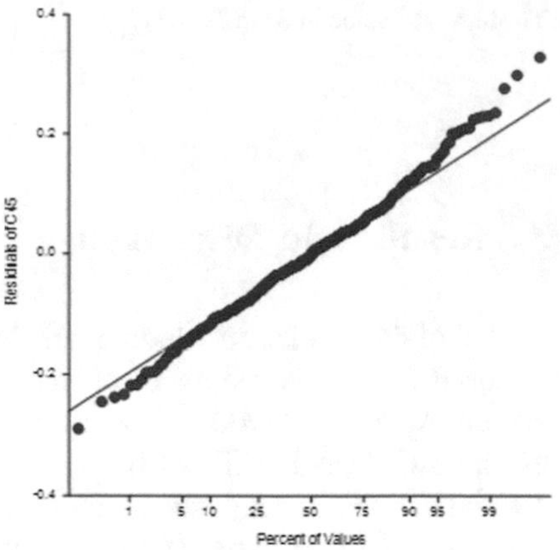

Fig. 3 Normal probability plot of residual of MM-2 output

Our study concludes that the performance of MM-2 is better than MM-1 to differentiate the carcinoma, fibrosis, and TB tissues. Despite satisfactory performance in the characterization of fibrosis, suspicious of TB, carcinoma from the healthy cases, the accuracy of the mathematical model always depends on the amount of dataset and the type of feature set.

Acknowledgements This research work was supported by Santhosham Chest Hospital, Chennai. We wish to thank Dr. Roy Santhosham and the technician of this hospital for giving us their support for the study.

References

1. http://www.who.int/mediacentre/factsheets/fs104/en/
2. World Health Organization: WHO Report 2011 Global Tuberculosis Control, pp. 9–27. WHO Press, Geneva, Switzerland (2011)
3. Hopewell, P.C., Pai, M., Maher, D., Uplekar, M., Raviglione, M.C.: International standards for tuberculosis care. Lancet Infect. Dis. **6**(11), 710–725 (2006)
4. World Health Organization1: Causes of Death 2008 Summary Tables. Global Health Observatory Data Repository (2011)
5. Levin, D.C., Rao, V.M., Parker, L.: The recent downturn in utilization of CT: the start of a new trend? Am. Coll. Radiol. **9**, 795–798 (2012)
6. Chen, C., Lee, G.: Image segmentation using multiresolution wavelet analysis and expectation maximum (EM) algorithm for mammography. Int. J. Imaging Syst. Technol. **8**(5), 491–504 (1997)
7. Wang, T., Karayaiannis, N.: Detection of microcalcification in digital mammograms using wavelets. IEEE Trans. Med. Imaging **17**(4), 498–509 (1998)
8. Majid, A.S., de Paredes, E.S., Doherty, R.D., Sharma, N., Salvador, X.: Missed breast carcinoma: pitfalls and pearls. Radiogr. **23**, 881–895 (2003)
9. Christiyanni, I., et al.: Fast detection of masses in computer aided mammography. IEEE Signal computer aided mammography. IEEE Signal

10. Devan, L., Santosham, R., Hariharan, R.: Automated texture based characterization of fibrosis and carcinoma using low-dose lung CT images. Int. J. Imaging Syst. Technol. **24**(1), 39–44 (2014)
11. Chan, H.P., et al.: Computer-aided classification of mammographic masses and normal tissue: linear discriminant analysis in texture feature space. Phys. Med. Biol. **40**, 857–876
12. Haralick, R.M., Shanmugam, K., Dinstein, I.: Textural features for image classification. IEEE Trans. Syst. Man. Cybernetics SMC-3, 610–621 (1973)
13. Tang, X., et al.: Texture information in run-length matrices. IEEE Trans. Image Process. **7** (11), 1602–1609 (1998)
14. Chicklore, S., Goh, V., Siddique, M., Roy, A., Marsden, P.K., Cook, G.J.R.: Quantifying tumour heterogeneity in ^{18}F-FDG PET/CT imaging by texture analysis. Eur. J. Nucl. Med. Mol. Imaging **40**, 133–140 (2013)
15. Hajian-Tilaki, K.: Sample size estimation in diagnostic test studies of biomedical iinformatics. J. Biomed. Inform. **48**, 193–204 (2014)

A Data Hiding Scheme with Digital Authentication Using Parity Checkers

Ayan Chatterjee, Tanmoy Bera, Soumen Kr. Pati and Asit Kumar Das

Abstract Information security is very important to authorize users for Internet-based system. Steganography is an approach to make communication secure in high-density space and maintained among respective users for hidden communication of secret information. Alternatively, unauthorized user cannot realize the existence of authorized communication. In this paper, a data hiding scheme is proposed using the parity checker in frequency domain. In time of inserting the secret message, the protection of steganographic communication is increased by parity checker from statistical attacks due to indirect occurrence of information in stego-image. In addition, this procedure develops the authentication between *LSB* matrix of *DCT* values of cover image and secret key. This scheme maintains all the conditions of information security with high concreteness during wireless communication among authenticated users. Efficiency of the scheme is analyzed from statistical attacks with the parameters such as *Peak Signal-to-Noise Ratio*, *Mean Squared Error*, *Embedding capacity,* and protection.

A. Chatterjee (✉)
D.El.ED Section, Sarboday Public Academy, East Midnapore, West Bengal, India
e-mail: ayanchatterje2012@gmail.com

T. Bera
Department of Information Technology, Murshidabad College of Engineering & Technology, Berhampore, West Bengal, India
e-mail: tanmoybera06@gmail.com

S. Kr.Pati
Department of Computer Science and Engineering, St. Thomas College of Engineering and Technology, Kidderpore, India
e-mail: soumenkrpati@gmail.com

A. K. Das
Department of Computer Science and Technology, Indian Institute of Engineering Science and Technology, Shibpur, Howrah, India
e-mail: akdas@cs.iiests.ac.in

J. Nayak et al. (eds.), *Soft Computing in Data Analytics*,
Advances in Intelligent Systems and Computing 758,
https://doi.org/10.1007/978-981-13-0514-6_13

Keywords Information security • Digital authentication • Frequency domain steganography • Statistical attacks • Parity checker • Peak signal-to-noise ratio Mean squared error

1 Introduction

Recently, Internet is one of the vital medium of exchanging information among the authorized users [1–3]. In this context, one of the most common drawbacks is that this medium or channel is not very much secured and highly protected from unauthorized access of information [4–6]. But the information should be provided to the authenticated users for maintaining confidentiality, integrity, and availability. As well as information security must be required during secured wireless communication. To achieve such secure communication, various important approaches (like cryptography and steganography) are developed gradually in different decades.

1.1 General Cryptography Model

It is effective due to hiding the meaning of secret information at the time of unauthorized access [7]. A sample model of cryptography is shown in Fig. 1.

From Fig. 1, it is realized that the actual secret information is changed to another text using proper encryption technique and the encrypted data is sent to the receiver. At the receiving end, the secret data is decrypted using corresponding decryption algorithm and secret key(s) available at the receiver side. Here, the weight of secrecy depends on the secret key(s). In this study, the popular and highly effective encryption and decryption techniques [3, 7, 8] (like Data Encryption Standard (*DES*), Advanced Encryption Standard (*AES*), *RSA* etc.) are developed. These cryptographic schemes are effective and safe from various unauthorized attacks (e.g., Brute force attack, password guessing attack, replay attack, inter-session chosen plaintext attacks). But the existence of communication cannot be hidden to the unauthenticated users due to some limitations of these algorithms.

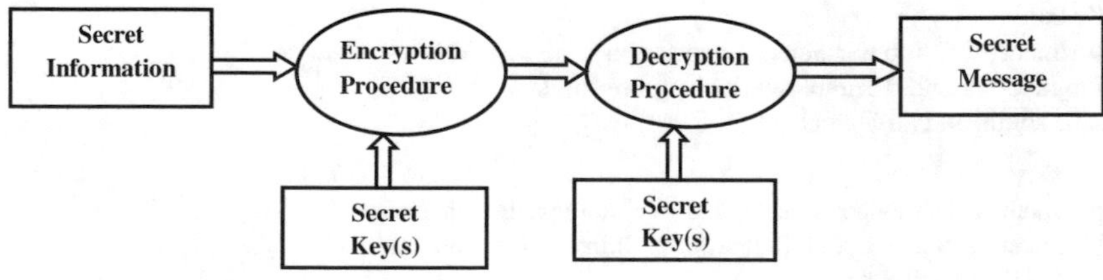

Fig. 1 A model of cryptography

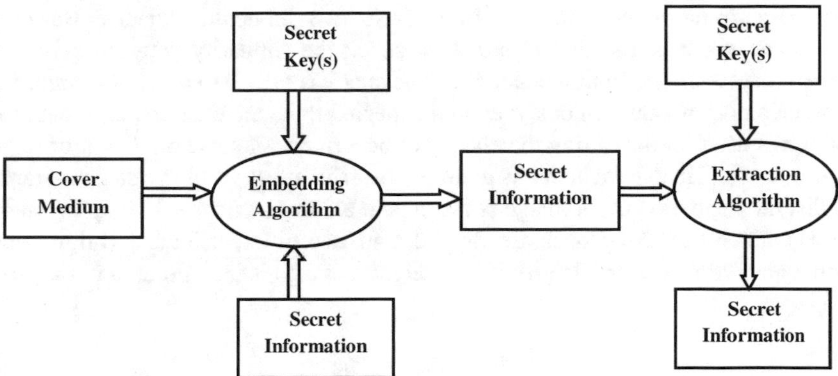

Fig. 2 A model of steganography

1.2 General Steganography Model

This approach is the next generation of cryptography and developed to reduce the limitations of cryptographic approaches, and the communication can be made hidden from unauthorized users or hackers. In this technology, the secret information is embedded in a cover medium (e.g., image, audio/video) with the minimum distortion [9–11]. More specifically, different types of pixels in cover mediums (e.g., *RGB*, entropy) are used to insert secret message in the particular cover medium. A model diagram of steganography approach is shown in Fig. 2.

There are two major types of steganographic approaches occurred in spatial domain and frequency domain: (i) In spatial domain steganography, the actual domain of a particular medium is used for inserting information and (ii) in frequency domain steganography, the actual domain is converted to another domain using some popular transformations (e.g., *DCT* (Discrete Cosine Transformation), *DWT* (Discrete Wavelet Transformation), *DFT* (Discrete Fourier Transformation)), and the transformed domain is used to insert the secret information [12, 13]. Generally, the distortion between actual cover medium and stego-medium is lesser in spatial domain steganography than frequency domain steganography. But in frequency domain, the hacking chance of secret information is very low due to use of secret key matrix and selection of key columns independently [14–16]. So, the data cannot be accessed by an unauthorized person directly by applying any type of statistical attacks (like chi-square and histogram).

1.3 Attacks in Steganography

The chi-square and histogram attacks are most popular and powerful attacks in steganography. There are very few techniques occurred [17] in both of spatial and frequency domains and are free from these two influential attacks. Basically, chi-square attack is a statistical attack to check the similarity between given and expected dataset. The central concept of this attack is fully based on the comparison between a pair of values in observed and expected frequencies. Here, a probabilistic concept is used for measuring the chance of occurrence of some hidden information in a cover file. Histogram attack is another powerful attack of image steganography. At first the histogram of an image is taken, and a different image is computed using Eq. (1). Then the *LSB* planes are flipped, and two different images (original and flipped) are compared to identify if any secret message is incorporated in the cover image.

$$D_{i,j} = I_{i+1,j} - I_{i,j} \tag{1}$$

where $I_{i,j}$ is the pixel value of stego-image and $D_{i,j}$ is the difference value of the matrix at position (i,j) and $(i+1,j)$. The matrix D is formed by computing the difference between two consecutive pixels (row-wise) of the stego-image.

In this paper, a new data hiding scheme is proposed using the concepts of parity checker in transformed image domain. In Sect. 2, some related works are described and analyzed with their speciality and drawbacks. The details of the proposed scheme followed by the algorithms of sender and receiver sides are illustrated in Sect. 3. The efficiency of the proposed scheme is analyzed with the help of *PSNR*, *MSE*, *RS* stego-analysis, chi-square attack, and histogram attack in Sect. 4. A conclusion of the proposed scheme is drawn with mentioning the corresponding future scope in Sect. 5.

2 Related Work

In the paper [1], Tahir Ali et al. proposed a data hiding scheme using parity checker where the message bits are placed in the *LSB* positions of any one of the three basic color components (red, green, and blue). This method [1] is developed on the basis of parity checker of these three components of color image of 24 bits, but the drawback of this technique is that it is applicable only for the color images. Another approach of data hiding is developed by Rajkumar et al. [2] where Pseudo Random Number Generator (*PRNG*) is used to locate pixel positions to use parity checker and a secret key is used for specification of *PRNG* among sender and receiver. So, this technique [2] is very much efficient for data hiding purpose in all types of images, but no authentication is provided beside the insertion of secret message. To reduce the problems regarding security conditions (confidentiality, integrity and availability of information among authorized users) in a single methodology and independency of types of pixels (like *RGB*, entropy), the proposed methodology is developed and it is illustrated with proper algorithms in the next section.

3 Data Hiding Scheme

In this section, the full procedures of data embedding at sender side and data extraction at the receiver side are illustrated step-wise. The concept of parity checker is used for both data embedding and data extraction purposes. A predefined secret key matrix $[R]_{4\times4}$ with elements 0 or 1 is used to maintain parity checking procedure. Another fact is that a digital authentication is provided using parity checker in a different combination with the key matrix. The data embedding and data extraction processes with digital authentication are discussed in the following subsections.

3.1 Data Embedding

In this process at sender side, a cover image (*I*) and secret message $(a_1a_2\ldots a_n)$ are taken as input. A secret key matrix $[R]_{4\times4}$, two secret key columns k_1, k_2, and a standard quantization matrix are set between sender and receiver before communication. Initially, entropy pixel values of the cover image (*I*) are divided into a finite number of (4 × 4) blocks, and each block is taken as the matrix *B*. Then 2-*DDCT* is applied on each block separately, and quantization is performed using predefined standard quantization matrix. After that, the matrix *R* is developed by taking the *LSB*'s of transformed values of each block *B*. Then the secret message $(a_1a_2\ldots a_n)$ is made to 4 *l* number of bits. More specifically, the number is changed when $n=4l$. But, $(4l-n)$ number of 0s are inserted at the *MSB* positions when $(n<4l)$ and the new shaped message becomes $(A_1A_2\ldots A_{4l})$. Then the new message $(A_1A_2\ldots A_{4l})$ is divided into 4-bit segmentation $(m_1m_2m_3m_4)$ sequentially. Each 4 bits of the dataset is inserted into each block *B* using a specific parity checker with the elements of secret key matrix (*R*). The elements of two secret columns k_1 and k_2 of each row of R and all the elements of corresponding rows(r) are used for inserting data using parity checker. Alternatively, the parity bit corresponding to each message bit $(m_{i=1,2,3,4})$ is set with the elements $(R_{ik1}, R_{ik2}, r_{i1}, r_{i2}, r_{i3}, r_{i4}: i=1,2,3,4)$ in each block. If the message bit $m_i=1$, then an odd parity is developed with the previously mentioned six elements. More specifically, if there is already odd parity occurred, then no change is required and if there is an even parity occurred, then one bit (R_{ik1} or R_{ik2}) is changed. Similarly, if message bit $(m_i=0)$, then an even parity is developed with previously mentioned six elements. Therefore, if required, at most one bit of R_{ik1} or R_{ik2} is changed. A digital authentication is provided by developing a parity checker with the elements of $\{R_{i1}, R_{i2}, R_{i3}, R_{i4}, r_{i1}, r_{i2}, r_{i3}, r_{i4}: i=1,2,3,4\}$ in each block. Actually an odd/even parity (as defined previously among authorized sender and receiver) is developed corresponding to all the eight elements of each row of *R* and *r* combinedly. In other words, according to requirement, at most one bit is changed among the values of two columns (other than k_1 and k_2) of each row(r) of *R*. Similar operation is performed in all the blocks. Then 2-*D* inverse *DCT* is applied in each block, and when merging all the blocks, a stego-image (*S*) is prepared.

3.2 Data Extraction

The stego-image (*S*) is taken as input at receiver side. The other parameters at this stage are secret keys that are taken from secret key matrix (*R*), two secret columns k_1, k_2, and the standard quantization matrix. Firstly, entropy pixel values of the stego-image (*S*) are divided into (4 × 4) pixel blocks, and each block is taken as B'. After that, 2-*DDCT* is applied in each block and quantization is performed on each block separately using the predefined standard quantization matrix. Then corresponding *LSB* matrix block (R') is developed by taking *LSB*'s of transformed values of each (B'). The originality of the stego-image is checked with the secret key matrix (*R*) and each *LSB* block (R') using parity checker. The even/odd parity (as defined before communication) is checked with the elements of each row of (R') and the elements of corresponding row of (*R*). More specifically, the parity is checked in each block of (R') with the eight elements of $\{R_{i1}, R_{i2}, R_{i3}, R_{i4}, r_{i1}, r_{i2}, r_{i3}, r_{i4}: i = 1, 2, 3, 4\}$. If this parity condition is satisfied in all rows of each block in (R'), then it is proved that the stego-image is original. Otherwise, it is realized that the stego-image is distorted by some external noise. Then, parity checker is used with the six elements of each row together with the block (R') and *R*. Alternatively, the parity is checked in each block with the elements of $\{R'_{ik1}, R'_{ik2}, r_{i1}, r_{i2}, r_{i3}, r_{i4}: i = 1, 2, 3, 4\}$. Mathematically, the parity checker is used for each message bit $(m_{i=1,2,3,4})$ according to Eq. (4), for $i = 1, 2, 3, 4$.

$$m_i = Parity(R'_{ik1}, R'_{ik2}, r_{i1}, r_{i2}, r_{i3}, r_{i4}) \tag{2}$$

If odd parity occurs in mentioned six elements in each stego-block using Eq. (2), then message bit $m_i = 1$ but if an even parity occurs, then message bit $m_i = 0$. In this way, the message bits are obtained sequentially from the blocks and the message is built up by concatenating message bits serially.

3.3 Sender and Receiver Side Procedure

The sender side algorithm is given bellow:

Algorithm 1: *Sender Side*
Input: Cover Image *(I)*,
Secret message $(a_1 a_2 \ldots ..a_n)$
Output: Stego-image (*S*)
Secret keys: (i) A 4×4 matrix with elements 0 and (*R*)
(ii) Two secret key columns k_1 and k_2,where$1 \leq k_1, k_2 \leq 4$
(iii) A standard quantization matrix

Procedure:
Step 1: Divide the entropy pixel values of cover image (*I*) into l number of (4×4) blocks and each block is considered as *B*.
Step 2: Apply 2-*DDCT* and perform quantization operation using a standard matrix on each *B*.
Step 3: Using the *LSB*'s of each block *B*, make corresponding *LSB* matrix *R*.
Step 4: Make secret message $(a_1 a_2 \ldots ..a_n)$ into $4l$ number of bits (i.e., insert a finite number of zeros at the front of message if required) and make it as $(A_1 A_2 \ldots \ldots A_{4l})$.
Step 5: Divide the message $(A_1 A_2 \ldots \ldots A_{4l})$ into 4-bits segmentation $(m_1 m_2 m_3 m_4)$ sequentially.
Step 6:In each block of *R* of order 4, insert 4-bits segmented message $(m_1 m_2 m_3 m_4)$ maintaining the following rule:
Make a parity checker for message bit (m_i), with the elements of$\{R_{ik1}, R_{ik2}, r_{i1}, r_{i2}, r_{i3}, r_{i4}\}$ as defined below:

I. **If**$(m_i = 1)$, **then**make a odd parity with mentioned six elements by at most one*LSB* distortion of the elements of R_{ik1} and R_{ik2}.

II. **If**$(m_i = 0)$, **then** make an even parity with mentioned six elements by at most one*LSB* distortion of the elements of R_{ik1} and R_{ik2}.

Step 7: Make an even/odd parity (explain in Sect. 3.1) with the elements ofeach row of *R* and all elements of each row *r*, (i.e. change at most one*LSB* among two elementsof each row of *R* (column positions other than k_1 and k_2) according to requirement).
Step 8: Apply 2-*D* inverse *DCT* on each block of *R*.
Step 9: Merge all the blocks sequentially and stego-image (*S*) is obtained.
Step 10: Stop.

Algorithm 2: *Receiver Side*
Input: Stego-image (*S*)
Output: Secret message $(a_1 a_2 \ldots ..a_n)$
Secret keys: (i) A 4×4 matrix with elements 0 and (*r*)
(ii) Two secret key columns k_1 and k_2,where,$1 \leq k_1, k_2 \leq 4$
(iii) A standard quantization matrix

Procedure:
Step 1: Divide the entropy pixel values of stego-image(*S*) into l number of (4×4) blocks and each block is considered as (*B* .
Step 2: Apply 2-*DDCT*operation and perform quantization using the defined standard matrix on each *B* .
Step 3: Using the *LSB*'s of each block (, make corresponding *LSB* matrix (*R* .
Step4: Check even/odd parity (explain in Sect. 3.2) with the elements ofeach row of (*R* and all elements of each row(*r*).
If the parity satisfies in all rows and all the blocks of*R* **then** go to *Step 5*.
Else Reject the stego-image for further data extraction procedure.
Step 5: Observe parity bit of each block of (*R* together with the secret matrix (*R*)(i.e. check parity ofthe six positions of{ for extracting data from each row of (*R* and *R*sequentially.

I. **If**(odd parity occurs)**then** message bit $(m_i = 1)$

II. **If**(even parity occurs)**then** message bit $(m_i = 0)$

Step 6: Concatenate the message bits sequentially obtained from all the blocks and get the secret message
Step 7: Stop.

4 Experimental Results and Performance Evaluation

The proposed data hiding scheme is implemented and tested on different images with various secret information of different lengths. All the selected images are *JPEG* in nature, and the size of each image is (512 × 512). The basic parameters (*PSNR* [7] and *MSE* [7]) are used to evaluate the efficiency of the proposed scheme by measuring the quality of the stego-image. The *PSNR* and MSE are defined using Eq. (3) and Eq. (4), respectively.

$$PSNR = 20\log_{10}\left(\frac{2^b - 1}{\sqrt{MSE}}\right) \tag{3}$$

$$MSE = \frac{1}{mn}\sum_{i=0}^{m-1}\sum_{j=0}^{n-1}\|I(i,j) - K(i,j)\|^2 \tag{4}$$

where I and K are the actual cover image and the image after insertion of secret message (stego-image), respectively. The *MSE* is used to determine the change of cover image after insertion of secret information.

The lower *MSE* and the higher *PSNR* achieved by the proposed method indicate that the efficiency of the proposed method is better compared to other methods [18, 19]. Mathematically, it is observed that the *PSNR* is inversely related to the *MSE*. Therefore, only the *PSNR* value is considered in the paper. Another important parameter *Embedding capacity* [20] is used to measure the efficiency of the methods. High embedding capacity indicates that the secret information with large bit streams is embedded into the cover mediums. The proposed scheme is applied in six popular selected *JPEG* images. The obtained *PSNR* and *Embedding capacity* values of the proposed and some other popular methods [7, 20] are shown in Table 1 and Table 2, respectively. The cover images (namely A, B, C, D, E, and F) [Standard image set of MATLAB (version 7)] and the corresponding stego-images (namely A1, B1, C1, D1, E1, and F1) are shown in Fig. 3 and Fig. 4, respectively.

Table 1 *PSNR* value of the proposed and compared methods for six images

Images	Size	PSNR		
		LSB-DCT [7]	DCT coefficient modification [20]	Proposed method
A	512 × 512	37.39	43.47	45.62
B	512 × 512	36.74	42.78	43.72
C	512 × 512	35.78	42.03	43.12
D	512 × 512	36.82	42.93	44.17
E	512 × 512	38.49	44.58	46.67
F	512 × 512	41.47	46.57	49.65

Table 2 *Embedding capacity* value of the proposed and other methods

Images	Size	Embedding capacity		
		LSB-DCT [7]	DCT coefficient modification [20]	Proposed approach
A	512 × 512	354,762	458,746	458,748
B	512 × 512	352,753	**457,732**	457,732
C	512 × 512	347,256	452,185	452,187
D	512 × 512	354,764	458,521	458,527
E	512 × 512	332,458	445,453	445,467
F	512 × 512	333,765	**446,689**	446,689

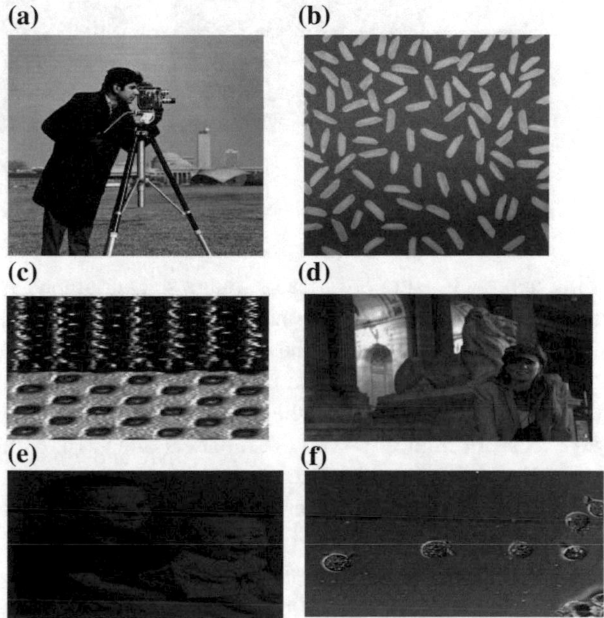

Fig. 3 Cover images (namely A, B, C, D, E, and F)

From Table 1, it is observed that the proposed method gives the best result in terms of higher *PSNR* value compare to other methods described in [7, 20] for all images.

Similarly, in Table 2, both the method [20] and proposed method give the best results for images B and F in terms of higher *Embedding capacity* value (indicated by bolt font) but for all other images, the proposed method gives the better results.

An important observation here is that, in various types of statistical analysis of the experiment using the concept of parity checker, the data is not embedded directly to the *LSB* positions of pixels. So, the data cannot be extracted and unauthorized access is not allowed by directly applying the chi-square attack and histogram attack.

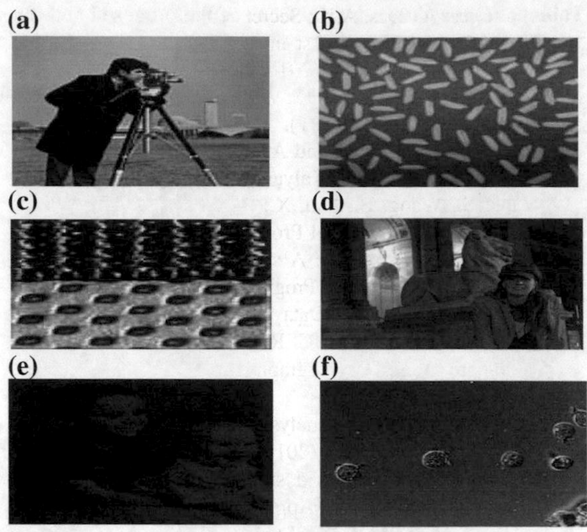

Fig. 4 Stego-images (namely A1, B1, C1, D1, E1, and F1)

5 Conclusion

A new data hiding approach using the idea of image steganography is presented in this paper. The model of parity checker increases the security level (e.g., confidentiality, integrity, and availability) at the time of wireless communication. The uniqueness of this approach is that the data bits are not inserted directly in the *LSB* positions of the cover image. As a result, the secret information cannot be retrieved by unauthorized access through any type of statistical attack (like chi-square, histogram). From another view, the use of secret key matrix and secret key columns increases the integrity level of security over the compared schemes. At a glance, the embedding of data and providing the digital authentication in a single system enhances the quality of this particular steganographic approach. Here, *DCT* is used to transform to the actual domain. But, other transformations, such as *DFT*, *DWT,* can be used to transform the domain and these may be used in future. Here, the proposed approach is applied only in images. But the method may be uniformly applicable for other mediums like audio, video. Use of the proposed approach in various cover mediums with different lengths of secret information may be taken as future extension of the paper.

References

1. Ali, T., Doegar, A.: A novel approach of LSB based steganography using parity checker. Int. J. Adv. Res. Comput. Sci. Software Eng. **5**(1) (2015). ISSN: 2277 128X
2. Rajkumar, A., et al.: New steganography method for gray level images using parity checker. Int. J. Comput. Appl. **11**(11) (2010)

3. Chatterjee, A., Das, A.K.: Secret communication combining cryptography and steganography. Paper presented at the 1st international conference on advanced computing and intelligent engineering, 21–23 Dec 2016, Bhubaneswar, India (2016)
4. Khamrui, A., Mandal, J.K.: A genetic algorithm based steganography using discrete cosine transformation (GASDCT). In: International Conference on Computational Intelligence: Modelling Techniques and Applications (CIMTA 2013), vol. 10, pp. 105–111. ELSEVIER, University of Kalyani, Kalyani, Procedia Technology (2013)
5. Wang, S., Wang, B., Niu, X.: A secure steganography method on genetic algorithm. J. Inf. Hiding Multimed. Signal Process **1**(1), 28–35 (2010)
6. Nosrati, M., Karimi, R.: A survey on usage of genetic algorithms in recent steganography researches. World Appl Program. **2**(03), 206–210 (2012)
7. Singla, D., Syal, R.: Data security using LSB & DCT steganography in images. Int. J. Comput. Eng. Res. (IJCER) **2**(2), 359–364 (2012)
8. Pachghare, V.K.: Cryptography and Information Security. PHI Learning Private Limited, New Delhi (2011)
9. Walia, E., Jain, P.: An analysis of LSB & DCT based steganography. Glob. J. Comput. Sci. Technol. **10**(01), 04–08 (2010)
10. Gokul, M., et al.: Hybrid steganography using visual cryptography and LSB encryption method. Int. J. Comput. Appl. **59**, (14), 05–08 (2012)
11. Soleimanpour, M.: A novel technique for steganography method based on improved genetic algorithm optimization in spatial domain. Iranian J. Electr. Electron. Eng. **09**(02), 67–75 (2013)
12. Mishra, S., Bali, S.: Public cryptography using genetic algorithm. Int. J. Recent Technol. Eng. (IJRTE) **02**(03), 150–154 (2013)
13. Mehndiratta, A.: Data hiding system using cryptography & steganography: a comprehensive modern investigation. Int. Res. J. Eng. Technol. **02**(01), 397–403 2015
14. Zamani, M., et al.: A genetic-algorithm-based approach for audio steganography. Int. J. Comput. Electr. Autom. Control Inf. Eng. **03**(06), 1562–1565 (2009)
15. Rajyaguru, M.H.: CRYSTOGRAPHY-combination of cryptography and steganography with rapidly changing keys. Int. J. Emerg. Technol. Adv. Eng. **02**(10), 329–332 (2012)
16. Nehru, G., Dhar, P.: A detailed look of audio steganography techniques using LSB & genetic algorithm approach. Int.l J. Comput. Sci. (IJCSI) **09**(01), 402–406 (2012)
17. Jain, R., Kumar, N.: Efficient data hiding scheme using lossless data compression and imege steganography. Int. J. Eng. Sci. Technol. (IJEST) **4**(8), 3908–3915 (2012)
18. Diwesh, D., et al., AAKRITI (ed.): An image and data encryption-decryption tool. Int. J. Comput. Sci. Inf. Technol. Res. **03**(02), 264–268 (2015)
19. Roy, S., Venkateswaran, P.: Online payment system using steganography and visual cryptography. In: Students Conference on Electrical, Electronics and Computer Science. IEEE (2014)
20. Lin, Y.-K.: A data hiding scheme based upon DCT coefficient modification. Sci. Direct Comput. Stand. Interfaces **36**, 855–862 (2014)

Graph-Based Text Summarization Using Modified TextRank

Chirantana Mallick, Ajit Kumar Das, Madhurima Dutta, Asit Kumar Das and Apurba Sarkar

Abstract Nowadays, the efficient access of enormous amounts of information has become more difficult due to the rapid growth of the Internet. To manage the vast information, we need efficient and effective methods and tools. In this paper, a graph-based text summarization method has been described which captures the aboutness of a text document. The method has been developed using modified TextRank computed based on the concept of PageRank defined for each page in the Web pages. The proposed method constructs a graph with sentences as the nodes and similarity between two sentences as the weight of the edge between them. Modified inverse sentence frequency-cosine similarity is used to give different weightage to different words in the sentence, whereas traditional cosine similarity treats the words equally. The graph is made sparse and partitioned into different clusters with the assumption that the sentences within a cluster are similar to each other and sentences of different cluster represent their dissimilarity. The performance evaluation of proposed summarization technique shows the effectiveness of the method.

Keywords Extractive summarization · Single-document source · Sentence index · Similarity graph · PageRank · ROUGE

C. Mallick (✉) · A. K. Das (✉) · M. Dutta · A. K. Das · A. Sarkar
Department of Computer Science and Technology, Indian Institute of Engineering Science and Technology, Shibpur, Howrah, India
e-mail: chirantana9@gmail.com

A. K. Das
e-mail: writetoajit@yahoo.com

M. Dutta
e-mail: madhurima.pg2016@cs.iiests.ac.in

A. K. Das
e-mail: akdas@cs.iiests.ac.in

A. Sarkar
e-mail: as.besu@gmail.com

J. Nayak et al. (eds.), *Soft Computing in Data Analytics*, Advances in Intelligent Systems and Computing 758,
https://doi.org/10.1007/978-981-13-0514-6_14

1 Introduction

Automatic text summarization reduces a text of a document in order to create a summary that retains the crux of the original text. The variety of applications like summaries of newspaper articles, book, magazine, resume, music, film, and speech increases the importance of text summarization methods.

In general, text summarization approaches can be categorized into two types: extractive summarization [21] and abstractive summarization [8]. Extractive summarization creates the summary from phrases or sentences in the source document(s) and abstractive summarization expresses the ideas in the source document using different words. Abstractive summarization is more efficient than extractive summarization as it generates humanlike summaries. Text summarization is also classified as indicative [7] and informative [17]. Indicative summary identifies the topics of the document which is mainly used for quick categorization and the informative summary elaborates these topics and includes conclusions, suggestions, and recommendations according to the reader's interest. Informative summarization technique is used for making generic [5] and query-oriented [20] summary. Generic summaries try to cover more information with preserving the generic topic of the given document and it provides author's view. Query-oriented summaries generate a query set that reflects user's interest. Background summary assumes reader's prior knowledge is poor and it is meant to be short and attention grabbing. Just-the-news summary [22] provides recent updated news according to users' interest. Single-document summary [10] takes sentences from the document itself, whereas multi-document summary [4] makes a summary by fusing sentences from different documents. Topic-oriented summarization [6] is made based on users' enthusiasm and the information extracted from the given document related to some specific topic. Centrality summarization [14] measures the centrality of a sentence in a given document. The centrality of a sentence is defined in terms of word centrality and the word centrality in a vector space is defined based on the centroid of the document cluster. Similarly, centroid-based summarization [16] is performed on centroid of sentences of the given document. The central sentences contain more words from the centroid of the cluster to measure how closely the sentence corresponds to the centroid of the cluster.

The proposed work is developed on extractive summarization technique. Extractive summarization contains three distinguished independent tasks: intermediate representation of the document, scoring sentences based on different parameters, strategies for selection of summary sentences.

A. Intermediate Representation: For any simplest summarizer, intermediate representation of the text to be summarized is done to identify the important content. It is an essential step. In topic representation approaches, the original text document is converted into an intermediate representation. Intermediate representation [14] includes frequency, TF-IDF [19] approach, and topic-word approach [6]. The topic representation of topic-word approaches consists of a simple list of words and their corresponding weights where higher weighted words are considered as indicative topic words. The topic signature words for latent semantic analysis are decided based on the patterns of word co-occurrence and the weights of each pattern. In graph models [2], such as LexRank [3], the entire input text document is represented as a graph based on the interrelated sentences. In Bayesian topic models, the document is represented as a combination of topics where each topic is presented by means of word probabilities (weights) for that topic. In indicator representation approaches, the input text document is represented as a list of indicators of such as sentence length, presence of certain phrases pertaining to the theme of the documet, location in the document.

B. Score Sentences: After converting the input text document into intermediate representation, score (or weight) is assigned to each sentence of the given document to identify the important sentences. The weight of each sentence is determined by examining the different parameters used in the method. Machine learning techniques are most commonly used to determine indicator weights. In LexRank, the score of sentence is determined from the graphical representation [12] of the text.

C. Selecting Summary Sentences: This is the final and last step to construct a relevant summary. Generally, it selects the best combination of sentences that gives the important information of the original text and the desired summary is the combination of top *n* important sentence. The optimal and best collection of sentences is selected to maximize overall importance, minimize redundancy, or maximize coherence in global selection procedure.

The summarization approach of Salton et al. [18] uses degree scores to extract the most important sentences of a document. Moens et al. [13] partition a text document into different topical regions using cosine similarity between the sentences. Zha [23] uses mutual reinforcement principal and a bigraph from the set of terms to the set of sentences to reduce a solution for the singular vectors of the transition matrix of the bipartite graph. Mihalcea and Tarau (2004) [3] proposed a centrality algorithm on weighted graphs for single-document summarization by computing eigenvectors. In our proposed method, to determine the similarity between two sentences, inverse sentence frequency modified cosine (isf-modified-cosine) similarity is measured to provide weights of the edges in the graph. In this paper, a graph-based method is used to extract the most important sentences from the given text, sufficient to provide the overall idea about the text document. Here, the overall centrality of a sentence is computed based on its similarity to other sentences. This approach mainly determines the similarity between the sentences based on the modified PageRank algorithm [12], named here as modified TextRank algorithm. The remaining part of the paper is organized as follows: Detailed procedure of the proposed method is described in Sect. 2. In this section, we have explained the process of extraction of the text document as input, the preprocessing of the raw input and summarization technique using TextRank. Subsequently, entire evaluation of the obtained summary using ROUGE method and experimental results is provided in Sect. 3. Finally, Sect. 4 holds future work and the final conclusion of the paper.

2 Proposed Methodology

With this paper, we have tried to present a modified graph-based approach on achieving extractive summary of a text document. Calculating similarity between pair of sentences, instead of the regular TF-IDF, isf-weight-based cosine similarity is used that provides the interesting results when tested with news articles. Figure 1 represents the system architecture of the overall proposed methodology.

2.1 Preprocessing

Several approaches exist to extract information from the Web. Structured data is preferably extracted through an API over Web screaping. Unfortunately not all Web sites provide an API because they do not want the users to extract data in structured way. We can perform Web scrapping to transform unstructured data in HTML or XML format into structured data (database or spreadsheet). In this paper, we have used Beautiful Soup [1] for scrapping a Web page. Beautiful Soup is a Python [15] library for extracting body of HTML and XML files. It also defines a class for auto detecting the encoding of an HTML or XML document, and converting it to Unicode. After extracting structured data, first we perform removal of stop words from the raw data. Next, we tokenize the given text into sentences and assign sentence index according to the input sequences of the sentences in the document. Finally, tokenize each sentence into a collection of words which is required for graphical representation.

2.2 TextRank-Based Summarization

In the proposed method, we have considered that the sentences of the document are equivalent to Web pages in the PageRank system [12]. The probability of going from sentence *A* to sentence *B* is equal to the similarity between the pair of sentences. The modified TextRank proposed in this paper uses the intuition behind the PageRank

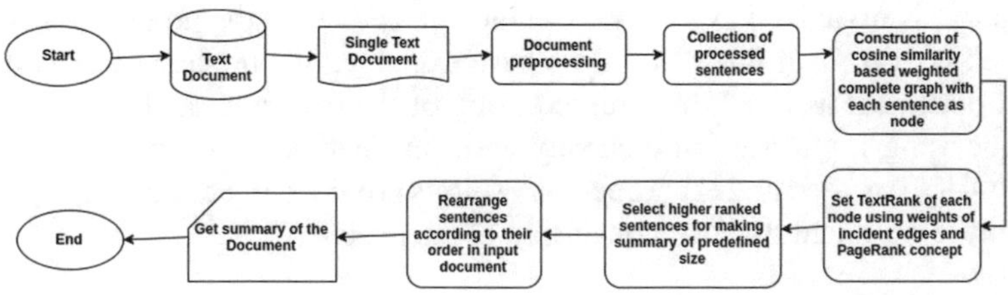

Fig. 1 System architecture

algorithm to rank sentences using which we can select the most important sentences from the input text document. In PageRank, important Web pages are linked with other important Web pages. Similarly, in our approach we assume that the important sentences are linked (similar) to other important sentences of the input document. In this model, at first to identify the sentences, we have transferred the input document D into sentences $\{s_1, s_2, \ldots, s_n\}$ using NLTK package, i.e., $D = \{s_1, s_2, \ldots, s_n\}$. We have assigned index for each of the sentence according to the input sequence of the sentences in the document. Index values are significantly used for proper ordering of the summarized sentences to make the summary meaningful. In the next step, each sentence is tokenized into a set of words. To define similarity among the sentences, we represent each sentence of the given text document as a word vector. Next, a complete weighted graph $G = (V, E, W)$ of the input document is built, where V is the set of vertices, each of which corresponds to a sentence represented by the word vector. E is the set of edges between every pair of vertices. W is the set of weights assign to the edges of the graph G. The weight w associated to edge $(u, v) \in E$ is assigned using following logic:

(i) Suppose $S(u) = \{w_u^1, w_u^2, \ldots, w_u^s\}$ is the sentence corresponds to the vertex u and $S(v) = \{w_v^1, w_v^2, \ldots, w_v^t\}$ is the sentence corresponds to the vertex v. Here, we have considered term frequency (*tf*) and inverse sentence frequency (*isf*) for each word in the sentence. For an example, $tf_w(S(u))$ is the term frequency of word w in $S(u)$ which gives the number of occurrences of the word w in the sentence. Similarly, $isf_w(D)$ is the inverse sentence frequency of the word w in the input document D which is defined by Eq. (1).

$$isf(w, D) = \log \frac{\|D\|}{\|\{S \in D : w \in S\}\|} \tag{1}$$

where, $\|\{S \in D : w \in S\}\|$ is the number of sentences in which word w appears, and $\|D\|$ is the number of sentences present in the input document.

(ii) Now, the isf-modified-cosine similarity is used to measure the similarity between every pair of sentences and it is used as weight w of edge (u, v) using Eq. (2).

$$w(u, v) = \frac{\Sigma_{\forall x \in S(u) \cup S(v)} tf_x(S(u)) tf_x(S(v)) (isf_x(D))^2}{\sqrt{\Sigma_{\forall y \in S(u)} (tf_y(S(u)) isf_y(D))^2} \times \sqrt{\Sigma_{\forall z \in S(v)} (tf_z(S(v)) isf_z(D))^2}} \tag{2}$$

Traditional cosine similarity only takes care of the term frequency of the corresponding words in the text document, and in case of cosine similarity, the dimension should be same for all sentence vectors, whereas isf-modified-cosine similarity takes care of the different level of importance for the corresponding words in the sentences and also considers different length of the sentence in the document.

Thus, we get a complete weighted graph which is made sparse by removing the edges having weight less than the threshold (t), set as the average weight of all the edges in the graph. This graph represents the similarity graph of the input document. For weight assignment, score is assigned to every sentence corresponding to every node of the graph G. In this proposed method, we initialise the TextRank score of each node of the graph by the average weight of the edges incident to it for giving importance to the weights associated with the edges(E), whereas in traditional PageRank the initial PageRank value of each node is set to $\frac{1}{T}$, where T is the total number of nodes in the graph. Next, the TextRank score is updated via modified TextRank as defined in Eq. (3).

$$TR(S(u)) = \frac{d}{T} + (1-d) * \Sigma_{v \in adj(u)} \frac{TR(S(v))}{deg(S(v))} \tag{3}$$

where $TR(S(u))$ is the text rank of sentence S corresponds to the node u, $TR(S(v))$ is the text rank of the sentence S corresponds to the node v such as $v \in adj(u)$, $deg(S(v))$ is the degree of the node v corresponding to the sentence S, T is the total number of nodes present in the graph, and d is a "damping factor." Here, we have considered the value of d as 0.15. Finally for summarization, we have selected top n scored sentences and rearranged those n sentences according to the sentence index which we have assigned at first step. The n output sentences construct the summary of our proposed model. Using the pseudocode of the proposed Algorithm (2.1), we can extract important sentences from a given input document.

Algorithm 2.1: MODIFIEDTEXTRANKSUMMARY(D, n)

Input: Extracted text from the target document D and size of summary = n (say).
Output: Summary of the input document.
1. Encode D into D' using UTF-8 format;
2. D' is tokenized into individual sentences (S_i) using the NLTK library;
3. Initialize *index_value* to zero for the sentence index;
4. **for each** $S_i \in D'$ in order of appearance in encoded D **do**
   4.1. Remove the stop words from the sentence;
   4.2. Increase the *index_value* by 1;
   4.3. Assign the *index_value* to the sentence;
5. **for each** processed tokenized sentence $S_i \in D'$ **do**
   Take word count vector $v_i = \{w_1, w_2, \ldots, w_{t_i}\}$;
6. Build a graph $G = (V, E, W)$ where V = set of vertices corresponding to sentences represented by word count vector, E = set of edges and W = set of weights associated with each edge $(u, v) \in E$ computed by Eq. (2);
7. Remove edges with weight less than the average weight considering all edges of G;
8. $\forall v \in V$ Initialize TextRank by the average weight of the edges incident to it;
9. Modify TextRank of every node $v \in V$ using Eq. (3) based on the initial TextRank;
10. Arrange the vertices of the graph in descending order of their TextRank score;
11. Take first n vertices from the sorted list;
12. Put in summary the sentences associated to the selected n vertices in order of the *index_value*;
return (*summary*)

3 Experimental Result

Since early 2000s, Recall-Oriented Understudy for Gisting Evaluation (ROUGE) [9] is widely used for performance evaluation of summarization techniques. Some of the ROUGE methods, i.e., ROUGE_N, ROUGE_L, ROUGE_W and ROUGES_U are

Table 1 Rouge values of the proposed method and traditional TF-IDF method with respect to ground truth

Method	Rouge values of the proposed method						Rouge values of traditional TF-IDF method					
Length	10%			15%			10%			15%		
Metric	Recall	Precision	f-score	Recall	Precision	f-score	Recall	Precision	f-score	Recall	Precision	f-score
Rouge-1	1.00000	0.72632	0.84147	0.68807	0.78947	0.73529	0.77215	0.64211	0.70115	0.72895	0.72895	0.72895
Rouge-2	0.98529	0.71277	0.82716	0.62963	0.72340	0.67327	0.67949	0.56383	0.60624	0.60963	0.60963	0.60963
Rouge-L	1.00000	0.72632	0.84147	0.59813	0.68817	0.64000	0.75494	0.63258	0.68956	0.74682	0.74682	0.74682
Length	20%			25%			20%			25%		
Metric	Recall	Precision	f-score	Recall	Precision	f-score	Recall	Precision	f-score	Recall	Precision	f-score
Rouge-1	0.68807	1.00000	0.73529	0.55232	1.00000	0.70561	0.62114	1.00000	0.70529	0.43379	1.00000	0.60510
Rouge-2	0.62963	0.95819	0.67327	0.55232	0.96809	0.58333	0.61232	0.94872	0.70459	0.41743	0.90590	0.58333
Rouge-L	0.63303	1.00000	0.67647	0.55232	1.00000	0.70561	0.62140	1.00000	0.66236	0.40092	1.00000	0.60510

commonly used to measure the performance. ROUGE_N is the N-gram ($N \geq 1$) recollection between a system summary and human-generated or reference summaries. It is used to estimate the fluency of summaries. ROUGE_N is computed using Eq. (4). Here, the value of N is considered as 1 and 2. ROUGE_1 and ROUGE_2 denote the overlap of 1-gram and bi-grams between the system and sample summaries, respectively.

$$Rouge\_N = \frac{\Sigma_{s \epsilon refsum} \Sigma_{gram_n \epsilon S} count_{match}(gram_n)}{\Sigma_{s \epsilon refsum} \Sigma_{gram_n \epsilon S} count(gram_n)} \tag{4}$$

Rouge_L is used to identify the longest co-occurring in sequence n-grams automatically. Suppose we assume that A is the set of sentences of the reference summary and B is the set of sentences of the candidate summary represented by the sequence of words and LCS-based F score (F_{lcs}) indicates the similarity between A (of length m) and B (of length n) according to Eqs. (5), (6), (7).

$$R_{lcs} = \frac{LCS(A, B)}{n} \tag{5}$$

$$P_{lcs} = \frac{LCS(A, B)}{m} \tag{6}$$

$$F_{lcs} = \frac{(1 + \beta^2) R_{lcs} P_{lcs}}{(R_{lcs} + \beta^2 P_{lcs})} \tag{7}$$

where LCS (A, B) denotes the length of the LCS of A and B and $\beta = P_{lcs}/R_{lcs}$. We have used the news articles generated from the BBC News Feed. The results of our evaluation and traditional TF-IDF are given below in Table 1. The unbiased ground truth or the reference summaries of the news articles are obtained from our fellow peers and research scholars having different field expertise. We have considered 10%, 15%, 20%, and 25% sentences of the whole document as the summary size and the performance of the summaries are evaluated using Eqs. (5), (6), and (7) as listed in Table 1. The values demonstrate the effectiveness of the proposed method in compared to the traditional TF-IDF.

4 Conclusion and Future Work

In this paper, we have discussed various methods of abstractive summarization as well as extractive summarization and different other types of summarization. We can get better view of important sentences by constructing the similarity graph of sentences using isf-modified-cosine similarity in comparison to the centroid approach. We also have tried to make use of more of the information from the graph representation of the input document and got better results in many cases. The results of applying these methods on extractive summarization are proving to be effective. Our proposed algorithm does not take care of anaphora [11] resolution problem. We can include it to obtain more informative summary. It may be possible that more than one similar type of sentences with high score is selected for the summary. To eliminate this kind of duplication, we may consider any kind of clustering algorithm and merge it with our proposed graph-based algorithm. In this proposed method, we have considered the weight of the graph only in the initialization of the TextRank score for every sentences correspond to the vertices of the graph. It may be possible to improve our result by considering the weights in the modification of the TextRank defined in Eq. (3). As an extension of the paper, we will compare our method with many more existing effective text summarization methods with different performance measurement metrics.

References

1. Beautifulsoup documentation. https://www.crummy.com/software/BeautifulSoup/bs4/doc/ (2017). Accessed 30 Nov 2017
2. Dutta, S., Ghatak, S., Roy, M., Ghosh, S., Das, A.K.: A graph based clustering technique for tweet summarization. In: 2015 4th International Conference on Reliability, Infocom Technologies and Optimization (ICRITO) (Trends and Future Directions), pp. 1–6. IEEE (2015)
3. Erkan, G., Radev, D.R.: Lexrank: graph-based lexical centrality as salience in text summarization. J. Artif. Intell. Res. **22**, 457–479 (2004)
4. Goldstein, J., Mittal, V., Carbonell, J., Kantrowitz, M.: Multi-document summarization by sentence extraction. In: Proceedings of the 2000 NAACL-ANLPWorkshop on Automatic Summarization, vol. 4, pp. 40–48. Association for Computational Linguistics (2000)
5. Gong, Y., Liu, X.: Generic text summarization using relevance measure and latent semantic analysis. In: Proceedings of the 24th Annual International ACM SIGIR Conference on Research and Development in Information Retrieval, pp. 19–25. ACM (2001)
6. Harabagiu, S., Lacatusu, F.: Topic themes for multi-document summarization. In: Proceedings of the 28th Annual International ACM SIGIR Conference on Research and Development in Information Retrieval, pp. 202–209. ACM (2005)
7. Kan, M.-Y., McKeown, K.R., Klavans, J.L.: Applying natural language generation to indicative summarization. In: Proceedings of the 8th European workshop on Natural Language Generation, vol. 8, pp. 1–9. Association for Computational Linguistics (2001)
8. Khan, A., Salim, N.: A review on abstractive summarization methods. J. Theor. Appl. Inf. Technol. **59**(1), 64–72 (2014)
9. Lin, C.-Y.: Rouge: a package for automatic evaluation of summaries. In: Text Summarization Branches Out: Proceedings of the ACL-04 Workshop, Barcelona, Spain, vol. 8 (2004)
10. Litvak, M., Last, M.: Graph-based keyword extraction for single-document summarization. In: Proceedings of the workshop on Multi-source Multilingual Information Extraction and Summarization, pp. 17–24. Association for Computational Linguistics (2008)
11. Loaiciga Sanchez, S.: Pronominal anaphora and verbal tenses in machine translation. Ph.D. thesis, University of Geneva (2017)
12. Mihalcea, R.: Graph-based ranking algorithms for sentence extraction, applied to text summarization. In: Proceedings of the ACL 2004 on Interactive Poster and Demonstration Sessions, pp. 20. Association for Computational Linguistics (2004)
13. Moens, M.-F., Uyttendaele, C., Dumortier, J.: Abstracting of legal cases: the potential of clustering based on the selection of representative objects. J. Assoc. Inf. Sci. Technol. **50**(2), 151 (1999)
14. Nenkova, A., McKeown, K.: A survey of text summarization techniques. Mining Text Data, pp. 43–76 (2012)

15. Python 2.7.14 documentation. https://docs.python.org/2/index.html (2017). Accessed 30 Nov 2017
16. Radev, D.R., Jing, H., Styś, M., Tam, D.: Centroid-based summarization of multiple documents. Inf. Process. Manag. **40**(6), 919–938 (2004)
17. Saggion, H., Lapalme, G.: Generating indicative-informative summaries with sumum. Comput. linguist. **28**(4), 497–526 (2002)
18. Salton, G., Singhal, A., Mitra, M., Buckley, C.: Automatic text structuring and summarization. Inf. Process. Manag. **33**(2), 193–207 (1997)
19. Seki, Y.: Sentence extraction by tf/idf and position weighting from newspaper articles (2002)
20. Tang, J., Yao, L., Chen, D.: Multi-topic based query-oriented summarization. In: Proceedings of the 2009 SIAM International Conference on Data Mining, pp. 1148–1159. SIAM (2009)
21. Wong, K.-F., Wu, M., Li, W.: Extractive summarization using supervised and semi-supervised learning. In: Proceedings of the 22nd International Conference on Computational Linguistics, vol. 1, pp. 985–992. Association for Computational Linguistics (2008)
22. Yeh, J.-Y., Ke, H.-R., Yang, W.-P., Meng, I.-H.: Text summarization using a trainable summarizer and latent semantic analysis. Inf. process. Manag. **41**(1), 75–95 (2005)
23. Zha, H.: Generic summarization and keyphrase extraction using mutual reinforcement principle and sentence clustering. In: Proceedings of the 25th Annual International ACM SIGIR Conference on Research and Development in Information Retrieval, pp. 113–120. ACM (2002)

A New Static Cost-Effective Parameter for Interconnection Networks of Massively Parallel Computer Systems

M. M. Hafizur Rahman, Mohammed N. M. Ali, Adamu Abubakar Ibrahim, Dhiren K. Behera, Yasuyuki Miura and Yasushi Inoguchi

Abstract The increasing of the signaling technology motivated the research community to find alternative solutions to replace the sequential computing systems. These systems reached their limits, and it became an infeasible choice to manage the computational grand challenges. Thereby, many types of research have been conducted to improve a system with special characteristics able to cope with the new technology. Massively parallel computer (MPC) systems have been emerged to solve the complex computing challenges in parallel and concurrently. The performance of these systems is affected by the structure of the underlying interconnection network. Therefore, many designs of interconnection networks topologies have been presented looking for an optimal one. In this paper, we present the architecture of MMN which is a hierarchical interconnection network. The static and dynamic performance parameters of MMN have been evaluated in previous studies, and it provided good results compared to the conventional interconnection network topologies. This paper will focus on evaluating and

M. M. Hafizur Rahman (✉)
College Computer Science and Information Technology, King Faisal University, Hofuf, Saudi Arabia
e-mail: rahmanjaist@gmail.com

M. N. M. Ali · A. A. Ibrahim
Department of Computer Science, KICT, IIUM, Kuala Lumpur, Malaysia
e-mail: moh.ali.exe@gmail.com

A. A. Ibrahim
e-mail: adamu@iium.edu.my

D. K. Behera
Indira Gandhi Institute of Technology, Sarang, Dhenkanal, Odisha, India
e-mail: dkb_igit@rediffmail.com

Y. Miura
Graduate School of Technology, SIT, Fujisawa, Kanagawa, Japan
e-mail: miu@info.shonan-it.ac.jp

Y. Inoguchi
School of IS, JAIST, Asahidai 1-1, Nomi-Shi 923-1292, Ishikawa, Japan
e-mail: inoguchi@jaist.ac.jp

J. Nayak et al. (eds.), *Soft Computing in Data Analytics*,
Advances in Intelligent Systems and Computing 758,
https://doi.org/10.1007/978-981-13-0514-6_15

determining the static cost-effective parameter of MMN which is the product of the relation between the node degree, network diameter, wiring complexity, and the total number of nodes in each network.

Keywords Massively parallel computer (MPC) · Hierarchical interconnection network (HIN) · Midimew-connected mesh network (MMN) · Packing density Static cost-effective analysis

1 Introduction

The great advancements in computer technology participated in richness the usage of the multicomputer systems in manipulating many of grand computational challenges. Utilizing several numbers of processors simultaneously will be useful in increasing the execution speed of a computer system. However, solving a computational problem in parallel pattern by using group of processors required the processors to communicate each other in order to share the date between them. In addition, to avoid the wasting of the computational power in connecting these processors, it is necessary for these processors to communicate efficiently. Therefore, using the interconnection networks in connecting the processing elements within the system is a time-effective way to maintain the proficiently of these systems. Moreover, these networks became a dominant factor in determining both the cost and the performance of the system. Thus, it is particularly important to design a proficient and cost-effective interconnection network. The research community steered several research work to design such interconnection networks by maintaining the effectivity of these networks [1]. Many kinds of studies also have been steered on the conventional interconnection networks revealed poor performance when the network size increased. Therefore, hierarchical interconnection networks (HINs) have been proposed; HIN is a cost-effective way to connect several network topologies. These topologies are integrated hierarchy to build the higher levels of the system [2, 3]. In addition, HIN maintains a system with good performance even with tens thousands or millions of nodes. Thus, using HIN to replace the conventional interconnection networks in building the future generation of massively parallel computer (MPC) systems is a credible solution to decrease the cost and to improve the performance of these systems [3].

Billions of transistors will be integrated in one system to compose tens to hundreds of intellectual property (IP) cores [4, 5]. The proficient cooperation between the IP cores is able to deliver good and rich services through utilizing of the available resources. The network-on-chip (NoC) is a suitable architecture which is able to provide accommodations for high number of cores; in addition, it guarantees the communication and data transfer between these cores. NoC has been emerged as an alternative solution to solve the performance limitations of long interconnect and to replace the bus architectures. Moreover, NoC plays main role in solving the global wire delay problem by easing the integration of high number of IP cores in a single system-on-chip (SoC). Therefore, NoC has been supported by the industry sector to be the popular choice in designing the on-chip interconnects for the multiprocessor system-on-chip (MPSoCs) [4, 6]. The performance and the modularity of the system are increased by using a network to replace the global wiring structure. NoC optimizes the electrical properties of the global wires; in addition, it makes it easy to be controlled. Controlling the electrical parameters is reducing the power dissipation by enabling the use of aggressive signaling circuits. The power dissipation will be reduced a factor of ten and increase propagation rate by three times. The usage of the wires will be more efficient if the wiring resources have been shared between many communication flows; when one client is idle, the network resources can be used by other clients [7, 8]. Massively parallel computer (MPC) systems are consisting of either thousands or millions of processors, and the recent advancements in very-large-scale integration (VLSI) and the network-on-chip (NoC) technology led to multiprocessor systems in three dimensions [9].

Two-dimensional a wraparound mesh network known as the midimew network; one of these dimensions is a diagonal wraparound, and the other one is a tori connected. A midimew-connected mesh network (MMN) is a hierarchical interconnection network which is containing several basic modules (BMs). The basic module (BM) of MMN is a two-dimensional mesh network. The BMs interconnected hierarchically to build the higher level networks in a midimew fashion [10]. The emphasis of this paper is on evaluating the static cost-effective parameter of MMN by using the relation between some of the static performance parameters including node degree, diameter, wiring complexity, and the total number of nodes. Section 2 is the description of the structure of MMN, and in Sect. 3, we will evaluate the static cost-effective parameter of this network, and Sect. 4 is the conclusion of this work.

2 The Structure of MMN

Minimal distance mesh with wraparound links (midimew) is a wraparound mesh network with two dimensions; these dimensions are a diagonally connected wraparound and tori connected. The perpendicular direction has been considered as tori connected. However, the horizontal direction is assumed as a diagonal connection. MMN composed of several clusters called basic modules (BMs); each basic module is a two-dimensional mesh network. The connection between these BMs is a hierarchical connection, in order to construct the higher levels network of MMN. The BM of MMN indicates the Level-1 network. Each level of this network considered as a sub-net module of the higher level.

2.1 Basic Module of MMN

The basic module of this network has either one or two exterior free nodes. The purpose of using such nodes is to be used in interconnecting the higher levels of MMN. In addition, in purpose to connect the sensibly adjacent BMs, extra links are used in these free ports. Moreover, these BMs are connected in hierarchy fashion to create the higher levels of MMN. The basic module of MMN composed of $2^m \times 2^m$ nodes, while the first 2^m represents the number of the rows and the second rcpresents the columns. Therefore, the basic module of MMN contains a 2^{2m} number of nodes, where m is a positive integer number. In order to achieve enhanced granularity, we have considered $m = 2$.

Figure 1 portrays the design of the (4×4) BM of MMN. The $2^m \times 2^m$ BM contains a number of 2^{m+2} free ports; these ports are used for the higher level interconnection. Therefore, the (4×4) basic module contains 16 free ports as it clear from Fig. 1. It is interesting to see that from Fig. 1, each free port defined according to Level $(1 \leq L \leq 5)$, direction (vertical, horizontal or diagonal), and link direction (incoming or outgoing). The bidirectional links are created by tying the incoming links with the corresponding outgoing links. As we mentioned earlier, the free ports are used to interconnect the higher levels of MMN. The communication links could be either unidirectional or bidirectional link. The unidirectional links are used to transmit and to receive the packets through fixed links. Therefore, packets will be sent through transmitting links and packets will be received through

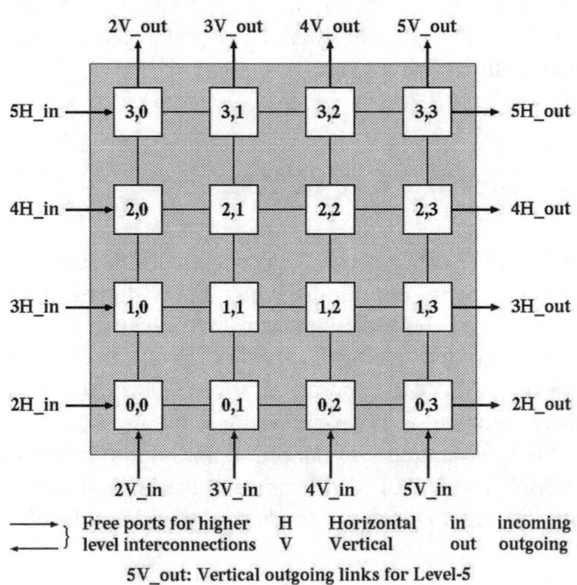

Fig. 1 Basic module (BM) of MMN

receiving links. In contrast, the bidirectional connection is using the same link to transmit and to receive the packets.

2.2 The Higher Level of MMN Network

The basic modules (BMs) of MMN interconnected through the free ports to construct the higher levels of this network. In addition, the higher level of MMN is a midimew network. Earlier in this paper, we have mentioned MMN has two directions; the connection of the perpendicular direction is toroidal; however, the horizontal one is considered a diagonal wraparound connection. Furthermore, the higher level network is using the immediately lower level network as a sub-net module. By considering $(m = 2)$, MMN needs $2^{(2 \times 2)} = 16$ BMs to be interconnected to build a Level-2 network. Similarly, Level-3 of MMN needs 16 sub-net module of Level-2 to be constructed. The number of the free ports and their associated links can be determined from $4 \times (2^q) = 2^{q+2}$ ports. These ports are separated as; $2(2^q)$ ports are for the vertical toroidal interconnection, and $2(2^q)$ is used by the neighboring BMs to connect horizontally and also for the diagonal wraparound connection of the end-to-end basic modules. The inter-level connectivity represented by *q;* where $q \in \{0, 1, \ldots, m\}$. In this paper, we considered that the minimal inter-level connectivity can be achieved when $q = 0$. However, the maximum inter-level connectivity achieved when $q = m$.

The interconnection of the higher level of MMN has been illustrated in Fig. 2 by considering $m = 2$ and $q = 0$. Thus, by using $q = 0$; the number of the free ports will be $4(2^0) = 4$. Therefore, we can divide the wires which are using for the higher level interconnection of MMN into two parts. These links will be 2 for the perpendicular toroidal connection, and the remaining 2 will be used for the horizontal and diagonal connection. In addition, the usage of these links will be for incoming and outgoing links. On the other hand, tying the outgoing and incoming wires of two sensibly neighboring nodes will help in creating the bidirectional connection. This situation is illustrated in Fig. 2, and it shows that an $MMN(m, L, q)$ is created of $2^m \times 2^m$ basic modules, where L is the level of the hierarchy, and q represent the inter-level connectivity. The highest level of the hierarchy of MMN can be calculated based on a $2^m \times 2^m$ basic module from $L_{max} = 2^{m-q} + 1$. by assuming $m = 2$ and $q = 0$, $L_{max} = 2^{2-0} + 1 = 5$, which mean that the maximum level of the hierarchy of MMN is Level-5 network. Furthermore, the number of nodes in each Level can be calculated from $N = 2^{2mL}$. Thus, by considering L_{max}, the highest level of the hierarchical connection, the supreme number of nodes which we can connect by $MMN(m, L, q)$ can be determined from $N = 2^{2m\left(2^{m-q+1}\right)}$. By considering the minimum inter-level connectivity, and by assuming $m = 2$, the highest number of nodes will be $N = 2^{2m(2m-q+1)} = 2^{2 \times 2\left(2^{2-0+1}\right)} = 2^{20} = 1,048,576$. This indicates that more than 1 million nodes can be interconnected in case of using MMN to build the new generation of massively parallel computer (MPC) systems.

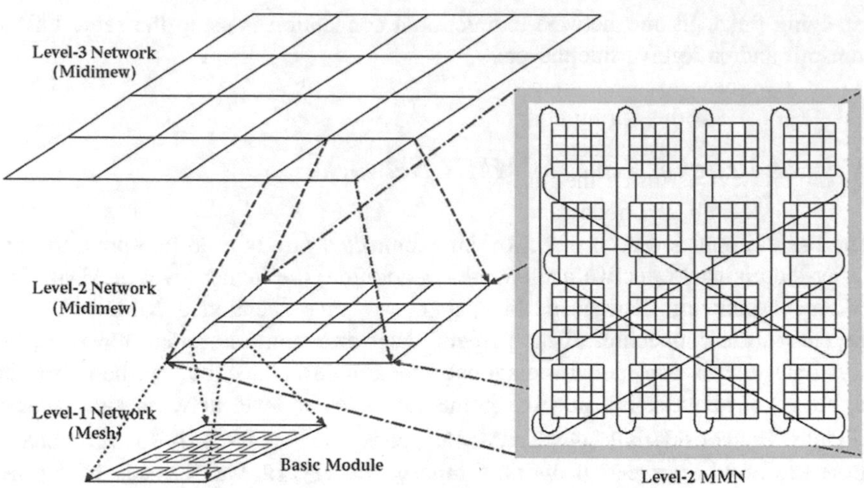

Fig. 2 Higher level of MMN

3 Static Cost-Effective Analysis of MMN

The interconnection network is the backbone of the massively parallel computer (MPC) systems, and it affects widely the performance of these systems. In addition, it has a crucial impact in either increasing or decreasing the cost of these systems [11]. Many of static parameters are used as a test to measure the effectivity of the topology of the interconnection network. Based on this test, the scientists determine the efficiency of these topologies compared to the other networks. In previous studies, we have evaluated the static performance of MMN in terms of node degree, diameter, average distance, wiring complexity, and many of these parameters. MMN shows good results compared to other networks such as mesh, torus, and TESH networks.

In this paper, we will use the relation between some of the static network parameters to evaluate and determine the static cost-effective parameter of MMN. Some of these parameters are related to the network cost and have the main role in either increasing or decreasing the cost, and other parameters have a serious effect on either improving or degrading the network performance. Parameters such as network diameter which have a crucial influence on the network performance are defined as the maximum number of nodes the packet needs to traverse to reach from the source node to the destination node by selecting the short route. Short diameter indicates good performance, while the long diameter degrades the network efficiency [12]. In contrast, parameters such as node degree and wiring complexity have the main role in increasing and decreasing the cost of the network. The node degree is defined as the maximum number of links emanating from this node; in addition, it has a great influence on the network cost. The node degree also is a measure of the node's I/O complexity [13]. On the other hand, the network wiring complexity affects the network cost by increasing or decreasing the number of interconnection links inside the network. In addition, it defined as the total number of wires we need to use to connect the nodes to each other. The node degree has a crucial effect on the wiring complexity of the interconnection network [14]. The number of nodes also affects the network cost by increasing the number of computational devices inside the interconnection network. In this paper, we will consider MMN with 256 nodes and we will compare it to other networks having the same number of nodes.

The static cost-effective analysis for each network can be calculated by determining the relation between particular static parameters; these parameters include the total number of nodes, the node degree, the network diameter, and the wiring complexity. The relation between these nodes will provide the actual cost of each network. In this paper, we will use the previously mentioned parameters to investigate the static cost effectivity of MMN and we will compare it to other networks such as 2D-mesh, 2D-torus, and TESH. The cost-effective analysis of each network can be derived from Eq. (1).

$$Static\ Cost-effective\ analysis = \frac{Node\ Degree \times Wiring\ complexity}{Diameter \times Total\ number\ of\ Nodes} \quad (1)$$

Table 1 provides a comparison between different networks including MMN. These networks have the same number of nodes in order to create a fair comparison. In this paper, we will consider each network with 256 nodes. From Table 1, we noticed that the node degree of each network is equal 4. In addition, Table 1 shows the network diameter of MMN is lower than that of 2D-mesh, and TESH networks. However, it is slightly higher than that of 2D-torus network. As we mentioned earlier, the short diameter improve the network performance. From Table 1, also we can see the wiring complexity of mesh and torus is higher than that of both MMN and TESH networks. Lower wiring complexity indicates network with less cost.

The static cost-effective analysis parameter has been calculated in Table 1 for these networks. It is clear that the cost-effective analysis parameter of MMN is higher than that of both 2D-mesh and TESH networks. In contrast, it is a little bit lower than that of 2D-torus network. Network with higher static cost-effective parameter is achieving better network performance. Thus, MMN is superior over 2D-mesh and TESH networks and approximately having a close value to torus network. Table 1 shows some facts, the network diameter, and the wiring

Table 1 Comparison between different networks

Network	Number of nodes	Degree	Diameter	Complexity	Static cost-effective analysis
2D-Mesh	256	4	30	480	0.25
2D-Torus	256	4	16	512	0.5
TESH	256	4	21	416	0.309
MMN	256	4	17	416	0.382

complexity of 2D-mesh is extremely higher than that of MMN; in addition, MMN achieved better network diameter compared to TESH network. As a result, MMN achieved the superiority over 2D-mesh and TESH networks. On the other hand, the network diameter of 2D-torus is a little bit lower than that of MMN. Nevertheless, the wiring complexity of MMN is sharply lower than that of 2D-torus network. As a result, a little bit higher diameter of MMN achieved a less cost network with good performance. Therefore, MMN is a good choice over 2D-torus network and it could be an excellent selection for the next generations of the MPC systems.

4 Conclusions

The design and the architecture of MMN have been discussed in this paper. In addition, the static network parameters of MMN have been evaluated and provided in Table 1. Moreover, we have compared the results of MMN to those results of 2D-mesh, 2D-torus, and TESH networks. MMN has a simple architecture, due to the combination of the mesh and midimew networks. In addition, it is a distinguished network about the other networks which have been used in the evaluation process through this paper. Moreover, the diagonal wrap links provided an effective distance parameter of MMN. In this paper, we focused on analyzing the static cost-effective parameter of MMN and this parameter has been calculated from the relation between some of the static network parameters; these parameters are node degree, diameter, wiring complexity, and the total number of nodes. MMN achieved lower cost comparing to 2D-torus network. On the other hand, the static cost-effective of MMN was greater than that of 2D-mesh and TESH networks and little bit lower than that of 2D-torus network. Network with high static cost-effective parameter is achieving better performance than the others. The future work on the MMN will be as: (1) measuring the fault tolerant and the reliability of MMN [15] and (2) using the optical links as an alternative to replace the long electrical links [16].

References

1. Cheng, S.Y., Chuang, J.H.: Varietal hypercube-a new interconnection network topology for large-scale multicomputer. In: Proceeding of the International Conference on Parallel and Distributed Systems, pp. 703–708. IEEE (1994)
2. Al Faisal, F., Rahman, M.M., Inoguchi, Y.: Topological analysis of the low-powered 3D-TESH network. IEICE Tech. Rep. **115**(399), 143–148 (2016)
3. Awal, M.R., Rahman, M.M.H., Akhand, M.A.H.: A new hierarchical interconnection network for future generation parallel computer. In: Proceeding of the 16th International Conference on Computer and Information Technology (ICCIT), pp. 314–319. IEEE (2013)

4. Anagnostopoulos, I., Bartzas, A., Vourkas, I., Soudris, D.: Node resource management for DSP applications on 3D network-on-chip architecture. In: Proceeding of the 16th International Conference on Digital Signal Processing, pp. 1–6. IEEE (2009)
5. Ali, M., Welzl, M., Zwicknagl, M.: Networks on chips: scalable interconnects for future systems on chips. In: Proceeding of the 4th European Conference on Circuits and Systems for Communications, ECCSC. pp. 240–245. IEEE (2008)
6. Goossens, K., Dielissen, J., Radulescu, A.: Æthereal network on chip: concepts, architectures, and implementations. IEEE Des. Test Comput. **22**(5), 414–421 (2005)
7. Dally, W.J., Towles, B.: Route packets, not wires: on-chip interconnection networks. In: Proceedings of Design Automation Conference, pp. 684–689. IEEE (2001)
8. Bjerregaard, T., Mahadevan, S.A.: survey of research and practices of network-on-chip. ACM Comput. Surv. (CSUR) **38**(1), 1
9. Miura, Y., Kaneko, M., Rahman M.M.H., Watanabe, S.: Adaptive Routing Algorithms and Implementation for TESH Network (2013)
10. Rahman, M.M.H., Ali, M.N.M., Nor, R.M., Sembok, T.M.T., Akhand, M.A.H.: Time-cost effective factor of a midimew connected mesh network. In: Proceeding of the 6th International Conference on Information and Communication Technology for the Muslim World (ICT4 M), pp. 264–268. IEEE (2016)
11. Kim, J., Dally, W.J., Scott, S., Abts, D.: Technology-driven, highly-scalable dragonfly topology. In: ACM SIGARCH Computer Architecture News, vol. 36, no. 3, pp. 77–88. IEEE Computer Society (2008)
12. Rahman, M.H., Inoguchi, Y., Yukinori, S.A.T.O., Horiguchi, S.: TTN: a high performance hierarchical interconnection network for massively parallel computers. IEICE Trans. Inf. Syst. **92**(5), 1062–1078 (2009)
13. Rahman, M.H., Horiguchi, S.: HTN: a new hierarchical interconnection network for massively parallel computers. IEICE Trans. Inf. Syst. **86**(9), 1479–1486 (2003)
14. Ali, M.N.M., Rahman, M.M.H., Nor, R.M., Sembok, T.M.B.T.: A high radix hierarchical interconnection network for network-on-chip. In: Recent Advances in Information and Communication Technology, pp. 245–254. Springer International Publishing (2016)
15. Mohanty, S.P., Ray, B.B., Patro, S.N., Tripathy, A.R.: Topological properties of a new fault tolerant interconnection network for parallel computer. In: Proceeding of the International Conference on Information Technology (ICIT'08). pp. 36–40. IEEE (2008)
16. Xiao, L., Wang, K.: Reliable opto-electronic hybrid interconnection network. In: Proceeding of the International Symposium on Parallel Architectures, Algorithms, and Networks (I-SPAN08). pp. 239–244. IEEE (2008)

MapReduce Accelerated Signature-Based Intrusion Detection Mechanism (IDM) with Pattern Matching Mechanism

Chinta Someswara Rao and K. Butchi Raju

Abstract Network protection was rendered an ultimatum due to expeditious growth in the pace of wired Internet and also due to the constantly increasing number of attacks. Discovery of dubious activities and prevention of their malicious impact can be effectively done by intrusion detection systems (IDS). Due to the pattern matching operation, existing signature-based IDS will bear more searching time and memory. A diligent system to truncate the overhead is the need of the hour. Matching the pattern through parallelizing a matching algorithm on a GPU using MapReduce is the objective of this research. This paper attempts to parallelize a pattern matching technique under the MapReduce framework. A speedup of four times is achieved using the GPU implementations in juxtaposing to the CPU under MapReduce framework.

Keywords Information security · Intrusion detection systems
Pattern matching · MapReduce · GPU · CPU

1 Introduction

Human life has become much affable due to the advent of digital devices and services. A series of obvious hindrances and skeptical behaviors enhances as the connectivity between various services increases. For different services, various strong attack detection processes are required [1].

IDS is the most prominent technique that will be used to inspect network payloads of packets for malicious threats. Intrusion signatures are required for IDS. Usage of software or hardware is the most customary proposal of several

C. S. Rao (✉)
Department of CSE, SRKR Engineering College, Bhimavaram, AP, India
e-mail: chinta.someswararao@gmail.com

K. Butchi Raju
Department of CSE, GRIET, Hyderabad, TS, India
e-mail: butchiraju.katari@gmail.com

J. Nayak et al. (eds.), *Soft Computing in Data Analytics*,
Advances in Intelligent Systems and Computing 758,
https://doi.org/10.1007/978-981-13-0514-6_16

researchers for the enhanced IDS implementations [2, 3]. For scanning the system files and malware signature by entreating the software-based IDS pattern matching algorithm. The system results are measured by means of searching time, memory usage, etc. These measures improvement of the efficiency of software solution is relatedly indispensable. This can be achieved by GPU-based computing under MapReduce framework. It is the most important component for achieving this by GPU.

In this paper, we propose a GPU-based MapReduce process with pattern matching techniques. Section 2 furnishes the related literature. Section 3 provides the proposed methodology of the techniques. The results pertaining to experiment and assessment of metrics are furnished in Sect. 4. Section 5 embellishes conclusion and the recommendation for future work.

2 Literature Survey

A lot number of software and hardware solutions for IDS are submitted, with an objective of aggravating the superior performance and throughput rates

Kharbutli et al. [4] proposed three parallelized algorithms for IDS. In this, authors split the input text and shared among the processing units along with count variable. Finally, all the results from processing units are collected.

Su et al. [5] proposed a multi-core with AC algorithm for IDS. Here also, input text splits into different equal parts and shared among the processing units along with count variable by following the overlap concept.

Holtz et al. [6] pool up the Internet traffic from different distributed IDSs related to the divergent locations of the network. MapReduce was the tool used to analyze the data in the network.

Hu et al. [7] proposed GPU-based Aho–Corasik algorithm for IDS in Gnort. Their experimental results have shown that their proposed techniques are very well for IDS.

Xu et al. [8] proposed a new proposal with impMASM for IDS. In this, authors split the input text into different equal parts and spread among the available processors along with pattern count under GPU environment.

Hung et al. [9] proposed GPU based approach for IDS with that will support of different memory architecture. In this also, authors split the text and spread among the processors along with pattern count. For implementation, authors used the multi-threading concept and hashing.

3 Methodology

Following are the details of the different preliminaries of the mathematical background used in the algorithm and the technical details of the frameworks used to implement it.

Hadoop [10] is a popular Apache open-source Java-based implementation platform that provides tools for parallel processing of vast amounts of data using the MapReduce paradigm [11] in a distributed manner majorly on gigantic clusters in a reliable and fault-tolerant fashion. MapReduce is a programming model for processing large datasets, and it crumbles a computational task into two steps such that the entire process is automatically parallelized and executed on a large cluster of computational elements. The user only needs to specify map and reduce functions.

3.1 The MapReduce Architecture

A typical MapReduce cloud environment consists in master node, worker nodes, and a distributed file system access through the environment. The master node acts as a frontend to the cloud environment, receiving jobs and distributing computing loads and the necessary data among the worker nodes. The worker nodes are common compute nodes, which actually perform the desired computation on the provided data and send their outputs back to the master node. This research document extends an attempt to throw light on the MapReduce implementation of the Hadoop project, which already includes the MapReduce processing environment, a distributed file system (Hadoop file system), several API spanning many languages, and accessory management services. A Hadoop environment mainly consists in the same components depicted in Fig. 4. In order to coordinate the interaction between the master node, the several worker nodes, the distributed file system, and other components that may exist in the cloud, Hadoop provides a set of background control services that run on each node, providing the master node with information about all the worker nodes and the other cloud components.

A HDFS environment is mainly composed by NameNodes and DataNodes, which are loosely equivalent to master nodes and worker nodes from a MapReduce perspective. When a file is stored in an HDFS volume, it is divided into several slices that are subsequently spread across different nodes that together provide physical storage resources for the volume.

Due to their similarity to the master and worker nodes in a MapReduce environment, the NameNode and DataNodes in a Hadoop setting are usually associated to the master node and the worker nodes, respectively. However, it is possible to set up an HDFS environment with more than one NameNode for backup and high availability support. The NameNode is responsible for maintaining a central catalog of the directory structures and files stored in each volume, which contains pointers to the location (i.e., DataNode address) of the slices of each file that are spread across the volume. It also handles all volume operations, such as read and write, acting as a frontend to the volume. The DataNodes are responsible for actually storing data, keeping slices of the files stored in the volume, providing access to the data that is stored locally and providing control information about the status of this data. The NameNode polls DataNodes in order to obtain information about data

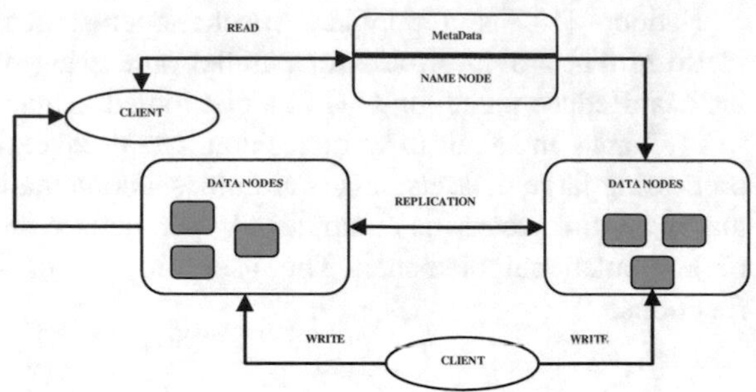

Fig. 1 Architecture of the HDFS file system

integrity and performs data slice replication tasks to distribute different slices of files across different DataNode, providing protection against failures of individual nodes. The general architecture of an HDFS environment is depicted in Fig. 1.

3.2 The Hadoop MapReduce Architecture

In a Hadoop environment, the MapReduce framework is strongly coupled to the HDFS file system by several control services in order to form an scalable and resilient cloud infrastructure. The master node is mainly composed by the Job-Tracker and NameNode services. JobTracker is a service that determines the worker nodes that have access to the necessary data and that are available to receive tasks. Based on the information collected from individual worker nodes, it allocates tasks to proper worker nodes. JobTracker also collects information on each running and scheduled task from the worker nodes, which it uses to allocate nodes and provide user with statistical and managerial information on the overall cloud status. NameNode simply acts as a NameNode for the Hadoop File System distributed file system. Client MapReduce applications access the NameNode service in order to perform tasks such as add, copy, move, or delete files stored in HDFS volumes.

The rest of the cluster worker nodes run other two services that compose the processing environment and data storage architecture, namely TaskTracker and DataNode. TaskTracker receives tasks (i.e., Map, Reduce, and Shuffle) from the JobTracker in the master node and reports back to the master node the status of currently running and scheduled tasks. It is has a set of virtual slots, which determine the number of tasks that it may accept and process. DataNode acts as a HDFS DataNode and manages local data storage of the slices allocated to each subordinate node by the sovereign node. When the master node first receives a job, JobTracker queries the DataNode in each worker node to determine which node already has the necessary data stored locally and then allocates a slot with the TaskTracker running in the given node. If it does not find a node that already has the data, it allocates the first empty slot it finds in any node. Usually, a file stored in a HDFS volume is replicated to several worker nodes through their respective

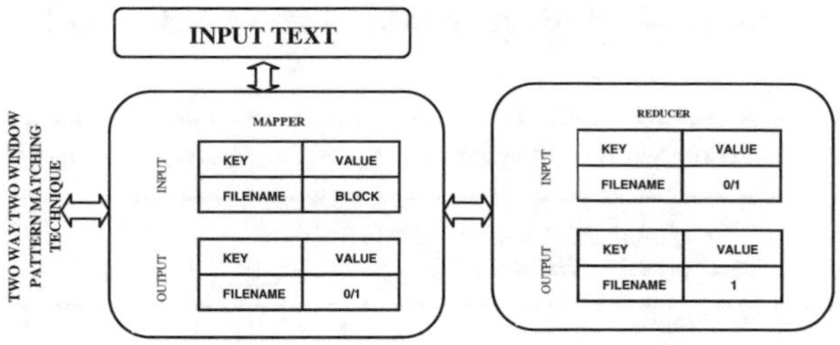

Fig. 2 Proposed MapReduce approach

DataNodes. When the TaskTracker receives a new job (consisting of the MapReduce task to be run and the description of the necessary dataset), it spawns a child process to execute it and then reports back on its status to the JobTracker.

3.3 Proposed Technique

The vivid particulars related to MapReduce implementation of IDS (the approach is shown in Fig. 2) are presented in this section.

Step 1:	In this step, data from HDFS in the block of 100 MB size are read by the custom input format. Here, *Key* is the file name and *block* is the value that will contain the several lines
Step 2:	The map() will compute the {key, value} pairs for different input data and transfer these into GPU which will perform the IDS operation
Step 3:	The data are transferred to HDFS cache. The map() will check for extension and pass the data and signatures to GPU. Here, different signatures are used for different extensions
Step 4:	The map() collects the occurrence count and position of the virus signatures in data from GPU
Step 5:	The searching results are returned to reduce with {key, value} pairs; here, filename is the key, and value is one or zero
Step 6:	The reducer() checks the values for 1 in {key, value} pairs and that will be returned to HDFS

Algorithm 1 describes the actual pattern matching process which is used in the MapReduce approach.

ALGORITHM 1: TWO WAY TWO WINDOW PATTERN MATCHING APPROCH	
Input: Text string 'T' of length 'n' and Pattern 'P' of length 'm'. **Output:** All occurrences of pattern in text string.	

STEP 1	get.CUDAProperties()
	start=0, end=T.length();
	threshold= T.length()/2;
	if (end - start > threshold){mid = (start + end) >>> 1; }
	left_window_limits=>left_window_start=0 &left_window_end=mid+m-1;
	right_window_limits=>right_window_start=T.length()-1&right_window_end=mid;

STEP 2	for
	i_left_window←left_window_starttoleft_window_end&&i_right_window←right_window_starttoi_right_window_enddo
	begin
	count=search(T, patterns[],left_window_limits)
	count=search(T,patterns[],right_window_limits)
	end for;

STEP 3	***int search(***Char[] T, Char[] P, intlimits)
	begin
	j=limits
	intcomp_pos=0;
	while (j >=0 && T_i==P_j){j--; i++;}
	if (j == -1){count++;}
	return count;
	end search;

4 Experimental Setup and Result Analysis

The proposed method is deployed under i7 processor, Windows 7 Ultimate operating system, and NVIDIA GPU. A total of 150,000 virus signatures were downloaded from Clam AV site [12], and proposed system implementation is performed in Java. The experimentation juxtaposed the time consumed by both CPU and GPU in the implemented pattern matching algorithm called *two-way two window pattern matching approach,* and time comparison results of this are shown in Table 1. Figure 3 shows the analysis between file size and searching time CPU under

Table 1 Performance statistics

	Searching time in minutes			Searching time in minutes	
File size (MB)	CPU under MapReduce	GPU under MapReduce	File size (MB)	CPU under MapReduce	GPU under MapReduce
1000	2.5	1.5	6000	15	10
2000	5	2.5	7000	18	13
3000	7.5	4	8000	22	18
4000	10	6	9000	26	21
5000	12.5	7	10,000	30	26

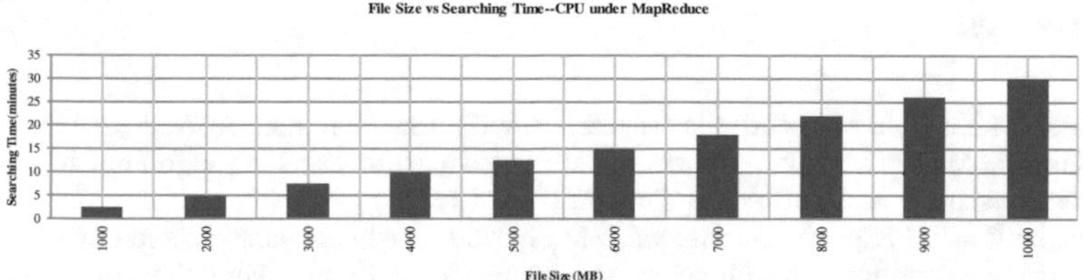

Fig. 3 File size versus searching time graph—CPU under MapReduce

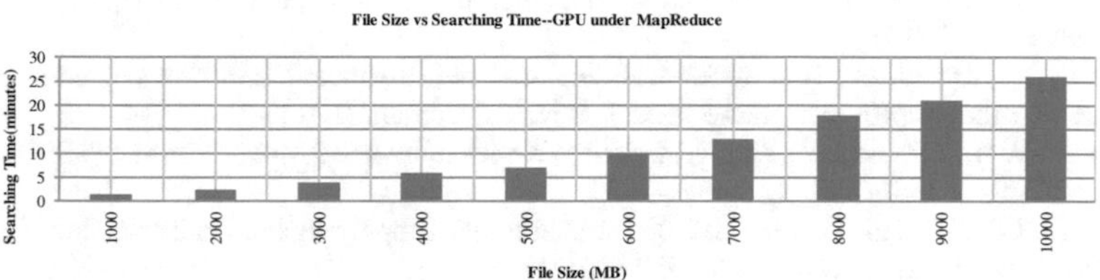

Fig. 4 File size versus searching time graph—GPU under MapReduce

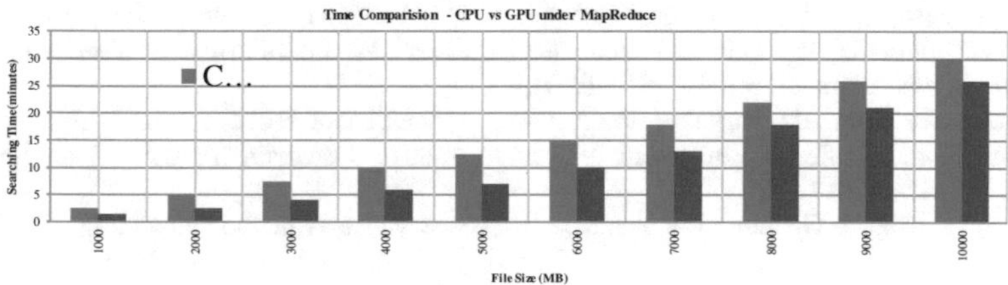

Fig. 5 Time comparison graph—CPU versus GPU under MapReduce

MapReduce; Fig. 4 shows the analysis between file size and searching time GPU under MapReduce, and Fig. 5 indicates the comparison graph between CPU and GPU under MapReduce.

5 Conclusions

Reduction of searching time through the usage of the GPU is the preliminary outcome that was observed through the proposed technique. Renowned two-way two window pattern matching was implemented to search for signatures in the content file. A towering gap between CPU and GPU times was observed while testing the large amount of data for testing. The next level of parallelism is achieved through the GPU in MapReduce process. In this paper, we are conducting the experiments on real-world data set. From the results, we observed that this GPU in MapReduce process will achieve milestones, especially for IDS.

References

1. Vardi, M.Y.: Cyber insecurity and cyber libertarianism. Commun. ACM **5** (2017)
2. Aldwairi, M., Ekailan, N.: Hybrid multithreaded pattern matching algorithm for intrusion detections systems. J. Inf. Assur. Secur. 512–521 (2011)
3. Kharbutli, M., Mughrabi, A., Aldwairi, M.: Function and data parallelization of Wu-Manber pattern matching for intrusion detection systems. Netw. Protoc. Algorithms 46–61 (2012)
4. Xu, K., Cui, W., Hu, Y., Guo, L.: Bit-parallel multiple approximate string matching based on GPU. Proced. Comput. Sci. pp. 523–529, 2013
5. Su, X., Ji, Z., Lia, A.: A parallel AC algorithm based on SPMD for intrusion detection system. In: Proceedings of the 2nd International Conference on Computer Science and Electronics Engineering (2013)
6. Holtz, M., David, B., Junior, R.: Building scalable distributed intrusion detection systems based on the MapReduce framework. J. Rev. Telecomun. (2011)
7. Hu, L., Wei, Z., Wang, F., Zhang, X., Zhao, K.: An efficient AC algorithm with GPU. Proced. Eng. 4249–4253 (2012)
8. Xu, K., Cui, W., Hu, Y., Guo, L.: Bit-parallel multiple approximate string matching based on GPU. Proced. Comput. Sci. 523–529 (2013)
9. Hung, C.L., Lin, C.Y., Wang, H., Chang, C.Y.: Efficient packet pattern matching for gigabit network intrusion detection using GPUs. In: International Conference on Embedded Software and Systems (HPCC-ICESS), pp. 1612–1617 (2012)
10. Rodrigues, R.A., Lima Filho, L.A., Gonçalves, G.S., Mialaret, L.F., da Cunha, A.M., Dias, LA.:Integrating NoSQL, relational database, and the hadoop ecosystem in an interdisciplinary project involving big data and credit card transactions. In: Information Technology-New Generations. Springer, pp. 443–451 (2018)
11. Zhang, L., Tang, B.: Parka, A Parallel Implementation of BLAST with MapReduce. In: International Conference on Intelligent and Interactive Systems and Applications, Springer, pp. 185–191 (2017)
12. ClamAV project: Clamav virus database. http://www.clamav.net/download.html

Ensemble of Artificial Bee Colony Optimization and Random Forest Technique for Feature Selection and Classification of Protein Function Family Prediction

Ranjani Rani Rangasamy and Ramyachitra Duraisamy

Abstract Protein function prediction is a prevalent technique in bioinformatics and computational biology. Even now, the computation of function prediction is an impudent task to provide efficient and statistically significant accurate results. In this work, the optimization approach and the machine learning method were proposed to predict the function families of a protein using the sequences regardless of its similarity. It is denoted as Prot-RF (ABC) (predicting protein family using random forest with artificial bee colony). The features of the protein sequences are selected using the ABC method, and they are classified using the random forest classifier. The Uniprot and PDB benchmark databases have been utilized to assess the proposed Prot-RF (ABC) method against the other well-known existing methods such as SVM-Prot, K-nearest neighbor, AdaBoost, probabilistic neural network, Naïve Bayes, random forest, and J48. The classification accuracy results of the proposed Prot-RF (ABC) method outperform the other remaining existing methods.

Keywords Protein sequence · Feature extraction · Feature selection Artificial bee colony · Random forest classification

1 Introduction

A large volume of uncharacterized protein sequence data has been increasing day by day which makes the existing methods for protein function prediction a challenging task. To progress any novel biological process, it is very significant to know

R. R. Rangasamy (✉) · R. Duraisamy
Department of Computer Science, Bharathiar University, Coimbatore, India
e-mail: ranjaniRSR91@gmail.com

R. Duraisamy
e-mail: jaichitra1@yahoo.co.in

J. Nayak et al. (eds.), *Soft Computing in Data Analytics*,
Advances in Intelligent Systems and Computing 758,
https://doi.org/10.1007/978-981-13-0514-6_17

the functionality of a protein. An empirically based function prediction of a protein involves huge time and human labor to analyze a solitary gene or a protein [1]. Thus, the computational approaches have been evolved to remove the disadvantages of the empirical methods. The function of a protein can be identified using numerous techniques such as clustering of a sequence, similarity of a sequence, fusion of a gene, interaction of a protein, analysis of an evolutionary relationship, detection of a remote homology of a protein, and classification of a protein function family [2].

The protein function that has null sequence homolog of known function is difficult to allocate the function depending on the sequence similarity. So, there is a need to explore the technique to predict the function of a protein that is not based on the sequence similarity. A protein functional family is a cluster of proteins with certain category of molecular functions which engrossed in explicit biological process demarcated by the Gene Ontology [3]. Many machine learning methods have been used to predict the protein function such as support vector machine [4–7], random forest [1, 2], KNN [1, 8], PNN [2], Naive Bayes [9], AdaBoost [10], and J48 [7]. Many of the protein family functional approaches have built the models for more largely defined functional families. Still there is a need to develop the novel techniques to explore the models with many other diverse functional families.

Every protein sequence has its own features, and those features have to be extracted to identify its functional family. Many number of online tools are available for the feature extraction from protein sequence. Also, the amino acid composition (AAC), dipeptide composition, and pseudoamino acid composition (PseAA) [4] techniques were also employed. Among them, the dipeptide composition technique extracts the better features but the dimensionality of this is larger in size which may lead to the decline of the classification accuracy, and it is time-consuming process. Many of the features are redundant and noisy in nature. Thus, for minimizing the features, the selection specific features are obtained by employing the optimization algorithms which predict the protein functional families.

By using the selected features, the random forest classification algorithm has been employed to classify the protein functional families depending on their sequence derivative structural and physiochemical properties. The proteins identified to be in functional family and outside the functional family are employed to train a classification model which distinguishes the particular sequenced derivative features for classifying the proteins either into or outside the family. The common structural and physiochemical properties are shared by the proteins with particular functional family that is identified by the classification model by producing the adequate various training datasets. This method was assessed based on the 192 protein families from Uniprot which was identified by the SVM-Prot [6].

Thus, in this paper, an ensemble technique termed ABC for feature selection along with the random forest method (Prot-RF) for classification is employed to predict various protein functional families of a protein using protein sequences regardless of its similarity. The remaining sections of the paper are as follows: Sect. 2 describes the methodology of the projected approach to extract and select the features of the sequences and classify the protein functional family classification. Section 3 discusses the experimental results of the projected method, and the Sect. 4 illustrates the implementation and discussion of the proposed work. To conclude finally, Sect. 5 describes the inference and the future enhancement of the research work.

2 Methodology

The complete workflow of the protein functional family classification is shown in Fig. 1. Initially, the raw protein sequence taken from the PDB and Uniprot has been given as an input. Using the dipeptide composition method [4], the sequence features are extracted. The selection of the relevant features is accomplished by the use of artificial bee colony optimization approach. Next, the datasets are arranged into different training and testing set with both positive and negative samples. Positive samples are defined as the sequences belong to the particular protein function and the negative samples are those which they do not belong to the specific protein function. The selected training set data has been applied to the Prot-RF classifier to distinguish the protein functional families from the sequence. Finally, the classification accuracy of the predicted functional families is displayed as a result.

The feature of the protein sequences has been extracted using the dipeptide composition technique. Some of the common features extracted from the optimization algorithm are hydrophobicity, composition of amino acid, atom count, residue count, etc. As it is an existing method [4], its results were not displayed due to the space constraint.

2.1 Feature Extraction

The ABC approach is one of the most popular evolutionary techniques for feature selection. The cautious selection of the features may reflect the performance of the

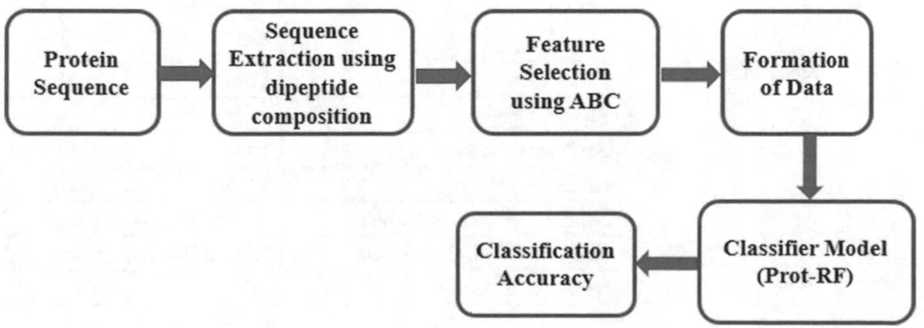

Fig. 1 Work flow of protein functional family classification

prediction accuracy. ABC algorithm was developed by inspiring the foraging behavior of the swarm of honeybees. It consists of three phases as employed bees, onlooker bees, and scout bees. The entire bee colony is separated into two clusters in which first half is for employed bees and second part is for onlooker bees. Scout bees are the one whose food source has been uninhibited. The location of the food source has been represented as a solution, and the nectar amount of the food source is considered to be the fitness value [11]. Subsequently ABC Algorithm employs the exploitative process resourcefully to converge minima and explorative process to deliver satisfactory varieties in the population for a given colony size. As the protein functional family prediction is a classification problem, here the classification accuracy has been considered as a fitness value. To obtain the optimal result for feature selection, the trial-and-error approach with the parameter values of the artificial bee colony optimization method has been employed. The better optimal features have been selected when using the following parameter values in Table 1.

2.2 Classification of the Protein Functional Family

Utilizing an exclusive set of features extracted by dipeptide composition and selected using an artificial bee colony optimization technique, the random forest classification technique has been employed to distinguish the protein functional family.

The Prot-RF model classifies a sequence into protein functional families depending on the sequence features which are selected by artificial bee colony optimization method. The random forest classification algorithm which utilizes a group of decision trees in which $p(x, \Theta_n)\ n - 1, 2, 3\ldots k$ where Θ_n are the identical dispersed independent random vectors and every tree adds a vote for the utmost prevalent class at input x that was established by Azad [12]. The specialty of this combination is that each decision tree is constructed from a parameter of random vectors. Among various approaches for classification, the random forest had proven to be efficient at building the ensemble of diverse trees using the bagging or random subspace methods which results in an efficient classification accuracy [12].

Table 1 Values of parameters

Parameters	Values
Swarm size	100
Number of iterations	100
Limit	50–500
Number of onlookers bees	50% of the swarm size
Number of employed bees	50% of the swarm size
Number of scouts	1

```
    Initialize the parameters: Max Iteration, swarm size, limit.
    Initialize the nectar of the food.
    Assess the nectar of the food by computing the fitness (Accuracy of classification
using Prot-RF).
    Iterations ←1
    While Iteration<Max Iteration Do.
    Produce new solutions using employed bees.
    Assess new solutions by computing fitness.
    Apply greedy operation for selection.
    Compute probability scores using fitness function.
    Produce new solutions using onlooker bees depending on the probability of
nectar.
    Assess new solutions by computing fitness
    Apply greedy operation for selection.
    Regulate uncontrolled solutions and produce new solutions arbitrarily using
scouts.
    Memorize the best solution found so far.
    Iteration ←Iteration+1
    End While
    Return best solution (Predictive features).
    Train the Prot-RF classifier using selected features.
    Classify protein functional families using Prot-RF.
Return classification accuracy.
```

Fig. 2 Pseudocode for Prot-RF (ABC) method

Proteins known to be inside and outside the functional family are utilized to train the classifier model, and the classification precision has also been increased.

The tenfold cross-validation technique is employed to assess the random forest algorithm. The values of sensitivity and the accuracy were low with the greater difference with specificity values. This has happened because of the imbalanced data. To conquer the problem of imbalanced dataset, the training data has to be rebuilt by oversampling the minority class and undersampling the majority class [12]. The general pseudocode of Prot-RF (ABC) method is depicted in Fig. 2.

3 Experimental Results

To assess the performance of the projected Prot-RF (ABC) method, most popular seven classification methods such as Naïve Bayes, K-nearest neighbor, probability neural network, J48, AdaBoost, support vector machine, and random forest were used.

3.1 Dataset Description

The benchmark dataset for protein sequences is taken from Protein Data Bank and UniProt database. In Prot-RF (ABC), the data for protein functional families was built from the Pfam Domain families, GPCRDB, NucleaRDB, BRENDA, TCDB, LGICdb, Gene Ontology, etc. [13]. The functional families are taken from broad different multiple sources to train the model with various diverse families.

3.2 Performance Measures

To measure the performance of the projected and existing methods of protein functional family classification, the following measures have been employed such as sensitivity (SE), specificity (SP), and the accuracy (AC). These measures are defined in given Eqs. 1–3.

$$\text{Sensitivity} = \frac{TP}{TP + FN} \tag{1}$$

$$\text{Specificity} = \frac{TN}{TN + FP} \tag{2}$$

$$\text{Accuracy} = \frac{TP + TP}{TP + FN + FP + TN} \tag{3}$$

where TP is the total quantity of true positive, FP is the total quantity of false positives, TN is the total quantity of true negatives, and finally FN is the total quantity of false negatives.

The feature selection approach has been employed using various optimization methods such as genetic algorithm, ant colony optimization algorithm, particle swarm optimization algorithm, and also with artificial bee colony optimization. The swarm size of all the optimization methods is in same size initially. After various trial-and-error methods with various parameters of iterations and populations, it has been observed that the ABC method selects best and novel features which is used for classifying the protein functional families accurately. The list of top seven features extracted from the optimization algorithm is hydrophobicity, composition of amino acid, atom count, residue count, polarizability, solvent accessibility, molecular weight.

After the feature selection, the classification accuracy of the protein functional family has been identified through this research work and the average classification accuracy value for different classifiers after obtaining the better optimal feature selection from ABC approach has been depicted in Table 2 and its chart representation in Fig. 3.

Table 2 Comparisons of the classification accuracy of the various classifiers

Function family	NB	J48	AdaBoost	KNN	PNN	SVM	RF	Prot-RF (ABC)
Chlorophyll biosynthesis	75	70	79	85	83	84	83	86
Coat proteins	76	80	84	86	88	97	80	84
Immune response	82	85	87	88	90	97	83	86
Lectin	71	74	76	78	79	83	76	81
mRNA-binding proteins	62	72	75	77	82	94	71	79
RNA-binding proteins	60	68	73	77	83	87	79	97
Zinc-binding	73	76	76	83	80	90	70	94
Metal binding	62	64	66	66	69	75	63	65

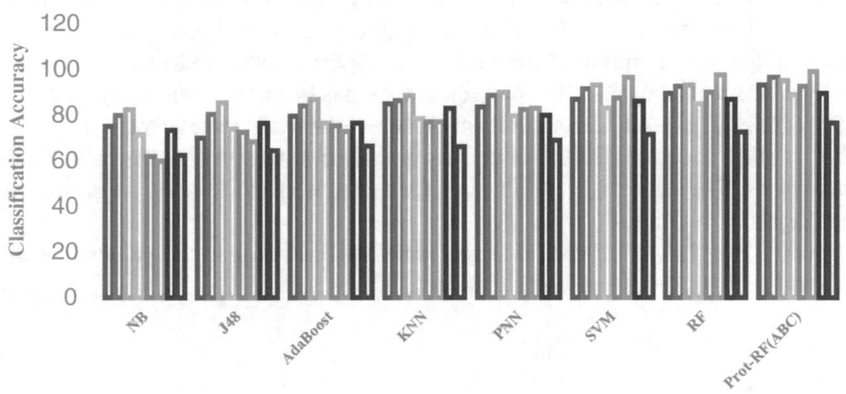

Fig. 3 Comparison of classification accuracy among the existing and proposed methods

4 Implementation and Discussion

This implementation was run in 2.00 GHz Intel CPU with 8 GB of memory and running on Windows 10. Initially, the sequence features are extracted by employing the dipeptide composition method. It comprises of a lot of irrelevant and redundant features. Thus, the features are selected using the ABC optimization method which has three phases such as employed bee, onlooker bee, and scout bee that search for the relevant features. The impact of feature extraction and selection leads to the prediction of numerous new functions of a protein sequence. After selecting the sequence features, they are classified as a protein functional family using

Table 3 Statistical significance among proposed and existing methods

Methods	NB	J48	AdaBoost	KNN	PNN	SVM	RF	Prot-RF(ABC)
NB	0	0.091	0.061	0.041	0.011	0.012	0.009	0.006
J48	0.091	0	0.057	0.052	0.031	0.024	0.018	0.009
AdaBoost	0.061	0.057	0	0.029	0.014	0.012	0.011	0.005
KNN	0.041	0.052	0.029	0	0.048	0.037	0.024	0.010
PNN	0.011	0.031	0.014	0.048	0	0.052	0.034	0.042
SVM	0.012	0.024	0.012	0.037	0.052	0	0.009	0.007
RF	0.009	0.018	0.011	0.024	0.009	0.009	0	0.006
Prot-RF (ABC)	0.006	0.009	0.005	0.010	0.007	0.006		0

the random forest classification technique. In the feature selection process, when the number of iterations increased, the probability of getting the most relevant features also increased. All the feature selection techniques and the classification algorithm work depending on the characteristics of dataset. In this work, among the various other existing methods, the Prot-RF (ABC) algorithm works for a large number of features and randomization concept well. That directs to the higher accuracy than all other. Thus, the Prot-RF (ABC) is capable of protein functional family prediction and it has minimum false rates with respect to other existing methods.

The statistical analysis of the projected work has been proficient using Wilcoxon signed rank test to verify the significant results of the projected method compared to other existing methods. The P value less than 0.05 has been considered as significant values. The statistical significance of the proposed and existing algorithms has been depicted in Table 3.

From Table 3, it is inferred that the projected method has the values less than 0.05 and it is proved to be significant in nature.

5 Conclusion

In this work, the ensemble of artificial bee colony algorithm with random forest for feature selection and classification of protein functional family has been accomplished. It has three phases; initially, the protein sequence features are extracted. Secondly, the features are selected, and finally, the classification of the protein functional family has been predicted. By investigating the diverse experimentations on the various datasets from PDB and UniProt, the outcomes proved that Prot-RF (ABC) method outperforms the other existing methods in the classification accuracy. Also, the statistical significance of the projected method has also checked. In future, many other novel functional families can be predicted and also various other methods and hybridization of them can be employed for feature selection and classification.

References

1. Lee, B.J., Shin, M.S., Oh, Y.J., Oh, H.S., Ryu, K.H.: Identification of protein functions using a machine-learning approach based on sequence-derived properties. Prot Sci. **7** (2009)
2. Tiwari, A.K., Srivastava, R.: A survey of computational intelligence techniques in protein function prediction. Int. J. prot. (2014)
3. Ong, S.A., Lin, H.H., Chen, Y.Z., Li, Z.R., Cao, Z.: Efficacy of different protein descriptors in predicting protein functional families. BMC Bio. **8** (2007)
4. Naveed, M., Khan, A.U.: GPCR-MPredictor: multi-level prediction of G protein-coupled receptors using genetic ensemble. Ami, Aci. **42** (2012)
5. Cai, C.Z., Han, L.Y., Ji, Z.L., Chen, X., Chen, Y.Z.: SVM-Prot: web-based support vector machine software for functional classification of a protein from its primary sequence. Nucl Aci Res. **31** (2003)
6. Li, Y.H, Xu, J.Y., Tao, L., Li, X.F., Li, S., Zeng, X., et al.: SVM-Prot 2016: A Web-Server for Machine Learning Prediction of Protein Functional Families from Sequence Irrespective of Similarity. PLoS ONE **11** (2016)
7. Cai, Y., Liao, Z., Ju, Y., Liu, J., Mao, Y., Liu, X.: Resistance gene identification from Larimichthys crocea with machine learning techniques. Sci Rep. **6** (2016)
8. Gao, Q.B., Wang, Z.Z.: Classification of G protein-coupled receptors at four levels. Prot. Eng. Design Sel. **19** (2006)
9. Lou, W., Wang, X., Chen, F., Chen, Y., Jiang, B., Zhang, H.: Sequence based prediction of DNA-binding proteins based on hybrid feature selection using random forest and Gaussian naive Bayes. PLoS ONE **9** (2014)
10. Gu, Q., Ding, Y.S., Zhang, T.L.: Prediction of G-protein coupled receptor classes in low homology using chous pseudo amino acid composition with approximate entropy and hydrophobicity patterns. Prot. Pept. Lett. **17** (2010)
11. Kaswan, K.S., Choudhary, S., Sharma, K.: Applications of artificial bee colony optimization technique: survey. In: Proceedings in 2nd International Conference on Computing for Sustainable Gloal Development (2015)
12. Azad, V.S.: Feature based protein function prediction by using random forest. Int. J. Eng. Res. Manag. Technol. **4** (2015)
13. Horn, F., Vriend, G., Cohen, F.E.: Collecting and harvesting biological data: the GPCRDB and NucleaRDB information systems. Nucl. Aci. Res. **29** (2001)

An Effective Path Planning of a Mobile Robot

S. Pattanayak, S. C. Sahoo and B. B. Choudhury

Abstract Recent advances in mobile robot path planning are turns into a prevalence research field. This paper proffers a metaheuristic approach to optimize the mobile robot path length by adopting particle swarm optimization (PSO) algorithm. This approach reckons the curtail path length between staring and goal point for the mobile robot without any physical contact to the obstacles. A static environment with known obstacles position is designed for evaluation of path length by this method. Total nine numbers of obstacles are taken into consideration for this study. The program for PSO optimization approach was written using MATLAB software.

Keywords Mobile robot · Path planning · PSO

1 Introduction

Development of new advanced industries and their requirement of continuous production, working in hazardous situation, and unattended machining operation limit the working of human beings. Therefore, it is necessary to develop a robot that can be controlled through a cellular phone/laptop/remote controller. The path in which robot reaches its destination is a challenging task for the designer. So the path is selected in such a way that the collision with obstacles can be completely avoided also the path length should be as small as possible. Path planning proce-

S. Pattanayak
Department of Production Engineering, Indira Gandhi Institute of Technology, Sarang, Dhenkanal 759146, Odisha, India
e-mail: suvranshupattanayak@gmail.com

S. C. Sahoo · B. B. Choudhury (✉)
Department of Mechanical Engineering, Indira Gandhi Institute of Technology, Sarang 759146, Odisha, India
e-mail: bbcigit@gmail.com

S. C. Sahoo
e-mail: sc007sahoo@gmail.com

J. Nayak et al. (eds.), *Soft Computing in Data Analytics*,
Advances in Intelligent Systems and Computing 758,
https://doi.org/10.1007/978-981-13-0514-6_18

dure is conducted in two environments such as static and dynamic [1]. In static environment, a complete information about surrounding is available with the robot. But in dynamic environment, the position of obstacles changes over time. There are some factors upon which the path selection depends such as shortest route, less cost, and minimum time required to move. Different soft computing approaches are followed for selection of optimal path.

Chaari et al. [1] investigated the capabilities of Tabu search (TS) to estimate the global path and compared the outcomes with genetic algorithm (GA) approach. It was reported that TS approach generates nearly optimal solution with reduction in execution time than GA approach. A modified A* (A star) algorithm was implemented by Duchon et al. [2] in case of mobile robot for optimization of computational time and path length. These modifications include Basic theta*, Phi*, Rectangular Symmetry Reduction (RSR) and Jump Point Search (JPS). The basic theta* algorithm was found appropriate for optimized path length. Hossen et al. [3] proposed a modified hybrid fuzzy approach for multi-sensors (large number of inputs) attached mobile robot navigation. This technique provides satisfactory accuracy in mobile robot navigation. Particle swarm optimization (PSO) approach was applied by Yusof et al. [4] for path planning and determined the shortest route which requires less cost. The path planning was done by considering some predetermined waypoints. This approach has an aim to provide mobility to the visually impaired people without any help of a caretaker. Improved gravitation search algorithm (IGSA) was proposed by Das et al. [5] to optimize the trajectory path for multi-robot. The author compared the outcomes of this approach with GSA and PSO. It was found that IGSA performs better in determination of optimized path as compared to GSA and PSO. Yakoubi and Laskri [6] developed a modified method using GA to simplify the path planning of the coverage region (PPCR) for a vacuum cleaner robot. Different sensors were used to access the robot through every part of the environment by avoiding obstacles. Adaptive neuro-fuzzy inference system (ANFIS) was employed by Pandey et al. [7] for mobile robot navigation and obstacle avoidance in its path using unknown static environments. The forward obstacles were exposed using sensors like ultrasonic and infrared range. Li and Choi [8] adopted the fuzzy logic concept to plan the path for the mobile robot in dynamic environment. The position and distance of obstacles were recognized by ultrasonic sensor. It was reported that the proposed method provides a finer way with obstacles shirking and quicker travel time. Ashoori et al. [9] explored a strategy for installation of wireless sensor networks (WSN), i.e., global positioning system (GPS) on mobile robot path planning. This approach provides better performance at a least consumed cost by providing the GPS in some nodes instead of all nodes. Rashid and Zain [10] develop a modular navigation technique by collecting signal from RFID tags and RFID reader. RFID reader communicates with RFID tags to determine the robot position. Mutib et al. [11] find the applicability of SLAM gathered data in a mobile robotic system for implementation of autonomous navigation system. The author also introduced a neuro-fuzzy framework to mix visual data with AI tools. A fuzzy control system was developed by Rulong et al. [12] to illustrate the operation of the robot. This developed approach was found effective while determining the path length for multi-robot system. Abdellatif and Montasser [13] developed and implemented two algorithms using C language for controlling the motion of the robot. First algorithm stops/start the motion according to the appearance/disappearance of obstacles. The second algorithm provides maneuvering behavior to escape from obstacles. The author also mounted four ultrasonic range sensors on the three-wheeled chassis to detect the obstacles. Zohaib et al. [14] made a comparison among a number of obstacle avoidance algorithms and discussed their advantages and drawbacks to select the best efficient algorithm for obstacle avoidance. It was concluded that the vector field histogram (VFH) algorithm takes less time to determine the shortest route among other algorithms. Shojaeipour et al. [15] developed a method using webcam and laser pointer to identify and measure the distance of obstacles from the robot. So that it can traverse to its target location. The program for this method was written using image processing toolbox of MATLAB.

In this paper, an attempt has been made to figure out the shortest route through which the robot reaches its destination without any obstruction. The environment considered in this study is static in nature with complete information about the position of obstacles. PSO soft computing approach was followed to estimate the optimal path length between starting and endpoint, for the mobile robot. The programming for PSO algorithm is written using MATLAB software.

2 Experimental Setup

An autonomous mobile robot is developed (as shown in Fig. 1) for estimation of path length from starting to goal point. Each components of mobile robot is mentioned in Fig. 1. It can run by avoiding all the obstacles present in the environment without any human involvement, i.e., static obstacles.

The environment condition used to clinch the path length while moving from one point to another is presented in Table 1. This environment condition have nine numbers of obstacles presented as pink circle in Fig. 2. The coordinates of starting and goal point are set as (1, 1) and (12, 14) meter respectively. The size and position of each obstacle are presented in Table 2. The size and position of the obstacle remain same for every test to be conducted.

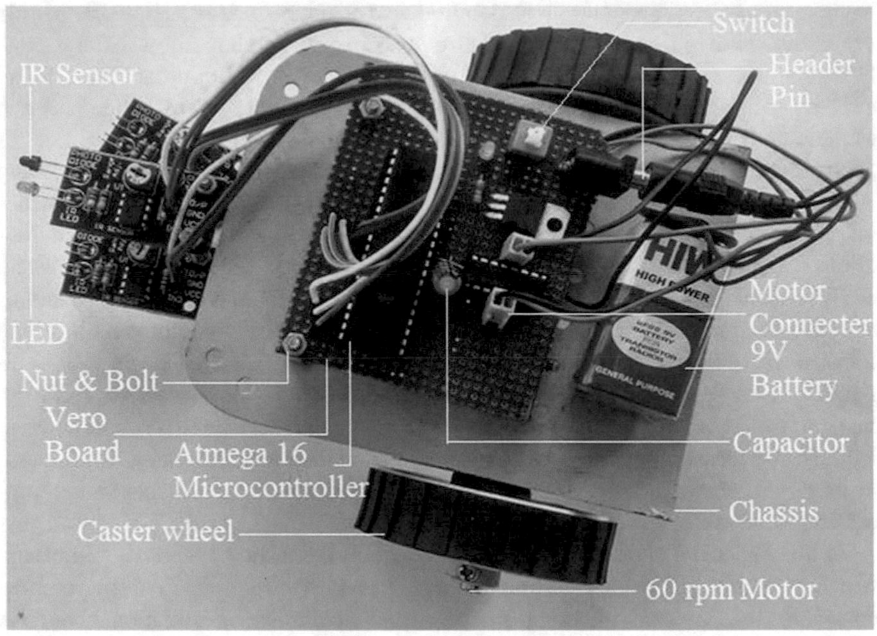

Fig. 1 Photographic view of mobile robot

Table 1 Environment condition

Sl. no	No. of static obstacles	Coordinates of starting point	Coordinates of endpoint
1	9	(1,1)	(12,14)

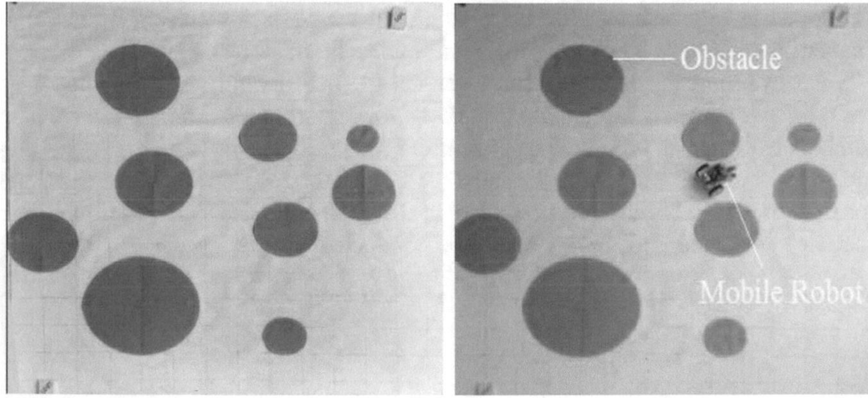

Fig. 2 Pictorial view of environment

Table 2 Size and position of the obstacles

Obstacle number	Coordinates of X-axis in (m)	Coordinates of Y-axis in (m)	Radius of the circle in (m)
Obstacle 1	4.0	3.5	1.8
Obstacle 2	4.5	8.2	1.2
Obstacle 3	8.0	10.0	0.9
Obstacle 4	8.5	6.5	1.0
Obstacle 5	11.0	10.0	0.5
Obstacle 6	8.5	2.5	0.7
Obstacle 7	4.0	12.0	1.3
Obstacle 8	11.0	8.0	1.0
Obstacle 9	1.0	6.0	1.1

3 Algorithm

PSO is a soft computing approach that depends upon a few programming commands to achieve the global optimum result. PSO approach holds good for the environment, which have outright knowledge about obstacles presents. In every iteration, the particle updates their position and velocity according to the following formula:

$$\mathrm{prtposij} = \mathrm{prtposij} - 1 + \mathrm{prtvelij} \tag{1}$$

$$\mathrm{prtvelij} = [\mathrm{w.\ prtposij} - 1 + \ \mathrm{c1r1}\,(\mathrm{pbestij} - 1 - \mathrm{prtposij} - 1) + \mathrm{c2r2}(\mathrm{gbesti} - 1 - \mathrm{prtposij} - 1)] \tag{2}$$

where

prtposij Position of *j*th particle in *i*th iteration
prtvelij Velocity of *j*th particle in *i*th iteration
pbestij Best position of *j*th particle till *i*th iteration
gbesti Best position within the swarm
c_1, c_2 Acceleration coefficient
r_1, r^2 Random numbers between 0 and 1 (varies at each iteration)
w Inertia weight (constant)

The fixed parameters used in PSO algorithm, for evaluation of the global best (gbest) value are listed in Table 3.

Table 3 PSO parameters

Sl. no	Parameter name	Parameter value
1	Number of particles (i)	(10–10000)
2	Number of iteration (it)	20
3	Acceleration coefficient (c1, c2)	c1 = c2 = 1.5
4	Inertia weight (w)	1

4 Results

Equations (1) and (2) are used for amending the position and velocity of the particle in every iteration. Then a checking is done to know whether the particles are lies inside the obstacles or not. If so, then cancel the swarm and go for further updating. If not, then enumerate the global best (gbest) and particle best (pbest) value of the particles. Lastly, it checks in case that the termination condition is satisfied or not. If satisfied it stops, contrarily goes for further forecasting of the fitness function. So here the fitness function is vital.

A number of tests have been carried out to figure out the shortest path between coordinate (1, 1) and (12, 14) using PSO algorithms. Figure 3 shows the shortest path length for the mobile robot in a static environment. The path length in every iteration for the best results is presented in Fig. 4.

The small red circles in Fig. 3 represent variables, and the line represents travel trajectory for the mobile robot starting from beginning to goal point. This figure also shows the optimal path. In every iteration, the path length is updated in Fig. 4. It shows the path length for the mobile robot in every iteration in which the number of iteration is taken in x-axis and the length of the path in y-axis. A number of tests have done to check the shortest length. Table 4 shows the path length of the mobile robot after every test. The optimal path length is found to be 19.3069 m from (Fig. 4 and Table 4).

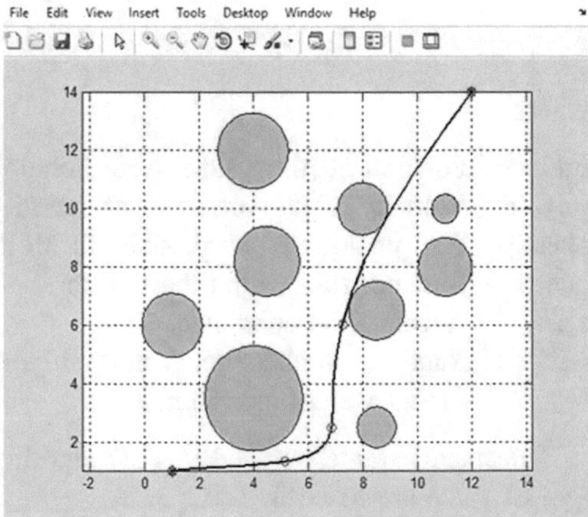

Fig. 3 Shortest path using PSO

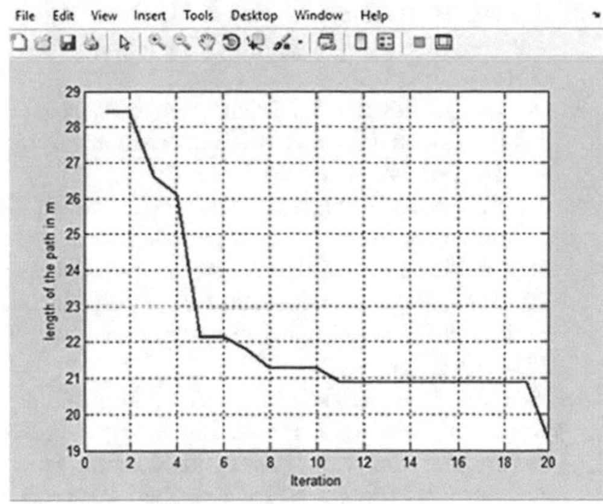

Fig. 4 Length of path in every iteration

Table 4 Path length obtained by using PSO

Number of tests	Length of the path in 'm'
1	20.5418
2	20.4238
3	20.2214
4	20.348
5	23.5941
6	20.4357
7	20.034
8	19.3069
9	20.6774
10	20.2422

5 Conclusion

In the present work, PSO algorithm is adopted for trajectory path planning of mobile robot to identify the collision-free optimal path in the static environment. The simulation results show that the PSO approach is quite simple and easy to implement, also quickly determine the path length for an autonomous mobile robot than the other existing algorithm. The shortest optimal path length is found to be 19.3069 m among all the tests.

References

1. Chaari, I., Koubaa, A., Ammar, A., Trigui, S., Youssef, H.: On the adequacy of tabu search for global robot path planning problem in grid environments. In: 5th International Conference on Ambient Systems, Networks and Technologies, vol. 32, pp. 604–613 (2014)

2. Duchon, F., Babinec, A., Kajan, M., Beno, P., Florek, M., Fico, T., Jurisica, L.: Path planning with modified a star algorithm for a mobile robot. Modell. Mech. Mechatron. Syst. **96**, 59–69 (2014)
3. Hossen, J., Sayeed, S., Iqbal, A.K.M.P.: A modified hybrid fuzzy controller for real-time mobile robot navigation. IEEE International Symposium on Robotics and Intelligent Sensors, vol. 76, pp. 449–454 (2015)
4. Yusof, T.S.T., Toha, S.F., Yusof, H.M.: Path planning for visually impaired people in an unfamiliar environment using particle swarm optimization. In: IEEE International Symposium on Robotics and Intelligent Sensors, vol. 76, pp. 80–86 (2015)
5. Das, P.K., Behera, H.S., Jena, P.K., Panigrahi, B.K.: Multi-robot path planning in a dynamic environment using improved gravitational search algorithm. J. Electr. Syst. Inf. Technol. **3**, 295–313 (2016)
6. Yakoubi, M.A., Laskri, M.T.: The path planning of cleaner robot for coverage region using Genetic algorithms. J. Innov. Dig. Ecosyst. **3**, 37–43 (2016)
7. Pandey, A., Kumar, S., Pandey, K.K., Parhi, D.R.: Mobile robot navigation in unknown static environments using ANFIS controller. Perspectives Sci. **8**, 421–423 (2016)
8. Li, X., Choi, B.J.: Design of obstacle avoidance system for mobile robot using fuzzy logic systems. Int. J. Smart Home **7**, 321–328 (2013)
9. Ashoori, E., Babagoli, I., Alipour, S.: A new method for localization of wireless sensor networks based on path planning of mobile robots. Adv. Sci. Technol. Res. J. **9**, 10–17 (2015)
10. Rashid, M.M., Zain, F.M.: Development of omni directional mobile robot navigation system using RFID for multiple objects. In: AASRI Conference on Modelling, Identification and Control, vol. 3, pp. 474–480 (2012)
11. Mutib, K.A., Mattar, E., Alsulaiman, M.: Implementation of fuzzy decision based mobile robot navigation using stereo vision. In: International Conference on Soft Computing and Software Engineering (SCSE), vol. 62, pp. 143–150 (2015)
12. Rulong, X., Qiang, W., Lei, S., Lei, C.: Design of multi-robot path planning system based on hierarchical fuzzy control. Adv. Control Eng. Inf. Sci. **15**, 235–239 (2011)
13. Abdellatif, M., Montasser, O.A.: Using ultrasonic range sensors to control a mobile robot in obstacle avoidance behavior. In: International Conference on Systemic, Cybernetics and Informatics (2001)
14. Zohaib, M., Pasha, M., Riaz, R.A., Javaid, N., Khan, R.D.: Control Strategies for Mobile Robot With Obstacle Avoidance (2013)
15. Shojaeipour, S., Haris, S.M., Shojaeipour, A., Zakaria, M.K.: Robot path obstacle locator using webcam and laser emitter. Phys. Procedia **5**, pp. 187–192 (2010)

Inverse Kinematics Solution of a 6-DOF Industrial Robot

Kshitish K. Dash, B. B. Choudhury and S. K. Senapati

Abstract A vital part of many industrial robot manipulators is to reach required position and orientation of end effectors so as to complete the pre-defined task. To get this, one should have knowledge of kinematics, i.e. inverse kinematics (IK). Though inverse kinematics never gives a closed form solution, it is too difficult to solve such problem of an industrial robot. There are so many analytical and other simulation methods which are adopted to solve this IK problem for our 6-DOF industrial pick and place robot. In this paper, artificial neural networks (ANN) are used and simulated by using MATLAB.

Keywords Industrial robot · Inverse kinematics · ANN

1 Introduction

Kinematics of robot indicates the analytic behaviour of the movement of robot manipulator. By taking appropriate kinematics models of an industrial robot, the kinematic behaviour, i.e. inverse kinematics and forward kinematics, can be analysed. These two spaces utilized as a part of kinematics demonstrating are known as Cartesian space and Quaternion space. The alteration among two Cartesian coordinate takes place in form of rotation and a translation as soon in Fig. 1. So many methods are adopted to solve forward kinematics and inverse kinematics problems of an industrial pick and place robot. out of different method Jacobian matrix and Denvit-Hertenberg theory is useful for analytic solution of straight forward kinematics and Screw theory is useful for inverse kinematic solution.

K. K. Dash (✉) · B. B. Choudhury · S. K. Senapati
Indira Gandhi Institute of Technology, Sarang, Dhenkanal 759146, Odisha, India
e-mail: kdash76@gmail.com

B. B. Choudhury
e-mail: bbcigit@gmail.com

S. K. Senapati
e-mail: sukantasenapati63@gmail.com

J. Nayak et al. (eds.), *Soft Computing in Data Analytics*,
Advances in Intelligent Systems and Computing 758,
https://doi.org/10.1007/978-981-13-0514-6_19

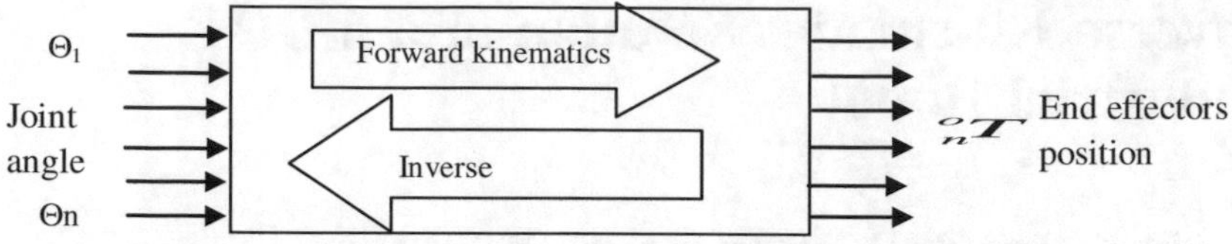

Fig. 1 Schematic portrayal of inverse and forward kinematics

Denavit–Hartenberg (D-H) gives common alteration between two joints, and it requires four parameters. The parameters used here are known as the D-H parameters. The kinematics of robot ordered in form of forward kinematics and inverse kinematics. Forward kinematics issue is simple, and also, there is no complication deriving the equations. Therefore, there is dependably a forward kinematics arrangement of a controller. Inverse kinematics is a considerably more troublesome issue than forward kinematics. The solution of the backwards kinematics issue is computationally extensive where some have solution, some have no solution, some have redundancy problem takes quite a while in the continuous control of controllers.

2 Literature Review

Li et al. [1] studied 6-R robot for virtual reality and analysed the virtual reality. The simulation is done by using EON software. The inverse kinematics of this robot is simulating by MATLAB, and the result is validating for the virtual reality using both the software. Lazar et al. [2] developed a robot manipulator control by using visual serving. The paper describes how integrating reference and image prediction novel architecture are used for the prediction camera velocity and time variation consider by interaction matrix. A simulation is done to improve its efficiency for a 6-DOF manipulator. Chen et al. [3] developed a paper to increase the inverse kinematics behaviour of a 6-DOF robot. The two coordinates are considered here as reference coordinate and tool coordinate. The screw motion calculated for each link based on reference coordinate. At last, a new algorithm is used to improve the efficiency of inverse kinematics. Duka [4] developed a paper inverse kinematics and forward kinematics of articulated robot. D-H theory is adopted to solve 4-axis articulated robot problem and considering the corresponding data inverse kinematics value also calculated. A comparison takes place between experimental data with calculated data for 4-axis articulated manipulator. Adrin et al. [5] developed a paper ANFIS-based inverse kinematics solution. They have discussed a 3-DOF planer manipulator which is considered to get effectiveness of this approach. The data obtained from forward kinematics is transfer for inverse kinematics and accepted the accuracy of the different joint angles. Dash et al. [6] developed a paper inverse kinematics of industrial using ANN. The end effectors' position is calculated by considering different joint angle. The multilayer neural network is used to train the data, and the analytical data is validate with ANN trained value. Chandra and Rolland [7] present 3RPR planer parallel manipulator where simulated annealing and relay collaborative met heuristics. The given method gave promise results. Henten et al. [8] devolved autonomous robot for harvesting. In this paper, the 3D image system implemented for controlling and to avoid collision-free motion during harvesting and 7-DOF manipulator was used to make more comfortable harvesting. Husty et al. [9] taken new algorithm to solve IK of 6-R manipulator. Here kinematics image considers to identify the displacement of each point, and the solution is made in two phase. In first phase, two joint angle value calculated algorithm by using and other four calculated by inverse kinematics equation. Srinivas et al. [10] accepting end-focuses for all segments of a multi-segment trunk are known and subtle elements are applying single-segment converse kinematics to each segment of the multi-segment trunk by adjusting for coming about changes in introduction. At last, an approach which registers per-area end-focuses given just a last segment endpoint gives an entire answer for the multi-segment converse kinematics issue. Iliukhina et al. [11] devolved modelling 5-DOF manipulator where the piece of research went for making mechanical controller controlled by methods for brain–computer interface for enhancing family unit confidence of people with incapacities and growing the extent of their movement. The mechanical controller gives plausibility of self-satisfaction in fundamental family works, and it shows a numerical model of the kinematics of the automated controller. Khuntia et al. [12] modelled a heuristics allocation of multitask robot. Here it has been created thinking about the earth, the framework parameters and the robots' capacities. An answer calculation has been created and executed to acquire the outcomes.

3 Inverse Kinematics

Backward kinematics is the inverse of forward kinematics. The position of end effectors calculated, after determining all joint angle position. The inverse kinematics solves the problem like end effectors position, what are the relating joint positions. In contrast with the forward kinematics issue, the arrangement of the converse issue is not generally conjugal. The end effectors position can be reached in several configurations according to position vectors. Our main aim is to solve 6-DOF industrial robot using artificial neural networks (ANN) and simulated the solution by using MATLAB.

The inverse kinematics model is shown in Fig. 2 whose angle can be calculated like θ_1, θ_2, $\theta_3 \ldots = f - (p)$. The arrangement is figured in two stages, first uses an area vector from the wrist to the wrist. The vector takes into consideration the arrangement of the initial three primaries DOF that finish the worldwide movement.

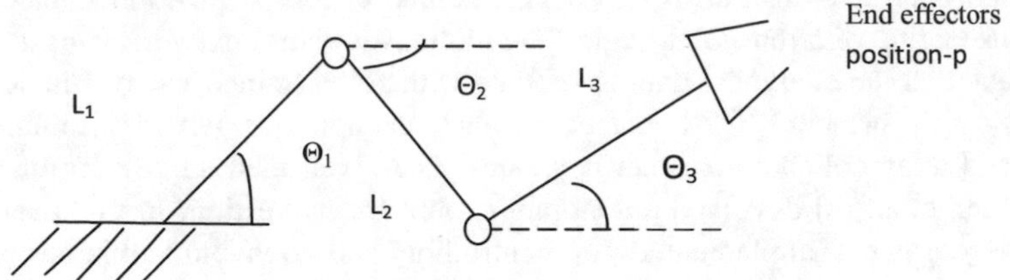

Fig. 2 Structure of 3 R arm

The last 3-DOF is discovered utilizing the computed estimations of the first 3-DOF and the introduction matrices T4, T5 and T6. The modelling is done here using D-H method for forward kinematics, and the values are obtained for end effectors with various corresponding joint angle position.

$$T = A_1A_2A_3A_4A_5A_6 \tag{1}$$

where T is the target position Θ_i is the joint variable.

4 Denavit–Hartenberg Theory

D-H parameters are one of the most useful theories to take care of forward kinematics issue of automated arms. The D-H formalization takes place by using only four parameters to describe the spatial relationship between successive link coordinate frames as shown in Fig. 3. The 6-DOF robot solution made by introducing two constraints to the placement of those frames: The axis x_i is perpendicular to the axis $z_{i-1,}$ and the axis x_i intersects the axis z_{i-1}.

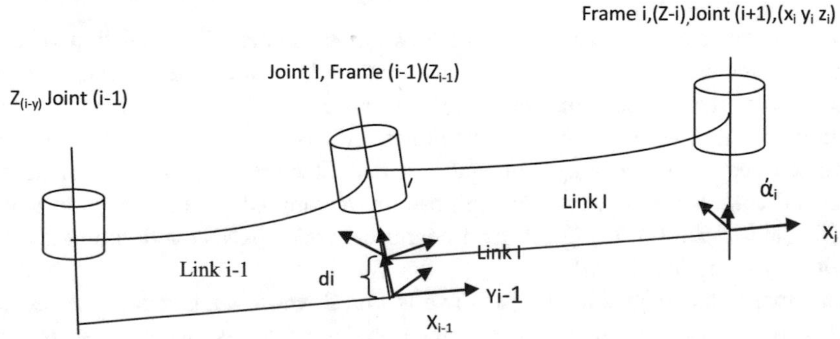

Fig. 3 D-H architecture

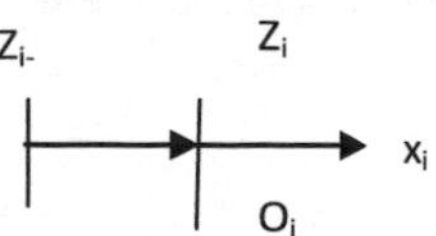

Fig. 4 D-H architecture for two axis

Frame i is rigidly attached to link joint i + 1. The frame is assign as D-H convention, i.e. D-H$_1$ x_i perpendicular to z_{i-1}, D-H$_2$ x_i intersects the axis z_{i-1}

Assign Z-axis as axis of motion. If Z_i and Z_{i-1} don't intersect and not parallel each other then x_i act along the common normal from origin O_i which is the meeting point between Z_i and the general normal line.

From Fig. 4 if Z_i and Z_{i-1} are parallel and do not intersect, then O_i can be consider anywhere along Z_{i-1} axis. The y_i value can be obtain by cross product of two axis when both Z_i and Z_{i-1} are intersect each other, Generally cross product that two axis each other. To get the above value from the frame i − 1 to frame i, some necessary steps to be followed then after we can get the equation like as stated below

$$T_i^{i-1} = T_z(d_i)T_z(\theta_i).T_x(a_i)T_x(\alpha_i) \tag{2}$$

$$= \begin{bmatrix} 1 & 0 & 0 & 0 \\ 0 & 1 & 0 & 0 \\ 0 & 0 & 1 & d_i \\ 0 & 0 & 0 & 1 \end{bmatrix}$$

$$\begin{bmatrix} Cos\,\theta_i & -Sin\,\theta_i & 0 & 0 \\ Sin\,\theta_i & Cos\,\theta_i & 0 & 0 \\ 0 & 0 & 1 & 0 \\ 0 & 0 & 0 & 1 \end{bmatrix} \begin{bmatrix} 1 & 0 & 0 & a_i \\ 0 & 1 & 0 & 0 \\ 0 & 0 & 1 & 0 \\ 0 & 0 & 0 & 1 \end{bmatrix} \begin{bmatrix} 1 & 0 & 0 & 0 \\ 0 & Cos\,\alpha_1 & -Sin\,\alpha_i & 0 \\ 0 & Sin\,\alpha_i & Cos\,\alpha_i & o \\ 0 & 0 & 0 & 1 \end{bmatrix}$$

$$= \begin{bmatrix} Cos\,\theta_i & -Cos\,\alpha_i Sin\,\theta_i & Sin\,\alpha_i Sin\,\theta_i & a_i Cos\,\theta_i \\ Sin\,\theta_i & Cos\,\alpha_i Cos\,\theta_i & -Sin\,\alpha_i Sos\theta_i & a_i Sin\,\theta_i \\ 0 & Sin\,\alpha & Cos\,\alpha_i & d_i 0 \\ 0 & 0 & 0 & 1 \end{bmatrix}$$

By considering 6-DOF architecture of Aristo as shown in Fig. 5, the parameters are calculated as stated Table 1. After getting parameter by using D-H principle, the joint angle for different position for inverse kinematics can be calculated.

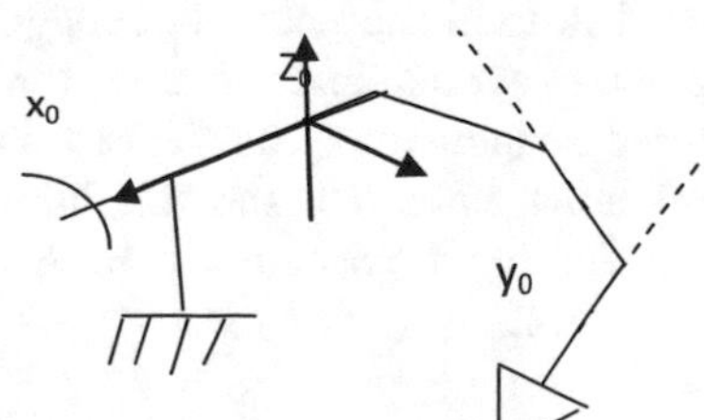

Fig. 5 6-DOF robot position

Table 1 D-H parameter value

DOF.	θi	A	ai	di
1	90	O	0	184
2	89.28	O	150	158
3	91.61	79.71	0	300
4	92.18	90	250	150
5	92.51	0	0	378.5
6	95.02	63.9	0	64.0

$$\theta_1 = \tan^{-}\left[\lambda q_y - d_2 q_x / \lambda q_x + d_2 q_y\right] \tag{3}$$

$$\theta_3 = \tan^{-}\left[\frac{q_x^2 + q_y^2 + q_z^2 - d_4^2 - a_2^2 - d_2^2}{\pm\sqrt{4d_4^2 a^2} - \left(q_x^2 + q_y^2 + q_z^2 - d_4^4 - a_2^2 - d_2^2\right)^2}\right] \tag{4}$$

$$\theta_2 = \tan^{-}\left[\frac{q_z(a_2 + d_4 s_3) - d_4 c_3(\pm\sqrt{q_x^2 + q_y^2 - d_z^2}}{q_z d_4 c_3 - (a + d_4 s_3)\left(\sqrt{q_x^2 + q_y^2 - d_2^2}\right)}\right] \tag{5}$$

$$\theta_4 = \tan^{-}\frac{C_1 a_y - S_1 q_x}{C_1 C_{23} q_x + S_1 C_{23} q_y - C_{23} q_z} \tag{6}$$

$$\theta_5 = \tan^{-}\left[\frac{(C_1 C_{23} C_4 - S_1 S_4) q_x + (S_1) C_{23} C_4 + C_1 S_4) q_Y - C_4 S1_{23} q_z}{C_1 S_{23} q_x + S_1 S_{23} q_y + C_{23} q_z}\right] \tag{7}$$

$$\theta_6 = \tan^{-}\left[\frac{-(S_1 C_4 + C_1 C_{23} S_4) nx + (-C_1 C_4 - S_1 C_{23} S_4) n_y + (S_4 s_{23}) n_z}{-(S_1 C_4 + C_1 C_{23} S_4) + C_1 C_4 - S_1 C_{23} S_4 + S_4 S_{23}}\right] \tag{8}$$

The above equation is used to calculate the joint angle value experimentally for different position of end effector.

5 Artificial Neural Network

The neural system design of this arrangement as appeared in Fig. 6 and the neurons are completely associated with this system. In this network, 18 nodes have been used to train the network. A sigmoid capacity is utilized as an exchange work among neurons, and there are three components of the system input. The initial three components speak to Cartesian position, and other six speak to the joint edge of various hubs. The showing informational index was set up keeping in mind that the end point position where ANN can be utilized for the inverse kinematics controller.

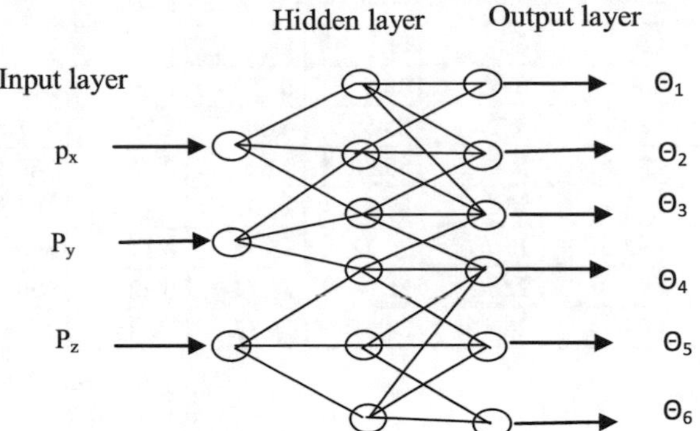

Fig. 6 Multilayer neural network

$$Ahi(t) = \sum_{k=1}^{i} WT_{ik}I_k + \sum_{k=1}^{m} Wh_{ik}f(Ah_k(t-1)); \tag{9}$$

The actuation work which is utilized as a part of concealed layer and the yield of the system is a weighted total of the shrouded unit output.

6 Results and Discussion

With reference to above equation and considering some equation from author [6], a number of joint angle value calculated which is shown below. From the table, it is observed that obtain joint angle value is varying within the maximum and minimum value of manipulator joint angle with a different position of end effector.

6.1 Experimental Data of Industrial Robot

See Table 2.

6.2 Specification of Industrial Robot

By taking number of experimental data, the neural network trained and obtained performance curve as shown in Fig. 7 (Tables 2 and 3).

Table 2 Joint angle and position of manipulator value

Θ_1	Θ_2	Θ_3	Θ_4	Θ_5	Θ_6	p_x	p_y	p_z
90	−89.98	90	0	90	0	0.30	378.88	393.84
89.85	−89.82	89.28	3.25	88.50	4.68	2.29	281.47	394.37
89.56	−89.78	91.61	5.25	85.00	7.84	2.56	380.14	394.74
89.20	−89.59	92.18	11.85	83.74	10.91	7.38	383.33	382.24
88.51	−89.22	92.51	17.17	79.71	15.46	8.11	390.57	375.52
86.45	−87.81	95.02	22.92	76.98	21.74	0.52	395.10	351.48
82.39	−85.56	95.44	26.56	63.93	24.37	29.85	414.98	336.66

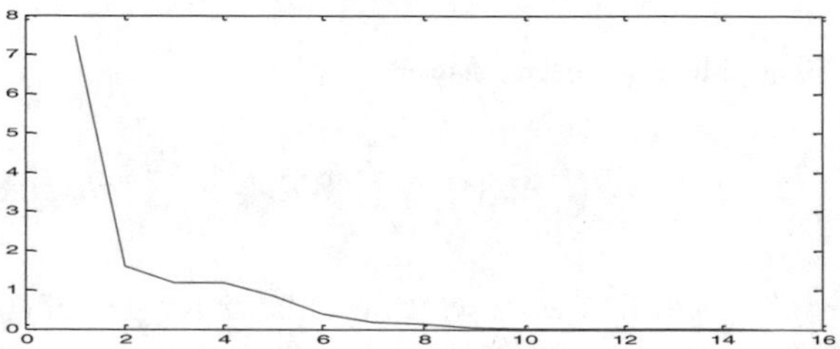

Fig. 7 Performance curve

Table 3 Joint angle specification

Minimum	Maximum	Home	Axis
−250°	90°	90°	Base
−90°	45°	90°	Shoulder
90°	−45°	−90°	Elbow
0°	340°	0°	Wrist
−90°	90°	90^0	Pitch
0°	340^0	0^0	Roll

The Levenberg–Marquardt (LM) algorithm is used to get the performance curve very fast. It is an iterative technique to get performance curve. The curve shows that the experimental value is nearly close to the theoretical value, and its errors are decrees with higher DOF.

The training rate is adapted for different epochs. Fig. 8 shows the change of the training rate decreasing with the number of epochs. By increasing the DOF, the joint angle value is closer with the train data.

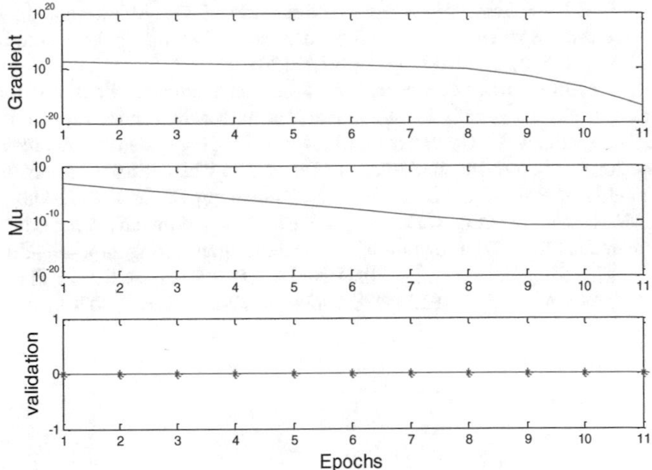

Fig. 8 Validation curve

7 Conclusion

From the analysis, it is observed that the calculated value of all joint angle of the industrial 6-DOF industrial robot is nearly equal to the experimental data. Out of different joint angle θ_1, θ_3, θ_4, θ_5 values are nearly equal to experimental data and θ_2, θ_6 values match 60% with experimental joint angle values. The trained value obtained by using neural network for the inverse kinematics gives approximate value after high training sample which is a big drawback of this paper.

References

1. Li, W.J., Song, Z.H., Zhu, Z.X., Mao, E.R.: Analysis and simulation of 6R Robot in virtual Reality. IFAC-PapersOnLine **49**(16), 426–430 (2016)
2. Lazar, C., Burlacu, A., Copot, C.: Predictive Control Architecture for Visual Servoing of Robot Manipulators (2011)
3. Chen, Q., Zhu, S., Zhang, X.: Improved inverse kinematics algorithm using screw theory for a six-dof robot manipulator. Int. J. Adv. Robot. Syst. (2015)
4. Duka, A.V.: Anfis Based Solution to the Inverse Kinematics of a 3-DOF Planar Manipulator. INTER-Eng-2014 (2014)
5. Parhi, D.R., Deepak, B.B.V.L., Nayak, D., Amrit, A.: Forward and inverse kinematic models for and articulated robotic manipulator. Int. J. Artif. Intell. Comput. Res. **4**(2) 103–109 (2012)
6. Dash, K.K., Choudhury, B.B., Khuntia, A.K., Biswal, B.B.: A Neural Network Based Inverse Kinematic Problem. IEEE. 978-1-4244-9477-4/11/$26.00 ©2011
7. Chandra, R., Rolland, L.: On solving the forward kinematics of 3RPR planar parallel manipulator using hybrid metaheuristics. Appl. Math. Comput. **217**(22), 8997–9008 (2011)

8. Van Henten, E.J., Schenk, E.J., van Willigenburg, L.G., Meuleman, J., Barreiro, P.: Collision-free inverse kinematics of the redundant seven-link manipulator used in a cucumber picking robot. Bio Syst. Eng. **106**(2), 112–124 (2010)
9. Husty, M.L., Pfurner, M., Schröcker, H.: A new and efficient algorithm for the inverse kinematics of a general serial 6R manipulator. Mech. Mach. Theory **42**(1), 66–81 (2007)
10. Srinivas, N., Matthew, A.C., Bryan, A.J., Ian, W.: A geometrical approach to inverse kinematics for continuum manipulators. In: Proceedings of IEEE/RSJ International Conference on Intelligent Robots and Systems, Nice, France, pp. 3565–3570 (2008)
11. Iliukhina, V.N., Mitkovskiib, K.B., Bizyanovaa, D.A., Akopyan, A.A.: The modeling of inverse kinematics for 5 DOF manipulator. Procedia Engineering **176**, 498–505 (2017)
12. Khuntia, A.K., Choudhury, B.B., Biswaland, B.B., Dash, K.K.: A Heuristics Based Multi-Robot Task Allocation. IEEE. 978-1-4244-9477-4/11/$26.00 ©2011

Dynamic Behaviour Analysis of an Industrial Robot Using FEM

Supriya Sahu and B. B. Choudhury

Abstract Industrial robots play a very important role in modern manufacturing industries, and great attention must be given to the design of robot structure to increase the productivity and quality of high-technology work environment. In this study, for industrial robot, the 3D model is developed by CATIA V5, and for the structural analysis, finite element method is used in ANSYS workbench. The design analysis in this paper consists of the investigation about deformation, shear stress, shear strain, and strain energy. Load is applied at the end effector, and the optimum values of parameters affecting the design of the structure are analysed in order to facilitate different structural design modifications.

Keywords ANSYS · CATIA V5 · FEA · Industrial robot
Design analysis

1 Introduction

Industrial robot is a six-axis articulated robotic arm having six different axes. Finite element method can be used for the static and dynamic behaviour analysis of any structure. Deformation, stress and strain analysis gives the idea about the weak part of the robot structure so that any design modifications if possible can be applied to get a better robotic structure.

Bhusnar and Sarawade [1] analysed the dynamic behaviour of structures like a rectangular plate, bolted lap joint and studied the natural frequencies and mode shapes using finite element analysis (FEA) and analytical methods. Chitte et al. [2]

S. Sahu (✉) · B. B. Choudhury
Deparment of Mechanical Engineering, I.G.I.T Sarang, Dhenkanal 759146,
Odisha, India
e-mail: supriyaigit24@gmail.com

B. B. Choudhury
e-mail: bbcigit@gmail.com

J. Nayak et al. (eds.), *Soft Computing in Data Analytics*,
Advances in Intelligent Systems and Computing 758,
https://doi.org/10.1007/978-981-13-0514-6_20

minimized the structural deformation of a six-axis robot manipulator in all the three x, y, and z directions by using calculus method on applied forces by means of which the stiffness can be improved. Ghiorghe [3] considered the criteria for minimizing the material used for building the robot structure. FEM has been used to determine the optimum value of design parameters. Jevan and Rao [4] analysed the structural parameters of industrial robot, and from the analysis, it has been found that the square-shaped robot arm vibrates less as compared to circular-shaped sustains. Manoj and Sambaiah [5] used finite element tool in the industrial robot for keeping the tasks in its accurate position with the desired force and torque in time. FEA tool is also used by Nor et al. [6] to find out high stress value to determine maximum deflection in low loader structure with modelling and simulation. Pachaiyappan et al. [7] used ANSYS, the commercially available analysis tool, to analyse various critical loads acting on the articulated robot. Ristea [8] designed the control system of a motorized robot structure, along with differentiating the material such as composite and aluminium. Sridhar et al. [9] studied about the behaviour of industrial robot structure and used finite element method to develop the model of robot for correctness in different loading conditions.

2 Modelling and Finite Element Analysis

2.1 Robot Modelling in CATIA V5

The six-axis industrial robot consists of several parts. The seven main parts which give rise to the complete robot are base, link, shoulder, elbow, pitch and roll. These members are having their individual range of motions. Base is the lowest part, which carries the total load of robot and is the supporting part. It is the first part whose modelling is done at the beginning using CATIA V5. Afterwards, modelling of other members is done one by one in sequence which is shown in Fig. 1.

After completion of modelling of all the parts, the assembly is done. Models of all the parts are imported to the environment of CATIA V5 for assembly modelling, and the assembly modelling is shown in Fig. 2.

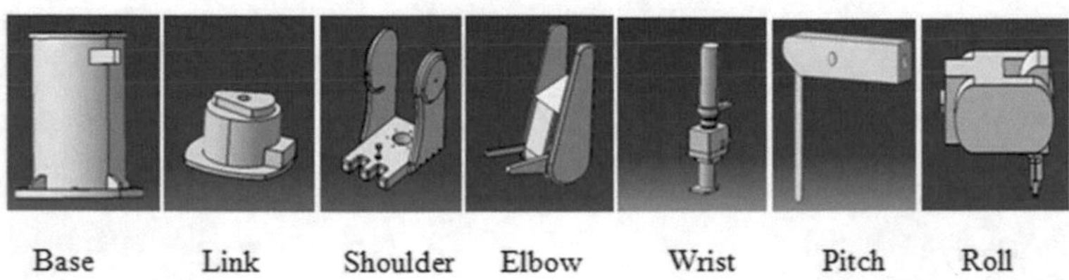

Fig. 1 Part modelling

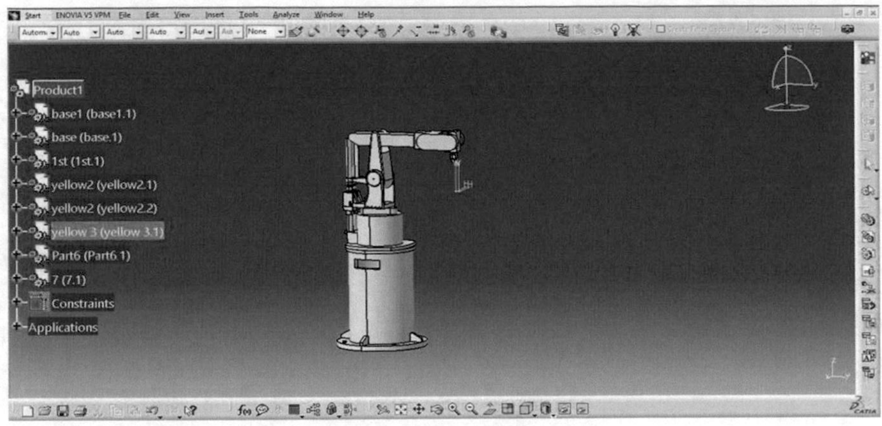

Fig. 2 Assembly modelling

2.2 *ANSYS Structural Analysis*

The solid modelling of 3D robot model is established using CATIA V5. Then, ANSYS workbench is used for structural analysis. From the library dialog window, a material has to be selected, and for this finite element analysis, the static structural is considered for the structural analysis. For the engineering data, structural steel is selected as the material of the six-axis industrial robot arm. All the constrained are added manually to the assembly parts, and all parts are considered as deformable bodies.

2.3 *Finite Element Analysis*

Before going for the FEA, first of all, the base of the structure requires the fixed constraint because during working with robot it has to be fixed to the floor. From all the directions, the base nodes are constrained and the structure is allowed to achieve its required degree of freedom which is shown in Fig. 3.

In order to define six axes of manipulator, the mash size is to be defined. For this finite element modelling and analysis, a mesh size of 0.01 mm has been chosen which give rise to a very good mesh design of the structure which is shown in Fig. 4.

As this type of robots is used for pick and place and verities of work in industries, load has to be applied at the end effectors which is the griper. In this analysis, twenty forces are applied starting from 5 to 100 N at the griper for analysis. The force applied to the griper is shown in Fig. 5.

Finally, all the necessary aspects of the robot model are then created such as constrains, loads, material and mesh. After importing all the data into the ANSYS,

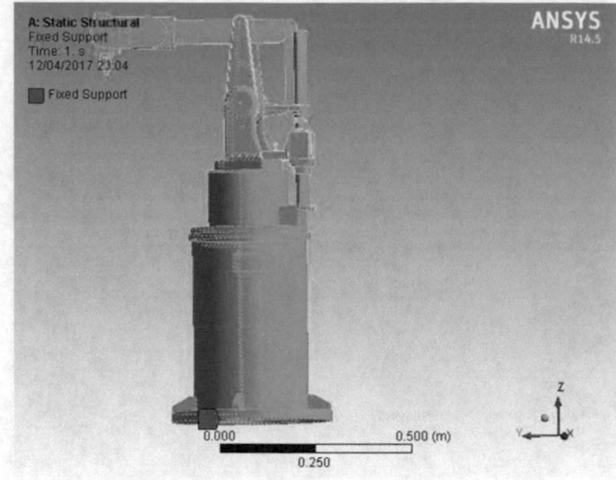

Fig. 3 Fixed base of robot model

Fig. 4 Finite element mesh of the model

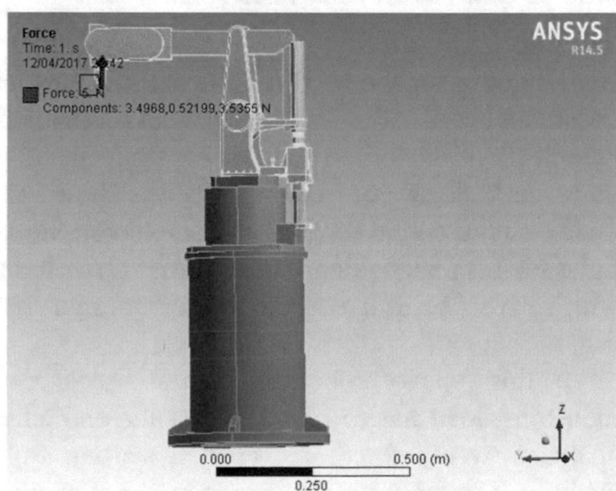

Fig. 5 Force applied at gripper

Fig. 6 Outline of the project

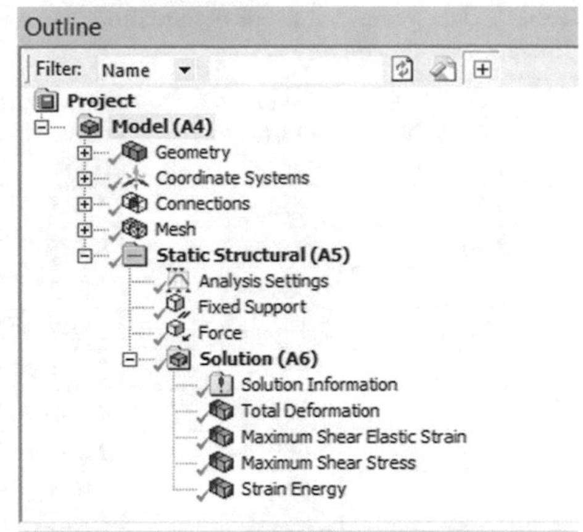

the ANSYS solver is used to get the results. The outline of the project is shown in Fig. 6.

3 Results

From the design analysis, the maximum values of all the parameters for the application of twenty number of forces are shown below in Table 1.

Total deformation of robot model structure for four different forces such as for 25, 50, 75 and 100 N that are applied among twenty forces is shown in Fig. 7. In this figure, it can see that at the left side there is a colour scale and the values of total deformation corresponding to that colour are shown. There are also five different colours such as dark blue, light blue, green, yellow and red which are seen in the robot model along with in the colour scale. The base of the structure is dark blue colour, then light blue, green, yellow and finally the griper is red in colour. The dark blue colour is seen at the base for minimum deformation, and red colour is seen at the griper for maximum deformation. These colours indicate the safety factor of the structure. Blue colour is the safe force corresponding to the load applied, and red colour is the unsafe force corresponding to the load applied to the structure. Among the total deformations shown for four forces, the deformation is maximum for the force of 100 N and the value is 0.0001054 mm.

Figure 8 shows the total deformation represented in graph for four different forces applied to the griper with four different colours where the total deformation is maximum for the force of 100 N.

Due to the application of external load, the member of the structure tries to shear off the material. Shear stress is the resisting force offered by the structure per unit area under an external force. This shearing action may lead to the breaking of the member. Shear stress is needed to find out the intensity of internal resisting force of

Table 1 Maximum values of deformation, shear stress, shear elastic strain and strain energy at different loading conditions

Sl no	Gripping load (N)	Maximum deformation (mm)	Maximum shear stress (Pa)	Maximum shear elastic strain (m/m)	Strain energy (J)
1	5	5.270e^{-6}	4.245e^{+005}	5.5152e^{-0065}	4.1274e^{-008}
2	10	1.054e–^{5}	8.485e^{+005}	1.103e-^{005}	1.651e^{-007}
3	15	1.581e–^{5}	1.2727e^{+006}	1.6546e^{-005}	3.7147e^{-007}
4	20	2.106e–^{5}	1.697e^{+006}	2.2061e^{-005}	6.6039e^{-007}
5	25	2.635e–^{5}	2.1212e^{+006}	2.7576e^{-005}	1.0319e^{-006}
6	30	3.162e–^{5}	2.5455e^{+006}	3.3091e^{-005}	1.4859e^{-006}
7	35	3.6891e–^{5}	2.9697e^{+006}	3.8607e^{-005}	2.0224e^{-006}
8	40	4.2161e–^{5}	3.3941e^{+006}	4.4122e^{-005}	2.6416e^{-006}
9	45	4.7431e–^{5}	3.8182e^{+006}	4.9637e^{-005}	3.3432e^{-006}
10	50	5.2701e^{-5}	4.2425e^{+006}	5.5152e^{-005}	4.127e^{-006}
11	55	5.7971e–^{5}	4.6667e^{+006}	6.0667e^{-005}	4.9942e^{-006}
12	60	6.3241e–^{5}	5.1212e^{+006}	6.5841e^{-005}	5.7251e^{-006}
13	65	6.8512e–^{5}	5.5152e^{+006}	7.1698e^{-005}	6.9754e^{-006}
14	70	7.3782e–^{5}	5.9395e^{+006}	7.7213e^{-005}	8.0898e^{-006}
15	75	7.9052e–^{5}	6.3637e^{+006}	8.2728e^{-005}	9.2867e^{-006}
16	80	8.4322e–^{5}	6.788e^{+006}	8.8244e^{-005}	1.0566e^{-005}
17	85	8.9592e^{-5}	7.2122e^{+006}	9.3759e-^{005}	1.1928e^{-005}
18	90	9.4862e–^{5}	7.6365e^{+006}	9.9274e^{-005}	1.3373e^{-005}
19	95	0.00010013	8.0607e^{+006}	1.0479e^{-004}	1.49e^{-005}
20	100	0.0001054	8.485e^{+006}	1.103e^{+004}	1.651e^{-005}

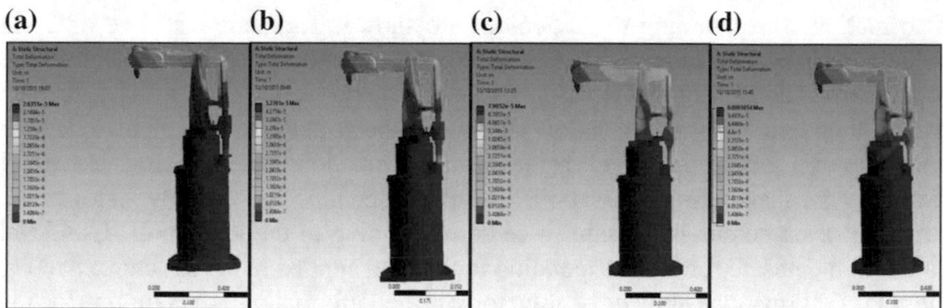

Fig. 7 Total deformation for load of **a** 25 N, **b** 50 N, **c** 75 N and **d** 100 N

the structure. Figure 9 shows the shear stress of the robot model structure at four different loads applied. From the colour scale of figure for griper load of 100 N, it is found that the shear stress is maximum and the value is 8.485e^{+006} Pa.

Shear stresses produced by the structure for the application of four different loads of 25, 50, 75 and 100 N are represented in the form of bars in the Fig. 10.

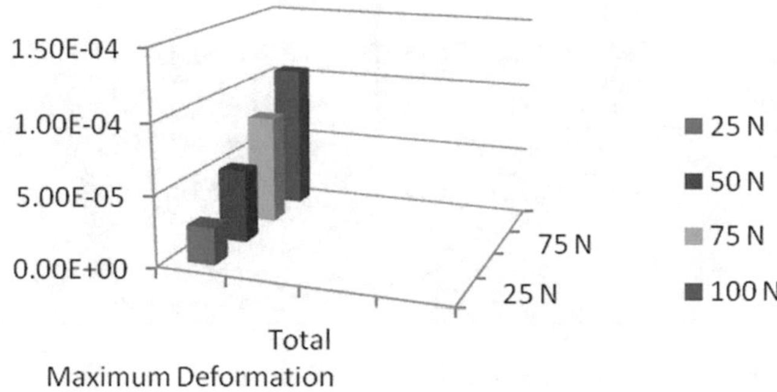

Fig. 8 Graph representing total deformation

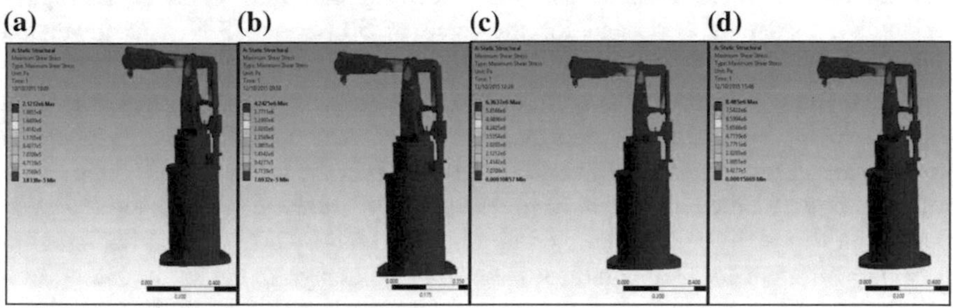

Fig. 9 Shear stress for load of **a** 25 N, **b** 50 N, **c** 75 N and **d** 100 N

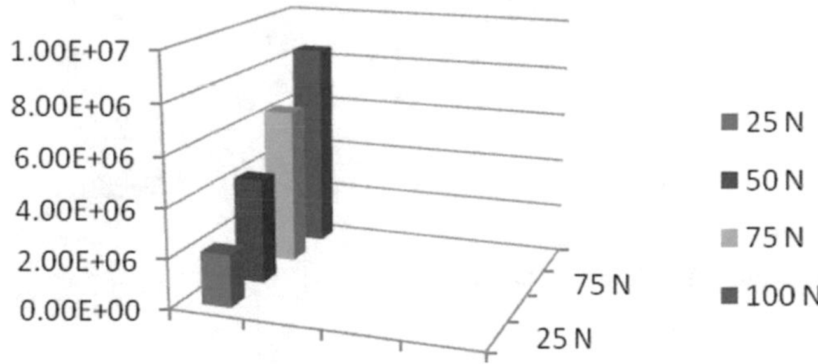

Fig. 10 Graph representing shear stress

Shear strain is also caused due to the application of external force which is defined as the change in dimension to the original dimension. In order to know the maximum change in dimension within the working limit of robot structure, the maximum shear strain is analysed. The maximum value of shear strain is obtained for Gripping load of 100 N, and the value is $1.103e^{-004\ \mathrm{m/m}}$, which is shown as red colour in the colour scale of Fig. 11.

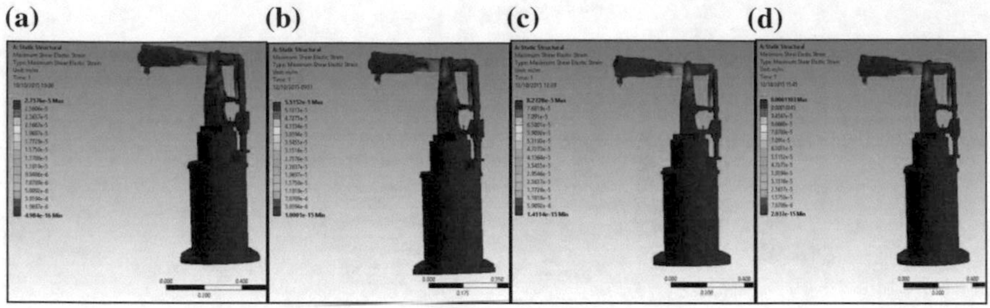

Fig. 11 Shear elastic strain for load of **a** 50 N, **b** 75 N and **d** 100 N

The graph representing shear elastic strain is shown in Fig. 12. From the graph, it is seen that the shear elastic strain is lowest for applied force of 25 N and highest for applied load of 100 N, and for the force of 50 N and 75 N, it is lower and higher, respectively.

Due to the application of force, strain is produced in the structures which are responsible for increasing the energy level of the structure. The energy observed during loading process is called as strain energy. It is the internal work done by the body and is very much important to find out in any structural analysis to predict the straight. In this analysis, the strain energy is found out by taking the robot structure in twenty different loading conditions. From these analyses, the strain energy only

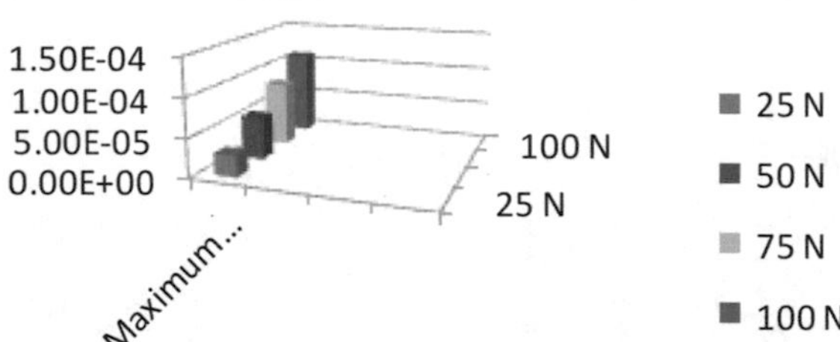

Fig. 12 Graph representing shear elastic strain

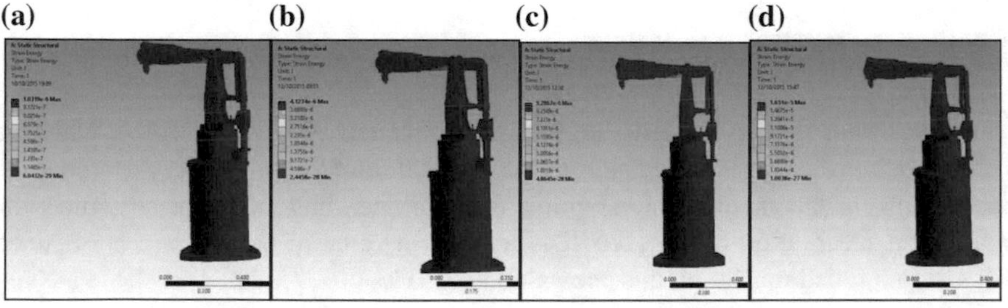

Fig. 13 Strain energy for load of **a** 25 N, **b** 50 N, **c** 75 N and **d** 100 N

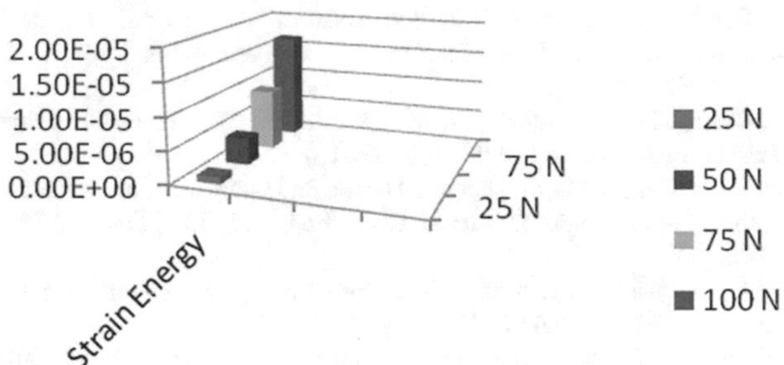

Fig. 14 Graph representing strain energy

for four loads applied is shown in Fig. 13. The highest value of strain energy is seen for the force of 100 N, and the value is $1.651e^{-005}$ j, which is shown in the colour scale of Fig. 13d. And the minimum value is found for the force of 25 N, and the value is $4.1274e^{-008}$ j.

The graphical representation of strain energy is shown in Fig. 14, where it is clearly seen that when the value of applied force increases the strain energy increases to a higher value.

4 Conclusions

The finite element analysis of the robot structure by taking the mesh size of 0.01 mm can be used to the best for the investigation of different parameters. For the simulation, the implementation of model with finite element is considerably good, because robot structural design is important to carry out so as to satisfy the customer's specific requirements and to decrease manufacturing cost and to increase quality of product. Complex engineering problems based on design can be solved by ANSYS structural analysis software which optimizes the product design and reduces the costs of physical testing.

References

1. Bhusnar, M.S., Sarawade, S.: Modal analysis of rectangular plate with lap joints to find natural frequencies and mode shapes. IOSR J. Mech. Civil Eng. (IOSRJMCE), 06–14(2016). e-ISSN: 2278-1684, p-ISSN: 2320-334XPP
2. Chitte, P.G., Bansode, S.S., Rathod, S.S., Motgi, N.S.: Structural and Irrational Analysis of Six axis ARISTO robot using ANSYS. IJIRT **3**(1) (2016). ISSN: 2349-6002
3. Ghiorghe, A.: Optimization design for the structure of an RRR type industrial robot. U.P.B. Sci. Bull. Ser. D **72**(4) (2010)
4. Jeevan, Rao, A.N.: Modeling and analysis of robot ARM using ANSYS. Int. J. Sci. Eng. Technol. Res. **04**(33), 6692–6697 (2015). ISSN 2319-8885

5. Manoj, K.K., Sambaiah, Ch.: Modelling and study of motion control system for motorized robot arm using mat lab and analysis of the arm by using ansys. Int. J. Mech. Ind. Eng. (IJMIE) **2**(1) (2012). ISSN No. 2231-6477
6. Nor, M.A.M., Rashid, H., Mahyuddin, W.M.F.W., Azlan, M., Mahmud, J.: Stress analysis of a low loader chassis. Procedia Eng., 995–1001 (2012)
7. Pachaiyappan, S., Balraj, M.M., Sridhar, T.: Design and analysis of an articulated robot arm for various industrial applications. J. Mech. Civil Eng., 42–53 (2014). e-ISSN: 2278-1684, p-ISSN: 2320–3334X
8. Ristea, A.: FEA Analysis for Frequency Behavior of Industrial Robot's Mechanical Elements, Sinaia, Romania, pp. 26–28 (2011)
9. Sridhar, C., Doukasa, J., Pandremenosa, P., Stavropoulosa, P., Foteinopoulosa, G.: On an empirical investigation of the structural behavior of robots. Procedia CIRP **3**, 501–506 (2012)

Design of Adaptive Frequency Reconfigurable Antenna for MIMO Applications

P. Manikandan, P. Sivakumar and C. Swedheetha

Abstract In recent wireless communication systems, the claim on multi-functional antennas is increasing. To meet with this demand, MIMO system has been developed which increases channel capacity without sacrificing additional spectrum or transmission power. This system uses multiple antennas at both transmitting and receiving ends. This paper explains the design of an antenna with adaptive frequency reconfigurability for MIMO applications. In addition to this to achieve miniaturization and multi-resonance, fractal concept that possesses the properties of self-similarity and space filling has been employed. The prototype of proposed antenna is simulated with the aid of ADS simulation software.

Keywords Reconfigurable antenna · MIMO · Axial ratio · Microstrip feed Impedance matching

1 Introduction

In modern era, the generation of antennas cannot be restricted to a certain function, but it has the tendency to alter their functionality depending on the execution necessities which are known as reconfigurable antennas. Related to broadband antennas, reconfigurable antennas provide numerous compensations, such as compact size, comparable radiation patterns for all required frequency bands, and frequency selectivity which aids in decreasing contrary effects like co-site interference and congestion [1].

P. Manikandan · P. Sivakumar · C. Swedheetha (✉)
Department of ECE, Kalasalingam University, Anand Nagar, Krishnankoil, Srivilliputtur 626126, Tamil Nadu, India
e-mail: swedhaece08@gmail.com; swedheetha.c@klu.ac.in

P. Manikandan
e-mail: maanip85@gmail.com

P. Sivakumar
e-mail: siva@klu.ac.in

J. Nayak et al. (eds.), *Soft Computing in Data Analytics*, Advances in Intelligent Systems and Computing 758,
https://doi.org/10.1007/978-981-13-0514-6_21

Modern research on reconfigurable antennas has produced other possibilities. In [2–4], various switching techniques are employed to change the current path and afford frequency reconfiguring ability. The rapid development of wireless technology throughout the past years has managed to increase in demand for miniaturized, cost-effective, and multiband antennas for use in wireless communication systems. By incorporating, the fractal structures over the patch will give the required miniaturization and its multiband characteristics. Normally, the application of fractal structures in antennas tends to miniaturize its physical size by varying its electrical length and yields multiband performance in the required resonating frequencies [5–8]. The generations of fractal arrangements have an iterative technique, to vary its current distribution path [9, 10].

In [11], the antenna is built on a dielectric substrate by using probe fed center strip with parasitic strips on both sides; each one is connected with two switched conditions. Multiple-input and multiple-output (MIMO) communication systems play a significant role in the IEEE 802.11n standard and 4G technologies. In MIMO applications, to provide extra functionalities 'n' antennas are desirable, whereas each antenna has been focused in different directions to cover the consistent area (space diversity). Each antenna is resonating at different frequencies. Due to the existence of more number of antennas, the area/cost/complexity is increased.

These 'n' numbers of antennas can be replaced with a single fractal incorporated antenna by applying the properties of self-similarities and iterative methods. An optimal method to implement the proposed concept is described in this paper.

This paper introduces a new reconfigurable patch antenna design, based on switches mounted at different slots inside the structure by using strip conductors. By switching these strip conductors, a change in the functionality of the antenna in the appropriate frequencies of operation is obtained. The radiation patterns for different configuration of switches are very similar.

Frequency reconfigurability is achieved by introducing parasitic elements on either side of the previously designed structure. By varying the length of parasitic elements, the concept of director–reflector method is applied, and the frequency reconfigurability is achieved.

1.1 Antenna Design Configuration

Antenna design configuration is based on the requirement of MIMO applications. The proposed antenna is basically a type of patch antenna. The substrate chosen for the proposed structure is FR4 substrate with a permittivity (ε_r) of 4.6 and loss tangent of 0.002. The structure is considered to be a non-magnetic, and hence, the value of permeability is 1. The thickness of the substrate layer is taken as 1.6 mm. FR represents the flame-retardant material, which is self-extinguishing substance. The standard thickness of the PCB board made of FR4 substrate is given as 1.6 mm.

The geometrical structure chosen for this proposed antenna is equilateral triangle of side 10 mm. The feeding technique has chosen to be via feed. The diameter of the hole is made as 0.5 mm. The length of the transmission line or the feed line length is given by 9.2 mm. Impedance narrates the relationship between the voltage and current at the input to the antenna. The real part of the antenna input impedance corresponds to power level that is either radiated away or absorbed within the antenna. The imaginary part of the impedance corresponds to power level that is accumulated in the near-field region. This is said to be the non-radiated power. An antenna with real input impedance having zero imaginary part is said to be antenna at the resonant condition. Note that the impedance of an antenna will vary with frequency.

The standard impedance when designing for transmission line applications is used to be as 50 Ω. While designing an antenna over the PCB, the traces need to be with controlled impedance, which is to be compensated for transmission line effects. If the impedance of the antenna trace (load) is matched with the output impedance of the source, it reduces the opportunity of reflections which would show the way to standing waves.

The triangular structure shown in Fig. 1 was found to have a resonant frequency of 8.063 GHz with −10.166 dB as return loss. In order to miniaturize the proposed structure, a slot was introduced at the centre. The resulting structure is none other than the Sierpinski fractal structure.

The structure shown in Fig. 2 is found to have a resonant frequency of 6.7 GHz with a return loss of −14.421 dB. This is how downshift in frequency is obtained. This explains the miniaturization property of the designed antenna. In addition to this, slots are made near the vertices of the triangle, and strip conductors are used as the switches for various switching actions. In order to generate reconfigurable

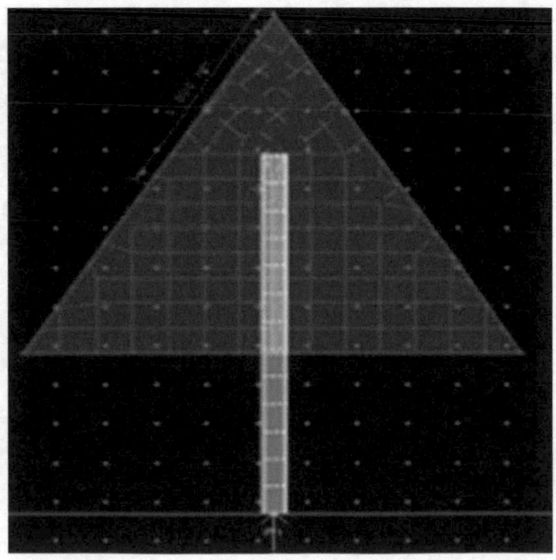

Fig. 1 Triangular structure without slots

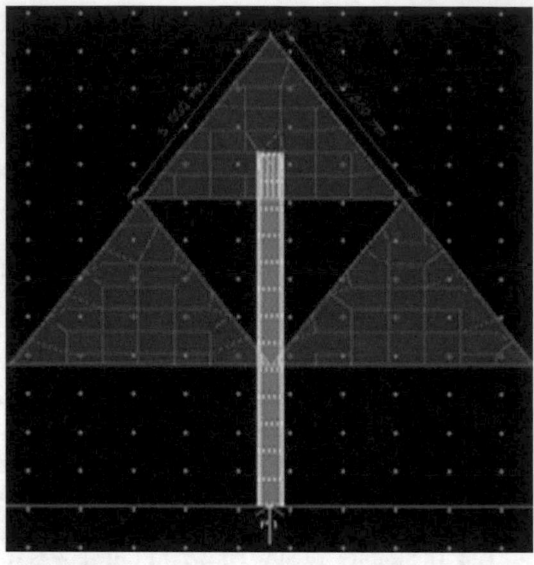

Fig. 2 Triangular patch antenna with slot at the centre

circuitry, various kinds of switching methods have been used which comprises of PIN diodes, RF MEMS switches, GaAs field-effect transistor (FET) switches. This switching mechanism is implemented to achieve frequency reconfigurability. This is shown in Fig. 3. It is found that the frequency remained the same (6.7 GHz) and there was better result regarding the reflection loss (−25.022 dB). These ‘n’ numbers of antennas can be replaced with a single fractal antenna by applying the properties of iterative approach.

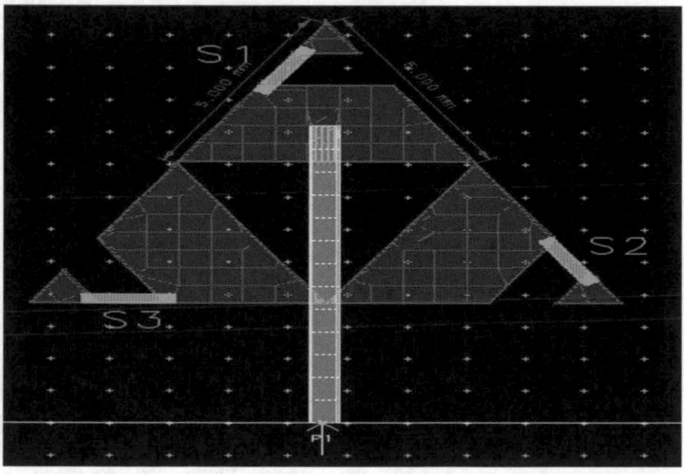

Fig. 3 Design with slot at the centre and also near the vertices

2 Design Methodology

Micostrip line feed is used in this design to match the input impedance of 50 Ω. Slot is mounted on the triangular patch and it is taken for simulation. To obtain the property of reconfigurability, different switching conditions (on and off conditions) are maintained using the strip conductors. Initially to verify this property, the triangular structure with and without slot in it is analyzed and simulated. We infer that there is a downshift in frequency. It is shown in Table 1.

By implementing the switch near the vertices as shown in Fig. 3, for the design in Fig. 3 the results obtained for different cases are tabulated in Table 2.

3 Passive Radiator as Fractal Yagi–Uda System

To achieve the pattern reconfigurability, the antenna structure shown in Fig. 3 is further extended by placing two rectangular strips on either side of the design. The resultant design is shown in Fig. 4. These rectangular strips behave like parasitic elements.

The dimensions of the rectangular strips placed on either side are similar. The dimensions which yield better results are found to be as follows. The length of the strip is found to be 0.5 λ = 22.35 mm. The switches 1, 2, 3, and 4 are used to vary the length of the rectangular strip from 0.5 λ = 22.35 mm to 0.2 λ = 8.94 mm.

Table 1 Comparison of structure with and without slot

A simple triangle without any slots (no strip conductors placed)		With slot only at the centre (first iteration of Sierpinski structure)	
Frequency (GHz)	Retur loss (dB)	Frequency (GHz)	Return loss (dB)
8.063	−10.166	6.700	−14.421

Table 2 Comparison of various switching status

Sl. no	Switch status			With slots both at the centre and near the vertices	
	S1	S2	S3	Frequency (GHz)	Return loss (dB)
1	ON	ON	ON	6.700	−25.022
2	ON	ON	OFF	4.895	−24.958
3	ON	OFF	ON	7.319	−17.568
4	OFF	ON	ON	7.345	−17.334
5	OFF	OFF	OFF	7.396	−14.044
6	OFF	OFF	ON	7.203	−12.185
7	ON	OFF	OFF	7.577	−27.391
8	OFF	ON	OFF	6.820	−44.955

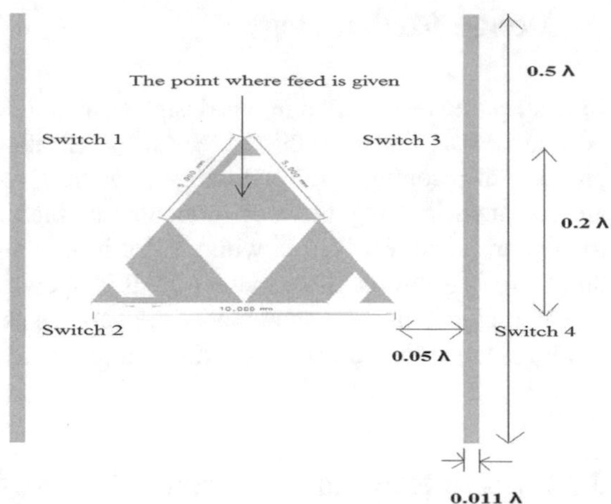

Fig. 4 Final design of the proposed idea

The width of the rectangular strip conductor was found to be 0.11 λ = 0.5 mm. The rectangular strip conductor was placed at 0.05 λ = 2.235 mm from the triangular structure.

When two rectangular slots on the left are shorted with strip conductors, the slots at the right parasitic element are open (as shown in Fig. 5). The element at the left acts as radiator and one at the right acts as director. Conversely, when the rectangular slots at the left parasitic element are closed and the gaps at the right element are open (as shown in Fig. 6), then the element at the right acts as reflector and the left-side element acts as director. The antenna much behaves like Yagi–Uda antenna placed on a dielectric substrate backed by a finite ground plane [5].

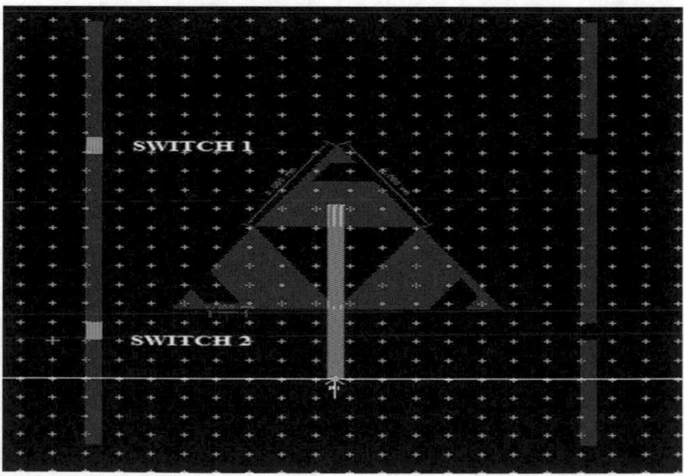

Fig. 5 Reflector–director mode

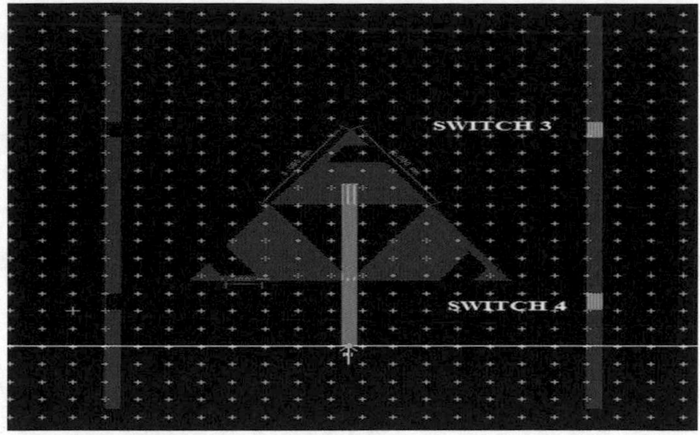

Fig. 6 Director–reflector mode

3.1 *Results and Discussion*

The design structure as shown in Fig. 4, We have analyzed with different switching conditions and observed that each case resonating at different frequencies. Figures 7, 8, 9, 10, 11, 12, 13, and 14 illustrate that the reflection coefficients at various resonating frequencies occurring due to diverse switching conditions. With the single structure, different switching conditions have analyzed which enables us to implement the reconfigurable property.

Fig. 7 Frequency for case 1

Case 1: S1=on, S2=on, S3=on.

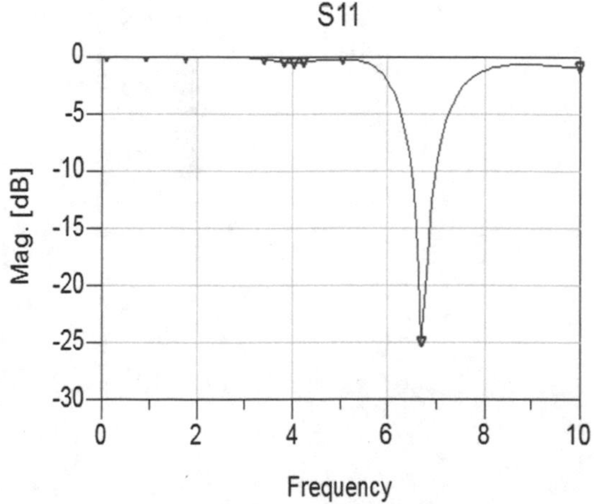

The different status and frequency obtained are illustrated below.

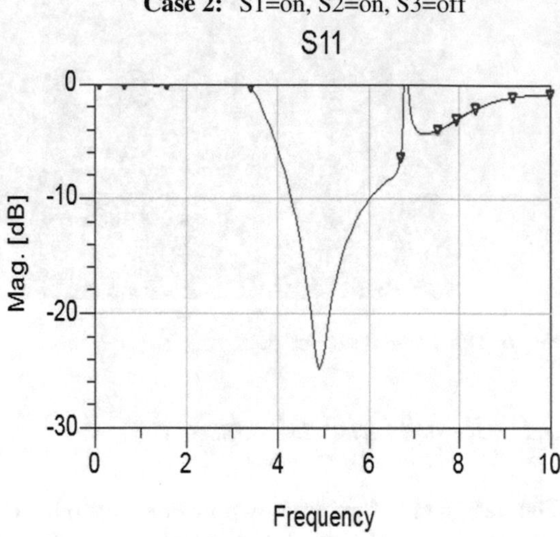

Fig. 8 Frequency for case 2

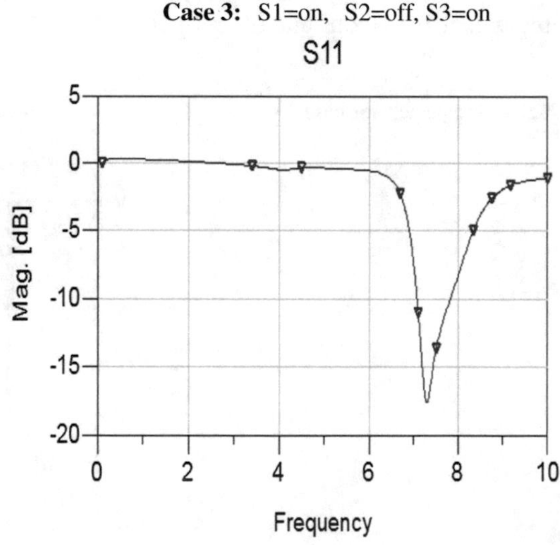

Fig. 9 Frequency for case 3

Fig. 10 Frequency for case 4

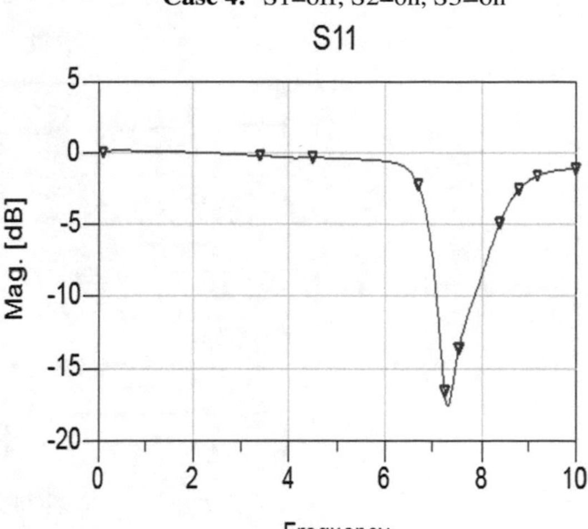

Fig. 11 Frequency for case 5

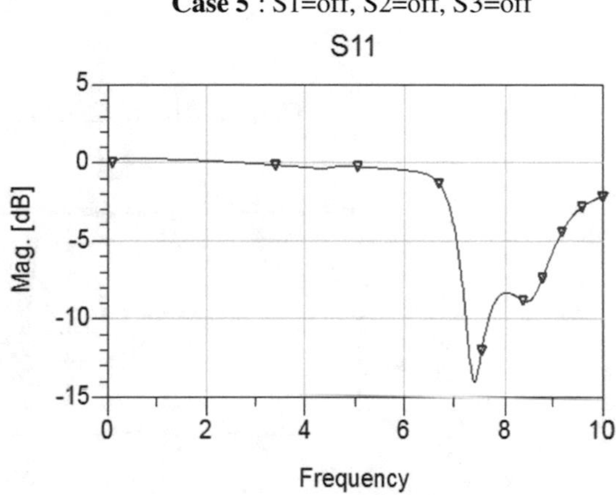

Fig. 12 Frequency for case 6

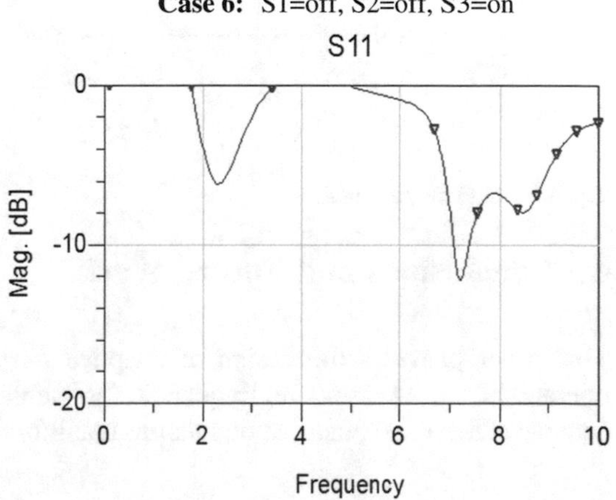

Case 7: S1=on, S2=off, S3=off

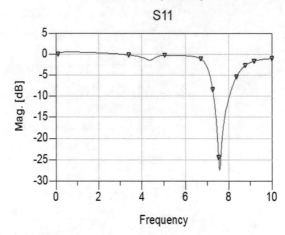

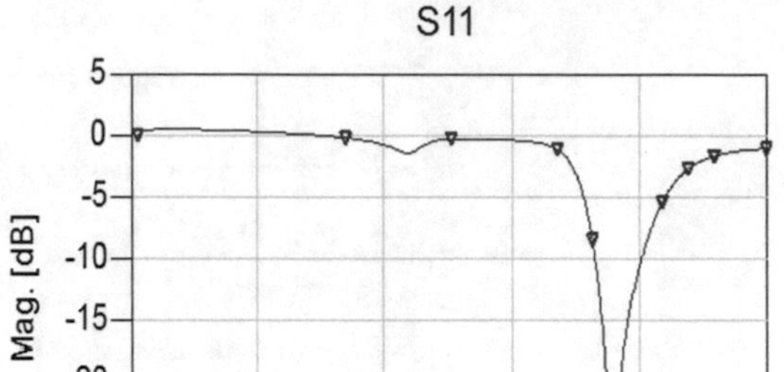

Fig. 13 Frequency for case 7

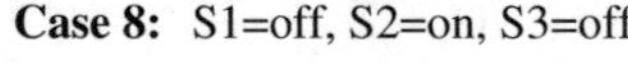

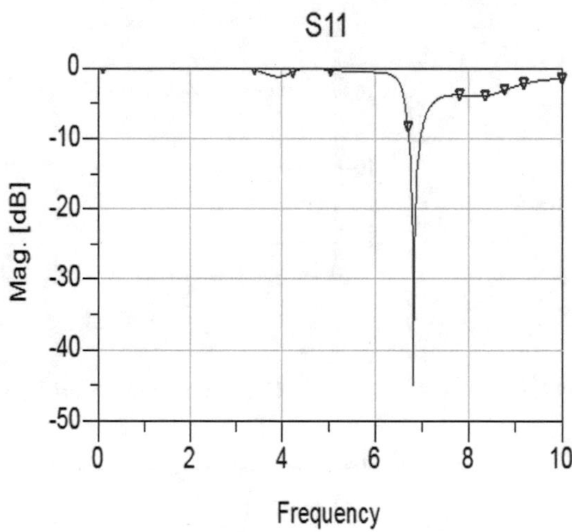

Fig. 14 Frequency for case 8

4 Conclusions and Future Work

This paper provides the design of adaptive frequency reconfigurable antenna to operate in the appropriate frequencies to achieve considerable bandwidth. To achieve pattern and polarization reconfigurability, we are proceeding with its fractal parasitic approach. The development of the proposed radiating element and the prototype measurement is under progress.

References

1. Jahromi, M., Falahati, A., Edwards, R.: Bandwidth and impedance matching enhancement of fractal monopole antennas using compact grounded co-planar waveguide. IEEE Trans. Antennas Propag. **59**(7) (2011)
2. Vinoy, K.J., Jose, K.A., Varadan, V.K., Varadan, V.V.: Hilbert curve fractal antennas with reconfigurable characteristics. In: Proceedings of the EEE, MTT-S International Microwave Symposium Digest, vol. 1, pp. 381–384 (2001)
3. Huff, G.H., Feng, J., Zhang, S., Bernhard, J.T.: A novel radiation pattern and frequency reconfigurable single turn square spiral microstrip antenna. IEEE Microw. Wirel. Compon. Lett. **13**, 57–59 (2003)
4. Hirasawa, K., Fujimoto, K.: On electronically-beam-controllable dipole antenna. In: Proceedings of the IEEE Antennas and Propagation Society International Symposium, vol. 18, pp. 692–695 (1980)
5. Wahid, P.F., Ali, M.A., DeLoach, B.C.: A reconfigurable Yagi antenna for wireless communications. Microw. Opt. Tech. Lett. **38**(2), 140–141 (2003)
6. Zhou, Y., Adve, R.S., Hum, S.V.: Design and evaluation of pattern reconfigurable antennas for MIMO applications. IEEE Trans. Antennas Propag. **62**, 1084–1092 (2014)
7. Werner, D.H., Ganguly, S.: An overview of fractal antenna engineering research. IEEE Antennas Propag. Mag. **45**(1), 38–57 (2003)
8. Lizzi, L., Azaro, R., Oliveri, G., Massa, A.: Printed UWB antenna operating over multiple mobile wireless standards. IEEE Antennas Wirel. Propag. Lett. **10** (2011)
9. Xu, H., Wang, G., Yang, X., Chen, X.: Compact, low return-loss, and sharp-rejection UWB filter using Sierpinski Carpet slot in a metamaterial transmission line. Int. J. Appl. Electromagn. Mech., 253–262 (2011)
10. Baldonado, M., Chang, C.-C.K., Gravano, L., Paepcke, A.: The stanford digital library metadata architecture. Int. J. Digit. Libr. **1**, 108–121 (1997)
11. Vinoy, K.J., Abraham, J.K., Varadan, V.K.: On the relationship between fractal dimension and the performance of multi-resonant dipole antennas using Koch curves. IEEE Trans. Antennas Propag. **51**(9), 2296–2303 (2003)

Online Training of Discriminative Parameter for Object Tracking-by-Detection in a Video

Vijay Kumar Sharma, Bibhudendra Acharya and K. K. Mahapatra

Abstract In this chapter, an online training algorithm to update a discriminative parameter vector is proposed. The initial discriminative parameter is obtained by training an SVM in the first video frame only. The positive example for SVM training is the initial target object, while the negative examples are cropped at some distance away from the target object. In the successive video frames, the parameter vector is updated based on the similarity score between the parameter vector and the vector corresponding to tracked object. The similarity score is measured using a Gaussian kernel. The learned parameter is used to construct a likelihood model. Using particle filter framework, a number of target candidates are cropped. The tracked object in each successive frame is the target candidate corresponding to the highest likelihood value.

Keywords Visual object tracking · Computer vision · HCI · SVM

1 Introduction

The purpose of visual object tracking is to estimate the location as well as the size of an object in the successive frames of a video. Visual object tracking has several applications in computer vision. Some important applications include human–computer interaction (HCI), gesture recognition, surveillance and action recognition.

Target appearance, motion model, and target localization are the three main components in visual object tracking [1]. The target appearance model is the

V. K. Sharma · B. Acharya (✉)
National Institute of Technology, Raipur, Raipur 492010, CG, India
e-mail: bacharya.etc@nitrr.ac.in

V. K. Sharma
e-mail: vijay4247@gmail.com

K. K. Mahapatra
National Institute of Technology, Rourkela, Rourkela 769008, Odisha, India
e-mail: kkm@nitrkl.ac.in

J. Nayak et al. (eds.), *Soft Computing in Data Analytics*,
Advances in Intelligent Systems and Computing 758,
https://doi.org/10.1007/978-981-13-0514-6_22

most important one. It is because the target appearance varies in each frame of a video sequence and the appearance model must be able to capture these variations efficiently without significant degradation of the model. Without online learning of an appearance model, the task to establish the object correspondence in successive video frames cannot be possible. The target appearance may vary in the form of pose change, deformation, motion blur, and illumination change. The target appearance may also undergo scale change and partial occlusion. Coping with the target clutter is also a challenging task.

There are two methods to form an appearance model. The first method utilizes the visual features such as edges, histogram, intensity for the appearance construction. The second method is to use statistical learning technique to learn a mathematical model of the target appearance. Generative and discriminative are the two mathematical models for the target appearance.

The generative model is composed of target objects which are tracked in successive video frames. It does not contain the information about the background [2–7]. Discriminative appearance model is composed of information from both target as well as its surrounding background. In the discriminative object tracking, the purpose is to construct a discriminative classifier to separate the object from the background. The important discriminative trackers are based on kernelized correlation filter (KCF) in [8, 9], multiple instance learning (MIL) in [10, 11], and support vector machine (SVM) in [12–14]. Kalal et al. in [15] formed discriminative classifier using similarity measure between patches. Hare et al. in [13] presented a tracker based on structured output SVM (STRUCK) framework.

If we have a learned classifier parameter $\mathbf{w}$ in the explicit form, then the classification task is carried out in linear space, and hence, it is computationally efficient than the one which uses nonlinear kernels (e.g., STRUCT in [13]). Therefore, Ning et al. in [14] proposed dual linear structured SVM (DLSSVM)-based method to get the classifier parameter $\mathbf{w}$ in the explicit form for fast learning and detection task.

In this chapter, we also learn the explicit classifier parameter using a simple update strategy. In the first video frame, the discriminative parameter is obtained by training SVM in linear space. In the successive video frame, the parameter is updated with the tracked object instance based on the similarity between the parameter vector and the vector corresponding to the tracked instance. From the current video frame, a negative example which is closest to the separating boundary is also added to the parameter vector to respect the constraint of SVM formulation in dual space. Using the learned parameter vector, a likelihood model is constructed to obtain the tracked instance corresponding to maximum likelihood.

The remaining part of this chapter is organized as follows. Discriminative appearance learning using SVM is given in Sect. 2. The proposed parameter learning method is discussed in Sect. 3. In Sect. 4, target candidates generation and likelihood construction are presented. Experimental results are given in Sect. 5. Finally, Sect. 6 concludes the chapter.

2 Discriminative Classifier Using Support Vector Machine

The SVM solves the following optimization problem in order to get the large margin separating hyperplane ($\mathbf{w}^\top\mathbf{x} + b = 0$, $\mathbf{w} \in \mathbb{R}^d$ and $b \in \mathbb{R}$) [16],

$$\min_{\mathbf{w},\xi} \frac{1}{2}\mathbf{w}^\top\mathbf{w} + C\sum_{i=1}^{n} \xi_i \tag{1}$$
$$\text{subject to } y_i(\mathbf{w}^\top\mathbf{x}_i + b) \geq 1 - \xi_i,$$
$$\xi_i \geq 0, i = 1, ..., n$$

where, $\{\mathbf{x}_i\}_{i=1}^n$ is training set with each $\mathbf{x}_i \in \mathbb{R}^d$ and labels $\mathbf{y}_i \in \{-1, +1\}$, $\mathbf{w}$ is the parameter vector, variable b represents distance of the hyperplane from the origin, ξ_i are slack variables, and C is the regularization constant. The formulation in Eq. (1) is called as soft margin SVM. For linear kernel ($\mathbf{x}^\top\mathbf{x}$), the normal weight vector $\mathbf{w}$ is expressed as linear sum of support vectors, given by Eq. (2) as,

$$\mathbf{w} = \sum_{i=1}^{nSV} \alpha_i y_i \mathbf{x}_i \tag{2}$$

where nSV is the number of support vectors ($nSV << n$). If $\alpha_i > 0$, the corresponding i-th example ($\mathbf{x}_i$) is called the support vector. The classification function $f(\mathbf{x})$ for a test example $\mathbf{x}$ is given by Eq. (3) as,

$$f(\mathbf{x}) = \mathbf{w}^\top\mathbf{x} + b \tag{3}$$

3 Proposed Discriminative Model Based on SVM Parameter Update

The SVM parameter needs to be updated with each tracked instance so as to carry the appearance variations of the target object. Most of the time, the object appearance varies very slowly from one frame to the next. Sometimes, however, the appearance variation may be high in the two consecutive frames. This happens especially when the tracked target is not accurate. If an inaccurate target is considered for appearance update, the overall appearance may get corrupted. This will further cause the inaccurate target tracking. To get rid of this problem, our strategy is to add the newly tracked examples based on their similarity with the discriminative parameter $\mathbf{w}$.

Let $\mathbf{w}_{t-1} \in \mathbb{R}^d$ be the updated parameter and $\mathbf{x}_{t-1}^{Tr} \in \mathbb{R}^d$ be the tracked instance in the $(t-1)$-th video frame. We update the parameter vector from $\mathbf{w}_{t-1}$ to $\mathbf{w}_t$ so as to correctly classify an example $\mathbf{x}_t^i$ from the t-th frame. Our update method is given in Algorithm 1. We first compute the similarity score (s_p) between parameter vector and the tracked instance given by Eq. (4) as,

$$s_p = \mathcal{K}(\widehat{\mathbf{w}}_{t-1}, \widehat{\mathbf{x}}^{Tr}_{t-1}) \tag{4}$$

where $\mathcal{K}$ is a kernel function and hat symbol on any vector $\mathbf{x}$ (i.e., $\widehat{\mathbf{x}}$) represents the vector with unit norm. Then the parameter is updated with the tracked instance based on similarity score as,

$$\mathbf{w}_{t-1} \leftarrow \mathbf{w}_{t-1} + (s_p \times \frac{\widehat{\mathbf{x}}^{Tr}_{t-1}}{C_1}) \tag{5}$$

where C_1 is a constant which is greater than 1. To further update the parameter with the negative instance, we select a negative instance $\mathbf{x}^-_{t-1,j}$ out of nn instances $X^-_{t-1} = \{\mathbf{x}^-_{t-1,i}\}^{nn}_{i=1}$ corresponding to the highest (classification) score cs_i with the parameter $\mathbf{w}_{t-1}$, where cs_i is computed by dot product Eq. (6) as,

$$cs_i =< \widehat{\mathbf{w}}_{t-1}, \widehat{\mathbf{x}}^-_{t-1,i} >, \; i = 1, ..., nn \tag{6}$$

We then obtain the updated parameter vector $\mathbf{w}_t$ using the following equation,

$$\mathbf{w}_t = \mathbf{w}_{t-1} - (s_p \times \frac{\widehat{\mathbf{x}}^-_{t-1,j}}{C_1}) \tag{7}$$

From Eqs. (5) and (7), it is clear that the positive and negative examples are added using the same factor which is s_p/C_1. This ensures that the constraint $\sum^n_{i=1} \alpha_i y_i = 0$ in the dual space is respected, where α_i are Lagrange multipliers [17].

Algorithm 1 Proposed online learning of parameter vector

Input: SVM parameter in the $(t-1)$-th frame, $\mathbf{w}_{t-1}$;Tracked target in the $(t-1)$-th frame, $\mathbf{x}^{Tr}_{t-1}$; Negative example set in the $(t-1)$-th frame, $X^-_{t-1} = \{\mathbf{x}^-_{t-1,i}\}^{nn}_{i=1}$; Constant C_1

1: Compute the similarity $s_p = \mathcal{K}(\widehat{\mathbf{w}}_{t-1}, \widehat{\mathbf{x}}^{Tr}_{t-1})$

2: Update $\mathbf{w}_{t-1}$ as $\mathbf{w}_{t-1} \leftarrow \mathbf{w}_{t-1} + (s_p \times \frac{\widehat{\mathbf{x}}^{Tr}_{t-1}}{C_1})$

3: Out of nn negative examples, find a negative example $\mathbf{x}^-_{t-1,j}$ corresponding to the highest (classification) score computed using $cs_i =< \widehat{\mathbf{w}}_{t-1}, \widehat{\mathbf{x}}^-_{t-1,i} >, \; i = 1, ..., nn$

4: Obtain the updated parameter as, $\mathbf{w}_t = \mathbf{w}_{t-1} - (s_p \times \frac{\widehat{\mathbf{x}}^-_{t-1,j}}{C_1})$

Output: Learned parameter vector $\mathbf{w}_t$

4 Likelihood Construction

In particle filter framework, a state vector $\mathbf{s}_t$ at time t is represented by six parameters of affine transformation, $(x_t, y_t, \theta_t, s_t, \alpha_t, \phi_t)$, corresponding to (x, y) translations, rotation angle, scaling, aspect ratio, and skew, respectively [2]. The parameters of the state are modeled independently using a Gaussian distribution around its previous state as [2],

$$p(\mathbf{s}_t|\mathbf{s}_{t-1}) = \mathcal{N}(\mathbf{s}_t; \mathbf{s}_{t-1}, \boldsymbol{\Psi}) \quad (8)$$

where $\boldsymbol{\Psi} = (\sigma_x^2, \sigma_y^2, \sigma_\theta^2, \sigma_s^2, \sigma_\alpha^2, \sigma_\phi^2)$ represents the variances of the parameters. In each video frame, n target candidates $\mathbf{x}_t^i, i = 1, ..., n$ are cropped corresponding to n particles $\mathbf{s}_t^i, i = 1, ..., n$. The likelihood model to compute the score (classification) for target candidates is given by,

$$p(\mathbf{x}_t^i|\mathbf{s}_t^i) = \mathcal{K}(\widehat{\mathbf{w}}_t, \widehat{\mathbf{x}}_t^i) \quad (9)$$

The tracked target $\mathbf{x}_t^{Tr}$ among $\mathbf{x}_t^i$ in the current (t-th) frame is the one which has the highest likelihood according to Eq. (9). However, in the initial 18 frames, the likelihood model is composed of product of the score $\mathcal{K}(\widehat{\mathbf{w}}_t, \widehat{\mathbf{x}}_t^i)$ and the score corresponding to reconstruction error in eigenbasis vector-based subspace (as in [2]).

5 Experimental Results

We performed the simulation on a notebook computer with core-i7, 8 GB of RAM, and Windows 10 OS. The parameter vector in first video frame is obtained by training SVM using LIBSVM software [16]. We used about 350 negative examples and 200 positive examples (similar copies) for training with LIBSVM. The number of target candidates generated in each frame is 600. A unique standard deviations ([12, 12, 0.03, 0.0, 0.0, 0.0]) is used for state dynamic model (in Eq. (8)) for all the videos. The minimum sample distance (after first video frame) for considering a sample

Table 1 Comparative average center error and average overlap rate

Video	Average center error			Average overlap rate		
	DLSSVM in [14]	Scale-DLSSVM in [14]	Proposed	DLSSVM in [14]	Scale-DLSSVM in [14]	Proposed
Car4	18.6	3.8	2.7	0.467	0.681	0.878
CarDark	1.4	1.3	1.1	0.858	0.857	0.857
Coke11	8.7	6.3	13.4	0.729	0.775	0.545
Dudek	12.6	10.2	11.0	0.728	0.793	0.780
Faceocc	15.9	17.7	13.0	0.756	0.724	0.795
Fleetface	23.4	23.0	23.9	0.609	0.619	0.534
MountainBike	7.4	6.7	8.8	0.727	0.661	0.681
Jogging	3.1	4.2	4.4	0.749	0.710	0.795
Singer1	15.4	5.2	3.2	0.350	0.551	0.844
Twinnings	7.2	6.5	2.5	0.747	0.644	0.724

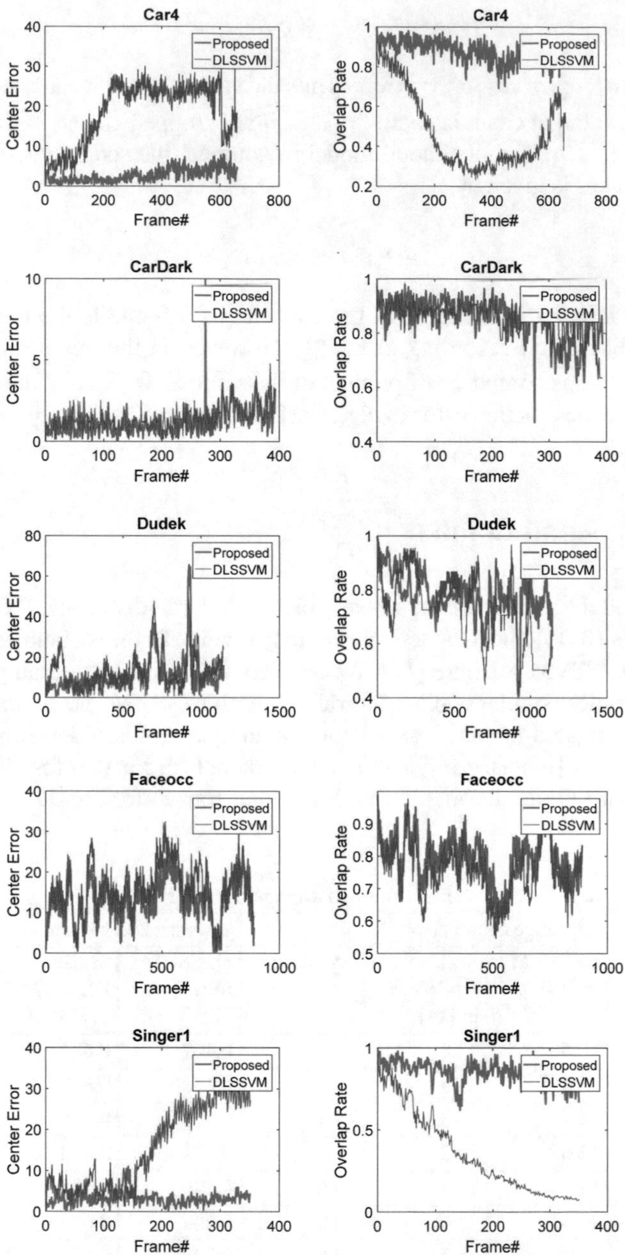

Fig. 1 Center error and overlap rate in Car4, CarDark, Dudek, Faceocc, and Singer1 video sequences

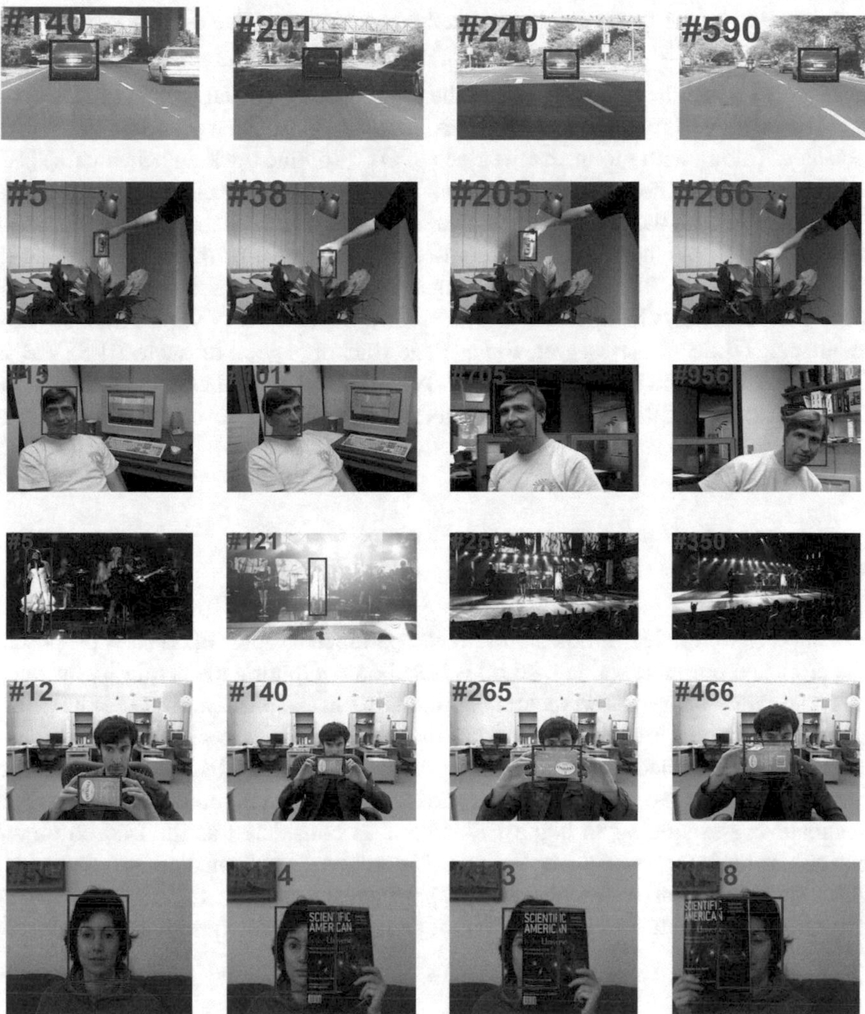

Fig. 2 Tracked rectangles in (top to bottom) Car4, Coke11, Dudek, Singer1, Twinnings, and Faceocc video sequences (the information about video sequences is given in online tracking benchmark (OTB)

as negative example is $(10 + (\frac{width}{20}))$, where *width* is the target width. All cropped examples and candidates are of size 32×32. We used Gaussian kernel with sigma value 1 to compute the similarity score s_p as well as the likelihood in Eq. (9). We set $C_1 = 10$, and therefore s_p/C_1 is less than 1. The initial value of C to train the SVM in the first video frame is 1. This ensures that the dual space constraint $0 \leq \alpha_i \leq C$ is respected [17].

We evaluated the tracking performances in terms of center location error and overlap rate. The center location error is the Euclidean distance between the ground truth location of the target and the tracked target location. The overlap rate is given by, $OR = \frac{area(B_T \bigcap B_G)}{area(B_T \bigcup B_G)}$, where B_T and B_G, respectively, are the bounding box of the tracked object and the bounding box of the ground truth. The DLSSVM tracker code in [14] is also used to perform tracking evaluation on the same platform. The video sequences (along with ground truth) given in OTB (online tracking benchmark) [18] have been used for the tracking evaluation. The average center error and the average overlap rate of both the trackers are shown in Table 1. Figure 1 shows the center error and the overlap rate across the entire video frames. Some of the tracked instances are shown in Fig. 2. The average tracking speed of the proposed tracker is 12 FPS while that of DLSSVM tracker is 9.5 FPS. The Scale-DLSSVM performs slightly better than DLSSVM. However, the average tracking speed of Scale-DLSSVM is 7.5 FPS, which is lower than that of DLSSVM method as well as proposed method. The performance of the proposed method can be further improved by using features as in [14] instead of raw pixels.

6 Conclusion

A method to online learn the discriminative parameter vector of SVM is proposed. The proposed online learning method is based on computing the similarity between the parameter vector and the vector corresponding to the positive (tracked) instance. A negative example is also used to update the parameter so as to satisfy the dual space constraint. The learned parameter vector is used to construct a likelihood model. The target candidates are cropped using the particle filter-based motion model. The target candidate corresponding to highest likelihood is considered as the tracked object instance in the successive video frames. The tracking performances are compared with a state-of-the-art video tracker. The performances can be further improved by integrating some better features instead of simple raw pixels-based features used in this work.

References

1. Li, X., Hu, W., Shen, C., Zhang, Z., Dick, A., Hengel, A.V.D.: A survey of appearance models in visual object tracking. ACM Trans. Intell. Syst. Technol. (TIST) **4**(4), 58 (2013)
2. Ross, D.A., Lim, J., Lin, R.-S., Yang, M.-H.: Incremental learning for robust visual tracking. Int. J. Comput. Vis. **77**(1–3), 125–141 (2008)
3. Cruz-Mota, J., Bierlaire, M., Thiran, J.: Sample and pixel weighting strategies for robust incremental visual tracking. IEEE Trans. Circuits Syst. Video Technol. **23**(5), 898–911 (2013)
4. Mei, X., Ling, H.: Robust visual tracking using l1 minimization. In: 2009 IEEE 12th International Conference on Computer Vision, pp. 1436–1443. IEEE (2009)
5. Mei, X., Ling, H., Wu, Y., Blasch, E., Bai, L.: Minimum error bounded efficient l1 tracker with occlusion detection. In: 2011 IEEE Conference on Computer Vision and Pattern Recognition (CVPR), pp. 1257–1264. IEEE (2011)

6. Bai, T., Li, Y.F.: Robust visual tracking with structured sparse representation appearance model. Pattern Recognit. **45**(6), 2390–2404 (2012)
7. Wang, D., Lu, H., Yang, M.-H.: Online object tracking with sparse prototypes. IEEE Trans. Image Process. **22**(1), 314–325 (2013)
8. Henriques, J., Caseiro, R., Martins, P., Batista, J.: High-speed tracking with kernelized correlation filters. IEEE Trans. Pattern Anal. Mach. Intell. **37**(3), 583–596 (2015)
9. Liu, T., Wang, G., Yang, Q.: Real-time part-based visual tracking via adaptive correlation filters. In: Proceedings of the IEEE Conference on Computer Vision and Pattern Recognition, pp. 4902–4912 (2015)
10. Babenko, B., Yang, M.-H., Belongie, S.: Robust object tracking with online multiple instance learning. IEEE Trans. Pattern Anal. Mach. Intell. **33**(8), 1619–1632 (2011)
11. Zhang, K., Zhang, L., Yang, M.-H.: Real-time object tracking via online discriminative feature selection. IEEE Trans. Image Process. **22**(12), 4664–4677 (2013)
12. Avidan, S.: Support vector tracking. IEEE Trans. pattern Anal. Mach. Intell. **26**(8), 1064–1072 (2004)
13. Hare, S., Saffari, A., Torr, P.H.: Struck: structured output tracking with kernels. In: 2011 IEEE International Conference on Computer Vision (ICCV), pp. 263–270. IEEE (2011)
14. Ning, J., Yang, J., Jiang, S., Zhang, L., Yang, M.-H.: Object tracking via dual linear structured SVM and explicit feature map. In: Proceedings of the IEEE Conference on Computer Vision and Pattern Recognition, pp. 4266–4274 (2016)
15. Kalal, Z., Mikolajczyk, K., Matas, J.: Tracking-learning-detection. IEEE Trans. Pattern Anal. Mach. Intell. **34**(7), 1409–1422 (2012)
16. Chang, C.-C., Lin, C.-J.: LIBSVM: a library for support vector machines. ACM Trans. Intell. Syst. Technol. (TIST) **2**(3), 27 (2011)
17. Platt, J.: Sequential minimal optimization: a fast algorithm for training support vector machines. Technical report (1998)
18. Wu, Yi., Lim, J., Yang, M-H.: Online object tracking: A benchmark. In: 2013 IEEE Conference on Computer Vision and Pattern Recognition (CVPR), pp. 2411–2418 (2013)

Fractional IMC-Based AGC for Interconnected Power System via Its Reduced Model Using Genetic Algorithm

Idamakanti Kasireddy and Arun Kumar Singh

Abstract In this paper, fractional IMC-based decentralized load frequency control (LFC) for multi-area power system is studied. At first, for decentralized controller design, the tie line power flow between the areas is assumed to be zero. Then, the transfer function models of each area have been reduced to lower order factional transfer function using genetic algorithm through step error minimization method. Based on the obtained reduced model, the controller has been designed. At last, the controller is equipped with multi-area power system to test the performance using MATLAB. The simulation results show that the proposed controller can minimize the load fluctuations and modelling errors effect on frequency and tie line power flow.

Keywords Automatic generation control · Multi-area power system IMC · Fractional calculus · Genetic algorithm · Step error minimization

1 Introduction

The power system is studied in terms of generation, transmission and distribution systems. In generation system, all synchronous generators are operated at specified frequency to meet the required load. The imbalance between the generation MW power and load MW power causes the deviation in frequency of the power system. Frequency deviation may be small or large based on the mismatch. There are two types of mismatches, i.e. small mismatches due to random load fluctuations and large deviations due to large generator or power plant tripping out, faults, etc.

I. Kasireddy (✉) · A. K. Singh
EEE Department, NIT Jamshedpur, Jamshedpur 831014, Jharkhand, India
e-mail: 2015rsee002@nitjsr.ac.in

A. K. Singh
e-mail: aksingh.ee@nitjsr.ac.in

J. Nayak et al. (eds.), *Soft Computing in Data Analytics*,
Advances in Intelligent Systems and Computing 758,
https://doi.org/10.1007/978-981-13-0514-6_23

Thus, the objective of load frequency control (LFC) is to minimize frequency deviation of the power system [1–3]. This can be achieved by adopting a secondary controller in addition to the governor. From literature [1–3], the conventional controller is used as secondary. To set the parameters of the controller, the power system is modelled and simulated. In this work, the modelling of power system is carried out through the fractional order differential equations.

The fractional order dynamic system is represented by differential equations where the powers of derivatives are any real or complex numbers. The concept of fractional order study is mainly used in the area of control, mathematics and physics [4]. Recently, the concept of fractional calculus and its applications in control system have increased significantly. The precision of modelling is achieved using the concept of fractional calculus. In view of above fact, integer operators of traditional control methods have been replaced by concept of fractional calculus [5, 6].

As per secondary controller for LFC, many modern controllers are available like sliding mode control (SMC) [7], decentralized SMC LFC [8], fuzzy gain scheduling of PI controller [9] and adaptive controller technique [10]. It is noted that power system parameters may change due to ageing of system, replacement and repair of system components and modelling errors. As a result, the problem to design optimum controller becomes a challenging task. A robust controller designed is inert to system parameter changes. Hence, a good robust controller design is needed to take care of parameter uncertainties as well as load disturbance.

Many robust control methods are available in literature for disturbance rejection and parameter uncertainty for LFC [11–13]. Vast research is going on internal model control (IMC) which is one of the robust control methods. Both set-point tracking and load disturbance rejection are achieved using two-degree-of-freedom IMC (TDF IMC) [12]. Therefore, IMC controller is an ideal choice for secondary control of the power system. Saxena [14] proposed a TDF IMC for LFC using Pade's and Routh's approximation which motivated us to consider the fractional IMC-PID controller as a secondary controller and is designed using a reduced fractional model of a system.

2 Multi-area Interconnected Power System

2.1 Decentralized LFC

Generally, the power system is subjected to small perturbation in load. Hence, the multi-area power system linear model is sufficient for decentralized controller design, which is shown in Fig. 1 [13]. The dynamic models of governor, turbine and power system are given as (1)

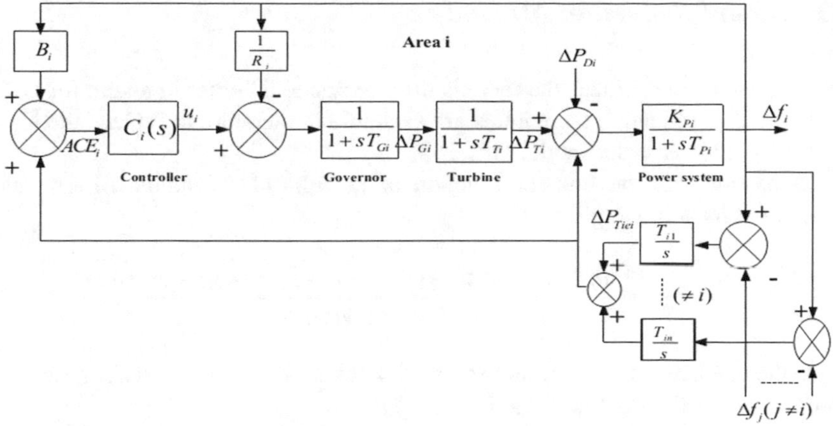

Fig. 1 Block diagram of multi-area linear power system model

$$G_G(s) = \frac{1}{T_G s + 1},\ G_T(s) = \frac{1}{T_T s + 1},\ G_P(s) = \frac{K_P}{T_P s + 1} \tag{1}$$

The objective of LFC problem requires that not only to minimize the frequency deviation ($\Delta f_i(s)$) but also to minimize tie line power deviation ($\Delta P_{tie}(s)$). Hence, the area control error (ACE) is utilized as a feedback signal to controller to fulfil the objective, which is defined as (2)

$$ACE_i = \Delta P_{tie} + B_i \Delta f_i \tag{2}$$

The control output $u_i(s)$ for area i is in the form (3)

$$u_i(s) = -K_i ACE_i \tag{3}$$

As per definition of decentralized controller [13], the controller should have only local measurements. In this, $\Delta f_i(s)$ is local measurement of ith area and $\Delta P_{tie}(s)$ is not a local measurement. Hence, decentralized controller is tuned assuming that the tie line power flow is zero. Therefore, the local control output becomes (4)

$$u_i(s) = -K_i B_i \Delta f_i \tag{4}$$

Hence, the transfer function from $\Delta f_i(s)$ to $u_i(s)$ can be obtained as (5)

$$G_1(s) = \frac{K_P}{T_P T_T T_G s^3 + (T_P T_T + T_T T_G + T_G T_P)s^2 + (T_P + T_T + T_G)s + (1 + K_P/R)} \tag{5}$$

2.2 Model Reduction Method

The proposed work utilizes the models that are fractional order in nature for design of filter. Many reduction techniques are available in literature to replace the higher system by a lower order model [15, 16].

Let us consider the transfer function of an nth-order dynamic SISO system represented by following Eq. (6)

$$T(s) = \frac{a_{n-1}s^{n-1} + a_{n-2}s^{n-2} + \cdots + a_2 s^2 + a_1 s + a_0}{b_n s^n + \cdots + b_1 s + b_0} \tag{6}$$

where the coefficients a_{n-1},b_n and so on are known. And let the transfer function of rth-order reduced model is given by Eq. (7)

$$T^{'}(s) = \frac{c_{r-1}s^{r-1} + c_{r-2}s^{r-2} + \cdots + c_1 s + c_0}{d_r s^r + \cdots + d_1 s + d_0} \tag{7}$$

where $r < n$, the coefficients c_{r-1},d_r and so on are to be determined by using MOR technique as follows.

The system $T(s)$ whose coefficients are known from the chosen system is reduced to $T^{'}(s)$ which has variables, i.e. c_{r-1}, d_r and so on as coefficients. The solution to these variables is found through step error minimization technique. This is as follows.

Step 1: A step input is given to system $T(s)$ and output is obtained.
Step 2: Similarly, a step input is given to $T^{'}(s)$ system constructed by assigning probable coefficients and output is obtained.
Step 3: The outputs of both the system are compared, and their change in error is passed to a performance index which determines the performance of allocated coefficients in representing original system. The chosen performance index is ITSE.
Step 4: Above steps keep on repeating until a optimal solution is achieved. The generation of these coefficients is handled by heuristic algorithms like genetic algorithm (GA) [17, 18].

Here, ITSE is the integral time multiplied squared error performance index and is denoted by J, which is defined as (8)

$$J = ITSE = \int_0^{t_{sim}} (y(=t) - y_r(t))^2 \cdot tdt \int_0^{t_{sim}} (e(t))^2 \cdot tdt \tag{8}$$

where $y(t)$ is the unit step response of full order system, i.e. for $T(s)$, and $y_r(t)$ is the unit step response of reduced order system, i.e. for $T^{'}(s)$. In the present case, $t_{sim} = 10$ s. Figure 2 shows the step error minimization method.

From (5), $G_1(s)$ is simplified as (9)

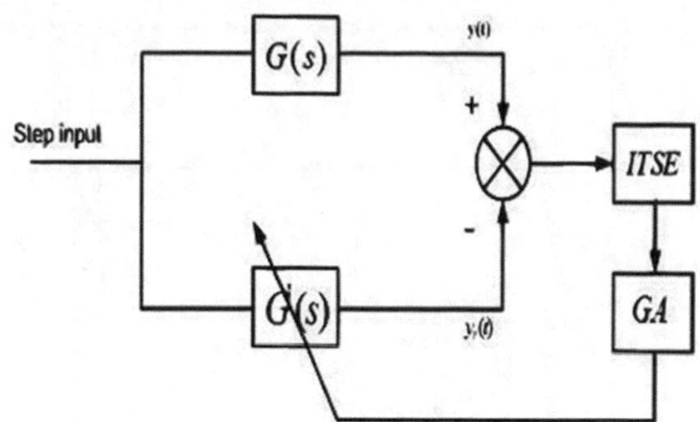

Fig. 2 Step error minimization method

$$T(s) = G_1(s) = \frac{250}{s^3 + 15.88s^2 + 42.46s + 106.2} \tag{9}$$

with $T_P = 20$ s, $T_T = 0.3$ s, $T_G = 0.08$ s, R = 2.4, $K_P = 120$.

Equation (9) represents higher order integer model which is converted to fractional model and is assumed to be $T^{'}(s)$.

Let us consider fractional SOPDT reduced model $T^{'}(s)$ given by (10)

$$T^{'}(s) = \frac{K_1 e^{-Ls}}{s^b + ps^c + q} \tag{10}$$

where the order of $T^{'}(s)$ is less than the $T(s)$ and is in fractional form.

This is done through MATLAB and Simulink, and obtained parameters of $T^{'}(s)$ are given as

$$K_1 = 16.385, L = 0.095, b = 1.897, c = 0.962, p = 1.954, q = 7.031$$

Thus, the obtained reduced fractional model $T^{'}(s)$ is obtained as (11)

$$T^{'}(s) = \frac{16.385e^{-0.095s}}{s^{1.897} + 1.954s^{0.962} + 7.031} \tag{11}$$

where the order of $T^{'}(s)$ is less than the $T(s)$ and is in fractional form.

The step responses of full order, proposed, Pade's approximation and Routh's approximation models are compared and shown in Fig. 3. The ITSE of proposed, Routh's and Pade's methods are 0.001, 0.0583 and 0.026, respectively. Hence, one can conclude that the response of the proposed method is almost closer to the original model.

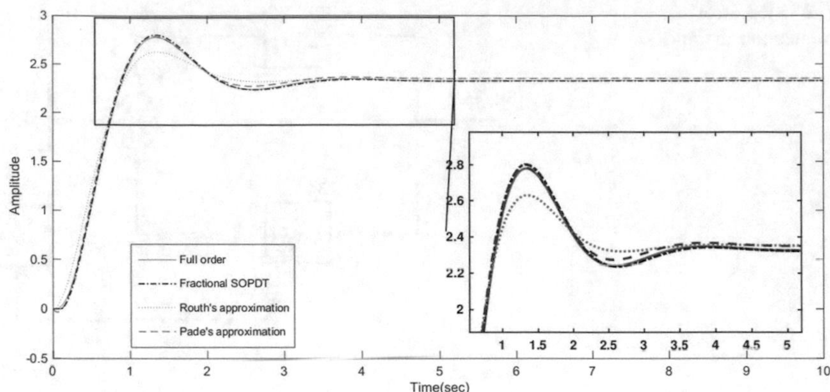

Fig. 3 Step responses of full order model, fractional SOPDT, Routh's and Pade's approximation

3 IMC-Based AGC/LFC

The block diagram of IMC structure [19, 20] is shown in Fig. 4, where $C_{IMC}(s)$ is the controller, $P(s)$ is the power plant, and $P_m(s)$ is the predictive plant.

Steps for IMC controller design [20, 21] are as follows.

Step 1:

The plant model can be written as (12)

$$P_m = P_m^+ P_m^- \tag{12}$$

where P_m^- = invertible part,

P_m^+ = non-invertible part, like time delay, zeros in right side of S-plane.

Step 2:

The IMC controller is (13)

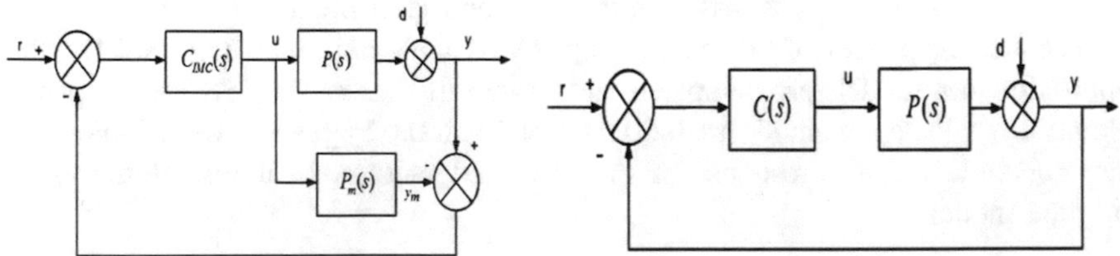

Fig. 4 Block diagram of internal model control and its equivalent conventional feedback control

$$C_{IMC}(s) = \frac{1}{P_m^-} f(s) \tag{13}$$

where $f(s)$ is low-pass filter, which has the form (14)

$$f(s) = \frac{1}{(1 + \tau_c s^{\lambda+1})^r} \tag{14}$$

where τ_c is the desired closed-loop time constant, and r is the positive integer, $r \geq 1$, which are chosen such that $C_{IMC}(s)$ is physically realizable. Here, r is taken as 1 for proper transfer function.

FIMC controller is designed for fractional SOPDT and is given by (11) via method discussed below.

Consider the system [20] given by (15)

$$P_m(s) = \frac{ke^{-\theta s}}{a_2 s^\beta + a_1 s^\alpha + 1}, \beta > \alpha \tag{15}$$

where $P_m(s)$ = Fractional SOPDT system, θ = delay,

α: $0.5 \leq \alpha \leq 1.5$, β: $1.5 \leq \beta \leq 2.5$, $a_2 = \tau^2$ and $a_1 = 2\zeta\tau$.

Using (12), the invertible part of $P_m(s)$ is given by (16)

$$P_m^-(s) = \frac{k}{a_2 s^\beta + a_1 s^\alpha + 1} \tag{16}$$

From (13), the fractional IMC controller is given by (17)

$$C_{IMC}(s) = \frac{a_2 s^\beta + a_1 s^\alpha + 1}{k} \frac{1}{(1 + \tau_c s^{\lambda+1})} \tag{17}$$

The conventional controller $C(s)$ is evaluated and given by (18)

$$C(s) = \frac{a_2 s^\beta + a_1 s^\alpha + 1}{k[(1 + \tau_c s^{\lambda+1}) - (1 - \theta s)]} \tag{18}$$

which can be simplified as (19)

$$C(s) = \frac{a_2 s^\beta + a_1 s^\alpha + 1}{k(\tau_c s^{\lambda+1} + \theta s)} \tag{19}$$

where $e^{-\theta s}$ is approximated as $(1 - \theta s)$ using Taylor expansion [20, 21].

Again, $C(s)$ is decomposed into FIMC-PID filter via technique discussed below. Multiplying and dividing RHS of (19) by $s^{-\alpha}$, the resultant is given by (20)

$$C(s) = \frac{a_1 + \frac{1}{s^\alpha} + a_2 s^{\beta-\alpha}}{\theta k s^{-\alpha}(s + (\tau_c/\theta)s^{\lambda+1})} \tag{20}$$

Equation (20) can be rearranged as (21)

$$C(s) = \left[\frac{s^{1-\alpha}}{1 + (\tau_c/\theta)s^\lambda}\right]\left[\frac{a_1}{k\theta}(1 + \frac{1}{a_1 s^\alpha} + \frac{a_2}{a_1}s^{\beta-\alpha})\right] \tag{21}$$

Now substituting $a_2 = \tau^2$ and $a_1 = 2\zeta\tau$ in (21), it becomes (22)

$$C(s) = \left[\frac{s^{1-\alpha}}{1 + (\tau_c/\theta)s^\lambda}\right]\left[\frac{2\zeta\tau}{k\theta}(1 + \frac{1}{2\zeta\tau s^\alpha} + \frac{\tau}{2\zeta}s^{\beta-\alpha})\right] \tag{22}$$

From above equations, the following parameters are obtained.

$\tau = 0.3768$, $\theta = 0.095$, k = 2.3303, $\beta = 1.897$
$\alpha = 0.962, \zeta = 0.3551$

Substituting above values in (22), the fractional IMC-PID filter for single area power system is obtained as (23)

$$C(s) = \frac{s^{0.038}}{1 + 10.526\tau_c s^\lambda} 1.21(1 + 3.736 s^{-0.962} + 0.5305 s^{0.935}) \tag{23}$$

4 Illustrative Cases

4.1 A Two-Area Power System Case

Based on overshoot, undershoot and settling time, best λ and τ_c are obtained and given as

$\lambda = 0.22$ and $\tau_c = 0.02$.
The resulting FIMC-PID controller is given as (24)

$$C(s) = \frac{s^{0.038}}{1 + 0.22 s^{0.02}} 1.21(1 + 3.736 s^{-0.962} + 0.5305 s^{0.935}) \tag{24}$$

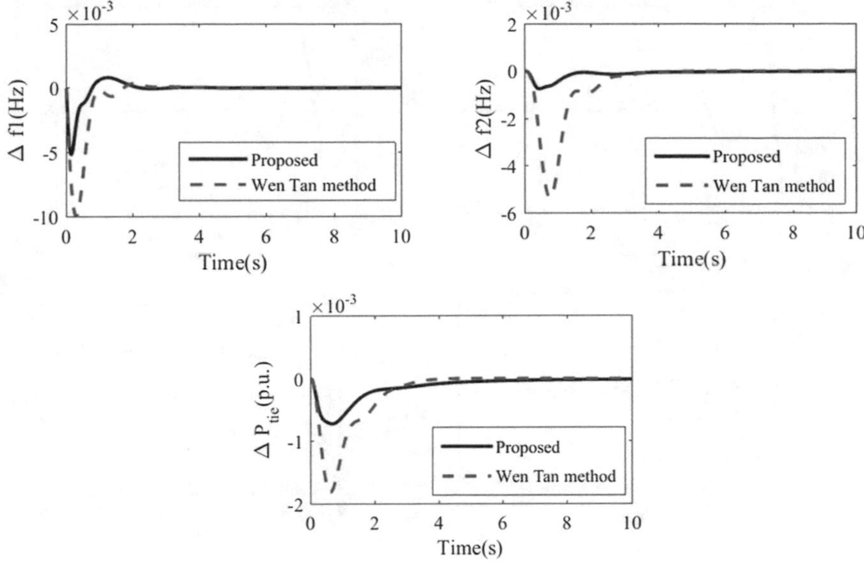

Fig. 5 Two-area power system responses when $\Delta P_d(s) = 0.01$ is applied in area1

To evaluate the performance of controller, a step load disturbance $\Delta P_d(s) = 0.01$(1% load change) is applied to a two-area non-reheated power system. The frequency deviations in area1 and area2 and deviation in tie line of the proposed method are compared with Wen Tan method [11], which is shown in Fig. 5. It is clear from figures that the deviations of the power system for the proposed controller due to load disturbance are diminished compared to Wen Tan method.

4.2 A Three-Area Power System Case

Based on overshoot, undershoot and settling time, best λ and τ_c are obtained and given as

$\lambda = 0.6$ and $\tau_c = 0.05$.

The resulting FIMC-PID controller is given as (25)

$$C(s) = \frac{s^{0.038}}{1 + 0.6s^{0.05}} 1.21(1 + 3.736s^{-0.962} + 0.5305s^{0.935}) \quad (25)$$

To evaluate the performance of controller, a step load disturbance $\Delta P_d(s) = 0.3$ (30% load change) is applied to a three-area non-reheated power system. The frequency deviations in area1, area2 and area3 and deviation in tie line of the

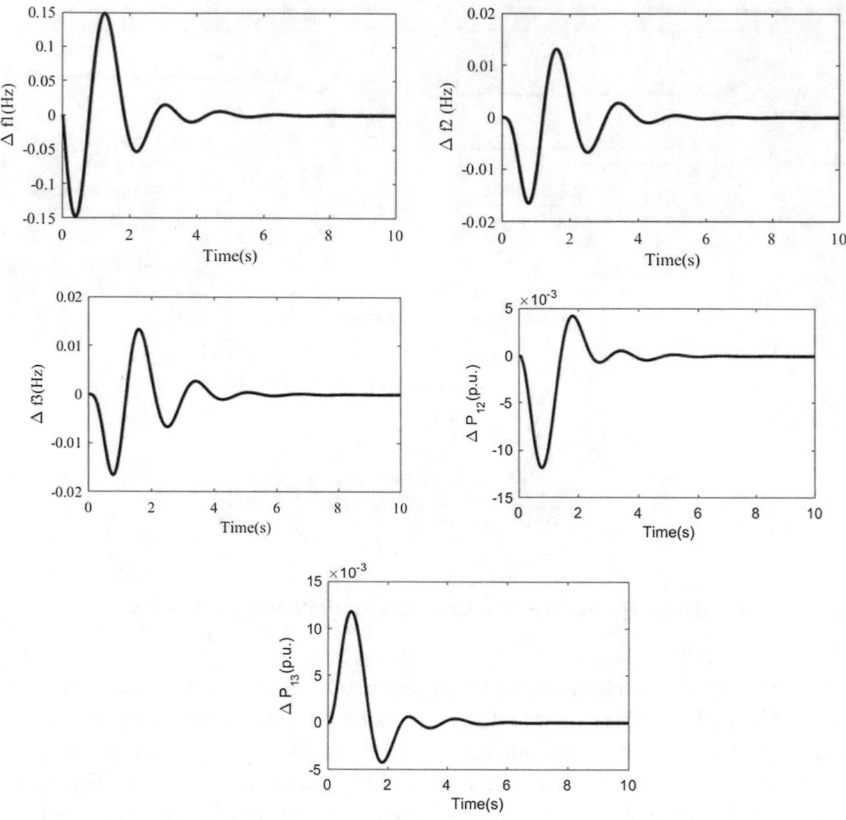

Fig. 6 Three-area power system responses when $\Delta P_d(s) = 0.3$ is applied in area1

proposed method are shown in Fig. 6. It is clear from figures that the deviations of the power system for the proposed controller due to load disturbance are diminished.

5 Conclusion

In the present scenario, a decentralized LFC technique is needed to mitigate the effect of load disturbances, system parameter variations and system modelling error on power system. In this work, a model reduction method is used to design an IMC-PID controller for power system. The controller consists of fractional filter and fractional order PID. The tuning parameters, time constant τ_c and non-integer λ are chosen based on settling time and overshoot/undershoot. The MATLAB simulation results showed that the proposed controller can minimize the load fluctuation and modelling errors effect on frequency and tie line power flow.

References

1. Kundur, P.: Power System Stability and Control. TMH 8th reprint, New Delhi (2009)
2. Elgerd, O.I.: Electric Energy Systems Theory. An Introduction. TMH, New Delhi (1983)
3. Manabe, S.: Early development of fractional order control. In: ASME 2003 International Design Engineering Technical Conferences and Computers Information in Engineering Conference. American Society of Mechanical Engineers, pp. 609–616 (2003)
4. Yu, W., Chen, Y., Luo, Y., Pi, Y.: Frequency domain modelling and control of fractional-order system for permanent magnet synchronous motor velocity servo system. IET Control Theory Appl. **10**(2), 136–143 (2016)
5. Pan, I., Das, S.: Fractional Order AGC for distributed energy resources using robust optimization. IEEE Trans. Smart Grid **7**(5), 2175–2186 (2016)
6. Vrdoljak, K., Peri, N., Petrovi, I.: Sliding mode based load-frequency control in power systems. Electr. Power Syst. Res. **80**(5), 514–527 (2010)
7. Mi, Y., Fu, Y., Wang, C., Wang, P.: Decentralized sliding mode load frequency control for multi-area power systems. IEEE Trans. Power Syst. **28**(4), 4301–4309 (2013)
8. Chang, C., Fu, W.: Area load frequency control using fuzzy gain scheduling of pi controllers. Electr. Power Syst. Res. **42**(2), 145–152 (1997)
9. Pan, C.T., Liaw, C.M.: An adaptive controller for power system load-frequency control. IEEE Trans. Power Syst. **4**(1), 122–128 (1989)
10. Tan, W., Xu, Z.: Robust analysis and design of load frequency controller for power systems. Electr. Power Syst. Res. **79**(5), 846–853 (2009)
11. Tan, Wen: Unified tuning of PID load frequency controller for power systems via IMC. IEEE Trans. Power Syst. **25**(1), 341–350 (2010)
12. Tan, W.: Decentralized load frequency controller analysis and tuning for multi-area power systems. Energy Convers. Manag. 2015–2023 (2011)
13. Saxena, S., Hote, Y.V.: Load frequency control in power systems via internal model control scheme and model-order reduction. IEEE Trans. Power Syst. **28**(3), 2749–2757 (2013)
14. Hutton, B.F.M.: Routh approximations for reducing order of linear, time-invariant systems. IEEE Trans. Autom. Control **20**, 329–337 (1975)
15. Saaki, I., Babu, C.K.R.P.C., Prasad, D.S.: Integral Square Error Minimization Technique for Linear Multi Input and Multi Output Systems, pp. 1–5 (2011)
16. Holland, J.: Adaptation in Natural and Artificial Systems: an Introductory Analysis with Applications to Biology, Control, and Artificial Intelligence. University of Michigan Press (1975)
17. Ismail, M., Moghavvemi, M., Mahlia, T.: Genetic algorithm based optimization on modeling and design of hybrid renewable energy systems. Energy Convers. Manag. **85**, 120–130 (2014)
18. Morari, M., Zafiriou, E.: Robust Process Control. Prentice Hall (1989)
19. Nasir, A.W., Singh, A.K.: IMC based fractional order controller for non-minimum phase system. In: 2015 Annual IEEE India Conference (INDICON), Dec 2015, pp. 1–6 (2015)
20. Bettayeb, M., Mansouri, R.: Fractional IMC-PID-filter controllers design for non integer order systems. J. Process Control **24**(4), 261–271 (2014)
21. Kasireddy, I., Nasir A.W., Signh A.K.: IMC based controller design for automatic generation control of multi area power system via simplified decoupling. Int. J. Control. Autom. Syst. **16** (2018)

RAID-6 Code Variants for Recovery of a Failed Disk

M. P. Ramkumar, N. Balaji, G. S. R. Emil Selvan and R. Jeya Rohini

Abstract With the increasing demand for capacity, speed, and reliability in large-scale storage systems, a mechanism should exist to ensure the data availability. Though there exist kinds of erasure code implementations in RAID-6, maximum distance separable (MDS) codes provide simple yet better way of data protection and recovery mechanism in the course of a disk failure. RAID-6 is preferred due to the capability of fault tolerance against simultaneous two disk failures. In addition to the provisioning of fault tolerance against disk failures, it is also necessary to concentrate on recovery to avoid data unavailability. Even though the RAID-6 supports two disk failures, when a number of disk failures are more than its parity, data will be lost. Hence, it is important to address the single disk failure and recover the failed disk at the earliest. The early recovery of a failed disk (i.e., recovery time) depends on the number of overlapping blocks. The hybrid code achieves the optimal recovery time than the other categories by consuming 22% of reused blocks.

Keywords Data availability · RAID-6 · MDS codes · Fault tolerance
Disk failure · Hybrid code · Overlapping blocks · Disk recovery

---

M. P. Ramkumar (✉) · G. S. R. Emil Selvan
Department of Computer Science and Engineering, Thiagarajar College of Engineering, Madurai 625015, India
e-mail: ramkumarmp1412@gmail.com

G. S. R. Emil Selvan
e-mail: emil@aol.in

N. Balaji
K.L.N. College of Information Technology, Pottapalayam, India
e-mail: balajin@klncit.edu.in

R. Jeya Rohini
K.L.N. College of Engineering, Pottapalayam 630612, India
e-mail: rjreee2008@gmail.com

J. Nayak et al. (eds.), *Soft Computing in Data Analytics*,
Advances in Intelligent Systems and Computing 758,
https://doi.org/10.1007/978-981-13-0514-6_24

1 Introduction

In the present scenario, there is an increase of data stored in storage systems which lead to the need for reliable and highly available storage systems. Redundant Array of Independent Disks (RAID) is a key technology used in critical servers, data centers, and large-scale storage systems. The RAID ensures the amalgamation of numerous hard disks into a consolidated logical resource that focuses on the data availability, reliability, and high performance at an acceptable cost. In recent times, researchers on parallel storage have focused on RAID-6 due to the increase in the probability of disk failure [1] in massive storage systems. The recent investigations on storage arrays have publicized the need for storage system's reliability. In a computationally intensive data center with thousands of hard drives, a number of disk failures are very common. Hence, the ability of the storage system to tolerate the disk failure as well as early recovery from failure is the need of the hour.

1.1 MDS Codes

Erasure coding is a data protection technique [2], where the blocks are encoded for parity generation. The objective of erasure coding is to reconstruct the corrupted data from the available disks and hence ensure the data availability [3].

Erasure codes that meet information theory's Singleton bound are called MDS codes. It involves only simple Exclusive OR (XOR) computation, which is more efficient than Reed–Solomon concerning computational complexity [4].

2 Literature Review

RAID-6 codes are classified as horizontal, vertical, and hybrid codes. The classification of codes is based on the position of parity blocks in the code array structures.

In the horizontal category, the latter two disks are spared to hold parity and the remaining disks are committed to storing data [5]. The horizontal codes are such as Conventional code, Reed–Solomon coding, EVENODD, Liberation Codes, Star,

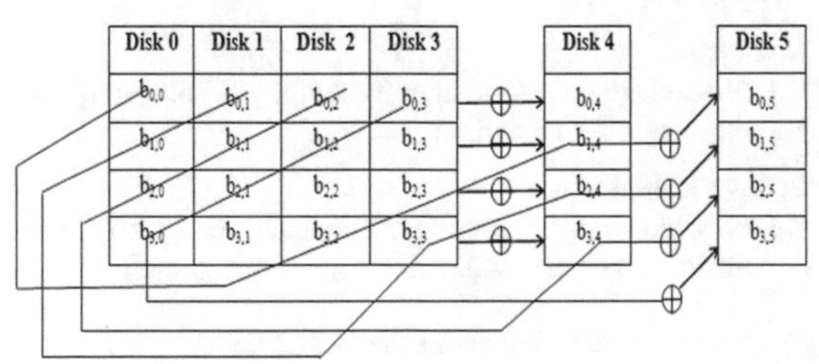

Fig. 1 RDP coding array structure

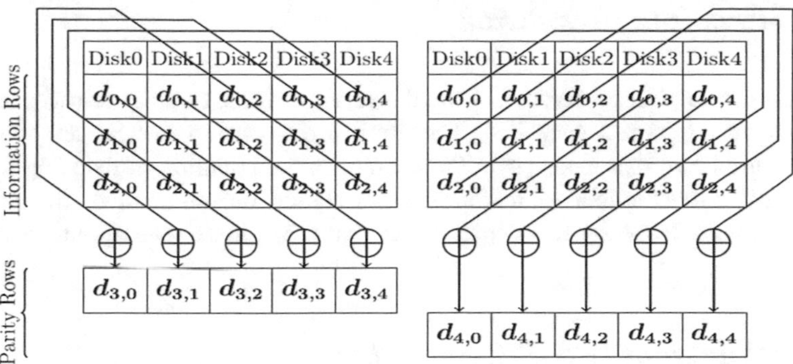

Fig. 2 X-Code array structure (Source: Xu S. et al. IEEE transactions on computers 2014)

and RDP. Among the codes, the RDP is taken into consideration for the performance evaluation. The coding structure of RDP contains a (p − 1) rows by (p + 1) columns, where p is a prime number greater than two. The row and diagonal parities are obtained [5] by Exclusive OR (XOR) summation of all the symbols in that stripe and the diagonal, respectively, as shown in Fig. 1.

Certain data blocks are not included in the diagonal parity calculation, and the corresponding diagonal is known as missing diagonal. From Fig. 1, the blocks along the missing diagonal ($b_{1,\ 3}$, $b_{2,\ 2}$, $b_{3,\ 1}$) are reconstructed using row parity.

In vertical codes, the last stripes are dedicated to storing the parity blocks that spread across all the disks. The codes such as X-Code and P-Code [6] are considered as vertical. The X-Code is considered for the performance evaluation under the vertical codes. X-Code for a 'p' disk array is illustrated in a p × p matrix. Data blocks occupy the first p – 2 rows and the parity blocks reside in last two rows [7, 8], which are computed along the diagonals, as shown in Fig. 2. The coding structure is named as X-Code, as its geometrical representation [9] is similar to the alphabet "X."

Due to the increasing storage demands, the capacity of the drives increases day by day, which results in disk failure more frequently [10]. The studies reveal that 98.08% of successful recoveries were from single disk failure [11]. Though parity provides data protection against disk failure, in the course of failed disks are more in number than the parity disks, data will be eventually lost. Hence, to guarantee the reliability of the storage system, a mechanism is required to recover and rebuild a failed disk at the earliest [12].

3 Challenges During the Recovery of a Failed Disk

The problems described below are the major challenges faced during the recovery of a failed disk.

3.1 *Minimizing Disk Reads*

The data blocks will be accessed from the survival disks, in order to regenerate the lost data during its recovery [13]. The survival disks read overhead are controlled by the number of disk reads [14]. To accomplish the intention of reducing the disk reads, either the number of reading blocks should be minimal, or the recovery process should be designed in such a way that it has more overlapping blocks.

3.2 *Minimizing Disk Recovery Time*

During recovery, if read requests are evenly disseminated across the survival drives, there will be a major improvement in the recovery time [15, 16]. If a protruding block is read for the first time, then it is read from the disk and stored it in the memory. The access of overlapping blocks directed to memory instead of from disk for the consecutive times. Since the time taken for reading from memory is less than that of from disk [11], it reduces the disk recovery time.

4 Proposed: Recovery of a Failed Disk Using Hybrid Codes

Hybrid codes are the one that combines the advantages [8] of horizontal and vertical codes. The HDP code, H-Code, RDOR-RDP, and PDRR are the examples of hybrid codes. From the hybrid category, the PDRR [17] is considered for the evaluation.

4.1 *Parity Declustering-Based RDOR-RDP (PDRR) Approach*

The problem with the RDP code (horizontal code) is, whenever a data block is modified, the respective row and diagonal parities should be recomputed. The recomputation process leads to writing overhead to the dedicated parity disks [18]. Hence, the parity blocks should be dispersed across all the disks in the array.

Even though the X-Code (vertical code) ensures the parity distribution across the entire disks, the reconstruction of a failed disk leads to computational overhead due to diagonal and anti-diagonal parities. Moreover, the number of overlapping symbols will be less due to a different kind of parity calculation. This restriction makes X-Code as unfeasible in the storage environment.

Disk0	Disk1	Disk2	Disk3	Disk4	Disk5
$B_{0,0}$	$B_{0,1}$	$B_{0,2}$	$B_{0,3}$	P	Q
$B_{1,0}$	$B_{1,1}$	$B_{1,2}$	$B_{1,3}$	Q	P
$B_{2,0}$	$B_{2,1}$	$B_{2,2}$	P	$B_{2,4}$	Q
$B_{3,0}$	$B_{3,1}$	$B_{3,2}$	Q	$B_{3,4}$	P

Fig. 3 PDRR implementation of stripe set 1

The PDRR is a hybrid approach that distributes the parity across the disks and stripes. Hence, the data modification in a disk will not be an overhead for specific disks. PDRR is a $(p - 1) \times (p + 1)$ array that addresses the single disk failure by using both the row and diagonal parities. The kind of parity combination leads to an increase in overlapping blocks [19]. The array layout of the stripe set 1 is shown in Fig. 3.

Disks are labeled from disk 0 through 5. The data blocks are mentioned as $B_{i,\ j}$ (block in the *i*th row of the *j*th disk) and the row and diagonal parities blocks are mentioned as 'P' or 'Q,' respectively.

5 Implementation

The RAID-6 environment and its code variants are implemented in Disk system Simulator Version 4.0 (DiskSim) [20]. It contains separate modules for disk specification, drivers, controllers, and buses. Table 1 indexing the disk array simulation parameters used in the implementation.

The metrics such as disk read optimality and recovery time are evaluated using the DiskSim for the RDP, X-Code, and PDRR codes, and the results are compared.

Table 1 Simulation parameters

Parameter	Value
Disk capacity	146.8 GB
Disk rotational speed	15,000 RPM
Average rotational latency	2 ms
Internal data transfer rate	960–1607 mb/s
Number of data surfaces	4
Number of cylinders	72,170
Number of disks	8, 12, 14, 18, 20
Block size	64 KB

5.1 Disk Read Optimality

Disk read optimality refers to the reduced number of blocks accessed for recovering the failed disk [14, 21]. The RDP, X-Code, and PDRR approach attempts to reduce the reads by increasing overlapping blocks as much as possible.

From Table 2, it is evident that the PDRR code contains more overlapping blocks for the disks 0, 2, and 5 than the remaining codes, and for the disk 1, the quantity of overlapping blocks is same as RDP. Since it is considered only for the stripe set 1 alone, the PDRR is not outperformed for the disks 3 and 4 when compared with the horizontal codes. Moreover, the PDRR is a balanced one; consideration of all the stripe sets together may yield better results for the disks 3 and 4 also. As an overall, the PDRR (hybrid code) achieves the highest average of overlapping symbols than RDP (horizontal), and X-Code (vertical) and the corresponding graph are shown in Fig. 4.

5.2 Recovery Time

Recovery time includes the time to regenerate the failed blocks and then writing it into the spare disk. Recovery time is calculated for the RAID-6 code categories by

Table 2 Amount of overlapping blocks for the disks 0–5 for three code categories

RAID-6 code	Amount of overlapping blocks (%)						
	Disk 0	Disk 1	Disk 2	Disk 3	Disk 4	Disk 5	Average
RDP	18.75	25	0	25	25	0	15.63
X-Code	13.33	13.33	13.33	13.33	13.33	13.33	13.33
PDRR	31.25	25	18.75	20	20	18.75	22.29

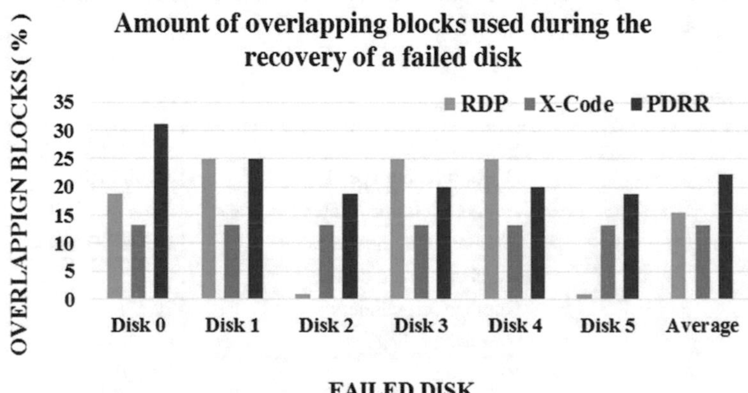

Fig. 4 Amount of overlapping blocks utilized by RDP, X-Code, and PDRR codes

Table 3 Recovery time of RDP, X-Code, and PDRR codes

Failed disk	Recovery approach	Number of disks				
		P + 1 = 8	P + 1 = 12	P + 1 = 14	P + 1 = 18	P + 1 = 20
		Recovery time (s)				
Disk 1	RDP	2484	5611	6663	8738	10248
	X-Code	2660	6008	7135	9357	10973
	PDRR	2102	4748	5638	7394	8671

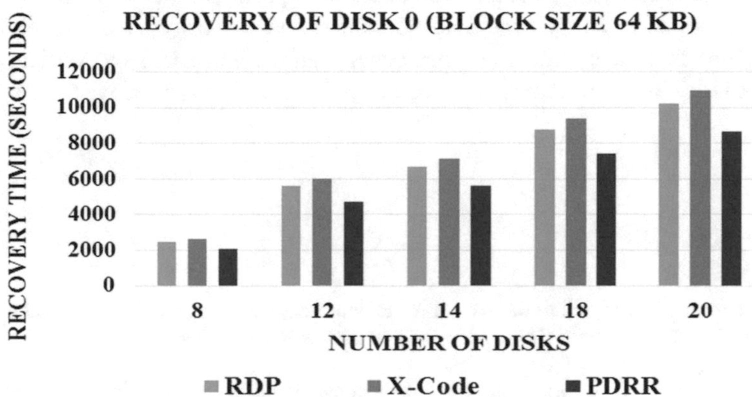

Fig. 5 Recovery time of RDP, X-Code, and PDRR codes

varying the number of disks, and the results are presented in Table 3. The corresponding graph is shown in Fig. 5.

The findings of the results are given below:

1. The amount of overlapping blocks used for recovering the failed disk is augmented for the PDRR (hybrid code) and compared with the RDP (horizontal) and X-Code (vertical code). More reused blocks designate that the recovery process can still be improved in real-time storage array scenario.
2. The hybrid code reduces the recovery time to get back from a single disk failure and is compared with the horizontal and vertical codes. When the number of disks changes from 8 to 20, the hybrid code always outdoes than the horizontal and vertical codes. This implies hybrid code scales well as the number of disks increases.

6 Summary

The RAID-6 codes discussed above emphasize the recovery of a failed disk in optimal time. Time to recover from a disk failure plays a crucial role in determining storage system's performance. The RDP code does not possess the parity distribution property [21]. The storage system architects strive to find an enhanced recovery scheme aimed at recovering the failed disk in optimal recovery time. The PDRR is a hybrid category of RAID-6 code that addresses the single disk failure. Since the hybrid codes use both the row and diagonal parities for the reconstruction of a failed disk, it achieves disk read optimality by retaining more overlapping blocks in the memory. Rather than interpreting 100% of blocks from the surviving disks, the hybrid code (PDRR) makes use of 22.29% overlapping blocks as an average, whereas the horizontal code (RDP) uses 15.6% and vertical code (X-Code) has just 13.33%. Moreover, the hybrid code is a balanced onc; it achieves optimal recovery time than horizontal and vertical codes. The simulated results show that the hybrid code does better than the horizontal and vertical codes related to disk read optimality and recovery time, during the recovery of a failed disk.

References

1. Pinheiro, E., Weber, W.D., Barroso, L.A.: Failure trends in a large disk drive population. In: FAST-2007: 5th Usenix Conference on File and Storage Technologies, vol. 7, pp. 17–23 (2007)
2. Guan, S., Kan, H., Wen, J., Xia, S.: A new construction of exact-repair MSR codes using linearly dependent vectors. IEEE Commun. Lett. (2017)
3. Zhijie, Huang, Jiang, Hong, Zhou, Ke, Zhao, Yuhong, Wang, Chong: Lowest density MDS array codes of distance 3. IEEE Commun. Lett. **19**, 1670–1673 (2015)
4. Zhijie, Huang, Jiang, Hong, Zhou, Ke, Wang, Chong, Zhao, Yuhong: XI-Code: a family of practical lowest density MDS array codes of distance 4. IEEE Trans. Commun. **64**, 2707–2718 (2016)
5. Xie, P., Huang, J., Cao, Q., Xie, C.: Balanced P-Code: a raid-6 code to support highly balanced I/O's for disk arrays. In: IEEE International Conference on Networking Architecture and Storage, pp.133–137 (2014)
6. Xu, S., Li, R., Lee, P.P.C., Zhu, Y., Xiang, L., Xu, Y., Lui, J.C.S.: Single disk failure recovery for x-code-based parallel storage systems. IEEE Trans. Comput. **63**, 995–1007 (2014)
7. Guangyan, Zhang, Guiyong, Wu, Yu, Lu, Jie, Wu, Zheng, Weimin: XSCALE: Online X-Code RAID-6 scaling using lightweight data reorganization. IEEE Trans. Parallel Distrib. Syst. **27**, 3687–3700 (2016)
8. Xu, L., Bruck, J.: X-code: MDS array codes with optimal encoding. IEEE Trans. Inf. Theory **45**, 272–276 (1999)
9. Wang, Y., Yin, X., Wang, X.: MDR codes: a new class of RAID-6 codes with optimal rebuilding and encoding. IEEE J. Sel. Areas Commun. **32**, 1008–1018 (2014)
10. Jin, C., Feng, D., Jiang, H., Tian, L.: A comprehensive study on raid-6 codes: Horizontal vs. vertical. In: IEEE International Conference on Networking, Architecture and Storage, pp. 102–111 (2011)
11. Xiang, L., Xu, Y., Lui, J., Chang, Q., Pan, Y., Li, R.: A hybrid approach to failed disk recovery using RAID-6 codes: algorithms and performance evaluation. ACM Trans. Storage. **7**, 11 (2011)
12. Schroeder, B., Gibson, GA.: A disk failures in the real world: what does an MTTF of 1,000,000 hours mean to you?, In Proceedings of the 5th USENIX Conference on File and Storage Technologies (FAST'07), USENIX Association, Berkeley, USA, vol 7, pp. 1–16 (2007)

13. Xie, P., Huang, J., Cao, Q., Qin, X., Xie, C.: A new non-MDS RAID-6 code to support fast reconstruction and balanced I/O's. Comput. J. **58**, 1811–1825 (2015)
14. Fu, Y., Shu, J., Luo, X., Shen, Z., Hu, Q.: Short code: an efficient raid-6 mds code for optimizing degraded reads and partial stripe writes. IEEE Trans. Comput. **66**, 127–137 (2017)
15. Wu, C., Wan, S., He, X., Cao, Q., Xie, C.: H-Code: A hybrid MDS array code to optimize partial stripe writes in RAID-6. In: IEEE International Conference on Parallel& Distributed Processing Symposium, pp. 782–793 (2011)
16. Wu, C., He, X., Wu, G., Wan, S., Liu, X., Cao, Q., Xie, C.: HDP code: a horizontal-diagonal parity code to optimize I/O load balancing in RAID-6. In: IEEE/IFIP 41st International Conference on Dependable Systems & Networks (DSN), pp. 209–220 (2011)
17. Ramkumar, M.P., Narayanan, B., Emil Selvan, G.S.R., Ragapriya, M.: Single disk recovery and load balancing using parity declustering. J. Comput. Theor. Nanosci. **14**, 545–550 (2017)
18. Wan, S.: Code-m: a non-MDS erasure code scheme to support fast recovery from up to two-disk failures in storage systems. in: IEEE/IFIP International Conference on Dependable Systems & Networks, pp. 51–60 (2010)
19. Jin, C., Jiang, H., Feng, D., Tian, L.: P-Code: a new RAID-6 code with optimal Properties. In: Proceedings of the 23rd International Conference on Supercomputing, pp. 360–369 (2009)
20. Bucy, J., Schindler, J., Schlosser, S., Ganger, G.: The disk sim simulation environment (v4.0). http://www.pdl.cmu.edu/disksim/ (2008)
21. Fu, Y., Shu, J.: D-code: an efficient raid-6 code to optimize i/o loads and read performance. In: IEEE International Symposium on Parallel and Distributed Processing, pp. 603–612 (2015)

Smart Agriculture System in India Using Internet of Things

Rama Krushna Das, Manisha Panda and Sweta Shree Dash

Abstract Agriculture plays a major role in the nation's economic contribution. The source of income and livelihood of around 70% of Indian population depends on agriculture. Harvest and post-harvest losses due to many reasons lead to heavy financial burden to farmers as well as Government. Though many agriculture security measures, like crop insurance, are provided to farmers by Government, still the farmers' suicide is a common phenomenon. To minimize the crop loss during harvest or post-harvest, this paper proposes the smart agriculture system with the help of Internet of Things (IoT). With the use of several sensors and Raspberry Pi, different models are proposed for supervision of soil moisture and pests, building intelligent seeds' corporation and efficient food corporation of India.

Keywords IoT · Humidity sensor · Moisture sensor · Odour sensor/E-nose Motion sensor · Pest sensor · FCI · Seed corporation · Agriculture Microcontroller · Raspberry Pi 3

1 Introduction

In India, agriculture is the main source of livelihood for the majority of the population, as it provides us with the required food grains and other needed raw materials. The agriculture system also includes irrigation, which is the main issue faced by the developing countries like India. Due to lack of proper irrigation

R. K. Das (✉)
National Informatics Centre, Berhampur 760004, India
e-mail: ramdash@yahoo.com

M. Panda
Berhampur University, Berhampur 760001, India
e-mail: manishapanda2013sai@gmail.com

S. S. Dash
Institute of Technical Education and Research, Bhubaneswar 751030, India
e-mail: sweta.soa@gmail.com

J. Nayak et al. (eds.), *Soft Computing in Data Analytics*,
Advances in Intelligent Systems and Computing 758,
https://doi.org/10.1007/978-981-13-0514-6_25

facility, farmers totally depend on a good monsoon. India has an isotropic climatic condition, but the climate change due to global warming results in erratic monsoon, and hence, the use of agricultural resources remains halfway in its implementation [1]. The agriculture system in India is facing a huge loss of nearly about 15% due to heavy industrialization and other service sectors. The farmers do not get proper price for their products due to lack of good storage and marketing system in their locality. The use of improper pesticides without experts' advice gives rise to crop and financial loss. Generally, the farmers go for traditional crops without soil testing and selecting the best crop suitable for their soil. The seeds are not properly stored by farmers or the Government agencies, which leads to bad quality of seeds for future use. The post-harvesting food grain storage system by the farmers and Government agencies is not up to the mark, resulting in heavy loss of food grains. In this paper, the authors propose smart agriculture system, which is e-build with the help of the recent emerging science and technology. In this paper, we have integrated IoT to create several smart agricultural models.

IoT is a system which allows several interactions of sensors for the in-depth knowledge that too without any manual labour. The use of IoT is to automate things and to analyse things for better efficiency by integrating it with a system. They enhance the accuracy and reduce manual labour. IoT makes use of the current and developing technology for sensing, networking, and robotics. So with help of IoT and other parameters such as smart GPS, cloud storage, Web integration and other devices, the agriculture system can be converted into a smart one. It can also prove that the use of IoT in place of manual labour can improve the efficiency and harvest cost [2]. Hence, it is high time for the implementation of modern science and technology in the agriculture sector for increasing the profit and converting it to smart agriculture.

2 Scope and Contribution

This research area mostly focuses on "Farmers' suicides in India": the ongoing hot topic for many years which earned a bad name for our agriculture system. Certain application of IoT, smart GPS, Web integration, etc., is integrated here in this paper with agriculture for better yield and less loss [3]. Some applications are briefly explained in this paper to convert a normal agriculture system to a smart one with stepwise procedures and models. First of all, several sensors such as moisture sensor, humidity sensor, soil type sensor, pest sensor, etc., are in use. With help of these sensors, different values of the parameters are measured. The soil samples are analysed by the sensors to identify the soil type and different harmful pests in the sample. After that, the farmer is suggested for applying pesticides and to take several steps in order to overcome the difficulty. For moisture control, moisture sensors and humidity sensors are connected deep in the soil to measure its percentage and the sensor is connected to an alarm so that it can beep or glow after the limit goes below the threshold limit. Secondly, the farmers and Government face frequent issues in the seeds godown. Mostly, the seeds in the seed godown are wasted due to the increase in humidity. This leaves the seeds useless, and farmers face a huge loss as they strive hard for getting germination from these seeds. So, here is a proposal to connect a humidity sensor in the seeds godown. If the humidity in the godown increases, then the sensor senses it and triggers an alarm to the persons in charge of the godown to do the needful. Alternatively, it can be done without any manual labour, and the sensor can automatically switch on the dehumidifier and switch off it after the humidity sensor senses that the humidity is normal and will not affect the seeds any more. Thirdly, the Food Corporation of India (FCI) faces several losses due to the wastage of food grain in warehouses due to mice and moisture disorder. So to reduce it, we can implement IoT and connect several sensors such as motion sensors and moisture sensors to detect and overcome the problem.

3 Literature Review

The present-day challenges in agriculture, such as decreasing soil quality, increasing loss of food grains in FCI and eradication of endangered species and their seeds, want an urgent attention and prompts for smart agriculture with latest state-of-the-art technology. In [4], a smart irrigation system is developed using soil temperature and moisture sensor. The proposed system mostly conserves water by enabling irrigation of the field during the required time and also minimizes the human intervention. The main advantage of the system is that it can be applied for agricultural practices where the key indicators are low population and semi-arid climate. This can be very useful for start-up farmers, who had just initiated their work and want their work to be done at a low cost. In reference [5], the authors explained the implementation of the automated drip irrigation system. This automated system was found to be very low cost and practicable for optimizing water resources in the agriculture sector. This automated system can be used in places with low rainfall and minimum irrigated water supply to improve the sustainability in the cultivation process. In reference [6], the authors presented a hierarchical soil measuring wireless sensor network. This includes several sensor nodes, relay nodes and base nodes. The operating frequency of the system is found to be 433 MHz range, which is possible for the transmission of data from beneath the soil. The affordable cost and negligible consumption of power make it the most suitable for large-scale implementation. In reference [7], the transfer from a conservative irrigation system to an electronically controllable system for individual control of irrigation sprinklers was discussed. Here, the use of GPS was made for the continuous monitoring of navigations. The authors also presented details about the wireless communications between the sensors connected in-field and programmable logic controller for its control. Another implementation of Bluetooth wireless technology was used in this paper which offers a plug-and-play communication module. This Bluetooth wireless technology also has an advantage of saving a good amount of time and cost by using easily available sensors and controllers using serial communication ports.

4 Internet of Things

The Internet of Things (IoT) is an embedded technology, where physical devices (also referred to as “connected devices” and “smart devices”) are interconnected through software, sensors, actuators and network connectivity for collecting and exchanging information among each other [8]. The reality and advantage of IoT are its effectively endless opportunities it offers. Most of the applications of IoT are out of our thinking and understanding. The reason that IoT is the hot topic nowadays is that it certainly opens the door to a lot of opportunities but also to many challenges. So, agriculture is one among these opportunities. Agriculture needs modernization as the farmers are being neglected the most. Their whole expenses and livelihood depend solely on the agricultural produce. Use of IoT can make the agriculture system more efficient and profitable [9].

5 Architectural Models

The proposed models of agriculture to make it smart, affordable and profitable are described below:

5.1 Supervision of Soil Moisture and Pests

The key foundation to a food system is a healthy soil. A healthy soil is responsible indirectly for the production of healthy crops which gives better nourishment to the peoples who consume the same. So in order to maintain a healthy soil, a farmer needs to put its utmost effort before the farming benign [10]. The sources that provide nutrients to plants are the organic matter and minerals. Organic matter mostly consists of plant or animal waste that is returning to the soil after it goes under the process of decomposition. In addition to providing nutrients and habitat to organisms living in the soil, the organic matter also binds soil particles and improves the water retaining capacity of soil. Thus, the most important challenge is to recognize soil management practices that will increase the soil organic matter formation and moisture retention [11]. This will indirectly increase the yield capability and hence enhance the profit for farmers in a short duration of time. So, the moisture and pests in the soil should be controlled with the help of sensors as it cannot be controlled by direct human intervention [12]. We propose use of IoT principles for detecting the friendly and harmful pests for the soil so that we can

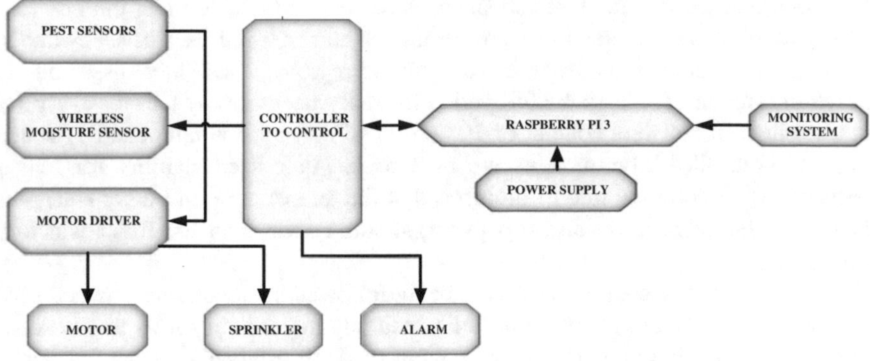

Fig. 1 Model for supervision of soil moisture and pests

suggest the farmers for applying pesticides and needed majors measures to be taken. Similarly, soil moisture control also plays an important role in the quality of soil for crop management which can be eliminated using IoT.

As shown in Fig. 1, a Raspberry Pi 3, microcontroller, pest sensor, wireless moisture sensor and a monitoring system are connected. This Raspberry Pi 3 connects the different sensors. After sensing, the Raspberry Pi 3 sends a signal to the monitoring system to send an alarm message to the farmers. If the wireless motion sensor and temperature sensor sense the moisture and temperature below the threshold value, then it starts the motor for spraying water which is called as sprinkler irrigation. Sprinkler irrigation is a method of spraying water into air which is allowed to fall on ground just like natural rainfall. Then again the sensor senses the temperature and moisture; if it is above the threshold, then the Raspberry Pi 3 sends signal to microcontroller to stop the motor. If some malfunctioning happens, then the alarm signal bccps continuously until it is solved. The wireless pest sensors are mounted under the soil (such as low-power image sensors, acoustic sensors, etc.) and used to recognize the harmful pest for the crop. Then, the data is sent through Raspberry Pi 3 to the monitoring system for further transmission to the department of agriculture from where they suggest the remedial measures.

5.2 *Intelligent Seeds Corporation*

Proper seed storage is encouraged for the conservation of its potency and liveliness. Seeds have a lifespan starting from a few years to hundreds of years. This lifespan of seeds is dependent on the storage conditions and the species. Generally, seeds stored in cool and dry conditions live longer whereas the seeds stored under wet and warm conditions have a shorter lifespan. Mostly in several parts of the world, the agricultural seeds are stored in open places called bins. In bins, the environmental conditions are the ambient conditions which result in losses of seeds due to low lifespan, degraded quality in hot and humid regions. It also face losses due to insects and rodents [13]. As the life period for seeds decreases by half for every 1% increase in seed moisture content or 10 °F (~5 °C) increase in temperature, sealed bins and controlled environments are used to maintain seed viability for longer periods [14]. A common rule of thumb is that the temperature (in Fahrenheit) plus the relative humidity in the air (in per cent) should be less than 100 for satisfactory seed storage.

Most of the above conditions are not being followed in Government-owned seed storage warehouses and godowns in India, resulting in producing low-quality seeds with less germination rate. These low-quality seeds distributed to farmers result in disappointment and heavy financial loss to farmers. Even some farmers lose one crop year due to these faulty supplied seeds. In the seeds storage godowns owned by seeds corporations, to control the humidity and rodents IoT techniques are proposed. As shown in Fig. 2, a Raspberry Pi 3, microcontroller, motion sensor, humidity sensor, alarm, camera and a monitoring system are connected. After receiving the signals, the Raspberry Pi 3 sends a signal to the monitoring system for sending an alarm message to the person in charge. If the motion sensor senses any motion, then it sends a signal to start the rodent repellent which kills if any rats or mouse are moving. Then after a given fixed time, the microcontroller triggers off the rodent repellent upon the instruction of Raspberry Pi 3. If someone makes theft, then also he/she can be trapped as the camera is also connected to the Raspberry Pi 3 to record once the motion sensor is activated. Similarly, the dehumidifier is turned on through the microcontroller via Raspberry Pi 3 once the humidity sensor senses humidity above a threshold value. Once the humidity becomes normal, the dehumidifier triggers off. After each step by Raspberry Pi 3, the user gets a message.

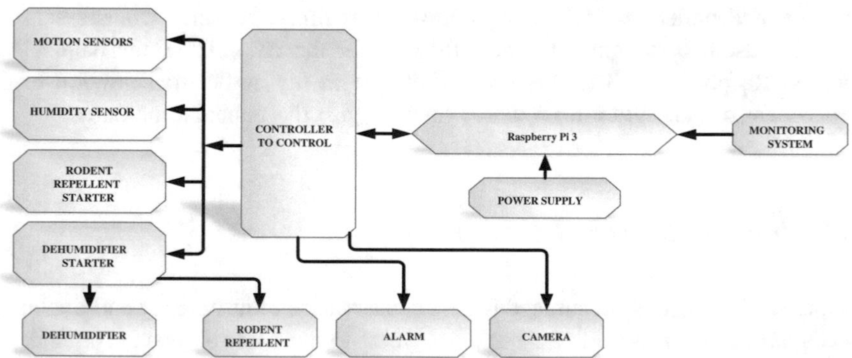

Fig. 2 Model for intelligent seed corporation

5.3 *Efficient Food Corporation of India*

For smooth supply and management of different food grains in India, FCI plays a pivotal role. Food grains procured from farmers through different Government channels are finally distributed through FCI to different warehouses and godowns, located in different geographical locations, as per their requirement. FCI has huge storage depots for buffer stock and to distribute the same through Public Distribution System (PDS) in different food security programmes to the common citizen.

During the long storage of food grains in FCI, there is heavy loss of food grains due to rodents, moisture, insects and old stock. So in order to reduce the said loss by FCI, IoT with several sensors is proposed to be installed in different stockyards, depots and warehouses. Motion sensor can be used to detect the rodents, humidity sensor to maintain humidity and the odour sensor/E-nose to detect if the old stock food grain emits bad odour or rotten. As shown in Fig. 3, a Raspberry Pi 3, microcontroller, humidity sensor, motion sensor, odour sensor/E-nose and a monitoring system are connected. This Raspberry Pi 3 connects to different sensors. After sensing, the Raspberry Pi 3 sends a signal to the monitoring system to send an alarm message to the person in charge. The dehumidifier is turned on through the microcontroller via Raspberry Pi 3 once the humidity sensor senses humidity above threshold value. If the motion sensor senses any motion, then it sends a signal to start the rodent repellent which drives out if any rats or mouse is moving. If someone makes theft, then also he/she can be trapped as the camera is also connected to the Raspberry Pi 3 to record once the motion sensor is activated. Then after a given fixed time, the microcontroller triggers off the rodent repellent upon the instruction of Raspberry Pi 3. The odour sensor is calibrated to sense the rotten smell or old stock smell, sends a message to the persons in charge and keeps the alarm on unless and until the cause of the smell is identified. This forces the FCI staff to remove the old stock or rotten food grains.

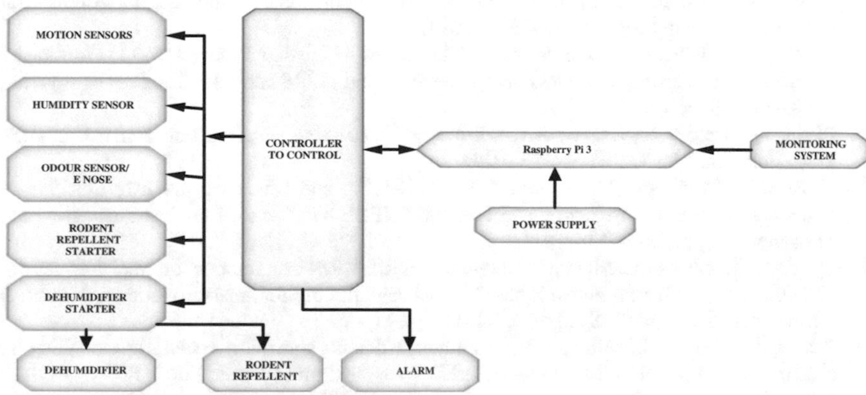

Fig. 3 Model for efficient Food Corporation of India

6 Conclusion

In India, the major problem in agriculture is the shortage of agriculture labourers and lack of interest of farmers due to less profit. Some states started using latest agricultural implements to compensate the labour problem and to increase profit margin. There is a very good scope of applying the latest state-of-the-art technology in agriculture to increase profit. The use of IoT in agriculture can improve the lives of Indian farmers. This paper explains a smart agriculture system with the above three proposed methodologies using IoT. Automation using IoT reduces manual labour and aims at controlling the financial loss. The proposed system enables a farmer to provide adequate information of applying pesticides which is mostly neglected by farmers. The wastage of seeds in godown and the vanishing of endangered seeds are reduced by using proposed model of intelligent seed storage. The loss due to moisture and rodents in FCI are minimized in a large scale by use of IoT. In the future, use of IoT can be enhanced in agriculture sector to increase the profit margin of Indian farmers and can guarantee decrease in their suicide rate.

References

1. Yoo, S., Kim, J., Kim, T., Ahn, S., Sung, J., Kim, D.: A2S: Automated agriculture system based on WSN. In: ISCE 2007. IEEE International Symposium on Consumer Electronics, Irving, TX, USA (2007)
2. Arampatzis, T., Lygeros, J., Manesis, S.: A survey of applications of wireless sensors and wireless sensor networks. In: 2005 IEEE International Symposium on Intelligent Control & 13th Mediterranean Conference on Control and Automation. Limassol, Cyprus, vol. 1–2, pp. 719–724 (2005)
3. IEEE: Wireless medium access control (MAC) and physical layer (PHY) specifications for lowrate wireless personal area networks (LR-WPANs). In: The Institute of Electrical and Electronics Engineers Inc.: New York, NY, USA, 2003
4. Nandurkar, S.R., Thool, V.R., Thool, R.C.: Design and development of precision agriculture system using wireless sensor network. In: IEEE International Conference on Automation, Control, Energy and Systems (ACES) (2014)
5. Joaquin, G., Juan Francisco V.-M, Alejandra N.-G., Miguel Angel P.-G.: Automated irrigation system using a wireless sensor network and GPRS module. IEEE Trans. Instrum. Meas. 0018–9456 (2013)
6. Wang, Q., Terzis, A., Szalay, A.: A novel soil measuring wireless sensor network. IEEE Trans. Instrum. Measur. 412–415 (2010)
7. Y. Kim, R. Evans and W. Iversen, “Remote Sensing and Control of an Irrigation System Using a Distributed Wireless Sensor Network”, IEEE Transactions on Instrumentation and Measurement, pp. 1379–1387, 2008
8. Tyagi, A., Reddy, A.A., Singh, J., Choudhari, S.R.: A low cost temperature moisture sensing unit with artificial neural network based signal conditioning for smart irrigation applications. Int. J. Smart Sens. Intell. Syst. **4**(1), 94–111 (2011)
9. Posada, J.F., Liou, J.J., Miller, R.N.: An automated data acquisition system for modelling the characteristic of a soil moisture sensor. IEEE Trans. Instrum. Measur. **40**(5), 836–841 (1991)
10. Mirabella, O., Brischetto, M.: A hybrid wired/wireless networking infrastructure for greenhouse management. IEEE Trans. Instrum. Meas. **60**(2), 398–407 (2011)

11. Vidya Devi, V., Meena Kumari, G.: Real-time automation and monitoring system for modernized agriculture. Int. J. Rev. Res. Appl. Sci. Eng. (IJRRASE) **3**(1), 7–12 (2013)
12. Kotamaki, N., Thessler, S., Koskiaho, J., Hannukkala, A.O., Huitu, H., Huttula, T., Havento, J., Jarvenpaa, M.: Wireless in-situ sensor network for agriculture and water monitoring on a river basin scale in Southern Finland evaluation from a data users perspective. Sensors **4**(9), 2862–2883. https://doi.org/10.3390/s90402862 (2009)
13. Liu, H., Meng, Z., Cui, S.: A wireless sensor network prototype for environmental monitoring in greenhouses. In: International Conference on Wireless Communications, Networking and Mobile Computing (WiCom 2007), Shangai, China; 21–25 Sept 2007
14. Baker, N.: ZigBee and bluetooth—Strengths and weaknesses for industrial applications. Comput. Control. Eng. **16**, 20–25 (2005)

A Multi-objective Flower Pollination Algorithm-Based Frequency Controller in Power System

Satya Dinesh Madasu, M. L. S. Sai Kumar and Arun Kumar Singh

Abstract This paper concentrates optimal design and performance analysis of proportional integral derivative (PID) controllers for automatic generation control (AGC) of an interconnected power system by using flower pollination algorithm. A two-area interconnected power system is considered for the design and analysis purpose. A different kind of approach is made to design a multi-objective function which contains weighted performance functions of system response. It is noticed that the performance of new objective optimized PID controller is better than the others mentioned in the literature. The objective function also includes performance response for various percentage of loads, so that obtained gain parameters are optimal for dynamic load conditions.

Keywords PID controller · Flower pollination algorithm · Performance functions · Interconnected power system

1 Introduction

The power system is a large-scale network consists of a number of generators interconnected through a transmission network. In this case, the amount of generated power is consumed at the same instant, any deviations from this will cause the imbalance of the network. Where frequency is one of the parameters of an AC network; its deviation is direct result of a load imbalance. So frequency of a power system

S. D. Madasu (✉)
Department of EEE, Gandhi Institute of Engineering & Technology,
Gunupur 765022, India
e-mail: msdinesh.nitjsr@gmail.com

M. L. S. Sai Kumar · A. K. Singh
Department of EEE, National Institute of Technology, Jamshedpur, Jamshedpur 831014, India
e-mail: saikumar.morla@gmail.com

A. K. Singh
e-mail: aksingh.ee@nitjsr.ac.in

J. Nayak et al. (eds.), *Soft Computing in Data Analytics*,
Advances in Intelligent Systems and Computing 758,
https://doi.org/10.1007/978-981-13-0514-6_26

is an important performance signal to the system operator for stability and security point of view [1]. The primary objective of the AGC is to regulate frequency to the specified nominal value and maintain the power exchanged via tie-line between the controller areas to the scheduled values by adjusting the generated power of specific generators in the areas. The combined effects of both the tie-line power and the system frequency deviation are generally treated as controlled output of AGC know as area control error (ACE). As the ACE is adjusted to zero by the AGC, both frequency and tie-line power errors will become zero [2, 3].

Some of the artificial techniques such as PSO, genetic algorithm (GA), fuzzy logic controller (FLC), and artificial neural network (ANN) have proposed for load frequency control areas [4–8]. Controller for the AGC can be developed in two ways. One of the self-tuning techniques with the help of neural network and fuzzy logic is adopted by group researchers and other follows a suitable optimization algorithm. Yeşil et al. [9] implemented fuzzy type controller on nonlinear system, but there is no specific mathematical formulation to decide the proper choice of fuzzy parameters. From the literature review, the enhancement of performance indices of the power system not only depending on control structure but also on artificial optimization techniques. So a new high-performance meta-heuristic algorithms are always welcome to solve the real-world problems in the various systems. To meet the real-world problems, it is proposed to design and formulate an optimization problem of control by the use of meta-heuristic algorithms. A two-area interconnected thermal power system along with nonlinearity is considered for the design and analysis purpose of optimal controller by flower pollination algorithm (FPA).

2 System Understudy

The primary objective of the automatic generation control (AGC) is to control the power system frequency to the specified nominal value for small perturbation in load. An interconnected power system with the thermal power plants is considered in Fig. 1. Each area is equipped with the non-reheat turbine and a governor modeled along with dead-band nonlinearity. These areas are connected through a tie-line, and the whole system is under investigation. The relevant parameters are given in Table 1 [10].

3 Outline of Flower Pollination Search Algorithm

FPA was developed by Xin-She Yang in 2012 [11] inspired by pollination of flowering plants. FPA along with multi-objective optimization function is utilized for controller design [12]. Flower pollination is an activity that involves the transfer of pollen among the flowers in self-pollination and cross-pollination ways. Self-pollination (or local pollination) is a biotic form which contributes 10% of pollination by non-pollinators, and cross-pollination (or global pollination) is an abiotic

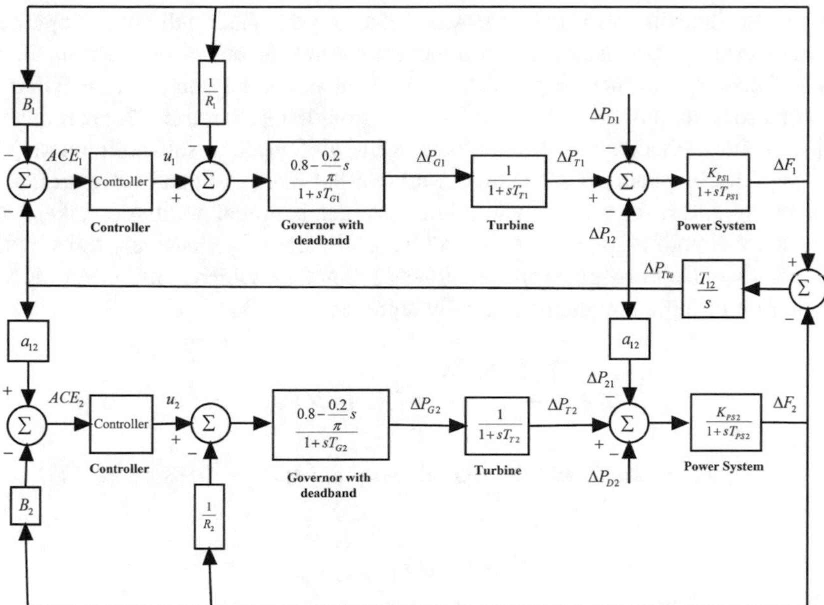

Fig. 1 Two-area interconnected thermal power system transfer function model

Table 1 Parameters of proposed model

Variables	Typical value
B_1, B_2	0.045 p.u. MW/Hz
R_1, R_2	2.4 Hz/p.u.
$T_{G1}, -T_{G2}$	0.2 s
T_{T1}, T_{T2}	0.3 s
K_{PS1}, K_{PS2}	120 Hz/p.u. MW
T_{PS1}, T_{PS2}	20 s
T_{12}	0.0707 p.u
a_{12}	−1 p.u

form which contributes 90% of pollination by pollinators like insects, birds, bats fauna. This phenomenon involves agents like pollinators that move from one flower to another flower exhibiting foraging behavior with a pollinator moving more frequently to certain flowers than others. The frequency of visit to a flower is indicated by term flower constancy, the proposed flower pollination algorithm deals with the selection size of population (N) and parameter (p) which help to decide the amount of self-pollination and cross-pollination to take place. The algorithm continues by initializing specified number of population, where each one contains a group of variables which are optimized by using the objective function. The proposed algorithm contains an indexing term called flower constancy for each individual population which determines how well their variables minimizes the objective function. On the bases of flower constancy, population are queued and best among them is found.

FPA proceeds through generation of new population based on the parameter p which decides whether this population is generated through self-pollination. This is carried out by generating a random variable between [0, 1] and comparing with parameter p, i.e., if the random variable is less than p, global pollination takes place (or) else local pollination occurs. For global pollination, agents would move with a different step size of length from one flower to another, which is mimicked by levy distribution of flight and mathematically represented as (1)

$$L \sim \frac{\lambda\Gamma(\lambda)sin(\frac{\pi\lambda}{2})}{\lambda}.\frac{1}{s^{1+\lambda}}, \ (s >> s_0 > 0). \tag{1}$$

The new population generated through global pollination is given by Eq. (2).

$$x_i^{t+1} = x_i^t + \gamma L(\lambda)(g_* - x_i^t) \tag{2}$$

The local pollination occurs within a small neighborhood of the current population. So its step size $'\epsilon'$ is taken from a uniform distribution. The mathematical expression for such an operation is expressed as (3)

$$x_i^{t+1} = x_i^t + \epsilon\ (x_j^t - x_k^t) \tag{3}$$

where x_j^t and x_k^t are pollen from different flowers of the same plant species.

Flower constancy for the new population is found in a similar manner as stated before. If new population flower constancy is better than the previous population, they are updated in the position of the previous one (or) else discarded. This process of generation and comparison will continue until the count reaches N. The best among the current population is found and declared as current global best. This process repeats for a maximum number of iterations as specified. The current global best is declared as best solution.

4 The Proposed Approach

AGC has come across many control strategies over the past decades, viz. proportional and integral (PI) controller and proportional, integral, and derivative (PID) controller in design of optimal controllers [13]. In this paper, PID controller utilized in the improvement of the dynamic performance of AGC in the two-area interconnected thermal power system. In this paper, FPA is applied to tune the parameters of proportional gain constant (K_P), integral gain constant (K_I), and derivative gain constant (K_D) for obtaining optimal values. The controllers in both the areas are considered to be identical so that $K_{P1} = K_{P2} = K_P$, $K_{I1} = K_{I2} = K_I$, and $K_{D1} = K_{D2} = K_D$. The inputs which deviates from nominal values fed to individual controller are the respective area control errors (ACE) which given by Eqs. 4 and 5

$$e_1(t) = ACE_1 = B_1 \Delta f_1 + \Delta P_{Tie} \tag{4}$$

$$e_2(t) = ACE_2 = B_2 \Delta f_2 + \Delta P_{Tie} \tag{5}$$

The controller outputs of the power system u_1 and u_2 along with PID structure are given by Eqs. 6 and 7

$$u_1 = K_{P1} ACE_1 + K_{I1} \int ACE_1 + K_{D1} \frac{dACE_1}{dt} \tag{6}$$

$$u_2 = K_{P2} ACE_2 + K_{I2} \int ACE_2 + K_{D2} \frac{dACE_2}{dt} \tag{7}$$

Based on the desire specification and constraints, the objective function is defined. This objective function customary tunes the parameters of PID controller which is based on performance criteria to obtain the optimal controller. Mainly, there are four kinds of performance criteria which are ITAE, ISE, ITSE, and IAE given by Eqs. 8–11.

$$J_1 = IAE = \int_0^{t_{sim}} [|\Delta f_1| + |\Delta f_2| + |\Delta P_{Tie}|] \cdot dt \tag{8}$$

$$J_2 = ISE = \int_0^{t_{sim}} (\Delta f_1)^2 + (\Delta f_2)^2 + (\Delta P_{Tie})^2 \cdot dt \tag{9}$$

$$J_3 = ITAE = \int_0^{t_{sim}} [|\Delta f_1| + |\Delta f_2| + |\Delta P_{Tie}|]^2 \cdot dt \tag{10}$$

$$J_4 = ITSE = \int_0^{t_{sim}} [(\Delta f_1)^2 + (\Delta f_2)^2 + (\Delta P_{Tie})^2] \cdot t \cdot dt \tag{11}$$

From [14], it is seen IAE criteria is opted, if the magnitude of response curve is < 1. Where as ISE criteria is prefered when it is > 1. In order to have better control of steady-state error, ITAE and ISE criteria are choosen depending on the magnitude and time of response curve.

In this paper, an attempt is made to construct a multi-objective function with four weighted performance criteria in Eq. (12). The instantaneous weights($\Delta\omega_1, \Delta\omega_2, \Delta\omega_3, \Delta\omega_4$) are chosen for small step of time whose magnitude depends on conditions, what was the response and which performance function is best suited at that point of step time. For the length of simulation time, instantaneous weights are added to give weights ($\omega_1, \omega_2, \omega_3, \omega_4$). These are multiplied with respective performance function, so that these act as extra penalty for a bad response which results in large cost value. Therefore optimization algorithm would reject such flower member tending to find a good solution (member) for given optimization function.

$$J_5 = \omega_1 \cdot ISE + \omega_2 \cdot ITSE + \omega_3 \cdot ITAE + \omega_4 \cdot IAE \tag{12}$$

In [15–18], the proposed objective function was based on fixed step load perturbation and the obtained controller parameters were optimal at fixed step load. But the system load is dynamic, so there is a requirement to design a controller that give optimal response for various load conditions.

In this paper, the objective function includes responses of various percentage step load changes, so the designed controller parameters give optimal response for most load disturbance.

5 Results and Discussions

In this paper, the proposed model which going to carried on under various testing conditions along with a FPA. This algorithm has been chosen as it had a parameter (p) which could control amount of local search and global search for given function.

The proposed objective function as stated above contains instantaneous weights which were chosen from a set of values [0.6, 0.2, 0.1, 0.1]. The best performance function instantaneous weight is assigned with highest value (0.6) whereas the least good performance function instantaneous weight is assigned with lowest value (0.1) for that instant of step time (0.01) depending on response. The instantaneous weights

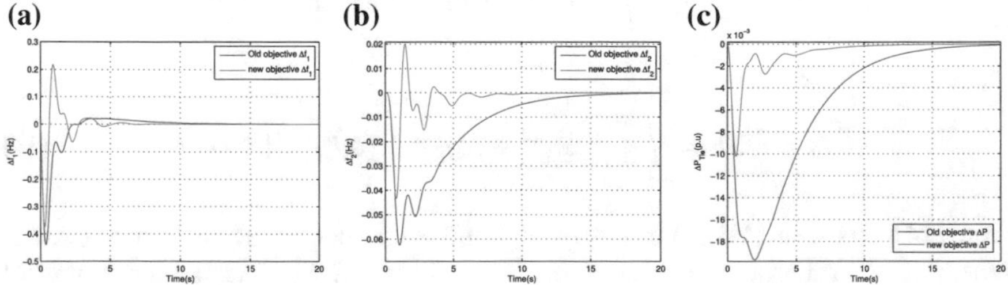

Fig. 2 Comparision of new and old objective functions **a** Δf_1 **b** Δf_2 and **c** Δp

are summed along simulation time and multiplied with performance function to give new objective function and gain values are tuned by using it. This tuned controller response for Δf_1, Δf_2, Δp_{tie} are compared with response of general objective function

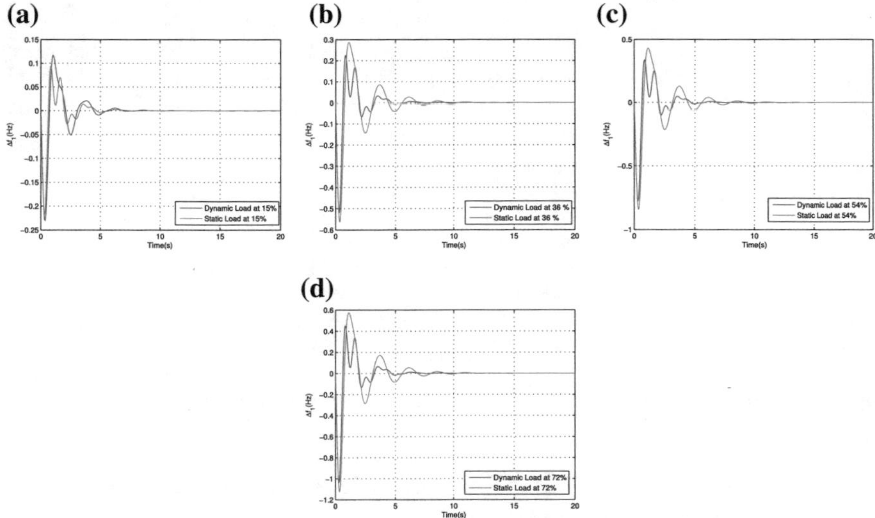

Fig. 3 Area -1 frequency deviation **a** At 15% load. **b** At 36% load. **c** At 54% load and **d** At 72% load

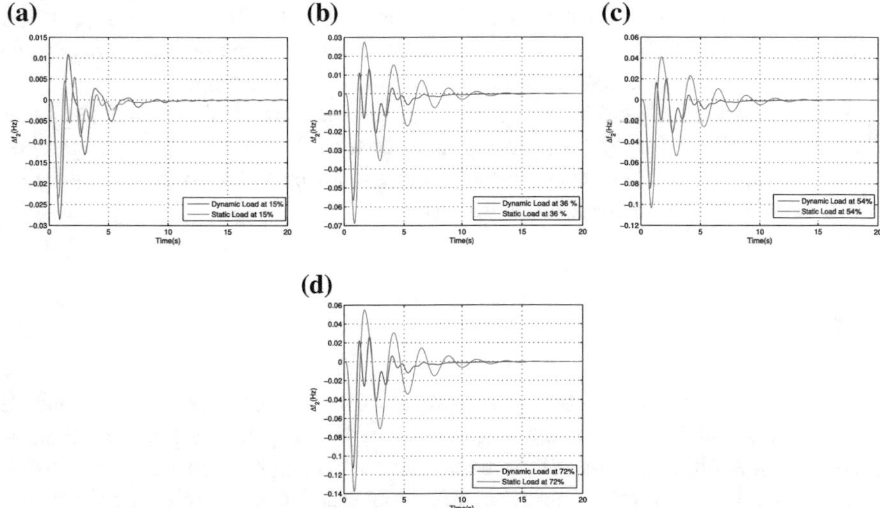

Fig. 4 Area -2 frequency deviation **a** At 15% load. **b** At 36% load. **c** At 54% load and **d** At 72% load

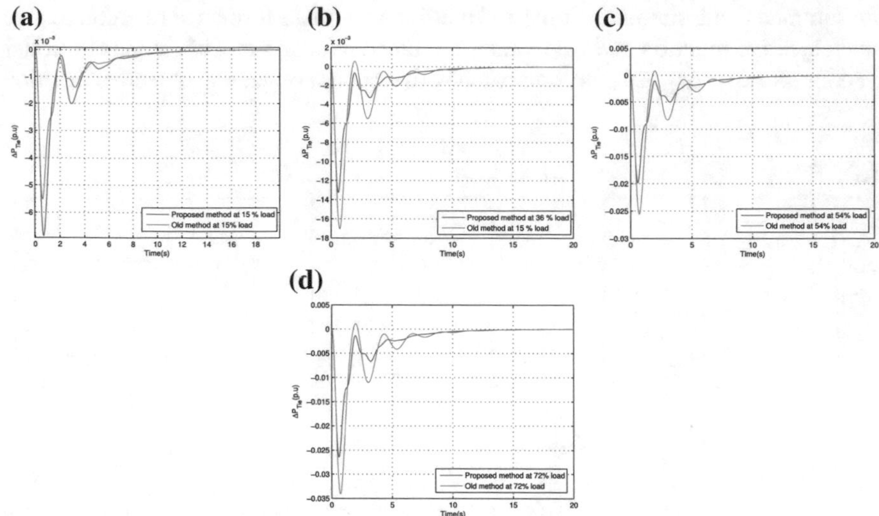

Fig. 5 Deviation of tie-line power **a** At 15% load. **b** At 36% load. **c** At 54% load and **d** At 72% load

with fixed weights [0.2, 0.2, 0.3, 0.3] are show Fig. 2. It is observed new objective proposed has a yielded a better performance.

The second aspect of proposed theory is to construct a objective function which includes performance for various load percentages $(1, 10, 30, 55, 65, 80, 90\%)$ to obtain tuned parameters for controller of considered system. This yielded gain parameters were optimal for any load change between $(1\ to\ 100\%)$. We have tested these gains for $15, 36, 54, 72\%$ load performance. We have also obtained gain parameters tuned for 15% fixed load and test for $36, 54, 72\%$ load changes.

The comparison of above two performance which are simulate and depict in Fig. 3, Fig. 4, Fig. 5 for Δf_1, Δf_2, Δp_{tie}, respectively. This shows parameters chosen from proposed method are optimal for most loads changes, so these could give optimal performance for dynamic loads also.

6 Conclusion

The study was carried out to design the optimal PID controller and also to analyze the proposed model at various loading conditions by using flower pollination algorithm (FPA). A different kind of approach is made to design a multi-objective function which contains weighted performance functions. These weights are functions of system response, such consideration showed an improvement in performance. The objective function also includes performance response for various percentage of loads, so that obtained gain parameters are optimal for different load conditions and this change could be observed in $\Delta f_1, \Delta f_2, \Delta p_{tie}$ responses.

7 Appendix 1

See Table 1.

References

1. Kundur, P.: Power System Stability and Control. TMH 8th reprint, New Delhi (2009)
2. Elgerd, O.I.: Electric Energy Systems Theory. An Introduction. TMH, New Delhi (1983)
3. Kothari, D.P., Nagrath, I.J.: Modern Power System Analysis. TMH 4th edn, New Delhi (2011)
4. Saikia, L.C., Nanda, J., Mishra, S.: Performance comparison of several classical controllers in AGC for multi-area interconnected thermal system. Int. J. Electr. Power Energy Syst. **33**(3), 394–401 (2011)
5. Parmar, K.P.S., Majhi, S., Kothari, D.P.: Load frequency control of a realistic power system with multi-source power generation. Int. J. Electr. Power Energy Syst. **42**(1), 426–433 (2012)
6. Ibraheem, I., Kumar, P., Kothari, D.: Recent philosophies of automatic generation control strategies in power systems. IEEE Trans. Power Syst. **20**(1), 346–357 (2005)
7. Ghoshal, S.P.: Optimizations of PID gains by particle swarm optimizations in fuzzy based automatic generation control. Electr. Power Syst. Res. **72**(3)
8. Golpra, H., Bevrani, H., Golpra, H.: Application of GA optimization for automatic generation control design in an interconnected power system. Energy Convers. Manag. **52**(5), 2247–2255 (2011)
9. Yeşil, E., Gzelkaya, M., Eksin, I.: Self tuning fuzzy PID type load and frequency controller. Energy Convers. Manag. **45**(3), 377–390 (2004)
10. Gozde, H., Taplamacioglu, M.C.: Automatic generation control application with craziness based particle swarm optimization in a thermal power system. Int. J. Electr. Power Energy Syst. **33**(1), 8–16 (2011)
11. Yang, X.-S.: Flower Pollination Algorithm for Global Optimization. Springer, Berlin Heidelberg (2012)
12. Yang, X.-S., Karamanoglu, M., He, X.: Multi-objective flower algorithm for optimization. Procedia Comput. Sci. **18**, 861–868 (2013)
13. Kumar, J., Ng, K.-H., Sheble, G.: AGC simulator for price-based operation. I. A model. IEEE Trans. Power Syst. **12**(2), 527–532 (1997)
14. Wade, M., Johnson, M.: Towards automatic real-time controller tuning and robustness. In: 38th IAS Annual Meeting. Conference Record of the Industry Applications Conference, 2003, vol. 1, pp. 352–359 (2003)
15. Mohanty, B., Panda, S., Hota, P.K.: Differential evolution algorithm based automatic generation control for interconnected power systems with non-linearity. Alexandria Eng. J. **53**(3), 537–552 (2014)
16. Sahu, R.K., Panda, S., Padhan, S.: A hybrid firefly algorithm and pattern search technique for automatic generation control of multi area power systems. Int. J. Electr. Power Energy Syst. **64**, 9–23 (2015)
17. Sahu, R.K., Panda, S., Padhan, S.: Optimal gravitational search algorithm for automatic generation control of interconnected power systems. Ain Shams Eng. J. **5**(3), 721–733 (2014)
18. Padhan, S., Sahu, R.K., Panda, S.: Automatic generation control with thyristor controlled series compensator including superconducting magnetic energy storage units. Ain Shams Eng. J. **5**(3), 759–774 (2014)

Satya Dinesh Madasu received M.Tech. degree from National Institute of Technology, Jamshedpur, Jharkhand, India, in 2010. He submitted the Ph.D thesis in the Department of Electrical and Electronics Engineering, NIT Jamshedpur. He is currently working as an Assistant Professor in the Department of Electrical and Electronics Engineering, Gandhi Institute of Engineering and Technology, Gunupur, Odisha, India. His areas of research interest are Application of Control System in Power System and Non-Conventional Energy.

M. L. S. Sai Kumar received B.Tech. degree from Anil Neerukonda Institute of Technology and Sciences, Vizag, India, in 2013. He received M.Tech. degree from National Institute of Technology, Jamshedpur, Jharkhand, India, in 2015. He is currently working toward Ph.D. degree in the Department of Electrical and Electronics Engineering, NIT Jamshedpur, Jharkhand, India. His areas of research interest are Automatic Generation Control and Power System Protection.

Arun Kumar Singh received B.Sc. degree from the Regional Institute of Technology, Kurukshetra, Haryana, India, M.Tech. degree from the IIT(BHU), Varanasi, UP, India, and Ph.D. degree from the Indian Institute of Technology, Kharagpur, India. He has been a Professor in the Department of Electrical and Electronics Engineering at National Institute of Technology, Jamshedpur, since 1985. His areas of research interest are Control System, Control System Application in Power System and Non-Conventional Energy.

Determinants of Farmers' Decision-Making to Accept Crop Insurance: A Multinomial Logit Model Approach

Partha Mukhopadhyay, Madhabendra Sinha and Partha Pratim Sengupta

Abstract With the developing trends of commercialization of agriculture, the magnitude of loss due to unfavorable eventualities is increasing and the farmers committed suicides. Burdwan district of West Bengal, India, becomes news headline because of farmer suicide. This study tries to underscore the elements which impact marginal farmers' decision to accept Pradhan Mantri Fasal Bima Yojona (PMFBY). We collect primary data from 701 farmers in the study over period of from February 2017 to October 2017. Seven estimators, namely education, age, cultivating land area, self-help groups (SHGs), marital condition, gender, and income, are considered, and multinomial logistic regression was used to analyze the data. Results show that gender coefficient is negative which indicates female farmer are prone to agro insurance (access to PMFBY) than male farmers. Results indicate that farmers' PMFBY and income are significant to the situation that one applies if others apply and get benefit.

Keywords Marginal farmers · Adoption decisions · Crop insurance Self-help groups · Risk attitude · Decision-making · India Pradhan Mantri Fasal Bima Yojona · Multinomial logistic model

---

P. Mukhopadhyay (✉) · M. Sinha · P. P. Sengupta
Department of Humanities and Social Sciences, National Institute of Technology Durgapur, Durgapur 741302, West Bengal, India
e-mail: pmukherjee400@gmail.com

M. Sinha
e-mail: madhabendras@gmail.com

P. P. Sengupta
e-mail: pps42003@yahoo.com

J. Nayak et al. (eds.), *Soft Computing in Data Analytics*,
Advances in Intelligent Systems and Computing758,
https://doi.org/10.1007/978-981-13-0514-6_27

1 Introduction

This paper explores the pattern of access to smallholder farmers' decision to accept the crop insurance Pradhan Mantri Fasal Bima Yojona (PMFBY) at Burdwan district, West Bengal, India [1]. It is renamed by West Bengal government as "Bangla Fasal Bima Yojona." Those farmers who have already a crop loan account through the Government are automatically eligible for this scheme. Agricultural production is subjected to large variations because of risk and uncertain factors completely beyond the farmer's control. The adoption of risk management strategies is affected by risk perceptions of farmers, their attitude toward risk, the characteristics of farm and farm household, and farmers' access to publically offered services including agricultural credit and information. These factors complicate management decisions in agricultural production and affect farmer's decision in production and marketing [8].

"Crop insurance" is the risk management tool. It is essential for farmers. Farmers are enabling to get financing and also survive in cataclysmic weather by crop insurance. Young farmers are exposed to risk as they are more leveraged than more established farmers. To eradicate the issues of risk of farmers, Hon'ble Prime Minister of India launched the new scheme PMFBY on January 13, 2016. In association with the respective state Governments, this scheme will be implemented in every state of India. This plan will be actualized in each condition of India, in relationship with separate State Governments [2]. The plan will be managed under the Agriculture Ministry and Farmers' Welfare, Government of India.

2 Brief Scenarios with Literature Survey

The farm sector is essential component to any Government. The common troubles of agricultural sector are that public policies are low and unstable incomes. As a public policy implementing of crop insurances seeks to minimize above issues. We have no control to the nature and occurrence of risky events. Besides bad weather, it is obvious that the higher premium rates discourage willingness to accept crop insurance. The imperfect information such as tendency of tempts of ethics and expected insurance payments exceed the value of efforts disheartened of private agencies in crop insurance market. Therefore, the private insurance companies of the regional nature may not be interested to penetrate the market with crop insurance. Higher agricultural production may be obtained by crop insurance. In case of failure of crops, the Government is exempted from large expenses for agricultural credit, relief, and credit loans. Despite crop insurance, depending on the yield of the field, it protects the amount of debt. It creates better opportunities for marginal and small farmers for institutional credit. Failure of crops or drought, in case of repayment of debts, is restored to the credit and thus reduces the loan repayment costs of loans and reducing excess and default loans.

Promoting the risk-diffusing mechanisms will encourage farmers to increase their income by adopting risky options. The crop seed benefit varies according to the nature and degree protection provided by the scheme. The "Priolker Committee" had prepared a budding format of crop insurance after thorough analysis, when India was a newly established nation with a huge but stagnating agricultural environment; Ghosh and Yadav [7]. In fact, the committee wanted to present a proposal with little success with the Central Government in India. It was a model of Dandekar by which a "homogeneous area based crop insurance" scheme that revives the idea, and even today, in India, the crop insurance is largely in size [5]. Risk-averse nature is found in farmers [4]. In addition, a poor farmer's penalty is too high [10]. Jodha [9] finds that the uncertainty of agriculture has shone on agricultural investment, which allocates non-formal assets. He also found that government loan companies are ill-equipped to reduce the risk of Indian farmers because they do not pay the cost of the drought-affected farmers.

Ahsan et al. [3] proved that under the risk, any rational and risk-averse person will endow less in input than a risk-free condition, so that the risk may only strengthen stability in crop productivity. The risk of agriculture is very costly for the nation to move forward in globalization era. In the early 1990s, it is seen that the banking system will not be beneficial for the agricultural loan and in short run it moves against the interest of the farming sector [6]. Ramkumar et al. [11] described that if the policy changes, then the moneylenders who charge high interest rates are returned. It is understood that India was less risky in agricultural parameter than three decades ago. Withdrawal of Government subsidies also raises the risk level of costly production and commercial input providers, raising the magnitude of risk of farmers.

3 Data and Methodology

The objective of this research is to highlight the basic disparities among the farmers through their decision-making process. On the other hand, observing the diversities of farmers according to selected indicators by using multinomial logit model constitutes a base for apprehending the risky behaviors of paddy producers in the region according to research results. Finally, the ultimate aim of the research can be exposed as to guide agricultural policy makers by using research results for preparing the background for more efficient future agricultural improvement or rural development projects. Therefore, it is intended to enable developing different strategies for farmers who embrace distinctive attitudes against agricultural insurances.

What are the factors responsible for smallholder farmers' decision to accept Crop Insurance (PMFBY)? In this regard, the specific objectives of the study are following:

- Farmers' awareness of agricultural insurance-taking;
- Examine the past attitude of farmers toward insurance-taking;
- Determine the willingness of farmers to accept insurance for agriculture in future;
- Examine the factors/constraints that may discourage farmers' confidence in taking agricultural insurance.

To fulfill our objectives, the basic null hypothesis to be tested in our present study can be represented as following:

- *Accept of agricultural insurance is not influenced by education, age, gender, marital status, self help group, ownership of land size and income of a farmer in case of India particularly in case of PMFBY.*

3.1 Study Area and Collection of Primary Data

For its huge production of paddy, Burdwan district is called "Granary of West Bengal." Total geographical area of the district is 7024 km^2. Total population as per 2011 census is as 7,723,663. Burdwan district of West Bengal, India, ranks seventh position population-wise, eleventh rank in the area-wise, and fifth on the basis of density of population in West Bengal. It is situated between 23° 53′ N′ and 22° 56″ north latitude and between 88° 25′ E′ and 86° 48′ E′ east longitude.

There are two types of paddy produces here, Aus and Aman. Out of total 277 Gram Panchayats, in the District of Burdwan, Primary data, which was mainly cross-sectional, was collected from 701 farmers randomly from 216 Gram Panchayats. We conducted the survey over the period from February 2017 to October 2017 just after one year launching of PMFBY. Independent variables included in the questionnaire were: farmer's age, gender, education, marital status, income, whether self-help group member or not, and cultivated land area. The decision factor of accepting agricultural insurance (PMFBY) by the farmer is considered as dependent variable (Y_i). Details of dependent and independent variables are described in Table 1.

The originality of this research as explained before is the use of multinomial logistic model for evaluating the decision-making process of farmers of Burdwan district. Thus, this chapter aims to display the methodology of the research while giving detailed information of the model, dependent and independent variables. According to the results, it is intended to figure out the validity of the research hypotheses and understand the impacts of social and economic factors on their decision-making and risk analysis process of farmers.

Table 1 Description of variables

Variable	Variable description	Influential factors
PMFBY (Y) (dependable variable)	Smallholder farmers' decision to accept crop insurance PMFBY (dependent variable) which takes the value of: 0 = if the farmers do not apply for crop insurance, 1 = for Apply, 2 = for Apply if others Apply.	
Age	Age of farmer (in years)	Individual factors
Education	Education (year of schooling)	
Married	Marital Status (1 = married, 0 = unmarried)	
Land	Size of Land (Bigha)	Asset
Income	Farmer's monthly non-farm income (rupees),	
SHG	Member of SHG (1 = member, 0 = non-member)	Decisive factors

Source Authors' own representation

3.2 Empirical Analysis

Multinomial logistic model is the main methodology of this study. The primary objective of this study is to highlight most risky attitude and decision-making process of farmers of Burdwan. Farmers were asked whether they have accepted the PMFBY and adopt innovative and modern agricultural practices or not. Their responses can be divided into three categories. Therefore, the multinational logistic model, including the three-class response variable, has been chosen as the most appropriate regression model for explaining the relationships between the responsive variable and explanatory variables. Three different categories of response or dependent variable are as follows:

- I never accept nor apply any crop insurance ("NOT APPLY" group).
- I apply crop insurance (PMFBY) and adopt innovative and modern agricultural practices only if other farmers apply and explain their satisfaction ("APPLY IF OTHERS APPLY" group).
- I have accept agricultural insurance (PMFBY) and adopt innovative and modern agricultural practices ("APPLY" group).

Agricultural insurance, primarily acceptance of PMFBY by Indian farmers is the main aim of this study. Only marginal and small farmers are considered in this study as it is observed that suicide rate is more for these groups only. To comprehend the basic components of farmers of Burdwan, we segregate the response variables into three distinct classes to conceive the complexity of decision-making process of farmers.

First group is refuses as they are not well aware about insurance and sustain their own way. They perceive agricultural insurance and innovative and modern agricultural practices as a risk, and they simply avoid risk. Second group prefers to accept PMFBY with innovative and modern agricultural practices only if other farmers apply and explain their satisfaction. They are more open and ready to bear risk. The last group of farmers ready to accept PMFBY and adopt innovative and modern agricultural practices introduced by ministry of agriculture. They are simply risk takers. After explaining the three different categories of the dependent or response variable Y, the general view of the multinomial logistic model used in this research is given as follows:

$$Y = \beta_0 + \beta_1 X_1 + \beta_2 X_2 + \beta_3 X_3 + \beta_4 X_4 + \beta_5 X_5 + \beta_6 X_6 + \beta_7 X_7 + \varepsilon$$

Y is the dependent or response variable (in our study, it is PMFBY) as expressed above. The explanatory variables used in the model are given below.

- X_1 is the age of the farmer;
- X_2 is the cultivated land size for individual small and marginal farmer;
- X3 is the gender;
- X4 is the marital status;
- X5 is educational level (years);
- X6 is self-help group member;
- X7 is income of an individual farmer.

3.3 *Description of Respondents*

Table 2 describes the status of decision taken by the farmers by doing the application for PMFBY. Out of 701 respondents, 467 are males and 234 are females. Out of 701, only 218 are accepting or promising for crop insurance out of which 144 are male and 74 are female.

Table 2 Case processing summary

Type of crop insurance	Group	Category as used	Number of farmers	Marginal percentage (%)
PMFBY	Not Apply	0	481	68.60
	Apply	1	109	15.50
	Apply if others Apply	2	111	15.90

Source SPSS output estimated by the Authors

4 Results and Discussions

Multinomial logistic regression has been used as the main methodology in the scope of this research. The impacts of social and economic factors of the paddy producers have been examined in order to understand the risky behaviors. Survey is chosen to be the major method for gathering the data from the research area. Goodness-of-fit tests: The multinomial logistic regression procedure reports Pearson and Deviance goodness-of-fit statistics. Model fitting information: A likelihood ratio test shows whether the model fits the data better than a null model. In our model, these values are shown in Tables 3 and 4.

Table 3 Goodness-of-fit

	Chi-square	df	Sig.
Pearson	460.526	452	0.381
Deviance	376.799	452	0.996

Source SPSS output estimated by the Authors

Table 4 Parameter estimation

Number of obs = 701					
LR chi^2 (14) = 595.64					
Prob > chi^2 = 0.0000					
Pseudo R^2 = 0.1116					
Log likelihood = −2371.6487					
PMFBY		Coef.	Std. Err.	z	P > |z|
0 (Not Apply group)		(Base outcome)			
1 (Apply group)	education	0.1251553	0.06453	1.94	0.052
	age	0.544234	0.0537749	10.12	0.000
	gender	−1.621944	0.1015247	−15.98	0.000
	married	−0.6403421	0.1424597	−4.49	0.000
	SHG	0.1549549	0.0860285	1.8	0.072
	land	0.1323865	0.0683932	1.94	0.053
	income	0.0717416	0.0656262	1.09	0.274
	_cons	−2.718938	0.2959628	−9.19	0.000
2 (Apply if others Apply group)	education	0.0945641	0.1174901	0.8	0.421
	age	0.6122606	0.1043777	5.87	0.000
	gender	−1.680616	0.1939104	−8.67	0.000
	married	−1.082159	0.3128265	−3.46	0.001
	SHG	0.1905167	0.161231	1.18	0.237
	land	0.0008435	0.1298624	−0.01	0.995
	income	0.4034891	0.1199532	3.36	0.001
	_cons	−4.689556	0.577431	−8.12	0.000

Source STATA output estimated by the Authors

It can be concluded that crop insurance has a positive impact on their decision-making process of farmers of Burdwan. However, the coefficient of crop insurance (PMFBY) is low because primary analysis shows that only 31% of farmers were accepting or promising for crop insurance. We consider significant when P value < 0.05. In the STATA output in "apply if others apply group," only age, gender and married are significant, which means besides age, married couple are prone to crop insurance and gender indicates female farmers are more inclined to crop insurance. In "apply group," only age, gender, married, and income are significant. More income group inclined to crop insurance.

4.1 *Factors Influencing the Acceptance of the Crop Insurance (PMFBY)*

In the focus interview, a number of farmers did not know about Bima Yojona, i.e., crop insurance policy. Moreover, most of them could not produce relevant papers of land where they cultivated. Ignorance, illiteracy, lack of awareness from farmers' side and lack of proper communication from Government are behind the constraints of the confidence of farmers.

5 Conclusion

The process of decision-making in agricultural activity is increasing with its importance. The main objective of this study is to investigate the risk assessment of insurance during the investigation of the decision-makers of the grain producers of Burdwan district. Therefore, it is intended to put forward social and economic issues to acceptance of PMFBY to paddy producers. It can be concluded that crop insurance has a positive impact on their decision-making process of farmers of Burdwan. However, the coefficient of crop insurance (PMFBY) is low because primary analysis shows that only 31% of farmers were accepting or processing the crop insurance. Again, farmers of Burdwan district are encouraged to further their learning/education about crop insurance because crop insurance reduces production or yield risk. In this study, we specifically investigate the factors that affect farmers' adoption of crop insurance, while taking into account the potential for simultaneous adoption and/or correlation among the adoption decisions using multinomial logistic approaches.

Because of the three-dimensional structures of the dependent variables in the model as "Apply," "If others Apply then Apply," and "Not Apply," it is believed that this study will maintain the valuable insight of the rural project in near future particularly in India. Using a multinomial logistic regression, we find that crop insurance adoption decisions are indeed correlated (even after controlling for observable factors). Using these research methods and response of farmers according to their feedbacks for agricultural insurance by using the multinomial logistic method is to be helpful for choosing the most appropriate regions for paddy-producing zone. Furthermore, our analysis suggests that the decision to adopt crop insurance (PMFBY) positively influences the decision to adopt the other tools. These results suggest that producers may increase their income by minimizing risk to accept agro insurance.

Acknowledgements Corresponding author is thankful to the Ethics Committee formed by the Department of Humanities and Social Science, NIT, Durgapur, for monitoring and conducting the primary survey for his doctoral research study regarding issues on Pradhan Mantri Fasal Bima Yojona on 701 farmers located at various rural areas of Burdwan district of West Bengal, India, over the period of February 2017 to October 2017. The survey provides the primary data used in the present paper.

References

1. Agriculture Insurance Company of India Ltd. www.aicofindia.org
2. Agricultural Statistics at a Glance: Agricultural Statistics Division, Department of Agriculture and Co-operation, Ministry of Agriculture, GOI, New Delhi (2017)
3. Ashan, M.S., Ali, A.G., Kurian, N.J.: Toward a theory of agricultural insurance. Am. J. Agric. Econ. **64**(3), 520–529 (1982)
4. Binswanger, H.: Attitudes toward risk—experimental measurement in rural India, World Bank Documents & Reports. Am. J. Agric. Econ. (1980)
5. Dandekar, Crop Insurance in India-A Review, 1976–77 to 1984–85, Economic & Political Weekly, vol. 20, issue 25–26 (1985)
6. Economic Survey (2007–2016): Ministry of Finance, Government of India, New Delhi
7. Ghosh, N., Yadav, S.S.: Problems and Prospects of Crop Insurance: Reviewing Agricultural Risk and NAIS in India (2008)
8. Hardaker, J.B., Lien, G., Huirne, R.B.M., Anderson, J.R.: Coping with risk in agriculture. Applied Decision Analysis, 3rd edn, pp. 107–126 (2015)
9. Jodha: Public Policies towards Formal Crop Insurance, pp. 20–23 (1978)
10. Mellor, J.W.: Production economics and the modernization of traditional agricultures. **13**(1), 25–34 (1969)
11. Ramkumar, M., Kumaraswamy, K., Mohanraj, R.: Environmental Management of River Basin Ecosystems, pp. 1–20 (2008)

A k-NN-Based Approach Using MapReduce for Meta-path Classification in Heterogeneous Information Networks

Sadhana Kodali, Madhavi Dabbiru, B. Thirumala Rao and U. Kartheek Chandra Patnaik

Abstract Classification of the nodes along with the interconnected semantic edges in a Heterogeneous Information Network (HIN) has a lot of significance in identifying the class labels which involves the application of knowledge and dissemination of knowledge from one node to the other. In this paper, the authors applied PathSim similarity measure for finding k-nearest neighbors along with the use of the well-known MapReduce paradigm to classify the meta-paths in a Heterogeneous Information Network. Applying MapReduce simplified the classification approach which deals with huge data present in the Heterogeneous Information Networks. Experiments were carried out on movie theater dataset, and the results are accurate and successful.

Keywords Classification · k-NN · MapReduce-based classification Meta-path

---

Sadhana Kodali (✉) · B. Thirumala Rao
Department of CSE, Koneru Lakshmaiah Education Foundation,
Vaddeswaram 522502, India
e-mail: sadhanaphd@gmail.com; sadhanalendicse@gmail.com

B. Thirumala Rao
e-mail: drbtrao@kluniversity.in

Madhavi Dabbiru
Department of CSE, Dr. L. Bullayya College of Engineering for Women,
Visakhapatnam 530016, Andhra Pradesh, India
e-mail: drlbcse@gmail.com

U. Kartheek Chandra Patnaik
Department of CSE, Lendi Institute of Engineering & Technology, Jonnada,
Vizianagaram 535005, India
e-mail: ukcp.lendi@gmail.com

J. Nayak et al. (eds.), *Soft Computing in Data Analytics*,
Advances in Intelligent Systems and Computing 758,
https://doi.org/10.1007/978-981-13-0514-6_28

1 Introduction

In today's world of interconnectivity between objects, a semantic network is formed which is called the Heterogeneous Information Network [1]. The autonomous objects present in the information network are treated as nodes which may be distributed, and the semantic relationships between these nodes are treated as edges. A social network is a best example of a heterogeneous network [2]. Observing and traversing the semantic paths connected between the nodes of a heterogeneous network can evolve many new and interesting things like the drug discovery, finding new friends, prediction of links, movie recommendations, item recommendations which are the major applications of the Heterogeneous Information Networks. To classify these nodes which are labeled is a challenging task. The existing algorithms like RankClass, HnetMine were based on path-based classification. The k-NN algorithm is one of the best approaches for classifying the data which is distributed. In this paper, the k-NN algorithm is applied to classify the meta-path which is formed by the traversal of the path between the objects using a similarity measure proposed in PathSim [3]. In this paper, the authors propose the approach to classify the nodes in the meta-path using k-NN with PathSim similarity measure, and later, they applied the MapReduce [4] to identify the nodes which are similarly close to each other.

1.1 Example of Heterogeneous Information Network

A Heterogeneous Information Network (HIN) is the semantic connectivity between the autonomous objects. The traversal of the semantic paths over a heterogeneous network gives a meta-path. Using a meta-path as an input, the commuting matrix is computed. The commuting matrix is calculated between two different kinds of attributes as in [5]. Attributes of category 1 are treated as row and attributes of category 2 are treated as columns. If there is a mapping present between the two attributes, then the matrix will have a value one; otherwise, the value is zero. Using this commuting matrix, the k-NN measure is applied with PathSim similarity calculation given as Eq. 1.

$$S(x:y) = 2\left(P_{x\to y:}\, P_{x\to y}\varepsilon P\right) / \left|\left(P_{x\to x:}\, P_{x\to x}\varepsilon P\right) + \left|\left(P_{y\to y:}\, P_{y\to y}\varepsilon P\right)\right| \qquad (1)$$

After which the input is given to the map task followed by the reduce task to observe the relevance between the classified entities. The rest of the paper is organized as follows: Sect. 2 narrates related work, Sect. 3 details the proposed approach, Sect. 4 discusses experimentation, and Sect. 5 concludes the paper followed by references.

2 Related Work

Heterogeneous Information Networks are formed by the interaction between different kind of objects and the semantic relationship between them. A number of classification algorithms are proposed for classification of HIN like the HetPath-Mine, Graffiti, and RankClass. But the well-known supervised learning-based classification using k-NN is more popular, and the use of this k-NN with the application of MapReducc is an excellent solution to classify huge networks which involve interaction among heterogeneous objects with semantic relationships.

2.1 Similarity Calculation

The network can be considered as a graph where the objects are the nodes and the connecting relationship between these objects is the edge. When these objects are traversed via a meaningful relationship, a meta-path is formed. The meta-paths can be more in number. The k-NN algorithm is applied to these meta-paths which is a graph-based k-NN, the similarity between the nodes is computed by using the PathSim similarity measure, and the weight factor is replaced with the similarity measure computed. To compute this similarity measure between nodes, the commuting matrix is formulated. The values presented in the commuting matrix are the weights between these nodes. The following example can be considered: if a person watches a movie in a theater, the meta-paths can be obtained in this manner: person —movie—theater. The commuting matrix can be formed between the person and theaters based on the number of times he visits the theater. This is shown in Fig. 1a.

The similarity between P1 and P2 are computed as 2 * (2 * 2 + 3 * 4)/(2 * 2 + 3 * 3) + (2 * 2 + 4 * 4) = 0.9696. The similarity scores in the same way are calculated between each person which measures how similar they visit the theaters as shown in Fig. 1b.

Fig. 1 **a** Commuting matrix, **b** similarity scores

Person\Theatre	T1	T2	T3
P1	2	3	0
P2	2	4	1
P3	1	2	3

(a) Commuting Matrix

Person\Similarity Score	PathSim Similarity
P1,P2	0.969
P1,P3	0.888
P2,P3	0.742

(b) Similarity scores

3 Approach

The Heterogeneous Information Network is traversed to find out all the possible meta-paths in it. After all the meta-paths are obtained, the k-NN algorithm is computed with PathSim similarity calculation.

Table 1 represents the algorithm which takes the graph which contains the labeled and unlabeled nodes from which the meta-paths are identified, and a commuting matrix is computed using the PathSim similarity measure. The approach is depicted in Fig. 2. The calculated similarities are given as an input to the MapReduce task. Table 2 shows the working of the MapReduce task in which we also identified the k-nearest neighbors, and the output is written to a part-r-00000 file.

The input to the Mapper is a file which contains the Manhattan distance of the similarity scores between two objects. The similarity score is computed using the PathSim similarity measure. The Manhattan distance is used to compute the distance between two similar points which is given by $dist\ (x_i: xj) = \Sigma|x_i - x_j|$. The Mapper writes a set of <*Key, Value*> pairs where *Key* contains the names of the attributes and *Value* is the list of similarity scores which are computed using the distance metric. The reducer compares a test value with every other value in the list, and with the specified *k*-value, the nearest neighbors are identified. The k-NN classifies the given similarity scores which are obtained as an output in the part-r-00000 file which is stored onto HDFS.

Table 1 Algorithm for graph-based k-NN using PathSim similarity construct

k-NN algorithm to classify nodes in HIN
Step 1:Input $X = X_l \cup X_{ul}$ Step 2: Construct the commuting matrix Step 3: Apply the PathSim similarity measure using Eq. 1. $S(x{:}y) = 2(P_{x \to y:}\ P_{x \to y} \in P)/|(P_{x \to x:}\ P_{x \to x} \in P) + |(P_{y \to y:}\ P_{y \to y} \in P)|$ Step 4: Write the similarity measures to a file which should serve as an input to MapReduce

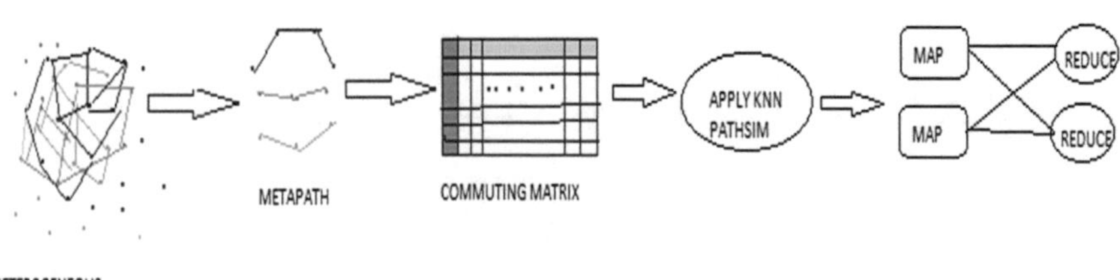

Fig. 2 Approach for k-NN classification using MapReduce

Table 2 Algorithm using MapReduce to obtain classification

k-NN algorithm to classify nodes in HIN
Step 1: Input file containing names of the nodes and corresponding similarities Step 2: Use the Map task to get the corresponding <key, Value> pairs Step 3: specify the similarity ranges to make partitions Step 4: The last reducer receives the unlabeled nodes and their similarities

4 Experimentation

The experimental results are tested on a single node Hadoop machine with 8 GB RAM. The training datasets are stored onto the HDFS. The experiment is initially preceded with 30 records then increased to 120, 191 and then slowly to 781 records. For all the above-mentioned, the map and reduce tasks were completed 100% and generated the k-neighbors with the specified *k*-value. The below figure shows an example of the result obtained after generating the Hadoop jar file and using the dataset stored on HDFS. We took the training data of the theater dataset and classified three labels.

Figure 3 shows the final execution after the MapReduce which is run on a pseudo-distributed environment. The runtimes for different dataset sizes are represented in a graph as shown in Fig. 4.

The x-axis represents the size of the dataset, and y-axis represents the runtime (in milliseconds). As the experiment is conducted using MapReduce, the execution times will not vary considerably.

The benchmark data is taken from movie lens by using the CSV file downloaded from https://grouplens.org/datasets/movielens/ [6] which consists of 138,000 users and 27,000 movies with their ratings given on various movies. By experimenting on this dataset, we find the nearest neighbor who gives similar rating. The MapReduce task is run on the dataset, and we identified the top five ratings given by the user for every movie for the movie ids given in the dataset. The figure depicts the part-r-00000 file.

Fig. 3 MapReduce execution results

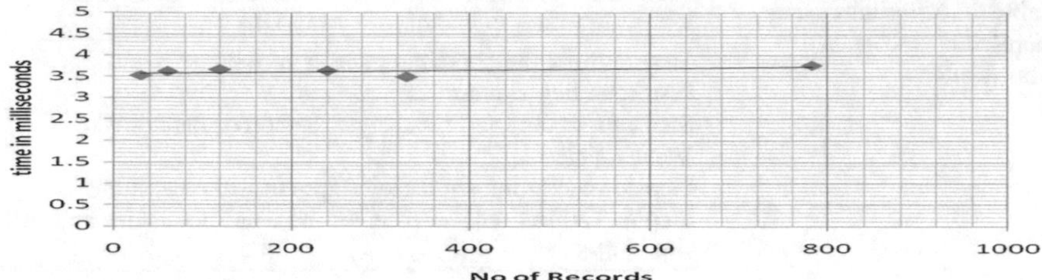

Fig. 4 Graph showing the runtime analysis

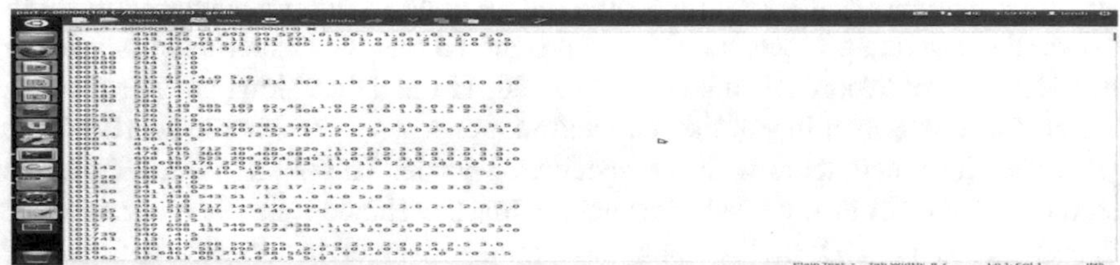

Fig. 5 Final output obtained on benchmark data

Synthetic data consists of 780 records generated by the authors. In the training dataset, we have considered person versus number of times the movie watched in nearby theaters. In the benchmark data, the person versus the movie ratings is considered. The ratings are given as inputs to the MapReduce task where the k-NN classifier is applied to identify the first five ratings of every movie present in the dataset. The final output is represented in Fig. 5. The execution times are shown in a graph, and the entire tuples in the dataset took just 4.55 ms to execute on a stand-alone machine where Hadoop is installed. The accuracy of the classification results is then computed in the subsection below.

4.1 Classification Accuracy

The classification accuracy is given as the number of correct predictions made divided by the number of predictions. The percentage of classification accuracy for our datasets has turned to be 100% for all the data records. The classification accuracy for the synthetic data is depicted in Table 3.

Table 4 represents the classification accuracy of benchmark data. All the jobs submitted to the MapReduce are done and the top five neighbors for every unique movie id present in the data are generated. Figure 6 represents the execution times plotted for the various sizes of datasets.

Table 3 Classification accuracy

Dataset size	Accuracy (%)
30	100
60	100
120	100
240	100
480	100
781	100

Table 4 Accuracy for benchmark datasets

Dataset size	Accuracy (%)
5000	90
10,000	94
20,000	92
30,000	95
40,000	92
65,536	98

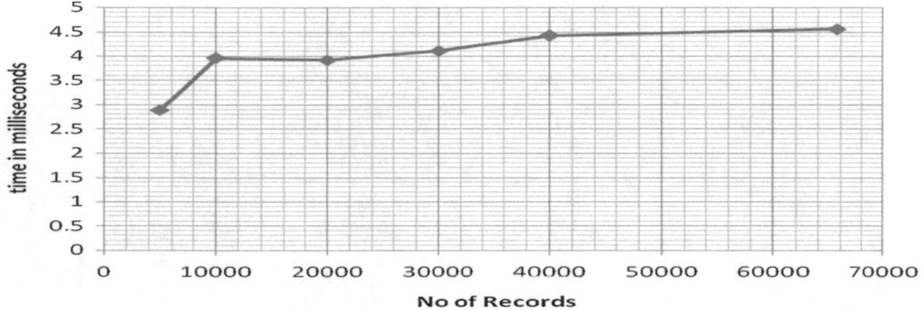

Fig. 6 Execution times for 65,536 records

5 Conclusions

In this paper, the authors suggested the approach of k-NN-based classification using meta-path similarity and the use of MapReduce to classify the Heterogeneous Information Networks. As the object heterogeneity increases, it is difficult to classify the labeled and unlabeled entities, but with this proposed approach, any kind of HIN input can be given and applied the basic classification approach with the support of MapReduce to handle big graphs which are nothing but the big data obtained from the HIN. The experiments are conducted on a stand-alone machine on which Hadoop is installed in pseudo-distributed mode. Our future work is to observe the distributed data as most of the HINs have the nodes scattered in a distributed manner by running the MapReduce task on parallel machines.

References

1. Shi, C., Li, Y., Zhang, J., Sun, Y., Philip, S.Y.: A survey of heterogeneous information network analysis. J. Latex Class Files **14**(8) (2015)
2. Han, J., Sun, Y.: Mining heterogeneous networks principles and methodologies (2012)
3. Sun, Y., Han, J., Yan, X., Yu, P.S., Wu, T.: PathSim: meta path based TopK similarity search in heterogeneous information networks. Proc. VLDB Endow. **4**(11) (2011)
4. Dean, J., Ghemawat, S.: MapReduce: simplified data processing on large clusters. OSDI 11 Mar 2004
5. Kodali, S., Dabbiru, M., Meduri, K.: Constraint based approach for minging heterogeneous information networks. In: Proceedings of 6th IEEE IACC 2016, 27–28 Feb 2016
6. https://grouplens.org/datasets/movielens/

Design of Fuzzy Logic-Based Controller for Nonlinear Power Systems

Rekha and A. K. Singh

Abstract The power system problem is essentially nonlinear in nature. To implement different control techniques, the nonlinear model has been converted into linear one by using direct feedback linearization (DFL) method. The design of DFL-PID controller through fuzzy logic based on heuristic knowledge has been considered in this paper which reduces the number of input variables for the controller. The fuzzy-based P, PI, PD, and PID controllers have been developed, and a comparative analysis has been carried out.

Keywords Linear control · Reactance · Linearization · Nonlinear system

1 Introduction

Concept of fuzzy logic has been given by Zadeh [1] in 1965. Fuzzy logic controller (FLC) is used as a controller in a power system as a fuzzy power system stabilizer [2–4]. In recent years, fuzzy logic has emerged as an effective tool in various power system applications [5–9] such as in power system stabilizer, motor control. The application of such control techniques has been motivated due to the following reasons: (a) improved robustness over the conventional linear control algorithm, (b) simple control design for difficult system models, and (c) simplified implementation. The parameters of the power system model are dependent on the operating condition, as operating condition changes and variation in plant parameters occurs. The variations in plant parameters are due to uncertain parameters such as damping ratio, transient and subtransient reactance.

Rekha (✉)
NIT Jamshedpur, Jamshedpur, India
e-mail: rchy72@gmail.com

A. K. Singh
Electrical & Electronics Engineering Department, NIT Jamshedpur, Jamshedpur, India
e-mail: aksnitjsr@gmail.com

J. Nayak et al. (eds.), *Soft Computing in Data Analytics*,
Advances in Intelligent Systems and Computing 758,
https://doi.org/10.1007/978-981-13-0514-6_29

2 Problem Statement

The dynamic behavior of the power system changes due to fault. Under this condition, the system may lose stability and voltage regulation at generator terminal. As the system conditions shift, the operating point at which the controller was optimized changes. For this reason, the developed fuzzy controller is taken as a better alternative.

The power system linearized model has been presented with the help of equations through (1)–(3) as given in appendix, where $v_f(t)$ is the controller output which acts as input to the system to be controlled in Eq. (3). $v_f(t)$ is realized here by using fuzzy logic controller. The detail of the process of linearization has been shown in Ref. [11].

3 Fuzzy Logic-Based Power System Control

Fuzzy-based PID controller is analyzed here to achieve the desired performance. The controller is designed with an objective to control two quantities of the generator: (i) the load angle δ and (ii) the speed ω with reference to δ_o and ω_o (i.e., the pre-fault power angle and speed of generator, respectively). The controller has to be adaptive to uncertain parameters throughout the simulation. The schematic block diagram representation for adaptive controlled system is shown in Fig. 1.

4 Fuzzy Logic Controller Design

Fuzzy logic controller has emerged as one of the effective tools [10] to control the complex systems. Figure 2 shows the stages of fuzzy controller design which mainly consists of blocks like fuzzification, defuzzification, inference mechanism, and rule base [11]. Two inputs' error and change in error are fed to the fuzzy

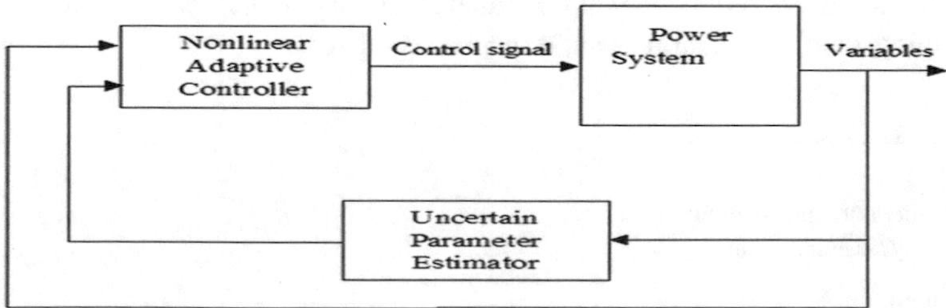

Fig. 1 Schematic representation of nonlinear adaptive control

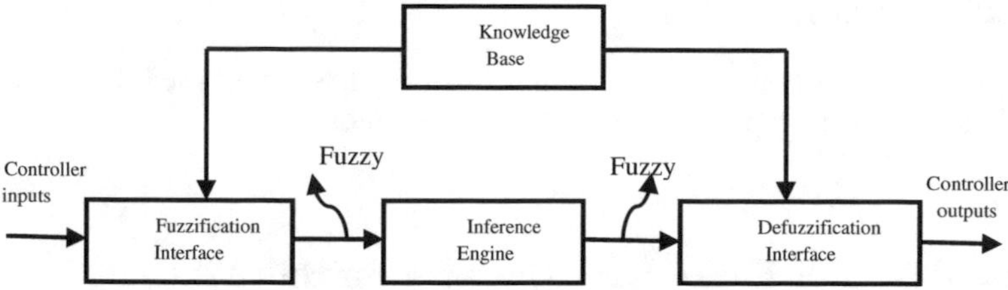

Fig. 2 Basic structure of fuzzy control systems

controller. For the problem, the two inputs are identified as speed and change in speed, and the fuzzy output acts as controlled input to the system.

For designing fuzzy controller for excitation control, speed deviation and acceleration are the two inputs and output is the control signal which is fed to the system. The steps followed for designing the fuzzy logic controller have been discussed below:

Step 1: To define input limits of fuzzy system

The ranges of speed deviation [NB….PB] and the acceleration deviation [NB…. PB] are set.

Step 2: To develop the membership function

The ranges of variables are decomposed into five sets of fuzzy regions: NB, NS, ZO, PS, and PB. The variables are defined for the speed deviation, acceleration, and output based on the heuristic knowledge.

Step 3: To develop the decision table

The decision table is prepared, considering the view that when a fault occurs in power system for a short time, the transient stability of the machine is subjected to a large impact, which causes significant reduction in machine terminal voltage, speed deviation, and acceleration. The five-member fuzzy system consists of 25 entries denoting each entry as a rule. The fuzzy rules with all 25 entries are shown in Table 1.

Table 1 Rules for fuzzy logic controller

Speed → ↓ Acceleration	NB	NS	ZO	PS	PB
NB	NB	NB	NS	NS	ZO
NS	NB	NS	NS	ZO	PS
ZO	NS	NS	ZO	PS	PS
PS	NS	ZO	PS	PS	PB
PB	ZO	PS	PS	PB	PB

Step 4: To defuzzify

The center of gravity type of defuzzification has been used here, and the defuzzified output is applied to the generator excitation.

5 Simulation of Fuzzy Logic-Based Controller

Two inputs and one output have been considered for implementing fuzzy PI or PD controller. The linguistic variables used here are listed as: NB, NS, ZO, PS, PB, and for simplification, triangular membership functions (M.Fs) have been considered.

The linguistic variables show the relation between the inputs and output of fuzzy controller. The two inputs speed and acceleration result in 25 rules. The typical rules are having the following structure:

Rule 1: If speed deviation is negative small (NS) and acceleration is positive small (PS), then output is zero (ZO).
Rule 2: If speed deviation is positive small (PS) and acceleration is negative big (NB), then output is negative small (NS).

The location of the fault and the system parameters considered for carrying out the simulation are assumed as in appendix, and three-phase short-circuit fault has been considered for analysis.

5.1 *Simulation Model for Fuzzy Logic Controller*

For implementing fuzzy logic controller discussed in previous section, modeling of the system has been carried out by using MATLAB/Simulink software package in the continuous time domain. The system has been divided into three subsystems: pre-fault, fault, and post-fault depending upon the time at which fault occurred and when it has been cleared. The transition between these subsystems is carried out by using switching block and using clock with it for enabling and disabling. Figure 3 shows the main simulation model for fuzzy-based controller design.

5.2 *Realization of Different Fuzzy Controllers*

One input (change in speed) and one output have been considered for implementing fuzzy P controller. For fuzzy PI or PD controller, two inputs considered here are: (i) change in speed and (ii) acceleration. Realization of fuzzy PID controller can be carried out by two ways: first one by considering three inputs for obtaining P, I, and D action, and the second consists of fuzzy controllers PI and PD combined together

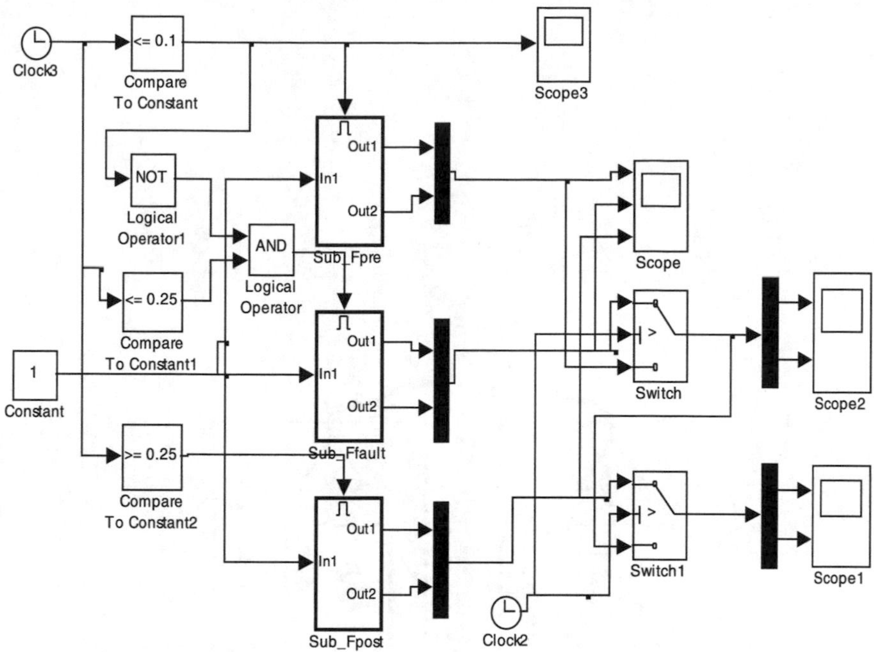

Fig. 3 Simulink model for single-machine infinite bus (SMIB) power system with fuzzy logic controller implemented

to obtain PID action. The realization of fuzzy PID controller has been done with the help of configuration shown in Fig. 4, which consists of two fuzzy blocks as: PI and PD controller. The detail of the components present in the PI or PD controller has been shown in Fig. 4. The dotted block with 'fuzzy PI controller' is used for PI controller implementation, and the dotted block with 'fuzzy PD controller' is used for PD controller whereas in case of PID controller design, both the blocks are active.

5.3 *Simulation Results for FLC*

The permanent type of fault has been considered for carrying out the simulation, and the operating points of the system for power angle and terminal voltage are assumed as: $\delta_o = 47°$ and $V_t = 1.0\,\text{p.u}$ [12]. For the system discussed in previous subsection, simulation has been carried out for different types of fuzzy PID controllers. The results obtained for above-mentioned controllers have been presented with the help of simulation results.

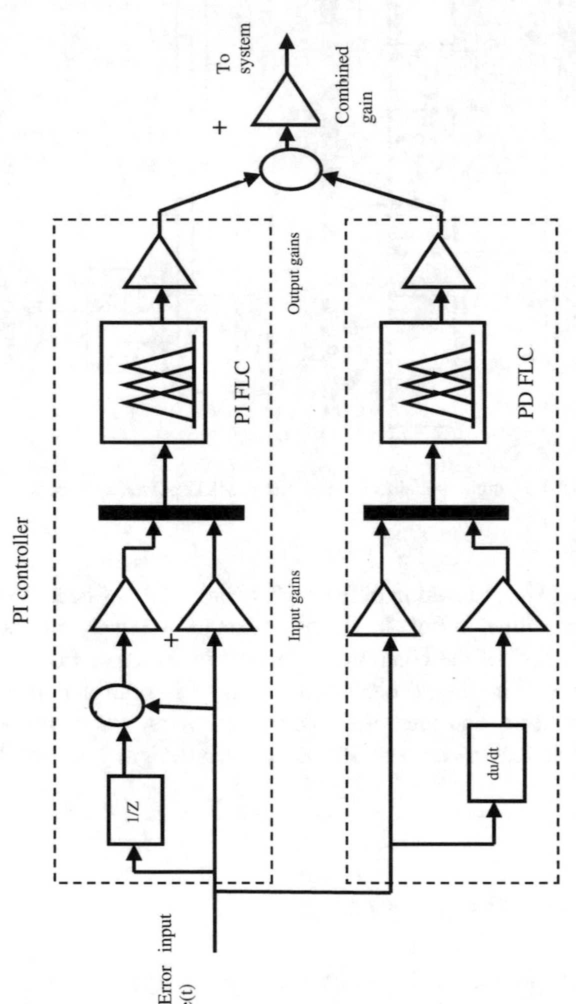

Fig. 4 Realization of fuzzy PID controller with PI and PD controllers

5.3.1 Fuzzy P Controller

The fuzzy P controller has been designed for the system under discussion using MATLAB/Simulink software. Single input (change in error) and single output (controlled output) are considered for designing the controller. The input and output gains of the fuzzy are tuned by trial and error approach. The response of FLC system for input gain ranging from 1.0 to 0.1 has been computed and shown in Fig. 5. The following inferences for P-controlled response could be drawn:

(1) For gain value of unity, the system is stabilized but there is steady-state error for both power angle and terminal voltage.
(2) Overall performance of P controller is not satisfactory due to the fact that the steady-state error in both the terminal voltage and load angle are high or having a poor regulation.

5.3.2 Fuzzy PI Controller

The fuzzy PI controller has been realized with two inputs, one for proportional (P) action and other for integral (I) action, and has been presented in Fig. 6. The following inferences could be drawn for PI controller:

(1) It has been observed that for output gain value of 0.55, the steady-state error for terminal voltage has reduced but at the same time load angle has stabilized below the pre-fault value of $\delta_o = 47°$.
(2) As the output gain of fuzzy PI controller is reduced below 0.55, both terminal voltage and power angle stabilize at value lower than initial operating point.

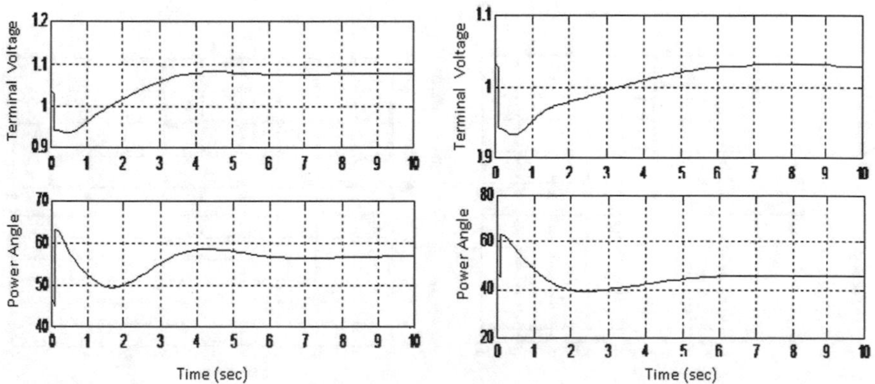

Fig. 5 Terminal voltage and power angle response for fuzzy P controller (input gain: 0.1 and output gain: 1 and 0.4, respectively)

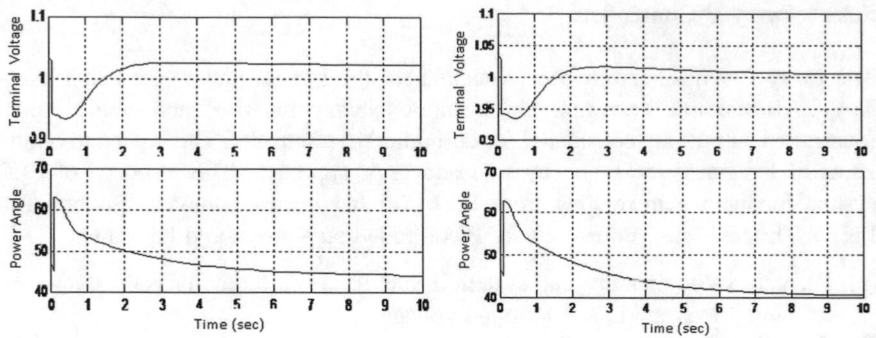

Fig. 6 Terminal voltage and power angle response (output gain: 0.6 and 0.55, respectively)

5.3.3 Fuzzy PD Controller

Fuzzy PD controller has been designed for the system under discussion, and the response with input and output for PD controller has been shown in Fig. 7.

(1) By increasing the output gain, the system does not reach the required operating points.
(2) Terminal voltage has stabilized to the pre-fault value of 1.0 p.u but at the cost of power angle being regulated at value lower than initial operating point.

Thus, it can be inferred that PI or PD controller alone is not sufficient to regulate both the power angle and terminal voltage. To overcome this demerit, it has been suggested to design fuzzy PID controller with the objective to achieve both regulations.

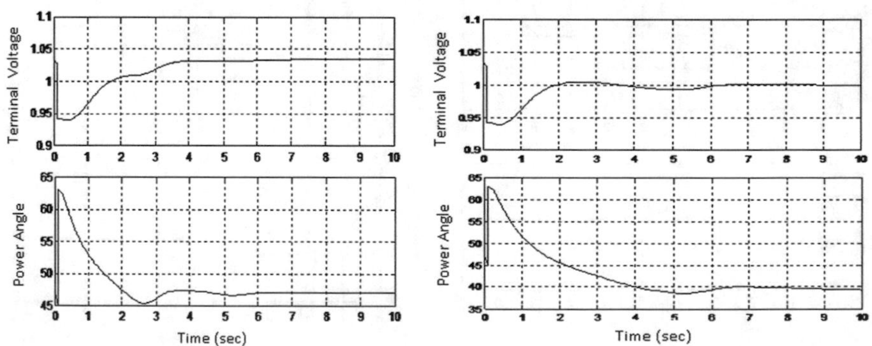

Fig. 7 Terminal voltage and power angle response for fuzzy PD controller (input gain: 0.1; 0.1, output gain: 1.01; input gain: 0.1 0.05, output gain: 1.01, respectively)

5.3.4 Fuzzy PID Controller

In this controller, two fuzzy controllers, PI and PD, have been combined to obtain the PID controller. The realization of fuzzy PID controller has been shown in Fig. 4. The input and output gains of PI and PD controllers, designed in earlier subsection, have been considered here for designing the PID controller, and only the output gain of combined controller has been tuned by trial and error with gain varying from unity to 0.5. Table 2 shows the gain values for PI and PD controllers.

The responses for fuzzy PID controller system have been presented in Fig. 8 for output gain of 1 and 0.5, respectively. By analyzing the response, it has been observed that:

(1) Stabilized output for both terminal voltage and power angle has been achieved for combined fuzzy output gain of unity. As the gain is reduced, the steady-state error in both terminal voltage and power angle reduces the gain value up to 0.5.
(2) To regulate both terminal voltage and power angle, along with fuzzy controller some optimization technique is required so that optimal controller gain values can be obtained.

Based on the results obtained for different fuzzy controllers discussed above, the inferences can be presented as shown in Table 3. It is observed that PI controller is able to regulate power angle, and PD controller is regulating the terminal voltage for the power system problem considered. Thus, PID controller has been able to combine the advantages of both PI and PD controllers.

Table 2 Gain values for PI and PD controllers

Type of controller	Input gain values	Output gain values
PI controller	Input1: 1; input2: 1	Output: 0.55
PD controller	Input1: 0.1; input2: 0.05	Output: 1.01

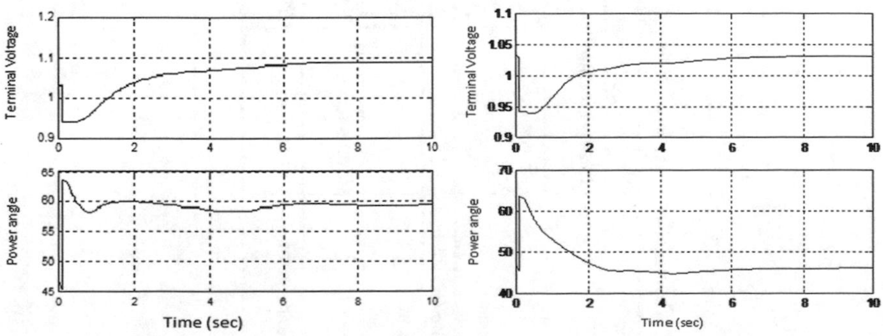

Fig. 8 Terminal voltage and power angle response for fuzzy PID controller (output gain: 1 and 0.5, respectively)

Table 3 Comparative analysis of different schemes of fuzzy controllers

Variation of parameter	Type of controller							
	Fuzzy P controller		Fuzzy PI controller		Fuzzy PD controller		Fuzzy PID controller	
	Power angle (δ)	Terminal voltage (V_t)	(δ)	(V_t)	(δ)	(V_t)	(δ)	(V_t)
Rise time (tr) (s)	0.25	0.25	0.25	0.25	0.25	0.25c	0.25	0.25
Settling time (ts) (s)	5	4.5	7	7	6	6	2.5	6
Overshoot (Mp)	Present	Present	Present	Present	Present	Approximately zero	Present	Present
Steady-state error (ess)	Present	8%	Present	2.5%	Approximately zero	Approximately zero	Approximately zero	2%

6 Conclusion

Fuzzy logic controller design for power system undergoing fault has been considered and successfully developed here. The parameters such as torque and line reactance change when a fault occurs in the system. This change in parameter has been taken into account by using switching block of MATLAB which transfers the control from one subsystem to another.

It is observed that the uncontrolled system is unstable, and by using fuzzy logic controller, system stability is improved but regulation is not proper. Thus to overcome this drawback, it is proposed to design PID controller using some optimization technique in place of fuzzy controller for the system under consideration and the controller parameters for the designed PID controller have been optimized by using PSO algorithm in foregoing chapter.

Appendix

The simulation data in this paper has been taken from Wang et al. paper [12].

$$\Delta\delta(t) = \omega(t) \tag{1}$$

$$\dot{\omega}(t) = \frac{-D}{H}\omega(t) - \frac{\omega_s}{H}\Delta P_e(t) \tag{2}$$

$$\Delta\dot{P}_e(t) = -\frac{1}{T'_{do}}\Delta P_e(t) + \frac{1}{T'_{do}}v_f(t) \tag{3}$$

References

1. Zadeh, L.: Fuzzy sets. Inform. Contr. **8**(3), 338–353 (1965)
2. Lin, Y.: Systematic approach for the design of a fuzzy power system stabilizer. In: International Conference on PowerCon 2004, vol. 1, pp. 747–752, Nov 2004
3. Roosta, A., Khorsand, H., Nayeripour, M.: Design and analysis of fuzzy power system stabilizer, Innovative smart grid technologies conf. Europe (ISGT Europe), 2010, pp. 1–7. IEEEPES, Oct 2010
4. Taher, A., Shemshad, S.A.: Design of robust fuzzy logic power system stabilizer. World Acad. Sci. Eng. Technol. **27** (2007)
5. Hoang, P., Tomsovic, K.: Design and analysis of an adaptive fuzzy power system stabilizer. IEEE Trans. Energy Convers. **11**(2) (1996)
6. El-Hawary, M.E.: Electric Power Applications Of Fuzzy Systems (1988)
7. Leung, F.H.F., Lam, H.K., Ling, S.H., Tam, P.K.S.: Optimal and stable fuzzy controllers for nonlinear systems based on an improved GA. IEEE Trans. Ind. Electron. **51**(1), 172–182 (2004)

8. Wong, C.C., Chen, C.C.: A GA-based method for constructing fuzzy system directly from numerical data. IEEE Trans. Syst. Manag. Cybern. B **30**, 904–911 (2000)
9. Pierre, G.: Fuzzy logic applied to motor control. IEEE Trans. Ind. Appl. **32**(1), 51–56 (1996)
10. Ziegler, J.G., Nichols, N.B.: Optimum setting for automatic controllers. Trans. ASME **64**(11), 759–65 (1942)
11. Rekha, Stabilizing PID controller for a class of nonlinear power systems, Ph. D thesis, 2017
12. Wang, Y., Hill, D.J., Gao, L., Middleton, R.H.: Transient stability enhancement and voltage regulation of power systems. IEEE Trans. Power Syst. **8**(2), 620–627 (1993)

Emotion Recognition Using Skew Gaussian Mixture Model for Brain–Computer Interaction

N. Murali Krishna, J. Sirisha Devi, P. Vijaya Bhaskar Reddy and Siva Prasad Nandyala

Abstract Over the past two decades, the research on EEG-based brain–computer interaction (BCI) made the BCI systems better and better, but building a robust BCI system for practical use is still a goal to achieve. To identify emotions of immobilized individuals, neuroscan machines like encephalography (EEG) is utilized. It uses the physiological signals available from EEG data extracted from the brain signals of immobilized persons and tries to determine the emotions, but these results vary from machine to machine, and there exists no standardization process which can identify the feelings of the brain diseased persons accurately. In this paper, a novel method is proposed, Skew Gaussian Mixture Model (SGMM) to have a complete emotion recognition system which can identify emotions more accurately in a noisy environment from both the healthy individuals and sick persons. The results of the proposed system surpassed the accuracy rates of traditional systems.

Keywords Emotion recognition · Electroencephalography (EEG)
Skew Gaussian Mixture Model · Brain–computer interaction

N. Murali Krishna
Department of CSE, Vignan Institute of Technology and Science, JNTU (H), Hyderabad 508284, India
e-mail: muralinamana@gmail.com

J. Sirisha Devi (✉)
Department of CSE, Institute of Aeronautical Engineering, JNTU (H), Hyderabad 500043, India
e-mail: musiri.cse21@gmail.com; siri.cse21@gmail.com

P. Vijaya Bhaskar Reddy
Department of CSE, CMR Engineering College, JNTU (H), Hyderabad 501401, India
e-mail: bhskr.dwh@gmail.com

S. P. Nandyala
DSP Specialist at Tata Elxsi, Bengaluru, Karnataka, India
e-mail: speech.nitw@gmail.com

J. Nayak et al. (eds.), *Soft Computing in Data Analytics*,
Advances in Intelligent Systems and Computing 758,
https://doi.org/10.1007/978-981-13-0514-6_30

1 Introduction

Brain–computer interface (BCI) is a communication system in which an individual can send messages or commands to an external device (e.g., a computer) without using the brain's muscular or peripheral nerves pathways. A person's intention to control or to communicate initiates brain activity, and the patterns from those brain activities can be detected from electrophysiological signals. Patients suffering from high spinal cord injuries (HSCI), amyotrophic lateral sclerosis (ALS), brainstems stoke, cerebral palsy, or other neural disorders find difficulties in communication and neural prosthetics. BCI aims at resolving the difficulties of such patients and raise their standard of living. It also presents a new supplementary way of communication and control for people with no disabilities. There are different invasive and noninvasive electrophysiological signal recording methods that are being used to detect brain activities. Electroencephalography (EEG) signals are the most studied type of signal to detect brain activities because of its noninvasive nature, portable, low setup, and equipment cost, and it is relatively easier to record in almost every kind of environment. Noninvasive methods include magnetoencephalography (MEG), positron emission tomography (PET), functional magnetic resonance imaging (fMRI), and optimal imaging [1]. These methods are more suitable for people with severe disorders.

In the following paper, Sect. 2 shows a detailed study on emotion recognition using electroencephalography, and Sect. 3 analyzes the gap in the literature. In Sect. 4, the proposed Skew Gaussian Mixture Model is presented, followed by the results in Sect. 5 and conclusion in Sect. 6.

2 Literature Survey

Electroencephalography (EEG) is a well-known method for analyzing and recording electrical patterns from the brain by using electrodes placed on the scalp. In 1875, Richard Caton for the very first time recorded the electrical patterns from an animal's brain. Han Berger recorded human brain electrical activity in 1924. In the start, EEG signal's application for communication was not realized and EEG was used to examine the neurological disorder in people by analyzing the patient's brain activity in the laboratories. Over the time, people have realized that EEG signals can also be used to read the intentions of a person for command or control by analyzing the EEG signals without using the peripheral and/or muscular channels. Although it has significant importance for people with disabilities, the communication application of EEG signals was not given enough importance because of three main reasons. First, experts at that time were not ready to work with the complex spatial and electrical geometry of the brain function and with the variation of trial-to-trial variations in the EEG recording for same subject. Second, the required technology to record EEG in real time was not either present or was only available in laboratories because of the high equipment cost. Third, the EEG-based communication has limited command or message transfer rate over that time.

For scalp-recorded EEG, usually Ag–AgCl electrodes are used. The values of the signals can be found in the range of 0.5–100 μV [2]. Other techniques are also being used for recording brain activities—magnetoencephalography (MEG), positron emission tomography (PET), functional magnetic resonance imaging (fMRI), and optimal imaging. Although these methods provide more precise mean of communication, these methods are invasive in nature, are technically more demanding for normal user, have longer time constants, not suitable for rapid communication, have expensive equipment and high setup cost, and specific environment is required for their recording. Because of these reasons, EEG is gaining popularity in the area of human–computer interaction.

EEG recording methods are of two types: invasive and noninvasive. The EEG signals recorded from the scalp of the brain are termed as noninvasive. The EEG signals can also be recorded by using epidural, subdural, and intra-cortical electrodes. In electrocorticography (ECoG) technique, the EEG signals are recorded by placing electrodes on the cortical surface, and in local field potential (LFP) techniques, the electrodes are placed inside the brain for capturing EEG signals [3]. These invasive techniques offer high topological and spatial resolution. These techniques can follow the individual's neurotic activity and better frequency bandwidth. Despite the fact that these methods can offer more precise and rapid source of communication and control than the scalp-recorded EEG, these methods are not used because of the invasive nature, they are generally used for people with severe disabilities.

3 Gap Analysis

The developments in the area of cognitive neuroscience have been instigated latest advancements in the area of brain–computer interaction in particular to brain image technology. The electroencephalography (EEG) [4], functional magnetic imaging (fMRI), and magnetoencephalography (MEG) are mainly used for brain analysis; among these technologies, EEG can be acquired cheaply and it is noninvasive which helps to identify the functional correlations in the brain [5, 6]. EEG plays a very important as well as vital role in the recognition of the emotions effectively from the immobilized persons. The emotions generated from the mobilized persons can be easily interpreted, and various methodologies exist in the literature to identify the emotions of the individual [7].

The hurdle one finds when recording EEG signals is most of the electrodes are not tightly contained within the scalp; therefore, many environmental, electromagnetic (EM), and other surrounding sources contribute to the high noise-to-signal ratio of EEG signals. EEG signals also have non-class-related artifacts, which include technical, environmental, electromyography (EMG), electrooculography

(EOG), and other non-CNS artifacts. These artifacts affect the classification of the feature set. The important aspect of a BCI system is, therefore, to maximize the accuracy by enhancing Gaussian Mixture Model.

4 Proposed Model: Skew Gaussian Mixture Model

Emotional signal is a collection of regions of several emotions. In each emotion region, the emotion data is quantized by EEG signal as it is influenced by random factors like vision, brightness, and contrast [8]. It is necessary to assume that the EEG signal in every emotion region follows a skew normal distribution. Skew normal distributions are shown in Fig. 1.

The mean EEG signal intensity of an emotion region is

$$E(y) = \mu + \sqrt{\frac{2}{\pi}} \cdot \delta(y) \tag{1}$$

The variance is

$$\mathrm{var}(y) = \left\{1 - \frac{2}{\pi} \cdot \delta^2(\lambda)\right\} \cdot \sigma^2 \tag{2}$$

where $\delta(y) = \frac{\lambda}{\sqrt{1+\lambda^2}}$

The moments are given by

$$\mu = m_1 - a_1\left(\frac{m_3}{b_1}\right)^{\frac{1}{3}}, \sigma^2 = m_2 - a_1^2\left(\frac{m_3}{b_1}\right)^{\frac{2}{3}}, \delta(\lambda) = \left\{a_1^2 + m_2\left(\frac{b_1}{m_3}\right)^{\frac{2}{3}}\right\}^{-\frac{1}{2}} \tag{3}$$

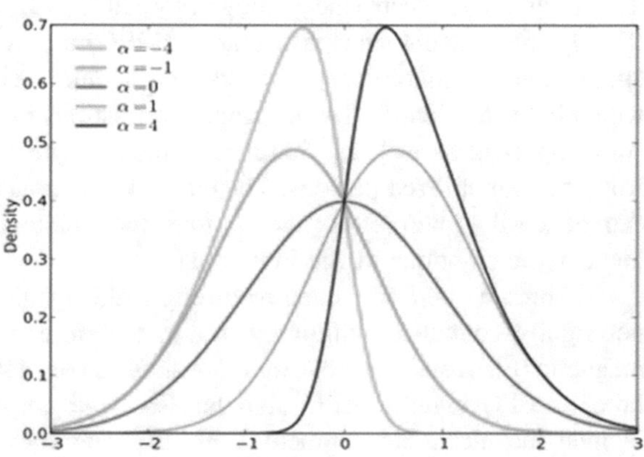

Fig. 1 Skew normal distributions

where $a_1 = \sqrt{\frac{2}{\pi}}; b_1 = \left(\frac{4}{\pi - 1}\right)a_1;\quad m_1 = n^{-1}\sum_{i=1}^{n} y_{i};\quad m_2 = \frac{\sum_{i=1}^{n}(y_i - \overline{y_i})^2}{n-1}$

$$m_3 = \frac{\sum_{i=1}^{n}(y_i - \overline{y_i})^3}{n-1}$$

Since the entire EEG sample is a collection of regions which are characterized by skew normal variants, we assume that the EEG samples intensities in the EEG region follow a k-component finite skew normal distribution and its probability density function is of the form

$$h(x) = \sum_{i=1}^{k} \alpha_i g_i\left(y_i|\mu_i, \sigma_i^2, \lambda\right) \tag{4}$$

where λ is the skewness parameter and k is the number of regions, μ_i, σ_i^2 and $g_i(y_i|\theta)$ are the mean, variance, and PDF of an EEG signal in the *i*th region, respectively.

4.1 Parameters Estimation

The likelihood function of the observations $x_1, x_2, x_3, \ldots . x_n$ drawn from EEG signal is given by

$$L(\theta) = \prod_{s=1}^{N}\left(\sum_{i=1}^{k} \alpha_i\, g_i(Z_s, \theta)\right) \tag{5}$$

where $\theta = (\mu_i, \sigma_i^2, \alpha_i = 1, 2, 3\ldots k)$ is the parameter set.

$$L(\theta) = \sum_{s=1}^{N}\left[\sum_{i=1}^{k}\left[\prod_{i=1}^{N}\left(\sum_{i=1}^{k}\alpha_i\left[\frac{2}{\sigma}\cdot\frac{1}{\sqrt{2\pi}}\cdot e^{-\frac{1}{2}\left(\frac{y-\mu}{\sigma}\right)^2}\Phi\left(\lambda\left(\frac{y-\mu}{\sigma}\right)\right)\right]\right)\right]\right] \tag{6}$$

4.2 E-Step

With initial parameter vector $\theta^{(0)}$, expectation value of $log\, L(\theta)$ is calculated with the given data 'y' as

$$Q\left(\theta; \theta^{(0)}\right) = \sum_{s=1}^{N}\log\left(\sum_{i=1}^{k}\alpha_i\, g_i(y_s; \theta)\right) \tag{7}$$

Given the initial parameters $\theta^{(1)}$, one can compute the density of any observation y_s, as

$$h(y_s, \theta) = \prod_{i=1}^{N} \left(\sum_{i=1}^{k} \alpha_i \left[\frac{2}{\sigma} \cdot \frac{1}{\sqrt{2\pi}} \cdot e^{-\frac{1}{2}\left(\frac{y-\mu}{\sigma}\right)^2} \Phi\left(\lambda\left(\frac{y-\mu}{\sigma}\right)\right) \right] \right) \tag{8}$$

where, $\Phi\left(\lambda\left(\frac{y-\mu}{\sigma}\right)\right) = \int_{-\infty}^{\lambda\left(\frac{y-\mu}{\sigma}\right)} \frac{1}{\sqrt{2\pi}} \cdot e^{-\frac{t^2}{2}} dt$ and the conditional probability of any observation y_i to belong to the segment 'k' is

$$t_k\left(y_s, \theta^{(1)}\right) = \frac{\alpha_k^{(1)} g_i\left(y_s, \theta^{(1)}\right)}{h\left(y_s, \theta^{(1)}\right)} = \frac{\alpha_k^{(1)} g_k\left(y_s, \theta^{(1)}\right)}{\sum_{i=1}^{k} \alpha_k^{(1)} g_i\left(y_s, \theta^{(1)}\right)} \tag{9}$$

where $h\left(y, \theta^{(1)}\right) = \sum_{i=1}^{k} \alpha_i^{(1)} g\left(y_s, \theta^{(1)}\right)$ from Eq. (10), evaluating the expectation value of $L(\theta)$ is

$$Q\left(\theta; \theta^{(1)}\right) = E_{\theta^{(1)}}\{\text{Log } L(\theta)/\bar{y}\} \tag{10}$$

From the heuristic arguments of Jeff A. Blimes (1997), we get

$$Q\left(\theta; \theta^{(1)}\right) = \sum_{i=1}^{k} \sum_{s=1}^{N} \left\{ t_i\left(y_i; \theta^{(1)}\right) (\log g_i(y; \theta) + \log \alpha_i) \right\} \tag{11}$$

But

$$g_i(y, \theta) = \frac{2}{\sigma} \cdot \frac{1}{\sqrt{2\pi}} \cdot e^{-\frac{1}{2}\left(\frac{y-\mu}{\sigma}\right)^2} \Phi\left(\lambda\left(\frac{y-\mu}{\sigma}\right)\right) \tag{12}$$

Substituting the value of $(log g_i(y, \theta))$, we have

$$Q\left(\theta; \theta^{(1)}\right) = \sum_{i=1}^{k} \sum_{s=1}^{N} \left\{ t_i\left(y_i; \theta^{(1)}\right) \left(\log\left(\frac{2}{\sigma} \cdot \frac{1}{\sqrt{2\pi}} \cdot e^{-\frac{1}{2}\left(\frac{y-\mu}{\sigma}\right)^2} \Phi\left(\lambda\left(\frac{y-\mu}{\sigma}\right)\right) \right) + \log \alpha_i \right) \right\} \tag{13}$$

4.3 M-Step

For constrained maxima by applying standard solution method, under the additional condition $\sum_{i=1}^{k} \alpha_i = 1$, for the segment weight α_k the maximum likelihood estimation problem can be solved. The first-order Lagrange-type function

$$L = \left[\log L\left(\theta^{(l)}\right) + \beta\left(1 - \sum_{i=1}^{k} \alpha_i^{(l)}\right) \right] \tag{14}$$

where β is a Lagrange multiplier. To update the estimate of α_k^l, we have to go to the next level, i.e., α_k^{l+1}. Taking the sum of all over $K = 1, 2 \ldots k$, and considering that for every y_S, the sum overall $t_k\left(y_s, \theta^{(l)}\right)$ must be equal to ONE, due to the probability character of $t_i\left(y_s, \theta^{(l)}\right)$, we get

$$\alpha_k^{(l+1)} = \frac{1}{N} \sum_{s=1}^{N} \left[t_k\left(y_s, \theta^{(l)}\right) \right] \tag{15}$$

From Eq. (12), we get

$$\alpha_k^{(l+1)} = \frac{1}{N} \sum_{s=1}^{N} \frac{\alpha_k^{(l)} g_i\left(y_s, \theta^{(l)}\right)}{h\left(y_s, \theta^{(l)}\right)} \tag{16}$$

4.4 Updating the Parameter *μ*

For updating the parameter μ_k, $k = 1, 2 \ldots K$, consider derivative of $Q\left(\theta; \theta^{(l)}\right)$ with regard to μ_k equated to 0

we have

$$Q\left(\theta; \theta^{(l)}\right) = E\left[\frac{\partial \log L(\theta; \theta^{(l)})}{\partial \mu}\right] \tag{17}$$

Since μ_i appears only in one region $i = 1, 2, 3 \ldots .k$ (regions), the value of μ can be updated as

$$\mu^{(l+1)} = y + \sigma^{2^{(l)}} + \frac{1}{\int_{-\infty}^{\alpha^{(l)}\left(\frac{y-\mu^{(l)}}{\sigma^{(l)}}\right)} e^{-\frac{1}{2}\left(\frac{t-\mu^{(l)}}{\sigma^{(l)}}\right)^2} dt} + \int_{-\infty}^{\alpha^{(l)}\left(\frac{y-\mu^{(l)}}{\sigma^{(l)}}\right)} \left(t - \mu^{(l)}\right) e^{-\frac{1}{2}\left(\frac{t-\mu^{(l)}}{\sigma^{(l)}}\right)^2} dt$$
$$- \sigma^{(l)} \alpha^{(l)} e^{\frac{\left[\left(\alpha^{(l)} + \sigma^{(l)}\right)\mu^{(l)} - \alpha^{(l)} y\right]^2}{2\sigma^{4^{(l)}}}} \tag{18}$$

4.5 Updating the Parameter σ^2

Consider

$$\frac{\partial}{\partial \sigma^2} Q\left(\theta; \theta^{(l)}\right) = 0 \tag{19}$$

The updated equation for σ is

$$\sigma^{(l+1)} = 1 / \left[\begin{array}{c} \frac{\left(y-\mu^{(l)}\right)^2}{\sigma^{3(l)}} + \frac{1}{\int_{-\infty}^{\alpha^{(l)}\left(\frac{y-\mu^{(l)}}{\sigma^{(l)}}\right)} e^{-\frac{1}{2}\left(\frac{t-\mu^{(l)}}{\sigma^{(l)}}\right)^2} dt} + \\ \int_{-\infty}^{\alpha^{(l)}\left(\frac{y-\mu^{(l)}}{\sigma^{(l)}}\right)} \frac{\left(t-\mu^{(l)}\right)^2}{\sigma^{3(l)}} e^{-\frac{1}{2}\left(\frac{t-\mu^{(l)}}{\sigma^{(l)}}\right)^2} dt + \\ \alpha^{(l)}\left(\frac{\mu^{(l)}-y}{\sigma^{2(l)}}\right) e^{\frac{\left[\left(\alpha^{(l)}+\sigma^{(l)}\right)\mu^{(l)}-\alpha^{(l)}y\right]^2}{2\sigma^{4(l)}}} \end{array} \right] \tag{20}$$

4.6 Updating the Parameter α

To update α, we equate $Q\left(\theta; \theta^{(l)}\right)$ with respect to α to zero. Therefore, the updated equation for α is

$$\alpha^{(l+1)} = \frac{\sqrt{2}\sigma^{2(l)}}{\mu^{(l)} - y} \left[\log\left(\int_{-\infty}^{\alpha^{(l)}\left(\frac{y-\mu^{(l)}}{\sigma^{(l)}}\right)} e^{-\frac{1}{2}\left(\frac{t-\mu^{(l)}}{\sigma^{(l)}}\right)^2} dt \right) - \log\left(\frac{y-\mu^{(l)}}{\sigma^{(l)}}\right) \right]^{\frac{1}{2}} - \frac{\sigma^{(l)}\mu^{(l)}}{\mu^{(l)} - y} \tag{21}$$

5 Experimentation and Results

Fp1, FpZ, Fp2, F7, F3, F2, F4, and F8 electrodes are utilized to recognize the emotions. The brain signals from the diseased persons are extracted using NeuroScan device with 64 electrodes (EEG cap), and the signals are preprocessed to minimize the noise, and the amplitude signals are extracted which are normalized into different ranges basing on the rhythm, for dimensionality reduction. For experimentation, DEAP database and PhysioBank by Gerwin Schalk were used. EEG cap is placed to the head, and an adhesive gel is filled in electrodes for decreasing the impedance. It is essential to have low impedance to obtain the noise-free emotion values. After looking at the low impedance levels in all electrodes, emotion values are extracted from the subject. Results of the experimentation are tabulated in Table 1.

Table 1 Emotion recognition (%) of immobilized persons and mobilized persons

Data acquisition method: Electroencephalography (EEG)	Happy	Sad	Angry	Neutral
Mobilized persons	90	86	92	82
Immobilized persons	82	81	89	79

6 Conclusion

Electroencephalography plays a very important as well as a vital role in the recognition of the thoughts effectively from the immobilized persons. These thoughts (or) emotions play a fundamental or major role in understanding the inherent feelings of the other person. Emotions act as a stimulus that evaluates the other persons based on the experiences. The proposed system is highly useful to immobilized persons who cannot express their feelings through speech signal. In future, research can be extended toward designing of BCI system with post-processing stage. Signal enhancement and feature extraction can also be termed as feature construction. A BCI system needs to be robust enough so that it can be used in practical applications.

References

1. Crosson, B., et al.: Functional imaging and related techniques: an introduction for rehabilitation researchers. J. Rehabil. Res. Dev. (JRRD) **47**(2), vii–xxxiii (2010)
2. Hartwigsen, G., Roman Siebner, H., Stippich, C.: Preoperative functional magnetic resonance imaging (fMRI) and transcranial magnetic stimulation (TMS). Curr. Med. Imaging Rev. **6**(4), 220–231 (2010)
3. Sakkalis, V.: Review of advanced techniques for the estimation of brain connectivity measured with EEG/MEG. Comput. Biol. Med. **41**, 1110–1117 (2011)
4. Sitaram, R., et.al.: Closed-loop brain training: the science of neurofeedback. Nat. Rev. Neurosci. **18**, 86–100. https://doi.org/10.1038/nrn.2016.164 (2017). Published online 22 Dec 2016
5. Devi, J.S., Yarramalle, S., Nandyala, S.P.: Speaker emotion recognition based on speech features and classification techniques. I. J. Comput. Netw. Inf. Secur. **7**, 61–77 (2014)
6. Wang, X.-W., Nie, D., Lu, B.-L.: Emotional state classification from EEG data using machine learning approach. Neurocomputing **129**, 94–106 (2014)
7. Krishna, N.M., Devi, J.S., Yarramalle, S.: A novel approach for effective emotion recognition using double truncated Gaussian mixture model and EEG. I. J. Intell. Syst. Appl. **6**, 33–42 (2017)
8. Rivet, B., Souloumiac, A., Attina, V., Gibert, G.: xDAWN algorithm to enhance evoked potentials: application to brain—computer interface. IEEE Trans. Biomed. Eng. **56**, 2035–2043 (2009)

Fault Detection and Classification of Multi-location and Evolving Faults in Double-Circuit Transmission Line Using ANN

V. Ashok, A. Yadav and Vinod Kumar Naik

Abstract This paper elaborated a new relaying scheme for fault detection and classification of shunt faults comprising of multi-location faults and evolving faults which are more severe in double-circuit transmission line. In these work, two DWT-ANN modules have been proposed to detect and classify multi-location faults and evolving faults using voltage and current signals of either end of a double-circuit transmission line. A 400 kV transmission system of Chhattisgarh state has been modeled in RSCAD software environment of real-time digital simulator (RTDS) to replicate various real-time fault scenarios and further analysis has been done in MATLAB software. The results demonstrated that all shunt faults are properly detected/classified in less than 10 ms. The distinctiveness of the DWT-ANN module is that it accurately detects/discriminates using one-terminal measurements/data only and it avoids maloperation of three-phase transmission line of another healthy circuit also, which has not been reported yet simultaneously.

Keywords Double-circuit line · Evolving faults · Multi-location faults ANN · RSCAD · RTDS

1 Introduction

The transmission lines are the most vulnerable element to fault among various power system elements. Fault diagnoses, fault classification along with faulty phase identification, are very important aspects of the protective system in order to prevent the revenue loss due to unplanned down time because of faults. There are

V. Ashok · A. Yadav (✉) · Vinod Kumar Naik
Department of Electrical Engineering, NIT Raipur, Raipur 492010, CG, India
e-mail: anamikajugnu4@gmail.com; yadav.ele@nitrr.ac.in

V. Ashok
e-mail: ashokjntuk@gmail.com

Vinod Kumar Naik
e-mail: forvinod.naik@gmail.com

J. Nayak et al. (eds.), *Soft Computing in Data Analytics*,
Advances in Intelligent Systems and Computing 758,
https://doi.org/10.1007/978-981-13-0514-6_31

various reasons due which a fault occurs in the transmission line such as lightning, thunderstorms, switching over voltage, falling of tree, birds, insulation failure due to contaminations and aging. Generally, transmission lines' faults are congregated such as shunt faults, multi-location faults, and evolving faults. The multi-location fault is defined 'as ground faults which occur in diverse phases of one circuit at different locations at same inception time' [1]. Moreover, "evolving faults are ground faults which initially occurs in one phase and dispersed to another healthy phase subsequently with few cycle delays" [2]. These multi-location faults and evolving faults have extra intricate characteristics consequently influence the efficacy of the distance protection scheme since the multi-location fault entails ground faults at diverse locations in dissimilar phases and the evolving fault starts with one type of fault and propagates into another type of fault. Thus, during the multi-location fault and evolving fault; the detection of actual faulty phase(s) thereby issuing a trip signal to respective circuit breaker is very complex task as compare to a common shunt fault.

2 Background Details and Related Work

There are several research papers on transmission line fault detection and classification using various techniques; a detailed view of transmission line protection including detection, classification, location, and direction estimation using neural network has been reported in [3]. A neural network-based high-speed fault classifier using neural network was proposed in [4]. Fast fault detection and classification based on a modular methodology with Kohonen-type neural network in power system were explained in [4]. An incorporation of supervised and unsupervised ANN for fault classification in parallel transmission lines was narrated in [5]. Wavelet transform-based neural network for fault detection and classification was explained in [6], and comparisons of different methods based on wavelet transform and ANN were reported in [7]. A comparative study of single and modular ANN-based fault detector and classifier is reported in [8]. By considering different operating/switching modes of double-circuit lines, Kohonen neural network was elucidated for fault detection in [9]. Detection and classification of an intercircuit faults and cross-country faults using neural network have been done in [10]. Nevertheless, all the above-reported techniques does not consider the detection and classification of multi-location faults and evolving faults in a double-circuit transmission line which is more crucial and complex in nature. While designing a relaying scheme, it is very important to consider the detection and discrimination of faulty phase of multi-location faults and evolving faults to avoid maloperation. To deal with the multi-location faults and evolving faults, a hybrid method based on discrete wavelet transform and artificial neural network has been proposed to detect the fault accurately and identify the faulty phase(s) and classify the fault type. In this work, the Bergeron model of the transmission line has been modeled in RSCAD/RTDS environment.

3 Power System Network Under Study

An existing 400 kV, 50 Hz double-circuit three-phase transmission system of Chhattisgarh state is shown in Fig. 1 comprising of consisting of two generating stations (Station-I with 4 × 500 MW and Station-II with 3 × 210 MW), one load (Load-1 with P = 1938.5 MW, Q = 20 MVAr) connected at bus-1 (KSTPS/NTPC) and bus-2 (Bhilai/Khedamara) is considered as a Thevenin's equivalent grid (400 kV). At bus-2, two loads are connected: Load-2 with P = 1347 MW, Q = 365 MVAr and Load-3 with P = 1800 MW, Q = 432 MVAr. Also, two line reactors are connected to absorb the reactive power in the line. In this paper, a double-circuit transmission line of 198 km length connected between bus-1 (KSTPS/NTPC) and bus-2 (Bhilai/Khedamara) has been considered to perform simulation studies for different types of shunt faults. The proposed distance-relaying scheme is employed at bus-1 (KSTPS/NTPC) and different types of faults are simulated using fault breakers.

4 Proposed Relaying Algorithm for Fault Detection/ Classification

The proposed relaying scheme deals with detection and classification of faults which are severe in double-circuit transmission lines. Therefore, two DWT-ANN modules have been designed, one for detection and another for classification. Using the DWT-ANN modules, detection and classification of all types of shunt faults can be done by using simply one-terminal measurement. The simulations are performed in RSCAD/RTDS environment at different fault conditions for evaluating the efficacy of the proposed relaying scheme is illustrated in Fig. 2. The deviation of the measured voltages and currents signals at one terminal of the line in the time sphere is very distinctive and explicit under numerous fault conditions.

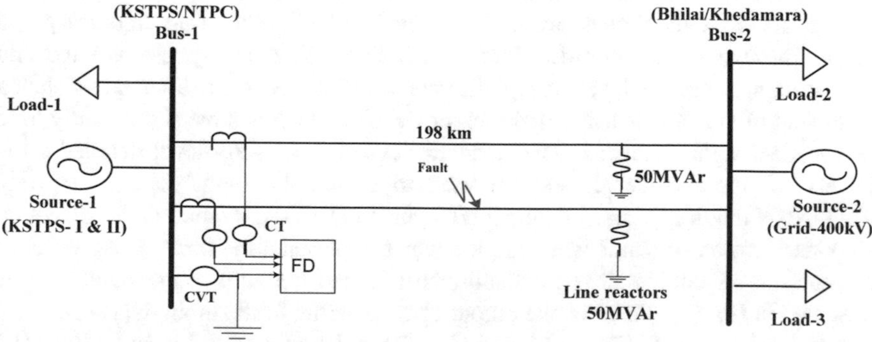

Fig. 1 400 kV double-circuit three-phase transmission system

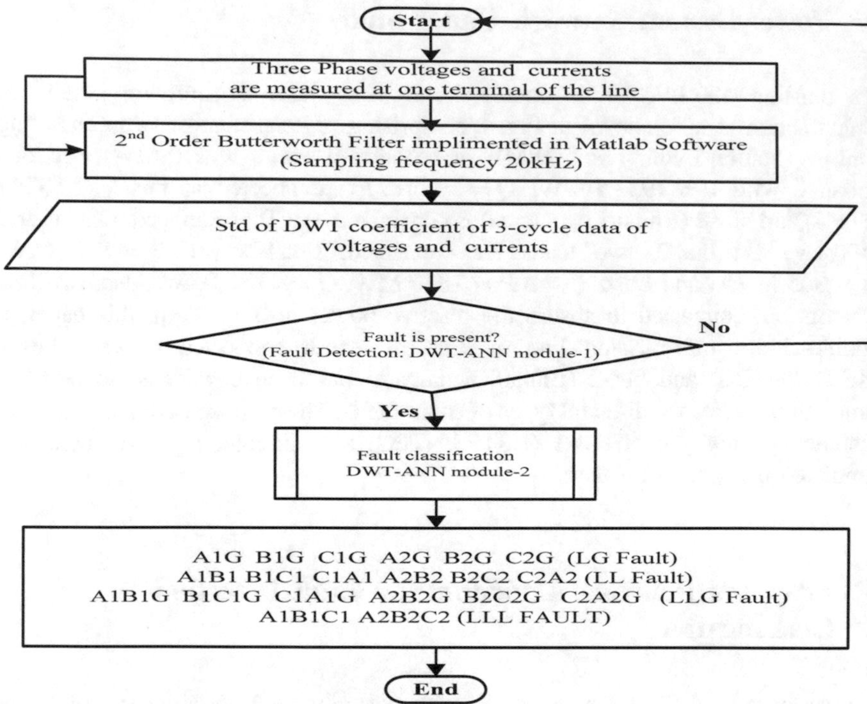

Fig. 2 Flowchart of the proposed relaying scheme

Designing of DWT-ANN-based relaying algorithm is based on the deviances in current and voltage signals of post-fault condition. When faults occur, it causes rise and drop in faulty phase(s) currents and voltages, respectively. As the functioning of neural network be determined by features input and output data set so, it is necessary to preprocess and abstract the constructive features to design the data set to train the DWT-ANN module. In this paper, input data set is generated by performing extensive simulation studies on existing 400 kV, 50 Hz double-circuit transmission line of 198 km span in RSCAD/RTDS environment. Moreover, the voltage and current signals are collected with 20 kHz sampling frequency and passed through a second-order Butterworth filter. The three-phase voltage and current signals are analyzed using discrete wavelet transform to obtain detailed coefficient of the signals using db4 mother wavelet which is most significantly used for transient signal analysis. The standard deviations of sixth-level detailed coefficients for three cycle signals are used to design the input data set to train DWT-ANN module. Therefore, the total inputs to DWT-ANN module for detection and classification of shunt faults are ten which represents the three-phase voltages and currents of double-circuit transmission line and the zero sequence current as presented in Eq. (1). There is one output corresponding to the input data set of fault detection module of DWT-ANN as per Eq. (2) which will be ‘0,’ if no fault, or ‘1,’ if there is fault.

$$X = [Ia1, Ib1, Ic1, Va1, Vb1, Vc1, Ig1, Ia2, Ib2, Ic2] \tag{1}$$

$$Y = [FD] \tag{2}$$

In addition to the fault detection, the fault type is determined by the fault classification DWT-ANN module. It has ten input vectors similar to the fault detection module which is shown in '(1).' Total seven outputs were considered as output vectors for determining the faulty phase classification (A1, B1, C1, G, A2, B2, C2) as shown in '(3).'

$$Y2 = [A1, B1, C1, G1, A2, B2, C2] \tag{3}$$

Faults comprising of multi-location shunt faults and evolving faults were simulated at different fault locations (between 1–197 km) with wide-ranging fault resistance (0–100 Ω) and fault inception angles (0–360°). From extensive simulation studies, it has been perceived that the DWT-ANN module is trained with the Levenberg–Marquardt algorithm confers superior results instead of back-propagation algorithm. Total fault cases taken for fault identification/detection module are 417. The neural network architecture for fault detection (DWT-ANN module-1) having three hidden layers with 'tansig' activation function and one output layer with 'purlin' function (10-10-30-40-1) minimized the mean square error (mse) to a decisive value of 0.00000989. Total number of fault cases have been taken for fault classification module are 417 and target/output data set has been designed analogous to the input cases. The DWT-ANN network-based fault classifier (DWT-ANN module-2) is trained using three hidden layers with 'tansig' activation function and one output layer with 'purlin' activation function (10-20-40-60-7), and the mean square error was reduced to 0.000000968 in 328 epochs.

5 Results and Discussions

The efficacy of the DWT-ANN module for fault detection and fault classification is tested for unknown faults' scenarios which have not been used earlier. Testing of DWT-ANN module was performed for common shunt faults, multi-location faults and evolving faults with different fault locations (L_f = 1–197 km), fault inception angles (Φ_f = 0–360°), and fault resistances (R_f = 0–100 Ω). The effect of various fault parameters on proposed DWT-ANN module has been reported hereunder in detail. For testing of trained ANN, a moving window of standard deviations of sixth-level detailed coefficients of current and voltage signals is used to get faulty phase detection time.

Table 1 Response of DWT-ANN module for variation of fault inception angle

Fault inception angle (Degree/Time)		Fault resistance (Ω)	Fault detection time (ms)	Fault phase identification/ classification time (ms)						
Fault 1	Fault 2			A1	B1	C1	G	A2	B2	C2
Multi-location faults										
AG at 4 km, Ø = 0°	BG at 7 km, Ø = 0°	5	1.0	1.0	1.0	–	1.0	–	–	–
AG at 4 km, Ø = 10°	CG at 13 km, Ø = 10°	5	1.0	1.0	–	1.0	1.0	–	–	–
AG at 4 km, Ø = 20°	BCG at 94 km, Ø = 20°	5	1.0	1	1.0	1.0	1.0	–	–	–
BG at 14 km, Ø = 30°	AG at 17 km, Ø = 30°	20	1.0	1.0	1.0	–	1.0	–	–	–
BG at 14 km, Ø = 40°	CG at 23 km, Ø = 40°	20	1.0	–	1.0	1.0	1.0	–	–	–
BG at 14 km, Ø = 50°	ACG at 104 km, Ø = 50°	20	1.0	4.0	1.0	1.0	1.0	–	–	–
CG at 24 km, Ø = 60°	AG at 27 km, Ø = 60°	70	1.0	1.0	–	1.0	1.0	–	–	–
CG at 24 km, Ø = 70°	BG at 33 km, Ø = 70°	70	1.0	–	1.0	1.0	1.0	–	–	–
CG at 24 km, Ø = 80°	ABG at 114 km, Ø = 80°	70	1.0	2.0	1.0	1.0	1.0	–	–	–
Evolving faults										
AG at 7 km, Øt1 = 1 s	BG at 7 km, Øt2 = 1.010 s	10	1.0	1.0	1.0	–	1.0	–	–	–
AG at 27 km, Øt1 = 1.001 s	CG at 27 km, Øt2 = 1.021 s	10	1.0	1.0		2.0	1.0	–	–	–
AG at 47 km, Øt1 = 1.003 s	BCG at 47 km, Øt2 = 1.033 s	10	1.0	1.0	1.0	1.0	1.0	–	–	–
BG at 67 km, Øt1 = 1.005 s	AG at 67 km, Øt2 = 1.0450 s	50	1.0	1.0	1.0	–	1.0	–	–	–
BG at 87 km, Øt1 = 1.006 s	CG at 87 km, Øt2 = 1.056 s	50	1.0	–	1.0	1.0	1.0	–	–	–
BG at 107 km, Øt1 = 1.008 s	ACG at 107 km, Øt2 = 1.068 s	50	1.0	1.0	1.0	–	1.0	–	–	–
CG at 127 km, Øt1 = 1.016 s	**AG at 127 km, Øt2 = 1.086 s**	**90**	**1.0**	**1.0**		**1.0**	**1.0**	**–**	**–**	**–**
CG at 147 km, Øt1 = 1.011 s	BG at 147 km, Øt2 = 1.091 s	90	1.0		1.0	1.0	3.0	–	–	–
CG at 167 km, Øt1 = 1.013 s	ABG at 167 km, Øt2 = 1.103 s	90	2.0	4.0	2.0	3.0	8.0	–	–	–

5.1 *Response of DWT-ANN Module for Variation of Fault Inception Angle*

In a real-time scenario, fault can occur at any instance in the line or at any inception angle which is measured from the zero crossing of the respective phase voltage. An extensive simulation studies have been performed by creating faults at different locations with different fault inception angles and different fault resistances. The effect of fault inception angle has been considered for 0°, 90°, and 270°. Test results of fault detector and classifier module for various multi-location faults and evolving faults are stated in Table 1. From Table 1, it can be noticed that the efficacy of the relaying scheme is not influenced by variation of different fault parameters.

From the above Fig. 3, it shows the currents and voltages signals of double-circuit transmission line during multi-location faults condition; A1G fault occurred at 34 km and C1G fault occurred at 43 km from relaying point at the same time (1.025 s.) and $R_f = 50\ \Omega$ and $Ø_f = 100°$. It can be clearly seen from the Fig. 3a that during multi-location fault, the magnitude of fault current in the faulty phases of circuit-1 is different; however, the proposed relaying scheme properly identified/classified the faulty phases and involvement ground as depicted in Fig. 3d, e.

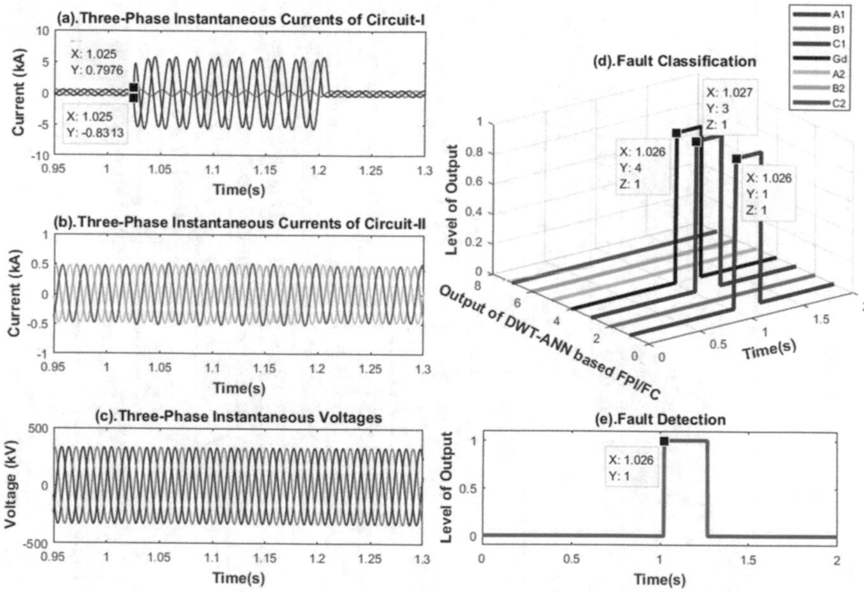

Fig. 3 Response of DWT-ANN-based FPI/FC in case of multi-location fault: A1G at $L_{f1} = 34$ km, C1G at $L_{f2} = 43$ km, $R_f = 50\ \Omega$, $\Theta_f = 100°$. (**a**)Three-phase instantaneous currents of ckt-1. (**b**) Three-phase instantaneous currents of ckt-2. (**c**) Three-phase instantaneous voltages at bus-1. (**d**) Response of fault classifier. (**e**) Response of fault detector

Table 2 Response of DWT-ANN module for variation of fault resistance

Fault resistance (ohm)	Fault inception angle (Degree/Time)		Fault detection time (ms)	Fault phase identification/classification time (ms)						
	Fault 1	Fault 2		A1	B1	C1	G	A1	B1	C1
Multi-location faults										
45	AG at 34 km, Ø = 100°	BG at 37 km, Ø = 100°	1.0	1.0	1.0	–	1.0	–	–	–
50	**AG at 34 km, Ø = 100°**	**CG at 43 km, Ø = 100°**	**1.0**	**1.0**	**–**	**1.0**	**1.0**	**–**	**–**	**–**
55	AG at 34 km, Ø = 100°	BCG at 124 km, Ø = 100°	1.0	1.0	1.0	1.0	1.0	–	–	–
60	BG at 44 km, Ø = 130°	AG at 47 km, Ø = 130°	1.0	1.0	1.0	–	1.0	–	–	–
65	BG at 44 km, Ø = 130°	CG at 47 km, Ø = 130°	1.0	–	1.0	1.0	1.0	–	–	–
70	BG at 44 km, Ø = 130°	ACG at 47 km, Ø = 130°	1.0	1.0	1.0	2.0	1.0	–	–	–
75	CG at 54 km, Ø = 160°	AG at 57 km, Ø = 160°	1.0	1.0	–	2.0	1.0	–	–	–
80	CG at 54 km, Ø = 160°	BG at 63 km, Ø = 160°	1.0	–	1.0	2.0	1.0	–	–	–
85	CG at 54 km, Ø = 160°	ABG at 144 km, Ø = 160°	1.0	1.0	1.0	1.0	1.0	–	–	–
Evolving faults										
0	AG at 7 km, Øt1 = 1.001 s	BG at 7 km, Øt2 = 1.011 s	1.0	1.0	1.0	–	1.0	–	–	–
10	AG at 7 km, Øt1 = 1.001 s	CG at 7 km, Øt2 = 1.011 s	1.0	1.0		1.0	1.0	–	–	–
20	AG at 7 km, Øt1 = 1.001 s	BCG at 7 km, Øt2 = 1.011 s	1.0	1.0	1.0	2.0	1.0	–	–	–
30	BG at 87 km, Øt1 = 1.003 s	AG at 87 km, Øt2 = 1.023 s	1.0	1.0	1.0	–	1.0	–	–	–
40	BG at 87 km, Øt1 = 1.003 s	CG at87 km, Øt2 = 1.023 s	1.0	–	1.0	2.0	1.0	–	–	–
50	BG at 87 km, Øt1 = 1.003 s	ACG at 87 km, Øt2 = 1.023 s	1.0	2.0	1.0	1.0	1.0	–	–	–
60	CG at 147 km, Øt1 = 1.005 s	AG at147 km, Øt2 = 1.035 s	2.0	2.0	–	2.0	10.0	–	–	–
70	CG at 147 km, Øt1 = 1.005 s	BG at147 km, Øt2 = 1.035 s	1.0	–	1.0	2.0	10.0	–	–	–
80	CG at 147 km, Øt1 = 1.005 s	ABG at 147 km, Øt2 = 1.035 s	1.0	1.0	1.0	2.0	10.0	–	–	–

5.2 *Response of DWT-ANN Module for Variation of Fault Resistance*

In general, phase to ground faults may occur with fault resistance and this fault resistance is caused by arcs and tower grounding. Mostly, this fault resistance is in the range of 10–20 Ω, but some of the cases it various to very higher value with high fault resistances, the fault current flow at relaying point is very small compared to other fault conditions and it may lead to maloperation of the relay. Therefore, it is very important to investigate the impact of fault resistance on proposed relaying scheme where the selectivity and reliability are more concerned in relaying scheme. In spite of this, different types of shunt faults have been considered to evaluate the effectiveness of proposed relaying scheme by changing the fault resistance from 0 to 100 Ω and the test results are depicted in Table 2. Figure 4 elucidates the response of DWT-ANN-based FPI/FC in case of evolving fault: CG fault occurred

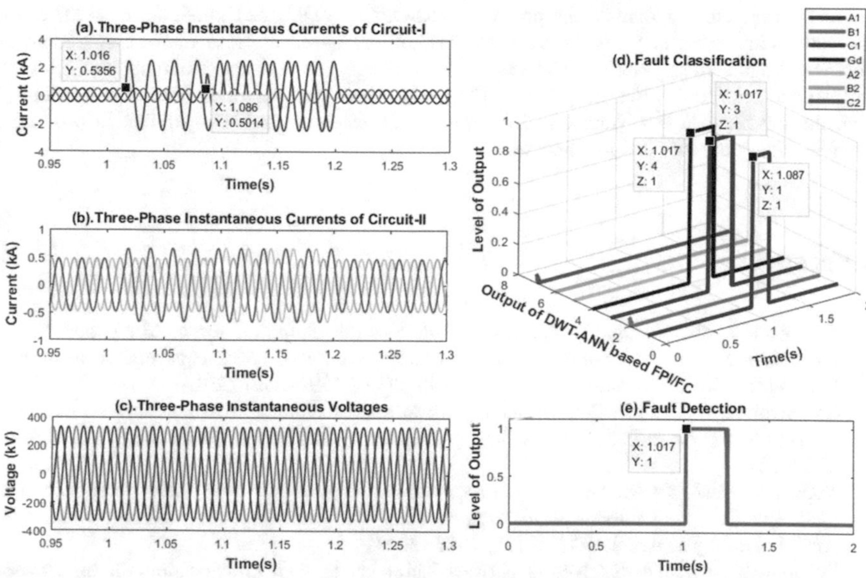

Fig. 4 Response of DWT-ANN-based FPI/FC in case of evolving fault: CG fault occurred at 1.016 s after 70 ms (3.5 cycles delay) converted to CAG fault with Lf = 127 km, Rf = 90 Ω, ϴf = 90°: **(a)** Three-phase instantaneous currents of ckt-1, **(b)** Three-phase instantaneous currents of ckt-2, **(c)** Three-phase instantaneous voltages at bus-1, **(d)** Response of fault classifier, **(e)** Response of fault detector

at 1.016 s after 70 ms (3.5 cycles delay) converted to CAG fault with $L_f = 127$ km, $R_f = 90\Omega$, $\Theta_f = 90°$. This evolving fault is discriminated by the different detection time of the fault phases C1 at 1.017 s and A1 at 1.087 s as depicted in Fig. 3 (d).

6 Conclusions

In this paper, DWT-ANN-based fault detection and classification approach is reported for multi-location faults, evolving and normal shunt faults. The scheme remains stable during extensive variations in fault type, fault location, fault inception angle, fault resistance. The method detects the fault type and faulty phase using one-terminal/single-end data. The fault detection is very fast 1–2 ms in most of the fault cases tested. The proposed scheme has been verified in real-time digital simulator (RTDS).

Acknowledgements The authors acknowledge the financial support of Central Power Research Institute, Bangalore, for funding the project no. RSOP/2016/TR/1/22032016, dated: 19.07.2016. The authors are thankful to the Head of the institution as well as Head of the Department of Electrical Engineering, National Institute of Technology, Raipur, for providing the research facilities to carry this research project. The authors are grateful to the local power utility (Chhattisgarh State Power Transmission Company Limited) for their cooperation in providing valuable data to execute this research work.

References

1. Swetapadma, A., Yadav, A.: Improved fault location algorithm for multi-location faults, transforming faults and shunt faults in thyristor controlled series capacitor compensated transmission line. IET Gener. Transm. Distrib. **9**(13), 1597–1607 (2015)
2. Swetapadma, A., Yadav, A.: All shunt fault location including cross-country and evolving faults in transmission lines without fault type classification. Electr. Power Syst. Res. **123**, 1–12 (2015)
3. Yadav, A., Dash, Y.: An overview of transmission line protection by artificial neural network: fault detection, fault classification, fault location, and fault direction discrimination. Adv. Artif. Neural Syst. Article ID 230382, 1–20 (2014)
4. Dalstein, T., Kulicke, B.: Neural network approach to fault classification for high speed protective relaying. IEEE Trans. Power Deliv. **10**, 1002–1011 (1995)
5. Agarwal, R.K., Xuan, Q.Y., Dunn, R.W., Johns, A.T., Bennett, A.: A novel fault classification technique of double-circuit lines based on a combined unsupervised/supervised neural network. IEEE Trans. Power Deliv. **14**, 1250–1256 (1999)
6. Silva, K.M., Souza, B.A., Brito, N.S.D.: Fault detection and classification in transmission lines based on wavelet transform and ANN. IEEE Trans. Power Deliv. **21**, 2058–2063 (2006)
7. Mahanty, R.N., Dutta Gupta, P.B.: Comparison of fault classification methods based on wavelet analysis and ANN. J. Electr. Power Compon. Syst. **34**, 47–60 (2006)
8. Yadav, A.: Comparison of single and modular ANN based fault detector and classifier for double-circuit transmission lines. Int. J. Eng. Sci. Technol. **4**, 122–136 (2012)

9. Skok, S., Marusic, A., Tesnjak, S., Pevik, L.: Double-circuit line adaptive protection based on kohonen neural network considering different operation and switching modes. Larg. Eng. Syst. Conf. Power Eng. **2**, 153–157 (2002)
10. Jain, A., Thoke, A.S., Patel, R.N., Koley, E.: Inter-circuit and cross-country fault detection and classification using artificial neural network, INDICON. In: 2010 Annual IEEE, pp. 1156–1161 (2010)

Shear Force Analysis and Modeling for Discharge Estimation Using Numerical and GP for Compound Channels

Alok Adhikari, Nibedita Adhikari and K. C. Patra

Abstract Experiments are conducted with a compound channel model framed in the laboratory to assess the shear stress between main channel and floodplain interfaces for smooth and rough floodplains which practically counts for the discharge assessment. Results are analyzed for compound channel with different rough floodplains. Modified method for calculation of shear force (*SF*) is suggested through numerical model development. Again, genetic programming technique is used to assess the % *SF* at different boundary conditions. Finally, methods of other researchers are compared with the suggested models and found that suggested models gave minimum error than other findings.

Keywords Genetic programming · Boundary shear · Shear force
Apparent shear force

1 Introduction

In hydrology, a reliable assessment for discharge of a compound river section encompassing a shallow main channel along with one or two flanking floodplains has been the subject of extensive research. The prediction helps in overall design, operation, and maintenance of streams and its periphery. To achieve this, the quantum of flow is the primary information to be predicted. According to researchers, the behavior of the flow in compound channels is complex as it

A. Adhikari (✉)
Department of Civil Engineering, NIT Rourkela, Rourkela 769014, India
e-mail: write2aa04@gmail.com

N. Adhikari
Biju Pattnaik University of Technology, Rourkela 769015, India
e-mail: drnibeditaadhikari@gmail.com

K. C. Patra
Department of Civil Engineering, NIT Rourkela, Rourkela 769008, India
e-mail: kcpatra@nitrkl.ac.in

J. Nayak et al. (eds.), *Soft Computing in Data Analytics*,
Advances in Intelligent Systems and Computing 758,
https://doi.org/10.1007/978-981-13-0514-6_32

depends upon various hydraulic and geometric parameters. The overall conveyance of the compound channels depends upon various parameters like the hydraulics radius, cross-sectional area, wetted perimeter, bed slope, roughness, shear, depth ratio, width ratio, sinuosity, viscosity, gravitational acceleration, momentum transfer mechanism, and other [1, 2].

Many investigations have been made to explore the phenomenon of momentum shift for a compound channel from one section to other. Flow irrespective of its stage encounters shear starting from different boundaries to shear taking place even between adjacent flow layers. The most prominent interaction takes place in between the boundary between main channel and floodplain, and the transfer takes the form of apparent shear stress. Studies relating to transfer of momentum across main channel and floodplain with varied roughness depicting common vegetation growth in floodplains are not much considered under research. Myers in 1978 described transmission of momentum within main channel–floodplain boundary as apparent shear stress [3]. It has been proven that the flow capacity reduces in a channel due to momentum transfer across the boundaries [4].

Knight et al. studied distribution of boundary shear force for a rough compound channel. They experimented with rough floodplains keeping main channel smooth and analyzed the effect of differential roughness between them [5]. They emphasized the computation of total shear force percentage carried by the floodplain (% SF_{fp}) through Eq. 1,

$$\%\mathrm{SF}_{\mathrm{fp}} = 48(\alpha - 0.8)^{0.289} 2\beta^{1/m}\left(1 + 1.02\sqrt{\beta}\log\gamma\right) \tag{1}$$

where m can be computed from $m = \frac{1}{0.75e^{0.38\alpha}}$, $\alpha = B/b$ is the dimensionless width ratio; B is width of compound channel, and b is the width of main channel base; $\beta = (H-h)/H$ is the relative depth, H is the depth of compound channel flow in the main channel, and h is the depth of main channel, Υ being the ratio of roughness coefficient of floodplain (n_{fp}) and that of main channel (n_{mc}) to that of the main channel sections [5, 9]. Equation 1 was developed for width ratio $\alpha = 4$ and γ in the range of 1–3.

Equation 1 is found to be complex, for which Khatua and Patra [6] modified it for compound channel with differential roughness having width ratio in the range 2–4,

$$\%\mathrm{SF}_{\mathrm{fp}} = 1.23\beta^{0.1833}(38\ \ln\alpha + 3.6262)\left(1 + 1.02\sqrt{\beta}\log\gamma\right) \tag{2}$$

Khatua et al. (2012) derived an equation for percentage of shear force as a function of floodplain area [7]. The method is derived for width ratio of 6.67 with uniform roughness for which the methods of Eqs. 1 and 2 predicted errors in the range of 70%. The percentage of shear force in floodplain is expressed as,

$$\%SF_{fp} = 4.105\left[\frac{100\beta(\alpha-1)}{1+\beta(\alpha-1)}\right]^{0.6917} \quad (3)$$

It is observed that although Khatua et al. predict well for the smooth channel, it provides large errors for channels having differential roughness of main channel and floodplain. Therefore, there is a need for an improvement in the model to successfully predict $\%SF_{fp}$ for such channels.

2 Experimental Arrangements

Experiments are conducted in a straight compound channel with symmetrical floodplains measuring 12 m × 2 m × 0.6 m in the Hydraulics Engineering laboratory of the National Institute of Technology Rourkela, India. Fabrication of channel is made with Perspex sheet of 0.006 m having a uniform Manning's n value to be 0.01. A trapezoidal channel is glued with chemicals and is installed inside a tilting flume. The channel has the width ratio (α) as 15.75 and the aspect ratio (δ) of 1.5. Water supply to the channel is made from a sump with a re-circulating system through one overhead tank (Fig. 1). The experimental channel section is shown in Fig. 2. Objective of this work is to know the effect of roughness in the floodplains on the flow behavior during high floods. Therefore, observations are carried out in one run with same Perspex sheet roughness in the floodplain (smooth) and roughened floodplains to other types for different runs. In the floodplains, different roughening materials are used to provide the effect of vegetation. For roughening, a synthetic mat is used in the floodplain surfaces having spikes 12 mm long 1.5 mm width with 72 spikes per square inch (Rough I). In another observation, wire mesh is used for roughening the surfaces in the floodplains (Rough II). In the third case, wire mesh in main channel with crushed stone at floodplains (Rough III) is used while in the fourth case smooth main channel with crushed stone in floodplains is used (Rough IV). Wire mesh used is having mesh opening size of 3 mm × 3 mm with wire diameter of 0.4 mm. Crushed stones used for roughening having equivalent sand roughness of 3.39 mm.

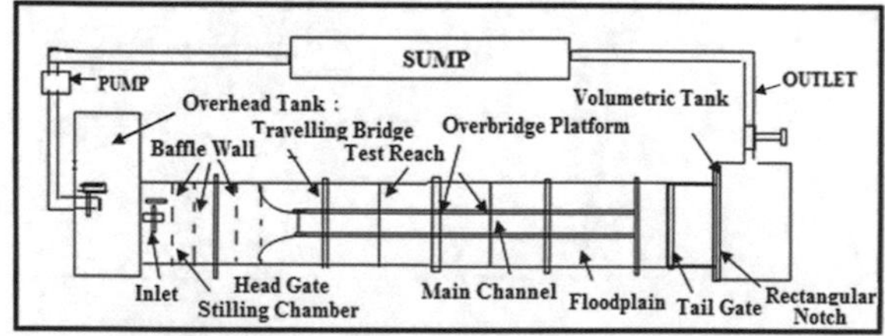

Fig. 1 Schematic drawing of experimental system

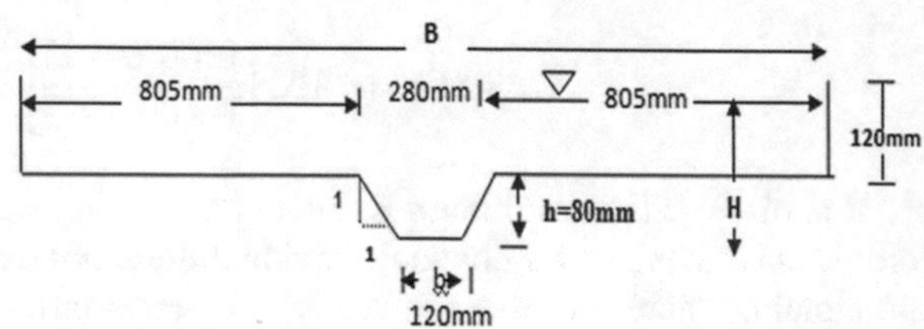

Fig. 2 Straight compound channel section

3 Boundary Shear Stress Assessments

Using the Preston tube technique, differential velocities are measured along the wetted perimeter which is converted to the boundary shear stress using Patel's Equation [6]. Boundary shear measurements are carried out at the entire cross section for relative depths (β) = 0.13, 0.2, 0.3, and 0.34. The shear stress profiles along the entire width of the channel are shown in Fig. 3 to all the seven sides of the channel, namely the floodplain walls (F-1, F-7), floodplain bed (F-2, F-6), main channel wall (F-3, F-5), and the main channel bed (F-4) and its mirroring.

Many researchers observed and reported that the distribution of boundary shear stress to be non-uniform over the wetted perimeter [5–11]. Study of boundary shear in rough compound channel with roughness variations with width ratio, α value more than 10, is very exceptional.

4 Allocation of Boundary Shear Force

The measured experimental boundary shear stress is integrated over the wetted perimeter to get the shear force at the different regions across the cross section for each depth of flow from one floodplain to the other and can be represented as,

Shear force in floodplain sidewall,

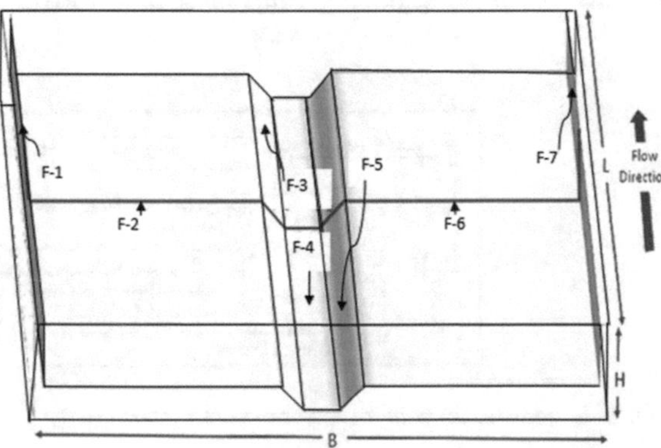

Fig. 3 Geometry showing the notations of each boundary for boundary shears' distribution

$$(\mathrm{F}-1)=\int_{F-1}\tau dp \quad (4)$$

Shear force in floodplain bed,

$$(\mathrm{F}-2)=\int_{F-2}\tau dp \quad (5)$$

Shear force in main channel wall,

$$(\mathrm{F}-3)=\int_{F-3}\tau dp \quad (6)$$

Shear force in the main channel bed,

$$(\mathrm{F}-4)=\int_{F-4}\tau dp \quad (7)$$

The total shear force can be represented as,

$$\mathrm{SF_T}=2\mathrm{SF_{F-1}}+2\mathrm{SF_{F-2}}+2\mathrm{SF_{F-3}}+\mathrm{SF_{F-4}} \quad (8)$$

Theoretically,

$$\mathrm{SF_T}=\rho gAS \quad (9)$$

Here, *SF* represents shear force, τ is the distribution of shear stress along the boundaries, p represents the respective wetted perimeter, and SF_{T} represents the total shear force. The total *SF* is computed using Eq. 8. The actual value is compared with the average of total theoretical *SF* by taking the mean of each section from Eq. 9. The error found between the values is less than 10% and is distributed proportionately among the bed and walls.

The floodplain sidewall region (i.e., F-1) is observed to increase its share of shear force with the increase of flow depth for both the cases of floodplain roughness. For same relative depth, the floodplains with roughened materials contribute more shear force. Similar observation is deduced for the main channel sidewall and the bed (i.e., F-3 and F-4, respectively). The *SF* carried by the floodplain bed (F-2) decreases with the relative depth to compensate the increase in the other regions. The main channel sidewalls are noticed to share higher shear force than the main channel bed.

Table 1 depicts the distribution of *SF*, the percentage of sharing of normalized *SF* and the averaged total theoretical *SF* for straight smooth compound channel with β varying from 0.13 to 0.4, whereas Table 2 shows the sharing of shear force

Table 1 % *SF* per length at different sections for straight smooth compound channel

Relative depth	Flow depth (m)	SF_T(Exp.) (N/m^2)	SF_T(Actual) (N/m^2)	SF_{F-1}as (% per m) of SF_T	SF_{F-2}as (% per m) of SF_T	SF_{F-3}as (% per m) of SF_T	SF_{F-4} as (% per m) of SF_T
$\beta = 0.13$	0.092	1.712	0.996	17.128	16.127	13.033	3.712
$\beta = 0.2$	0.100	2.380	1.420	9.993	20.081	15.861	4.065
$\beta = 0.23$	0.104	2.605	1.640	12.760	15.191	17.582	4.466
$\beta = 0.27$	0.110	1.986	1.920	14.740	17.342	12.553	5.366
$\beta = 0.32$	0.117	2.544	2.268	9.016	16.294	18.657	6.033
$\beta = 0.4$	0.132	2.000	3.016	12.041	15.455	16.483	6.021

Table 2 % *SF* per length at different sections with rough floodplain

Relative depth	Flow depth (m)	SF_T(Exp.) (N/m^2)	SF_T(Actual) (N/m^2)	SF_{F-1}as (% per m) of SF_T	SF_{F-2}as (% per m) of SF_T	SF_{F-3}as (% per m) of SF_T	SF_{F-4} as (% per m) of SF_T
$\beta = 0.2$	0.100	2.201	1.420	14.689	17.485	14.025	3.801
$\beta = 0.23$	0.104	2.276	1.640	14.689	17.372	14.039	3.901
$\beta = 0.27$	0.110	1.906	1.918	12.544	14.325	17.874	5.257
$\beta = 0.3$	0.115	2.438	2.168	12.123	14.274	17.334	6.269
$\beta = 0.33$	0.120	2.547	2.418	10.724	14.775	18.706	5.795
$\beta = 0.36$	0.125	1.995	2.666	12.742	15.170	16.179	5.910
$\beta = 0.38$	0.130	2.025	2.916	12.680	16.043	15.431	5.846

distribution throughout the channel sections for straight compound channel having rough floodplain with varying β from 0.2 to 0.38.

The *%SF* per unit length in the floodplain side wall is observed to be more in the case of rough channels as compared to the similar value of relative depth in smooth channel. At less relative depth, the *%SF* per unit length for the floodplain in rough channel is bit more than smooth channel, whereas at higher relative depth it becomes more or less constant. At the main channel-side slope, *%SF* per unit length is more in the case of rough channels. Similar observation is also seen in the case of main channel bed.

5 Boundary Shear Model Development

To assess the boundary shear distribution, the component-wise *SF* along the wetted perimeter (SF_i) of floodplain is obtained by multiplying the attenuated shear stress on each point with appropriate wetted perimeter element of floodplain, and then all are summed. The resulting value is doubled due to mirroring to provide total shear force of the floodplains (SF_{fp}) and the % SF. The % SF carried by the floodplains by the experimental analysis conducted for different channel roughnesses in the laboratory and their corresponding flow depths are recorded for analysis. It is observed that $\%SF_{fp}$ is directly proportional to the relative depth of flow in all the channel types whereas is inversely proportional to the differential roughness, γ. This may be due to the surface roughness, which retards the velocity of the flow on floodplain resulting in the decrease in SF. Simultaneously, it increases the velocity in the main channel.

Table 4 illustrates the predicted values of % SF_{fp} by Eqs. (1), (2), and (3) [6]. To improve the results derived by Khatua et al. a multiplication factor for % SF_{fp} between the actual and derived values from Eq. (3) with variation of differential roughness is found for all the data sets. An exponential relation is observed and the modified equation for Eq. (3) is expressed as,

$$\%\mathrm{SF}_{\mathrm{fp}} = 4.105\left[\frac{100\beta(\alpha-1)}{1+\beta(\alpha-1)}\right]^{0.6917}\left(1-3.22\ln\gamma e^{-1.441\beta}\right) \quad (10)$$

Equations. (1), (2), (3), and (10) are used to determine the $\%SF_{fp}$ and the percentage of error for the different experimental data sets of NITR. The developed model is observed to predict acceptably better results for all the channels with respect to the other models.

6 Apparent Shear Stress

The apparent shear forces (ASF) act on the imaginary interface of the compound section. The momentum transfer between the floodplains and the main channel is evaluated by the percentage of floodplain shear. It plays a vital role in selecting the interface while estimating the discharge. The compound section discharge is obtained by dividing the channel into hydraulically homogeneous regions. The planes are originating from the junction of the floodplain and main channel. Thus, it can be considered that the floodplain region is moving separately from the main channel. The assumed interfaces may be (1) vertical, (2) horizontal, or (3) diagonal. The different division methods are illustrated in Fig. 4 [6].

Because the boundary shear stress carried by the compound section (ρgAS_f) is equal to 100%, where A is the total cross section of the compound channel, the % SF carried by the main channel surfaces can be calculated as;

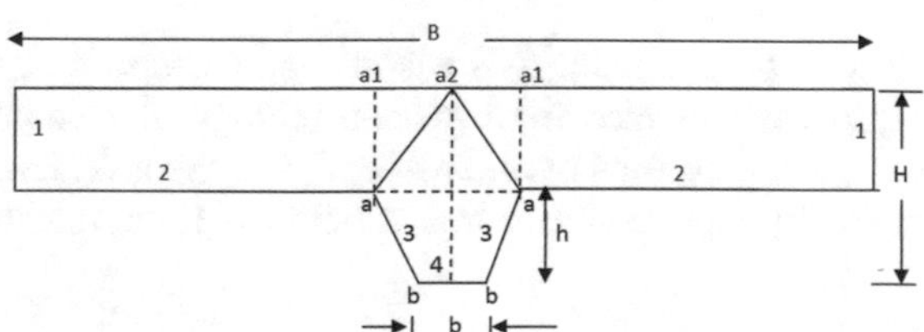

Fig. 4 Division methods in compound channel

$$100\frac{\int \tau dpA_{mc}}{\rho gAS_f} = \%\left[SF_{(3)} + SF_{(4)}\right] = 100\frac{\rho gA_{mc}S_f}{\rho gAS} - 100\frac{ASF_{ip}}{\rho gAS_f} \tag{11}$$

But $\%\left[SF_{(3)} + SF_{(4)}\right] = 100 - \%\left[SF_{(1)} + SF_{(2)}\right]$; and $100\frac{ASF_{ip}}{\rho gAS_f} = \%$ SF on the assumed interface. Substituting the values, the *ASF* on the interface plane can be calculated as

$$\%ASF_{ip} = 100\frac{A_{mc}}{A} - \{100 - \%[SF_1 + SF_2]\} \tag{12}$$

where $\% ASF_{ip}$ = percentage of shear force in the interface plane.

For vertical interface between the boundary of the floodplain and main channel shown by the lines aa1 in Fig. 4, the value of A_{mc} is the area marked by a1abbaa1, which when substituted in Eq. (11), yields $\%ASF_V$. Similarly, for horizontal or diagonal interfaces, A_{mc} is estimated from the area marked as aabb or a_2abbaa_2, respectively, in Fig. 4. This *ASF* is expressed as percentages of the total channel shear force using the following equations for vertical (Eq. 13), horizontal (Eq. 14), and diagonal (Eq. 15) interfaces.

$$\%ASF_V = \frac{50}{[(\alpha - 1)\beta + 1]} - \frac{1}{2}\{100 - \%[SF_1 + SF_2]\} \tag{13}$$

$$\%ASF_H = \frac{100(1 - \beta)}{[(\alpha - 1)\beta + 1]} - \{100 - \%[SF_1 + SF_2]\} \tag{14}$$

$$\%ASF_D = \frac{25(2 - \beta)}{[(\alpha - 1)\beta + 1]} - \frac{1}{2}\{100 - \%[SF_1 + SF_2]\} \tag{15}$$

% *ASF* for the three assumed interface planes for the present experimental dataset series of NITR is shown in Tables 4 (column 6 and 12) and 5 (column 6 and 12). The table compares the measured shear force percentages carried by the floodplains in each case along with the computed values. It is observed that the average error is minimum for the present experimental model for smooth and all the three cases of floodplain roughness. Again, it is concluded that the diagonal division method is a more appropriate method of channel division.

7 Genetic Programming

Genetic programming (GP) is an optimization tool which uses machine learning approach motivated from general biology. It randomly generates a population of computer programs in order to optimize the task. Here, the computer programs are internally represented as tree structures. Then, mutation and crossover are carried

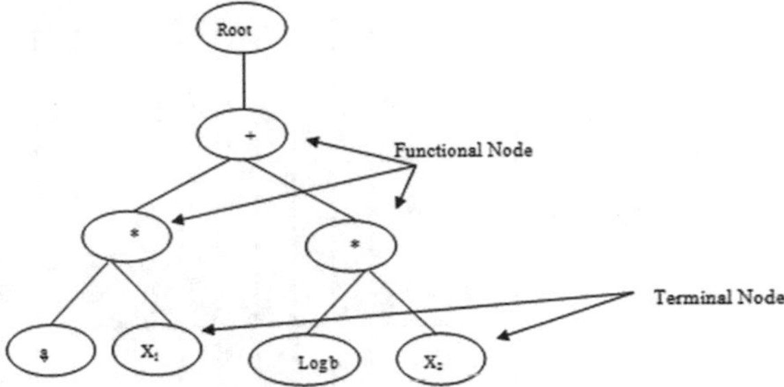

Fig. 5 GP tree for expression $ax_1 + Log\ b\ x_2$

out on best performing trees to find a new population. This process is iterated until the population contains the programs solve the task well. The mathematical models derived in GP are more compact and robust [12–14].

For the current research, the open-source MATLAB toolbox [15] called GPTIPS is used as it is developed for specific purpose of performing symbolic regression. It employs a unique type of symbolic regression called multigene symbolic regression that evolves linear combinations of nonlinear transformations of the input variables [14]. GP model is described through a tree structure which is composed of nodes. Nodes are either members of a functional set or terminal set. A functional set may include arithmetic operators (+, −, *, /), mathematical functions (sin, cos, cot, or log), Boolean operators (AND, OR, NOT, etc), logical expressions like IF, THEN, and ELSE or any other suitable functions defined by the users, whereas the terminal set includes variables (like x_1, x_2, x_3) and/or constants for example 1, 2, 3, and 4. The functions and terminals are arbitrarily chosen to construct a GP tree with a root node and the branches implicitly describe a mathematical equation consisting of function nodes which end with terminal nodes as shown in Fig. 5 for the expression, $y = aX_1 + \log bX_2$.

8 Shear Force Prediction Using GPTIPS

From previous studies, it is observed that different hydraulic parameters play crucial role for fluctuation of discharge data. Using artificial neural networks with stage as single input to predict discharge as the output, different network models are derived earlier. It is observed the model equations derived are much complex. Therefore, different common hydraulic characteristics are taken as input for calculation of *SF* for predicting discharge using GP.

Table 3 Description of GP models for different data sets

Model	Training data (Initial population)	Testing data	No. of generations	Tournament size	D_{max}	G_{max}	Function node set
Smooth	120	20	77	4	3	5	+,-,*, plog, square
Rough	28	10	0	4	4	5	

Table 4 Calculation sheet for $\%S_{fp}$, $\%ASF_V$ and average error

Channel type	$\%SF_{fp}$						$\%ASF_V$					
	Observed	Equation (1)	Equation (2)	Equation (3)	Equation (10) new model	Equation (16) GP model	Observed	Equation (1)	Equation (2)	Equation (3)	Equation (10) new model	Equation (16) GP model
1	2	3	4	5	6	7	8	9	10	11	12	13
Smooth channel	89.85	104.40	91.15	73.64	73.64	74.35	32.98	40.25	33.63	24.87	24.87	25.23
	90.31	104.56	99.26	81.10	81.10	74.06	28.80	35.93	33.27	24.20	24.20	20.68
	86.59	104.62	102.13	83.21	83.21	73.70	25.32	34.34	33.10	23.64	23.64	18.88
	89.73	104.67	105.06	85.12	85.12	73.12	25.25	32.72	32.92	22.94	22.94	16.95
	85.85	104.72	107.95	86.77	86.77	72.25	21.71	31.15	32.76	22.17	22.17	14.91
	86.49	104.80	112.39	88.92	88.92	69.86	19.65	28.81	32.60	20.87	20.87	11.34
Average error								**33.87**	**33.05**	**23.11**	**23.11**	**18.00**
Rough floodplain	90.19	106.16	100.77	82.34	66.03	74.06	28.74	36.72	34.03	24.81	16.66	20.67
	90.13	106.34	103.81	84.59	68.49	73.69	27.10	35.20	33.94	24.32	16.27	18.88
	85.25	106.54	106.93	86.63	70.88	73.11	23.01	33.65	33.85	23.70	15.83	16.94
	84.78	106.68	109.21	87.98	72.55	71.66	21.59	32.54	33.81	23.19	15.48	15.03
	84.90	106.80	111.14	89.04	73.93	70.74	20.67	31.62	33.79	22.73	15.18	13.59
	86.45	106.91	112.81	89.88	75.09	69.68	20.61	30.84	33.79	22.32	14.93	12.22
	87.50	107.01	114.26	90.58	76.07	68.41	20.42	30.17	33.80	21.96	14.70	10.87
Average error								**32.96**	**33.86**	**23.29**	**15.58**	**15.46**

Table 5 Calculation sheet for $\%ASF_{H}$, $\%ASF_{D}$ and average error

Channel type	$\%ASF_H$						$\%ASF_D$					
	Observed	Equation (1)	Equation (2)	Equation (3)	Equation (10) new model	Equation (16) GP model	Observed	Equation (1)	Equation (2)	Equation (3)	Equation (10) new model	Equation (16) GP model
1	2	3	4	5	6	7	8	9	10	11	12	13
Smooth channel	53.20	67.75	54.50	36.99	36.99	37.70	29.79	37.06	30.44	21.68	21.68	22.04
	40.15	54.41	49.10	30.94	30.94	23.91	24.44	31.56	28.91	19.83	19.83	16.31
	31.48	49.51	47.02	28.11	28.11	18.59	20.53	29.55	28.30	18.84	18.84	14.09
	29.58	44.52	44.91	24.97	24.97	12.97	20.02	27.49	27.69	17.71	17.71	11.72
	20.80	39.67	42.90	21.71	21.71	7.20	16.06	25.49	27.11	16.51	16.51	9.26
	14.14	32.45	40.04	16.58	16.58	-2.48	13.36	22.52	26.31	14.58	14.58	5.05
Average error		**48.05**	**46.41**	**26.55**	**26.55**	**16.31**		**28.95**	**28.13**	**18.19**	**18.19**	**13.08**
Rough floodplain	40.03	56.00	50.61	32.18	15.88	23.90	24.38	32.36	29.67	20.45	12.30	16.31
	35.02	51.23	48.71	29.48	13.38	18.58	22.30	30.41	29.14	19.53	11.48	14.08
	25.10	46.39	46.78	26.48	10.73	12.96	17.78	28.42	28.62	18.47	10.60	11.71
	21.00	42.90	45.43	24.20	8.77	7.88	16.05	27.00	28.26	17.65	9.93	9.48
	18.09	40.00	44.34	22.23	7.13	3.94	14.86	25.81	27.98	16.92	9.37	7.78
	17.09	37.55	43.45	20.52	5.72	0.31	14.58	24.81	27.75	16.29	8.89	6.19
	15.94	35.45	42.71	19.02	4.52	-3.15	14.19	23.95	27.58	15.73	8.48	4.65
Average error		**44.22**	**46.00**	**24.87**	**9.45**	**9.20**		**27.54**	**28.43**	**17.86**	**10.15**	**10.03**

9 GPTIPS Run Settings

A GPTIPS run with the following settings is performed for all five models discussed above. Initially, a set of GP trees, as per the data sets of each model, are randomly generated using various functions and terminals assigned. The fitness criteria calculate the objective function, which determines the best population. In the mating pool, the reproduction process creates a new population at each generation through crossover and mutation of the selected GP trees. Then, the existing population gets replaced by new population. This entire process gets repeated until the termination criterion, which can be either a threshold fitness value or maximum number of generations, whichever is earlier is satisfied. The best GP model (Table 3) based on its fitness value is selected in order to minimise the root-mean-squared prediction error on the testing data.

Besides, the data collected from the present experimental setup data sets from flood facility at HR Wallingford, UK, and the reported data from other investigators are taken for analysis [3, 9, 16, 17]. The GP model derived Eq. 16 after several runs with minimum bias and maximum number of parameters.

$$\begin{aligned}\%SF_{fp} = {} & 3.445\,\alpha + 12.5\mathrm{plog}(\mathrm{plog}(\alpha(S_0+\beta)) - 3.445\gamma^2\beta^2 + 14.19\alpha(S_0-\beta) \\ & + 70.2\mathrm{plog}(\alpha(2S_0+\beta)) + 42.68\end{aligned} \tag{16}$$

where α is width ratio, S_0 the bed slope, γ the differential roughness between floodplain to main channel, and β the depth ratio. Percentages of *ASF* for the three assumed interface planes for the present experimental dataset series of NITR for GP model are shown in Tables 4 (column 7 and 13) and 5 (column 7 and 13) using Eq. 16. It is observed that GP model even better performed than the suggested experimental model with least average error.

10 Conclusions

As reported earlier, for computation of discharge of a stream flow with assessment of boundary shear or apparent shear stress, diagonal division method is proved to be the best and using this numerical model is suggested. Again using genetic programming, much better prediction is observed in case of smooth and rough floodplains with minimum average error for different cases, which can be used for similar studies with more number of runs of flow or data sets.

References

1. Moharana, S. and Khatua, K. K.: In: Murillo (ed.) Prediction of Discharge in a Meandering Channel using ANFIS, River Flow, pp. 1049–1055 (2012)
2. Mohanty, P.K., Khatua, K.K.: Estimation of discharge and its distribution in compound channels. J. Hydrodyn. **26**(1), 144–154 (2014)
3. Myers, W.R.C.: Velocity and discharge in compound channels. J. Hydraul. Eng. ASCE **113** (6), 753–766 (1987)
4. Rajratnam, N., Ahmadi, R.M.: Interaction between main channel and flood plain flows. J. Hydraul. Eng. ASCE **105** (HY5), 573–588 (1979)
5. Knight, D.W., Hamed, M.E.: Boundary shear in symmetrical compound channels. J. Hydraul. Eng. ASCE **110** (19217), 1412–1430 (1984)
6. Khatua, K.K., Patra, K.C.: Boundary shear stress distribution in compound open channel flow. J. Hydr. Eng. ISH **12**(3), 39–55 (2007)
7. Khatua, K.K., Patra, K.C., Mohanty, P.K.: Stage discharge prediction for straight and smooth compound channels with wide floodplains. J. Hydraul. Eng. ASCE **138**(1), 93–99 (2012)
8. Khatua, K.K., Patra, K.C., Jha, R.: Apparent shear stress in compound channels. J. Hydraul. Res. (ISH) Spec. Issue Taylor & Francis **16**(3), 1–14 (2010)
9. Knight, D.W., Demetriou, J.D.: Floodplain and main channel flow interaction. J. Hydraul. Eng. ASCE **109**(8), 1073–1092 (1983)
10. Kean, J.W., Smith, J.D.: Calculation of Stage-Discharge relations for gravel bedded channels. J. Geophys. Res. **115**, F03020 (2009). https://doi.org/10.1029/2009JF001398
11. Knight, D.W., Tang, X., Sterling, M., Shiono, K., McGahey, C.: Solving open channel flow problems with a simple lateral distribution model. River Flow **1**, 41–48 (2010)
12. Azamathulla, H.Md., Aminuddin, Ab.G.: Genetic programming to predict river pipeline scour. J. Pipeline Syt. Eng. Practi ACSE **1**(3), 127–132 (2010)
13. Muduli, P.K., Das, M.R., Samui, P., Das, S.K.: Uplift capacity of suction caisson in clay using artificial intelligence techniques. Marine Georesources Geotechnol. **31**, 375–390 (2013)
14. Searson, D.P., David, E.L., Mark, J.W.: GPTIPS: an open source genetic programming toolbox for multigene symbolic regression. In: Proceedings of International Conference of Engineering and Computer Scientists, IMECS2010, vol. I, 7–19 Mar 2010
15. Demuth, H., Beale, M.: Neural Network Toolbox User`s Guide, the Math Works Inc (2004). www.mathworks.com
16. Atabay, S.A., Knight, D.W.: The influence of floodplain width on the stage discharge relationship for compound channels, river flow. In: Proceeding of International Conference on Fluvial Hydraulics, Louvail- La-NEUVE, Belgium, vol. 1, pp. 197–204 (2002)
17. Rezaei, B.: Overbank Flow in Compound Channels with Prismatic and Non-prismatic Floodplains. University of Birmingham, Diss (2006)

A Software Approach for Mitigation of DoS Attacks on SDN's (Software-Defined Networks)

Trupti Lotlikar and Deven Shah

Abstract Software-defined networking (SDN) is a network technology that aims to make the network more flexible and centralized. The main aim of the architecture is decoupling of network and control plane which enables the network control to be programmed and hence the forwarding devices are abstracted from higher application services. However, before this technology evolves on a large scale, it is important to understand the vulnerabilities associated with it. The agenda for this paper is to study the different types of attacks on the three layers of the OpenFlow protocol and the possible mitigation strategies to reduce the impact of those attacks on the network, implement SDN network which consists assorted topologies connecting numerous hosts, switches, and controllers. The implemented framework monitors the various events occurring in the network, identifies malicious events causing DoS attack, and mitigates the same. Through implementation, we elaborate different ways of making SDN more secure.

Keywords SDN · OpenFlow · DoS attacks · Mininet · PuTTY

1 Introduction

Considering the traditional approach to networking, which implements network devices like switch, router to deliver the packets, which impacts in a lot of administrative overhead and hence a really good chance of configuring at least one of these devices incorrectly. So this problem with traditional configuration approaches in the data centers is at the heart of software-defined networking to make it easier, to make it more consistent in order to enact changes in the network

T. Lotlikar (✉)
Terna College of Engineering/Fr.CRIT, Navi Mumbai 400703, India
e-mail: trupslotlikar@gmail.com

D. Shah
TCET, Kandivli, Mumbai 400101, India
e-mail: sir.deven@gmail.com

J. Nayak et al. (eds.), *Soft Computing in Data Analytics*,
Advances in Intelligent Systems and Computing 758,
https://doi.org/10.1007/978-981-13-0514-6_33

environment that we need. Software-defined networking pulls the intelligence away from the hardware itself. This authorizes the network operators with more agility to make their networks programmable [1]. The basic of software-defined networking technology is about breaking out the control plane of operation from the data plane as shown in Figs. 1 and 2.

Now the data plane is the forwarding intelligence for the device, and this is implemented typically in application specific integrated circuit. So we have our decision-making process or control of the device and we separate this from the raw forwarding power. Thus, we can control security on the control plane and in a separate distinct manner from how we secure data plane traffic. We can optimize control plane functions in a separate manner from how we optimize data plane functions. However, SDN's main benefits are control logic centralization and network computing capability which introduce new threats and attack planes [2]. This paper discusses threats and attacks associated with data forwarding layer, control layer, and application layer of OpenFlow protocol along with the countermeasures to mitigate the effect of those attacks on the network. Software-defined networking pulls the intelligence away from the hardware itself. It empowers network operators with traffic. We can optimize control plane functions in a separate manner from how we optimize data plane functions [3].

The remaining paper is organized as follows. Section 2 discusses the different types of attacks. Section 3 gives network experiment setup. Section 4 shows the DoS attack. Section 5 gives solution to reduce the effect of the attacks on SDN, and Sect. 6 gives the conclusion.

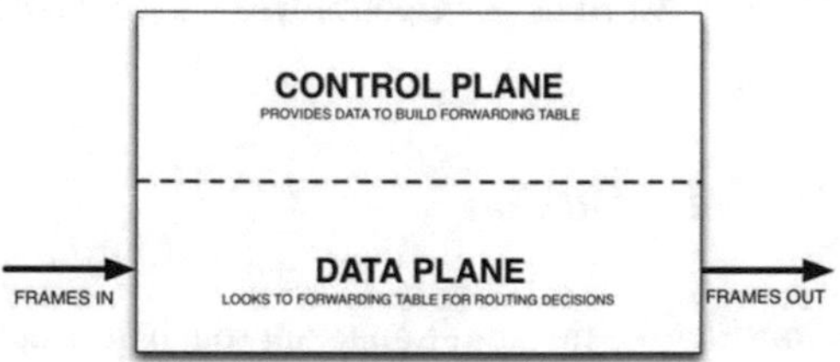

Fig. 1 Traditional network

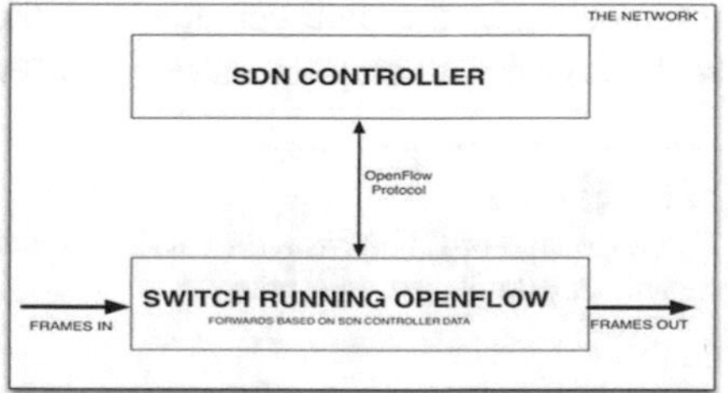

Fig. 2 SDN architecture

2 Types of Attacks

With increasing popularity of SDN, it has become important to understand the security issues and vulnerabilities of it before any large-scale evolvement. As per the OpenFlow architecture, we have divided the types of attacks on the basis of three layers of the protocol, viz. data forwarding layer, control layer, and application layer. By making controller centralized, SDN opens door of opportunities to an attacker to make network weak. We concentrate more on the attacks that performed on data plane layer and control plane layer as these layers hold high probability of getting attacked [4].

I. Attack on control plane layer

- Attack on controller's bandwidth
  Controller handles the packet flow between the hosts and takes decision on whether to forward or to drop the packets accordingly. There can be number of users who communicate through network. As a result, this may lead to the inefficiency of the controller's bandwidth which makes it vulnerable to DoS attack [5].

II. Attack on data plane layer

- Flow table overflow
  The switches used in SDN reside at data plane. Every switch has one flow table to store all information regarding the packet flow. By forwarding multiple packets at one time, flow table gets overflowed; that is, how the attack occurs at data plane. The proposed framework tries to perform an attack on the control plane and data plane of SDN and prevents the same [6].

3 Network Experimental Setup

A. System specification

- CPU: Intel(R) Core(TM) i5-4200U CPU@1.60 GHz
- Primary Memory: 16GiB (2 × 8GiB SODIMM DDR3 Synchronous 1600 MHz (0.6 ns))
- Hard disk: 200GiB

B. Required software

- Virtual box: VirtualBox all additional operating systems to be installed on it, as a Guest OS, and run in a virtual environment. We use VirtualBox 5.1.4 to implement the framework.

- Mininet: Mininet is a software emulator for prototyping a large network on a single machine. We are using Mininet 2.2.1 for Ubuntu 14.04 LTS—32 bit.
- PuTTY: PuTTY is a free and open-source terminal emulator for Windows operating system. Since we are using windows platform, we need PuTTY.
- Xming: We need Xming to display graphics such as an image from the remote Linux server (VirtualBox). We are using Xming-6-9-0-31(2.2 MB).
- Wireshark: We need this software for analyzing network problems, monitor usage and gather statistics. We are using Wireshark version 2.2.4.
- Hping: Hping is a command-line-oriented packet assembler/analyzer.
- MiniEdit: We required MiniEdit to create the network and expcriment the same.

C. Network setup

To design a framework for software-defined network (SDN), we have to create networking environment using Mininet [7]. We use MiniEdit to create different topologies. There can be different topologies in one network. We have experimented few topologies which are as follows:

I. Minimal Topology

Minimal topology is the basic topology for any network. In this, minimum two hosts are capable of communicating with each other via switch. To create the topology shown in Fig. 3, we use following command from PuTTY:

$$\$sudomn - topo\,minimal$$

II. Linear topology

Linear topology, which we have implemented, is the one which has several switches connected in linear fashion with separate hosts and a controller to manage whole network. To create the topology shown in Fig. 4, following command is used:

$$\$sudomn - topo\,linear, 4$$

Fig. 3 Minimal topology

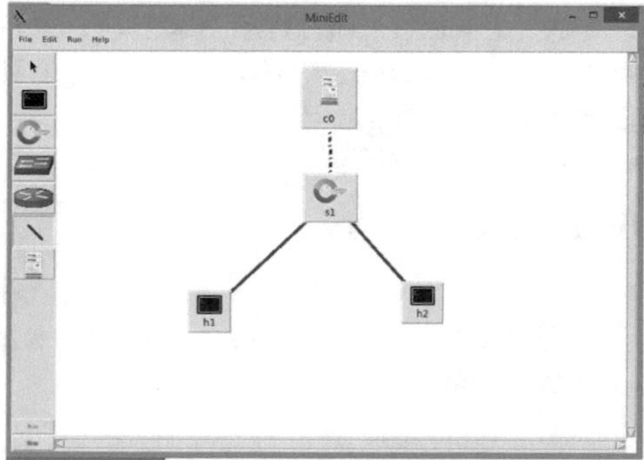

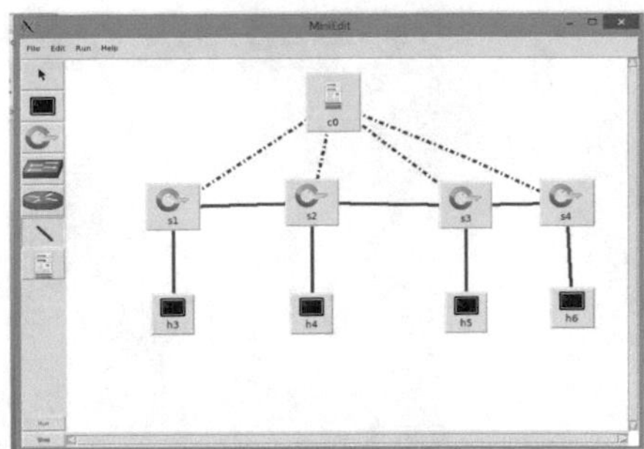

Fig. 4 Linear topology

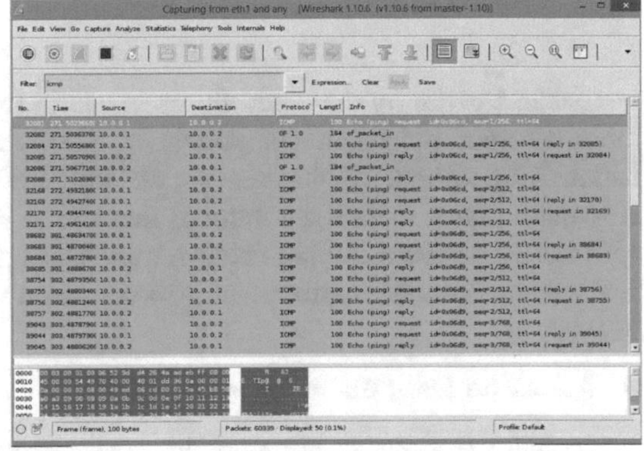

Fig. 5 Snapshot of Wireshark implementation

Monitoring and Analysing Network

As topologies have been designed, the urge of monitoring and filtering packet flows direct us to the Wireshark software. Wireshark provides the interface from which we can analyze the actions taken by each host connected in network. To monitor the network traffic in minimal topology we must use following command:

$$\$sudowireshark\& > /dev/null\&$$

After executing the above command from PuTTy,

Create one minimal topology by using command which we have discussed earlier. As we create topology, Wireshark starts monitoring the packet flow. Same procedure is followed for the linear topology. Figure 5 shows Wireshark analysis

4 Implementation of DOS Attack

A DoS attack is an attempt to make a system or server unavailable for legitimate users and, finally, to take the service down. As discussed earlier in Sect. 1, control plane and data plane makes the SDN less secure. At data plane, Switch performs the packet forwarding function as per the directions provided by controller. All the

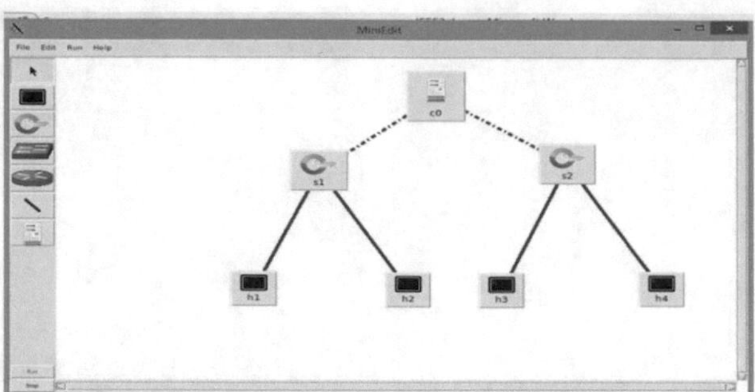

Fig. 6 Sample topology

instructions given by the controller is stored in the flow table of switch. The flow table becomes the major component of SDN as it has plan of action which is followed by switch. So attackers choose the flow table to perform DoS attack [8]. At control plane, there is a controller that acts as a master in master–slave network architecture. It controls whole network and takes decisions accordingly. Bandwidth is the major issue at control plane which gives a way to an attacker to get into the network [9].

- **Attack on Data plane layer**

To attain DoS attack at data plane layer, one can flood the switch by sending multiple requests at a time. The attack can be initiated using Hping3 (packet generator) from Mininet.

In minimal topology, host1 (h1) can send number of packets using Hping command as follows:

$$\text{hping3} - \text{S} - \text{flood} - \text{V destIP}$$

where -S set SYN flag,–flood send packets as fast as possible, -V verbose mode.

Hping command floods the switch with number of requests such that host2(h2) will not be able to create connection with switch. We have experimented DoS attack on one more topology whose structure is as shown in Fig. 6.

In above scenario, if h1 pings h2 using Hping command, only switch (s1) gets attacked. Otherwise, if h1 pings h3 using Hping command both the switches will be freezed [10].

5 Solution

Except for unique and unusual cases, all IP addresses that are attempting to access the internet from inside the network should bear an address, i.e., assigned to LAN. Inversely, traffic from the outside of the Internet should not claim a source address that is part of internal network [11]. Attacker tries to break these facts in order to attack the system. Using illegitimate IPs or by modifying IP field of packets, attacker enters in network and do unlawful use of network.

To address this issue, we can use Ingress–Egress policy. These policies are filtering techniques; with the help of these techniques we can limit the DoS attack [12].

- **Ingress Policy**

Ingress filtering is a technique which ensures that incoming packets are legitimate. In this technique, we check the IP field of packets and drop the packet which is having less than predefined limit required to send lesser amount of data. Using Ingress policy, we can block the illegitimate incoming packets [13].

- **Egress policy**

Egress filtering is a practice of monitoring and potentially restricting the flow of information outbound from one network to another [14].

For implementing Ingress–Egress policy, we have created one topology with two domains A and B as shown in Fig. 7:

We have programmed the controllers using python such that controllers can mitigate the attack by limiting the bandwidth between the target and the attacker node [15].

In order to start simulation, we have to keep both the controllers running in background using following commands:

1. Run custom controller 1 (for domain A) as shown in Fig. 8.

sudo python./ryu/bin/ryu-manager–ofp-tcp-listen-port 6633 ryu/ryu/app/Controller1.py

2. Run custom controller 2 (for domain B) as shown in Fig. 9.

sudo python./ryu/bin/ryu-manager–ofp-tcp-listen-port 6634 ryu/ryu/app/Controller2.py

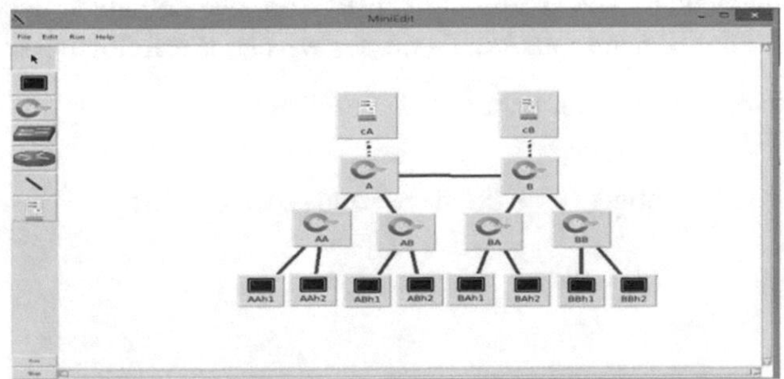

Fig. 7 Custom topology

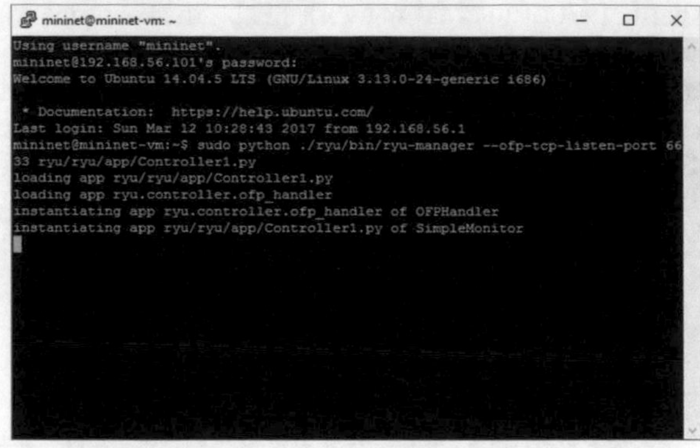

Fig. 8 Snapshot of running controller 1

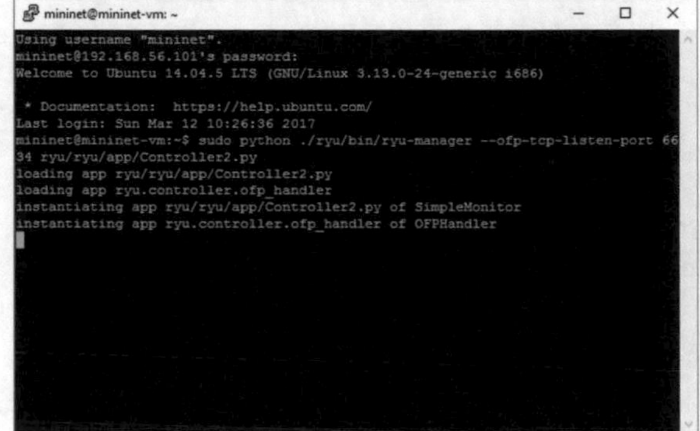

Fig. 9 Snapshot of running controller 2

Our next step is to create the topology in MiniEdit and run that file using following command:

sudo python customTopology.py

The attacks can be performed in two ways, i.e., within domain or in between two domains as per custom topology. The attacks performed within domain are known as intra-domain attacks. The attacks performed from one domain to another domain are known as inter-domain attacks. To do so, we can use following commands:

1. To start inter-domain attack on BBh1

hping3 – –flood – –udp AAh1&BBh1

2. To start intra-domain attack

$$\text{hping3} - -\text{flood} - -\text{udp AAh1\&ABh1}$$

$$\text{hping3} - -\text{flood} - -\text{udp AAh2\&ABh2}$$

We can monitor that even after flooding packets using Hping3 (packet generator) controller prevents the attack.

6 Conclusion

We have discussed the concept of SDN. Our main agenda was to discover the loopholes of SDN and to make SDN networks more secure. We can conclude that the control plane layer and data plane layer of SDN are vulnerable to DoS attacks. As there are major security risks in design of SDN, we have tried to protect and give the solutions with the implementations to make SDN more secure. We hope that the research will trigger discussions in the SDN community around issues related to security and dependability, to serve as a catalyzer of joint efforts in these critical issues.

References

1. Kreutz, D., Fernando, M.V., Ramos, P.V.: Towards secure and dependable software defined networks published. In: IEEE Transaction y. University of Lisbon, Portugal, Aug 2013
2. Wang, H., Xu, L., Gu, G.: OF-GUARD: a dos attack prevention extension in software defined network SDN. In: Proceedings of the 45th Annual IEEE/IFIP International Conference on Dependable Systems and Networks. Texas University (2015)
3. Kandoi, R., Antikainen, M.: Denial of service attacks in Openflow SDN networks IEEE/IFIP, department of computer science In: Finland in International Conference on Integrated Network Management, May 2015
4. Braga, R., Mota, E., Passito, l: Lightweight DDoS flooding attack detection using NOX/ Openflow. In: 35th Annul IEEE Conference on Local Computer Networks, Denver, Colorado (2010)
5. Sezer, S., Scott-Hayward, S.: Pushpinder Kaur Chouhan: implementation challenges for software defined networks published. IEEE Commun. J. (2013)
6. Douligeris, C., Mitrokotsa, A.: DDOS attacks and defense mechanisms: classification and state-of-the-art. Comput. Netw. **44**, 643–666 (2004)
7. Prete, L.R., Schweitzer, C.M., Shinoda, A.A., Santos de Oliveira, R.L.: Simulation in an SDN Network: Scenario using Pox Controer. IEEE (2014)
8. Hakiri, A., Gokhale, A., Berthou, P.: Software defined networking: challenges and research opportunities for future. Internet Comput. Netw. J. **75**, Part A (2014)
9. Tariq, U., Hong, M., Lhee, K.-S.: A Comprehensive Categorization of DDoS Attack and DDoS Defenseense Techniques, in Advanced Data Mining and Applications, pp. 1025–1036. Springer, Heidelberg (2006)

10. Scott-Hayward, S., Natarajan, S., Sezer, S.: A survey of security in software defined networks. In: IEEE Communication Survey and Tutorials, vol 18, no. 1. First quarter (2016)
11. Scott-Hayward, S., O' Callaghan, G., Sezer, S.: SDN security: a survey. In: IEEE SDN for Future Networks and Services (SDN4FNS), Trento, Italy (2013)
12. http://searchitchannel.techtarget.com/tutorial/Establish-Ingress-and-Egress-address-filtering-policies
13. https://en.wikipedia.org/wiki/Ingress_filtering
14. https://en.wikipedia.org/wiki/Egress_filtering
15. Mininet. //http://www.mininet.org/accessed (2016). Accessed 15/11/2016
16. Zhang, P., Wang, H., Hu, C., Lin, C.: On denial of service attacks in software defined network published. In: IEEE Conference on Network Forensics and Surveillance for emerging networks November (2016)

Techno-economic Analysis of Different Combinations of an Existing Educational Institute-Based PV–Diesel–Grid Hybrid System in India

T. S. Kishore and S. Mahesh Kumar

Abstract In developing countries like India, there is a rapid increase in load growth and energy demand due to heavy industrial development, urbanization, and economy rise. Most of this demand is met through power generation from conventional sources which have adverse environmental impact. Renewable energy resources provide a good alternative but lack of continuous generation, high initial cost, complex control, conversion between AC and DC generation, grid integration, etc., are some of the drawbacks associated with renewable energy sources. To overcome these disadvantages and to efficiently utilize the available energy resources, a hybrid power system combining the advantages of individual power sources can be seen as a preferred option to generate and deliver power. In this paper, an attempt has been made to perform techno-economic analysis of a hybrid power system existing in an educational facility with conventional grid power supply, diesel generator, and photovoltaic generation and the results are summarized.

Keywords Conventional · Hybrid power system · Renewable energy
Techno-economics

1 Introduction

Electricity is the crucial part for industrialization, development, economic growth, and enhancement of excellence of life in society [1]. India is world's third largest producer and the fourth largest consumer of electricity. In order to reduce the use of fossil fuels, the most feasible way out is to increase our reliance on renewable energy sources (RES) [2]. The isolated rural areas can be electrified by RES as in

T. S. Kishore (✉) · S. Mahesh Kumar
GMR Institute of Technology, Rajam 532127, Andhra Pradesh, India
e-mail: srinivasakishoret@gmail.com

S. Mahesh Kumar
e-mail: sanapalamahesh92@gmail.com

J. Nayak et al. (eds.), *Soft Computing in Data Analytics*,
Advances in Intelligent Systems and Computing 758,
https://doi.org/10.1007/978-981-13-0514-6_34

many cases extension of grid to these areas is uneconomical [3, 4]. Apart from electrification of individual household by renewable energy, integrated renewable energy system (IRES), distributed generation, and hybrid power system (HPS) can also be developed to supply power. These require the information of constraints like existing technologies, available government policies, customer requirement, and resource limitations [5, 6]. In order to reduce the complexity of operation and increase efficiency, power system has changed from regulated or vertical mode of operation to deregulated mode of operation. There are many causes that fueled the concept of deregulation of the power industry. One major thought that prevailed during the early nineties raised questions about the performance of monopoly services. Due to initiation of deregulated power system many different types of power systems and subsystems came into existence, viz. smart grid, microgrid, distributed generation, IRES, HPS. Each system has its own benefits and also complexities in coordinated operation with main grid. In the present study, HPS is opted because it is the integration of the RES along with the conventional sources. The main purpose of the HPS is to combine multiple energy sources and/or storage devices which are complement of each other. Thus, higher efficiency can be achieved by taking the advantage of each individual energy source and/or device while overcoming their limitations [5]. In this paper, an attempt has been made to perform techno-economic analysis of a hybrid power system in an educational facility with conventional grid power supply, diesel generator, and photovoltaic generation. The results summarized provide information regarding load assessment, simulation studies, and economic evaluation of existing hybrid power system. These results are useful in understanding the operational techno-economics of the existing system, devise more efficient operation and maintenance strategies, which finally reduce the dependence on conventional sources for energy needs and increase renewable energy utilization.

2 Study Area

GMR Institute of Technology (GMRIT) is an academic institution catering to the needs of engineering education, situated in Rajam, a small remote town in Andhra Pradesh, state of India, having the geographical coordinates of 18.4665 N, 83.6608 E. GMRIT is established in the year 1997 by the GMR Varalakshmi Foundation, the Corporate Social Responsibility (CSR) wing of GMR Group. At inception, the institute has five blocks with a total connected load of nearly 700 kW. At present GMRIT has a total connected load of 1450 kW. The existing power supply system in GMRIT is a HPS. The choice of the site has been taken from the fact that it is the only educational institution having a HPS in Srikakulam district of Andhra Pradesh state with net metering method provided by the Eastern Power Distribution Company of Andhra Pradesh Limited (APEPDCL), supplying power through 11 kV bus interconnection.

3 Modeling of Hybrid Power System

Modeling of the HPS has been carried out based on assessment of load profile and system resources in HOMER software, which is an application developed by the National Renewable Energy Laboratory (NREL), USA [7].

3.1 Load Profile

The load profile of the present study area is characterized by annual electricity consumption of 5248 kWh/d, peak load of 501 kW, and a load factor of 43.7%. In this analysis, the day-to-day and time step to time step random variability of load is taken as zero. The energy requirements vary daily and monthly depending on several factors like work break hour, institute operating time, vacations, events, and seasonal changes are shown in Fig. 1 [8].

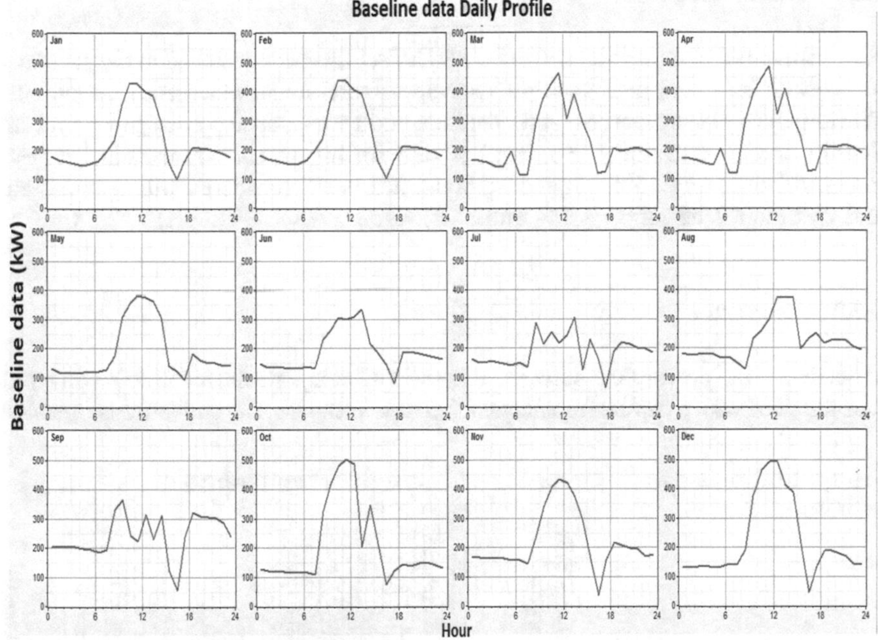

Fig. 1 Yearly load profile

3.2 *Resource Assessment*

The energy resource assessment has been made on the basis of data collected from institute database.

3.2.1 Solar Resource

Figure 2 shows the global horizontal radiation with daily radiation and clearness index. It can be observed that the global horizontal irradiation is maximum in the month of April. The total monthly average irradiation is 1867.221 kWh/m^2/d. The annual average irradiation is 5 kWh/m^2/d. A derating factor of 80% and ground reflectance of 20% were considered for analysis purpose. Based on the above technical data, a 700 kW PV plant has been installed with a capital (replacement) cost and O&M cost of $812405 and $2660, respectively. A lifespan of 20 years has been considered for PV panels without tracking devices. The PV arrays are installed with an inclination angle equal to the latitude of the site, i.e, 18.98°.

3.2.2 Diesel Resource

Two generators of rating 500 kVA and 380 kVA have been installed in the premises to generate base and peak load power respectively, under emergency conditions. The capital (replacement) cost and operation and maintenance cost of generators are $44873 and $29915 and $1.079 and $0.820 for higher and lower rating respectively. A lifetime of 15000 operating hours has been considered with a minimum and maximum load ratio of 30% and 80%, respectively.

3.2.3 Converter

To facilitate the power flow between AC and DC components, a 750 kW converter was installed with a capital (replacement) cost of $28046. The converter is operated

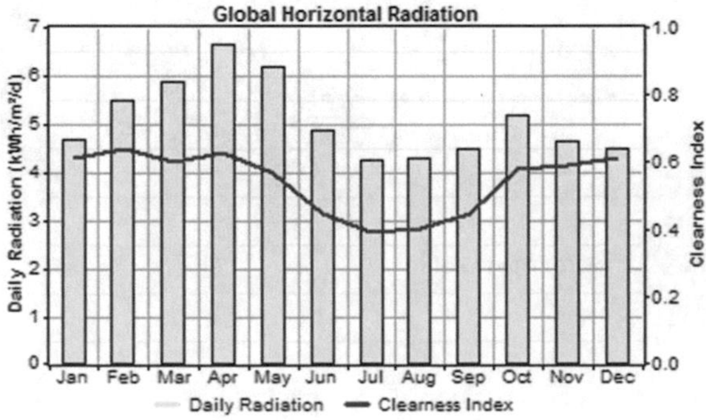

Fig. 2 Global horizontal radiation

at 100% capacity relative to inverter and it can also operate simultaneously with an AC generator.

3.2.4 Grid

Grid interconnection of the existing HPS is facilitated by DISCOM with different tariffs via. Normal tariff and time of day (TOD) tariff. Normal tariff will be charged by the DISCOM between 00:00 to 18:00 and 22:00 to 00:00 daily at a rate of 0.112 $/kWh, and TOD tariff will be charged daily from 18:00 to 22:00 at a rate of 0.130 $/kWh. The fixed demand charge per kVA will be charged at a rate of 5.55 $/kVA.

4 HPS Configuration

In this study, three different types of HPS configurations have been considered for analysis purposes. The models employ a combination of generating sources in each case and the results obtained are analyzed based on technical, economic, and environmental parameters such as cost of energy (CoE), net present cost (NPC), renewable fraction (RF), and carbon emissions.

4.1 Diesel Generator/Grid

Figure 3a illustrates the grid-connected diesel generator system. In this configuration, four combinations of DG capacities have been considered, i.e., 0, 380, 500, and 880 kW. The DG has been force scheduled for three hours per day.

4.2 Diesel Generator/PV

Figure 3b illustrates the diesel generator/PV configuration. As considered in configuration 1 in Sect. 4.1, four combinations of DG capacities have been considered,

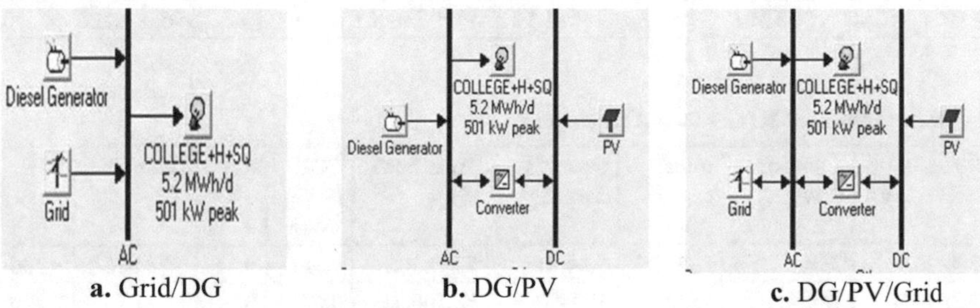

a. Grid/DG **b.** DG/PV **c.** DG/PV/Grid

Fig. 3 Proposed configurations

i.e., 380, 500, and 880 kW. However, the scheduling of the diesel generator is done in such a way that it is required to operate only when PV system is unable to generate power. For analysis purposes, 0 kW and 700 kW ratings are considered for PV system.

4.3 Diesel Generator/PV/Grid

Figure 3c illustrates the diesel generator/PV/grid configuration. As considered in configuration 1 in Sect. 4.1, four combinations of DG capacities have been considered, i.e., 380, 500, and 880 kW. PV capacities of 0 and 700 kW have been considered. Net metering method is functional in this configuration with other operational features of the grid remaining same as mentioned in Sect. 3.2.4. According to Andhra Pradesh State Electricity Regulatory Commission (APSERC), the sell back price/kWh of a system is $0.106818.

5 Techno-economic Evaluation and Results

In this section, techno-economic analysis has been done for the three configurations mentioned in Sect. 4. Techno-economic evolution incorporates optimization studies taking into consideration various combinations of available energy generating resources and gives various parameters as output. In this study, CoE has been considered to identify the best optimized operating schedule. The results of techno-economic evaluation have been presented in Tables 1, 2 and 3 below.

Table 1 Optimized results of configuration 1

Diesel (kW)	Grid (kW)	Initial capital ($)	Operating cost ($/yr)	Total NPC ($)	COE ($/kWh)	RF	Diesel (L)	Diesel Generator (hrs)
–	700	0	2,59,654	2,771,748	0.138	0	–	–
380	700	33,266	3,06,026	3,300,029	0.164	0	66,321	1,126
500	**700**	**48,224**	**3,21,494**	**3,480,096**	**0.172**	**0**	**87,265**	**1,126**
880	700	95,591	3,77,743	4,127,910	0.205	0	1,53,586	1,126

Table 2 Optimized results of configuration 2

PV	DG (kW)	Converter (kW)	Initial capital ($)	Operating cost ($/yr)	Total NPC ($)	COE ($/kWh)	RF	Diesel (L)	DG (hrs)
700	**500**	**750**	**885,324**	**6,54,413**	**7,871,041**	**0.391**	**0.44**	**6,78,855**	**8,200**
700	880	750	932,691	10,84,203	12,506,317	0.621	0.34	11,21,785	8,200
–	880	–	92,240	11,92,554	12,822,485	0.637	0	12,49,749	8,760

Table 3 Optimized results of configuration 3

PV (kW)	DG (kW)	Con (kW)	Grid (kW)	Initial capital ($)	Operating cost ($/yr)	Total NPC ($)	COE ($/kWh)	RF	DG (L)	DG (hrs)
700	–	500	700	834,453	1,57,577	2,516,551	0.125	0.51	–	–
700	–	750	700	843,802	1,56,830	2,517,922	0.125	0.51	–	–
–	–	–	700	0	2,59,654	2,771,748	0.138	0	–	–
–	–	250	700	825,105	1,85,329	2,803,455	0.139	0.48	–	–
700	380	500	700	864,368	2,03,613	3,037,888	0.151	0.51	66,321	1,126
700	380	750	700	873,717	2,02,865	3,039,260	0.151	0.51	66,321	1,126
700	500	500	700	879,326	2,18,473	3,211,476	0.159	0.5	87,265	1,126
700	**500**	**750**	**700**	**888,675**	**2,17,726**	**3,212,848**	**0.16**	**0.5**	**87,265**	**1,126**
700	380	250	700	855,020	2,31,365	3,324,792	0.165	0.47	66,321	1,126
–	500	–	700	48,224	3,20,887	3,473,617	0.172	0	87,265	1,126
700	500	250	700	869,978	2,46,226	3,498,380	0.174	0.47	87,265	1,126
700	880	500	700	926,693	2,65,528	3,761,147	0.187	0.48	1,53,586	1,126
700	880	750	700	936,042	2,64,781	3,762,519	0.187	0.48	1,53,586	1,126
–	880	–	700	95,591	3,67,942	4,023,288	0.2	0	1,53,586	1,126
700	880	250	700	917,345	2,93,281	4,048,051	0.201	0.45	1,53,586	1,126

6 Conclusion

In this paper, the techno-economic analysis of a HPS has been carried out by considering three configurations. From Table 1, it can be understood that out of all possible combination results, the result with 500 kW DG and 700 kW Grid is optimum. Considering the worst-case scenario of grid failure, the 500 kW DG is sufficient to supply the required load demand. If further, load is increased DG cannot meet the demand and load shedding has to be done for insignificant loads. Although 880 kW DG is available which can accommodate the load increase, it is not considered as optimal due the capital cost which is two times when compared to 500 kW DG capital cost. From Table 2, it can be understood that from the obtained three results, the result with 700 kW PV, 500 kW DG, and converter with 750 kW is considered as the optimum. During daytime, PV system is sufficient to meet the load. During night time, the load is approximately 300 kW. Therefore, DG of 500 kW is sufficient to meet the load. From Table 3, it can be observed that the result with 700 kW PV, 500 kW Diesel Generator, 700 kW grid, and converter with 750 kW is chosen as the optimized and feasible one. Considering the worst-case scenario of grid and PV system failure during daytime and only grid failure during night time, the 500 kW DG can meet the required load demand. An NPC of $3,480,096, $7,871,041, and $3,212,848 has been obtained for configurations 1, 2, and 3, respectively. Similarly, a COE of $0.172, $0.391, and $0.16 has been obtained for configurations 1, 2, and 3, respectively. A comparison of the above NPC and COE values for the three optimum configurations gives a basis for selecting the best configuration. From the analysis, it can be concluded that the third configuration, i.e., PV/grid/DG HPS configuration is best suitable for meeting the necessary load.

References

1. Khare, V., Nema, S., Baredar, P.: Status of solar wind renewable energy in India". Renew. Sustain. Energy Rev. **27**, 1–10 (2013)
2. Rhaman, M.: Hybrid renewable energy system for sustainable future of Bangladesh. Int J. Renew. Energy Res. **3**(4), 4–7 (2013)
3. Kishore, T.S., Singal, S.K.: Analysis of investment issues and transmission schemes for grid integration of remote renewable energy sources. Int. J. Renew. Energy. Res. **5**(2), 483–490 (2015)
4. Kishore, T.S., Singal, S.K.: Design considerations and performance evaluation of EHV transmission lines in India. J. Sci. Ind. Res. **74**(2), 117–122 (2015)
5. Upadhyay, S., Sharma, M.P.: A review on configurations, control and sizing methodologies of hybrid energy systems. Renew. Sustain. Energy Rev. **38**, 47–63 (2014)
6. Das, H.S., Dey, A., Wei, T. C., Yatim, A.H.M.: Feasibility analysis of standalone PV/wind/battery hybrid energy system for rural Bangladesh. Int. J. Renew. Energy Res. **6**(2) (2016)
7. Barzola, J., Espinoza, M., Cabrera, F.: Analysis of hybrid solar/wind/diesel renewable energy system for off-grid rural electrification. Int. J. Renew. Energy Res. **6**(3) (2016)
8. Bhatt, A., Sharma, M.P., Saini, R.P.: Feasibility and sensitivity analysis of an off-grid micro hydro/photovoltaic/biomass and biogas/diesel/battery hybrid energy system for a remote area in Uttarakhand state. India. Renew. Sustain. Energy Rev. **61**, 53–69 (2016)

MidClustpy: A Clustering Approach to Predict Coding Region in a Biological Sequence

Neeta Maitre and Manali Kshirsagar

Abstract Data mining can act like a medium to discover new avenues in bioinformatics rather than just a pattern recognition in the biological sequences. It is useful in the sequence analysis, and clustering can be used to reduce the total number of operating sequences to perform this analysis. Expressed sequence tags (ESTs) are the complimentary DNA sequences, shorter in size and instrumental in locating coding region in genomic sequences. Clustering of these ESTs requires basic computer knowledge for sequence analysis and its relevance in the field of biology. MidClustpy is an algorithm specifically designed to cluster ESTs based on the most accurate part in the sequence. The similarity search for locating coding region in a query sequence can be assisted by MidClustpy algorithm. The research paper is, thus, focussed on the effective use of expressed sequence tags using MidClustpy for prediction of coding region.

Keywords Data mining · Expressed sequence tags · Bioinformatics
Coding region · MidClustpy

1 Introduction

Clustering is a data mining technique which can be used efficiently in finding coding region in the genomic sequence. The most popular method in bioinformatics is Basic Local Alignment Search Tool (BLAST) from National Center for Biotechnology Information (NCBI). This method performs similarity searches for combinations of nucleotide and protein sequences. These similarity searches are mainly focussed on the basic sequences which are very large in size and need high

N. Maitre (✉)
G.H. Raisoni College of Engineering, Nagpur 440016, India
e-mail: neeta.maitre@gmail.com

M. Kshirsagar
Rajiv Gandhi College of Engineering and Research, Nagpur 441110, India
e-mail: manali_kshirsagar@yahoo.com

J. Nayak et al. (eds.), *Soft Computing in Data Analytics*,
Advances in Intelligent Systems and Computing 758,
https://doi.org/10.1007/978-981-13-0514-6_35

computations. Clustering of these sequences and then generation the cluster representatives for the similarity searches in the efficient way is hard to handle large volume of 29 biological datasets. DNA sequences clustered in such a fashion when used for similarity search are still found to be a cumbersome task. In case of locating coding region in a biological sequence, total number of similarity searches can be reduced if DNA sequences can be replaced by expressed sequence tags as they are smaller in size and contain only exonic region [1].

MidClustpy is an approach applied to the EST sequences in order to cluster them based on the most accurate part of EST and thus provides a novel way to provide prediction of coding region in the query sequence w.r.t. a species.

1.1 Background

- Data mining is the process of analyzing large volume of stored data. It refers to the extraction of information from large amount of available data. It helps in generating and finding useful patterns. Data mining, thus, needs preprocessing in the form of data cleaning, integration, and transformation of data. Tasks of data mining generally can be achieved in two ways [2]:

  - Supervised learning, and
  - Unsupervised learning

  Supervised data mining deals with inferring from labeled training dataset. Here, the training set is a pair of input and the desired output value which is known beforehand. Classification is the example of supervised data mining.

  Unsupervised data mining takes a form where no model is available. Clustering is an unsupervised data mining type. Here, an unsupervised clustering approach named MidClustpy is suggested to perform clustering of special type of biological sequences.
- Bioinformatics is a multidisciplinary field that can have a combination of molecular biology, genetics, computer science, mathematics, and statistics. It generally deals with biological sequences, majorly DNA sequences or protein sequences. Thus, sequence analysis plays a vital role in generating patterns, structures in bioinformatics.

1.2 Basics in Bioinformatics

Structure and function of an organism are defined by DNA which contains genetic information [3]. Genetic code of DNA forms proteins. Processes which are responsible for conversion of this genetic information can be of three types:

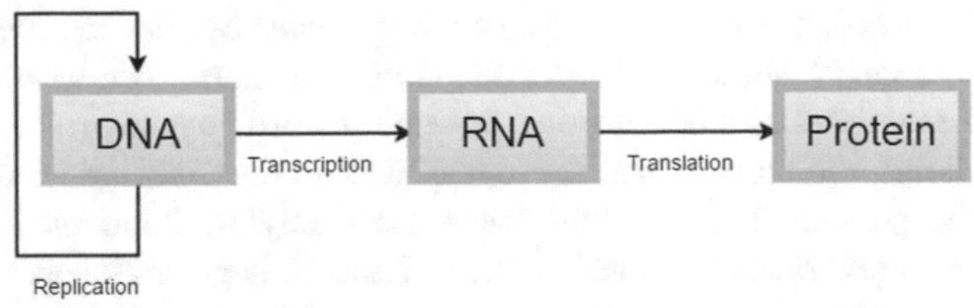

Fig. 1 Central dogma of molecular biology

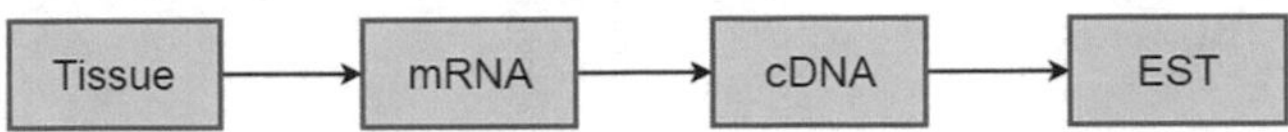

Fig. 2 EST formation

- Replication: Nucleic acid duplication to form identical copies.
- Transcription: A single-stranded RNA is transcribed by a gene-constituting DNA sequence.
- Translation: This process converts RNA into a protein-coding amino acid.

This constitutes the central dogma of molecular biology as shown in Fig. 1. Coding region is that region of a gene's DeoxyriboNucleic Acid (DNA) or RiboNucleic Acid (RNA) made up of exons which are responsible for protein codes. Ribosomal translation is the process which actually translates nucleotides (concatenated exons) in the coding region into amino acid.

Expressed sequence tags are the short subsequences of complimentary DNA (cDNA) [4]. cDNA is a copy of DNA that contains mRNA. ESTs are a valuable adjunct to whole genome sequencing as they facilitate gene identification [5]. Thus, ESTs are the sequences that contain exon region which may represent (partial) genes. The main advantage of using EST is its small size. Size of ESTs ranges between 200–500 base pairs. It is very "tiny" as compared to whole genome sequence. Size of genome is large due to the occurrence of exon region interrupted by intron region. The EST formation process is depicted in Fig. 2.

Exons generally represent coding region of DNA. Thus, ESTs can be utilized as a tool to find coding region.

UniGene is resource where the EST datasets are clustered. The source is developed and maintained by National Center for Biotechnology Information. Though UniGene is providing information about clustered ESTs, it is not available for the entire set of organisms. The EST sequences covered in UniGene are not exhaustive.

1.3 Basic Local Alignment Search Tool

Basic Local Alignment Search Tool (BLAST) from NCBI is very useful in sequence analysis and comparison of biological sequences. It provides a set of algorithms to find sequence similarity of major sequences. BLAST works on the principle of detection of local sequence alignment by finding best-matching sequence to the query sequence. Generally, alignment starts with word size mostly three. It then calculates number of times that query word appears in the sequence. It takes into account the close words those may lack in one match among three. The query score is generated which tells the nearest neighborhood dataset of query sequence. A computerized "picture" of BLAST search is generated as a result. This picture gives an idea about how many sequences show a comparison with given query sequence in the form of a "hit." Few BLAST algorithms are blastp, blastn which compare query sequence with protein database, nucleotide database, respectively [6].

2 Materials and Methods

Representative biological sequences in the form of EST (approximately 2000) were selected from NCBI Gossypium EST database. This is a comprehensive database of cDNA. All records chosen for this work were from Gossypium family (a cotton plant having 13 chromosomes). The number of ESTs available at NCBI is about 66000. The dataset contains five attributes as per the FASTA file format of EST data. This dataset is then processed by using MidClustpy algorithm.

Cotton is chosen as the material due to considerably large-sized genome and its economic importance in agriculture. Cotton has very high farm gate value and a promising research area in the field of agriculture blended with power of computer science. Cotton has variety of uses and is widely used. Many of the Asian countries grow cotton and have different species like Gossypium hirsutum, Gossypium raimondii, Gossypium arboreum, and Gossypium herbacium. The subsets of ESTs of these species are chosen and are mapped against the query sequences using proposed approach named MidClustpy.

MidClustpy uses an incremental algorithm to cluster EST sequences. The approach is focussed on the middle part of the EST sequences where the accuracy and the probability of coding region location are high. To achieve this, the process of splitting ESTs into three equal parts was computationally obtained. The middle part of these three segments is extracted from each sequence to process further as per MidClustpy algorithm. Original sequences are also preserved in a separate file to have comparative analysis of results obtained. The extracted dataset is subject to the clustering process.

The algorithm uses an incremental clustering mechanism. The sequences in the input file are first sorted in the descending order of their length. A distance matrix is created with these sequences and then the matrix is normalized as per the diagonal elements.

The first sequence in list forms the initial cluster. This sequence is added in the cluster_repr file. The remaining sequences in the file are then matched one by one with this sequence. The criteria for matching are similarity with the defined value of threshold. This threshold value is programmer-defined and can be changed through program.

The conditions those could be encountered are as follows:

- If the threshold condition is not met, a new cluster is formed.
- If the threshold is mapped to already existing cluster, the sequence is added as a candidate to the already existing cluster.

The algorithm for clustering is shown in the figure.

Here, the threshold is then considered as a mean value of normalized elements of the distance matrix generated. Similarly, the median of all values can also be considered as threshold parameter. But mean value gives better performance as compared to median.

The algorithm for clustering is shown in the Fig. 3.

The output of clustering is then given to a similarity checking module which uses a dynamic programming approach. The similarity metrics followed in this work are based on local alignment.

Local alignment gives better optimality of similarity as compared to the global alignment [7]. Coding region requires an ungapped match so also the EST data sequences rarely contain gaps. Optimality is considered as the main focus of sequence similarity and is taken care by doing pairwise alignment that inherently uses Smith–Waterman algorithm.

3 Implementation

MidClustpy algorithm is implemented in Python. The choice of using this language is made on the basis of higher code readability, comprehensive and large library of functionalities. It is easy to interpret, code and is highly extensible.

Biopython module is used for performing pairwise alignment of sequences, and following tasks are performed by using biopython [8]:

- To create high quality, reusable classes and modules.
- To parse almost all bioinformatics file formats like FASTA.
- To provide interface to common programs like BLAST.
- To perform indexing and dictionary interface which makes it easy to iterate record by record.

The overall implementation of the query system to find coding region is shown in Fig. 4.

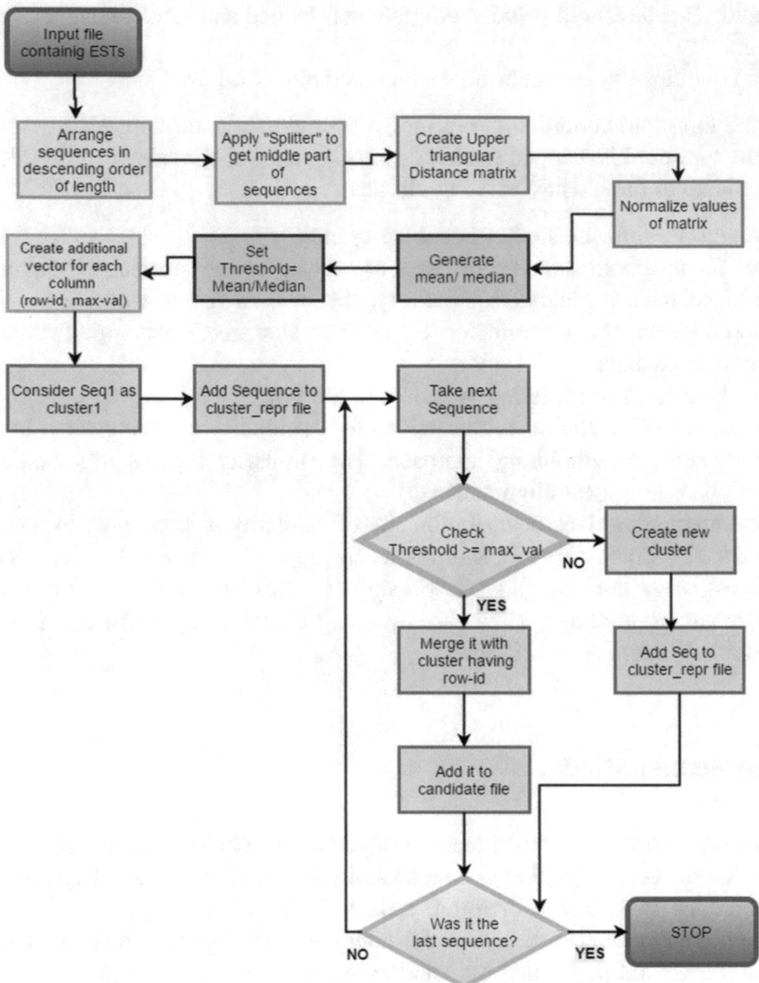

Fig. 3 Algorithm of MidClustpy

4 Result and Analysis

The dataset is used for performing comparison of similarity using MidClustpy of five cotton breeds. A dataset of about 500 sequences is clustered by taking variable number of sequences at a time separately for each breed. Clusters are formed and the time required for clustering is noted. Sampling of the results is done in order to get an overall picture of time taken for clustering. Similar results are generated in BLAST search also. Here, for this, a query sequence is matched with random EST datasets of cotton under consideration.

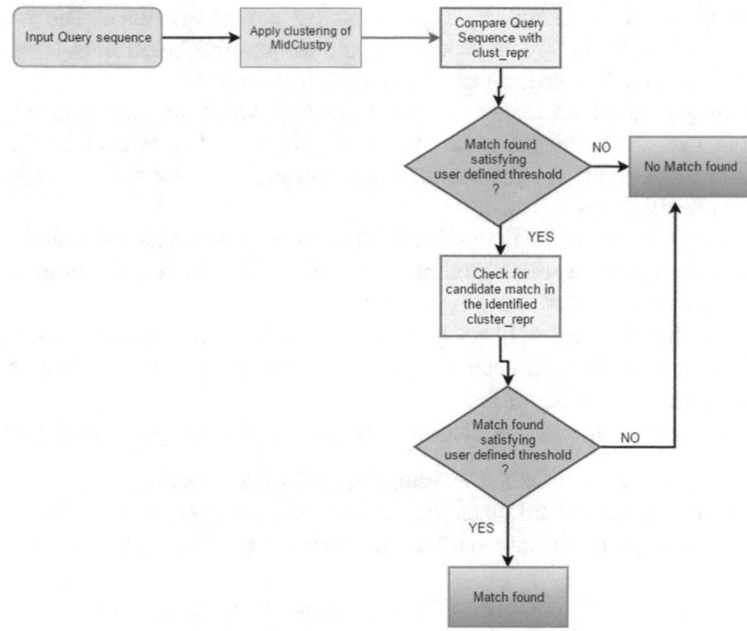

Fig. 4 Query system using MidClustpy

The performance of BLAST (online) varies for the number of base pairs in the query sequence. The higher is the number of base pairs, the performance on the lower side emerges through BLAST.

Thus, a standalone BLAST is downloaded and is used efficiently using Python modules. Standalone BLAST option is chosen in order to have uniformity in operating environment.

The results suggest that the memory requirement and time required for the complete EST sequence comparison are far greater than that of the middle portion of the ESTs. The analysis conducted on various datasets from Gossypium family ESTs, with mean and median as threshold, showed that the threshold is nearly same for same set of sequences. The values set for computing pairwise alignment may alter the results. So it is advised to analyze all the parameters during predictions separately. The use of Python, biopython, pandas is application-specific; other scripting languages like Perl can also be used.

Incremental algorithm suggested here requires two BLAST searches in order to get into the prediction. The choice is to be made among mean and median which deviates by 5%.

MidClustpy is proved to be effective as it inherently uses Smith–Waterman algorithm for generating similarity. This approach tries to give an optimal solution as compared to the heuristic approach followed in BLAST search. Number of comparisons made in the BLAST search in terms of number of records is definitely found to be lesser as it first deals with the cluster representatives given by

MidClustpy for a given threshold provided by the programmer. The recursive search is made with the identified cluster, and the optimal solution is obtained by matching the candidate sequences of the identified cluster.

MidClustpy gives an option to set threshold value programmatically. The noticeable fact about this threshold value is that the value remains in the range 36.5–39 throughout all the observations and also proved to be similar for mean or median, both the cases of run.

MidClustpy is coded in Python and hence is easy to understand, deploy, and execute in an operating environment. Number of clusters formed from a set of biological sequences is uniform in every run.

Pairwise alignment using biopython uses Smith–Waterman algorithm which is a dynamic programming approach to generate similarity and hence gives optimal results as compared to BLAST.

The summary of the results obtained for dataset of Gossypium is as follows:

- In 86% cases, the results are matching with BLAST results.
- In 92% cases, results are matching with the known query sequences.
- In 6% cases, query covers are not matching up with the top five elements of the hit table.
- Results of clustering complete EST sequences and clustering by MidClustpy are comparable, and only 5% deviation is seen in the result.

Few of the facts that are observed and are important from research perspective are as follows:

- A variation in time required to compare query sequence with the given dataset was observed for standalone BLAST implementation.
- The threshold value is found to lie in the range of 36.5 to 39 during the computation of distance matrix.
- Though standalone BLAST takes less time to compare sequences, MidClustpy gives optimal solution at every run.

5 Conclusions

MidClustpy is apt to its name as it clusters smartly expressed sequence tags by taking into account Python implementation and operates on the most accurate part of EST. The middle part being exon-rich has a high probability to get coding region. The results proved that use of expressed sequence tags is useful in getting coding region and ultimately helps to predict genes. MidClustpy can be very well applied to the biological sequences and could be instrumental in predicting genetically modified organisms with the help of gene sequences for the given organism.

References

1. Kshirsagar, M., et al.: Data mining strategy for gene prediction with special reference to cotton genome. Cotton Sci. **20** (2008)
2. Han, J., Pei, J., Kamber, M.: Data Mining: Concepts and Techniques. Elsevier (2011)
3. Lee, J.K., et al.: Data Mining in Genomics. Clin Lab. Med. (2009)
4. Shivashankar, H.N., et al.: A hitchhiker's guide to expressed sequence tag (EST) analysis. Brief. Bioinform. **8** (2006)
5. Parkinson, J., et al.: Making sense of EST sequences by CLOBBing them. BMC Bioinform. (2002)
6. Camacho, C., et al: BLAST Help. NCBI (2010)
7. Dubey, V.K.: Expressed Sequence Tags. Lecture notes in IIT Guwahati
8. Chang, J., Chapman, B., Friedberg, I., Hamelryck, T., De Hoon, M., Cock, P., Antao, T., Talevich, E.: Biopython Tutorial and Cookbook (2010)

Data Manipulation Technique in a Keyword-Based Text Retrieval System on Encrypted Data

Navjeet Kaur Randhawa and Prabhat Keshari Samantaray

Abstract Preserving the privacy of the data is necessary especially when outsourcing sensitive information to the cloud servers. Symmetric searchable encryption (SSE) is known to be an efficient solution for preserving data privacy and keyword-based search on encrypted data. In this work, we have constructed an efficient data manipulation method on encrypted data using SSE with clustered document collections. We have implemented various data operations such as insert, update, and delete operations over encrypted data and at the same time validate data integrity in the encrypted domain. Rigorous experiments have been executed to evaluate the performance of the system.

Keywords Encrypted search · Vector encryption · Document clustering
Encrypted index construction · Dynamic data manipulation

1 Introduction

SSE facilitates the efficient search mechanisms on encrypted domain using text keywords and at the same time allows to performing various data manipulation. Song et al. [1] formulated the first known SSE model, and subsequently, other SSE schemes were proposed including single-keyword based search [1–7], single-keyword ranked search [8, 9], and multi-keyword ranked search [10–16].

In this paper, we have implemented a data manipulation technique on clustered encrypted document collection using an index tree-based SSE model. The docu-

N. K. Randhawa
Integrated Test Range, Defence Research and Development Organization,
Balasore, India
e-mail: navjeetkaur30@gmail.com

P. K. Samantaray (✉)
Department of Computer Science and Engineering,
College of Engineering and Technology, Bhubaneswar 751003, India
e-mail: prabhat.samantray@gmail.com

J. Nayak et al. (eds.), *Soft Computing in Data Analytics*,
Advances in Intelligent Systems and Computing 758,
https://doi.org/10.1007/978-981-13-0514-6_36

ments are clustered using bisection k-means clustering [17] method. The index is constructed with vector space model, and the vectors are encrypted with secure k-NN computation [18]. We perform dynamic data manipulation on encrypted document collection followed by necessary update mechanism on the encrypted index tree to reflect the change.

2 Background

2.1 System Model

A general SSE system comprises of the data owner, authorized user, and the semi-trusted cloud server. The data owner creates the encrypted index tree I from the clustered document collection D and encrypted document collection E. The data users are the authorized users who perform search operation with encrypted queries called trapdoor (TDR) using the shared symmetric keys. The cloud server performs the search process in the encrypted index and finally delivers the most relevant references of the documents to the data users which are then decrypted by using the secret key. Figure 1 shows the general model of SSE.

2.2 Preliminaries

Vector Space Model: Vector space model is a method used for vectorization of textual data, and the TF X IDF rule [19] in this mode is used for calculating the

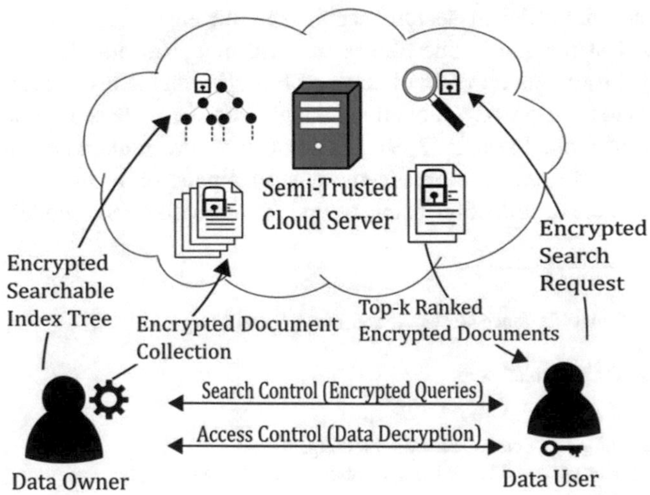

Fig. 1 A general view of an SSE model

relevance between document vectors and query vector. In this model, we use vector space model to construct the index and perform search and update operations.

Document Clustering: The document collection is clustered using bisecting k-means clustering [17] method. It is a hierarchical k-means clustering method which efficiently generates high quality clusters.

In this clustering method, we assume that a cluster having highest cluster size or lowest overall similarity (intra-cluster similarity) is chosen for the split. The cluster center is calculated from the equation given in Eq. (1):

$$C = \frac{1}{|D|} \sum_{d \epsilon S} d \tag{1}$$

where d is the vector of the document in the collection D where |D| is cardinality of the collection, and C represents the centroid and is calculated by taking the average value of the respective elements of document vectors in D.

During clustering process, we need to find the highly similar cluster for a document d and the similarity between a cluster center c with the document can be computed using cosine measure and it is given as in Eq. (2):

$$\begin{gathered} Sim(d, c) = cosine(d, c) \\ Sim(d, c) = d \cdot c / ||d|| \; ||c|| = d \cdot c / ||c|| \end{gathered} \tag{2}$$

where $||d|| = 1$, since the document vectors represented as normal vectors.

The overall similarity of a cluster that represents the cohesiveness of the cluster which can be calculated as in Eq. (3):

$$overall\,similarity = \frac{1}{|D|^2} \sum_{\substack{d \epsilon D \\ d' \epsilon D}} cosine\left(d', d\right) \tag{3}$$

Index Tree Construction: The searchable index tree in this model is adapted from keyword balanced binary (KBB) tree from Xia et al. [15], and we have implemented the index tree based on clustered documents in our previous paper [20]. Similar to KBB, the searchable index tree in this model stores vector V at each node i, and we denote it as V_i. A node in the index tree can be represented in 5-tuple:

$$i = \langle ID, V, Lch, Rch, DocID \rangle$$

where ID stores the unique identity value for the node i, and Lch and Rch store the reference to the left child node and the right child node, respectively, and the elements of index vector V_i are assigned to the values computed from its child nodes as given in Eq. (4).

$$V[r] = max\{i.Lch \rightarrow D[r], i.Rch \rightarrow D[r]\} \tag{4}$$

for each r = 1, 2, … m.

The leaf nodes are represented by clusters, and *DocID* field contains the list of documents in the cluster.

The cluster vector CV_j of cluster j that represents the leaf node of the tree is calculated as given in Eq. (5).

$$CV_j[r] = max\{d_1[r], d_2[r], \ldots d_s[r]\} \tag{5}$$

for each r = 1, 2 … m and s = cluster size of the cluster j.

Finally, index tree is built similar to *KBB* from the cluster vectors which are treated at leaf nodes. Algorithm 1 shows the index tree construction process.

Algorithm 1. *CreateIndexTree (D, z)*

1: *ClusterList = Create Cluster from Document Collection;*
2: Create cluster vector CV_j using Eq. (5) for each cluster in ClusterList.
3: For each cluster *j* in *ClusterList*, create leaf node having cluster vector CV_j and insert it into *NodeList*.
4: Construct parent nodes from the nodes in*NodeList* by using Eq. (4) until the root node is created [15].
5: Return the reference of the root node.

Search Process in Index Tree: The search operation is adapted from keyword balanced binary (KBB) tree [15] which was implemented in our previous paper [20]. In this process, the search follows a depth-first algorithm by choosing the child node having highest relevance score. The leaf nodes consist of cluster vector which represents that particular cluster. After reaching the leaf node, the search process takes linear approach within the cluster to find the most relevant document.

3 Index and Query Vector Encryption

The index and query vectors are encrypted using secure k-NN method [18]. In this method, the data administrator creates the symmetric secret key $S = \{S, M_1, M_2\}$, where S is a secret binary vector and M_1, M_2 are two Gaussian random invertible matrices. The vectors are encrypted by partitioning the given vector in two vectors on the basis of secret vector S. The index vector V_i is partitioned into $V_i^{'}$ and $V_i^{''}$ and its encrypted version is $EV_i = \{M_1^T V_i^{'}, M_2^T V_i^{''}\}$. Similarly, the query vector Q_i is partitioned into $Q^{'}$ and $Q^{''}$ and its encrypted version is $TDR = \{M_1^{-1} Q^{'}, M_2^{-1} Q^{''}\}$

where TDR is known as trapdoor. In our previous paper, we proposed two security schemes such as basic multi-keyword ranked search on clustered documents (BMRSCD) to address ciphertext threat model and enhanced multi-keyword ranked search on clustered documents (EMRSCD) to address known background threat model.

4 Data Manipulation in Encrypted Domain

The index tree needed to be updated when a document is inserted into or deleted from the document collection. The index vectors of the nodes are updated to reflect the changes in the document collection. To support the update operation in the index tree on the basis of insertion or deletion of a document in the collection, the following procedures are described.

$\{I_L^{'}, e_i\} \leftarrow CreateUpdateInfo(SK, T_L, i, updateType)$—The purpose of this algorithm is to create the update information regarding the nodes needed to be updated in the index tree. This algorithm returns a subset of nodes $I_L^{'}$ from the index tree T which needed to be updated and the encrypted document e_i that is inserted into or deleted from the document collection on the basis of the parameter *updateType* where $updateType \in \{Insert, Delete\}$. This algorithm is executed at the data administrator side and the update information is then sent to the remote cloud server to carry out update operation on the index tree. The update information is generated according to the parameter *updateType* and is described as follows.

If *updateType* is *insert,* then the data owner creates a document node i = <*GenUniqueID*(), *V*, null, null, *x*> for the document d_x where $i.V = NormTF_{d_x, k_r}$, where r = 1, … m. For the document d_x, the nearest cluster C_j is selected having highest similarity between its cluster center CC_j and document vector $d_x.V$ which is calculated from Eq. (2), and finally, the document d_x is assigned to the cluster C_j. The cluster center CC_j is updated as given in Eq. (1), and the cluster vector CV_j is recalculated as given in Eq. (5). Based on the updated cluster vector CV_j, the index vectors of the list of parent nodes P_L which constitutes a path between the parent of cluster C_j and the root node are updated by using Eq. (4). The nodes in P_L and the cluster C_j constitute a final list of nodes $T_L^{'}$ and encrypted with secret key *SK* to form encrypted list of nodes $I_L^{'}$. The document d_x is encrypted using the symmetric key to create encrypted documents e_x. The data owner finally uploads $I_L^{'}$ and e_x to the server to reflect the updates of the respective nodes in the index tree.

If *updateType* is *delete*, the data owner deletes the document d_x from the document collection. For the cluster C_j such that $d_x \in C_j$, the cluster center CC_j is updated by using Eq. (1) and the cluster vector CV_j is also updated by using Eq. (5). Similar to the insert operation, the index vector of the parent nodes in P_L are updated. The index vectors of P_L and the cluster C_j are encrypted with *SK* to create I_L'. The data owner uploads I_L', and the *DocID* of d_x is deleted from document collection from the cloud server to carry out the update process.

$\{I', E'\} \leftarrow ExecuteUpdate(I, E, updateType, I_L', e_x)$—This algorithm performs the update operation in encrypted index tree I on the cloud server. It replaces the index vectors of respective nodes with the index vector in I_L'. If *updateType* is *insert*, the document e_x is inserted into the document collection E and if *updateType* is *delete*, and the document e_x is deleted from the document collection E. Finally, the algorithm returns the updated encrypted index tree I' and updated encrypted document collection E'.

The dynamic update operation defined in [15] implemented the delete operation by replacing the elements in index vector of the specified leaf node with zeros instead of actually deleting that from the index tree and the parent nodes are updated accordingly. This procedure does not need to recreate the index tree, thus reducing the overall time for index tree construction. During insert operation, a new leaf node is created and it is added to the index tree by replacing an empty leaf node created by the previous delete operation. Although this insert method is efficient when an empty node is available for a new node, it needs to recreate the index tree when no empty nodes are present. In the proposed model, we implement the index from the clustered documents and the leaf of the index consists of a cluster of similar documents. In an insert operation, there is no need to find an empty node from previous delete operation to insert a new document. A cluster with high similarity is selected and the given document is assigned to it. In the case of a delete operation, there is no need to create empty nodes; the document node can be deleted from the cluster efficiently without affecting the index tree. This method completely eradicates the overhead of recreation of index tree; thus, it provides a faster implementation of the insert, delete, and update operation on large document collection. So, we may conclude the index tree generated from clustered document collection is almost static unless the data owner chooses to create new clusters.

5 Performance Analysis

The performance of the schemes is evaluated using the dataset built with text documents from BBC News archives [21]. In this section, we test the efficiency of index tree construction and search process.

5.1 Search and Update Efficiency

We have compared the search efficiency between the secure schemes defined in [15] and our proposed method. Time complexity of the search operation found out to be $O(\theta m log(z) + s)$, and for the update operation, it is $(m^2 log(z))$, where θ is

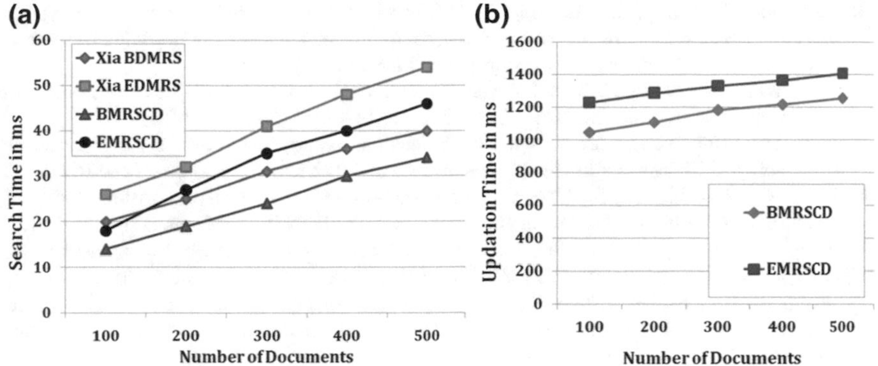

Fig. 2 Comparison of search operation: **a** BDMRS, BMRSCD, EDMRS, and EMRSCD with number of keywords in query 25. **b** The time to update a document in BMRSCD and EMRSCD with dictionary size 1000 and number of clusters 20

total cluster number that contains the resultant documents, k be the number of search result, z be the number of clusters in the index tree, n be the cardinality of document collection, and s be the size of the cluster and assuming $\theta < k$ and $\theta \ll n$ (Fig. 2).

6 Conclusions

We have constructed a data manipulation technique implemented with encrypted searchable index tree on the encrypted data. We have used our SSE model constructed from clustered document collection in order to carry out various data operation such as insert, delete, update, and search, and we can verify that this approach leads to better data management and efficient data manipulation process. The index structure tends to be more compact and resilient to any changes to the contents in document collection.

However, SSE schemes are still an open research area and there is lot of security and scalability issues that needed to be addressed. In the proposed scheme, if the data owner creates large number of clusters number of clusters, then it may lead to inefficient index construction and search operation. These issues must be taken into consideration when implementing in practice.

References

1. Song, D.X., Wagner, D., Perrig, A.: Practical techniques for searches on encrypted data. In: Proceedings IEEE Symposium on Security & Privacy, pp. 44–55 (2000)
2. Goh, E.-J.: Secure Indexes. IACR Cryptology ePrint Archive, vol. 2003, p. 216 (2003)

3. Curtmola, R., Garay, J., Kamara, S., Ostrovsky, R.: Searchable symmetric encryption: improved definitions and efficient constructions. In: Proceedings of the 13th ACM Conference Computer and Communications Security, pp. 79–88 (2006)
4. Li, J., Wang, Q., Wang, C., Cao, N., Ren, R., Lou, W.: Fuzzy keyword search over encrypted data in cloud computing. In: IEEE Proceedings INFOCOM, pp. 1–5 (2010)
5. Kuzu, M., Islam, M.S., Kantarcioglu, M.: Efficient similarity search over encrypted data. In: Proceedings IEEE 28th International Conference on Data Engineering, pp. 1156–1167 (2012)
6. Wang, C., Ren, K., Yu, S., Urs, K.M.R.: Achieving usable and privacy-assured similarity search over outsourced cloud data. In: Proceedings IEEE INFOCOM, pp. 451–459 (2012)
7. Wang, B., Yu, S., Lou, W., Hou, Y.T.: Privacy-preserving multikeyword fuzzy search over encrypted data in the cloud. In: Proceedings IEEE INFOCOM, pp. 2112–2120 (2014)
8. Wang, C., Cao, N., Li, J., Ren, K., Lou, W.: Secure ranked keyword search over encrypted cloud data. In: Proceedings IEEE 30th International Conference Distributed Computing Systems (ICDCS 2010), pp. 253–262 (2010)
9. Wang, C., Cao, N., Ren, K., Lou, W.: Enabling secure and efficient ranked keyword search over outsourced cloud data. IEEE Trans. Parallel Distrib. Syst. **23**(8), 1467–1479 (2012)
10. Cao, N., Wang, C., Li, M., Ren, K., Lou, W.: Privacy-preserving multi-keyword ranked search over encrypted cloud data. In: Proceedings IEEE INFOCOM, pp. 829–837 (2011)
11. Sun, W., Wang, B., Cao, N., Li, M., Lou, W., Hou, Y.T., Li, H.: Privacy-preserving multi-keyword text search in the cloud supporting similarity-based ranking. In: Proceedings 8th ACM SIGSAC Symposium on Information, Computer and Communications Security, pp. 71–82 (2013)
12. Orencik, C., Kantarcioglu, M., Savas, E.: A practical and secure multi-keyword search method over encrypted cloud data. In: Proceedings IEEE 6th International Conference on Cloud Computing, pp. 390–397 (2013)
13. Zhang, W., Xiao, S., Lin, Y., Zhou, T., Zhou, S.: Secure ranked multi-keyword search for multiple data owners in cloudcomputing. In: IEEE 44th Annual IEEE/IFIP International Conference on Dependable Systems and Networks (DSN), pp. 276–286 (2014)
14. Pang, H., Shen, J., Krishnan, R.: Privacy-preserving similarity-based text retrieval. ACM Trans. Internet Technol. **10**(1), 4 (2010)
15. Xia, Z., Wang, X., Sun, X., Wang, Q.: A secure and dynamic multi-keyword ranked search scheme over encrypted cloud data. IEEE Trans. Parallel Distrib. Syst. **27**(2), 340–352 (2016)
16. Zerr, S., Olmedilla, D., Nejdl, W., Siberski, W.: Zerber + r: Top-k retrieval from a confidential index. In: Proceedings of the 12th International Conference on Extending Database Technology: Advances in Database Technology, pp. 439–449 (2009)
17. Steinbach, M., Karypis, G., Kumar, V.: A comparison of document clustering techniques. In: KDD Workshop on Text Mining, vol. 400, no. 1, pp. 525–526, Aug 2000
18. Wong, W.K., Cheung, D.W.-L., Kao, B., Mamoulis, N.: Secure KNN computation on encrypted databases. In: Proceedings of ACM SIGMOD International Conference on Management of Data, pp. 139–152, June 2009
19. Manning, C.D., Raghavan, P., Schutze, H.: Introduction to Information Retrieval. UK, Cambridge University Press, Cambridge (2008)
20. Samantaray, P.K., Randhawa, N.K., Pati, S.L.: An efficient multi-keyword text search over outsourced encrypted cloud data with ranked results. In: Proceedings of the 4th International Conference on Computational Intelligence in Data Mining (2017)
21. BBC Datasets. http://mlg.ucd.ie/datasets/bbc.html

Implementation of Grasshopper Optimization Algorithm for Controlling a BLDC Motor Drive

Devendra Potnuru and Ayyarao S. L. V. Tummala

Abstract The paper implemented a recently developed grasshopper optimization algorithm for speed response improvement during transient and steady-state conditions for a BLDC motor drive. An objective function is formulated to reduce the integral square error in such a way that the gains of speed control (PID) are optimally tuned. To know the validity of the present approach, simulation experiments are conducted extensively to get the proper tuning of PID gains in MATLAB/Simulink and then the same gains can be used in off-line for the hardware implementation.

Keywords BLDC · Grasshopper optimization · Speed controller PID tuning

1 Introduction

The brushless DC motor is nowadays well known for its superiority due to its kind of the characteristics such as noise-free operation, flexible speed control, and high-speed range. However, the closed-loop speed control is necessary for some applications where in the speed should be constant irrespective of sudden changes of load and/or supply. Hence, the gain tuning of speed controller designed using PID controller is a tedious job. To reduce the time of gain tuning and improve the performance of the speed control in closed loop, many researchers developed various techniques. The Ziegler–Nichols method is a usually employed method for determining the PID gains in many applications. However, in high-performance

D. Potnuru (✉) · A. S. L. V.Tummala
GMR Instutute of Technology, Rajam 532127, India
e-mail: devendra.p@gmrit.org; devendra.p.07@gmail.com

A. S. L. V.Tummala
e-mail: ayyarao.tslv@gmrit.org

J. Nayak et al. (eds.), *Soft Computing in Data Analytics*,
Advances in Intelligent Systems and Computing 758,
https://doi.org/10.1007/978-981-13-0514-6_37

applications the accurate closed-loop speed control with good steady-state and transient behavior is very much important.

In [2], BAT algorithm is implemented for speed control of BLDC motor. Ye et al. [7] have been implanted with modified PSO and PID controller in. In [6]. Bacteria foraging and PSO optimization has been applied to speed control DC motor. Multiobjective Optimization has been applied for Controller Tuning of a TRMS Process in. Fuzzy-Neural-Network Self-Learning Control Methods for Brushless DC Motor Drives has been implemented in [4]. DE and PSO algorithms have been applied for PID controller tuning in [3].

However, tuning of PID speed controller using grasshopper algorithm for closed-loop speed controller for a BLDC motor has been not yet implemented in the existing literature. Hence, the present paper is mainly devoted to implementation of recently developed nature-inspired grasshopper optimization algorithm for tuning of the PID gains. Integral speed error is considered as objective function optimization in the present work. The extensive simulations using MATLAB/Simulink has resulted in optimum gains of PID, and they are in off-line for real-time speed control of BLDC motor in laboratory arrangement.

The paper is organized as follows: In Sect. 2, the operation of BLDC drive is described. Formulation objective function and PID control design are elaborated in Sect. 3. Later, the experimental results are described in Sect. 4 and conclusions are given in the last section.

2 The Drive Scheme of BLDC Motor

The BLDC motor drive operation mainly consists of three-phase inverter, BLDC motor, DSP controller and speed and position sensors. As the BLDC motor drive run in self-control mode, the inverter IGBTs are switched on based on rotor position. The DSP controller is responsible for generating the PWM pulses from the given input conditions. Then, the gate drivers are used to drive the output pulses of the DSP controller and turn on the IGBTs. In the present work BLDC motor is a tetra square wave type and with inbuilt hall sensors to sense the rotor position.

The block diagram for PID tuning for speed control of BLDC motor drive using grasshopper optimization algorithm is shown in Fig. 2. For more details of the drive scheme, one can refer [1].

First, the MATLAB simulation file is converted into DSP-enabled code and dump in the DSP processor (dSPACE DS 1103). The LEM makes current, and voltage sensors are used to sense the current and voltages. In the closed-loop speed control, the torque reference is generated based on the speed error. The output of the PID controller is scaled down using the motor torque constant to obtain the reference currents of the hysteresis current controller. The current controller generates the required PWM pulses to the inverter switches.

3 Grasshopper Optimization

Grasshopper is one kind of insects which damages the agriculture. The swarm behavior of the grasshopper is modeled to develop an efficient nature-inspired algorithm. The size of the grasshopper swam is very large than any other creature in nature. Most noted thing is that swarm behavior can be observed in nymph and adulthood. They jump and move like rolling and eat the vegetation in their way and become nightmare for the farmers. Later, they form very big swarm in the air and migrated over large distances [5]. Figure 1 shows the life cycle of swarm, Fig. 2 shows the primitive corrective patterns in swarm of grasshoppers, and Fig. 3 shows the primitive corrective patterns in swarm of grasshoppers.

The implementation algorithm for grasshopper optimization is given as below [5].

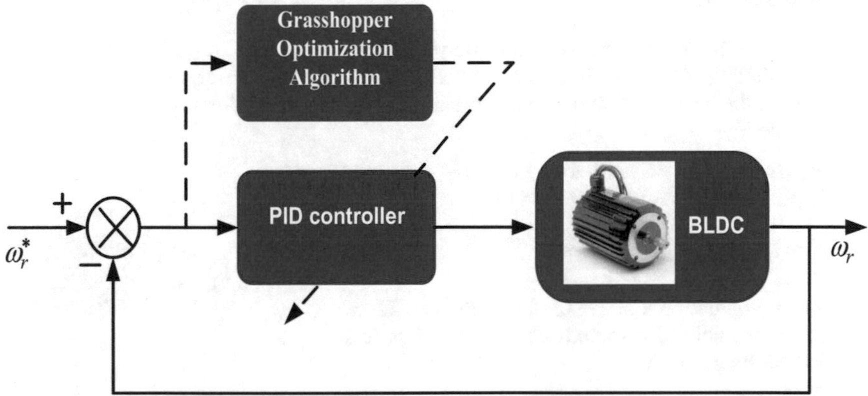

Fig. 1 Speed controller of BLDC motor using grasshopper algorithm

Fig. 2 Grasshopper and life cycle of grasshopper [5]

Fig. 3 Primitive corrective patterns in swarm of grasshoppers [5]

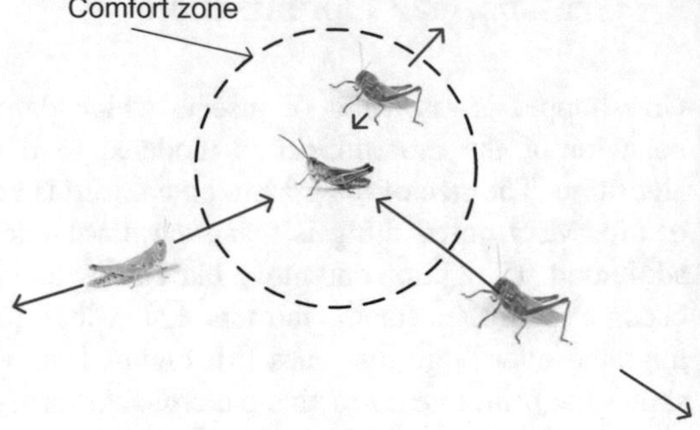

Grasshopper optimization Algorithm

Initialization

Initialize the population, Max-iteration λ, ζ_{max}, ζ_{min}
Calculate the objective function value from (ISE) for each solution
Obtain the best solution of X with minimum objective function -
-value F

While (i < λ) % main loop begins
Update the following equation (1)

$$\zeta = \zeta_{min} - i\frac{\zeta_{max} - \zeta_{min}}{\lambda} \tag{1}$$

for j < population size % to obtain the new X
obtain normalized distance between grasshoppers
update the position
j=j+1
end for loop
Replace X with the new one if the objective function is less than F
i=i+1
end % end main loop

---

4 Result Analyses

The closed-loop speed control using grasshopper algorithm has been extensively tested for two reference speeds. Firstly, the algorithm is run for the closed-loop speed control for number of times by keeping the proper ranges for the PID gains. Then finally, the optimal gains are selected for implementation of closed loop speed control. The typical convergence characteristic of fitness function is as shown in

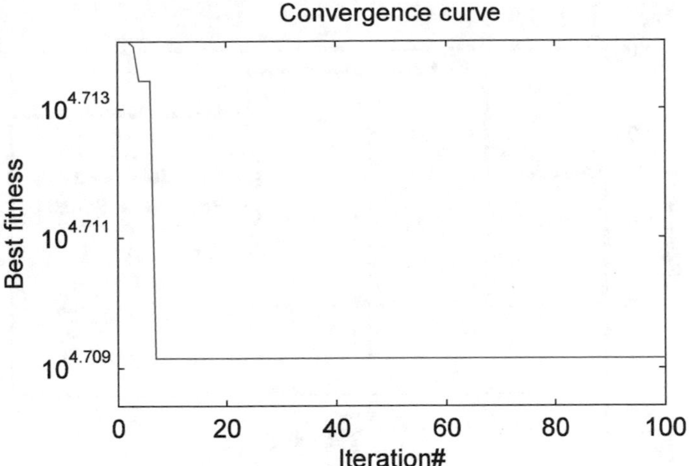

Fig. 4 Convergence curve of optimization

Fig. 4. The optimal PID gains for final selection are with kp = 23.418, ki = 13.8416, and kd = 0.001. These gains are applied to observe the dynamic performance of the drive. In first test case, reference speed consists of stepped speed command and second case is the combination of step and ramp speeds (hybrid speed reference).

Figure 5a shows the performance of the closed-loop speed control using stepped speed command, and Fig. 5b shows speed error with the stepped reference speed.

Figure 6a shows the performance of the closed loop speed control using hybrid speed reference, and corresponding speed error is shown in Fig. 6b.

One can observe that the closed-loop speed control performance is superior during transient and steady-state conditions. Speed error is very small during the steady-state conditions.

5 Conclusions

The closed-loop speed control of BLDC motor has been implemented successfully by using the PID gains obtained with the grasshopper optimization algorithm. It is observed that the closed-loop speed control performance is superior during transient and steady-state conditions. Speed error is very small during the steady-state conditions.

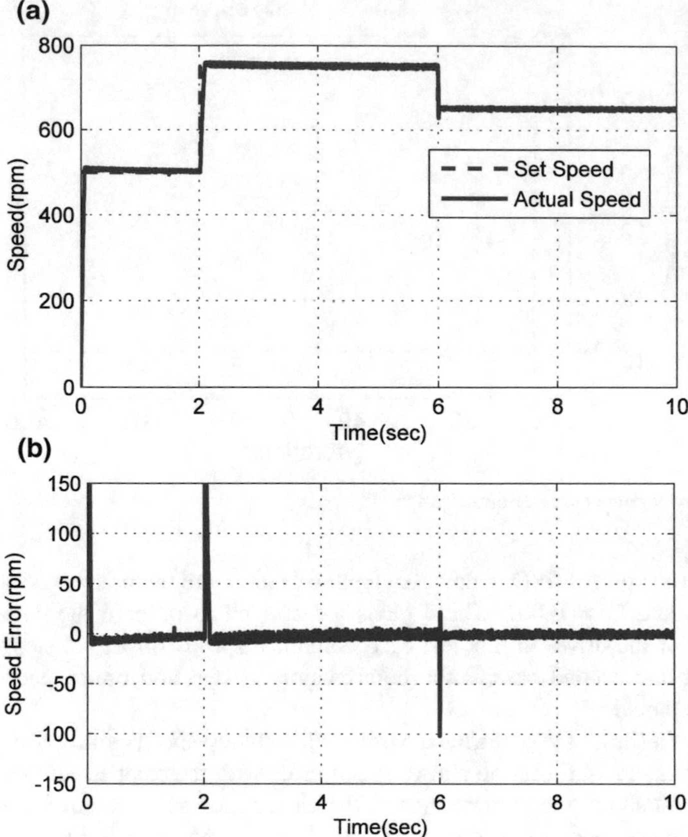

Fig. 5 **a** Closed-loop speed control using grasshopper algorithm. **b** Speed error of closed-loop speed control with stepped speed reference

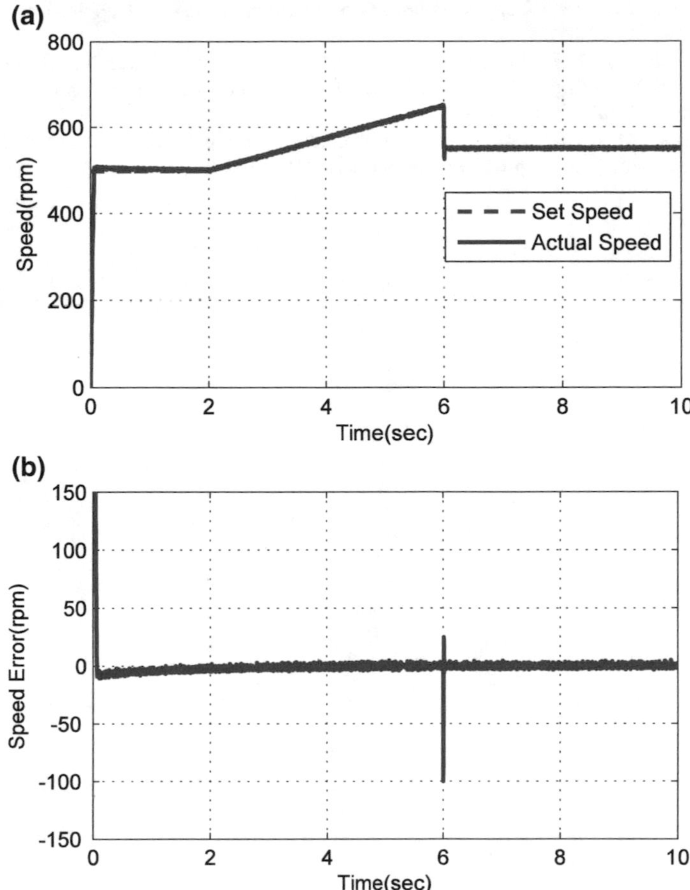

Fig. 6 **a** Closed-loop speed control hybrid step and ramp speed command. **b** Speed error of the closed-loop speed control for hybrid step and ramp speed command

References

1. Potnuru, D., et al.: Design and implementation methodology for rapid control prototyping of closed loop speed control for BLDC motor. J. Electr. Syst. Inf. Technol., 4–16 (2017)
2. Premkumar, K., Manikandan, B.V.: Bat algorithm optimized fuzzy PD based speed controller for brushless direct current motor. Eng. Sci. Technol. an Int. J. **19**(2), 818–840 (2016)
3. Puralachetty, M.M., Pamula, V.K.: Differential evolution and particle swarm optimization algorithms with two stage initialization for PID controller tuning in coupled tank liquid level system. In: ICARM 2016–2016 International Conference on Advanced Robotics and Mechatronics, pp. 507–511 (2016)
4. Rubaai, A., Young, P.: Hardware/software implementation of fuzzy-neural-network self-learning control methods for brushless DC motor drives. IEEE Trans. Ind. Appl. **52**(1), 414–424 (2016)

5. Saremi, S., et al.: Grasshopper optimisation algorithm: theory and application. Adv. Eng. Softw. **105**, 30–47 (2017)
6. Thangavelusamy, D., Ponnusamy, L.: Study and comparison of PI controller tuning techniques using bacteria foraging and bacteria foraging based particle swarm optimization. Int. J. Electr. Eng. **5**(5), 571–586 (2012)
7. Ye, Y., et al.: Position control of nonlinear hydraulic system using an improved PSO based PID controller. Mech. Syst. Signal Process. **83**, 241–259 (2017)

Economic Load Dispatch in Microgrids Using Real-Coded Genetic Algorithm

Gouthamkumar Nadakuditi, U. Mohan Rao, Venkateswararao Bathina and Srihari Pandi

Abstract This paper proposes a simple and heuristic-based solution method for solving economic load dispatch (ELD) in an islanded mode of microgrid. The economic dispatch for different combination of conventional and nonconventional sources is performed by using real-valued genetic algorithm (GA). This paper compares the total power generation cost for different combination of sources. Further, it considers the solar renewable energy credits (REC), provided by the government to encourage the usage of renewable energy sources in order to reduce the emission pollutants and to overcome the fuel availability problems. Thus, results show the integration of renewable sources and simultaneous consideration of solar credits gives more profit to the microgrid.

Keywords Microgrid · Economic load dispatch · Renewable energy
Genetic algorithm · Renewable energy credits

G. Nadakuditi · V. Bathina
Department of Electrical and Electronics Engineering,
V. R. Siddhartha Engineering College, Vijayawada 520007, AP, India
e-mail: gowthamkumar218@gmail.com

V. Bathina
e-mail: bvrao.eee@gmail.com

U. Mohan Rao (✉)
Department of Electrical and Electronics Engineering,
Lendi Institute of Engineering and Technology, Vizianagaram 533005, AP, India
e-mail: mohan13.nith@gmail.com

S. Pandi
Infosys Pvt. Limited, Hyderabad, India
e-mail: srihari.nith@gmail.com

J. Nayak et al. (eds.), *Soft Computing in Data Analytics*,
Advances in Intelligent Systems and Computing 758,
https://doi.org/10.1007/978-981-13-0514-6_38

1 Introduction

Enormous growth in electricity utilization causes need for massive amount of power production [1]. Due to insufficient availability of conventional energy sources, the necessity of renewable energy sources (RES) has been increased day by day. Hence, the whole world is looking toward the existence of abundant quantity of RES. Generally, microgrid encourages the usage of RES in power generation; therefore, it could be a better solution for our energy crisis [2]. Microgrids are designed and operated by a single consumer or a group of consumers for having an economic, clean, and reliable power supply in order to satisfy their load demand. Microgrid is normally considered as small-scale power producer to provide continuous power supply for small community. It is a combination of different loads and several microsources operating as one controlled unit to supply power as well as heat wherever it is needed. Since it is comprising of different low power generating sources that makes it more efficient and flexible. Microgrid can be operated in two modes. One is isolated or islanded mode, and another is grid-connected mode. In islanded mode, microgrid operation is independent from the utility grid, whereas in grid-connected mode, microgrid is connected to utility grid to avoid power supply outages because of mutual power exchange. It is very important to perform economic load dispatch (ELD) in microgrids because of various generations and different generating costs [3]. ELD is nothing but the minimization of total generation cost with the allocation of all generators to meet the load demand while satisfying the power balance and generation limits constraints. In this microgrid, CHP unit supplies both power and heat. But due to the presence of single CHP unit, the heat generated should be equal to the heat demand. So no need to perform heat dispatch. Therefore, only power dispatch is considered in this work [4]. Due to the burning of coal, thermal generating plants release large amount of emission pollutants such as NO_x, CO_2, and SO_2, which causes global warming. In order to get lower generation cost and minimal emission pollutants from the thermal units, the RES such as solar and wind has been in vogue along with the thermal plants. Generally, sources of renewable energy (RE) are random or erratic in nature. So it is difficult to treat them as regular or steady power producers. Therefore, due to this random nature, these sources cannot be included in ELD, and with some accuracy, these are assumed as negative loads [5].

The main contribution of this paper is to study the ELD in islanded mode of microgrid consisting of two RES, i.e., wind and solar energy generation, one combined heat and power (CHP) unit and two conventional thermal units. The simulations are performed and compared total generation costs with various combination of sources in microgrid by using the real-coded genetic algorithm (GA).

2 Problem Formulation

2.1 Objective Function

The objective function is a smooth quadratic cost function for each unit of ELD problem and can be represented in Eq. (1) as

$$min\ F_T = \sum_{i=1}^{3} \sum_{t=1}^{24} \left(a_i + b_i P_i + c_i P_i^2\right) \tag{1}$$

where F_T is the total fuel cost to minimize, P_i is the power generated by ith unit, and a_i, b_i, and c_i are the cost coefficients for ith unit.

2.2 Constraints

Power balance is represented as the equality constraint, and the generated output power of each generator must lie between its minimum and maximum limits treated as inequality constraint for the ELD problem, which are represented in Eq. (2).

$$\begin{cases} \sum_{1}^{3} P_i = P_D \\ P_i^{\min} \le P_i \le P_i^{\max} \end{cases} ; i = 1, 2, 3 \tag{2}$$

where P_D is the total load demand, $P_i^{\min}$ is the minimum power generation limit, and $P_i^{\max}$ is the maximum power generation limit of ith unit, respectively.

2.3 Wind Power Generation and Wind Cost Calculation

Wind turbine output power is calculated by using (3) obtained from [6].

$$P_w = \frac{1}{2}\left(\rho \times A \times u^3\right) \tag{3}$$

where P_w is wind power generation (kW), ρ is density of air which is taken as 1 kg/m^3, A is wind turbine swept area m^2, and u is velocity of wind (m/s). The total wind power generation cost function includes the total investment cost for the equipment and the operation and maintenance cost and depreciation of all the generating equipment, which is represented in Eq. (4) as follows:

$$\begin{cases} F(P_w) = \Omega \times I^p \times P_w + G^E \times P_w \\ \Omega = \frac{r}{\left(1-(1+r)^{-N}\right)} \end{cases} \quad (4)$$

where Ω is coefficient of annuitization (no units), r is rate of interest (for base case $r = 0.09$), N is investment lifetime (taken as $N = 20$ years), I^p is investment cost per kW installed power ($/kW), and G^E is O and M cost per kW generated power ($/kW).

2.4 Solar Power Generation and Solar Cost Calculation

The output power from the photovoltaic (PV) panels is calculated by using (5) adopted from [5].

$$E_t = 3.24 \times M_{PV}[1 - 0.0041 \times (T_t - 8)] \times S_t \quad (5)$$

where E_t is solar output power (kW), M_{PV} is capacity of generation from all PV panels, T_t is temperature, and S_t is solar radiation at time t (kW/m^2). Renewable energy credit includes the state tax and federal tax subsidies on solar panels. So the consideration of REC results in the reduction of the investment cost for solar. The new reduced cost of the investment due to the consideration of REC is calculated in Eq. (6) as follows:

$$I_p' = I_p - (I_p \times S_t) - [I_p \times (1 - S_t) \times F_t] \quad (6)$$

where I_p' is reduced cost of the investment ($/kW), I^p is actual cost of the investment ($/kW), S_t is state tax subsidy (%), and F_t is federal tax subsidy.

3 Solution Methodology

Genetic algorithm (GA) is a population-based algorithm developed by Goldberg from the inspiration of Darwin's evolution theory. GA works on a hope that the new generation will be superior to the old generation. In GA, initial population is generated randomly and it gets modified from one generation to next generation through the genetic operators such as selection, crossover, and mutation [7]. GA basic steps are as follows.

3.1 Initialization

Initialize the random population which has a set of chromosomes, where each chromosome represents one solution set. Typical value of the population size may vary in the range 30–200. Chromosome can be represented in Eq. (7) as follows [8]:

$$\text{Chromosome} = [\mathrm{P}_{1,1}\mathrm{P}_{2,1}\mathrm{P}_{1,2}\mathrm{P}_{2,2}, \ldots \mathrm{P}_{1,t}\mathrm{P}_{2,t}], \mathrm{t} = 1, 2, \ldots 24 \tag{7}$$

where $P_{1,1}$, $P_{1,2}$, and $P_{1,t}$ are power dispatched to CHP unit at 1st, 2nd, and tth hour, respectively.

3.2 Fitness Evaluation

After creating the initial random population, the fitness value is to be calculated for each chromosome. Chromosome which is having more fitness value has more chances to survive in the next generations. For ELD, the objective function is total generation cost minimization. So chromosome corresponds to least cost should have highest fitness value and chromosome corresponds to highest cost should have least fitness value. To imply this inverse relation, the fitness function can be represented in Eq. (8) as follows:

$$\begin{cases} F(i) = \frac{1}{1 + F_{obj}(i)} \\ F_{obj}(i) = \sum_{t=1}^{24} \sum_{i=1}^{3} F_i(P_i) + Penalty \end{cases} \tag{8}$$

where $F(i)$ is the fitness of ith chromosome and $F_{obj}(i)$ is the objective function value of ith chromosome. Penalty is added to the total cost of 24 h to eliminate the chromosome which corresponds to violation of generation limits. Later the evaluation of the fitness value, population should undergo selection, crossover, and mutation operators, respectively.

3.3 Selection

It is a process to pick up the parent chromosomes for crossover operation. There are so many selection methods, but mostly Roulette wheel selection is adopted from [9] by using the probability of fitness $P(i) = F(i) / \sum_{i=1}^{n} F(i)$ for ith chromosome out of n number of chromosomes.

3.4 Crossover and Mutation

Crossover is used to recombine the selected parent chromosomes to produce new pair of offsprings or child chromosomes [10]. The single-point crossover procedure is opted, and the typical value of crossover probability value is taken as 0.75. Mutation is necessary to make the chromosomes dissimilar in the population.

Single-point mutation procedure is adopted, and typical value of mutation probability value is taken as 0.01.

3.5 Termination

Stop the procedure if the total number of generations is reached. Later, performing the ELD for the three conventional generators, minimum cost of the conventional generation is obtained by substituting the dispatched power values in their corresponding cost equations. Then, the costs of RE generations are also obtained from their cost equations corresponding to wind generation and solar generation, respectively. Afterward, the total generation cost is found by adding the cost of conventional generation and RE generation.

4 Results and Discussion

The load demand for 24 h in a day is adopted from [5]. The cost function for CHP and two conventional generators is assumed as quadratic, and the coefficients of the quadratic cost functions for these generators are also adopted from [5]. The active power limits for these three generators are shown in Table 1.

The wind velocity data required to find out the wind power generation is adopted from [5]. It provides the hourly averaged wind velocity for 24 h in a day. Total 30 number of wind turbines are considered in the wind farm with a swept area of 3000 m^2, and each turbine is rated as 1 MW, respectively. The total wind power output is calculated for the duration of 24 h in a day. The investment cost per kW installed power (I^p) and O and M costs per kW generated power (G^E) are adopted from [5] as $1400 and 1.6 cents, respectively. The solar radiation data required to find out the solar power generation is adopted from [5]. It provides the hourly averaged solar radiation for 24 h in a day. Totally 1,40,000 PV panels are considered with a panel capacity of 350 W. The total solar power output is calculated by using (8) for the duration of 24 h in a day.

The ELD program is performed by using MATLAB-2013(a) software, and the total cost of generation for the below four scenarios in microgrid is compared. The population size, maximum number of iterations, crossover rate, and mutation rate have been selected for GA as follows: 20, 100, 0.75, and 0.01.

Table 1 Active power limits on generators

	CHP	Unit 1	Unit 2
P^{min} (MW)	0	0	30
P^{max} (MW)	160	200	220

Case 1—Considering all generations: This combination results in more total generation cost comparing with remaining all cases explained below, because the initial investment of solar equipment is very high. Therefore, solar generation cost becomes more, since solar cost equation includes the investment cost also. In this case, there is no REC for solar panels; hence, it leads to more cost. The result of power dispatch and corresponding total cost is shown in Table 2.

Case 2—Considering all generations except wind and solar: In this case, power generation is only from CHP and two thermal generators, so there is no amount of generation from RES such as solar and wind. Therefore, it does not give any loss or profit to the microgrid, but results in more cost compared with Case 3.

Case 3—Considering all generations except solar: In this case, solar generation is not considered and only wind generation is considered along with CHP and

Table 2 Economic dispatch considering all generations

Time (h)	CHP (MW)	Gen 1 (MW)	Gen 2 (MW)	Wind (MW)	Solar (MW)	Total cost ($/h)
1	45.19	42.95	50.17	1.69	0	6145.28
2	69.29	18.02	54.18	8.51	0	6380.93
3	30.37	35.68	79.71	9.24	0	6459.42
4	10.37	37.76	95.21	16.66	0	6569.31
5	89.07	2.90	65.83	7.20	0	6801.57
6	19.47	73.33	72.25	4.91	0.03	6845.83
7	45.78	37.92	70.38	14.65	6.27	7128.15
8	45.62	36.90	54.02	26.48	16.98	7602.21
9	40.66	77.77	46.65	20.87	24.05	8625.50
10	12.43	67.05	93.28	17.87	39.36	9728.92
11	46.35	113.06	60.23	12.94	7.41	8792.10
12	74.58	58.03	95.06	18.68	3.65	8773.38
13	26.23	86.83	80.67	14.34	31.94	9688.63
14	66.93	36.89	79.01	10.37	26.81	9018.92
15	49.09	37.32	95.26	8.25	10.08	7912.21
16	35.48	60.18	65.32	13.72	5.30	7209.52
17	13.38	66.63	76.97	3.45	9.57	7232.66
18	46.36	62.48	71.99	1.86	2.31	7274.20
19	75.67	67.53	56.04	0.75	0	7564.66
20	51.32	65.25	123.26	0.17	0	8538.37
21	45.28	95.17	84.40	0.15	0	8177.34
22	52.22	50.69	86.77	0.31	0	7314.35
23	52.83	25.01	81.08	1.07	0	6631.77
24	34.60	32.57	77.24	0.58	0	6276.05
Total cost/day						182691.29

conventional generators. Wind generation cost is very less compared with remaining all generations. So this combination gives profit to the microgrid.

Case 4—Considering all generations with REC for solar panels: In this scenario, all five generations and the REC for solar panels are also considered. Therefore, this combination of generations gives maximum profit to the microgrid. Here, 35% of state tax rebate and 50% of federal tax subsidy are considered from [5]. The power dispatch corresponding to this case is shown in Table 3.

The convergence characteristics of GA for four scenarios are shown in Fig. 1. Here, convergence characteristics for Case 1 and Case 4 are merged, due to the same power dispatch. But the total generation cost is less for Case 4 due to inclusion of REC for solar. The cost comparison among the four scenarios is shown in Table 4.

Table 3 Economic dispatch considering all generations with renewable energy credits for solar

Time (h)	CHP (MW)	Gen 1 (MW)	Gen 2 (MW)	Wind (MW)	Solar (MW)	Total cost ($/h)
1	45.19	42.95	50.17	1.69	0	6145.28
2	69.29	18.02	54.18	8.51	0	6380.93
3	30.37	35.68	79.71	9.24	0	6459.42
4	10.37	37.76	95.21	16.66	0	6569.31
5	89.07	2.90	65.83	7.20	0	6801.57
6	19.47	73.33	72.25	4.91	0.03	6844.56
7	45.78	37.92	70.38	14.65	6.27	6863.53
8	45.62	36.90	54.02	26.48	16.98	6885.36
9	40.66	77.77	46.65	20.87	24.05	7610.33
10	12.43	67.05	93.28	17.87	39.36	8067.51
11	46.35	113.06	60.23	12.94	7.41	8479.18
12	74.58	58.03	95.06	18.68	3.65	8619.45
13	26.23	86.83	80.67	14.34	31.94	8340.80
14	66.93	36.89	79.01	10.37	26.81	7887.33
15	49.09	37.32	95.26	8.25	10.08	7486.67
16	35.48	60.18	65.32	13.72	5.30	6985.62
17	13.38	66.63	76.97	3.45	9.57	6828.76
18	46.36	62.48	71.99	1.86	2.31	7176.88
19	75.67	67.53	56.04	0.75	0	7564.66
20	51.32	65.25	123.26	0.17	0	8538.37
21	45.28	95.17	84.40	0.15	0	8177.34
22	52.22	50.69	86.77	0.31	0	7314.35
23	52.83	25.01	81.08	1.07	0	6631.77
24	34.60	32.57	77.24	0.58	0	6276.05
Total cost/day						174935.03

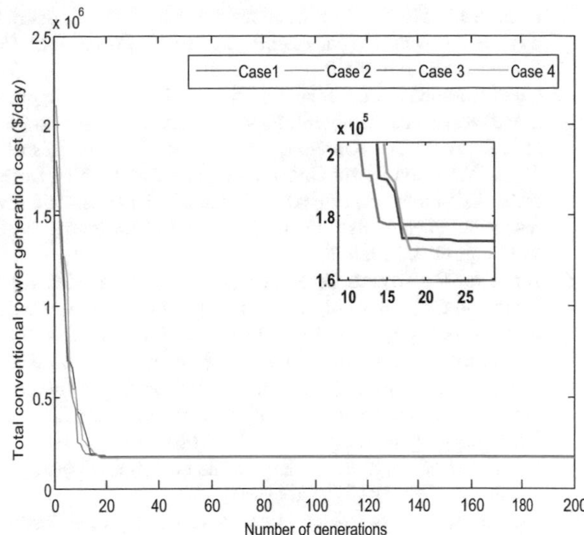

Fig. 1 Convergence characteristics of GA for all scenarios

Table 4 Cost comparison among different scenarios

Case no	Scenario	Total generation cost ($/day)
1	Considering all generations	182691.29
2	Considering all generations except wind and solar	176598.78
3	Considering all generations except solar	175548.30
4	Considering all generations with inclusion of REC for solar panels	174935.03

5 Conclusion

This paper proposed a heuristic-based real-coded genetic algorithm for solving ELD problem in microgrids by considering different combination of generation sources. From the above discussion and analysis, it can be concluded that the integration of RES in microgrid is a profitable solution if and only if REC for solar is considered. Otherwise, only integration of wind energy source is better than integration of both solar and wind energy sources to the microgrid.

References

1. Gouthamkumar, N., Sharma, V., Naresh, R.: Disruption based gravitational search algorithm for short term hydrothermal scheduling. Exp. Syst. Appl. **42**, 7000–7011 (2015)

2. Ahn, S.J., Moon, S.I.: Economic scheduling of distributed generators in a microgrid considering various constraints. In: IEEE Power and Energy Society General Meeting, Calgary, AB, July 2009
3. Gouthamkumar, N., Sharma, V., Naresh, R.: Application of nondominated sorting gravitational search algorithm with disruption operator for stochastic multiobjective short term hydrothermal scheduling. IET Gener. Transm. Distrib. **10**, 862–872 (2016)
4. Tong, W.: Wind Power Generation and Wind Turbine Design. WIT press, Hand book (2010)
5. Noel, A., Sindhu, S., Prajakta, M., Kashif, S.: Economic dispatch for a microgrid considering renewable energy cost functions. In: IEEE PES Innovative Smart Grid Technologies (ISGT), Washington, DC, Jan 2012
6. Norberto, F., Yosune, S., Morcos, R., Carlos, M., Cesar, D.: The use cost-generation curves for the analysis of wind electricity costs in Spain. Appl. Energy **88**, 733–740 (2011)
7. Kaur, A., Singh, H.P., Bharadwaj, A.: Analysis of economic load dispatch using genetic algorithm. Int. J. Appl. Innov. Eng. Manag. **3**, 240–246 (2014)
8. Venkateswarao, B., Nageshkumar, G.V.: A comparative study of bat and firefly algorithms for optimal placement and sizing of static var compensator for enhancement of voltage stability. Int. Energy Optim. Engine. **4**, 68–84 (2015)
9. Chen, P.H., Chang, H.C.: Large scale economic dispatch by genetic algorithm. IEEE Trans. Power Syst. **10**, 1919–1926 (1995)
10. Venkateswarao, B., Nageshkumar, G.V.: Optimal location of thyristor controlled series capacitor to enhance power transfer capability using firefly algorithm. Elec. Power Comp. Syst. **42**, 1541–1552 (2014)

Design and Simulation of a DC–DC Converter for Fuel Cell-Based Electric Vehicles with Closed-Loop Operation

J. S. V. Siva Kumar and P. Mallikarjunarao

Abstract Petroleum fuels produce more pollution and environment changes are also happening because of these fuels. Because of all these reasons, the whole world is concentrating on renewable energy sources. In that point of view, more research is going on fuel cell vehicles. These vehicles run with hydrogen gas rather than any foreign oil and with no harmful emission. But the output voltage of fuel cell is low. It is necessary to increase the output voltage of fuel cell to run an electric vehicle. This paper contains design and simulation of DC–DC boost converter with closed-loop operation. In this converter, DC voltage is converted with high voltage gain. The working principle of this converter is based on ZVS and ZCS switching methods. In the closed-loop operation, PID controller is used for getting better results. The simulation of this DC-DC to converter is done by using PSIM 9.0.4 software. Finally, the simulation results are discussed.

Keywords Fossil fuel · Fuel cell stack · High voltage gain
Electric vehicle · Hybrid boost DC–DC converter

1 Introduction

In the recent years, the energy consumption from the renewable energy resources is increasing compared to that of conventional sources due to the demand for clean power. According to the environmental summits, the fossil fuels might be completely exploited by the year 2050. So we have to reserve these sources for next generations and give importance to the renewable resources for generating electric energy. Renewable energy resources include photovoltaic systems, wind energy

---

J. S. V.Siva Kumar (✉)
G.M.R. Institute of Technology, Rajam 532127, Andhra Pradesh, India
e-mail: jsvsivakumar99@gmail.com

P. Mallikarjunarao
Andhra University College of Engineering, Visakhapatanam 530003, India
e-mail: electricalprofessor@gmail.com

J. Nayak et al. (eds.), *Soft Computing in Data Analytics*,
Advances in Intelligent Systems and Computing 758,
https://doi.org/10.1007/978-981-13-0514-6_39

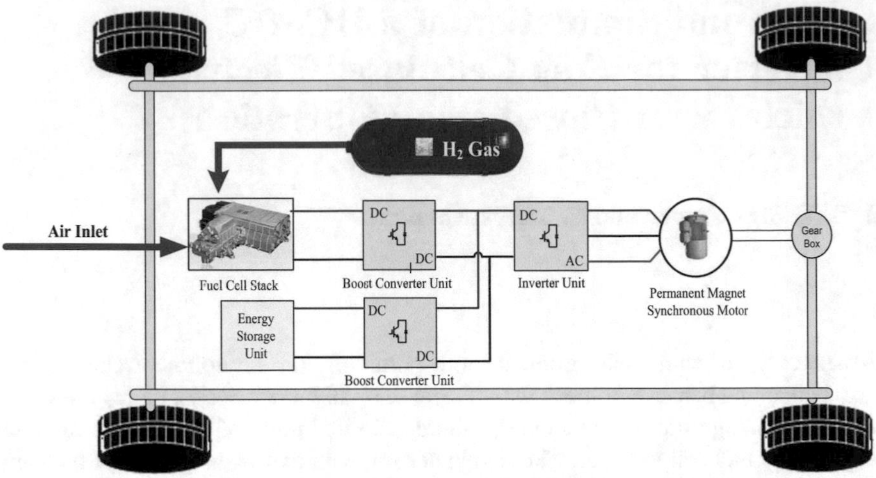

Fig. 1 Architecture of an electric vehicle powered with fuel cell

systems, fuel cells and others. Fuel cells are efficient, reliable and easy to generate on-board electricity for automobiles [1]. Block diagram of an electric vehicle powered with fuel cell is shown in Fig. 1. The output voltage of a single cell is not enough to connect directly to the alternate current utility system. Therefore, each cell is connected with others in series and parallel to obtain high-level DC voltage. However, with the above topology, the efficiency of the fuel cell gets deteriorated. So, the DC–DC converters with high voltage gain are used [2]. These converters boost the low voltages (23–68 V) to high voltages (300–500 V). The salient features of this converter are large transfer gain and high efficiency with low duty ratio. According to theoretical calculations, classical boost converters have high voltage gain with extreme duty ratio. However, functioning of the converters is devolved with high duty ratio because of low efficiency, reverse recovery and electromagnetic interference problems [3]. There are two types of the converters: 1. isolated converters and 2. non-isolated converters.

This paper introduces the closed-loop operation of hybrid isolated DC–DC boost converter with high voltage conversion ratio phenomenon for fuel cell-based electric vehicle applications. In the present paper because of the inductor with high-frequency transformer responsible to boost up the voltage. PID controller is used in closed-loop operation for getting better outputs.

2 Design of Fuel Cell

Fuel cell is a device that produces electricity by combining hydrogen and oxygen without any combustion and produce water and heat as by-products. Fuel cell has few moving parts so it requires less maintenance. A single cell will not produce sufficient amount of electricity, therefore, for a fuel cell stack, individual fuel cells are connected in series [4–6].

The different fuel cell technologies, their operating temperature, efficiency and advantages are explained in [1, 5]. Out of all fuel cell technologies, PEM fuel cell technology is most popular because of its low operating temperature range, smaller size, less start-up time and causes less corrosion. Simulation is done based on the equations from [1, 5] (Figs. 2, 3 and 4).

Table 1 Values used for design of boost converter

Parameter	Symbol	Value
Input voltage	V_{in}	12 V
Output voltage	V_{out}	300 V
Switching frequency	f_{sw}	100 kHz
Input boost inductors	L_1, L_2	36 uH
Snubber capacitors	C_1–C_{12}	0.5 pF
Turns ratio of transformer	1:k	1:2.5
Input capacitor	C_{in}	4.7 mF
Output filter capacitors	C_{01}–C_{04}	680 uF
Leakage inductance of T/F	L_{lk}	1.6 uF
Duty ratio	$\tilde{d}$	0.8
Current ripple	ΔI	1 A

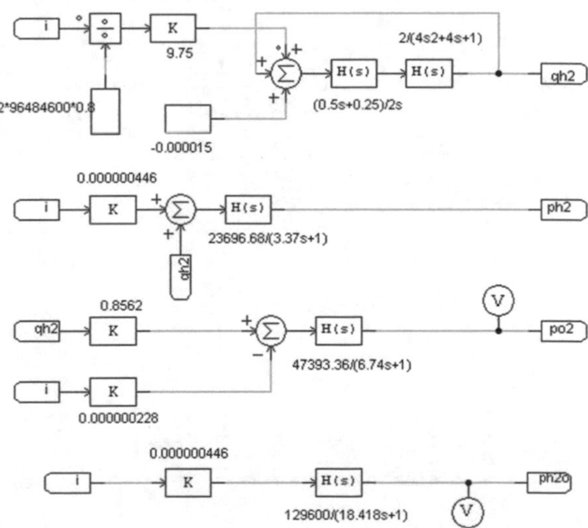

Fig. 2 Partial pressures of the gases

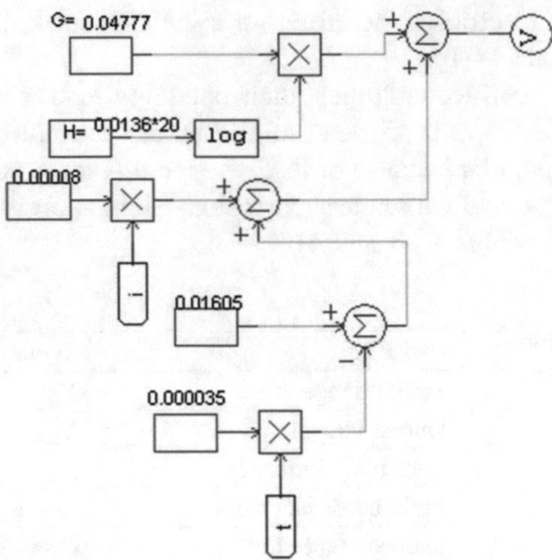

Fig. 3 Activation and ohmic losses of the fuel cell

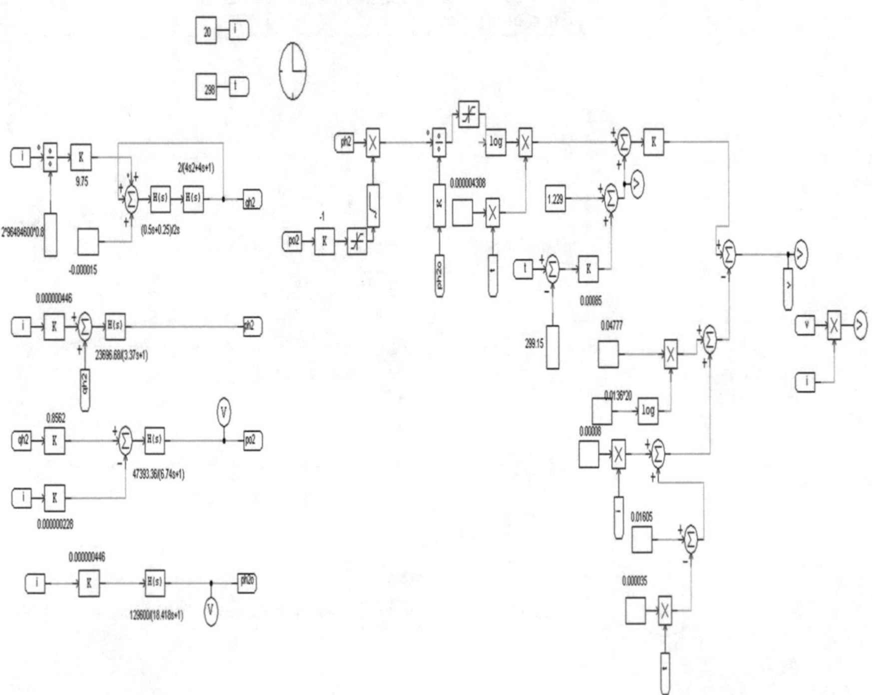

Fig. 4 Simulation model of the PEM fuel cell

3 Design of Hybrid Boost Converter

3.1 Analysis of Hybrid Topology

The suggested DC–DC converter comprises two parts with half-bridge voltage-doubler coupled in series at high voltage side and current-fed full-bridge coupled in parallel at low voltage side. The simulation diagram of the DC–DC converter is shown in Fig. 5. The suggested converter is an interleaved type which has more advantages than the single-cell converter because of high efficiency, high power handling capacity, reduced thermal requirements, reduced turns ratio of the transformer and switch ratings [7–10] (Figs. 6 and 7).

Merits of the proposed converter are that switching losses are reduced greatly because of zero current switching of primary side and zero voltage switching of secondary side; voltage across the primary side devices does not depend on duty cycle with varying input voltage. In case of failure of any one of the cell, it will supply 50% of the rated output power. It will satisfy bidirectional power flow that means in forward direction it acts as a current-fed converter and during regenerative breaking condition it will act as a voltage-fed converter and charges the energy storage system.

3.2 Operation and Analysis of DC–DC Converter

In this section, zero current switching of the primary side of the cells has been explained. Before turning on the off switches (diagonal pair) (e.g. S_2–S_3), the already turned on switches should be turned off (e.g. S_1–S_4). The total output voltage is the sum of two cells output (V_0). The output voltage of each cell is $V_0/2$.

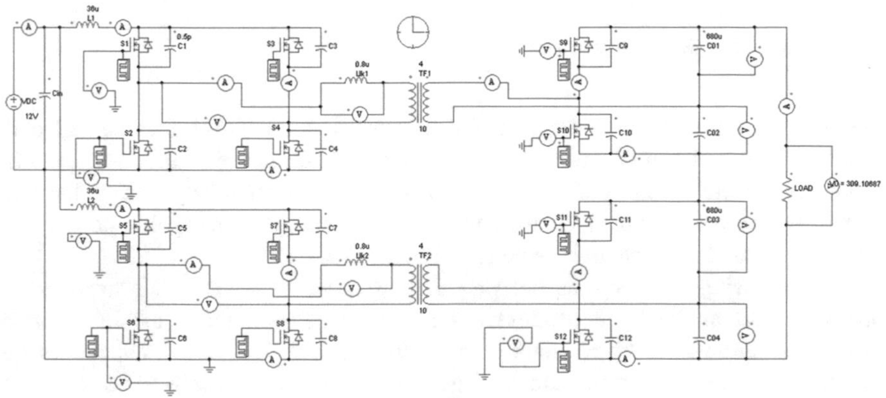

Fig. 5 Simulation model of the hybrid boost converter

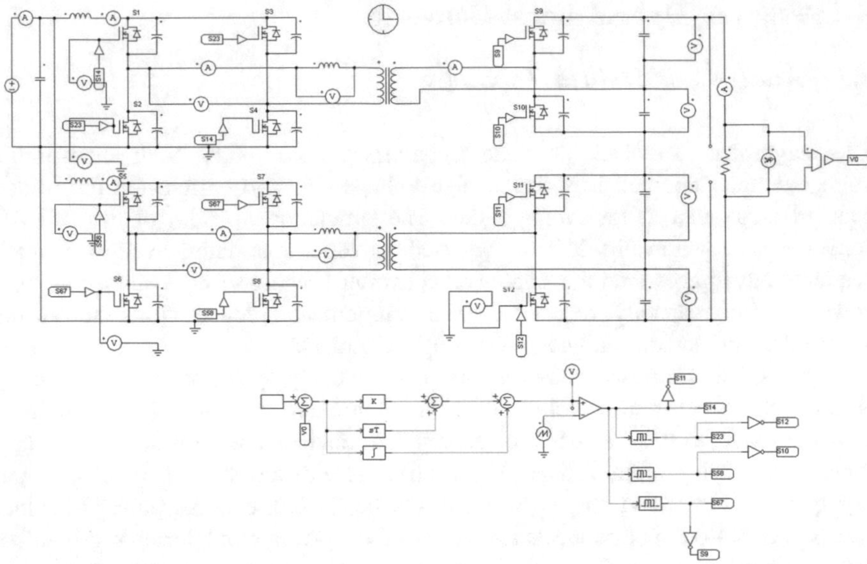

Fig. 6 Simulation model of the closed-loop operation of hybrid boost converter

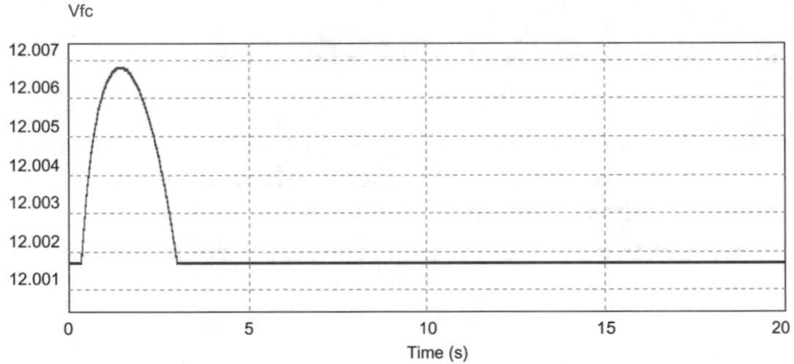

Fig. 7 Fuel cell stack output voltage

The secondary side voltage of each HF transformer is $V_0/4$. This voltage is reflected to primary side of transformer as $V_0/4n$, where n is turns ratio. This voltage will cause the current in one pair to increase and finally reach steady-state value and current in another pair to decrease and finally reach zero. During this period, the body diodes of the respective switches will conduct.

The following conventions are prepared in order to analyse the steady-state operation and analyse the proposed converter. 1. The input boost inductor is assumed as large to make constant current through it. 2. The output capacitors are assumed as large to make constant voltage across it. 3. By assuming all the switches as ideal. 4. Transformer magnetizing inductance is assumed as very large.

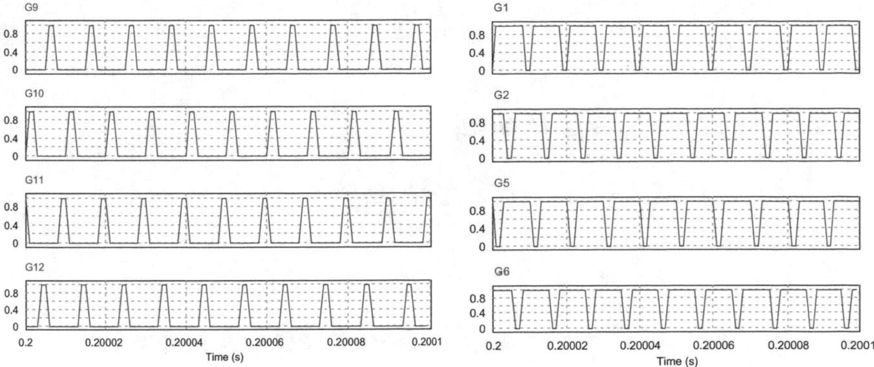

Fig. 8 Gating signals of all the switches of boost converter

In the suggested converter, diagonal switch pairs of cell1 (S_1, S_4) and (S_2, S_3) should work with same gate signals having phase shift of 180°. Similarly, switch pairs of cell2 (S_5, S_8) and (S_6, S_7) should be operated with identical gate signals having phase shift of 180°. The phase shift between gating signals of (S_1, S_4) in cell1 and (S_5, S_8) in cell2 is 90°. Similarly, the phase shift between gating signals of (S_2, S_3) in cell1 and (S_6, S_7) in cell2 is 90°. Duty ratio of all the switches should be maintained at more than 50%. The gate signals of both the side switches are shown in Fig. 8.

3.3 *Design of the Proposed Converter*

The design of the proposed converter is discussed in this section. The parameters in table 1 are required to design the converter.

(A) Boost Inductor Value:

While choosing the inductor, first the most basic parameter is the saturation current of the inductor. In the event that the saturation current of the inductor is not as much as the converter required peak current, at that point the converter would not be ready to supply the fundamental output power. By using the parameters in Table 1 the inductor value can be calculated according to Eq. (1).

$$L = \frac{V_{in} \cdot (\tilde{d} - 0.5)}{\Delta I \, . \, f_s} \tag{1}$$

B) Voltage Gain:

According to volt–sec balance on inductor, the voltage conversion ratio is given in Eq. 2

$$M = \frac{V_0}{V_{in}} \tag{2}$$

where output voltage V_0 is given in Eq. 3

$$V_0 = \frac{2 \cdot k \cdot V_{in}}{1 - \tilde{d}}. \tag{3}$$

3.4 Closed-Loop Operation of Hybrid Boost Converter

Based on the output voltage requirement, we will design time constants of PID controller for getting better output values as compared to the open-loop operation of hybrid converter as it is shown in Fig. 6. In this, we observed for the required output voltage between 200 and 300 V. The output of PID controller is compared with sawtooth waveform for generating the pulses. These pulses are given to the switches in hybrid boost converter for getting better output.

4 Results and Discussions

The suggested converter was simulated by using PSIM 9.0.4 software. The output voltage of fuel cell is shown in Fig. 7. With increase of temperature of the fuel cell stack, the no-load voltage, stack output voltage and output power are decreased. Input boost inductor currents i(L1) and i(L2) are separated from each other by 180° is shown in Fig. 9. Their ripple frequency is double the device switching frequency. Input current i is the vector addition of i(L1) and i(L2), and the output voltage of boost converter is shown in Fig. 10. The ripple frequency is four times the device switching frequency, so it reduces the input capacitor filter Cin requirement. The

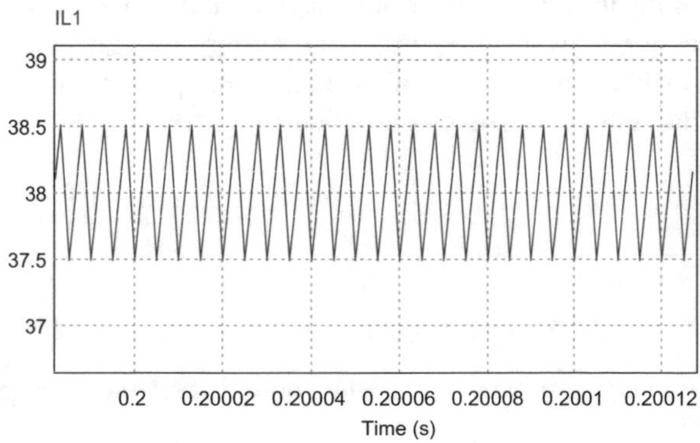

Fig. 9 Simulation waveform of input boost inductor currents $I(L_1)$

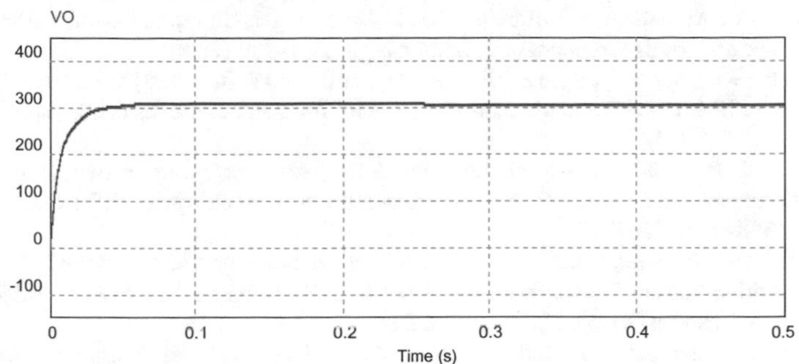

Fig. 10 Output voltage of boost converter

phase shift between voltage across C01 of cell1 and C03 of cell2 is 90°. Due to the cancellation effect, the output voltage contains lower ripple, which eliminates the filter requirement at output. The maximum voltage across the primary switches is very less, i.e. V0/4n.

5 Conclusion

The suggested converter is an isolation-based converter. The proposed converter operates as ZCS turn off of primary side devices and ZVS turn on of secondary side devices, due to which the switching losses of the converter are very low. Operating frequency of the converter is very high so that size of the converter reduces. With this interleaved design, power handling capacity increases. The major advantage with this converter is that the voltage across the power switches is very low. In this paper, the closed-loop operation of the hybrid boost converter is implemented and the performance with closed-loop operation is better than the open-loop operation. Because of all these reasons, this converter is suitable for fuel cell-based electric vehicle.

References

1. Yu, X., Starke, M.R., Tolbert, L.M., Ozpineci, B.: Fuel cell power conditioning for electric power applications IET Electr. Power Appl. **1**(5), 643–656 (2007)
2. Park, K.B., Kim, C.E., Moon, G.W., Youn, M.J.: Non-isolated high step-up converter based on boost integrated half-bridge converter. In: International Telecommunications Energy Conference, vol. 2, pp. 1–6 (2009)
3. Hsieh, Y.P., Chen, J.F., Liang, T.J., Yang, L.S.: Novel high step-up DC-DC converter with coupled-inductor and switched-capacitor techniques for a sustainable energy system. IEEE Trans. Power Electron. **26**(12), 3481–3490 (2011)

4. Salam, A.A., Mohamed, A., Hannan, M.A.: Modeling and simulation of a PEM fuel cell system under various temperature conditions, pp. 204–209 (2008)
5. Uzunoglu, M., Alam, M.S., Member, S.: Dynamic modeling, design, and simulation of a combined PEM fuel cell and ultracapacitor system for stand-alone residential applications, **21** (3), 767–775 (2006)
6. Prasanna, U.R., Rathore, A.K., Mazumder, S.K.: Novel zero-current-switching current-fed half-bridge isolated DC/DC converter for fuel-cell-based applications. IEEE Trans. Ind. Appl. **49**(4), 1658–1668 (2013)
7. Xuewei, P., Rathore, A.K.: Novel bidirectional snubberless naturally commutated soft-switching current-fed full-bridge isolated DC/DC converter for fuel cell vehicles. IEEE Trans. Ind. Electron. **61**(5), 2307–2315 (2014)
8. Xuewei, P., Member, S., Rathore, A.K., Member, S.: Novel interleaved bidirectional snubberless voltage doubler for fuel-cell vehicles, **28**(12), 5535–5546 (2013)
9. Siva Kumar, J.S.V., Satish, P.V.: Design and simulation of DC-DC converter with high voltage gain for fuel cell based vehicles, **8**(8), 565–574 (2017)
10. Zhang, Y., Sun, J.T., Wang, Y.F.: Hybrid boost three-level DC-DC converter with high voltage gain for photovoltaic generation systems. IEEE Trans. Power Electron. **28**(8), 3659–3664 (2013)

Improved Content-Based Image Retrieval with Multivariate Gaussian Distribution

B. Ramesh Naik and T. Venu Gopal

Abstract This paper presents an overview of content-based image retrieval based on integrating image features obtained from scale-invariant feature transform (SIFT) with multivariate Gaussian distribution for the efficient retrieval of images in the image database. The state-of-the-art methods for image retrieval and object recognition use SIFT and HoG to extract image visual features. Though these descriptors are helpful in a variety of applications, they exploit zero-order statistics as they only collect histogram features, and this lacks high descriptiveness of object features and quantization problem. The novel method is having each pixel of the object which is associated with multivariate Gaussian distribution and approximated new features in the locality of the region. The key issue of this approach lies in space of the multivariate Gaussian distribution which lies in Riemannian manifold. But the linear space is suitable domain to discriminate image feature vectors efficiently. With the basis of Lie group structure and Riemannian geometry, multiplication operations are determined on the manifold to embed Gaussian space into linear space and are referred to as log-Euclidean multivariate Gaussian descriptors. These descriptors determine distinctive low- and high-dimensional image features efficiently. The experiments were conducted on Caltech-101, WANG database to validate thoroughly this approach.

Keywords Covariance matrices · Feature extraction · Gaussian distribution Image matching · Vectors · SIFT

B. Ramesh Naik (✉)
Department of CSE, GITAM School of Technology,
GITAM (Deemed to be University), Bengaluru 561203, India
e-mail: rameshnaik3@gmail.com

T. Venu Gopal
Department of CSE, JNTUH College of Engineering, Jagityal,
Karimnagar 505501, India
e-mail: t_vgopal@rediffmail.com

J. Nayak et al. (eds.), *Soft Computing in Data Analytics*,
Advances in Intelligent Systems and Computing 758,
https://doi.org/10.1007/978-981-13-0514-6_40

1 Introduction

Image matching in content-based image retrieval (CBIR) gained significant importance in last 15 year where the relevant images are extracted to a query image on the basis of color, shape, texture, special locations, or derived visual features. A great research has been attracted to characterize local properties of the image in past 10 years. Diverse descriptors can be used for describing region of interest in the query and target images localized by region detectors, shape detectors, or color gradient detectors [1–3]. The challenging vision task in CBIR is to develop descriptors which are high distinctiveness; i.e., features are invariant to scale, rotation, and translation for general image retrieval tasks.

Content-based visual information retrieval (CBVIR) and (CBIR) [2, 4] wide-ranging applications for image matching in the computer vision task that is, searching for digital images in large databases using low level and derived features. The proposed approach presents novel function-valued descriptors that exploit the image visual features in higher dimensions, and hence, these are competent in many applications of vision tasks such as relevant image retrieval, recognizing objects, image classification [5]. Proposed approach first extracts features using popular local sampling descriptors such as SIFT or HoG, and then, new features are described using multivariate Gaussian distribution [3, 4]. The SIFT features are successful in a variety of applications, but they only take advantage of histogram frequencies, and there lacks a natural mechanism for high distinctiveness of images [6, 7]. Apart from this, discrete histograms frequently undergo quantization problem. Multivariate Gaussian (μ, $\sum$) can be used to represent image features in d-dimensional space N(d) [3, 5, 8] where Gaussian space is not a linear but a manifold. The space of the Gaussian can be equipped with Lie algebra by specifying operations in form of multiplication on manifold. The process of embedding Gaussian in linear space is described in detail in this paper. These features are again described using log-Euclidean multivariate Gaussian [8, 9]. The resulting descriptors are referred to as log-Euclidean multivariate Gaussian descriptors.

2 Introduction to Scale-Invariant Feature Transforms

The scale-invariant feature transform (SIFT) [3, 7, 10] is a promising approach which models low-dimensional visual features from fixed sized patches picked based on pixel information which can then be validated and matched to other features. This approach has a set of parameters, choice, and modification of the parameters that can be varied to improve the quality of the results. In the original paper by David Lowe [3, 7, 11], a set of parameters is explained with dissimilar images but is not clear whether or not the parameters are optional.

2.1 Scale-Space Extrema Detection

Finding scale-invariant locations is determined by scale function that looks for stable features across various scales. Convolution of Gaussian is used to create multiple octaves from blurred images. Dissimilarity among two successive images within an octave is referred as the difference of Gaussian which is described in [3]. The scale space is characterized by Gaussian function $L(x, y, \sigma) = G(x, y, \sigma) * I(x, y)$, where $G(x, y, \sigma)$ is a Gaussian scale variable given in [3], $I(x, y)$ is the input image, and the convolution operation is performed on $G(x, y, \sigma)$ and $I(x, y)$. Keypoint locations which are stable in the scale-space are determined by a variety of techniques. The difference of Gaussian $D(x, y, \sigma)$ between two images is computed, one with $L(x, y, \sigma)$ and other with k times $L(x, y, \sigma)$. $D(x, y, \sigma)$ is then given by $D(x, y, \sigma) = L(x, y, k\sigma) - L(x, y, \sigma)$. Local minimum and local maximum are identified by comparing eight neighbors at the same scale, and its nine neighbors above and below one scale from DoG [5]. Minimum or maximum of all these 26 points is referred as a local extrema.

2.2 Feature Localization

The number of features which are less important is eliminated in this stage. DoG function normally encompasses a high reaction at edges even when the location along the edge is weakly established and therefore sensitive to little amount of noise. The low contrast points can be rejected because they are responsive to little amount noise or are weakly confined along an edge. A Taylor series is expanded (up to the quadratic terms) to the DoG $D(x; y; \sigma)$ [3], so the origin is located at the sample point where x is offset from particular point, D and its derivatives described at each sample point. The location of the extremum $\hat{x}$ is determined by taking the derivative of D with respect to offset $x = (x; y; \sigma)^T$ and setting it to zero. Particularly, the extremum is closer to a dissimilar sample point [3, 10] when the offset $\hat{x}$ is greater than 0.5 for any given dimension.

2.2.1 Orientation Assignment

To validate rotation invariance, orientation is assigned to the each keypoint in gradient directions of its neighborhood. For each keypoint, image gradient directions in the keypoint neighborhood are calculated and added to 36 bins orientation histogram [3, 4]. The values those closer to the center have a higher effect on the resulting orientation as neighborhood values are weighted by Gaussian. Separate key orientation is used to represent each feature. The orientation is determined as followed: The Gaussian smoothed image L is selected by using keypoint scale, from gradient magnitude (m) shown in Eq. 1.

$$m(x,y) = \sqrt{(L(x+1,y) - L(x-1,y))^2 + (L(x,y+1) - L(x,y-1))^2}$$
$$Where\ orientation\ (\theta)\ is\ \theta(x,y) = \tan(L(x,y+1) - \frac{L(x,y-1)}{(L(x+1,y) - L(x-1,y))}) \quad (1)$$

For each keypoint, the orientation gradient is determined and highest peak is identified in the histogram. By using this peak as basis, any other peaks are identified which are falling within 80% highest peak to create a keypoints and multiple orientations are determined for each of this peaks.

2.2.2 Creating the Feature Descriptor

A feature descriptor is created with 128 dimensional vectors which describe pixel properties of area surrounding a keypoint [3, 5, 12]. Each region is divided into subregions of size 4 × 4 and creates histogram for each subregion; 16 bins are used to create 16 histograms. Trilinear interpolation is used to place features into histogram bins. Samples are Gaussian weighted with N(σ) = 1.5 region width.

3 Introduction to Riemannian Manifold

A manifold M is typically described from any topological space and looks like vicinity of Euclidean space with all of its usual topology. A topological space consists of set of points along with its neighborhoods for each point. In other words, each point in an n-dimensional space has a region which is homeomorphic to Euclidean space of n-dimension [13–15]. Topological spaces are said to be homeomorphic if there exist a continuous function that maps one topological space to another space and has a continuous inverse function. A three-layer structure is used to build Riemannian manifold. Topological layer typically describes the notion of continuity and convergence. Differential layer extends the notion of differentiability to the manifold, and Riemannian manifold describes rigid geometrical quantities such as distances, angles, and curvature on the manifold. A Riemannian manifold (M, *k*) is extended from differential manifold [14, 15] with a Riemannian metric *k* on tangent vectors by a local inner product on them which is shown in Eq. 2

$$\mathrm{k_x}(\dot{},\dot{}) : \mathrm{Q_xM} \times \mathrm{Q_xM} \to \mathrm{R}, \mathrm{x} \in \mathrm{M} \quad (2)$$

The Riemannian metric k is bilinear, symmetric, positive definite and is also C^∞ differentiable in *x* as shown in Eqs. 3, 4, 5, 6, respectively.

$$k_x\left(\sum_{i=1}^{n} u_i, \sum_{i=1}^{n} v_i\right) = \sum_{i=1}^{n}\sum_{j=1}^{n} k_x(u_i, v_j) \tag{3}$$

$$k_x(u, v) = k_x(v, u) \tag{4}$$

$$k_x(u, u) \geq 0 \tag{5}$$

$$k_x(u, u) \geq 0 \Leftrightarrow u = 0 \tag{6}$$

Riemannian metric k is defined for every u, $v \in Q_xM$, based on bilinearity of inner product as shown in Eq. 7.

$$k_x(v, u) = \sum_{i=1}^{n}\sum_{j=1}^{n} v_i u_j k(\partial i, \partial j) \tag{7}$$

In Eq. 7, k_x is obtained by $\{di\}_{i=1}^{n}$ of Q_xM where $\{di\}$ is inner product set between its basis elements, i.e., $\{k_x(\partial i, \partial j) : 1 \leq i, j \leq n\}$, where $k_x(\partial i, \partial j)$ is Gram matrix $[G(x)]_{ij}$ which is a symmetric and positive definite in nature and used to describe the metric g_x [6, 5]. Hence, $\sqrt{k_x(v, u)}$ is used as measure to describe tangent vectors $v \in Q_xM$ and their lengths; the curves γ: [a, b] → M and its lengths are defined by

$$L(\gamma) = \int_a^b \sqrt{k_x(\dot{\gamma}(t), \dot{\gamma}(t))dt} \tag{8}$$

In Eq. 8, γ represent a curve and $\dot{\gamma}$ (t) denotes velocity vector at time t; then, $d_k(x, y)$ is geodesic distance differentiating two points $x, y \in M$ which is obtained from Eq. 9:

$$d_k(x, y) = \inf_{\gamma \in \Gamma(x, y)} \int_a^b \sqrt{k_x(\dot{\gamma}(t), \dot{\gamma}(t))dt} \tag{9}$$

Set of curves which are connecting x and y points are represented by $\Gamma(x, y)$. The geodesic distance d_k is used as basis to describe geodesic curves and smallest curve achieving it. Usual requirement of a distance satisfied by geodesic distance is typically compatible with the structure of topological manifold. Contextually, if question is clear in the manifold, then the subscript is eliminated and only d is used to represent geodesic distance. In particular, if any geodesic curve $c(t), t \in [a, b]$ is comprehensively used to define for all its $t \in$ R, then manifold said to be geodesically complete.

4 Embedding Gaussian in Linear Space

4.1 Lie Group Algebra

Let a group G, multiplication operation on G with itself determines components which join any two components a and b in G i.e. G × G → G produces $(a \cdot b)$ *of G* which satisfies following characteristics [16, 17]:

i. For every a, b, and c in G, (a · b) · c = a · (b · c) is associative.
ii. For every a in G, a · e = e · a = a. where e identity element.
iii. There is a^{-1} for each element a in G; then, $a \cdot a^{-1} = a^{-1} \cdot a = e$
iv. Let a group G with elements (a, b), then G is commutative group if a · b = b · a, for all a, b.

Let a group G with its subset F, if F forms a group under the result of multiplication, then F said to be subgroup of G [18, 17]. The multiplication and inverse operations of G are smooth functions under Lie group algebra [19]. Lie group is locally equitable to linear space since it lies in differential manifold, and hence, locality of any component in any group can be efficiently determined by its tangent.

Let GL be a Lie group [16]. Let ·, ∘ be the multiplication operations on be Lie groups GL, GL′ respectively. A *Lie group* is said to *homomorphic* (φ : GL → GL′) and a smooth function if it satisfies $\varphi(a \cdot b) = \varphi(a) \circ \varphi(b)$ for all $a, b \in GL$.

4.2 Matrix Group

The square matrices are significant sources to form lie groups. Let a set A be with n × n real matrices; a Lie group is formed if A is invertible under matrix multiplication; then, it is said to be general linear group, and its subgroups are principle source to get matrix group [8]. Let A be a square matrix and its exponent be represented as exp (A). The series mapping is smooth or any X and exp(X) [18–20]. Let the matrix B, and Log on B is represented as *log*(B), is a matrix A such that exp (A) = B. Let the set of all real numbers R and its Lie algebra R^+. The log-Euclidean notation is formulated as *exp*: $R \rightarrow R^+ \; x \rightarrow$ *exp(x)*. This states that exponential function *exp* of any real number R and inverse log of R is smooth bijection which are said to be diffeomorphisms; i.e., we define

$$\otimes : R^+ X R^+ \rightarrow R^+, x \otimes y = \exp(\log(x) + \log(y)) \tag{10}$$

$$\odot : R X R^+ \rightarrow R^+, \lambda \odot x = (\lambda \log(x)) = x^\lambda \tag{11}$$

Above Eqs. 10, 11 specify that multiplications under Lie group R^+ are equipped into additions in the linear space (logarithm domain) hence said to be "Log-Euclidean."

C. Embedding of Gaussian in Linear space

The detailed process of embedding Gaussian in linear space is described by Peihua et al. [8, 9, 12, 19]. Let S(n + 1) be a set with $S_{t,X}$ where X belongs to upper triangular matrix with positive diagonals and t belongs to R^n. According to the properties of X, matrix exponentials the set S(n + 1) can be equipped to Lie group structure of the matrix group S^+(n + 1). Under matrix exponential, the mapping from S^+(n + 1) to its S(n + 1) is a diffeomorphism and is proved in the following Proposition 1 [5, 15, 19].

Proposition 1 Converge of $\exp : S(n+1)\, to\, S^+(n+1),\ S_{t,x} \rightarrow \exp(S_{t,X})$ is a smooth one to one and onto mapping, and its inverse mapping is also smooth as well.

Proof It is simple to prove that any $S_{t,\ X}$ in S(n + 1) is uniquely equipped into S^+(n + 1) under exponential function. Let exp of S(n + 1) is represented as exp(S(n + 1)) and logarithm of S^+(n + 1) is represented as log(S^+(n + 1)) according to the properties of matrix group S(n + 1) can determined as exp(log(S^+(n + 1)) = S(n + 1) [16, 19, 21]. Hence proved that exp : S(n + 1) → S^+(n + 1) is 1 to 1 and onto. So Log-Euclidean structure can be obtained from S^+(n + 1) which is determined in Proposition 2 [22].

Proposition 2 S^+(n + 1) under log-Euclidean operation. This theorem denotes that

$$\begin{aligned} &\otimes = S^+(n+1) \times S^+(n+1) \rightarrow S^+(n+1) \\ &\quad S1 \otimes S2 = \exp(\log(S1) + \log(S2)) \\ &\odot : R \times S^+(n+1) \rightarrow S^+(n+1) \\ &\lambda \odot S = \exp(\lambda \log(S)) = S^\lambda \end{aligned}$$

This states that $S^+(n+1)$ is a commutative under Lie group structure and $\log : S^+(n+1) \rightarrow S(n+1),\ S^+ \rightarrow \log(S)$ essentially satisfies isomorphism [19, 23]. In addition, $S^+(n+1)$ is representing a linear space with respect to the operators $\otimes$ and $\odot$. Therefore, this theorem states manifold specific structure of $S^+(n+1)$ is mapped to a liner space $S(n+1)$ with the basis of Lie group structure and matrix logarithm. The complete process of embedding is given in the Eq. 12:

$$N(\mu,\ \textstyle\sum)\Phi^{-1} S_{\mu,L^{-T}} \log\log\,(S_{\mu,L^{-T}}) \tag{12}$$

In Eq. 12, $\sum = L^{-T}L^{-1}\ and\, L^{-T}$ is n × n upper triangular matrices with positive diagonal entries and Cholesky decomposition $\sum^{-1} = LL^{-T}$.

5 Log-Euclidean Multivariate Gaussian Distribution

Let M be a given image, and first, the features of M are extracted through an interest point detector such as SIFT [3]. The process of extracting SIFT features is presented in Sect. 2. Let X = {A_1 : : : A_N} be the local features extracted from given image M using SIFT algorithm [19]. Multivariate Gaussian (μ, ∑) can be used to represent image features in d-dimensional space N(d) [8, 9, 19]. The multivariate Gaussian distribution [8] of a set of d-dimensional vectors X is calculated by Eq. 13, i.e.,

$$f(X|\xi, \sum) = (2\pi)^{-\frac{d}{2}}(\det K)^{\frac{1}{2}}e^{-(x-\xi)^T K(x-\xi)/2} \tag{13}$$

In Eq. 13, *det* is the determinant, ξ is the mean vector, and K is the covariance matrix with space of real symmetric positive semi-definite matrices [24].

The information about the variance and their correlation is depicted in the covariance matrix. Properties of the covariance matrix state that the covariance space is not closed under multiplication; thus, this space is said to be in Riemannian manifold. The multiplication on covariance matrix is possible if it lies within vector space in the form of points [5]. The majority of the general machine learning approaches states that covariance space which lies in Riemannian manifold can be transformed to vector space in the form of points prior to their use. Hence, log-Euclidean metric is adapted since covariance matrix is symmetric positive definite in nature. The covariance matrix encoded in Gaussian space is equipped in Lie algebra by defining multiplication operation on manifold [8, 18, 22] and is referred to as log-Euclidean multivariate Gaussian descriptors (LEMVG) since the covariance matrix is symmetric positive definite in nature; i.e., the descriptors obtained from SIFT features are mapped to Euclidean space by adapting log-Euclidean metric [7, 12] on it. The basic inspiration of this metric is to establish compatible relationship between these two spaces, i.e., Riemannian manifold [19] and the vector space. The process of mapping Riemannian manifold to Euclidean space is described as following:

(i) First covariance matrix Projected on vector space tangent to the Riemannian manifold as given in Eq. 14.

$$t_k = \log_p(K) = Q^{1/2}\log(Q^{-\frac{1}{2}}KQ^{\frac{1}{2}})Q^{\frac{1}{2}} \tag{14}$$

where log (Q) *is manifold specific* log *operator*

(ii) Then, the orthonormal coordinates of the projected vectors are extracted using Eq. 15.

$$Vec_p(t_K) = vec_I(Q^{-\frac{1}{2}}t_K Q^{\frac{1}{2}}) \tag{15}$$

(iii) While the vector operator on the tangent space at identity of a symmetric matrix Y is defined in Eq. 16:

$$Vec_I(Y) = \left[y_{11}\sqrt{2y_{12}}\sqrt{2y_{13}}y_{22}\sqrt{2y_{23}}\ldots y_{dd}\right] \quad (16)$$

(iv) According to the Eq. 16 substituting t_K in Eq. 14, the projection of K on the hyperplane tangent to Q is given in Eq. 17.

$$K = vec_I\left(\log(Q^{-\frac{1}{2}}KQ^{\frac{1}{2}})\right) \quad (17)$$

The projection point Q could influence the performance (distortion) of the projection since it is random in nature. The covariance matrix which is derived from SIFT descriptors is projected on a Euclidean space and concatenated to the mean vector to obtain the final descriptor.

6 Results and Analysis

We carry out experiments on the challenging Coral (Wang) Database and Caltech-101 database [25] to evaluate our descriptors and the components involved in LEMVG. The Coral (Wang) Database contains 10,000 images in total and divided into ten classes. We follow the standard measure average precision (AP) to validate accuracy over different categories of the images. Caltech-101 has about 9 K images spanned over 101 categories, consisting of diverse object poses, sizes, and under various lighting conditions [25]. The objective of log-Euclidean multivariate Gaussian descriptors (LEMVG) is to describe efficient local statistics of the images. The frequently used low-level visual features such as intensity, color, location, first- and second-order derivatives are calculated and contrasted at different scales. To extract image feature descriptors in different directions and at various scales, we explore various operators. Our experimental results are compared with the results obtained by linear discriminant analysis (LDA) which is generalization of Fisher's discriminant analysis over WANG and Caltech-101 databases [18, 19, 26]. LDA is a machine learning approach which efficiently discriminates image descriptors. The descriptors using LEMVG and LDA are computed with 16 × 16 patch size [8]. Table 1 presents the retrieval accuracy AP over Caltech-101 and WANG database which represents their performance against various combinations of raw features [18]. The covariance descriptors [8] computed which contains intensity, location (@1 in Table 1) [12]. The orientation histogram of edges (OHEs) [5] compute gradients which collect the zero-order statistics of the image (@2 in Table 1). The first-order derivative operator of Gaussian (FDOG) and second-order operators of Laplacian are combined to compute first- and second-order derivatives (@3 in Table 1). Finally, descriptors are computed by evaluating additional color, Gabor filters (@4 in Table 1). LEMVG descriptor accomplishes best performance over WANG database with combination of first-order and second-order derivatives. LEMVG outperforms LDA over

Table 1 Experimental results obtained using LDA and LEMVG over different operator on WANG database

No	Raw features	Results on WANG database		Results on Caltech-101 database	
		LDA (AP, %)	LEMVG (AP, %)	LDA (AP, %)	LEMVG (AP, %)
@1	Covariance descriptors (intensity, location)	48.53	52.21	46.64	51.54
@2	Orientation histogram of edges (eight bins)	46.62	49.92	44.81	46.89
@3	First- and second-order derivatives (Laplace)	48.37	52.68	47.34	50.65
@4	Gabor filters	50.83	51.93	48.63	52.03

Caltech-101 database by using Gabor filter. The retrieval results @2 and @3 indicate that the OHE and Gabor filter operators slightly improve performance over WANG database. We observe that in most cases LDA shown nearly similar accuracy to SIFT, and LEMVG is better performed over both of them by 4.2%, 2.2% for distinct and various scales, respectively.

Size of the patches and modeling them have greater impact on local configuration of the images and their local descriptors density [8, 5]. We have performed analysis at various patch sizes from 8×8 to 24×24. Image retrieval performances improved progressively as patch size reduces step by step to 12×12. This determines that local structures are more characteristic at finer scale to get better accuracy. But the results at small patch size lead to inadequate number of descriptors. Out of 101 categories from Caltech-101 image database, ten images are randomly chosen from each category and are used as queries. For each query, average precision of the image retrieval at each level of the recall is computed.

We have used 16×16 as stable patch whenever distinct scale LEMVG descriptors are computed. The retrieval results AP of our method LEMVG are presented in Table 1. With LEMVG, we further achieved average precision improvement of 4.2% on average.

7 Conclusion

We have proposed a novel approach log-Euclidean multivariate Gaussian descriptors to distinguish local, higher-dimensional statistics of images. We have determined a log-Euclidean measure to describe Gaussian into Euclidean space as Gaussian structure lies within manifold. Unlike SIFT or HoG which is described from Histogram quantities, the proposed descriptors are consistent and model higher-order analysis of images which are very much competitive in image retrieval task. Further, we have designated the Gaussian space can be smoothly mapped with a Euclidean space, which is equivalent to a linear space in terms of logarithm operation. These conclusions provide novel direction within the Gaussians in terms geometrical structure and Lie group algebra. Empirical evaluation demonstrates the efficiency of the resulting descriptors on both synthetic and real problems in image retrieval. Our further study focuses on the applications of these descriptors to other visions tasks in the real world.

References

1. Hiremath, P.S., Pujari, J.: Content based image retrieval using color, texture and shape features. In: International Conference on Advanced Computing and Communications, 2007. ADCOM 2007. IEEE (2007)
2. Venu Gopal, T., Ramesh Naik, B., Prasad, V.K.: Image retrieval using adapted Fourier descriptors. Int. J. Signal Imag. Syst. Eng. **3**(3), 188–194 (2010)
3. Lowe, D.: Distinctive image features from scale-invariant keypoints. Int. J. Comput. Vis. **60**, 91–110 (2004)
4. Persoon, E., Fu, K.S.: Shape discrimination using Fourier descriptors. IEEE Trans. Syst. Man Cybern. **21**(3), 170–179 (1997)
5. Dalal, N., Triggs, B.: Histograms of oriented gradients for human detection. In: Proceedings of the International Conference on Computer Vision and Pattern Recognition, pp. 886–893 (2005)
6. Arsigny, V., Fillard, P., Pennec, X., Ayache, N.: Geometric means in a novel vector space structure on symmetric positive-definite atrices. SIAM J. Matrix Anal., Appl (2006)
7. Lebanon, G.: Metric learning for text documents. IEEE Trans. Pattern Anal. Mach. Intell. **28** (4) (2006)
8. Li, P., Wang, Q., Zeng, H., Zhang, L.: Local log euclidean multivariate gaussian descriptors and its application to image classification. IEEE Trans. PAMI **39**(4), 803–817 (2017)
9. Naik, B.Ramesh, Venugopal, T.: Object recognition using log euclidean multivariate gaussian descriptors. Int. J. Appl. Eng. Res. **12**(14), 4130–4137 (2017)
10. Shi, H., Zhang, H., Li, G., Wang, X.: Stable embedding of Grassmann Manifold via Gaussian Random Matrices. IEEE Trans. Inf. Theory **61**(5), 2924–2924 (2015)
11. Ke, Y., Sukthankar, R.: PCA-SIFT: a more distinctive representation for local image descriptors. In: Proceedings of the IEEE Conference on Computer Vision and Pattern Recognition, pp. II–506 (2004)
12. Grana, C., et al.: UNIMORE at ImageCLEF 2013: Scalable Concept Image Annotation. CLEF (Working Notes) (2013)
13. Serra, G., Grana, C., Manfredi, M., Cucchiara, R.: Modeling local descriptors with multivariate Gaussians for object and scene recognition. In: Proceedings of the ACM International Conference on Multimedia, pp. 709–712 (2013)
14. Lebanon, G.: Riemannian geometry and statistical machine learning. Thesis, Language Technologies Institute, 2005
15. Huang, Y., Wu, Z., Wang, L., Tan, T.: Feature coding in image classification: a comprehensive study. IEEE Trans. Pattern Anal. Mach. Intell. **36**(3), 493–506 (2014)
16. Hall, B.C.: Lie groups, Lie algebras, and Representations: an elementary entroduction. Graduate Texts in Mathematics, vol. 222, 2nd edn. Springer. https://doi.org/10.1007/978-3-319-13467-3. ISBN 978-3319134666 ISSN 0072-5285. 2015
17. Skovgaard, L.T.: A Riemannian geometry of the multivariate normal model. Scand. J. Stat. **11** (4), 211–223 (1984)

18. Sánchez, J., Perronnin, F., Mensink, T., Verbeek, J.: Image classification with the Fisher vector: theory and practice. Int. J. Comput. Vis. **105**(3), 222–245 (2013)
19. Calvo, M., Oller, J.M.: A distance between multivariate normal distributions based in an embedding into the siegel group. J. Multivar. Anal. **35**(2), 223–242 (1990)
20. Gallier, J.: Logarithms and square roots of real matrices. CoRR (2013). arXiv:0805.0245
21. Cimpoi, M., Maji, S., Kokkinos, I., Mohamed, S., Vedaldi, A.: Describing textures in the wild. In: Proceedings of the IEEE Conference on Computer Vision and Pattern Recognition, pp. 3606–3613 (2014)
22. Garnett, R., Osborne, M.A., Hennig, P.: Active learning of linear embeddings for Gaussian processes. arXiv:1310.6740 (2013)
23. Donahue, J., Jia, Y., Vinyals, O., Hoffman, J., Zhang, N., Tzeng, E., Darrell, T.: DeCAF: a deep convolutional activation feature forgeneric visual recognition. In: Proceedings of the International Conference on Machine Learning, pp. 647–655 (2014)
24. van de Sande, K., Gevers, T., Snoek, C.: Evaluating color descriptors for object and scene recognition. IEEE Trans. Pattern Anal. Mach. Intell. **32**(9), 1582–1596 (2010)
25. Griffin, G., Holub, A., Perona, P.: The Caltech-256. Tech. Rep, California Institute of Technology (2007)
26. Lazebnik, S., Schmid, C., Ponce, J.: Beyond bags of features: spatial pyramid matching for recognizing natural scene categories. In: Proceedings of the IEEE Conference on Computer Vision and Pattern Recognition, pp. 2169–2178 (2006)

Performance Analysis of Embedded System for Data Acquisition on FPGA

B. Jhansi Rani, A. Ch. Sudhir and T. Vidhyavathi

Abstract Miniaturization is one of the major demands for the development of any modern embedded systems. This can be achieved by incorporating the entire design including processor on to a single FPGA. Field-programmable gate array is an IC (integrated circuit) which is programmable in the field after it gets manufactured either by the designer or by the customer and so it is called field programmable. The configuration of the FPGA is specified more generally by the HDL. The system developed on FPGA has the features as follows: standardization, modularization, and scalability. One of the applications is in an airborne SAR signal processing system. The main intention of this paper is to design and develop a data acquisition system using FPGA through designing and coding, verification, configuration of processor, and driver development.

Keywords Field-programmable gate array (FPGA) · Resistor–transistor logic (RTL) · UART · Hardware descriptive language (HDL)

1 Introduction

A processor along with a set of peripherals used for a specific purpose is termed as an embedded system. The data acquisition system typically is used to acquire real-time data and the physical parameters from sensors and is used to automate the

B. Jhansi Rani · A. Ch. Sudhir (✉)
GITAM (deemed to be) University, Visakhapatnam 530045, India
e-mail: sudhir.ach1@gmail.com

B. Jhansi Rani
e-mail: jhansiraniboggavarapu@gmail.com

T. Vidhyavathi
GVP College of Engineering, Visakhapatnam 530045, India
e-mail: thota.vidyavathi9@gmail.com

J. Nayak et al. (eds.), *Soft Computing in Data Analytics*,
Advances in Intelligent Systems and Computing 758,
https://doi.org/10.1007/978-981-13-0514-6_41

system. The components of the data acquisition system are ADC controller which converts physical parameters from sensors to digital values, UART which is the interference for serial communication for modems, printers, debugging, etc., SRAM memory controller which is a module manages the data flow to and from processor bus and memory.

The present-day electronic demands are reduction in size, power with high performance. This has paved the way to think about single-board solutions for electronic system. Due to the difference in the computing application, the idea of having a FPGA [1] with a processor meet all the computing requirements, moreover there is a tremendous growth in the processor industry with respect to increase in performance with less power. In this paper following are the major works involved: design, coding, and verification of ADC interface module, UART, and memory controller, configuration of processor, memory interface, ADC interface and other modules using embedded development kit, driver development and implementation and testing on FPGA.

2 Hardware Modeling

IP core is a reusable unit of logic or cell or a component developed using HDL languages like VHDL or Verilog. These are the basic building blocks of any ASIC or FPGA designs [2].

2.1 Types of Cores

There are three types of IP Cores:

a. Soft IP Core: These are typically referred to as synthesizable RTL codes delivered as VHDL or Verilog. User can modify the design by altering RTL code.
b. Firm IP Cores: These are gate-level netlist IPs. User cannot modify the design but can implement the design in various technology nodes or among different boundaries.
c. Hard IP Cores: These IPs are by nature layout level or very primitive low-level representation of the logic, which is foundry specific. So, user cannot alter or go with any foundry other than the implemented process.

Compilation and Simulation

A simulator and an analyzer are the components present in a typical VHDL system. The analyzer reads the design description and compiles it into a design library after validating the syntax and performing static semantic checks. The message processor [3] can automatically locate errors detected during compilation and reporting error, information, and warning messages for design problems such as connection and syntax errors. We introduce the notion of balance between the rates at which data is fetched from memory and accessed by the computation, combined with estimation from behavioral synthesis [4]. Simulation is the act of exercising a model of an actual component for analyzing its conduct under a given set of conditions and/or stimuli.

Synthesis

Synthesis is the automatic method of converting a design description to a lower-level circuit representation such as net list. A synthesizer is used for automatic hardware generation. To minimize and remove redundant logic, synthesizer selects appropriate logic, ensuring that the device logic sources are used as efficiently as possible for the target device architecture. Today converting register transfer-level descriptions to gate-level netlists can be done by current synthesis tools available [5].

Programming

The assembler module creates one or more programming object files (.bit) for a compiled project. The tool programmer uses UCF files together with standard hardware to program devices. In programming the device, a file is produced by the synthesis, optimization, and fitting software.

Optimization

Optimization is dependent on three things:

Boolean equations form, type of available resources, and user-applied synthesis devices (constraints) or automatic devices [5].

We can do better optimization of anything in the best way if it is in its simplest form. Generally, Boolean algebra finds its effective or practical use in the logic circuits simplification. Here, we convert the function of logic circuit into Boolean form and apply some rules of algebra to the equation, to simplify the equation, by reducing the number of terms or arithmetic operations and translating the equation back into circuit form so that we can achieve the circuit in its simplest form. This reduces the utilization of product term and the number of logic block input/outputs requirement for any expression given. Moreover, two-level size optimization using algebraic methods implementing a Boolean function using only two levels of gates —a level of AND gates followed by one OR gate—usually results in a circuit having minimum delay. RTL Code is implemented not to exceed two-level hierarchy.

3 Design Flow on FPGA

The design flow on FPGA shown in Fig. 1 initiates with the architecture and then moves to the HDL modeling which is a language used to describe the structure and behavior of electronic circuits and most commonly digital logic circuits and then to

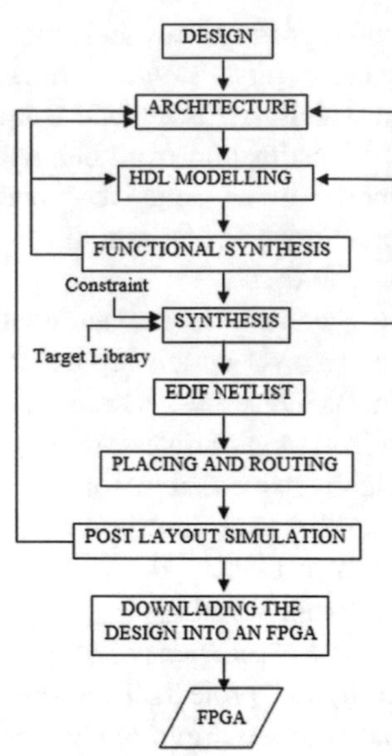

Fig. 1 Design flow on FPGA

functional simulation, an iterative process which requires multiple simulations to achieve the desired end functionality of the design. Then it moves to synthesis, where flow makes use of knowledge of the layout and timing of the target device in order to achieve the minimum area usage at the required speed and later to EDIF net list, which transfers information from one system to another and then to placing and routing by the tool provided by the FPGA vendor or another software manufacturer. Later, post-layout simulation will be done and then we download the design into an FPGA [6] (Fig. 1).

4 Implementation

The data acquisition system shown in Fig. 2 is mainly used to acquire analog input data and transmits the values periodically through UART. The system consists of a 32-bit processor along with peripherals like memory controller, UART, and ADC interface embedded in the FPGA. The processor communicates with the peripherals through a high-speed bus. The FPGA is interfaced to Flash/SRAM through memory controller, to RS232 transceiver through UART, and to ADC through interface (Fig. 2).

ADC Controller Development
ADC controller is shown in Fig. 3 an interface between processor bus and analog devices AD977a ADC. The hardware block diagram is shown in Fig. 3 . ADC is a module which converts real continuous analog voltages into digital data. This is

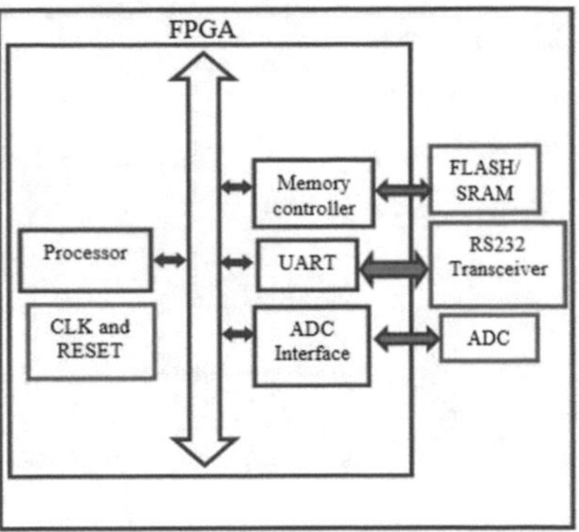

Fig. 2 Block diagram of embedded system for data acquisition system on FPGA

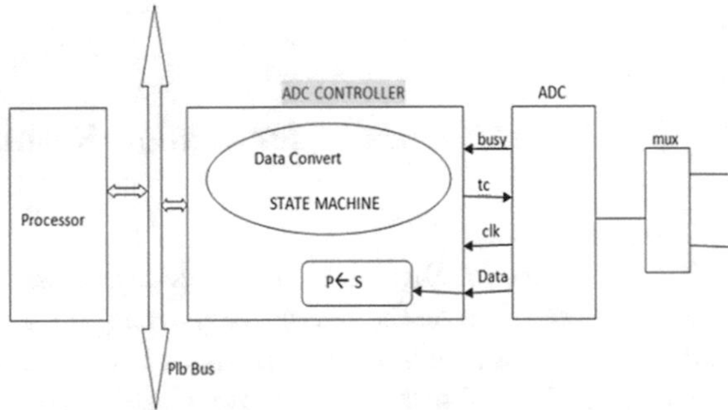

Fig. 3 ADC controller

used to acquire sensor data like temperature and pressure. ADC controller is the IP core which was developed using VHDL that will be an interfaced to the processor bus on one side and the ADC on the other side. Initially, processor gives a command to start the conversion for ADC. Then the ADC controller gives start convert to ADC if busy signal is not active then after getting the start convert signal. The ADC will give clock and serial data to ADC controller. The ADC controller will sample the serial data coming from ADC.

UART Specifications and Block Diagram
UART is a serial communication protocol which communicates at a data rate of 9600 bps, 115200 bps, and 2 Mbps. UART has two FIFOs, one for transmitter and second one for receiver of size 256 bytes; it also has a transmitter block in which it reads parallel data from FIFO and converts into serial data and also appends start bit of value 0 to the initial state and stop bit value 1 to the final bit (Fig. 4).

UART has two FIFOs, one for transmitter and second one for receiver of size 256 bytes; it also has a transmitter block in which it reads parallel data from FIFO and converts into serial data and also appends start bit of value 0 to the initial state and stop bit value 1 to the final bit.

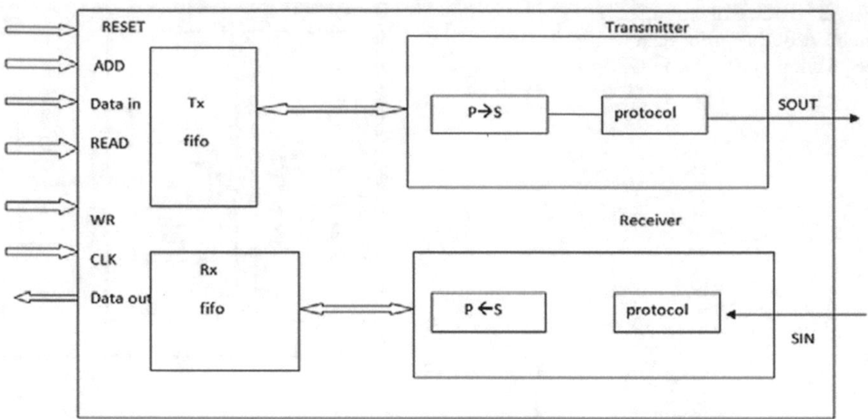

Fig. 4 Block diagram of UART

5 Implementation and Simulation Results

Simulation

The total functionality of UART is described in the result. Initially, transmitter schedules serial data on the line based on processor's command to transmit (Fig. 5). Subsequently, the receiver receives data on serial line and converts to parallel data according to UART protocol to acquire data as "00000001".

Transmitter

The above Fig. 6 shows the processor transmits the data_in—"00000001" to IP core register address "000". Then gives "start_tx to 'o'" for the IP to transmit UART serial protocol data on "sout" port—with starting with a start bit 0, next data "00000001" with LSB first and MSB last, then stop bit 1.

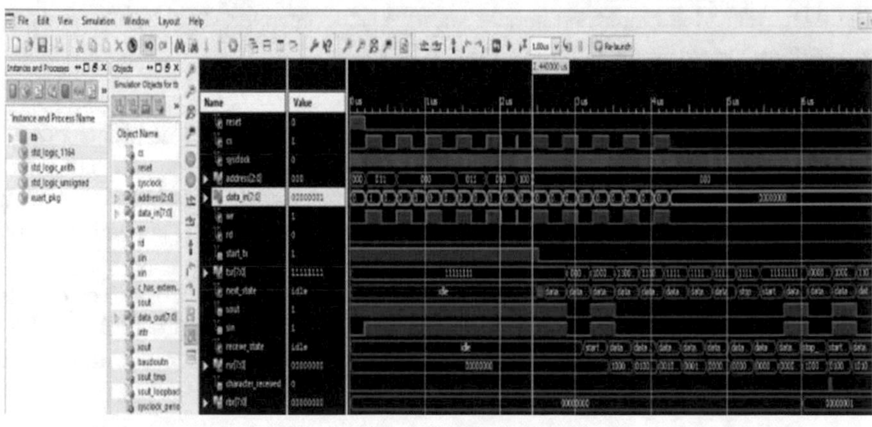

Fig. 5 Simulation

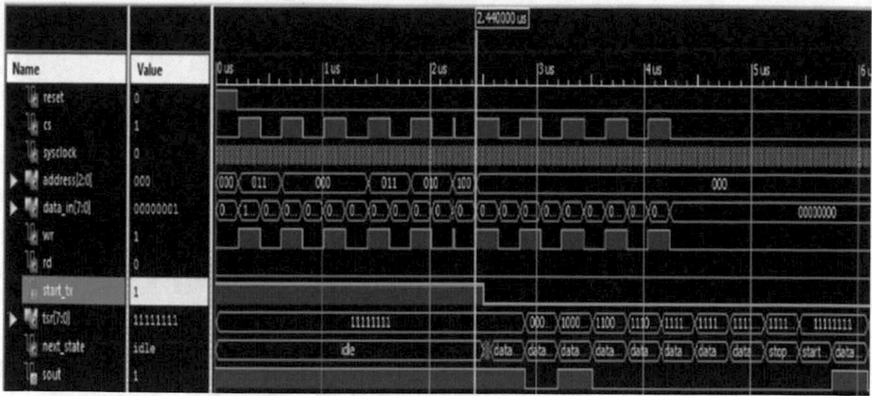

Fig. 6 Transmitter

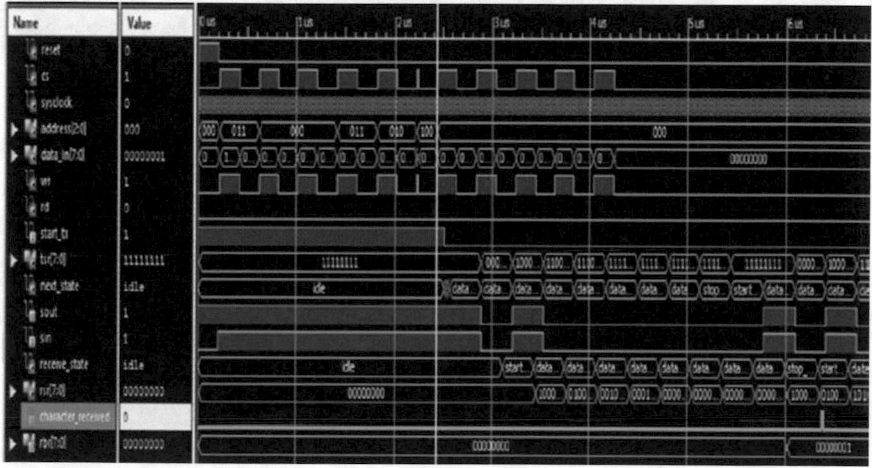

Fig. 7 Receiver

Receiver

The above Fig. 7 shows that the IP core always checks for UART serial protocol data on “sin” port—with a start bit 0, next data “any data” with LSB first and MSB last, then stop bit 1. Once the IP core receives all 8 bits of serial data, it alerts the processor with “character received” port. Then processor reads the data “00000001”.

6 Conclusion

Here the data acquisition system has been realized by processor from FPGA it is used. The developed IP cores like memory controller, UART, and ADC interface is verified using VHDL and the integrated IP Cores to processor using Xilinx EDK [7]. All the results like simulation and system levels are satisfactory.

References

1. Cheung, P.Y.K., Constantinides, G.A., de Sousa, J.T. (eds.): Field-programmable logic and applications. In: 13th International Conference, FPL 2003, Lisbon, Portugal, Sept 2003
2. Catthoor, F., Franssen, F., Wuytack, S., Nachtergaele, L., DeMan, H.: FPGA-based system on chip design for real-time applications in particle physics. In: IEEE Workshop on VLSI Signal Processing, La Jolla, CA, Oct 1994
3. Kulkami, D., Najjar, W., Rinker, R., Kurdah, F.: Fast area estimation to support compiler optimizations in FPGA-based reconfigurable systems. In: Proceedings of the IEEE Symposium on FPGAs for Custom Computing Machines (FCCM'02) (2002) (to appear)
4. Gokhale, M., Stone, J.: Automatic allocation of arrays to memories in FPGA processors with multiple memory banks. In: Proceedings of IEEE Symposium on FPGAs for Custom Computing Machines (FCCM'99), pp. 63–69. IEEE Computer Society Press, Los Alamitos, CA, Oct 1999
5. Miranda, M., Catthoor, F., Janssen, M., DeMan, H.: High-level address optimization and synthesis techniques for data-transfer intensive applications. IEEE Trans. Very Large Scale Integr. VLSI Syst. **6**(4), 677–686 (1998)
6. Padmanabhan, T.R., Bala Tripura Sundari, B.: Design Through Verilog HDL, WSE. IEEE Press (2004)
7. www.xilinx.com

Performance Comparison of FOA Optimized Several Classical Controllers for Multi-area Multi-sources Power System

B. V. S. Acharyulu, S. K. Swamy, Banaja Mohanty and P. K. Hota

Abstract This study extensively represents a practical power system using various sources in each control area connected through AC/DC transmission link. Each area includes reheat thermal, hydro and nuclear power plant using proper generation rate constraints. A new population-based algorithm, i.e. fruit fly optimization algorithm (FOA), is applied for tuning purpose. Initially, the parameters of integral (I) controller and parameters of HVDC are optimized with FOA algorithm employing integral time absolute error (ITAE) as an objective function, and to show supremacy of FOA technique, the system performances are compared with GA and PSO algorithms. Further, performances of different conventional controllers like integral-derivative (ID), proportional-integral (PI), proportional-integral-derivative (PID), integral-double-derivative (IDD) and proportional-integral-double-derivative (PIDD) are compared with FOA algorithm for the concerned system. The investigation reveals that PIDD controller tuned with FOA algorithm outperforms than other controllers. Additionally, sensitivity analysis is executed with parameters of the system and operative load conditions variation. It is observed from simulation outcomes that the optimum gains of the controller are robust enough to sustain wide variation in loading condition and system parameters.

Keywords Automatic generation control (AGC) · Fruit fly optimization algorithm (FOA) · High-voltage direct current (HVDC) link

B. V. S. Acharyulu (✉) · S. K. Swamy · B. Mohanty · P. K. Hota
Department of Electrical Engineering, Veer Surendra Sai University of Technology (VSSUT), Burla 768018, Odisha, India
e-mail: acharyulu201@yahoo.com

S. K. Swamy
e-mail: kumar.simhadri@gmail.com

B. Mohanty
e-mail: banaja.m@yahoo.com

P. K. Hota
e-mail: pkhota_ee@vssut.ac.in

J. Nayak et al. (eds.), *Soft Computing in Data Analytics*,
Advances in Intelligent Systems and Computing 758,
https://doi.org/10.1007/978-981-13-0514-6_42

1 Introduction

Previously, power system network is not so complex or complicated. Due to this, stability of the system was not much of concern. But, due to increase in intricacy of the present power system network serious distress about the voltage and frequency steadiness of the system. Automatic generation control (AGC) acting as a vibrant role in operation and control of complex interconnected power system [1].

Different controllers are implemented as supplementary controller in the system not only to maintain the frequency at scheduled value but also to settle the frequency deviation as quickly as possible. Classical controllers [2], state feedback controllers [3], fuzzy logic controllers, ANFIS controllers [4], sliding mode controller (SMC) [5], fractional order controller [6] are implemented for AGC system in last few years. Different controller has some shortcomings, like for fuzzy logic, ANN and ANFIS controller; a properly trained person is required to operate the controllers. SMC has the chattering problem, whereas proportional-integral-derivative (PID) controller is simple in construction, easy to understand, operate and tuning. For this reason, PID controller is established to be best controller for operating engineers. In [2], a number of classical controller's structures have implemented and their enactment has been evaluated for an AGC system. In [7], a multiplicity of generations' conventional as well as renewable energy has been well thought out in different control areas. In [8], BFOA optimized fuzzy IDD controller has been compared with several classical controllers for a hydro-thermal AGC system. But, PID controller is a fixed point controller. The environmental effects change the system parameters, which affect the controller performance. Due to this, the optimal controller parameter must be a robust one to sustain the system parameter variation. To achieve this, different meta-heuristic algorithms are applied to AGC system like BFOA [9], PSO [10], ABC [11], DE [12], BBO [13], TLBO [13], KHA [13], QOHS [14].

All the above-mentioned techniques are successfully implemented, and improvement in system performance is also furnished. However, some algorithm requires to initialize number of parameters over which performance of the algorithm depends. For some algorithm, computational time is more, execution process is complex, convergence rate is poor, etc. Fruit fly optimization algorithm (FOA) is a recently developed algorithm successfully applied in many engineering problems, and it is developed by Pan [15]. This algorithm is a parameter-free optimization algorithm; its computational time is less and easy to implement and understand.

The main aim of this paper

- FOA is considered to tune different controllers for two area system. To show the dominance of the recommended algorithm, the dynamic performance of the system is evaluated with GA, PSO algorithm for integral controller.
- Dynamic performances of different classical controllers are evaluated and better performance is acquired with PIDD controller.
- Sensitivity analysis is conducted for best controller for parameter variation and system loading condition.

2 System Modelling

A convincing power system shown in Fig. 1 taken for analysis is consisting of thermal, hydro and nuclear power plant [16]. PIDD controllers are separately considered for each unit. In this study, different controllers are taken for each unit. The transfer function of PIDD controller is given in Eq. (1), and for tuning with different algorithms, the objective function considered is given in Eq. (2).

$$PIDD = K_P + \frac{K_I}{S} + K_{DD}\,S^2 \tag{1}$$

$$J = \int_0^{t_{sim}} \left(|\Delta f_1|^2 + |\Delta f_2|^2 + |\Delta P_{tie}|^2\right) \cdot dt \tag{2}$$

Therefore, the design problem can be formulated as the optimization problem. This problem and conditions are given in Eqs. (3) and (4).

$$\text{Minimize } J \tag{3}$$

Subject to

$$K_{P\,\min} \le K_P \le K_{P\max} K_{I\,\min} \le K_I \le K_{I\max} \text{ and } K_{D\min} \le K_D \le K_{D\max} \tag{4}$$

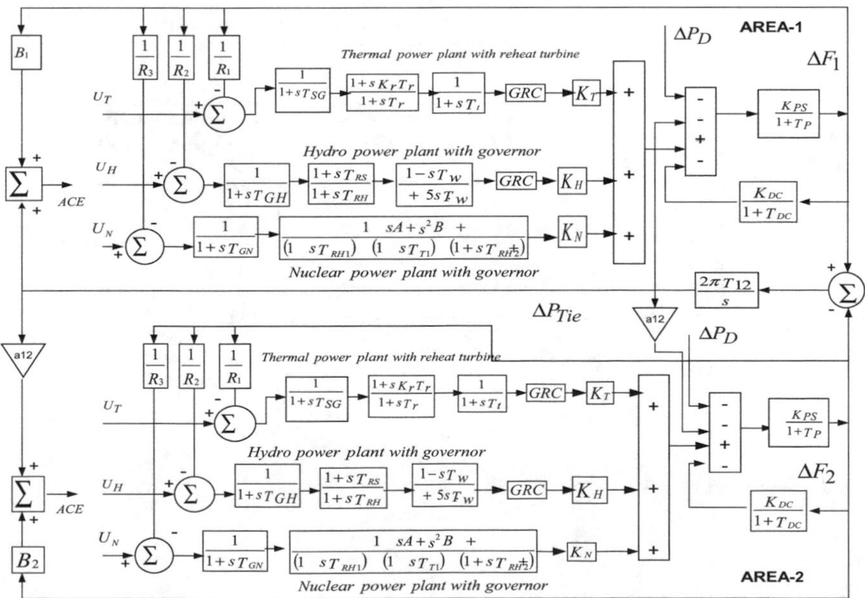

Fig. 1 Block diagram of multi-area multi-source interconnected power system with HVDC link

FOA is a process for locating global optimization considering the fruit fly's food searching behaviour. Considering the behaviour of sensing and perception, fruit fly is exceptional to another type of species, especially in osphresis and vision. Even from a distance of 40 km, it can smell food through osphresis organs. Then, after getting near to food location, exercise its sensitive vision to search food and fly in that direction [15]. FOA has been effectively employed in various fields such as load forecasting [16], phasor measure unit (PMU) placement [17], standalone hybrid PV-wind-diesel system sizing [18], PID controller tuned with modified FOA [19], scheduling and testing of semiconductor [20]. In this work, FOA technique is proposed for optimizing gains of different classical controller for AGC system. Steps of FOA are explained in [15] and mentioned in flow chart of Fig. 2.

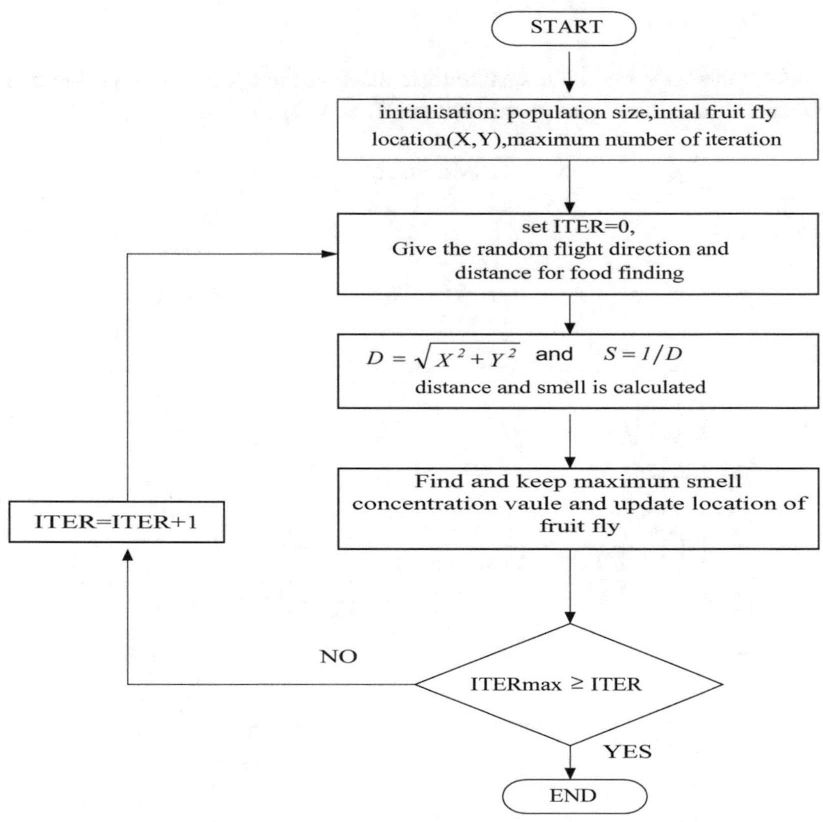

Fig. 2 Flow chart of FOA

Table 1 Gain values of integral controller and parameters of HVDC and performance analysis with different algorithms

Different algorithms	Optimized controller and HVDC parameters					ITAE	Settling time Ts (s)		
	KI_1	KI_2	KI_3	K_{DC}	T_{DC}		Δf_1	Δf_2	ΔP_{Tie}
GA	0.6378	0.6340	0.386	0.1994	0.8844	9.7336	45.77	44.82	37.18
PSO	0.2691	0.1655	0.5887	0.1975	0.6911	6.579	34.49	34.49	30.02
FOA	0.4561	0.5017	0.5954	0.1321	0.4973	6.4138	30.86	29.92	23.39

3 Result Analysis

Implementation of FOA algorithm

For this study, a population size and maximum number of iteration are taken as 100 for FOA technique. Similarly, parameters for GA method are 100, 50, 0.85 and 0.01 for population size, number of generations, crossover probability and mutation probability, respectively. The best controller and HVDC parameters found in the 50 runs are given in Table 1.

4 Analysis of Results

To illustrate the supremacy of FOA algorithm for optimizing controller parameters, the results of FOA algorithm are competed with the results obtained with GA and PSO algorithm, considering only I controller for the proposed system. As observed from Table 1 that with proposed FOA technique, the ITAE value as objective function is obtained as 6.4138, whereas with GA and PSO algorithm ITAE is 9.7335 and 6.579, respectively. The settling time of frequency deviation and tie-line power deviation with FOA optimized I controller is less compared to PSO and GA techniques. Time domain simulations are performed with 10% step load change in area-1 and are shown in Figs. 3 and 4. Results obtained with FOA optimized controller give best result as competed with other two algorithms considered in this paper. The unique algorithm converged to exact optimal solution with less time compared to other algorithms. Hence, FOA algorithm is considered for further optimization purpose of several controllers, i.e. ID/PI/PID/IDD/PIDD. The optimization was run for 50 times, and the optimum controller parameter among this is selected as the gain of different controller and HVDC parameters and is given in Table 2.

Different controllers are applied in order to explore its effect on performance of the realistic concern power system. A 10% step load increase is applied at area-1. The performances of the system as ITAE value and settling times of frequency and tie-line power deviations are mentioned in Table 3. As observed from Table 3, for same system, minimum ITAE value is achieved with PIDD controller (ITAE = 1.9885) compared to IDD controller (ITAE = 3.00), PID controller (ITAE = 4.916), PI

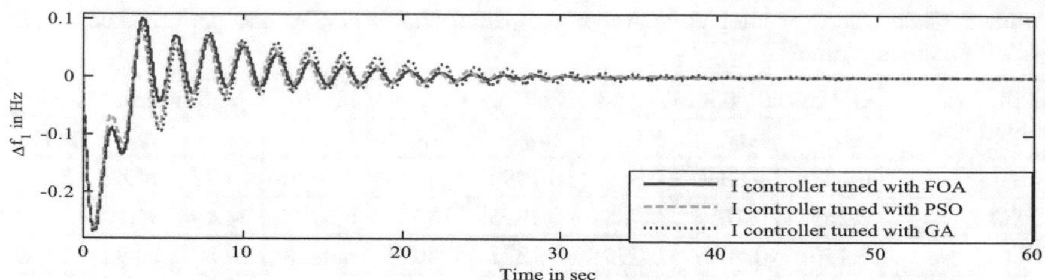

Fig. 3 Frequency deviation of area-1 for 10% step load change in area-1 with I controller

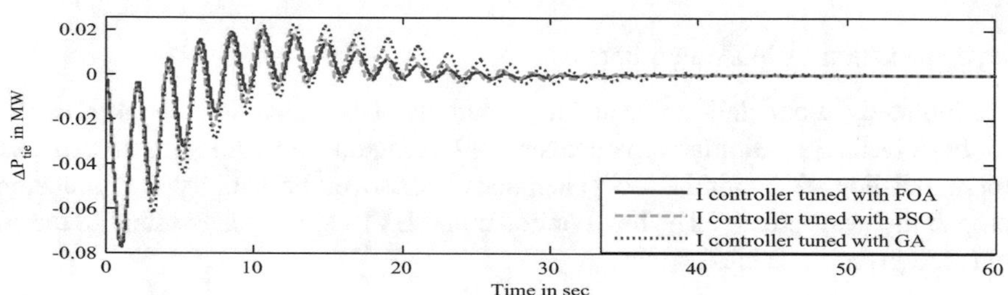

Fig. 4 Tie-line power deviation for 10% step load change in area-1 with I controller

Table 2 Gain values of different controllers and parameters of HVDC with FOA algorithm

Gain values	PI controller	ID controller	PID controller	IDD controller	PIDD controller
KP_1	0.8891		0.6565		0.1903
KP_2	0.7179		0.3099		0.1745
KP_3	0.9509		0.3316		0.0641
KI_1	0.4581	0.6403	0.7206	1.9938	1.1482
KI_2	1.2082	0.5047	0.1606	0.7179	0.6236
KI_3	0.6573	0.5945	0.6917	1.2505	1.7903
KD_1		0.0386	0.0285	0.0393	0.0835
KD_2		0.0292	0.0663	0.0076	0.0023
KD_3		0.0234	0.0602	0.0545	0.0640
K_{DC}	0.7187	0.2009	0.2882	0.5091	0.4631
T_{DC}	0.1092	0.5803	0.4007	0.2468	0.0270

controller (ITAE = 5.6482) and ID controller (ITAE = 6.1395). Hence, it can be established that PIDD controller outperforms from different controllers considered. Therefore, better system performance is accomplished with proposed FOA optimized PIDD controller compared to other controllers considering reduced settling times of frequency and tie-line power deviations. The above investigation confirmations that improved system performance are achieved with the proposed technique and PIDD controller.

Table 3 Performance analysis of different controllers

Performance index		ID controller	PI controller	PID controller	IDD controller	PIDD controller
ITAE		6.1395	5.6482	4.9169	3.000	1.9885
T_S (s)	Δf_1	23.73	26.55	21.11	20.72	17.61
	Δf_2	22.96	26.88	22.01	20.87	18.31
	ΔP_{Tie}	20.54	18.22	18.54	17.32	14.97

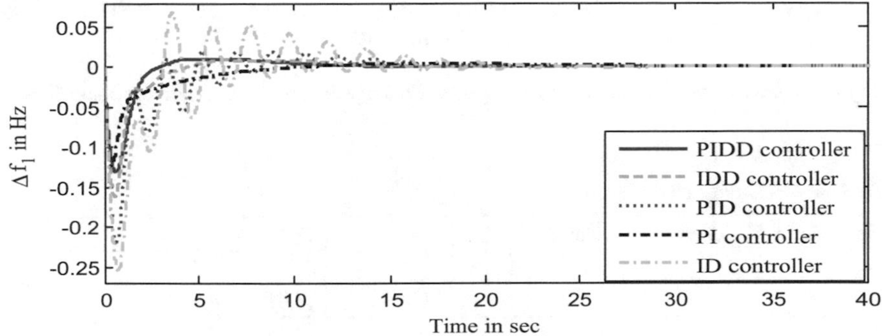

Fig. 5 Frequency deviation of area-1 for 10% load change in area-1 with different controllers

The system performances are shown in Figs. 5 and 6 for 10% step increase in load in area-1. Better dynamic performance is attained with the proposed FOA optimized PIDD controller compared to other controllers as seen from Figs. 5 and 6. Sensitivity analysis is accomplished with variation in system parameter and loading conditions in the range of +25% to −25% from their nominal values. Table 4 gives system performances under nominal and varied conditions. Acute investigation of Table 4 divulges that system performances vary within satisfactory ranges. So it can be established that, the proposed approach offers a satisfactory robust and stable control. The system performances with the varied conditions of loading T_{SG}, T_{GH}, T_{GN}, T_{RH}, T_{RS}, T_T, K_R, T_R and T_{12} are shown in Figs. 7, 8 and 9. It can be detected from Figs. 7, 8 and 9 that the system performances are negligibly effective with the variation of loading condition and system parameters. As observed, the controller parameters do not require to be retuned for varying system conditions. Hence, the proposed technique is proved to be robust one.

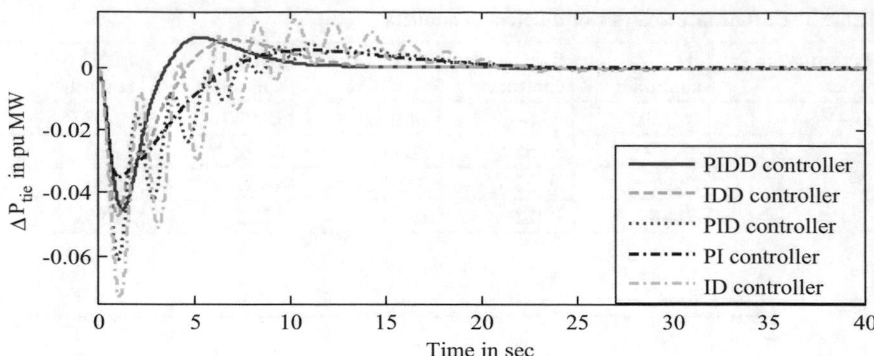

Fig. 6 Tie-line power deviation for 10% step load change in area-1 with different controllers

Table 4 Sensitivity analysis

Parameter variation	% change (%)	ITAE	T_S (s)		
			Δf_1	Δf_2	ΔP_{Tie}
Nominal		1.9885	17.61	18.31	14.97
Loading	+25	2.3032	18.25	18.75	14.92
	−25	2.3039	18.13	18.47	14.84
T_{12}	+25	2.2343	18.95	19.25	15.46
	−25	2.2602	18.16	18.54	14.99
T_{SG}	+25	2.2456	18.35	18.7	14.98
	−25	2.2447	18.36	18.71	14.99
T_{GN}	+25	2.2535	18.91	19.24	15.55
	−25	2.2375	18.39	18.75	15.02
T_{GH}	+25	2.2518	18.37	18.71	15
	−25	2.2376	18.77	19.13	15.4
T_T	+25	2.2376	18.21	18.5	14.84
	−25	2.2431	18.38	18.71	15.01
T_{RH}	+25	2.3347	18.27	18.58	14.39
	−25	2.1082	18.78	19.09	15.36
T_R	+25	2.3374	18.87	19.27	15.17
	−25	2.18	17.48	17.81	14.91

5 Conclusion

In the present analysis, FOA algorithm as new population-based optimization technique used to optimize hybrid interconnected AGC system and its tuning performance is analysed. To compare the tuning performance of FOA algorithm, the dynamic performance of the system with integral controller is studied and

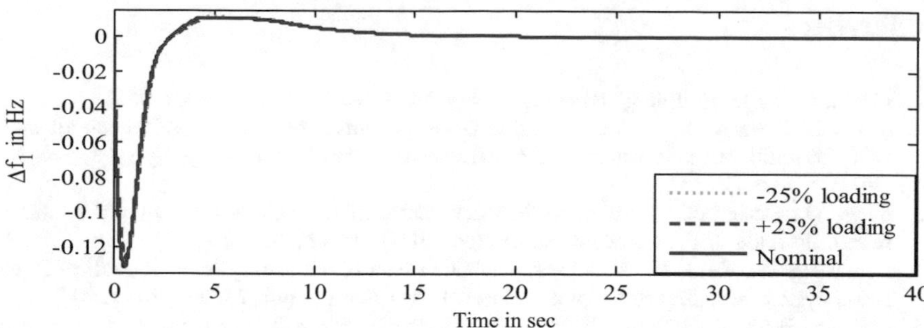

Fig. 7 Frequency deviation of area-1 for 10% step load change in area-1 with variation in loading

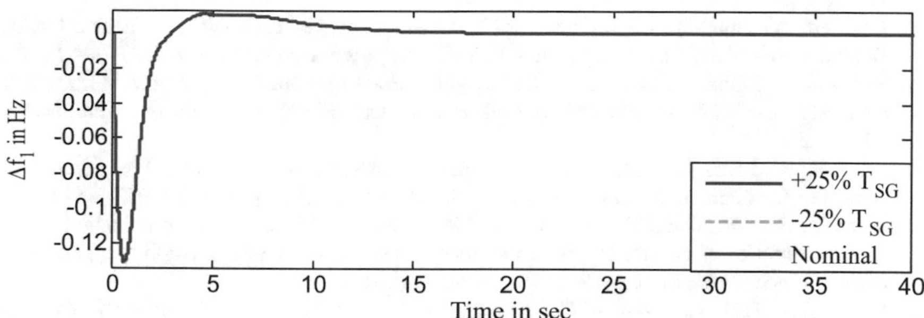

Fig. 8 Frequency deviation of area-1 for 10% step load change in area-1 with variation in T_{SG}

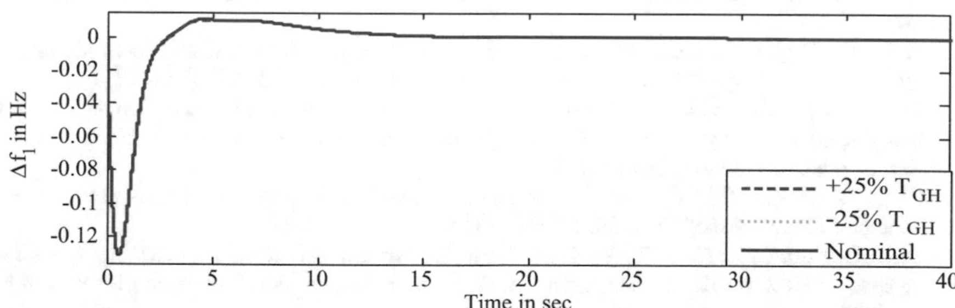

Fig. 9 Frequency deviation of area-1 for 10% step load change in area-1 with variation in T_{GH}

compared with GA and PSO algorithm. In the following studies, gains of ID/PI/PID/IDD/PIDD controllers are tuned optimally with FOA technique. The proposed PIDD controller optimally tuned with FOA algorithm is successfully applied to realistic power system, and its control performance is investigated. Then, sensitivity analysis validates the strength of the optimized controller parameters. It is detected that the recommended FOA-based PIDD controller offers better performance compared to other controllers.

References

1. Elgard, O.I.: Electric Energy Systems Theory. Mc Graw-Hill, New York (1982)
2. Saikia, L.C., Nanda, J., Mishra, S.: Performance comparison of several classical controllers in AGC for multi-area interconnected thermal system. Electr. Power Energy Syst. **33**, 394–401 (2011)
3. Singh, O.: Design of optimal state feedback controller for AGC using a hybrid stochastic search. In: 2008, IEEE Conference, Powercon, IIT Delhi (2008)
4. Demiroren, A., Yesil, E.: Automatic generation control with fuzzy logic controllers in the power system including SMES units. Electr. Power Energy Syst. **26**, 291–305 (2004)
5. Mohanty, B.: TLBO optimized sliding mode controller for multi-area multi-source nonlinear interconnected AGC system. Int. J. Electr. Power Energy Syst. **73**, 872–881 (2015)
6. Taher, S.A., Fini, M.H., Aliabadi, S.F.: Fractional order PID controller design for LFC in electric power systems using imperialist competitive algorithm. Int. J. Ain Shams Eng. **5**, 121–135 (2014)
7. Ibraheem, Nizamuddin, Bhatti, T.S.: AGC of two area power system interconnected by AC/DC links with diverse sources in each area. Electr. Power Energy Syst. 55:297–304 (2014)
8. Saikia, L.C., Sinha, N., Nanda, J.: Maiden application of bacterial foraging based fuzzy IDD controller in AGC of a multi-area hydrothermal system. Electr. Power Energy Syst. **45**, 98–106 (2013)
9. Ali, E.S., Abd-Elazim, S.M.: Bacteria foraging optimization algorithm based load frequency controller for interconnected power system. Electr. Power Energy Syst. **33**, 633–638 (2011)
10. Bhatt, P., Roy, R., Ghoshal, S.P.: GA/particle swarm intelligence based optimization of two specific varieties of controller devices applied to two-area multi-units automatic generation control. Electr. Power Energy Syst. **32**, 299–310 (2010)
11. Gozde, H., Taplamacioglu, M.C., Kocaarslan, I.: Comparative performance analysis of Artificial Bee Colony algorithm in automatic generation control for interconnected reheat thermal power system. Electr. Power Energy Syst. **42**, 167–178 (2012)
12. Rout, U.K., Sahu, R.K., Panda, S.: Design and analysis of differential evolution algorithm based automatic generation control for interconnected power system. Ain Shams Eng. J. **3**, 409–421 (2012)
13. Guha, D., Roy, P.K., Banerjee, S.: Quasi-oppositional symbiotic organism search algorithm applied to load frequency control. Int. J Swarm Evol. Comput. **33**, 46–67 (2017)
14. Nandi, M., Shiva, C.K., Mukherjee, V.: TCSC based automatic generation control of deregulated power system using quasi-oppositional harmony search algorithm. Int. J. Eng. Sci. Technol. **20**, 1380–1395 (2017)
15. TPan, W.: A new fruit fly optimization algorithm: taking the financial distress model as an example. Knowl.-Based Syst. **26**, 69–74 (2012)
16. Li, H.Z., Guo, S., Zhao, H.R., Su, C.B., Wang, B.: Annual electric load forecasting by a least squares support vector machine with a fruit fly optimization algorithm. Energies **5**, 4430–4445 (2012)
17. Ramachandran, B., Bellarmine, G.T.: Improving observability using optimal placement of phasor measurement units. Electr. Power Energy Syst. **56**, 55–63 (2014)
18. Zhao, J., Yuan, X.: Multi-objective optimization of stand-alone hybrid PV-wind-diesel-battery system using improved fruit fly optimization algorithm. Soft Comput. **20**, 2841–2853 (2016)
19. Li, C., Xu, S., Li, W., Hu, L.: A novel modified fly optimization algorithm for designing the self-tuning proportional integral derivative controller. J. Converg. Inf. Technol. **7**, 69–77 (2012)
20. Zheng, X.L., Wang, L., Wang, S.Y.: A novel fruit fly optimization algorithm for the semiconductor final testing scheduling problem. Knowl.-Based Syst. **57**, 95–103 (2014)

Integration of Goal Hierarchies Built Using Multiple OLAP Query Sessions

N. Parimala and Ranjeet Kumar Ranjan

Abstract A goal hierarchy consists of a strategic goal, decision goals, information goals and tasks. The goal hierarchy can provide a decision maker different ways in which the data can be analysed in order to achieve a strategic goal. In our earlier work, we built a goal hierarchy for a query session, using OLAP query recommendation technique. It is possible to build more than one goal hierarchy for a given strategic goal if multiple query sessions are used. In this paper, we propose an integration technique. It is a mixed approach where decision goals and information goals are integrated top-down and tasks are integrated bottom-up. The result of integration of multiple goal hierarchies is a new goal hierarchy with a unified view of analysis in order to identify the different perspectives of analysis in a single diagram.

Keywords Data warehousing · OLAP · Goal hierarchy · Integration of goal hierarchies · OLAP query session

1 Introduction

A data warehouse is a collection of a large amount of integrated data collected from various operational sources [1]. The data warehouse systems are used by decision makers to analyse data to help in taking decisions. Goal decomposition has been widely used for requirement engineering for software development as well as for data warehouse schema design. In the case of data warehousing, the goal decomposition techniques are used to identify the fact and dimensions [2–8]. It is expected

N. Parimala · R. K. Ranjan (✉)
School of Computer and Systems Sciences, Jawaharlal Nehru University,
New Delhi 110067, India
e-mail: ranjeetghitm@gmail.com

N. Parimala
e-mail: dr.parimala.n@gmail.com

J. Nayak et al. (eds.), *Soft Computing in Data Analytics*,
Advances in Intelligent Systems and Computing 758,
https://doi.org/10.1007/978-981-13-0514-6_43

that all the fact and dimension attributes to fulfil the strategic goals can be identified.

Suppose, a data warehouse schema has been arrived at by analysing the existing data sources [9–12] but not by analysing goals, the question is, whether a strategic goal can be achieved by the existing warehouse. The issue can be addressed by building the goal hierarchy diagram for a strategic goal using the data warehouse schema [13].

If a user uses only one session, then one goal hierarchy is created. The goal hierarchy may not have sufficient information to achieve a strategic goal. Also, the goal hierarchy may not cover all the parts of the data warehouse schema. To get more information in terms of tasks, information goals and decision goals, the user should build more goal hierarchies using other query sessions. The goal hierarchies built using multiple sessions may have same or distinct strategic goal.

If we have multiple goal hierarchies for a common strategic goal, then these should be combined to give a complete uniform view of the decomposition of the strategic goal. The combined goal hierarchy will cover more parts of the data warehouse schema. It will have more number of tasks, information goals and the decision goals. This can make the analysis tasks easier to achieve the strategic goal.

In this paper, we propose to integrate the different goal hierarchies for a given strategic goal into a single hierarchy. The overall approach for integration of two goal hierarchies is as follows.

First, all the tasks are disconnected from the information goals to which they are linked. The common strategic goal is identified. Next, the decision goals of both the goal hierarchies are integrated. The union of the decision goals is kept in the resulting goal hierarchy. The information goals with their respective links to the decision goals in the original hierarchies form part of the new integrated hierarchy. Next, the tasks are integrated after eliminating the redundant tasks. Finally, all the tasks are attached to the information goals.

The rest of this paper is organized as follows. In Sect. 2, the survey of integration of diagrams has been described. Section 3 explains the goal hierarchy. The manner in which two hierarchies are integrated is explained in Sect. 4. An illustrative example is shown in Sect. 5 followed by conclusion in Sect. 6.

2 Related Work

To the best of our knowledge, integration of goal hierarchies has not been attempted. Some researchers have proposed merging multiple diagrams or graphs to find a unique diagram. In [14], a semi-automated integration approach has been proposed to integrate multiple decision diagrams into a single concise and coherent decision diagram. The approach of merging decision diagram is based on the conditions attached to a node. In [15], an algorithm for connecting nodes from multiple disconnected structured knowledge graphs such as ontologies and taxonomies has been proposed. Here, a knowledge bridge is built between two nodes using domain knowledge. In our work, the nodes are goals or tasks represented using textual English phrases which are chosen by the user. In the goal hierarchy, the nodes are goals or tasks represented using textual English phrases which are chosen by the user. Thus, we propose a new graph integration technique which takes into account these aspects of a goal hierarchy.

3 Goal Hierarchy

The goal hierarchy consists of a strategic goal, decision goals, information goals and tasks [2, 13]. In order to build the goal hierarchy, the high-level goal is decomposed into low-level goals followed by the identification of actions or tasks.

The strategic goal has a high level of abstraction with very less information about the data required to analyse. In order to realise the strategic goal, some decisions have to be taken. So, the strategic goal can be decomposed into some decision goals, such as "*increase the number of stores*", "*decrease the sales price*", "*launch some promotions*". To take the decision, the decision makers have to analyse some relevant data which will give the idea about what decision he should take. To get more precise information for a decision goal, some information goals should be identified. For example, for the decision goal "*increase the number of stores*", the information goal can be derived as "*analyse the sales of stores*", "*analyse the store cost*", etc. The information goal mainly provides information about the measure attributes. The information about the dimension attributes is not provided adequately. The next step is the identification of tasks. Tasks correspond to the information requirements of the information goal. For example, for the information goal "*analyse the store cost*", the tasks can be "*get the store cost for all store type*", "*get the store cost for all size of store country wise*", etc.

We have used the notation of strategic rationale (SR) model to represent the goal hierarchies based on the Computation Independent Model (CIM) [2] which uses the notations of SR model of *i** framework [16, 17] to model the goals and information requirements for the data warehouse.

As brought out earlier, multiple goal hierarchies are built using multiple sessions. The goal hierarchies may have same or distinct strategic goals. Using two sessions, the goal hierarchies are built for a strategic goal "*Increase the sales*" as shown in Figs. 1 and 2.

3.1 Basic Definitions

In this section, we have defined the goal hierarchy diagram [13] and the terms used in this paper.

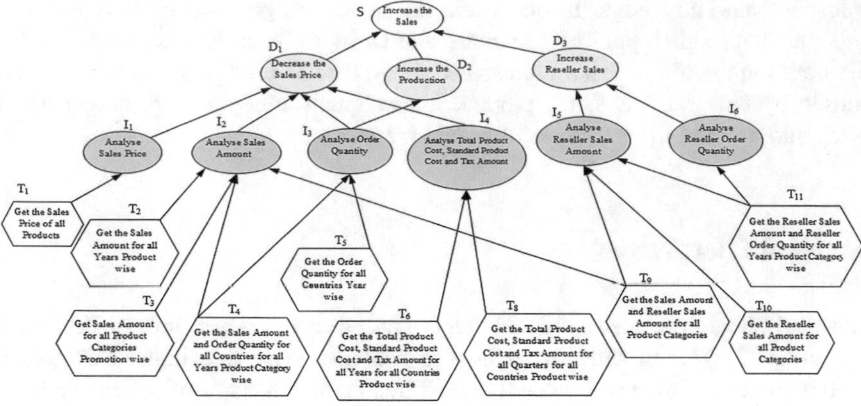

Fig. 1 First goal hierarchy G built for strategic goal "*Increase the sales*"

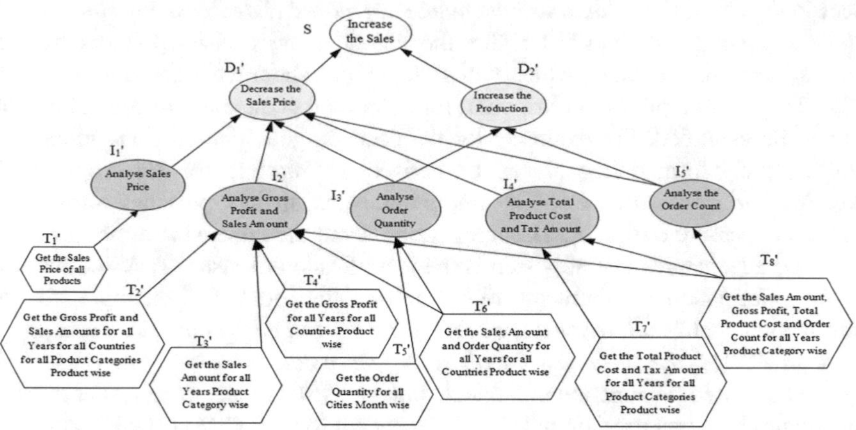

Fig. 2 Second goal hierarchy G' built for strategic goal "*Increase the sales*"

Definition 1 Goal hierarchy The goal hierarchy consists of four levels of nodes, and it can be denoted by a four tuple, $G = \langle S, Ds, Is, Ts \rangle$ where, S is the strategic goal for which the goal hierarchy is built, $Ds = \{D_1, D_2, \ldots, D_n\}$ is the set of decision goals, $Is = \{I_1, I_2, \ldots, I_k\}$ is the set of information goals, and $Ts = \{T_1, T_2, \ldots, T_m\}$.

Definition 2. Attribute set of a task The attribute set associated with a task is the set of business terms or schema terms. These are the measures and the dimensional attributes. The set is defined as $a\_set(T) = \{ms(T), ds(T)\}$, where $ms(T) = \{m_1, m_2, \ldots, m_i\}$ is a measure set and $ds(T) = \{d_1, d_2, \ldots, d_j\}$ is a dimension set of the task T.

Definition 3. Measure set of an information goal The measure attribute set associated with an information goal I is defined as $ms(I) = \{m_1, m_2, \ldots, m_n\}$.

4 Integration of Goal Hierarchies

Let $G = \langle S, Ds, Is, Ts\rangle$ and $G' = \langle S, Ds', Is', Ts'\rangle$ be the two goal hierarchies to be integrated and let the resulting goal hierarchy be $G_N = \langle S, Ds_N, Is_N, Ts_N\rangle$.

We adopt a mixed approach to integrate the hierarchies. The decision goals and information goals are integrated top-down. The tasks are integrated bottom-up. Subsequently, we link the tasks to the information goals.

4.1 *Identification of Strategic Goal*

As brought out earlier, the integration of two goal hierarchies can be done only if both the goal hierarchies are built for a common strategic goal. This common strategic goal becomes the strategic goal of the resulting goal hierarchy G_N.

4.2 *Integration of Decision Goals*

It is possible that the two decision goal sets, Ds and Ds', of the two goal hierarchies G and G' respectively have some common decision goals as well as some distinct decision goals. Ds_N, the union of decision goals ($Ds_N = Ds \cup Ds'$), forms the set of decision goals in the resulting goal hierarchy G_N. All the decision goals of Ds_N are connected to the strategic goal by preserving the links that existed in G and G'.

4.3 *Integration of Information Goals*

All the information goals of Is and Is' with the links to their respective decision goals are retained in the new set of information goals Is_N. Further, the information goals consisting of more than one measure are decomposed using AND decomposition technique [5] into the information goals with only one measure attribute. That is, if an information goal I has multiple measures, $ms(I) = \{m_1, m_2,\ldots, m_n\}$, the information goal I is decomposed into a set of new information goals $\{I_1, I_2,\ldots, I_n\}$ with $ms(I_i) = m_i$. The decomposed goals with one measure are connected to the composite information goal that was decomposed. The result of such decomposition is shown in Fig. 3.

At the end of the foregoing decomposition strategy, all the leaf information goals will have only a single measure.

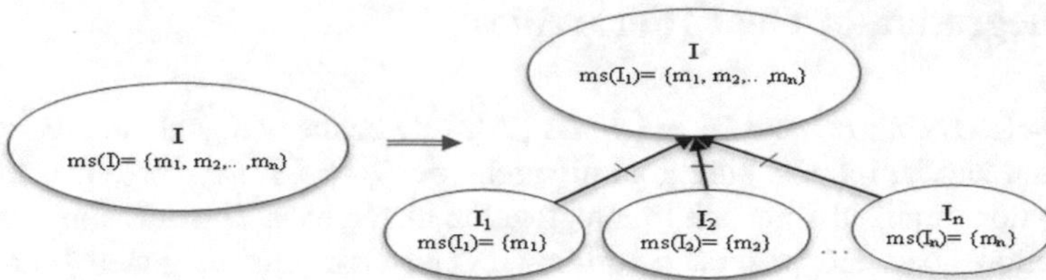

Fig. 3 Information goal decomposition

4.4 Guidelines to Combine Two Tasks

As brought out earlier, the tasks are integrated using a bottom-up approach. The tasks in the goal hierarchies have already been disconnected from their respective information goals. Tasks are combined in order to remove the redundant tasks. The combination of tasks is done after comparing the measure and dimensional attributes of the tasks.

Every task has a set of measures and a set of dimensional attributes associated with it. The measures in the two tasks may be identical, totally different or some measures may be common. Similarly, the dimensional attributes in the two tasks may be identical, different or some attributes may be common. In addition, the different attributes may belong to the same hierarchy.

The different possibilities that arise are discussed in this section. For each possibility, the tasks that are retained or going to be created follow the rules given below:

(1) If the measure and dimensional attributes associated with the tasks are same, then one of the two tasks is retained and other is deleted.
(2) If both tasks have same measure attribute sets and distinct dimensional attribute sets but the dimensional attributes belong to some common dimensional hierarchies, then a new task is created using the attributes of both the tasks.
(3) If the measure attributes associated with the tasks are distinct, then both tasks are retained in the resulting goal hierarchy.

The different possibilities are discussed below in detail. Towards this, a schema *SalesSummary* was generated using *SQL Server Data Tool* [18] and *AdventureWorks* database [19, 20]. From the dimension hierarchies, available with *AdventureWorks* database, relevant ones were also chosen while generating the schema.

Let the two tasks be T and T′. Let ms(T) and ms(T′) be the measure sets, ds(T) and ds(T′) be the dimension sets of task T and T′ respectively. The different possibilities that can arise between measure sets and dimension sets of two tasks are listed in Table 1 and Table 2 respectively.

Table 1 Relationships between measures

Measure Sl.no.	Condition	Example
M1 (Identical measures)	ms(T) ∩ ms(T′) = ms (T) = ms(T)′	ms(T) = {Sales Amount, Order Quantity}, ms(T′) = {Sales Amount, Order Quantity}
M2 (Different measures)	ms(T) ∩ ms(T′) = ϕ	ms(T) = {Sales Amount, Order Quantity}, ms(T′) = {Gross Profit, Tax Amount}
M3 (Some common measures)	ms(T) = {m_1, m_2, …m_i, …, m_n} ms(T′) = {m_1, m_2, …m_i, m_{i+1}'…, m_m'} ms(T) ∩ ms(T′) = {m_1, m_2, …, m_i}	ms(T) = {Sales Amount, Order Quantity}, ms(T′) = {Sales Amount, Tax Amount}
M4 (Subset/ Superset)	ms(T) ⊂ ms(T′)	ms(T) = {Sales Amount, Order Quantity}, ms(T′) = {Sales Amount, Order Quantity, Tax Amount}
	ms(T) ⊃ ms(T′)	ms(T) = {Sales Amount, Order Quantity, Tax Amount}, ms(T′) = {Sales Amount, Tax Amount}

Table 2 Relationships between dimensional attributes

Dimension Attributes Sl.no.	Condition	Example
D1 (Identical dimensional attributes)	ds(T) ∩ ds(T′) = ds (T) = ds(T′)	ds(T) = {Year, Country}, ds(T′) = {Year, Country}
D2 (Different dimensional attributes)	ds(T) ∩ ds(T′) = ϕ	ds(T) = {Year, Country}, ds(T′) = {Product, Promotion}
D3 (Some common dimensional attributes)	ds(T) = {d_1, d_2, …d_i, …, d_n} ds(T′) = {d_1, d_2, …d_i, d_{i+1}' …, d_m'} ds(T) ∩ ds(T′) = {d_1, d_2, …, d_i}	ds(T) = {Year, Country, Product}, ds(T′) = {Year, Country, Promotion}
D4 (Subset/Superset)	ds(T) ⊂ ds(T′)	ds(T) = {Year, Country, Product}, ds(T′) = {Year, Country}
	ds(T) ⊃ ds(T′)	ds(T) = {Year, Country}, ds(T′) = {Year, Country, Product}

The cross product of these two tables gives the total number of possibilities for creating new tasks. We consider each of these in turn. In some of the cases considered below, a new task may be created. We refer to this as T". In the rules given below to compare the tasks, MiDj refers to the serial number Mi of Table 1 and Dj of Table 2.

M1D1: In this case, since the measures as well as the dimensional attributes are identical in both the tasks, only one task is retained in the resulting goal hierarchy.

If ms(T) ∩ ms(T′) = ms(T) = ms(T′) and ds(T) ∩ ds(T′) = ds(T) = ds(T′), then only one of the tasks T and T′ is retained in the goal hierarchy.

M1D2: In this case, the measures in the two tasks are identical but the dimensional attributes are different. Different situations arise as far as dimensional attributes are concerned. These are considered below:

- **D2.1**: All the attributes of ds(T) participate in common hierarchies with the attributes of ds(T′).
- **D2.2**: One or more than one attribute of ds(T) is outside all common hierarchies that may exist between ds(T) and ds(T′) or vice versa.

We consider each of these in turn.

M1D2.1: In this case, a new task is created by combining both the tasks. Since the attributes of tasks T and T′ participate in common hierarchies, we pick the attribute that is at the lower level in the respective hierarchies. This is because, if the query corresponding to a task analyses data at a lower level, it is always possible to roll up in the hierarchy for analysis and arrive at a new task. The difference between the two tasks in only in their granularity and therefore, the tasks are not very significantly different. We could have chosen either the lower level attribute or the higher level attribute without losing the tasks that can be arrived at for a strategic goal. The new task is T" where ms(T") = ms(T) = ms(T′). T" will have all the dimensional attributes at the lower level of the common hierarchies of task T and T′

M1D2.2: If there is a non-common attribute, then the axis of analysis is different. Therefore, both the tasks T and T′ are retained in the resulting goal hierarchy. If ds (T) ∩ ds(T′) = ϕ, then both the tasks T and T′ are retained in the resulting goal hierarchy.

M1D3: The measures in the two tasks are identical but not all the dimensional attributes are identical. There are some common attributes and some different attributes in both the tasks. If ms(T) ∩ ms(T′) = ms(T) = ms(T′) and ds(T) ∩ ds (T′) ! = ds(T)! = ds(T′).

Let ds(T) = $\{d_1, d_2, \ldots d_i, \ldots, d_n\}$, ds(T′) = $\{d_1, d_2, \ldots d_i, d_{i+1}', \ldots, d_m'\}$ and ds (T) ∩ ds(T′) = $\{d_1, d_2, \ldots, d_i$. Due to the existence of non-common dimensional attributes in both the tasks, the following situations can arise:

- **D3.1**: All the non-common attributes of ds(T) participate in common hierarchies with the non-common attributes of ds(T′). That is, attributes d_{i+1}, …, d_n and d_{i+1}', …, d_m' participate in common hierarchies.
- **D3.2**: One or more than one non-common attribute of ds(T) is outside all common hierarchies that may exist between ds(T) and ds(T′) or vice versa. That is, one of the attributes d_{i+1}, …, d_n and d_{i+1}', …, d_m' is not taking part in any of the common hierarchies.

We consider each of these in turn.

M1D3.1: In this case, a new task T" is created by combining both the tasks. The common attributes are included in ds(T"). Since d_{i+1}, …, d_n of task T and d_{i+1}'…, d_m' of task T′ participate in common hierarchies, we pick the attribute that is at the lower level in the respective hierarchies. The reasoning is the same as that given for the rule M1D2.1.

M1D3.2: In this case, both the tasks T and T′ are retained in the resulting goal hierarchy following the reasoning given for rule M1D2.2.

M1D4: The set of dimensional attributes of one task is either a subset or superset of the set of attributes of the other task. If ds(T) ⊂ ds(T′) then, without loss of generality, we assume task T to be a subset of T′. Let ds(T) = {d_1, d_2, …, d_n}, ds(T ′) = {d_1, d_2, …d_n, d_{n+1}, …, d_m}. Here ds(T) ⊂ ds(T′). Because of the extra attributes in ds(T′), the following situations arise:

- **D4.1**: Each extra attribute of ds(T′) participates in some hierarchy of the attributes of ds(T). That is, every one of the attributes d_{n+1}, …, d_m is taking part in one of the hierarchies in which d_1, …, d_n participate.
- **D4.2**: The extra attributes of ds(T′) are outside the hierarchies of the attributes of ds(T). That is, at least one of the attributes d_{n+1}, …, d_m is not taking part in any of hierarchies in which d_1, …, d_n participate.

M1D4.1: We use the same argument as in M1D2.1 and pick up from the attributes d_1, d_2, …, d_m the ones which are at the lowest level in each of the hierarchies.

M1D4.2: Following the reasoning given in M1D2.2, both the tasks are retained.

M2D1, **M2D2**, **M2D3**, **M2D4**, **M2D5**: In these cases, the measures of the two tasks are completely different. In other words, the subject of analysis is different. Therefore, we consider the two tasks as different even if there may be similar dimensional attributes. Both the tasks are retained. If ms(T) ∩ ms(T′) = ϕ then, both the tasks T and T′ are retained in the resulting goal hierarchy.

M3D1, **M3D2**, **M3D3**, **M3D4**, **M3D5**: In these cases, the measures of the two tasks have some measures in common but both have some measures which are not common. Even here, since some of the subjects of analysis are different, the tasks are not combined. So, both the tasks are retained in the resulting goal hierarchy.

M4D1, **M4D2**, **M4D3**, **M4D4**, **M4D5**: In these cases, the set of measures of one task is either a superset or a subset of the set of measures of the other task. With the same reasoning as above, both the tasks are retained in the resulting goal hierarchy.

Summarizing then, only one task is retained if it falls under the rule M1D1. Both the tasks are retained in case of rules; M1D2.2, M1D3.2, M1D4.2, M2D1, M2D2, M2D3, M2D4, M2D5, M3D1, M3D2, M3D3, M3D4, M3D5, M4D1, M4D2, M4D3, M4D4 and M4D5. A new task is created for the rules; M1D2.1 and M1D3.1.

4.5 *Linking Tasks to Information Goals*

A task *T* is connected to an information goal *I* if $ms(I) \subset ms(T)$. If a task has more than one measure associated with it, then it is connected to as many information goals as there are measures in this task. An example of linking tasks to the information goals is shown in Fig. 4.

Using the steps detailed in Sect. 4.1–4.5, the two goal hierarchies can now be integrated.

5 Illustrative Example

Let the two goal hierarchies *G* and *G′* for the strategic goal S, "*Increase the sales*", be as shown in Figs. 2 and 3. The resulting goal hierarchy, after integration, is shown in Fig. 5. The integration steps are explained below.

First, the strategic goal "*Increase the sales*" which is same in both the goal hierarchies is created in the integrated goal hierarchy. Next, the decision goals D_1 and D_2 of *G* are identical to D_1', D_2' of *G′* respectively. The remaining decision goals are distinct. By applying the rule given in Sect. 4.2, the resulting goal

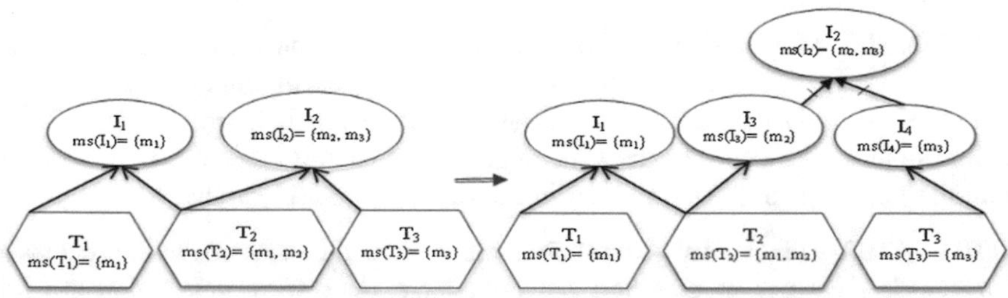

Fig. 4 Attaching tasks to information goals

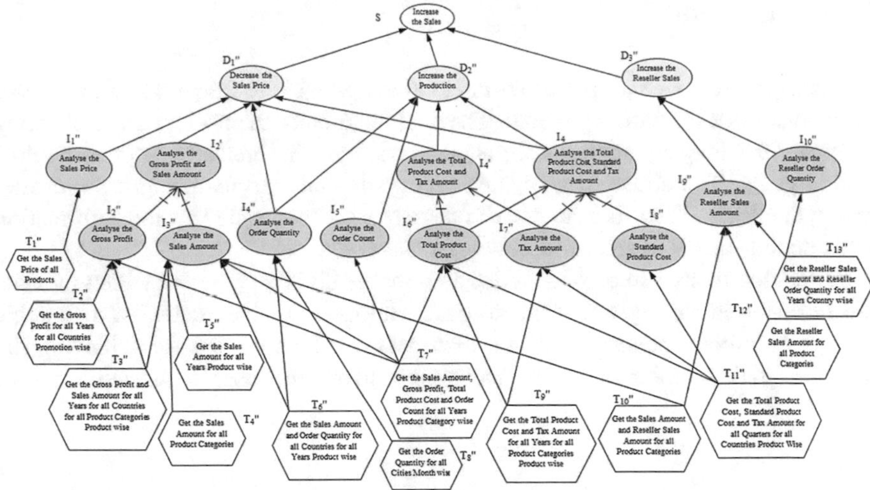

Fig. 5 Resulting goal hierarchy G_N after integration of G and G'

hierarchy G_N has three decision goals; "*Decrease the sales price*", "*Increase the productions*" and "*Increase the reseller sales*". All these decision goals are linked to the strategic goal.

Next, the information goals are integrated following the rule described in Sect. 4.3. Information goal I_1 of G and I_1' of G' is identical. Also I_3 and I_3' are identical. Remaining information goals are distinct. So, I_1, I_3 and remaining distinct information goals are retained in G_N. The composite goals are I_4, I_2' and I_4'. The composite information goal I_4 of G is decomposed into three information goals I_6'', I_7'' and I_8'' using AND decomposition links. Also the information goal I_2' of G is decomposed into I_2'' and I_3'' and I_4' is decomposed into I_6'' and I_7''. The identical information goals I_6'' and I_7'', after decomposition of I_4 and I_4', are represented only once and linked to I_4 and I_4'. In the resulting goal hierarchy G_N, all the leaf information goals are distinct and with only one measure attribute.

Next, the tasks are integrated after comparing the measures and dimensional attributes using the rules described in Sect. 4.4. In the goal hierarchy G, the tasks T_6 and T_8 are combined based on the rule M1D3.1 and the result is denoted as T_{11}'' in the resulting goal hierarchy G_N. The tasks T_1 of G and T_1' of G are identical so only one task is retained as T_1'' in G_N as the rule M1D1. The tasks T_2 and T_3' are combined in one task T_5'' using rule M1D3.1. The tasks T_4 and T_6' are also combined using same rule in the task T_6''. The tasks T_5 and T_5' are combined into one task T_8'' using rule M1D2.1. The tasks T_3, T_9, T_{10} and T_{11} of G are retained in G_N as T_4'', T_{10}'', T_{12}'' and T_{13}'' respectively. Similarly, the tasks T_2', T_4', T_7' and T_8' are retained in G_N as T_3'', T_2'', T_9'' and T_7'' respectively. Finally, the tasks are linked to the respective goal hierarchies based on the rule given in Sect. 4.5.

6 Conclusion

In this paper, we have proposed an integration approach to integrate multiple goal hierarchies built for a strategic goal. The goal hierarchies themselves are built using multiple OLAP query sessions. First, the tasks are disconnected from the information goals. The strategic goal, decision goals and information goals are integrated top-down. Then the tasks are integrated and connected to the information goals bottom-up. This creates a new goal hierarchy.

A goal hierarchy can provide a decision maker different ways in which the data can be analysed for a given strategic goal. Multiple goal hierarchies of a strategic goal give different perspectives in which data can be analysed for achieving the strategic goal. Combining these hierarchies gives the user a unified view of analysis.

References

1. Inmon, W.H.: Building the Data Warehouse. Wiley (2005)
2. Mazón, J., Pardillo, J., Trujillo, J.: A model-driven goal-oriented requirement engineering approach for data warehouses. In: International Conference on Conceptual (2007)
3. Liu, X., Xu, X.: Requirement analysis for data warehouses based on the tropos. In: 2011 3rd International Workshop on Intelligent Systems and Applications (ISA), pp. 1–5 (2011)
4. Giorgini, P., Rizzi, S., Garzetti, M.: Goal-oriented requirement analysis for data warehouse design. In: Proceedings of the 8th ACM international workshop on Data warehousing and OLAP-DOLAP. p. 47 (2005)
5. Giorgini, P., Rizzi, S., Garzetti, M.: GRAnD: A goal-oriented approach to requirement analysis in data warehouses. Decis. Support Syst. **45**, 4–21 (2008). https://doi.org/10.1016/j.dss.2006.12.001
6. Salinesi, C., Gam, I.: A requirement-driven approach for designing data warehouses. In: 12th Working Conference on Requirements Engineering: Foundation for Software Quality, REFSQ. pp. 179–192 (2006)
7. Prakash, N., Gosain, A.: Requirements Driven Data Warehouse Development. In: CAiSE Short Paper Proceedings (2003)
8. Castro, J., Kolp, M., Mylopoulos, J.: Towards requirements-driven information systems engineering: the Tropos project. Inf. Syst. **27**, 365–389 (2002)
9. Golfarelli, M., Maio, D., Rizzi, S.: Conceptual design of data warehouses from E/R schemes. In: Proceedings of the Thirty-First Hawaii International Conference on System Sciences, pp. 334–343 (1998)
10. Golfarelli, M., Maio, D., Rizzi, S.: The dimensional fact model: a conceptual model for data warehouses. Int. J. Coop. Inf. Syst. **7**, 215–247 (1998). https://doi.org/10.1142/S0218843098000118
11. Guo, Y., Guo, Y., Tang, S., Tang, S., Tong, Y., Tong, Y., Yang, D., Yang, D.: Triple-driven data modeling methodology in data warehousing: a case study. In Proceedings of the 9th ACM International Workshop on Data Warehousing and OLAP, pp. 9–66 (2006). https://doi.org/10.1145/1183512.1183524
12. Jukic, N., Nicholas, J.: A framework for requirement collection and definition process for data warehousing projects. In: 32nd International Conference on Information Technology Interfaces (ITI), pp. 187–192 (2010)

13. Ranjan, R.K., Parimala, N.: A bottom-up approach for creating goal hierarchy using olap query recommendation technique. Int. J. Bus. Inf. Syst. (2017)(In Press)
14. Eiter, T., Krennwallner, T., Redl, C.: Declarative merging of and reasoning about decision diagrams. In: Workshop on Constraint Based Methods for Bioinformatics (WCB 2011), Perugia, Italy. pp. 3–15 (2011)
15. Al-Mubaid, H., Bettayeb, S.: An Algorithm for Combining Graphs Based on Shared Knowledge. In: ISCA BICoB. pp. 137–142., Las Vegas, Nevada, USA (2012)
16. Yu, E.S.K.: Towards modelling and reasoning support for early-phase requirements engineering. In: Requirements Engineering, Proceedings of the Third IEEE International Symposium on. pp. 226–235 (1997)
17. Yu, E.: Modelling strategic relationships for process reengineering. Soc. Model. Requir. Eng. 11, (2011)
18. Microsoft SQL Server 2012 Analysis Services, https://www.microsoft.com/
19. AdventureWorks: AdventureWorksDW Database, https://msftdbprodsamples.codeplex.com/
20. Smith, B., Clay, C.: Microsoft SQL Server 2008 MDX Step by Step. Pearson Education (2009)

Energy-Aware Reliable Routing by Considering Current Residual Condition of Nodes in MANETs

Arshad Ahmad Khan Mohammad, Ali Mirza Mahmood and Srikanth Vemuru

Abstract Mobile ad hoc network design's goal is to provide Internet connectivity anywhere and anytime regardless of geographical location. Application of MANET includes disaster recovery, military, and environment monitoring. Resource-constrained environment of MANET makes its communication operations very much challenging. Moreover, network nodes are equipped with constrained batteries and it is much tough to replace or recharge the batteries during the mission. Thus, MANET requires an energy-efficient mechanism to address the constraints. We are achieving the energy efficiency through network layer, as MANET is infrastructureless peer-to-peer network. We develop a new reactive energy-efficient routing protocol based on knapsack mechanism. It selects the routing path based on the current residual condition of network nodes. Simulation results conclude that our proposed method is better in comparison with the existing works E-AODV, MRPC with respect to network lifetime and link stability.

Keywords MANETs · Energy · Buffer · Bottleneck node
Lifetime of network

1 Introduction

Wireless infrastructureless distributed network, such as MANET's design goal is to provide Internet access anywhere and anytime regardless of geographical locations. Mobile ad hoc network gains high attention due to its self-organization and

A. A. K. Mohammad (✉) · S. Vemuru
KL University, Andhra Pradesh, India
e-mail: ibnepathan@gmail.com

S. Vemuru
e-mail: vsrikanth@kluniversity.in

A. M. Mahmood
DMS.SVH Engineering College, Machilipatnam, India
e-mail: alimirza.md@gmail.com

J. Nayak et al. (eds.), *Soft Computing in Data Analytics*,
Advances in Intelligent Systems and Computing 758,
https://doi.org/10.1007/978-981-13-0514-6_44

self-maintenance characteristics. Nodes in a network are provided with network intelligence so that they can perform the task of routing. Communication between nodes happens directly if they are present in a communication range of one another otherwise they rely on intermediate nodes.

The major constraint in MANET is energy resource, as nodes in network are provided with constrained batteries. It is much tough to replace/recharge the batteries during the mission. Thus, energy-efficient routing mechanism in MANET is powerful considerable aspect. But it is challenging job in comparison with infrastructure-based network due to MANET characteristics like mobility, heterogeneity and dynamic network topology. Thus, in this work, we propose a novel energy-efficient routing protocol based on the current residual condition of network nodes with respect to energy and traffic, to avoid the node to become bottleneck.

2 Energy-Efficient Routing in MANET

Energy-efficient routing is prime considerable factor in mobile ad hoc network, as heterogeneous nodes present in a network act as router as well as host. Routing functionality consumes the energy of node. Work [1] analyzed that the power consumption for receiving and transmitting packets is about 800–1200 MW. To assist large substantial applications, routing protocol of MANET should be energy efficient. In the literature [2–16], various routing protocols have been designed based on energy. These protocols are majorly categorized into three types

1. Routing protocol to find reliable link route
2. Routing protocol to find minimum energy consumption route
3. Routing protocol to find higher energy nodes route

The energy-efficient routing protocol considering just minimum energy consumption route is not always reliable route but also needs to consider link reliability and current energy state of nodes. Work [17] demonstrates reliable routing based on presumed transmission count; this technique reduces the energy consumption during the communication and recovers the packet loss by retransmission. Algorithm proposed [18] is based on higher remaining energy of nodes to enhance the network lifetime; however, this technique does not contemplate the energy efficiency. Algorithms proposed in [2, 6, 7, 10, 13, 14] depend on transmission power of nodes but does not contemplate the energy consumption. In all above routing protocols, route choice metric is based on higher remaining energy node or minimum energy consumption route or minimum retransmission count route. All these routing protocols reserve the particular routing path for communication. However, these routing protocols create the situation of bottleneck node.

2.1 Bottleneck Node

MANET is a peer-to-peer multi-hop network; if communicating nodes do not lie in a radio communication range of each other, then they communicate with the help of intermediate nodes. Thus, each node present in a network is equipped with buffer (memory) to enable communication. This buffer space is used to hold the packets for small delay for proper synchronization between communicating entities. Figure 1 shows the buffer internal structure of node. Whenever packets arrive at intermediate node buffer, it encounters the following tasks:

1. The packet arrived at input interface is put inside the input buffer queue, and it moves forward to reach at the edge of the queue, where it waits for routing decision.
2. Decision module takes the packet from queue of input buffer and makes the decision about the packet according to underlying routing protocol specifications.
3. Packet is placed at queue of output buffer and upended till it arrives the output interface and stays for its turn based on MAC protocol specifications to transmit.

For instance, intermediate node (router) becomes bottleneck, if the packets arrived at the input interface are greater than its processing capacity, then queue at input buffer grows. If this queue is higher than its buffer size, then packets will drop from node. Figure 2 shows the bottleneck intermediate node at node 4. This paper's objective is to find the routing path to avoid the node to become bottleneck, extend the network lifetime, and provide the link stability. Route selection metric of our proposed routing protocol is current residual condition of node with respect to energy and buffer with the help of knapsack mechanism. Major contribution of the proposed mechanism is an analytical model for bottleneck node, calculation of current residual condition of nodes, priority assignment, and routing path initiation.

Input interface	i/p buffer ……………………o/p buffer	Output interface

Fig. 1 Node buffer internal structure

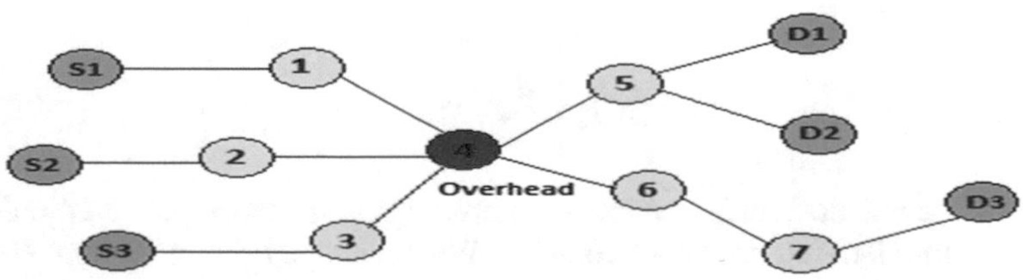

Fig. 2 Bottleneck intermediate node at node 4

2.2 Analytical Model for Bottleneck Node

In this section, we design an analytical model to predict the chance of node to become bottleneck in a given network condition with the help of poison probabilistic model. This prediction is helpful to control the traffic toward the intermediate node by estimating the average input queue size of input buffer and average waiting delay of packet inside the node buffer. Without loss of generality, we are considering the following conditions:

1. Consider energy consumption $(E_{\Delta t})$ of an active node during the time interval Δt time, then mean value of energy consumption (ϑ) of an active node is computed by Eq. 1.

$$\vartheta = (E_{\Delta t}/P_{\Delta t}) \tag{1}$$

where $P_{\Delta t}$ is number of packets processed (received and transmitted) by an active node during Δt time interval.

2. Probability of energy consumption due to one packet process from active node is $\vartheta\Delta t$, and the probability of active node to become an idle state is $1 - \vartheta\Delta t$, and these conditions are independent of one other during the time interval (Δt)

With the help of the above two conditions, we can derive an equation to compute exact amount of energy consumption through an active node in a given time interval with the help of poison probabilistic model, which given in Eq. 2.

$$P(\delta) = \left(e^{-\vartheta t}\right)\vartheta t^{\delta}/\delta! \tag{2}$$

where $'\delta'$ is the energy consumption of an active node and $'t'$ is a time interval, If μ is the packet arrival rate toward the input buffer of node and α is the active nodes packet-forwarding capacity in a provided time interval, then probability of average number of packets (n) present in a node buffer is calculated by Eq. 3.

$$P(n) = \frac{\mu^n}{\alpha}\left(1 + \frac{\mu}{\alpha}\right) \tag{3}$$

And probability of packet stay in an input buffer more than delay (t) seconds is calculated by Eq. 4.

$$P(t) = \frac{\mu}{\alpha}e^{-t(\mu - \alpha)} \tag{4}$$

We used computed analytical model to investigate the performance of multi-hop MANET in underlying network protocols. We considered the multi-hop MANET with 200 heterogeneous nodes geographically distributed in an area of 1500*1500 sq.m. All nodes are equipped with buffer and battery with capacity to hold 20 packets and 20 J of energy, respectively. We considered the packet arrival intervals as 0.004 s and 0.005 s. Then, we calculated the probability of average packet waiting time inside the node buffer and the average packet queue length at an active intermediate node buffer during communication, which is shown in Figs. 3, 4, 5 and 6. Figures are clearly indicating that packet drop from an active intermediate node occurs when packet arrival rate toward it is increased. This situation arises in network whenever an active node acts as a relay or intermediate node for more than one sources simultaneously. If the number of sources simultaneously transmits the packets to identical active intermediate node, then packet arrival rate increases. This in turns increases queue length and packet waiting time. Packets drop forms an active intermediate node in two conditions, i.e., if the queue size is more than its buffer size and if packet waiting time is more than TTL (lifetime) interval, and this situation in the network also leads the node to become bottleneck.

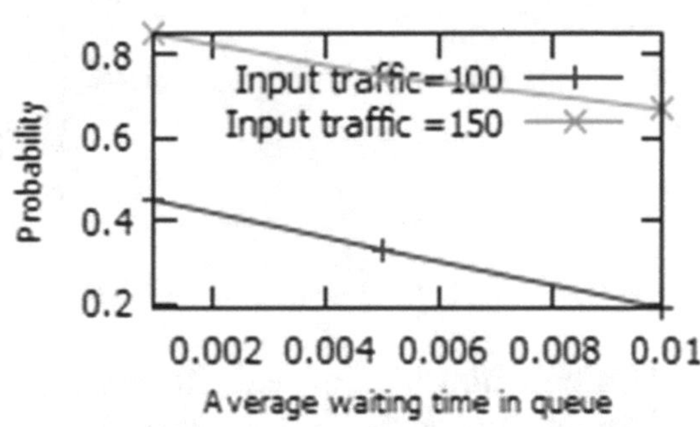

Fig. 3 Probability of packet waiting delay in buffer of node (Arrival rate = 0.005 s)

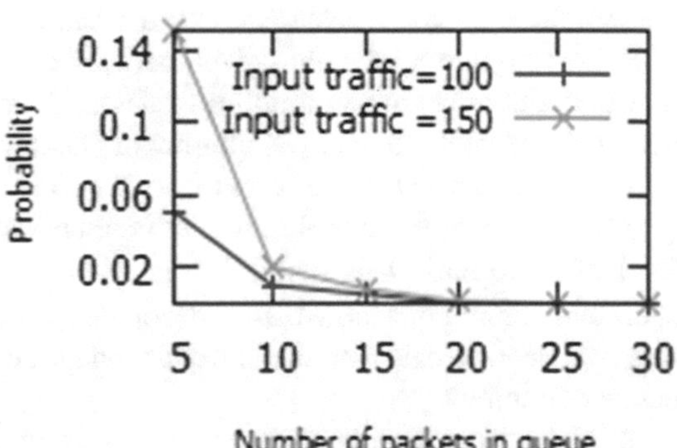

Fig. 4 Probability of queue length of buffer for (Arrival rate = 0.005 s)

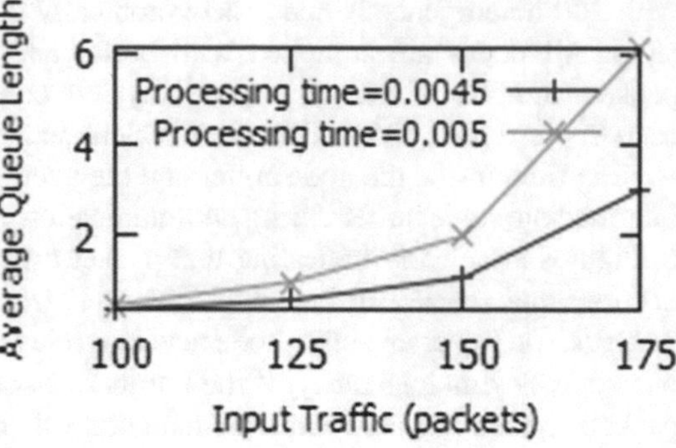

Fig. 5 Probability of packet waiting delay in buffer of node (Arrival rate = 0.004 s)

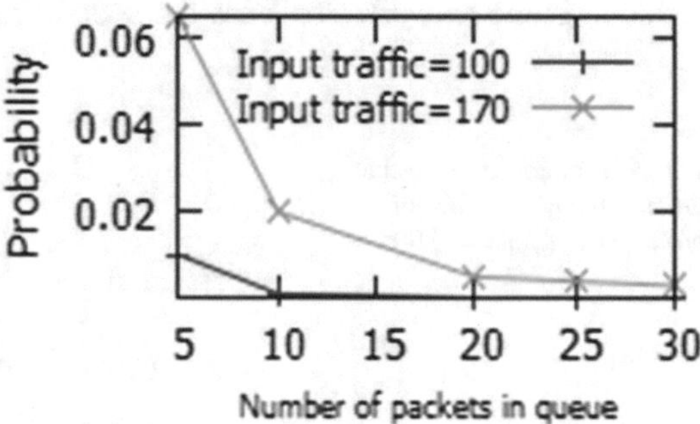

Fig. 6 Probability of queue length of buffer for (Arrival rate = 0.004 s)

3 Calculation of Current Residual Condition of Nodes Regarding with Energy and Buffer

In order to overcome the problem of bottleneck node and packet loss in a network, we introduce the route calculation metric called current residual condition of node (CR).To compute '*CR*' within given network conditions, we consider the multi-hop network with the number of mobile nodes equipped with battery capacity (E) joules and buffer capacity to hold (B) number of packets. We consider the communication in network occurring in the form of packets. For instance, consider the packet (P_i) $\{i = 1, 2, 3\ldots\}$ processing through an intermediate node $(\mathrm{I_n})$ by consuming energy $\left(\mathrm{E_p}\right)$ of node and waits inside the node buffer delay $\left(\mathrm{T_p}\right)$. In order to provide successful communication between communicating entities without packet dropping, we need to calculate the current residual condition of nodes by satisfying the below conditions.

All the communicating packets process through node at most energy (E) joules and buffer (B) bytes. Node must process as many as possible packets. Node should not process the part of the packet.

The metric current residual condition of node is computed by packet queue length and energy drain rate. Table 1 shows the notation used in our work. We consider the multi-hop MANET with multiple sources trying to transmit the packets through an active intermediate node (I_n) equipped with battery of energy (E) joules and buffer capacity of (B) bytes. The node I_n can process maximum of 'n' packets of multiple sources say $\{S_1, S_2, S_3, S_4, \ldots\}$ within its provided energy and buffer capacity. In order to achieve successful communication between communicating entities, the packets arriving at an intermediate node (I_n) must be processed by satisfying the conditions such as packets should either be completely processed or discarded but partial operation is not possible. We contemplate positive values of n topples as follows:

Maximum number of packets $\{P_1, P_2, P_3, P_4, P_5, P_6, \ldots\}$ needs to process via an intermediate node (I_n) within its maximum buffer size of (B) bytes, and each packet contains the size of (x) bytes.

Energy drain (E_d) occurs due to the process of single packet through an active intermediate node, and it is calculated by Eq. 5.

$$E_d = \left(E_r + E_p + E_t\right)/E \tag{5}$$

If 'n' number of packets process through an active intermediate node, then the energy drainage (E_{dn}) of an intermediate node (I_n) is calculated by Eq. 6.

$$E_{dn} = \left(\left(E_r + E_p + E_t\right) * n\right)/E \tag{6}$$

Time needed to process the packets from an active intermediate node is calculated by Eq. 7.

$$T_p = \frac{n_p * P_i}{T_o - T_i} \tag{7}$$

Table 1 Notation used to compute the current residual condition of the node

Notations	Description
(E)	Battery capacity, in joules
(E_d)	Energy drain
(I_n)	Intermediate node
T_p	Time to process one packet from an intermediate node
(CR)	Current residual condition
P_i	Processing packets $\{i = 1, 2 \ldots\}$
CR_{max}	Maximum threshold value
CR_{min}	Minimum threshold value
$L[0 \ldots. n, \quad 0 \ldots. E_d]$	Maximum packets as a subset of processing packets P_i $\{i = 1, 2 \ldots\}$ of energy drain at most E_d
$M[0 \ldots. n, \quad 0 \ldots. T_p]$	will operate maximum packets as a subset of processing packets and process time at most T_p

where n_p is the packet holding capacity of buffer, $(T_o - T_i)$ packet delay inside the buffer, and P_i processing packets $\{i = 1, 2 \ldots\}$. Energy drain and processing time required to processed packet via active intermediate node are given as E_d and T_p.

For instance, we need to find out the energy drain and processing time of packet of size $K \epsilon \{P_i\}$ to maximize $\sum P_i$, where $i \epsilon K$ with respect to $T_p(P_i) \leq T_P$ and $E_d(P_i) \leq E_d$.

In order to obtain the current residual condition of node with regarding energy and buffer capacity, one convenient way is to attempt all feasible process subject to K. We obtain it with the help of knapsack mechanism, which provides two multi-dimensional arrays with entries as follows:

$$L[0 \ldots n, 0 \ldots E_d] \quad \forall \quad 1 \leq P_i \leq n \, and \, 0 \leq E_d(P_i) \leq E_d$$

$$M\left[0 \ldots . n, 0 \ldots . T_p\right] \quad \forall \quad 1 \leq P_i \leq n \, and \, 0 \leq T_p(P_i) \leq E_d$$

Such that V[i, E] and V[i, B] will operate maximum packets as a subset of processing packets P_i $\{i = 1, 2 \ldots\}$ of energy drain and process time at most E_d and T_p. Array entries of V[i, E] and V[i, B] consist of maximum packets to process through provided active intermediate node. Array entries must avoid the following conditions:

$$L[0, E_d] = 0 \quad \forall 0 \leq E_d(P_i) \leq E_d \quad \text{and} \, M\left[0, T_p\right] = 0 \quad \forall \quad 0 \leq T_p(P_i) \leq E_d$$

$L[0, E_d] = -\infty \quad \forall \quad E_d(P_i) \leq 0,$ not acceptable and $M\left[0, T_p\right] = -\infty \quad \forall \quad T_p(P_i) \leq 0$, not acceptable.

Calculation of current residual condition of nodes with regarding energy and buffer is as follows:

$$\mathrm{L}[\mathrm{i}, E_d(P_i)] = \max(\mathrm{L}[\mathrm{i}-1, E_d(P_i)], \mathrm{L}_\mathrm{i} + \mathrm{L}[\mathrm{i}-1, E_d(P_i) - E_d(P_{i+1})]) \forall \, \text{and} \, 0 \leq E_d(P_i) \leq E_d$$

$$\mathrm{M}\left[\mathrm{i}, T_p(P_i)\right] = \max\left(\mathrm{M}\left[\mathrm{i}-1, T_p(P_i)\right], \mathrm{M}_\mathrm{i} + \mathrm{L}\left[\mathrm{i}-1, T_p(P_i) - T_p(P_{i+1})\right]\right) \forall \, \text{and} \, 0 \leq T_p(P_i) \leq T_p$$

To calculate desirable subset knapsack mechanism, [19] appends a Boolean auxiliary array Keep $[\mathrm{i}, E_d(P_i)]$ and Keep $\left[\mathrm{i}, T_p(P_i)\right]$, respectively. This Boolean array becomes one if active intermediate node processes the i_{th} packet from communication array Keep $[\mathrm{i}, E_d(P_i)]$ and Keep $\left[\mathrm{i}, T_p(P_i)\right]$; otherwise, it becomes zero. The algorithm to compute current residual condition of node is provided in algorithm 1.

4 Proposed Model

Computed current residual condition of node regarding energy and buffer acts as a route selection metric between communication entities. Every node present in a network must calculate its current residual condition whenever it becomes an active intermediate node between communicating entities with the help of algorithm 1. This metric is reactively maintained by the intermediate node same as existing AODV routing mechanism hop count value. We are considering threshold values as CR_{max} in idle conditions and CR_{min} in least conditions; node actual CR value is compared with threshold value to decide whether the node needs to participate in routing path or not by the following conditions.

If nodes computed current residual condition 'CR' is higher or equal to CR_{max}, then node gains higher priority and allows to participate in routing path.

If nodes computed current residual condition 'CR' is in between CR_{max} and CR_{Min}, then node gains medium priority and are used for backup of higher priority nodes and allows to participate in routing path.

If nodes computed current residual condition 'CR' is lower or equal to CR_{min}, then node gains lower priority and do not allow to participate in routing path.

Our routing protocol is based on AODV reactive routing protocol, but instead of metric hop count, we are using the metric current residual condition of nodes (CR).

Algorithm 1

Knapsack(l, E_d (Pi),n,E_d){
for (E_d (Pi) = 0 to Ed) S[0, E_d (Pi)] = 0;
for (i = 1 to n)
for (E_d (Pi) = 0 to Ed)
If ((E_d (Pi) $\leq$ Ed)and(l[i] + L[i-1,E_d- E_d (Pi)] > L[i-1, E_d (Pi)])){
L[i, E_d (Pi)] = l[i] + L[i-1, E_d- E_d (Pi)];
Keep [i, E_d (Pi)] = 1;}
else V [i, E_d (Pi)] = L [i-1, E_d (Pi)];
Keep [i, E_d (Pi)] = 0;}
K = E_d;
for (I = n down to 1)
If (keep [i, K] == 1) {
Output i;
K = K- E_d (Pi);
Return L[n, E_d];

5 Performance Evaluation

Performance evaluation of our proposed algorithm has been done with the help of NS-2 simulator by necessary extensions. We compare the performance of our proposed model with the existing energy-aware routing protocol [9, 20] in same network environment. Simulation parameters for performance evaluation are same as in Sect. 2.2. Communicating entity initiates its communication by CBR traffic.

We generated random traffic session between communicating entities with constant packet size of 512 bytes. Whenever node becomes an intermediate node, it must run our algorithm to calculate its current residual condition. This computed value decides whether node needs to participate in communication or not.

Our major objective is to examine the node's current residual condition regarding with energy and buffer, and based on computed value, node gets priority to participate in routing path or not. We further computed the lifetime of network, reliability of link with respect to packet delivery ratio and compared it with underlying energy-based routing protocols E-AODV [20], MRPC [9]. We considered lifetime of network as the time for first node in a network to exhaust its energy. Network lifetime greatly impacts on the network performance, majorly in MANET as it is peer-to-peer network, and communication is relaid on other nodes. Performance results presented in Figs. 7, 8, and 9 clearly demonstrated that proposed work performed better in comparison with existing protocols E-AODV and MRPC.

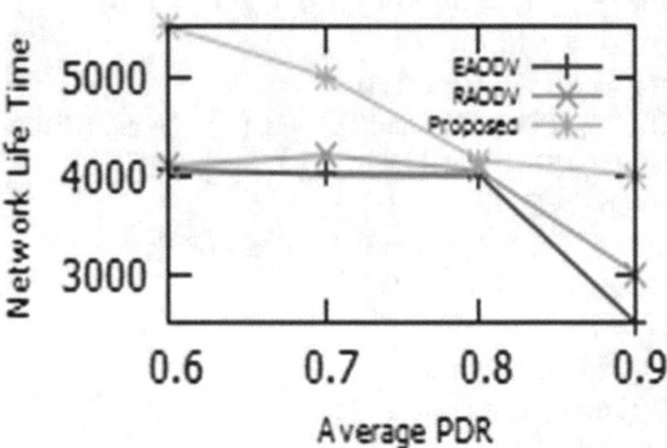

Fig. 7 Network lifetime versus PDR

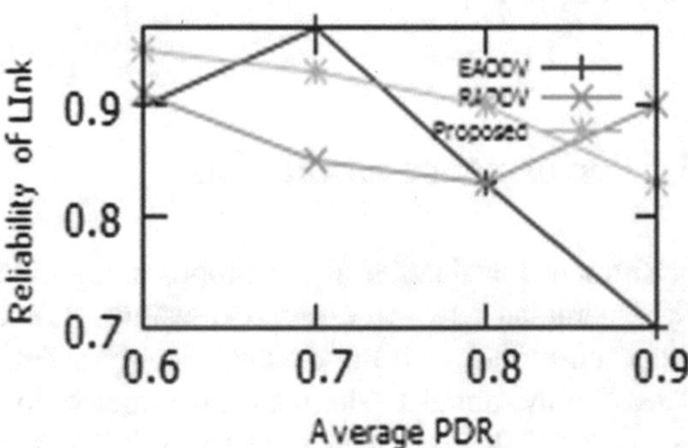

Fig. 8 Route reliability versus PDR

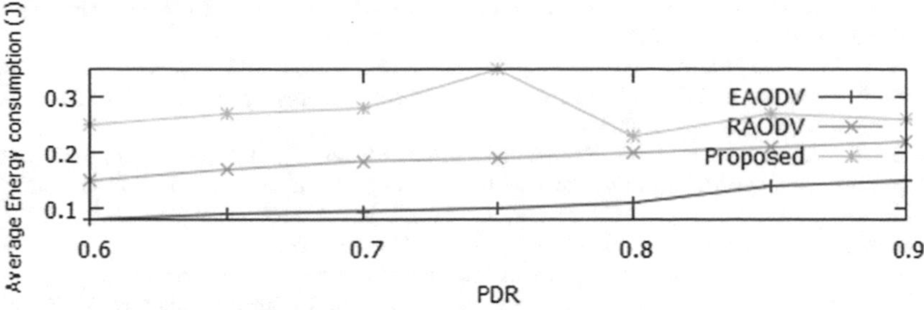

Fig. 9 Average energy consumption per route versus packet delivery ratio

6 Conclusion

Routing protocols design based on residual energy and energy consumption per route causes the intermediate node to bottleneck. Performance of network is greatly affected by bottleneck intermediate node. In this work, we designed an energy-efficient routing protocol to mitigate the node to become bottleneck by novel metric called 'current residual condition' of node regarding energy and buffer. Moreover, our proposed protocol extended the network lifetime and saved the energy of nodes. This metric provides the method to observe the present status of intermediate nodes in the network, so it could be used as proactive route calculation, and it is our future work.

References

1. Gautam, S., Kumar, R.: A Review of Energy-Aware Routing Protocols in MANETs. Int. J. Mod. Eng. Res. (IJMER) **2**(3), 1129–1133 (2012)
2. Mohammad, A.A.K., Mirza, A., Razzak, M.A.: Reactive energy aware routing selection based on knapsack algorithm (RER-SK). In: Emerging ICT for Bridging the Future-Proceedings of the 49th Annual Convention of the Computer Society of India CSI, vol. 2, pp. 289–298. Springer, Cham (2015)
3. Mohanoor, A.B., Radhakrishnan, S., Sarangan, V.: Online energy aware routing in wireless networks. Ad Hoc Netw. **7**(5), 918–931 (2009)
4. Vergados, D.J., Pantazis, N.A., Vergados, D.D.: Energy-efficient route selection strategies for wireless sensor networks. Mob. Netw. Appl. **13**(3–4), 285–296 (2008)
5. Nagy, A.S., El-Kadi, A.A., Mikhail, M.N.: Swarm congestion & power aware routing protocol for MANETs. In: 2008. CNSR 2008. 6th Annual Communication Networks and Services Research Conference, pp. 517–525. IEEE, May 2008
6. Li, X.Y., Chen, H., Shu, Y., Chu, X., Wu, Y.W.: Energy efficient routing with unreliable links in wireless networks. In: 2006 IEEE International Conference on Mobile Adhoc and Sensor Systems (MASS), pp. 160–169. IEEE, Oct 2006
7. Li, X.Y., Wang, Y., Chen, H., Chu, X., Wu, Y., Qi, Y.: Reliable and energy-efficient routing for static wireless ad hoc networks with unreliable links. IEEE Trans. Parallel Distrib. Syst. **20** (10), 1408–1421 (2009)

8. Park, J., Sahni, S.: Maximum Lifetime Routing in Wireless Sensor Networks. University of Florida, Computer & Information Science & Engineering (2005)
9. Kim, D., Garcia-Luna-Aceves, J.J., Obraczka, K., Cano, J.C., Manzoni, P.: Routing mechanisms for mobile ad hoc networks based on the energy drain rate. IEEE Trans. Mob. Comput. **2**(2), 161–173 (2003)
10. Gomez, J., Campbell, A.T., Naghshineh, M., Bisdikian, C.: PARO: supporting dynamic power controlled routing in wireless ad hoc networks. Wirel. Netw. **9**(5), 443–460 (2003)
11. De Couto, D.S., Aguayo, D., Bicket, J., Morris, R.: A high-throughput path metric for multi-hop wireless routing. Wirel. Netw. **11**(4), 419–434 (2005)
12. Misra, A., Banerjee, S.: MRPC: Maximizing network lifetime for reliable routing in wireless environments. In: 2002 Wireless Communications and Networking Conference, WCNC2002, vol. 2, pp. 800–806. IEEE, Mar 2002
13. Banerjee, S., Misra, A.: Minimum energy paths for reliable communication in multi-hop wireless networks. In: Proceedings of the 3rd ACM International Symposium on Mobile ad hoc Networking & Computing, pp. 146–156. ACM, June 2002
14. Singh, S., Raghavendra, C.S.: PAMAS—power aware multi-access protocol with signalling for ad hoc networks. ACM SIGCOMM Comput. Commun. Rev. **28**(3), 5–26 (1998)
15. Toh, C.K.: Maximum battery life routing to support ubiquitous mobile computing in wireless ad hoc networks. IEEE Commun. Mag. **39**(6), 138–147 (2001)
16. Singh, S., Woo, M., Raghavendra, C.S.: Power-aware routing in mobile ad hoc networks. In: Proceedings of the 4th annual ACM/IEEE International Conference on Mobile Computing and Networking, pp. 181–190. ACM, Oct 1998
17. Maleki, M., Dantu, K., & Pedram, M.: Power-aware source routing protocol for mobile ad hoc networks. In: Proceedings of the 2002 International Symposium on Low Power Electronics and Design, ISLPED'02, pp. 72–75. IEEE (2002)
18. Vazifehdan, J., Prasad, R.V., Niemegeers, I.: Energy-efficient reliable routing considering residual energy in wireless ad hoc networks. IEEE Trans. Mob. Comput. **13**(2), 434–447 (2014)
19. Martello, S., Toth, P.: Knapsack Problems. (1990) (catalog.enu.kz)
20. Karadge, P.S., Sankpal, S.V.: A performance comparison of energy efficient AODV protocols in mobile ad hoc networks. Int. J. Adv. Res. Comput. Commun. Eng. **2**(1), 2278–1021 (2013)

Cluster Head Selection by Optimized Ability to Restrict Packet Drop in Wireless Sensor Networks

Amairullah Khan Lodhi and Syed Abdul Sattar

Abstract Wireless sensor networks (WSNs) are comprised of spatially distributed autonomous nodes attached to the sensors to detect and maintain physical and environmental states. Energy efficiency is an important challenge in wireless sensor networks, in which the batteries are equipped with these sensors with limited amount of power and act as a power source, which having limited storage capacities. Thus, energy efficient routing techniques are required incorporate operations of wireless sensor networks to provide the network connectivity and routing of data with less energy consumption. Clustering in WSNs is greatest widespread mechanism in routing processes. Existing energy efficient clustering algorithms selects the cluster head based on energy status. However, these protocols cause the cluster head to become bottleneck and drops the packets due to insufficient buffer. Thus in this work, we propose a novel efficient metric to select the cluster head known as "optimized ability to restrict packet drop" to enhance the network lifetime. This metric provides the status of nodes with respect to energy and memory. Calculation of residual status of an intermediate node is done by knapsack algorithm. Performance of proposed work is analyzed by NS2, and the results show that our work outperforms in comparison with existing protocols.

Keywords Energy efficiency · Cluster head · Wireless sensor network Routing

A. K. Lodhi (✉)
Vignan's Foundation for Science Technology & Research, Guntur 522213, AP, India
e-mail: lak_resumes@yahoo.co.in

S. A. Sattar
NSAKCET, Hyderabad, India
e-mail: syedabdulsattar1965@gmail.com

J. Nayak et al. (eds.), *Soft Computing in Data Analytics*,
Advances in Intelligent Systems and Computing 758,
https://doi.org/10.1007/978-981-13-0514-6_45

1 Introduction

Wireless sensor networks are emerging as a powerful requirement in wireless technology [1]. These sensor networks expected to acquire thousands of nodes with sufficient sensing capabilities and limited communicating and computational energy allowing utilizing a sensor network in large scale. WSNs consist of smaller elements to monitor natural and physical states like temperature, humidity, pressure, motion, pollutants. Such networks can be deployed widely in various environments for civil, commercial, and military applications such as vehicle tracking, medical, surveillance, acoustic data gathering, climate, and habitat monitoring. Power, processing, and storage are the key drawbacks of WSN technology. Due to these limitations and specific sensor node architecture, there is an increasing necessity in secure and energy efficient communication protocols. An advanced technology like MEMS combined with low-cost, low-power DSPs and RF devices accelerates its viability of these inexpensive sensor networks. These sensors are comprised of radio transceivers, actual sensors, microcontrollers, and power supply. The sensing circuits collectively compute the ambient states associated with the environment surrounding the sensor and alter them into an electrical signal. Managing such a signal exhibits some properties about objects located and/or events happening in the sensor area. A sensor collects such data and transmits either directly or through a gateway toward a central node usually via radio transmitter.

Generally, these nodes are distributed throughout the space which needs to be monitored. They organize themselves in such a manner that they can collaborate easily with each other to complete the usual tasks through wireless communication medium. Wireless sensor networks [2] are provided with essential features such as dynamic network topology, self-organizing capabilities, node mobility, failure of a node, multi-hop/single hop routing, and short-range broadcast communication and large-scale deployment. Wireless sensors are flexible and scalable and hence are powerful to be deployed in many applications. These networks can organize themselves and do not require any fixed infrastructure to communicate, which leads them to be arranged in an ad hoc manner in hazardous locations and remote area without any existing physical shape. Therefore, computational node and communicating protocol operations must be energy efficient. Since the energy consumed during data transmission takes 70% of the total energy required, data transmission protocols are more important among these protocols.

In this work, we provide a comprehensive survey on energy efficient WSN protocols and propose a cluster head selection metric "optimized ability to restrict packet drop." Proposed work minimizes the energy consumption and increases the lifetime of wireless sensor networks by considering optimized ability to restrict packet drop with respect to energy and buffer. The remaining part of the paper is organized as follows. Second section provides the related work which presents survey of several routing protocols, Sect. 3 gives proposed work which explains the novel metric calculation algorithm in WSNs, Sect. 4 gives cluster head selection, Sect. 5 gives performance analysis of WSN routing protocols. Finally, Sect. 6 presents conclusion of this paper.

2 Literature Survey

Various researchers are adopting the methods to reduce energy consumption by studying the cause which effects energy consumption. Hence, till now various papers are being analyzed to minimize the energy requirement of these types of wireless sensor networks. Some of these are specified in this paper to consider both the consumption of energy and security of the protocol too. Since energy and security of nodes are always required in wireless sensor networks, various attacks are the main threat which can steal or alter the confidential information and can be destructive in emergency situations. We considered recent papers familiar to the latest issues in WSN technologies involving energy and security.

A genetic algorithm provides number of optimized cluster heads, due to which there are possibilities of local minimization. It can be improved further by stimulated annealing thereby resulting in global minimization. With the help of genetic algorithm, Roslin S. E. [1] created a hierarchical network for WSN. This network administers the network topology without affecting its characteristics. The network performance impact is studied for networks with various densities. A quantitative analysis of the methodologies proposed on an N-Tier with various densities proves that the two-tier architecture provides reduced consumption of energy in comparison with other architectures. Hence, a two-tier sensor network developed with GA can be utilized in any hazardous regions. Based on AHYMN approach, A. S. Uma Maheshwari [2] utilized a new method and represented a generic algorithm to form a sink node dynamically in sensor networks. Hence, it is more accurate and quicker to recognize the nodes with maximum node energy and to choose the cluster heads. The benefits of heterogeneous nodes are the longer network lifetime, network reliability improvement, and it decreases data transfer delay.

For effective selection of cluster heads and to improve HBO algorithm, Kiranpreet Kaur [3] proposed a cluster head selection protocol known as EDC-HBO. It considers the distance factor and energy as a parameter to enhance sink node selection. This is used to improve the lifetime and to reduce the power consumption as well. Results show that EDC-HBO is more energy efficient as compare to other protocols. As WSN is data redundant, our future work will be over designing and realizing routing protocol with optimal data aggregation. R. Aiyshwariya [4] proposed the hierarchical fizzy integral over the scheme in order to make energy efficiency to be a single entity to decide the cluster heads, which is the key innovation and advancement of classical protocols. Moreover, the new scheme effectively eliminates the unwanted data so that both the energy consumption and traffic are reduced. In terms of cost, this FMPDM is more beneficial than existing protocols. The simulation results show that the energy efficiency and network lifetime of FAHP are superior to the other classical protocols. When compared to FMPDM, it improves decision-making consistency and also increases the parameters of energy, throughput and lifetime efficiently thereby increasing robustness.

For improving LEACH by utilizing equally distributed cluster and reducing unequal cluster topology. Bin Deni Raj [5] introduced a new algorithm based on clustering for wireless sensor network with maximum lifetime known as EDR-LEACH. The new network protocol can be established upon the shortcoming of LEACH to test and amend them. Result shows it increases lifetime considerably. Nabil Ali Alrajeh [6] studied that our protocol cannot be transferred from wired networks to wireless environments; Because WSNs have special liability that does not obtain in wired networks. Hence, in this paper, some of the protocols are used to describe with some methods to determine where the packets can be checked. This includes new theoretical approach which observes that the author is able to enhance energy through energy saving sufficiently for resource-constrained and large networks.

3 Proposed Work

Selecting suitable cluster head in a WSN is a challenging task due to its characteristics and constrained properties. Cluster head should be effective in terms of energy and memory, as limited battery of node is major considerable issue. Available energy of node should be used to complete communication task as much as possible to prolong the network lifetime. In the literature, number of protocols developed to solve the issue of cluster head selection based on energy. However, these protocols cause the cluster node to drop the packets due to insufficient buffer. This situation causes the tremendous negative impact on network performance with respect to throughput and also causes the congestion in network. We argue that energy and memory both parameters are important to decide the node to become cluster head. In this paper, we design a mechanism to select cluster head based on the two parameters, i.e., energy and memory of node. We name the cluster selection metric as "*optimized ability to restrict packet drop*" with respect to memory and energy. This metric help to predict the node ability to restrict the packet drop from a node, and which interns help to prevent the packet loss, congestion, and also improve network lifetime. Without loss of generality, we are considering following conditions to compute the metric.

Every node in a network equipped with battery of "E" joules of energy to process 'P_E' packets and buffer 'B' capacity to hold the 'P_B' number of packets. During time interval 't', let one packet let 'i' through the node, it consumes finite amount of energy of node and utilizes the buffer size. Then the node parameters, remaining energy of node E_r and residual buffer of node B_r become as follows in Eqs. (1) and (2),

$$E_r = P_E - P_{ei}, P_E > P_{ei} \tag{1}$$

$$B_r = P_B - P_{bi}, P_B > P_{bi} \tag{2}$$

In what extent the node drop the packet is nothing but, the capacity of node to process the packets without dropping in an available energy and memory. With help of knapsack algorithm, we computed it by satisfying below conditions

1. Packets must be process within the given residual energy and memory of node.
2. Maximum packets must be processed by node.
3. Part of packet must not processed by node, and this condition will consider as illegal.

The metric "*optimized ability to restrict packet drop*" is calculated by residual energy and memory of intended node. The notation we are considered in our work shown in Table 1. We assume that the node N_i receives the packets from multiple sources within its residual energy and memory E_r and B_r, respectively. The nodes ability to restrict packet drop are up to P_n number of packets. In order to successful process of n packets from a given node without drop, the node must satisfy the below conditions

1. Packet must process completely from a node.
2. Partial process must not be happen, but packet can drop from an node completely.

We contemplate positive values '*n*' of topples as follows to compute the metric "optimized *ability to restrict packet drop*."

1. Let the ability to restrict the packet drop of a node is up to $\{P_1, P_2, P_3, P_3, P_4, \ldots, P_n\}$, every packet consumes P_{ei} of energy and P_{bi} buffer of node to process.
2. We need to calculate the capacity of node to restrict the packet drop in the manner of energy and memory, and also, we need to compute the maximum capacity of the node to process number of packets without dropping the packets.

Table 1 Notation used in our work

Notations	Description
E	Battery energy in joules
P_E	Number of packets for node
P_B	Number of packets for buffer
T	Time interval
E_r	Remaining/residual energy of node
B_r	Remaining/residual buffer of node
Pei	Individual energy node packet
Pbi	Individual buffer packet
ORmax	Maximum threshold value
ORmin	Minimum threshold value
B	Buffer of sensor node
P_n	Number of packets

For instance, we need to find out the capacity of node to process maximum packet of count $K\epsilon\{P_n\}$ to maximize the packets $\sum P_n$, where $n\,\epsilon\,K$ with respect to $E_r(P_n) \leq E$ and $B_r(P_n) \leq B$.

In order to compute the metric "*optimized ability to restrict packet drop*" with respect to buffer and memory, one expedient method is to attempt all the achievable processes subject to *K*. We computed this with the help of knapsack algorithm, which give two-dimensional array with following entries in Eqs. (3) and (4)

$$J[0\ldots n, 0\ldots E]\forall 1 \leq P_n \leq n\, and\, 0 \leq E_r(P_n) \leq E \tag{3}$$

$$L[0\ldots n, 0\ldots B]\forall 1 \leq P_n \leq n\, and\, 0 \leq B_r(P_n) \leq B \tag{4}$$

Such that J[i, E] and L[i, B] are the maximum packet processing capacity of node in a residual energy and memory without packet loss. The entries must not have the following conditions

$$J[0, E] = 0\,\forall\, 0 \leq E_r(P_n) \leq E \text{ and } L[0, B] = 0\,\forall\, 0 \leq B_r(P_n) \leq B$$

$$J[0, E] = -\infty \forall E_r(P_n) \leq 0, \text{ not acceptable and}$$

$$L\left[0, T_p\right] = -\infty \forall B_r(P_n) \leq 0, \text{ not acceptable}$$

To compute "*optimized ability to restrict packet drop*" *with respect to energy and memory is as follows*

$$J[\mathrm{i}, E] = \max(\mathrm{J}[\mathrm{i}-1, E_r(P_n)], \mathrm{L_i} + \mathrm{L}[\mathrm{i}-1, E_r(P_n) - E_r(P_{n+1})])\forall 1 \leq i \leq n \text{ and } 0 \leq E_r(P_n) \leq E \tag{5}$$

$$\mathrm{L}[\mathrm{i}, B] = \max(\mathrm{M}[\mathrm{i}-1, B_r(P_n)], \mathrm{M_i} + \mathrm{L}[\mathrm{i}-1, B_r(P_n) - B_r(P_{n+1})])\forall 1 \leq i \leq n \text{ and } 0 \leq B_r(P_n) \leq B \tag{6}$$

To find the desirable solution from the subset of knapsack algorithm [7, 8], we have to extend the auxiliary array $Keep[\mathrm{i}, E_r(P_n)]$ and $Keep[\mathrm{i}, B_r(P_n)]$ correspondingly. This Boolean array becomes one if and only if the node processing the *i*th packet does not drop; otherwise, it becomes zero.

4 Cluster Head Selection

Evaluated metric "*optimized ability to restrict packet drop*" with respect to energy and memory is used for cluster head selection. In the network initialization stage, every node present in a network must compute their "*optimized ability to restrict packet drop*" metric and also, it calculates reactively to obtain node status. We keep one threshold value to decide the cluster head known as OR_{max} in idle condition and OR_{min} in least conditions. Computed *OR* value is greater than the maximum threshold value, and then the node can consider for cluster head. Cluster head selection has done by the conditions given below.

1. Node becomes cluster head if evaluated "*optimized ability to restrict packet drop*" *OR* is higher than OR_{max}.
2. Node becomes as backup cluster head if evaluated value of "*optimized ability to restrict packet drop*" *OR* lies between OR_{max} and OR_{Min}.
3. Node cannot considered for cluster head if evaluated "*optimized ability to restrict packet drop*" *OR* is less than OR_{max}.

5 Performance Evaluation

Performance evaluation of proposed mechanism is carried out by NS-2 [9] with appropriate extension in existing libraries and compared with existing cluster head selection algorithms [2, 4] in identical network environment. In our simulation, we considered the communication distance of each node is about 250 m and single hop transmission delay is 10 ms. All sensor nodes arranged across the 1000 × 1000 square area. All nodes are equipped with battery of 10 J energy, IEEE 802.11 e network interface card, and memory. Send data is transmitted in the form of packets of size 512 bytes with consumption of transmission power is 600 mW, the receiving power is 300 mW of node. Each sensor node should run our cluster head selection mechanism and cluster head broadcast its beacons over two hops every 20 s.

Main intention toward our work is to determine the node's *optimized ability to restrict packet drop* within residual buffer and memory, and rendering to it decides the cluster head. We therefore evaluate the performance of network with respect to throughput, packet loss, and network lifetime and compare it with the existing-1 [2], existing-2 [4]. Figures 1, 2, and 3 clearly shows that our proposed mechanism is enhance the network performance and extend the network lifetime.

6 Conclusion

Clustering in WSNs is greatest widespread mechanism in routing processes. Cluster head selection is done based on only one parameter energy results in packet loss due to insufficient buffer. In this work, we consider two parameters energy and buffer to select the cluster head in network. We introduce the novel metric to select the cluster head called "optimized ability to restrict packet drop," which indicates the status of node regarding its status of buffer and energy. Moreover, our proposed

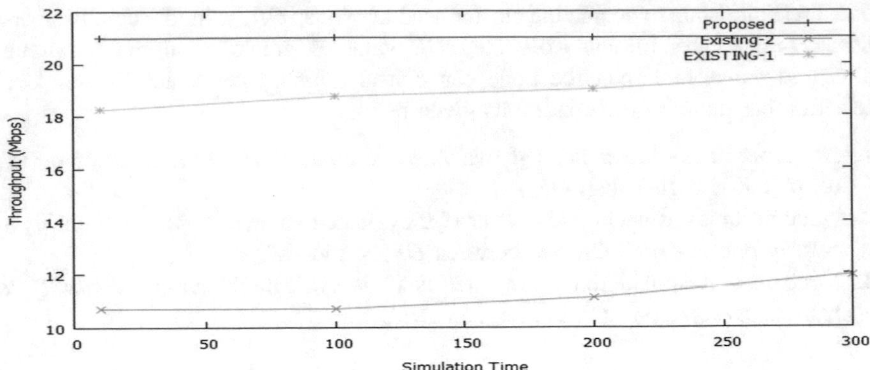

Fig. 1 Throughput versus simulation time comparison of proposed and existing works

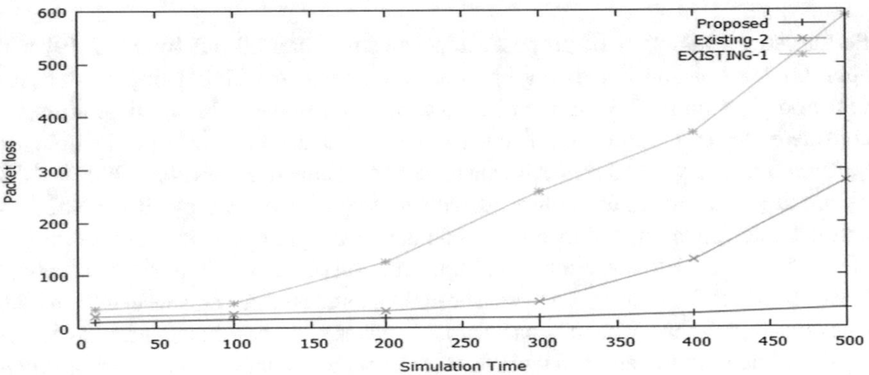

Fig. 2 Packet loss versus simulation time comparison of proposed and existing works

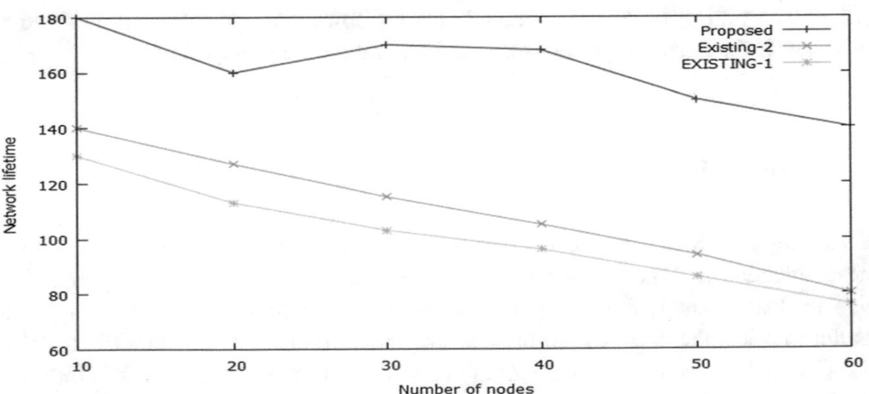

Fig. 3 Network lifetime versus simulation time comparison of proposed and existing works

protocol extended the network lifetime and save the energy of nodes. This metric provides the method to observe the present status of nodes in network, so it could be used as proactive cluster head selection and it is our future work.

References

1. Roslin, S.E.: Genetic algorithm based cluster head optimization using topology control for hazardous environment using WSN. In: 2015 International Conference on Innovations in Information, Embedded and Communication Systems (ICIIECS), 19 Mar 2015, pp. 1–7. IEEE
2. Maheswari, A.U., Pushpalatha, S.: Dynamic Cluster Head Selection based on Genetic Algorithm using Routing Approaches in WSN
3. Kaur, K., Singh, H.: Cluster head selection using honey bee optimization in wireless sensor network. Int. J. Adv. Res. Comput. Commun. Eng. **4**(5), 716–721 (2015)
4. Devi, R.A., Buvana, M.: Energy efficient cluster head selection scheme based on FMPDM for MANETs. Int. J. Innov. Res. Sci. Eng. Technol. **3** (2014)
5. Raj, E.D.: An efficient cluster head selection algorithm for wireless sensor networks-Edrleach. IOSR J. Comput. Eng. **2**(2), 39–44 (2012)
6. Alrajeh, N.A., Khan, S., Shams, B.: Intrusion detection systems in wireless sensor networks: a review. Int. J. Distrib. Sens. Netw. **9**(5), 167575 (2013)
7. Mohammad, A.A., Mirza, A., Razzak, M.A.: Reactive energy aware routing selection based on knapsack algorithm (RER-SK). In: Emerging ICT for bridging the future-proceedings of the 49th annual convention of the computer society of india CSI, vol. 2, pp. 289–298. Springer, Cham (2015)
8. Armstrong, R.D., Kung, D.S., Sinha, P., Zoltners, A.A.: A computational study of a multiple-choice knapsack algorithm. ACM Trans. Math. Softw. (TOMS) **9**(2), 184–198 (1983)
9. Issariyakul, T, Hossain, E.: Introduction to network simulator NS2. Springer Science & Business Media, 2 Dec 2011

FDI and Industry in Developed and Developing Countries: A Comparative Dynamic Panel Analysis

Madhabendra Sinha and Partha Pratim Sengupta

Abstract The paper investigates the sector-specific dynamic impacts of FDI inflow on industrial productivity on a comparative basis between developed and developing countries. Different studies observe both positive and negative impacts of FDI on economic growth. However, the existing evidences have not cleared the sector-wise impact of FDI, particularly in industrial sector separately in developing and developed economies. We explore and compare the dynamic relationship between FDI and industrial performances in two groups of countries using data from UNCTAD and World Bank over the period of 1970–2015 in the dynamic panel framework. The stochastic properties of variables are looked into by carrying out various panel unit root tests followed by the GMM estimation. Empirical findings imply that FDI inflows significantly promote industrial growth in both groups of countries; however, the relationship is bi-directional in developing countries and the degree of influence is higher in developed nations.

Keywords FDI · Industry · Developed countries · Developing countries Dynamic panel model · GMM estimation

1 Introduction

The study analyzes and explores the dynamic impact of inflows of foreign direct investment (FDI) on industrial production on a comparative basis between the groups of developing and developed economies. In the economic growth process the role of FDI has been a burning and one of the most debatable issues nowadays,

---

M. Sinha (✉) · P. P. Sengupta
Department of Humanities and Social Sciences,
National Institute of Technology Durgapur, Durgapur 713209, West Bengal, India
e-mail: madhabendras@gmail.com

J. Nayak et al. (eds.), *Soft Computing in Data Analytics*,
Advances in Intelligent Systems and Computing 758,
https://doi.org/10.1007/978-981-13-0514-6_46

particularly in transitional developing nations. Low base of technology of production is identified as a crucial factor impinging upon growth of the nations. In this regard inflows of FDI can alleviate the constraints to economic growth to some extents. FDI carries foreign capital with modern technologies, advanced managerial techniques and organizational structures as viewed by Prakash and Balakrishnan [14]. The current phase of globalization experiences the service led FDI across the world due to considerably higher return from investments in tertiary sector as compared to primary and secondary sectors. However, it is also observed that FDI flows to services sector are more volatile than other sectors and that is why trends of FDI inflows in developed countries are more erratic as maximum portion of FDI inflows to those countries is captured by services sector [16]. Approximately two-third of global FDI flows are concentrated in the services sector, whereas manufacturing sector has accounted for 26% of total flows of FDI in 2016. Five major industries within the manufacturing sector are chemical, electronics, motor vehicles, food and beverages and petroleum products, those have been accounting for approximately 70% of total flows of FDI in the sector.

UNCTAD [17] reported that global inflows of FDI cut down by about 2% to US $1.75 trillion in 2016. Inward flows of FDI in developing economies have been turning down by around 14%, and inflows of same to poor-structured underdeveloped nations have remained low with more volatile. Whereas FDI inflows to developed nations have increased further by 5% to US $1 trillion in 2016, after a significant growth in the preceding year. Though, the scenario in contexts of developing and developed nations was almost reverse over the globe during the previous year. The decrease in FDI inflows to European countries was more than its compensation by the modest growth of the inflows of FDI to North American nations and a substantial increase in same in other developed countries. The share of FDI inflows in developed countries has grown by a significant rate in 2016. The developing as well as high-income countries have been displaying extensive disparities in the method of industry-driven economic growth. During last four decades, both developing and developed countries have experienced the rising trends in industrial production and developed countries stood higher level than developing nations. However, the trends in developing countries overtook the trends in developed nations in 2012, and FDI inflow is considered as a significant factor in this matter, where China and some other nations play a vital role in the world economy [16]. Against this backdrop present study explores the dynamic relationship between FDI inflow and growth of industrial production for two groups of countries separately on a comparative basis empirically over a long period. The rest of the paper has been organised as follows. The next section briefly documents the existing studies with the present trends in FDI inflows and industrial performances in developed as well as developing economies. The data sources and methodological issues are properly discussed before analysing the empirical findings. The final section summarizes and concludes the paper.

2 Existing Literature and Present Trends of FDI and Industrial Performance

The area of the theoretical as well as empirical studies on FDI flows and economic growth performance has been so vast whereas comparing to this, the studies on the sector-specific analysis of the impact of FDI particularly in secondary sector are very few till now. We start the reviews of theoretical evidences ([3, 13, 15] etc.) by the models of endogenous growth, which consider a particular type of fund, as a source of profit, productivity gain. The basic idea is to contribute for creating the new technological and organizational knowledge through the research and development activities and innovations of human resources capital. This creation of knowledge brought by inflows of FDI has been rewarding the impact of the supply of the reduced capital and helping the nation for maintaining a good and sustained level of economic growth. Therefore, FDI inflow may influence the economic growth in host countries significantly in a positive sense by enhancing the total factor productivity through the spread of knowledge and externalise the technologies. Caves [6], Globerman [8] and Blomström [4] had been pioneering the empirical literature by analyzing the data on Australia, Canada and Mexico respectively. Subsequently their models of reproduction have been revised and expanded; however, the basic method still remains almost similar. The measurement of productivity spillovers from the multinational enterprises includes a variable of the foreign companies' infusion of proxies, usually counting as a part of employment of factors or sales in multinationals. The model of regression allows capturing the effect of inflow of FDI on productivity of home country's firms. Although the effect of FDI inflow on the economic development in the host country has been a theoretical evaluation, as reported in belief, especially in terms of the level of experience related to the organization, and there are some limitations to oppose it. The initial empirical findings confirm the positive relationship between inflow of FDI and productivity of local firms in the host countries. In this connection, Blomström and Wolf [5] had observed a positive and significant effect of inflow of foreign capital on local productivity in Mexico over the period of 1965–1982.

By investigating the dynamic role of FDI inflow in the manufacturing productivity in China over the period of 1988–1994, Démurger and Chen [7] found a positive correlation between the growth of manufacturing productivity and the presence of multinational investment in home industries. The positive influence of inward FDI flow on total factor productivity in Chinese industries has been documented by Liu and Wang [12]. By same kind of evaluation in case of Lithuanian industries over the period of 1996–2000, Javorcik [10] reported that the vertical links as derived from the contacts between international and domestic firms generates the highest benefits. So it can be viewed that impact of inflows of FDI varies across the nations as well as regions with different perceptions and this diversified impact may found to be more significant between developing and developed countries. In this context, one explorative study should be there regarding this issue.

Trends in FDI inflows and growth of production in industrial sectors across developing and developed nations in the world have not been homogeneous over the periods as Fig. 1 shows. UNCTAD has documented diversified and frequently changing scenarios in both developing and developed nations. Developing Asia has been hampered by the 15% contraction of FDI inflows and almost the same scenarios were there in Latin America, Caribbean islands and also in Africa. Flows to least developed countries have strongly retreated and to landlocked developing countries have been showing the trends of marginal falling. However, there was a 6% shrank of FDI flows to small island developing states. Figure 2 shows trends of industrial production in developed and developing group of countries. Developing

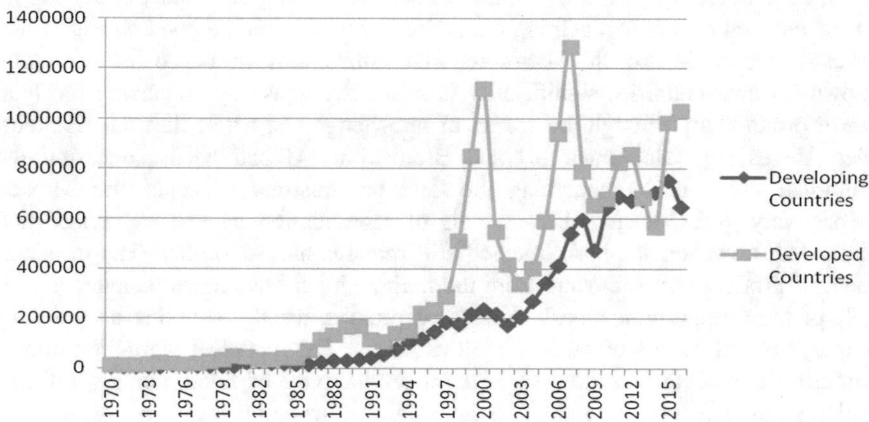

Fig. 1 Trends of FDI inflows in US $ million. *Source* UNCTAD stat data base, 2017

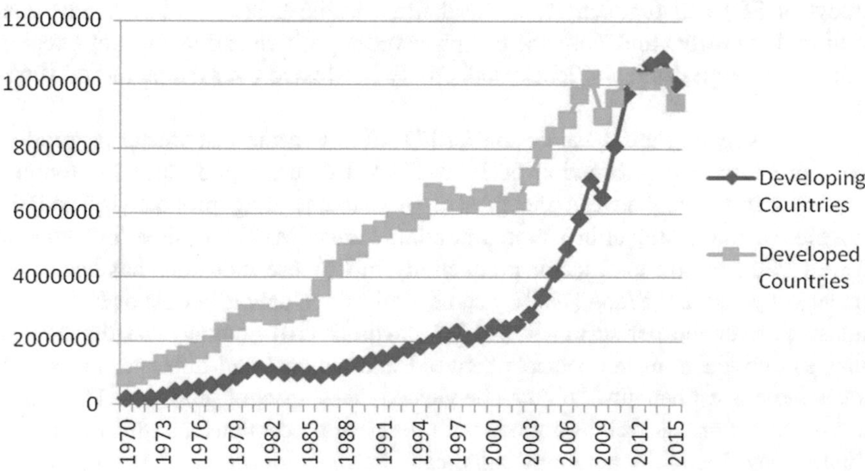

Fig. 2 Trends of industrial production in US $ million. *Source* UNCTAD stat data base, 2017

nations have experienced a sharply rising trend since 2000 onwards, and the trend for developing countries exceeded that for developed economies in 2012 and FDI should found to be an important factor in this regards in developing economies. Though the increased share of Asia was lesser than the declining shares of the same in American and European countries, the large size of the economy of developing Asia has been making its enough boosts for compensating the declining trends in other developing nations. Africa has experienced the sharply dropping share during 2000 onwards. Oceania was having the steadily decreasing share of manufacturing since the 1970s and at present its share is at the lowest level as compared to all other regions.

3 Data and Methodology

Two different panels are built up over the period of 1970–2015 for 50 developed and 54 developing countries, selected on the basis of World Bank classifications of the nations according to their income levels. UNCTAD STAT Data Base provides the country wise data on total FDI inflow (FDI) and utilizing various UNCTAD reports on the sector wise share of FDI, we calculate the data on FDI in industries. We collect the industrial production (INP) data measured by gross value addition from world development indicators (WDI) provided by World Bank. The data on gross domestic capital formation (DCF), international trade measured by export and import (TRD) and domestic demand measured by per capital GDP (PCG) are collected from WDI to control them in the regression models as the possible determinants of INP. Now to fulfil our basic objective we carry out the testing of the general null hypothesis as follows:

- *FDI inflow does not impact industrial production and the scenarios are indifferent in developing and developed countries.*

The generalized method of moments (GMM) estimation is used in dynamic panel framework for controlling the endogeneity. The panel data models are more efficient to control the heterogeneity at the individual level with the availability of more information. We employ the panel unit root tests developed by Levin et al. [11] and Im et al. [9] to check stochastic properties of variables. Augmented Dickey-Fuller (ADF) specification of panel unit root test is given in Eq. 1:

$$\Delta y_{it} = \rho y_{i,t-1} + \sum_{j=1}^{p_i} \eta_{ij} \Delta y_{i,t-j} + X_{it}^{'} \delta + \varepsilon_{it} \tag{1}$$

LLC test permits intercepts with residual variances; dynamic trends and order of autocorrelation; however, it needs the auto-generated time series with common sample size and autocorrelation coefficient (ρ). The lag order, varying across individuals, is selected on the basis of t-statistics of η_{ij} by permitting the utmost lag.

The estimate of ρ can be obtained after estimating the regression of Δy_{it} on $\Delta y_{i,t-j}$ and X_{it}. The common ρ criterion is the main weakness of the LLC test. But the IPS test considers the different ρ for all cross section units under heterogeneous panel. Arellano and Bond (1991) proposed GMM technique is the widely used estimator for fixed effect dynamic panel models, where, first we eliminate the fixed effects using first differences of the variables and then estimate the instrumental variables of the differenced equation. The validity of the instruments can be checked by Sargan test. The one period lag dynamic panel model can be presented as follows in Eq. 2:

$$y_{it} = \alpha_i + \theta_t + \beta y_{i,t-1} + x_{it}'\eta + \varepsilon_{it} \tag{2}$$

where α_i = fixed effect, θ_t = time dummy, x_{it} represents $(k - 1) \times 1$ exogenous variables' vector and $\varepsilon_{it} \sim N(0, \sigma^2)$ denotes random disturbance. Hausman test suggests that in panel data structure fixed effect model is more appropriate than random effect model. In order to eliminate the unobservable state-specific effects, we difference Eq. 2 and then it becomes:

$$\Delta y_{it} = \Delta\theta_t + \beta \Delta y_{i,t-1} + \Delta x_{it}\eta + \Delta\varepsilon_{it} \tag{3}$$

The lagged difference of the logarithm of the dependent variable is correlated with the difference of error term. To remove this kind of endogeneity in Eq. 3, instrumental-variables with lag as suggested by moment condition are to be used. The differenced components of endogenous explanatory variables should also be treated cautiously. The GMM estimation also involves for instruments specification, weighting matrix choosing and estimator determination.

4 Empirical Results

4.1 Findings from Developing Countries

To estimate the impact of FDI inflow on industrial performance, first, we employ the panel unit root tests developed by Levin et al. [11] and Im et al. [9]. The panel unit root test statistics are calculated for all underlying panels of selected developing nations. Variables' lag lengths are selected on the basis of minimum [1] information criterion (AIC) rule. Both the individual level effects and also linear trends as the exogenous variables have been incorporated in the estimated GMM equation. We find that all the variables are non-stationary at level in terms of LLC and IPS tests as Table 1 shows. But, the first differences of the variables become stationary as LLC and IPS statistics report. We use the first differenced equation of [2] prescribed GMM estimation for controlling the unobserved heterogeneities involved in the estimated model, where INP is the dependent variable and FDI is treated as the focussed independent regressor. DCF, TRD and PCG are considered

Table 1 Estimated panel unit root tests statistics

Series	LLC (2002)			IPS (2003)
	Level	First difference	Level	First difference
FDI	1.68	−6.15*	−1.65	−6.47*
INP	−1.26	−5.89*	−1.74	−5.86*
DCF	−2.03	−5.76*	−1.81	−5.41*
TRD	−1.81	−6.11*	−1.89	−6.69*
PCG	1.86	−5.98*	−1.83	−6.07*

*Represents the stationary at 5% level of significant

Source Authors' estimation using data from UNCTAD and World Bank

Table 2 GMM estimation of INP and FDI relationship in developing countries

Dependent variable: ΔINP (1, it)			
Total balanced panel observations: 2430			
Variable	Coefficient	t-Statistics	Probability
ΔINP (1, it-1)	0.17	17.23	0.0000
ΔFDI (it)	0.11	16.92	0.0001
ΔDCF (it)	0.12	11.73	0.0110
ΔTRD (it)	−0.09	13.11	0.0102
ΔPCG (it)	0.08	8.01	0.1193
J-Statistics	12.79 (0.0000)	Instrument rank	15

Source Authors' estimation using data from UNCTAD and World Bank

as the control instruments. The dependent variable in lagged form captures the dynamism of industrial performances. Table 2 represents the estimated model of the relationship between FDI and INP in the dynamic panel structure.

The J-statistic is nothing but the Sargan statistic and rank of instrument exceeds the number of coefficients in the estimated equation. Our findings refer that growth of FDI is significantly influencing the INP growth in selected developing countries. It is also observed that TRD negatively impacts INP, whereas DCF and PCG are found to be significant and insignificant determinants of INP respectively.

Various theoretical and empirical evidences refer that growth of industrial sector can also influence the FDI inflow in the home country. Thus we also estimate the GMM equation by considering FDI and INP as dependent and independent variables respectively, shown in Table 3; where INP, TRD and PCG are found to be significant determinants of FDI inflow in selected developing nations. So, it can be argued that the relationship between the growths of FDI and INP is bi-directional in case of developing countries.

Table 3 GMM estimation of FDI and INP relationship in developing countries

Dependent variable: ΔFDI (1, it)			
Total balanced panel observations: 2430			
Variable	Coefficient	t-Statistics	Probability
ΔFDI (1, it-1)	0.21	16.94	0.0000
ΔINP (it)	0.09	11.89	0.0101
ΔDCF (it)	−0.08	10.73	0.0110
ΔTRD (it)	0.12	14.16	0.0010
ΔPCG (it)	0.06	9.14	0.0129
J-Statistics	13.09 (0.0000)	Instrument rank	16

Source Authors' estimation using data from UNCTAD and World Bank

4.2 Findings from Developed Countries

Similarly in case of developed countries we employ LLC (2002) and IPS (2003) panel unit root tests. Again these are found that all the variables are non-stationary at levels but stationary at first differences as of LLC and IPS tests statistics report at Table 4.

Table 5 shows the estimated results of the [2] GMM estimation for selected developing countries considering INP and FDI as dependent and independent variables respectively; DCF, TRD and PCG as control instruments. The findings imply that growth of FDI is significantly influencing INP growth. We also observe that TRD and DCF have the significant positive impact on INP, whereas PCG is an insignificant determinant of INP. It must be noted that degree of influence of FDI on INP in developed countries is higher than that in developing economies. Based on the results of developing countries here also we employ the GMM estimation by considering FDI and INP as the dependent and independent variables respectively in case of developing countries. However, unlike the developing countries, growth of INP is not found to a significant promoter of FDI in developed countries. So the FDI and INP relationship is unidirectional in developed nations.

Table 4 Estimated statistics of panel unit root tests

Series	LLC (2002)		IPS (2003)	
	Level	First difference	Level	First difference
FDI	1.79	−5.98*	1.71	−6.16*
INP	−1.91	−6.12*	−1.69	−5.97*
DCF	−1.84	−6.08*	−2.11	−5.99*
TRD	−2.11	−6.54*	−2.06	−6.03*
PCG	−2.01	−6.21*	−1.99	−5.76*

*Represents the stationary at 5% level of significance.

Source Authors' estimation using data from UNCTAD and World Bank

Table 5 GMM estimation of INP and FDI relationship in developed countries

Dependent variable: ΔINP (1, it)			
Total balanced panel observations: 2250			
Variable	Coefficient	t-Statistics	Probability
ΔINP (1, it-1)	0.21	18.41	0.0001
ΔFDI (it)	0.19	17.88	0.0011
ΔDCF (it)	0.16	13.53	0.0121
ΔTRD (it)	0.11	14.78	0.0114
ΔPCG (it)	−0.16	7.92	0.1619
J-Statistics	14.08 (0.0000)	Instrument rank	16

Source Authors' estimation using data from UNCTAD and World Bank

5 Conclusion

The paper empirically examines dynamic impacts of FDI inflows in secondary sectors on overall production performances of industries in groups of selected developed and developing countries across the world on a comparative basis over the period of 1970–2015. The findings of the paper recommend that growth of FDI inflows has significant positive on the industrial productivity in both groups of nations and the impact of FDI of industry is greater in a positive sense in developed economies. We have also observed that international trade measured by export plus import of developing countries negatively influence the industrial growth in those countries but the result is completely reverse in case of developed economies. It must be pointed out that the FDI-industrial growth relationship is bi-directional in selected developing economies, whereas the relationship is unidirectional in the group of developed nations. Domestic demand measured by per capita GDP is found to be insignificant to promote the industrial sector in both types of countries. So our findings move against some conventional views of open economy macroeconomics and international trade.

References

1. Akaike, H.: Information theory and an extension of the maximum likelihood principle. In: Petrov, B.N., Csáki, F. (eds.) 2nd International Symposium on Information Theory, Budapest, Akadémiai Kiadó, pp. 267–281 (1973)
2. Arellano, M., Bond, S.: Some tests of specification for panel data: Monte Carlo evidence and an application to employment equations. Rev. Econ. Stud. **58**, 277–297 (1991)
3. Barro, R.J., Sala-i-Martin, X.: Economic Growth. MIT Press, Cambridge, MA (1995)
4. Blomström, M.: Foreign investment and productive efficiency: the case of Mexico. J. Ind. Econ. **35**, 97–112 (1986)

5. Blomström, M., Wolff, E.: Multinational corporations and productivity convergence in Mexico. In: Baumol, W. et al. (eds.) Convergence of Productivity: Cross-National Studies and Historical Evidence. Oxford University Press (1994)
6. Caves, R.E.: Causes of direct investment: foreign firms' shares in Canadian and United Kingdom manufacturing industries. Rev. Econ. Stat. **56**(3), 279–293 (1974)
7. Démurger, S., Chen, Y.: Croissance de la productivité dans l'industrie manufacturière chinoise: Le rôle de l'investissement direct étranger. Economie internationale, 2002/4 No. 92: 131–163 (2002)
8. Globerman, S.: Foreign direct investment and 'spillover' efficiency benefits in Canadian manufacturing industries. Can. J. Econ. **12**, 42–56 (1979)
9. Im, K.S., Pesaran, M.H., Shin, Y.: Testing unit roots in heterogeneous panels. J. Econometrics **115**, 53–74 (2000)
10. Javorcik, B.S.: Does foreign direct investment increase the productivity of domestic firms? In search of spillovers through backward linkages. Am. Econ. Rev. **94**(3), 605–627 (2004)
11. Levin, A., Lin, C.F., Chu, C.: Unit root tests in panel data: asymptotic and finite-sample properties. J. Econometrics **108**, 1–24 (2002)
12. Liu, X., Wang, C.: Does foreign direct investment facilitate technological progress? Evidence from Chinese industries. Res. Policy **32**, 945–953 (2003)
13. Lucas, R.: On the mechanics of economic development. J. Monetary Econ. **22**, 342–367 (1988)
14. Prakash, S., Balakrishnan, B.: Managerial approach to conceptualization of development and growth-convergence of macro to micro theory. Bus. Perspect. **8**(1)
15. Romer, P.: Idea gaps and object gaps in economic development. J. Monetary Econ. **32**(3), 543–573 (1993)
16. United Nations Conference on Trade and Development (UNCTAD): Investment and the digital economy, World investment report 2017, Geneva (2017)
17. United Nations Industrial Development Organization (UNIDO): The role of technology and innovation in inclusive and sustainable industrial development, Industrial development report 2016, Vienna (2016)

Speaker Identification from Mixture of Speech and Non-speech Audio Signal

Ghazaala Yasmin, Subrata Dhara, Rudrendu Mahindar and Asit Kumar Das

Abstract Separating speaker from an amalgam of multiple sounds is a challenging area in the domain of speech processing. Henceforth, it has been quickly led to the new area of development in the subfield of speech processing called speaker identification. The proposed work presents a new approach to catch this problem by using acoustic features of the audio signal. The mixture of speech and non-speech audio signal has got separated by using filtering algorithm followed by the recognition of the speech audio by extracting noteworthy acoustic features. A new feature has got implemented as part of contribution to the proposed work named del-MFCC. The computed features have been served for identification of speakers using different popular classifiers. The performance of the presented methodology has been compared with the existing related methods to express the usefulness of the proposed method.

Keywords Speech separation · Feature extraction · Pitch · Speaker identification · MFCC · Classification

G. Yasmin (✉)
CSE Department, St. Thomas' College of Engineering & Technology,
Kolkata 700023, India
e-mail: ghazaala.yasmin@gmail.com

S. Dhara · R. Mahindar
ECE Department, St. Thomas' College of Engineering & Technology,
Kolkata 700023, India
e-mail: subratadhara36@gmail.com

R. Mahindar
e-mail: rudrendumahindar@gmail.com

A. K. Das
CST Department, Indian Institute of Engineering Science and Technology,
Shibpur, Howrah 711103, India
e-mail: akdas@cs.iiests.ac.in

J. Nayak et al. (eds.), *Soft Computing in Data Analytics*,
Advances in Intelligent Systems and Computing 758,
https://doi.org/10.1007/978-981-13-0514-6_47

1 Introduction

Numerous levels of information can be gathered from the speech. It is also an informative resource to identify the speakers. Not only the speech but also music and voice have unique nature by which one speaker can discriminate from others. Human vocal structure has been configured as a discriminative species. It achieves extremely speaker-dependent characteristics. These discriminative features also reflect in case of the sounds containing other non-speech sounds like music, environmental sound. The unique feature in a speech of a particular speaker persists even if it is amalgamated with a mixture of other sounds. This certitude has been inspired us to generate a new system for identifying different speakers from a mixture of their speech with other non-speech sounds. Various researchers have worked with different audio processing and speech recognition [1] techniques. The contribution of some of the researchers has been presented in this section. Reynolds [1] focused on building a statistical speaker recognizer for speaker identification. Speaker verification was based on isolating a voice known to the speaker from voices unknown to the system. Mel-frequency cepstrum coefficients (MFCCs) and vector quantization approach were used for processing the input audio signal and speaker identification, respectively, by Dudeja and Kharbanda [2] to implement text-independent speech recognition system. Characteristic parameters were extracted from sound signal by using linear prediction cepstrum coefficients (LPCCs) and mel-frequency cepstrum coefficients (MFCCs) by Xu [3] to implement a text-dependent speaker recognition system. A speaker recognition algorithm was used by improving Linde–Buzo–Gray (LBG) algorithm and by training the samples using continuous left–right hidden Markov model (HMM). Iterative clustering approach was used in the paper [4] to implement text-independent speech recognition system. Reynolds et al. [5] used Gaussian mixture model–Universal background model (GMM-UBM) for developing text-independent speaker verification system. Spectral centroid magnitude (SCM) and spectral centroid frequency (SCF) were used in [6] to develop speaker recognition model. Differences of idiolects among different speakers were used as a basis of speaker recognition in [7] using likelihood ratio and speaker entropy. Speaker-independent speech recognition of isolated words was implemented by Paul and Prakash [8] using first three format frequencies of the vocal tract and the mean zero-crossing rate (ZCR) of the audio signal. Estimation of format frequencies was done by stimulating vocal tract by linear predictive coding (LPC) filter and calculating its resonant frequencies. Emotional state detection was done in [9] using speech processing. Automatic speaker recognition (ASR), used to verify a person's claimed identity automatically from voice, was developed in paper [10]. Automatic singer identification was described in [11] that involved continuous voice recognition techniques using different methods like support vector machine (SVM) for vocal/instruments boundary detection and Gaussian mixture model (GMM) for modelling the singers' voice. Mermelstein used local differences by any distance metric for speech recognition in [12]. Lartillot, Toiviainen and Eerola developed MIRToolbox [13], an immensely useful toolbox kit for music information retrieval. Different musical features that are associated with MIRToolbox are zero-crossing rate (ZCR), RMS energy, centroid, kurtosis, skewness, flatness, brightness, roughness, spectral flux, mel-frequency cepstral spectrum (MFCC) and its fluctuation, chromagram, pitch, tempo. They are used in tonality, rhythm, timbre and data analysis and can be applied for studying different acoustic features of music and emotions. Beigi proposed an audio source classification system [14] using Gaussian mixture model (GMM). Biometric voice deconstruction was developed by Mazaira-Fernandez, Álvarez-Marquina and Gómez-Vilda [15] for gender-dependent speaker recognition using Gaussian mixture model–Universal Background Model (GMM-UBM).

The remaining part of the work is organized as follows: proposed gender recognition and speaker identification extracting respective relevant features are described in Sect. 2. Section 3 depicts the experimental results and comparative performance analysis, and finally, conclusion and future scope of the work are discussed in Sect. 4.

2 Proposed Methodologies

There is a fine difference between speaker recognition and speaker identification. Speaker recognition deals to check whether a particular speaker exists in the database or not, whereas speaker identification matches a given speech with the speech contained in the sample database.

The workflow of the proposed speaker identification system is described in Fig. 1. The system consists of three major stages. The first stage performs the separation of speech signal from non-speech signal using centre channel separation technique. The second stage extracts the related features for gender recognition from speech signal which divides the speech signal basically into two files, namely female speech signal and male speech signal. Finally, from each speech file, some new features, relevant to speaker identification are extracted and the speakers (both female and male) are identified separately from the individual files.

2.1 Speech Separation

Here, filtering technique is applied to separate speech from non-speech in an amalgam of multiple sounds. By filter, here means the digital filter, which is a system that computes mathematical operations on the sampled signals and passes the signals with a particular range of frequencies while blocked the other signals. In case of voice signal, the lowest voice frequency is 300 Hz and the maximum frequency of human voice is 3.4 kHz. A filter can be designed which passes the signals which lie within a particular band of frequencies, and this type of filter is known as band-pass filter. So, if a band-pass filter with lower cut-off frequency

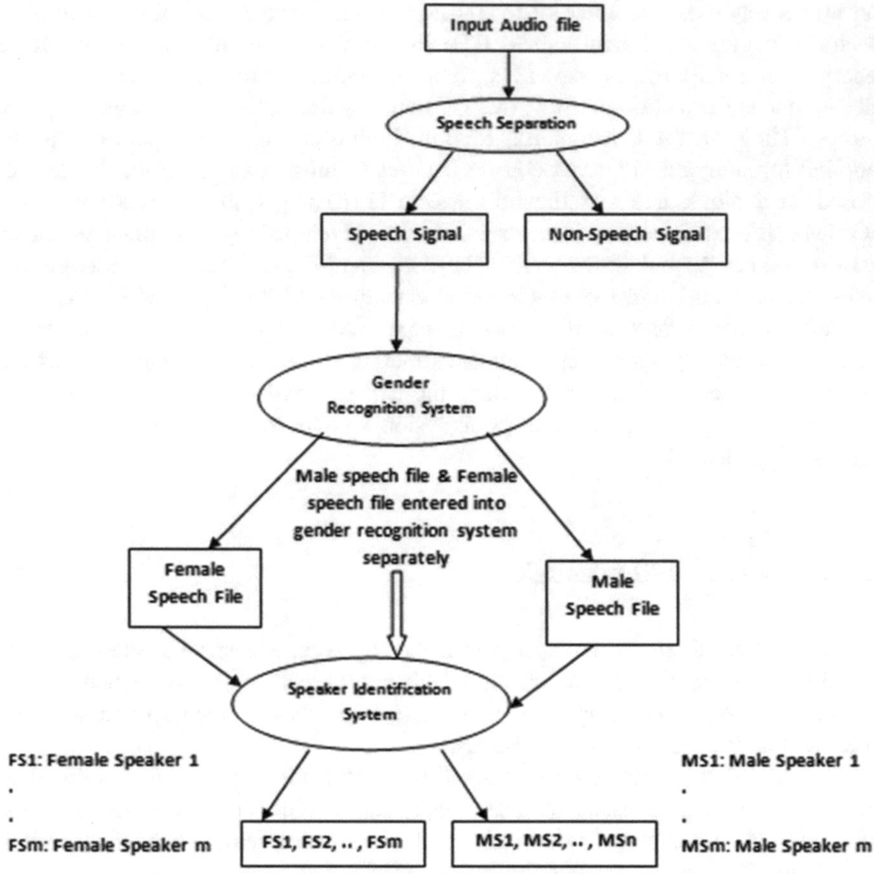

Fig. 1 Overall layout of the proposed speaker identification system

300 Hz and upper cut-off frequency 3.4 kHz is designed and a audio signal file is passed through it, then the signals with the above range are obtained which are mostly of voice signals. Thus, the voice is extracted out from the audio signal. For voice signal, the filter response should be as sharp as possible. So, the digital filter with very high order is applied here.

Figure 2a gives the audio signal where speech and non-speech parts are overlapped, and Fig. 2b represents the speech part separated from non-speech part of the audio signal represented by Fig. 2a.

In the paper, the speech signals are separated from non-speech signals using centre channel separation technique. Voice signals are symmetrical along x-axis and mostly lie in the centre portion of the channel. So, extracting out the centre portion, the voice signal can be separated from the audio signal. This is done by the centre channel separation method based on spatial analysis, and the separated voice signal is shown in Fig. 3, which is separated from the mixture of signal shown in

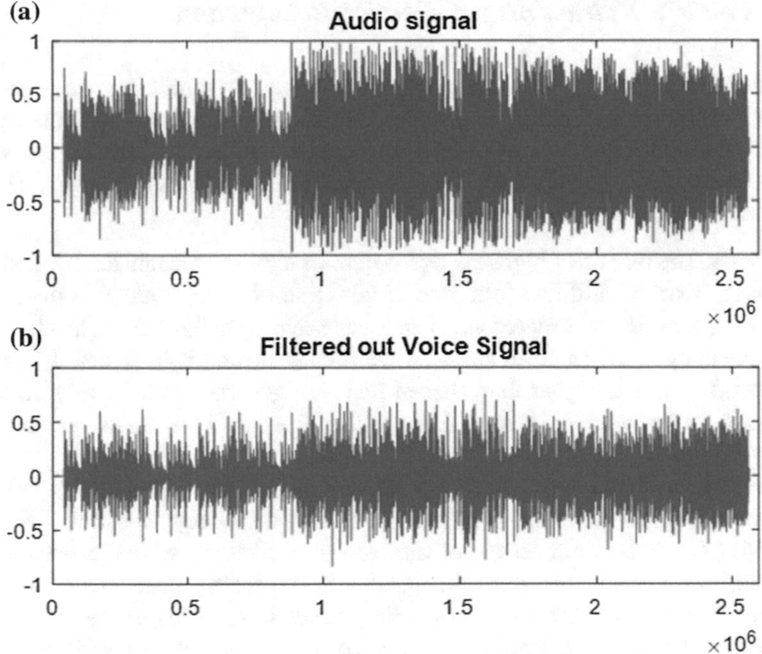

Fig. 2 **a** Speech signal overlapped with non-speech **b** separated speech

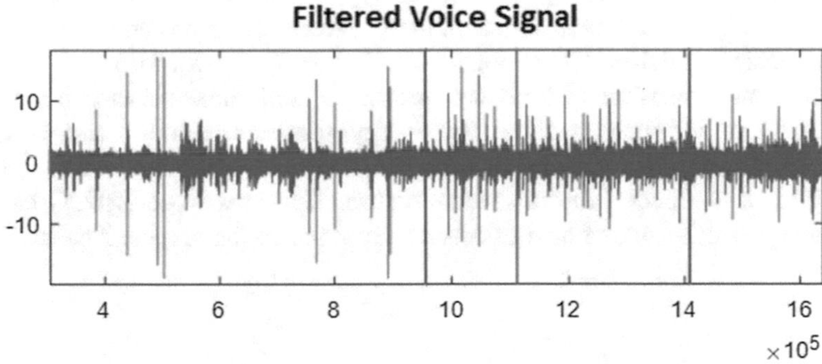

Fig. 3 Speech signal separated from Fig. 2 using centre channel separation

Fig. 2a. The output after the centre channel separation method contains a lot of noise along with the voice signal.

It has been observed that filter method contains very little amount of music where the centre channel separation method has noise with the voice signal. So, we conclude that the second method is better than the first one as it contains only the voice signal. In the proposed work, centre channel separation technique has been used for separating speech signal from the mixture of non-speech signal.

2.2 Feature Extraction for Gender Recognition

As described in Fig. 1, two different systems have been implemented; one is for gender recognition and other for speaker identification. Consequently, the features have got divided into two sets. One set contains the set of features for gender discrimination, and the other set is the features used to identify speakers. The first set consists of the following features:

(i) **Pitch**: The degree of highness or lowness of a tone is recognized by pitch. The pitch varies according to the rate of vibration of vocal folds. It is an auditory perception and is ordered on a musical scale from low to high. Higher the frequency of vibration of sound wave, higher is the pitch. Generally, pitch of female voice is higher than that of male. In general, voice of a typical adult male will have a fundamental frequency from 85 to 180 Hz, and that of a typical adult female from 165 to 255 Hz.

(ii) **MFCC and del-MFCC**: Mel-frequency inspection of speech is based on human perception experiments. The mel-frequency cepstral coefficient (MFCC) is a frequency domain acoustic feature, which represents the short-term power spectrum of a sound with respect to a linear cosine transform of a log power spectrum on a nonlinear Mel scale of frequency. This feature represented as a sequence of cepstral vector consists of 13 coefficients. From this feature vector, a new feature named as del-MFCC has got implemented for the proposed methodology, as shown in Fig. 4. MFCC is a 13-tuple feature vector, say $(X_1, X_2, X_3, \ldots, X_{12}, X_{13})$ and del-MFCC is the same dimensional feature vector $(X_1, \Delta X_1, \Delta X_2, \ldots, \Delta X_{12})$ obtained from MFCC so that the values of the first dimension for both the vectors are same and the value of i-th dimension of del-MFCC is obtained as $\Delta X_i = X_{i+1} - X_i$, for $i = 1, 2, \ldots, 12$. The first dimension of both the vectors is kept same because it gives the maximum information about the energy spectrum. Though rest of the 12 coefficients of MFCC is not showing the major change in their values, but their differences reflect a discriminative nature. That is why del-MFCC has been incorporated as one of the important feature set for the proposed methodology.

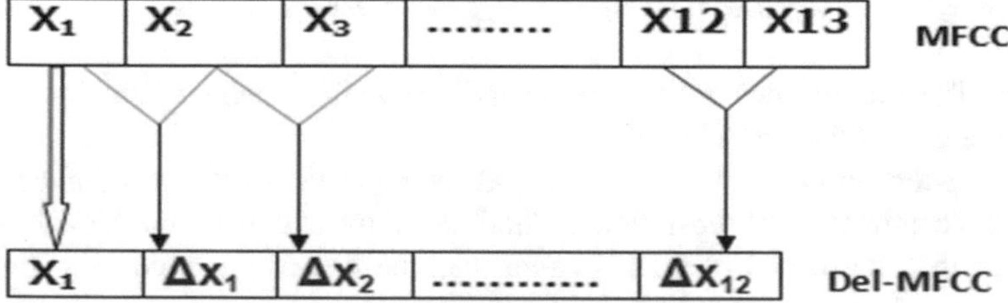

Fig. 4 Diagrammatical representation of MFCC and del-MFCC

2.3 Feature Extraction for Speaker Recognition

(i) **Roughness and Spectral Brightness**: Roughness exposes the amount of sensory dissonance at each successive moment throughout the signal. Roughness quantifies how rapid is the amplitude modulation of sound. High roughness gives rise to more harsh sound. An estimation of the total roughness is computed using the peaks of the spectrum, and the average of all the dissonance between all possible pairs of peaks is taken. The brightness shows the evolution of brightness throughout the picce of music. High values indicate moments in the music where most of the sound energy is on the high-frequency register, whereas low values indicate moments where most of the sound energy is on the low-frequency register.

(ii) **Chromagram**: The chromagram coefficient shows the distribution of energy along the 12 pitch classes. Spectral coefficients obtained from a short-time Fourier transform give the chroma features. Chroma representation signifies the intensity of each of the 12 distinct musical chroma of the octave accordingly with each time frame. The main notion of chroma features is to aggregate all spectral information that relates to a given pitch class into a single coefficient. Experimentally, it has been observed that the 12 chromagram coefficients have increased the accuracy of the result by around 20%. This is why this has been chosen to be better feature compared to the other features for the proposed work.

(iii) **Short-Time Energy (STE)**: The amplitude of a speech signal diversifies with time. Likewise, energy content also differs with time. In particular, the amplitude of unvoiced segment is generally much lower than the amplitude of voiced segments. STE accomplishes a convenient representation to reflect these amplitude variations. Since, energy of different speakers is a significant feature, henceforth it has been taken as a choice of feature for the methodology. STE can be expressed by Eq. (1).

$$E_n = \sum [s(m)w(n-m)]^2 \tag{1}$$

where $s(m)$ is a short-time speech segment obtained by passing the speech signal $x(n)$ through the window $w(n)$.

2.4 Feature Selection

In the domain of pattern recognition, it is always a better practice to reduce the dimension of feature set. Consequently, based on the domain knowledge, the relevant features are extracted, but combination of some features may make the other features redundant. So feature selection is an important step for selecting only the minimum number of important features without losing any information. Inspiring from this fact, we have reduced the feature set using correlation-based feature selection (CFS) algorithm. The prime goal of this algorithm is to find the correlation among the features. The correlated features have to be reduced to single value. In the proposed methodology, the feature set for gender recognition consists of 114 (pitch 88 + MFCC 13 + del-MFCC 13) features, which has been reduced to 14. This algorithm is further applied on the feature set extracted for speaker identification. For feature identification, 15 features have been computed (12 chromagram coefficients + 2 roughness and brightness + 1 STE). By using CFS, this extracted feature set has been reduced to nine. We have used Wcka tool [16] to execute the feature selection algorithm.

3 Experimental Results

The sample dataset collected for testing the proposed system contains audio speech files mixed with different music as well as other environmental sounds. Many song files have also been considered for testing the efficiency of the system. Fifteen different speakers' speech sample data have been assembled for the classification of speaker. Out of 15 speakers, eight are male speakers and remaining seven are female speakers. Many of the sample data contain multiple files of same speakers with a different speech. This has been done to make our dataset text independent. Total 150 files of audio have been considered for experimental purpose. Weka tool [16] has been chosen for the classification in this methodology.

Several popular classification techniques have been performed on the total extracted and selected features. The result mentioned in Table 1 is the accuracy of some of the classifiers for the testing dataset. The selected features using CFS have came up with better accuracy in most of the classifiers, and in few of the classifiers, total extracted features provide slightly better result. This shows the importance of feature selection algorithm for speaker identification.

The propounded method is also compared with the related existing methods [2, 4–6] in terms of accuracy of the classifiers. In most of the classifiers (out of eight

Table 1 Classification accuracy for male and female

Classifier	All features (114)	CFS (14)
Naïve Bayes'	96.88	**98.44**
SVM	81.36	**84.25**
Logistic	**98.80**	98.30
MLP	99.10	**99.20**
Random forest	92.15	**93.79**
PART	92.19	**93.07**
Adabooster	**95.31**	95.01
multiclass classifier (MCC)	89.06	**93.75**

Table 2 Comparative study

Methodology	Accuracy (%)
Revathi et al. [4]	86
Kua et al [6]	89
Dudeja and Kharbanda [2]	91
Reynolds et al [5]	87
Proposed work	**94**

considered classifiers), the proposed method gives better accuracy than the others. The average accuracies of all the methods are listed in Table 2.

4 Conclusion

A hierarchical approach for identification of the speaker has been carried out in the proposed work. The fascination of the proposed system is that, it has separated the speech signal from overlapped non-speech signal. This system is aiming to explore further with the new problem, where mixture of multiple speech would be separated in their component signal. The new feature set del-MFCC has been incorporated in the proposed work and observed the better accuracy of the system. The paper describes the importance of feature selection technique in the proposed method. In future, some other popular feature selection methods and novel feature selection method may be explored for selecting only the minimum relevant features from all probable relevant extracted features. The proposed speaker identification problem is language dependent which may be modified to adduce language-independent speaker identification from the mixture of multiple speeches.

References

1. Reynolds, D.A.: Automatic speaker recognition using Gaussian mixture speaker models (1995)
2. Dudeja, K., Kharbanda, A.: Applications of digital signal processing to speech recognition. Int. J. Res. **2**(5), 191–194 (2015)
3. Xu, H.H.: Text Dependent Speaker Recognition Study (2015)
4. Revathi, A., Ganapathy, R., Venkataramani, Y.: Text independent speaker recognition and speaker independent speech recognition using iterative clustering approach. Int. J. Comput. Sci. Inf. Technol. (IJCSIT) **1**(2), 30–42 (2009)
5. Reynolds, D.A., Quatieri, T.F., Dunn, R.B.: Speaker verification using adapted Gaussian mixture models. Digit. Signal Process. **10**(1–3), 19–41 (2000)
6. Kua, J.M.K., Thiruvaran, T., Nosratighods, M., Ambikairajah, E., Epps, J.: Investigation of spectral centroid magnitude and frequency for speaker recognition. In: Odyssey, p. 7 (2010)
7. Doddington, G.R.: Speaker recognition based on idiolectal differences between speakers. In: Interspeech, pp. 2521–2524 (2001)

8. Paul, D., Parekh, R.: Automated speech recognition of isolated words using neural networks. Int. J. Eng. Sci. Technol. (IJEST) **3**(6), 4993–5000 (2011)
9. Otero, P.L.: Improved strategies for speaker segmentation and emotional state detection (2015)
10. Campbell, J.P.: Speaker recognition: a tutorial. Proc. IEEE **85**(9), 1437–1462 (1997)
11. Atame, S., Shanthi Therese, S., Gedam, M.: A Survey on: Continuous Voice Recognition Techniques
12. Mermelstein, P.: Distance measures for speech recognition, psychological and instrumental. Pattern Recog. Artif. Intell. **116**, 374–388 (1976)
13. Lartillot, O., Toiviainen, P., Eerola, T.: A matlab toolbox for music information retrieval. In: Data Analysis, Machine Learning and Applications, pp. 261–268 (2008)
14. Beigi, H.: Audio source classification using speaker recognition techniques. World Wide Web (2011)
15. Mazaira-Fernandez, L.M., Álvarez-Marquina, A., Gómez-Vilda, P.: Improving speaker recognition by biometric voice deconstruction. Front. Bioeng. Biotechnol. **3** (2015)
16. Srivastava, S.: Weka: a tool for data preprocessing, classification, ensemble, clustering and association rule mining. Int. J. Comput. Appl. **88**(10) (2014)

Single-Sentence Compression Using SVM

Deepak Sahoo and Rakesh Chandra Balabantaray

Abstract Presenting a sentence in less number of words compared to its original one without changing the meaning is known as sentence compression. Most recent works on sentence compression models define the problem as an integer linear programming problem and solve it using an external ILP-solver which suffers from slow running time. In this paper, we have presented a machine learning approach to single-sentence compression. The sentence compression task is modeled as a two-class classification problem and used support vector machine to solve the problem. Different learning models are created using different types of kernel functions. Finally, it has been observed that RBF kernel gives good result compared to other kernel functions for this compression task of single sentence.

Keywords Compression · SVM · Kernel · Classification · Parsing

1 Introduction

Sentence compression is an active research area in last ten to twelve years due to various reasons like increasing penetration of smartphones, use of Internet through smartphone and increase in online shopping through smartphones. This situation demands small but complete informative sentence which fits in small screen of smartphone than large sentences. Sentence compression can be of two types single-sentence compression and multi-sentence compression. Research on both the types shares two common properties either they depend on syntax or they are supervised. In single-sentence compression, the sentence is represented in fewer words without compromising the original meaning of the sentence, whereas in multi-sentence compression [1] more than one sentence that are highly related (e.g., +ve/−ve view

D. Sahoo (✉) · R. C. Balabantaray
Department of CSE, IIIT-Bhubaneswar, Bhubaneswar 751003, Odisha, India
e-mail: c114005@iiit-bh.ac.in; deepsahoo@gmail.com

R. C. Balabantaray
e-mail: rakesh@iiit-bh.ac.in

J. Nayak et al. (eds.), *Soft Computing in Data Analytics*,
Advances in Intelligent Systems and Computing 758,
https://doi.org/10.1007/978-981-13-0514-6_48

regarding a product or two sentences having transition or anaphoric relationship and some common words) are joined to produce a new sentence whose size (no. of words) is less than the combined size (no. of words) of original sentences.

We worked on single-sentence compression where the new generated sentence is represented in less no. of words compared to original sentence without negotiating the meaning of original sentence. Let sentence Si belongs to document D. S_i contains a set of words. $S_i = W_1 \ldots W_i \ldots W_n$, i = 1, 2...n; n = Total no. of words in the sentences. Let after the compression, we get compressed sentence $CS_i = W_1 \ldots W_i \ldots Wm$, i = 1, 2,...m; m = total no. of words in compressed sentence and $m \leq n$.

Example 1 Original Sentence = Ram is a boy
Compressed Sentence = Ram is a boy
In this case m = n; there is no change. Here n = 4, m = 4.

Example 2 Original Sentence = Five people have been taken to hospital with minor injuries following a crash on the A17 near Seaford this morning
Compressed Sentence = Five people have been taken to hospital with minor injuries following a crash on the A17 near Seaford
In this case m<n. Here n = 20, m = 18.

The paper is organized as follows: Sect. 2 presents the related work. Detailed methodology is discussed in Sect. 3. Evaluation is discussed in Sect. 4, followed by the conclusion.

2 Related Work

Different systems have been introduced since its very first approach [2–5] based on syntactic information. Some research methods highly depend on parser to identify syntactic information, but supervised techniques do not depend completely on parser [6] or use syntactic information for feature extraction [7].

To produce grammatically correct compressed sentence, some researchers use techniques [8–12] that modify/rectify syntactic trees.

Some researchers are used rule-based approach [13] and language-based model [6, 14] for sentence compression. Some researchers combine sentence compression with summarization system and frame it as an integer linear programming (ILP) [15–17] and then solve the problem with an external IP-solver. But the issue with integer linear programming approach is that they suffer from slow run-time.

The framework is proposed by Wood-send and Lapata [18], Almeida and Martins [19] based on integer linear programming, but they have improved running time compared to previous ILP-based systems. Filippova et al. [20] proposed a supervised model that is based on local deletion decision in association with a recursive procedure to get most probable compression at every node in the tree. This approach takes less time compared to previous models.

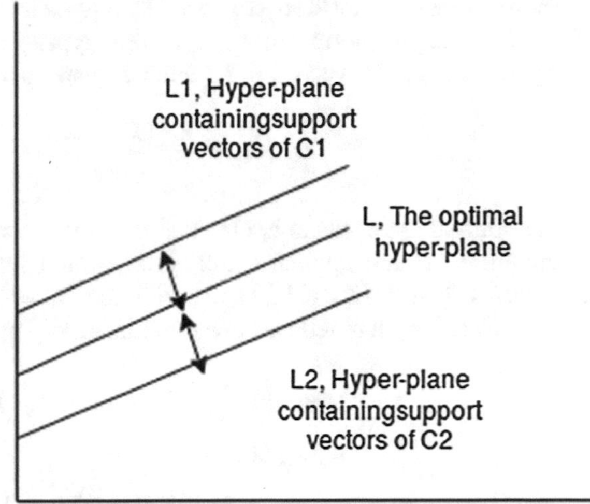

Fig. 1 Svm classification

Still there is scope to increase the efficiency (time complexity) and grammatical accuracy of the compression model. We framed the sentence compression task as a classification problem and solved the problem using machine learning technique.

3 Methodology

We formulated sentence compression as a two-class classification problem, where words present in the output sentence after compression belong to one class and words not present after compression belong to another class. Support vector machine (SVM) [21] is used for this classification task. The goal of SVM is to produce a model (based on the training data) which predicts the target values of the test data given only the test data attributes.

Given a training set $\{X_i, D_i\}_{i=1}^{n}$, where $x_i \in R^m$ and $D_i \in \{1, -1\}$. $D_i = 1$ if $xi \in C_1$, $D_i = -1$ if $x_i \in C_2$.

A new data x (class is unknown), the model determine the class to which it belongs.

Let the equation in Eq. 1 represents the hyperplane separating the two class of data.

$$W^T + b = 0 \tag{1}$$

Then, any data x_i belong to class C1 or C2 depends on the condition in Eqs. 2 and 3.

$$W^T x_i > 0,\ if\ x_i \in C_1\ i.e\ D_i = +1 \tag{2}$$

$$W^T x_i < 0,\ if\ x_i \in C_2\ i.e\ D_i = -1 \tag{3}$$

The distance between the hyperplanes L1 and L2 in Fig. 1 is equal to the projection of $X_1 - X_2$ along the normal to the separating hyper-plane. Lets say the distance is r. X_1 and X_2 are support vectors in L1 and L2, respectively. 'r' is defined as in Eq. 4.

$$r = \frac{W^T(X_1 - X_2)}{\|W\|} \tag{4}$$

The optimal hyper-plane is obtained when the distance between L1 and L2 is maximum. To obtain optimal hyper-plane minimizing $\|W\|$ over all possible W' s. Subject to constraints $W^T X_i + b >= 1$ for $D_i = +1$ and $W^T X_i + b <= 1$ for $D_i = -1$. The problem reduces to a minimization problem given below in Eq. 5.

$$\begin{aligned} &\text{Minimize } \frac{1}{2}\|W\|^2, Given\{X_i, D^i\}_{i=1}^n \\ &\text{Subject to} \\ &\qquad D_i(W^T x_i + b) >= 1. \end{aligned} \tag{5}$$

3.1 Training Data Creation

In machine learning, training data creation is a very important task. For this single-sentence compression task, the structure of the sentence plays very important role for the creation of training data. A sentence is formed by joining two or more clauses which are the basic building blocks of English grammar. Again clauses are made up of phrases; so sentence is a combination of phrases.

We have used Stanford parser to create the parse tree of the sentence; then, the parse tree is used to create the training data. Stanford Parser uses the Penn Treebank[1] tag set that gives well-defined levels to clause, phrase, and POS tags. We have used the first 1000 sentence released by Filippova and Altun (2013)[2] for training and test data creation.

To create the training data, original sentence is passed to the parser; parser generates the parse tree. For example, lets take a sentence "Pita restaurant chain Pita Pit has closed in Edwardsville," the parse tree generated by the parser is given in Fig. 2.

Each word in the parse tree can be represented as a sequence of clause or phrase at different levels and parts of speech. Clause or phrases at different level and parts of speech are taken as features to represent each word. Each clause, phrase, and part of speech given a distinct numeric value. Clause, phrases up to fifteenth level and part of speech are taken as features. So every feature in this training data is a categorical data.

---

[1] https://www.ling.upenn.edu/courses/Fall_2003/ling001/penn_treebank_pos.html.

[2] http://storage.googleapis.com/sentencecomp/compression-data.json.

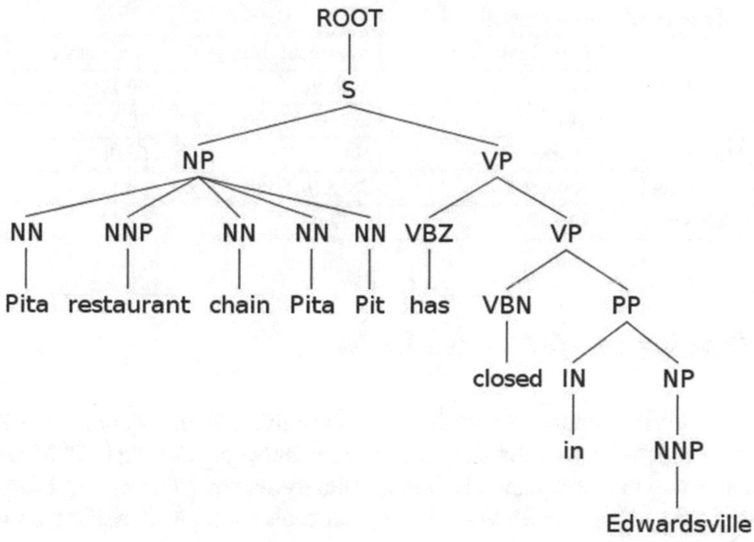

Fig. 2 Parsetree of sentence

Pita S NP NN
restaurant - S NP NNP
chain S NP NN
Pita S NP NN
Pit S NP NN
ha1s S VP VBZ
closed S VP VP VBN
in S VP VP PP IN
Edwardsville S VP VP PP NP NNP

The human-generated compression of the sentence is "Pita Pit has closed in Edwardsville." First three words Pita restaurant chain are removed from the original sentence after compression. Words to be retained, and words to be removed after compression are given the level 1 and −1, respectively.

Pita S NP NN -1
restaurant - S NP NNP -1
chain S NP NN -1
Pita S NP NN 1
Pit S NP NN 1
has S VP VBZ 1
closed S VP VP VBN 1
in S VP VP PP IN 1
Edwardsville S VP VP PP NP NNP 1

Table 1 Accuracy of different models

Model	Kernel function	Argument (if any)	Accuracy (%)
Rbf_Model_default	RBF	Default	69.19
Rbf_Model_best	RBF	C = 32, Gamma = 2	70.37
Linear_Model	Linear	NA	66.08
Polynomial_Model	Polynomial	d = 3 (default)	58.12
Sigmoid_Model	Sigmoid	Default	59.42

3.2 Training and Testing the Model

We have manually annotated nine hundred sixty-five sentences containing 20,404 words form which seven hundred seventy-five sentences containing 16,003 words are used for training the model and one hundred ninety sentence containing 4401 words for testing the system. LIBSVM [22] a popular tools for SVM classification is used for creating compression model. Five different models are created using five different kernel function with same set of training data. The different kernel functions used are RBF kernel, polynomial kernel, sigmoid kernel, and linear function. All the models are tested with same set of test data set, and the accuracy of each model is given in Table 1. Out of the five model, RBF kernel with cost/penalty of misclassification value C is 32 and gamma is 2.

4 Evaluation

For evaluation task, we have taken RBF model with best argument values as its accuracy is better than other models. Sentences from 1001 to 1100 are taken for evaluation released by Filippova and Altun (2013), but out of 100 sentence, 76 sentences are taken for evaluation as rest of the sentences have height more than 25. The model predicts whether a word in the original sentence will present in compressed sentence or not. All positively predicted words are concatenated to get the compressed sentence. In evaluation, we measure the quality of compression using two measures, compression rate, and informativeness. For both the measures, we strictly follow the grammatical correctness of the compressed sentence.

4.1 Informativeness

Informativeness is an important measure to measure the quality of compression. Gramatical structure of the compressed sentence is not compermised to mesure the informativeness. Each compressed sentence comes under one of the following

Table 2 Informativeness of compression

Total sentences	Category (A)	Category (B)	Category (F)	Total grammatical correct sentence (A and B)	Accuracy (%)
76	23	28	25	51	67.10

Table 3 Comparison of compression rate

Total sentences	Total correct sentences (A and B)	Total words of correct source sentence	Total words of compressed and correct source sentence	Total deleted words	Compression rate (%)
76	51	1145	705	440	38.42

categories based on its grammatically correctness and informativeness.

$$x = \begin{cases} A, \textit{If Grammatical and Informative} \\ B, \textit{If Grammatical and Informative but few extra words are} \\ \textit{present or absent} \\ F, \textit{If Ungrammatical} \end{cases} \tag{6}$$

We have taken seventy-six sentences that are not present in training and testing data. The target of words in a sentence is not known. Table 2. Given below gives the accuracy of the model in terms of grammatically correctness and informativeness of sentences. Except ungrammatical (Category 'F'), other two categories are taken as correct for accuracy measure.

4.2 Compression Rate

In compression rate, we need to find out how many words are removed after compression compared to original source sentence. In this measure, we have only considered the grammatically correct compressed sentences (Category (A) and Category (B)). Table 3 shows the compression rate of our system. Our system is compared with Filippova and Alfonseca [20], and it is given in Table 4. Source and compressed sentence of different category (A, B, C) is given in Tables 5, 6 and 7 in Appendix 1.

Table 4 Compression rate

	Compression rate (%)
TOP-DOWN	38.30
TOP-DOWN+NSS	38.10
Our System	38.42

5 Conclusion

In this paper, we have formulated the sentence compression as a two-class classification problem. Different training models are created using different kernel functions. Finally, it has been observed that RBF kernel with gamma value 2 and C value 32 gives better result than other kernel functions. Manual evaluation of system is done to measure the number of grammatical correct and informative sentences generated by model which is found to be 68% and compression rate is 38.42%. We are still working to further improve the quality of compression using other linguistic features and techniques.

Appendix 1

Table 5 Sentences that are grammatically correct and informative (category A)

Original sentence	Compressed sentence
1. Manchester United record signing Juan Mata needs more time to prove his mettle team mateWayne Rooney has said	1. Manchester United record signing Juan Mata needs more time to prove mettle team
2. Lehigh University President Alice P Gast the colleges 13th president in its 148 year history will step down on July 31 according to a posting on the university Website	2. Lehigh University President Alice P Gast the colleges 13th president in its will step down on July 31 a posting university website
3. The Fair Loans personal injury cash advance firms agree to reforms has proven to be a convenient and private process that is borrower friendly and enables people to cover sudden or emergency costs in a reasonable and manageable fashion	3. The Fair Loans injury cash advance firms agree has proven that enables people to cover sudden or emergency costs reasonable and manageable

Table 6 Sentences that are grammatically correct and Informative but few extra words (category B) given in brackets

Original sentence	Compressed sentence
1. Gia Allemand a reality TV star and girlfriend of NBA Pelicans player Ryan Anderson has died after being taken off life support at a hospital in New Orleans	1. Gia Allemand a reality TV star and girlfriend of NBA Pelicans Ryan Anderson has died (after being off)
2. A 20 year old Kissimmee man was killed Tuesday when his motorcycle crashed east of St Cloud	2. A 20 year old Kissimmee man was killed Tuesday (motorcycle crashed east of)
3. Fox has canceled comedy Hope after four seasons the network announced Monday	3. Fox has canceled comedy Raising Hope (after)

Table 7 Sentences that are grammatically incorrect (C)

Original sentence	Compressed sentence
1. Three people were injured in a two car crash on Route 20 in LafayetteSunday morning according to Onondaga County Sheriffs Deputies	1. Three people were injured in a two morning
2. The Gilmer City Council on Tuesday plans to consider canceling its May 10 council election since none of the incumbents drew achallenger during the filing period	2. The Gilmer City Council on plans to consider canceling
3. Los Angeles March 2 U2 frontman Bono reportedly takes his own bed on tour with him so that he doesnt aggravate his back which was operatedon in 2010	3. Los Angeles March 2 U2 reportedly takes with him so he doesnt, aggravate his back operated on in 2010

References

1. Ganesan, K., Zhai, C., Han, J.: Opinosis: a graph-based approach to abstractive summarization of highly redundant opinions. In: Proceedings of the 23rd International Conference on Computational Linguistics, pp. 340–348, Aug, 2010
2. Grefenstette, G.: Producing intelligent telegraphic text reduction to provide an audio scanning service for the blind. In: Working Notes of the Workshop on Intelligent Text Summarization, Palo Alto, Cal., pp. 111–117, 23 Mar 1998
3. Knight, K., Marcu, D.: Statistics-based summarization step one: Sentence compression. In: Proceedings of AAAI-00, pp. 703–710 (2000)
4. Knight, K., Marcu, D.: Summarization beyond sentence extraction: a probabilistic approach to sentence compression. Artificial Intelligence-139, pp. 703–710 (2002)
5. Jing, H., McKeown, K.: Cut and paste based text summarization. In: Proceedings of NAACL-00, pp. 178–185 (2000)
6. Clarke, J., Lapata, M.: Global inference for sentence compression: an integer linear programming approach. J. Artif. Intell. Res. **31**, 399–429 (2008)
7. McDonald, R.: Discriminative sentence compression with soft syntactic evidence. In: Proceedings of EACL-06, pp. 297–304 (2006)

8. Galley, M., McKeown, K.R.: Lexicalized Markov grammars for sentence compression. In: Proceedings of NAACL-HLT-07, pp. 180–187 (2007)
9. Filippova, K., Strube, M.: Dependency tree based sentence compression. In: Proceedings of INLG-08, pp. 25–32 (2008)
10. Cohn, T., Lapata, M.: Sentence compression as tree transduction. J. Artif. Intell. Res. **34**, 637–674 (2009)
11. Nomoto, T.: A comparison of model free versus model intensive approaches to sentence compression. In: Proceedings of EMNLP-09, pp. 391–399 (2009)
12. Wang, L., Raghavan, H., Castelli, V., Florian, R., Cardie, C.: A sentence compression based framework to query-focused multi-document summarization. In: Proceedings of ACL-13, pp. 1384–1394 (2013)
13. Dorr, B., Zajic, D., Schwartz, R.: Hedge trimmer: a parse-and-trim approach to head-line generation. In: Proceedings of the Text Summarization Workshop at HLT-NAACL-03, Edmonton, Alberta, Canada, pp. 1–8 (2003)
14. Hori, C., Furui, S., Malkin, R., Yu, H., Waibel, A.: A statistical approach to automatic speech summarization. EURASIP J. Appl. Signal Process. **2**, 128–139 (2003)
15. Martins, A.F.T., Smith, N. A.: Summarization with a joing model for sentence extraction and compression. In: ILP for NLP-09, pp. 1–9 (2009)
16. Berg-Kirkpatrick, T., Gillick, D., Klein, D.: Jointly learning to extract and compress. In: Proceedings of ACL-11 (2011)
17. Thadani, K., McKeown, K.: Sentence compression with joint structural inference. In: Proceedings of CoNLL-13, pp. 65–74 (2013)
18. Woodsend, K., Lapata, M.: Multiple aspect summarization using integer linear pro-gramming. In: Proceedings of EMNLP-12, pp. 233–243 (2012)
19. Almeida, M.B., Martins, A.F.T.: Fast and robust compressive summarization with dual decomposition and multi-task learning. In: Proceedings of ACL-13 (2013)
20. Filippova, K., Alfonseca, E.: Fast k-best sentence compression (2015). arXiv:1510.08418. Accessed 28 Oct 2015
21. Cortes, C., Vapnik, V.: Support-vector networks. Mach. Learn. **20**(3), 273–297 (1995)
22. Chang, C.C., Lin, C.J.: LIBSVM: a library for support vector machines. ACM Trans. Intell. Syst. Technol. (TIST) **2**(3), 27 (2011)

Differential Evaluation Base Gain Tune of Proportional–Integral–Derivative Controller for MLI Base-Integrated Wind Energy System with Multi-winding Transformer

L. V. Sureshkumar and U. Salma

Abstract Renewable energies play a major role in power networks and have several advantages: Wind energy generation is almost pollution free; similarly, fuel cells and photovoltaic cells have several advantages over the other conventional methods. Harmonics are interacting with advanced converters which affect both the distribution system equipment and the loads connected to it. The objective of this paper is to tune PID controller gains for a new stand-alone multi-winding transformer (MWT) connected to load through a multi-level converter (MLC), using differential evaluation algorithm for the minimization of the harmonics. A five-level diode-clamped multi-level inverter (MLI) to operate with advanced five-level switching schemes is used to control the variable wind power voltage with PID controller through the multi-winding transformer. The effect of MLC switching operation on the total harmonic distortions (THDs) is analyzed for different pulse width modulation (PWM) techniques like sinusoidal PWM, phase opposition disposition (POD), anti-phase opposition disposition (APOD), and phase disposition (PD). The results are presented and analyzed to ascertain the effectiveness of the best switching scheme for the minimization of the harmonics with DE algorithm.

Keywords Stand-alone integrated wind conversion system (SIWCS)
Multi-winding transformer (MWT) · Modular multi-level converter (MLC)
Differential evaluation (DE)

L. V. Sureshkumar (✉)
GMR Institute of Technology, Rajam 532127, India
e-mail: lvenkatasureshkumar@gmail.com

U. Salma
GITAM University, VSKP, Visakhapatnam 530045, India
e-mail: usalma123@gmail.com

J. Nayak et al. (eds.), *Soft Computing in Data Analytics*,
Advances in Intelligent Systems and Computing 758,
https://doi.org/10.1007/978-981-13-0514-6_49

1 Introduction

A large growth in power demand yields more power generation. A synchronization of the frequencies from renewable like wind, solar, hybrid, diesel generators affects their interfacing to the main grid. In the present wind energy system topologies, generators are integrated to the grid through the power electronic converters. Wind system with induction generator is relatively efficient, need low maintaining cost, robust construction, and is easy to control. Stability enhancement the stability of power system with high number of wind energy system penetration is carried [1–8]. The tuning of controller gains plays a vital role. Improvement in inverter voltage stability and enhancement of voltage profile can be obtained by appropriately tuning the controller gains using heuristic algorithms [9–13]. Multi-winding transformers transfer three phase to variable phases like twelve, eleven, nine, seven, and five. These transformers play a vital role in power generation and also in microgrid base-integrated wind energy systems connected to multi-winding transformer through MMC and MLC [13–18]. MLI and MMC are used in grid-connected integrated wind and solar power system. MLI or MMC is operated by modulation techniques like PD, POD, and APOD for smooth switching operation, increasing efficiency, and easy control of the system [10–19]. PID controllers are more popular in the electrical system because one can reduce rise time, overshoot, settling time, steady-state error and can improve stability. Controller gains are tuned by using optimization techniques like GA, PSO, DE [15–23]. In this paper, MLC base-integrated wind energy system with multi-winding transformer has been introduced, which converts nine phase to three phase. Three 3-phase induction generators are connected to star connection of nine-phase transformer primary and 3-phase delta connection secondary. The secondary transformer winding is connected to five-level diode-clamped converter through a universal bridge rectifier. The inverter output voltage is controlled with PID controller, and gains are tuned with DE to achieve low total harmonic distortion with multi-level switching methods.

2 Stand-alone Integrated Wind Energy System

The block diagram of the stand-alone wind energy conversion system integration is shown in Fig. 1. The electrical generator produces electrical energy using the kinetic energy of the wind. These wind turbines are either vertical or horizontal axis type. The generator used mostly in wind energy system is induction generator. The ability of the induction generator to produce useful power at variable rotor speed is more suitable for wind energy system, as speed is always a varying factor in these systems. These generators are also mechanically and electrically simpler and rugged in construction [3, 8].

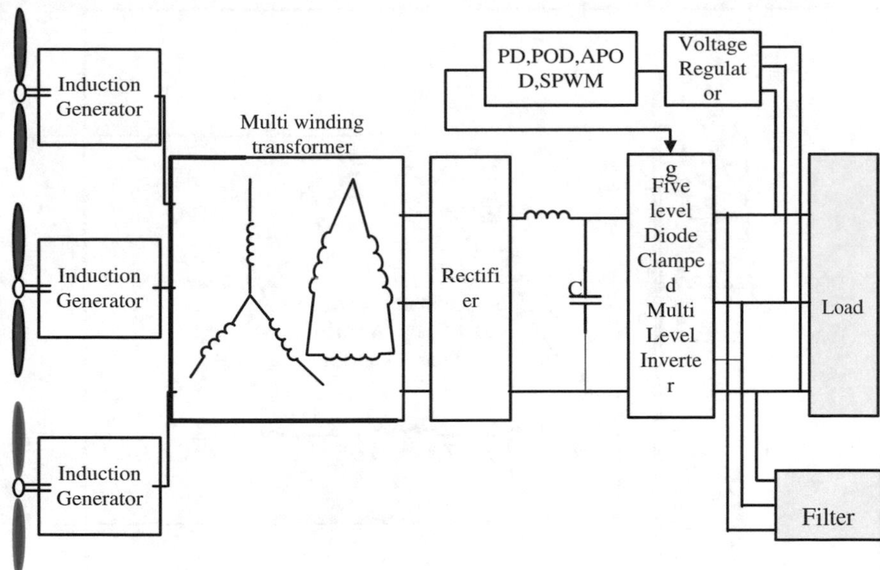

Fig. 1 Stand-alone integrated wind conversion system

The energy generated from the wind turbine is step-up by using a multi-winding transformer. As often failure of normal transformers occurs, multi-winding transformers are designed to overcome that. The operating principle of these multi-winding transformers is similar to the ordinary transformers, but they have several primary or secondary windings. They can be used for either step-up or step-down or both, and they can also be employed for single phase or for three phase. In Figs. 1 and 2, nine windings, three each from single induction generator, are connected to a multi-winding transformer. A converter system is used between the wind farm and load. A rectifier is used for conversion of AC to DC, and an RC filter is used to eliminate higher order harmonics. An inverter is used to convert the DC to AC and fed it to load. A diode-clamped MLI that reduces harmonics, ensures more sinusoidal wave, and improves efficiency can be a useful technique. By using various modulation techniques like SPWM, PD, POD, APOD, an output from the multi-level inverter can have less THD. Active filters are placed between the inverter and the load in order to mitigate the voltage sag, voltage ripples, an imbalance in the source voltage. These filters can also compensate reactive power and negative sequence currents. A feedback is given to the voltage regulator from the load. The voltage regulator is used to maintain the constant voltage level automatically at load. It gives the signal to the modulation block of the inverter by comparing the reference value and the actual value at load. Thus by using the feedback loop, desired constant value can be attained at any load.

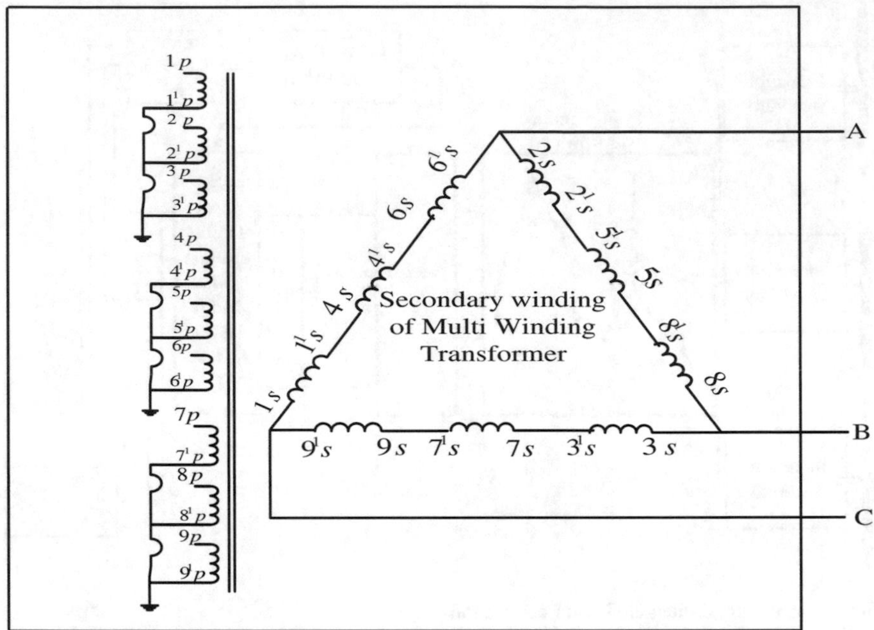

Fig. 2 Multi-winding transformer

3 Multi-winding Transformer

Although multi-winding transformer (MWT) is used for many applications, it is mainly used for galvanic isolation. In this paper, MWT is used for interconnection of stand-alone wind energy system. A multi-winding transformer generally consists of two to three 3-phase star winding for the primary side and one 3-phase delta winding for the secondary side. Each star-connected winding is connected to respective wind generators. From a wind farm where a large number of generators are installed, using the multi-wind energy systems, voltages developed in their primary are transferred to their secondary by the process of mutual induction [8, 13].

Multi-winding transformer model is a star-delta transformer; i.e., all the individual primaries are connected in star and individual secondary of delta connected. In the above circuit, 1p, 2p, 3p are connected to first generator output and 1'p, 2'p, 3 'p are grounded collectively so that it forms a star-connected system; similarly fourth, fifth, and sixth individual primaries are connected to the second generator output and for seventh, eighth, and ninth individual primaries third generator output is connected. Secondaries winding of the proposed multi-winding transformer is a delta connection formed such that 1st limb formed by connecting secondary of 1st, 4th and 6th and 2nd limb formed by secondaries of 2nd, 5th and 8th are connected in series and third limb by attaching of secondaries of 3rd, 7th and 9th. All these limbs are made parallel by connecting in series forms, 1s end links with 9's, 2s links with 6

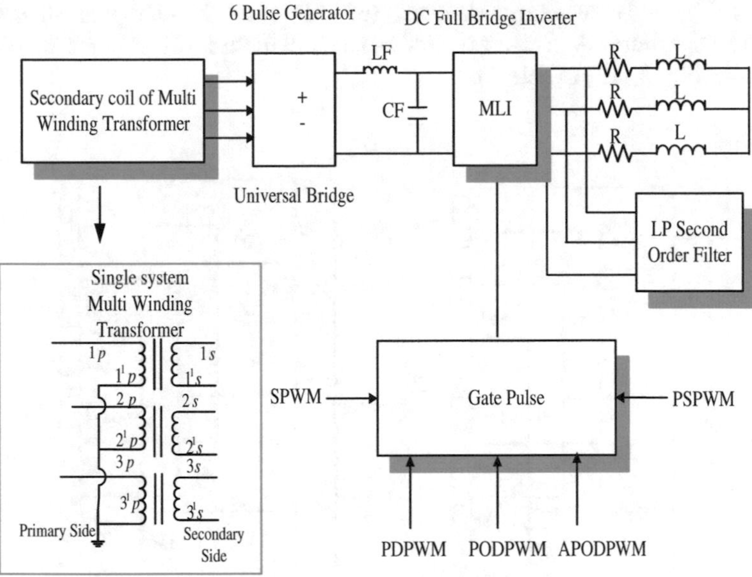

Fig. 3 Internal circuitry of distributed network from wind warm to load

's, and 3s links with 8's to form a delta connection. Three individual primary and individual secondary sides can be observed. It shows the single system of multi-winding transformer, where 1p and 1'p are first and second terminals of first individual primary winding, 1s and 1's are first and second terminals of first individual secondary winding, similarly 2p and 3p are first terminals of individual primaries, 2'p and 3'p are second terminals of individual primaries, 2 s and 3 s are first terminals of individual secondary, 2's and 3's are second terminals of secondary. Here, all the coils are electrically isolated but magnetically coupled to each other [8, 13]. The individual primaries can be connected in series where the higher primary supply voltage is required and in case of lower primary supply voltage requirement parallel connection of individuals is done; this is applied similarly to the secondaries shown in Fig. 2 and overall system block diagram also shown Fig. 3.

4 Diode-Clamped Multi-level Inverter

Diode-clamped inverter (DCI) is a classical inverter topology which is effectively suitable for medium voltage and high-power applications. The basic concept of this inverter is the diode acts like a clamping device to control DC voltage to get desired voltage. The usual 'm'-level diode-clamped inverter consists of (m − 1) capacitor on the DC bus to produce 'm' levels on the phase voltages. An m-level inverter leg requires 2(m − 1) switching devices, (m − 1) (m − 2) clamping diodes, and (m − 1) capacitors. A five-level diode-clamped multi-level inverter topology is studied in Fig. 4 [13] (Table 1).

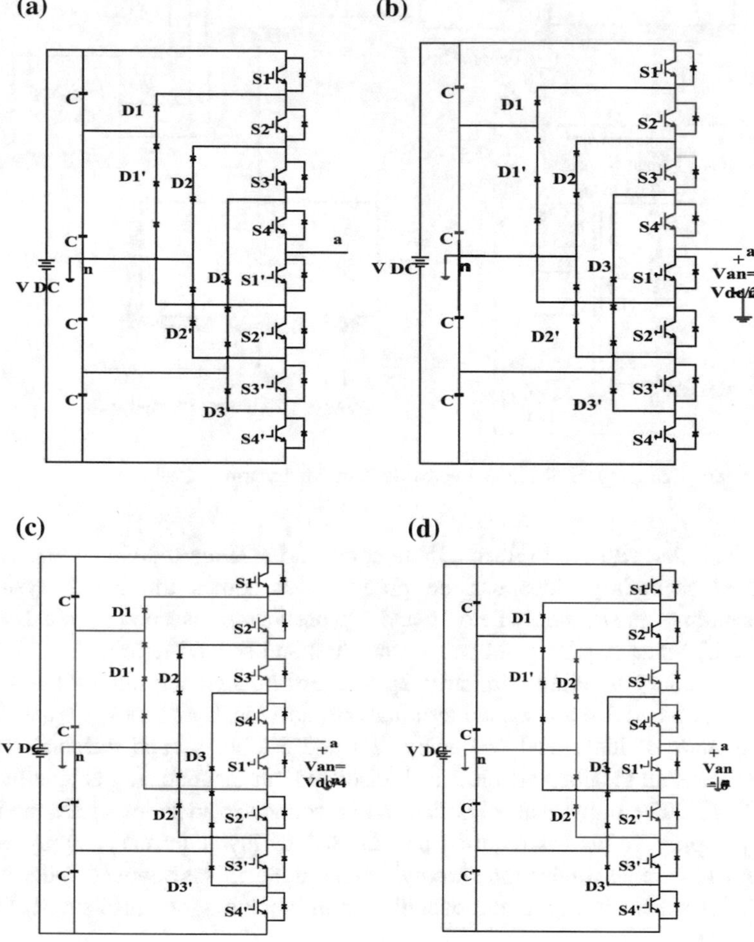

Fig. 4 **a** Circuit showing the path for +Vdc/2, **b** circuit showing the path +Vdc/4, **c** circuit showing the path +Vdc = 0, **d** circuit showing the path −Vdc/4

Table 1 Switching instants of diode-clamped MLI

Van states	S1	S2	S3	S4	S1′	S2′	S3′	S4′
Vdc/2	1	1	1	1	0	0	0	0
Vdc/4	0	1	1	1	1	0	0	0
0	0	0	1	1	1	1	0	0
−Vdc/4	0	0	0	1	1	1	1	0
−Vdc/2	0	0	0	0	1	1	1	1

5 Modulation Techniques

Mostly, the application of DCMI is constrained to three-level because voltage unbalancing problem in DC link arises as the levels increase. The three-level DCMLI also serves with problem of narrow pulse width and capacitor voltage unbalanced problem. This problem leads to inefficient performance of DCMLI which effects on power quality. By selecting proper modulation techniques, these problems can be reduced and one can achieve low THD, because proper switching of the devices reduces thc strcss on the switches and the voltage unbalancing.

Multi-level modulation method can be achieved by comparing a reference sinusoidal signal with a high-frequency triangular carrier signal. Multi-level modulation shown in Fig. 5 is further classified as space vector-based and voltage-level-based. The bus utilization is better for space vector PWM compared to sinusoidal PWM, but this method neither provides the solution for capacitor imbalance nor for narrow pulse width modulation. So, research is going on voltage-based algorithms in which multiple carrier PWM technique shows better outputs. Multiple carrier PWM methods for 78% linear modulation range give maximum RMS output voltage.

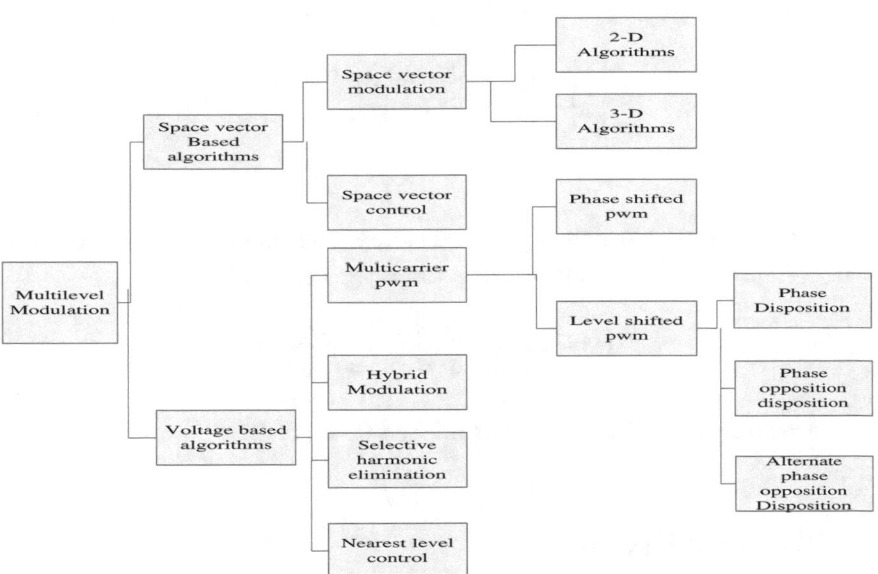

Fig. 5 Classification of multi-level modulation techniques

6 Differential Evolution Algorithm

First Step: Initial Population—Initial population is represented in a vector form x = [x1, x2,…xN]T, and xi represents real number. Differential evaluation identities a global minimum value given search space. DE initializes every parameter randomly and uniformly probability distributed within the search space shown in Eq. (1) [9, 22].

$$x_i^0 = x_{i(\min)} + rand \cdot \left(x_{i(\max)} - x_{i(\min)}\right) \tag{1}$$

Second Step: Mutation

The weight difference between two random vectors is added to donor vector $\left(x_{new,i}^{k+1}\right)$, and predetermined trial vector $\left(v_i^k\right)$, newly generated vector from the current (kth) population, three different parameter vectors xw1, xw2, and xw3 are randomly selected to create donor vector for ith target. A constant F is multiplied to difference of any two of the three vectors shown in Eq. (2).

$$x_{new,i}^k = x_{w3}^k + F \cdot \left(x_{w1}^k - x_{w2}^k\right) \tag{2}$$

Third Step: Crossover—Whether the new vector is to be recombined (or) not as shown in Eq. (3) [9, 22].

$$\begin{cases} v_i^k = x_{new,i}^k & \text{if rand } [0,\ 1] \leq CR \\ x_i^k & \text{Otherwise} \end{cases} \tag{3}$$

Fourth Step: Selection

The previous fitness function $\left(f\left(x_i^k\right)\right)$ and current generation $\left(f\left(v_i^k\right)\right)$ are compared with the corresponding vectors as shown in Eq. (4) [21, 22].

$$x_i^{k+1} = \begin{cases} v_i^k & \text{if} \left(f\left(v_i^k\right)\right) \leq \left(f\left(x_i^k\right)\right) \\ x_i^k & \text{if} \left(f\left(v_i^k\right)\right) > \left(f\left(x_i^k\right)\right) \end{cases} \tag{4}$$

Differential evaluation flowchart is referred from [19–22].

7 Simulation Results

Comparison between the APOD, POD, and PD modulation techniques is done with and without filters using MATLAB/simulink, and the results were given below. Anti-phase opposition disposition (APOD) method is applied. Fig. 6 shows output voltage waveforms and THD without using filter, while APOD is applied. THD in this is 23.17%, and the output voltage waveform is with more harmonic distortion.

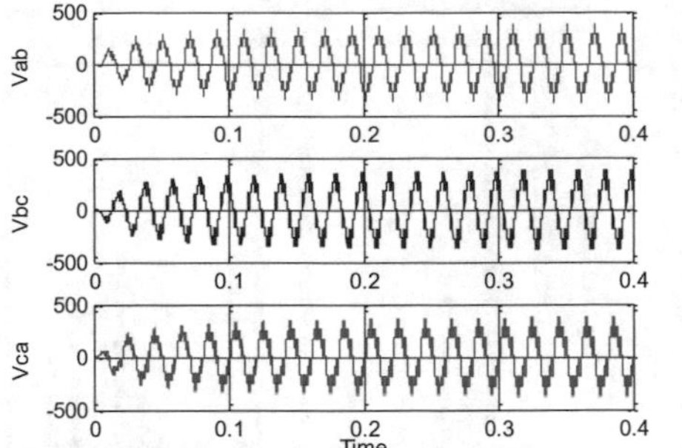

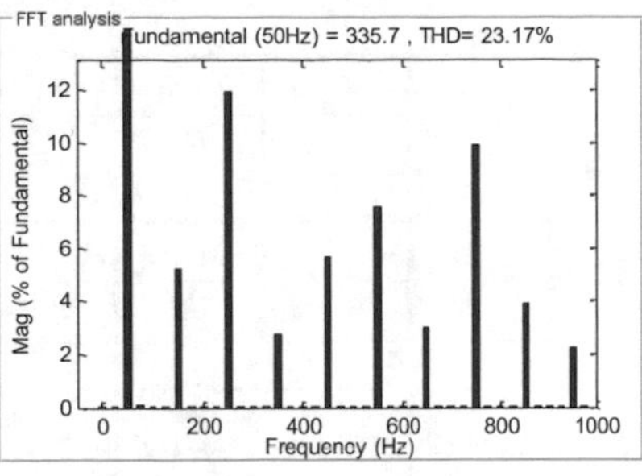

Fig. 6 Inverter output voltage and THD without filter using APOD

Anti-phase opposition disposition method is applied. Figure 7 shows output voltage waveforms and THD with using filter, while APOD is applied.

The THD value observed here is 2.96%, which is much less than compared to without filter application, and also the harmonic content in the output voltage waveform is decreased by using the filter. Phase opposition disposition (POD) method is applied. Figure 8 shows output voltage waveforms and THD without using filter, while POD is applied. THD in this is 19.79%, and the harmonic content in the output waveform is slightly decreased here compared to APOD (without filter) application but while using filter POD gives better THD value as THD value in POD with filter is 2.05%. Figure 9 shows the response with filter POD method.

The output voltage waveform and THD content while applying sinusoidal PWM without filter are shown in Fig. 10. THD value in this is 32.76%, and it is more compared to APOD and POD methods.

The output voltage waveform and THD content while applying SPWM with filter are given in Fig. 11. THD value in this is 18.31%, and it is more compared to APOD and POD methods, as with filter application of APOD gives 2.96% and POD gives 2.05%.

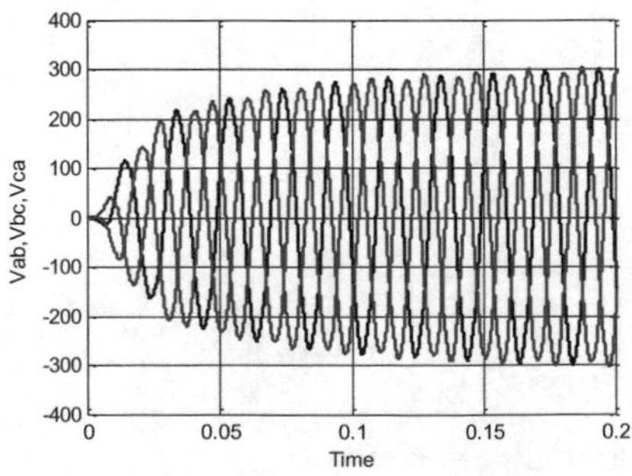

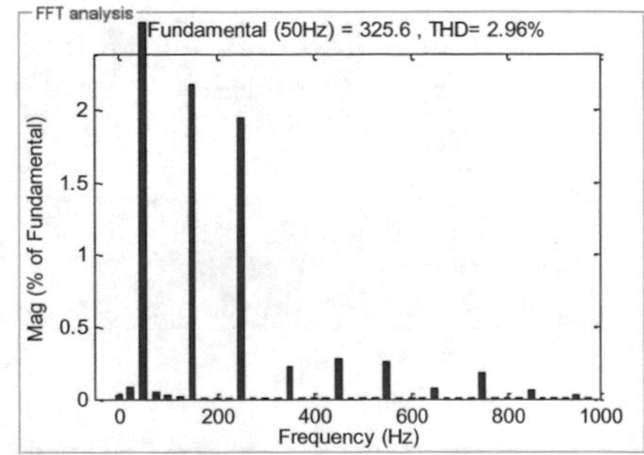

Fig. 7 Inverter output voltage and THD with filter using APOD

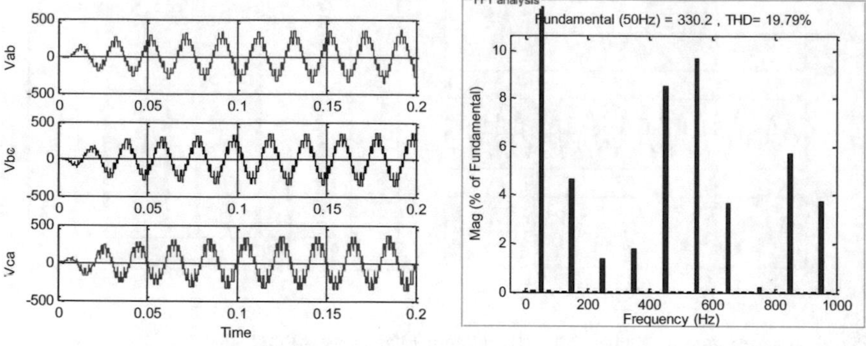

Fig. 8 Inverter output voltage and THD without filter using POD

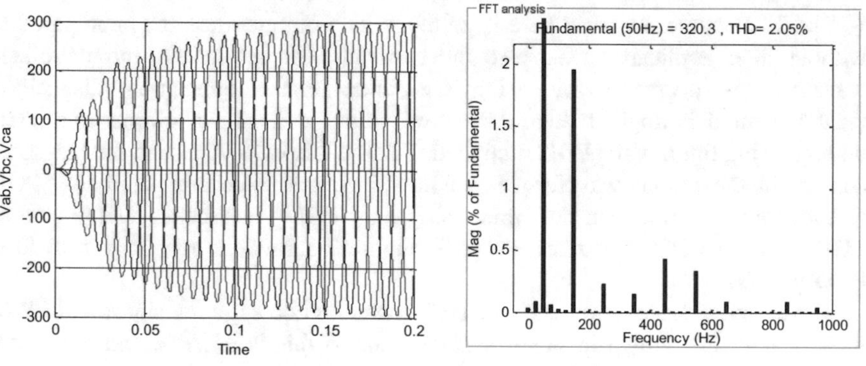

Fig. 9 Inverter output voltage and THD with filter using POD

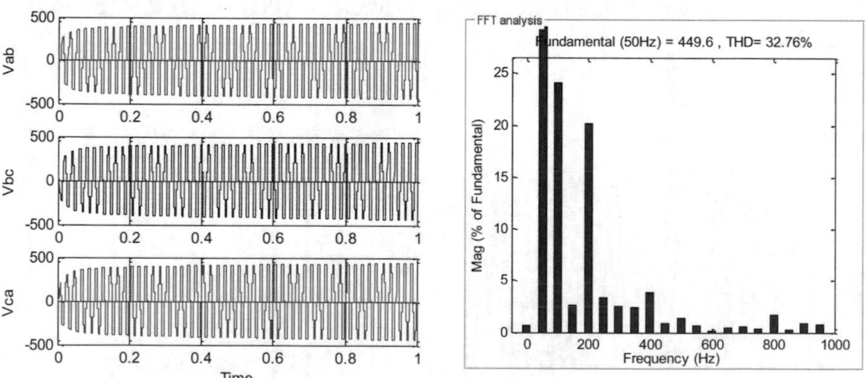

Fig. 10 Inverter output voltage and THD without filter using SPWM

Voltage waveform and THD content while applying PD without filter are shown in Fig. 12. THD value in this is 16.29%, and it is observed that is less when compared to all the three PWM techniques, i.e., APOD, POD methods.

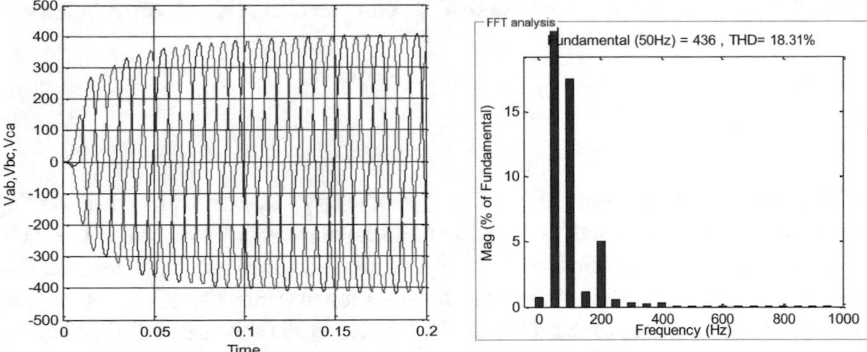

Fig. 11 Inverter output voltage and THD with filter using SPWM

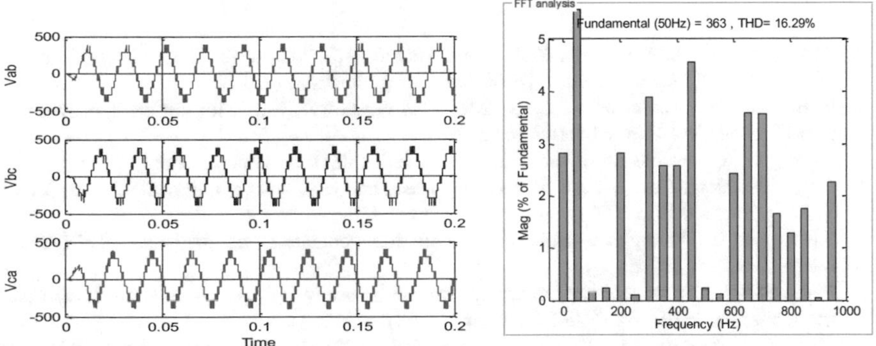

Fig. 12 Inverter output voltage and THD without filter using PD

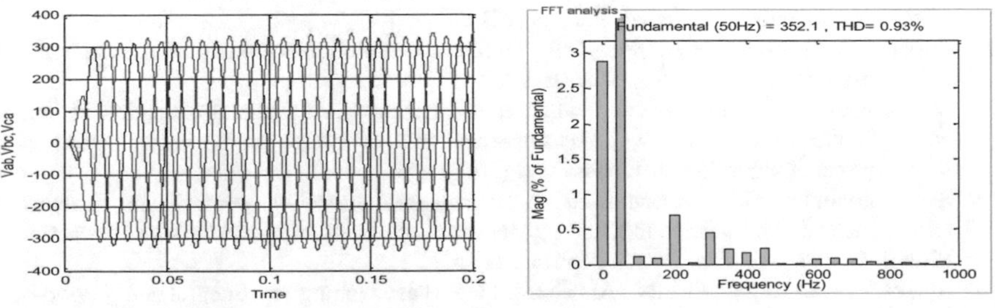

Fig. 13 Inverter output voltage and THD with filter using PD

The output voltage waveform and THD content while applying PD with filter are given in Fig. 13. THD value in this is 0.93% which is much less compared to APOD, POD, and SPWM methods, as with filter application of APOD gives 2.96%, POD gives 2.05%, and SPWM gives 18.39% THD. Also, the output voltage waveform observed here is more sinusoidal compared to other three techniques.

8 Conclusion

In this paper, a new stand-alone integrated wind energy system with multi-winding transformer is developed and THD of the multi-level inverter is observed. The THD is analyzed with and without filter by applying multi-level switching techniques like PD, APOD, and POD. PID controller gains are tuned using DE at each switching technique system. It was found that the THD is less in PD scheme of diode-clamped MLC base-integrated wind energy system with multi-winding transformer.

References

1. Abdullah, R., Rahim, N.Abd., Raihan, S., Ahmad, A.Z.: Five-level diode-clamped inverter with three-level boost converter. IEEE Trans. Ind. Electron. (2014)
2. El Ela, A.A.A., Abido, M.A., Spea, S.R.: Optimal power flow using differential evolution algorithm. Electric Power Syst. Res. (2010)
3. Afgan, N.: Renewable energy strategies for sustainable development. Energy (2007)
4. Ang, K., Chong, G., Li, Y.: PID control system analysis, design, and technology. Control Syst. Technol., IEEE (2005)
5. Bansal, R.C.: Three-phase self-excited induction generators: an overview. IEEE Trans. Energy Convers. (2005)
6. Bialasiewicz, J.T.: Renewable energy systems with photovoltaic power generators: operation and modeling. IEEE Trans. Ind. Electron. (2008)
7. Chen, Z., Guerrero, J.M., Blaabjerg, F.: A review of the state of the art of power electronics for wind turbines. IEEE Trans. Power Electron. (2009)
8. Mishra, S., Palu, I., Madichetty, S., Suresh Kumar, L.V.: Modelling of wind energy based microgrid system implementing MMC. Int. J. Energy (2016)
9. Das, D.C., Roy, A.K., Sinha, N.: PSO based frequency controller for wind-solar-diesel hybrid energy generation/energy storage system. In: Proceedings, International Conference on Energy, Automation and Signal, ICEAS (2011)
10. El Ela, A.A.A., Abido, M.A., Spea, S.R.: Differential evolution algorithm for optimal reactive power dispatch. Electric Power Syst. Res. (2011)
11. Sravan Kumar, B., Suryakalavathi, M., Nagesh Kumar, G.V.: Optimal allocation of unified power flow controller for power loss minimization and voltage profile improvement using harmony search algorithm. J. Electr. Eng. (2016)
12. Konstantopoulos, G.C., Alexandridis, A.T.: Full-scale modeling, control, and analysis of grid-connected wind turbine induction generators with back-to-back AC/DC/AC converters. IEEE J. Emerg. Sel. Top. Power Electron. (2014)
13. Kumar, L.V.S., Kumar, G.V.N., Anusha, D.: PSO based tuning of a integral and proportional integral controller for a closed loop stand alone multi wind energy system (2016)

14. Venkateswara Rao, B., Nagesh Kumar, G.V.: Firefly algorithm based optimal power flow for sizing of thyristor controlled series capacitor to enhance line based voltage stability. J. Electr. Eng. (2016)
15. Madichetty, S., Dasgupta, A., Suresh Kumar, L.V.: Application of modular multilevel converter for AGC in an interconnected power system. Int. J. Electr. Power Energy Syst. (2016)
16. Mei, J.: A new selective loop bias mapping phase disposition PWM with dynamic voltage balance capability for modular multilevel converter. IEEE Trans. Ind. Electron. (2014)
17. Mihet-Popa, L., Blaabjerg, F., Boldea, I.: Wind turbine generator modeling and simulation where rotational speed is the controlled variable. IEEE Trans. Ind. Appl. (2004)
18. Mishra, S., Palu, I., Madichctty, S., Surcsh Kumar, L.V.: Modelling of wind energy-based microgrid system implementing MMC. Int. J. Energy Res. (2016)
19. Oinoddin, S.: Three-phase to seven-phase power converting transformer. IEEE Trans. Energy Convers. (2012)
20. Odulio, C.M.F, Escoto, M.T.: Multiple winding transformer model for power supply applications in circuit simulations. In: World Congress on Engineering (2013)
21. Chen, Q.: A new model for multiple-winding transformer. In: Proceedings of Power Electronics Specialist Conference—PESC (1994)
22. Rodriguez, J., Lai, J.-S., Peng, F.Z.: Multilevel inverters: a survey of topologies, controls, and applications. IEEE Trans. Ind. Electron. (2002)
23. Suresh Kumar, L.V, Nagesh Kumar, G. V., Prasanna, P.S.: Differential evolution based tuning of proportional integral controller for modular multilevel converter STATCOM. In: Computational Intelligence in Data Mining, Jan 2016

Chemical Reaction Optimization: A Survey with Application and Challenges

Janmenjoy Nayak, Sura Paparao, Bighnaraj Naik, N. Seetayya, P. Pradeep, H. S. Behera and Danilo Pelusi

Abstract Both physical and chemical objects, properties, behaviours have remained a great inspiration for the optimization community to develop competitive algorithms in contrast to nature-inspired, swarm- and evolutionary-based algorithms. Although the number of developments in both the areas is only a few, still those algorithms are quite efficient to compete with other nature-inspired algorithms. In this paper, a brief review on a novel chemical reaction optimization algorithm is presented with its applications. The algorithm is based on certain chemical properties and has been a keen interest for solving various problems. Further, some future challenges are discussed for further improvements and applicability in other real-life problems.

J. Nayak (✉) · S. Paparao · N. Seetayya · P. Pradeep
Department of CSE, Sri Sivani College of Engineering, Srikakulam 532410, AP, India
e-mail: mailforjnayak@gmail.com

S. Paparao
e-mail: surapaparao@gmail.com

N. Seetayya
e-mail: seeth.n@gmail.com

P. Pradeep
e-mail: pradeepau.mtech@gmail.com

B. Naik
Department of Computer Application, Veer Surendra Sai University of Technology, Burla, Sambalpur 768018, Odisha, India
e-mail: mailtobnaik@gmail.com

H. S. Behera
Veer Surendra Sai University of Technology, Burla, Sambalpur 768018, Odisha, India
e-mail: mailtohsbehera@gmail.com

D. Pelusi
Communication Sciences, University of Teramo, Coste Sant'agostino Campus, Teramo, Italy
e-mail: dpelusi@unite.it

J. Nayak et al. (eds.), *Soft Computing in Data Analytics*,
Advances in Intelligent Systems and Computing 758,
https://doi.org/10.1007/978-981-13-0514-6_50

Keywords CRO · Chemical reaction optimization · Real-coded CRO

1 Introduction

God has created a beautiful gift for us called nature, where every possible creation occurs with a different intention. There are certain governing rules for each of the elements in nature, and possibly human being is the best findings than others. The principles of nature are quite different and extraordinary in solving any complex problems as like solving the problems in mathematical sciences with optimization techniques. Not only in the early days, but the present scenario of mathematics, science, engineering and computer science (especially advance computing) is quite dependent on optimization techniques. Optimization is nothing but solving the problems with two aspects like minimization and maximization with certain pre-defined constraints. The basic reason behind the popularity of optimization is the inefficiency of non-optimization-based techniques (classical methods) in solving complex problems. Moreover, finding accurate solutions, rounding and shaping the variables, satisfying the constraints, etc., are some of the added challenges for such methods. Although compared to such classical methods, some of the earlier developed optimization methods like constraint-based techniques are efficient, but they are based on the objective and some typical constraints. Prior to those, the most famous and frequently used optimization based on the concept of evolutionary theory and Darwin's principle is genetic algorithm (GA) that came into the picture and has been a successful history in almost all the diversified areas. Then, another effective algorithm based on the flocking behaviour of birds called particle swarm optimization (PSO) was developed by Kennedy et al. to solve large-scale complex problems. Likewise, thousands of algorithms based on the evolutionary and swarm behaviour were developed in recent years for meeting various challenges in different problem-solving environment. A brief history of such development (evolutionary-based, swarm-based, nature-based, physical- and chemical-based) is presented in Tables 1 and 2.

The statement of no free lunch theorem [1] states that 'there is no such algorithm exist which can solve all the problems better than some other algorithms'. Also, the algorithm, which is good at global maxima, is also good at some of the average points in the search space. So, always there will be a chance of overcoming one algorithm by the other at any point. As per the findings of optimization community, the performance of all the algorithms is mostly equivalent and also different in some specified problems only. Moreover, after the number of developments of such optimization techniques, still there is a huge scope to go ahead as the number of problem is more than the developed solutions. Especially, in the problems like NP-hard in complexity theory, more research on achievement of exact solutions rather approximate solutions is highly required. In 2010, Lam and Li [2] developed a new metaheuristic based on certain properties of chemical reactions called chemical reaction optimization (CRO), which has been widely acclaimed.

Table 1 Various swarm-based optimization algorithms

Algorithm name	Inspiration	Year of development	Name of the developer	Refs.
Bee system	Behaviour of bees	2001	Lucic and Teodorovic	[3]
Bacterial foraging	Social foraging behaviour of bacteria	2002	Passino	[4]
Fish school	Behaviour of fish swarm in searching of food	2002	Li et al.	[5]
Beehive	Communication in the hive of honey bees	2004	Wedde et al.	[6]
Virtual bees	Generation of virtual bees	2005	Yang	[7]
Glow swarm	Behaviour of glow swarms	2005	Krishnanand and Ghose	[8]
Bees algorithm	Mimics the foraging behaviour of honey bees	2005	Pham et al.	[9]
Artificial bee colony (ABC)	Cooperative behaviour of bee colonies	2005	Karaboga and Basturk	[10]
Virtual ant algorithm	Generation and behaviour of virtual ants	2006	Yang	[11]
Cat swarm	Behaviour of cats	2006	Chu et al.	[12]
Good lattice swarm	Number theory and particle swarm	2007	Su et al.	[13]
Monkey search	Behaviour of monkeys for searching food	2007	Mucherino and Seref	[14]
Hierarchical swarm model	Natural hierarchical complex system	2010	Chen et al.	[15]
Bumble bees	Collective behaviour of social insects	2011	Comellas and Martinez	[16]
Cuckoo search	Emulation of lifestyle of cuckoo birds	2011	Yang and Deb	[17]
Wolf search	Imitating the way of searching foods by wolves	2012	Tang et al.	[18]
Krill herd	Simulation of herding behaviours of krill individuals	2012	Gandomi and Alavi	[19]
Social spider algorithm	Cooperative behaviour of social spiders	2014	Cuevas and Cienfuegos	[20]
Antlion	Hunting mechanism of antlions	2015	Mirjalilli	[21]

Since its inception, it has outperformed many other swarm and evolutionary algorithms in solving various applications. However, as always there is a better chance for improvement and reaching at more appropriate solutions, a deep research is required for further enhancement of this algorithm. In this paper, a brief review on CRO has been conducted with its applicability in various research domains till date. Section 2 describes the preliminary concepts of CRO with all its components and their working modules. The applications of CRO in various

Table 2 Various bio-inspired optimization algorithms

Algorithm name	Inspiration	Year of development	Name of the developer	Refs.
Differential evolution	Evolutionary concept	1997	Storn and Price	[22]
Marriage in honey bees	Marriage process of honey bees	2001	Abbass	[23]
Gene expression	Character linear chromosomes composed of genes	2001	Ferreira	[24]
Artificial immune	Immune system of the human being	2002	de Castro and Leandro	[25]
Queen bee evolution	Reproduction process of bees	2003	Jung	[26]
Shuffled frog leaping	Communication among frogs	2003	Eusuff and Lansey	[27]
Invasive weed	Colonizing weeds	2006	Mehrabian and Lucas	[28]
Roach infestation	Behaviour of common cockroaches such as Periplaneta Americana	2008	Havens	[29]
Fish school	Behaviour of fish schools	2008	Lima et al.	[30]
Biogeography based	Principal of immigration and emigration of the species from one place to other	2008	Simon	[31]
Group search optimizer	Animals searching behaviour	2009	He et al.	[32]
Human-inspired algorithm	Intelligent search strategies of mountain climbers who use modern techniques (such as binoculars and cell phones) to effectively find the highest mountain in a given region	2009	Zhang et al.	[33]
Paddy field	Scattering seeds on the field	2009	Premaratne et al.	[34]
Termite colony	Intelligent behaviours and movement of termites	2010	Hedayatzadeh et al.	[35]
Brain storm	Human brainstorming process	2011	Shi	[36]
Eco-inspired evolutionary algorithm	Ecological concepts of habitats, relationships and ecological successions	2011	Parpinelli and Lopes	[37]
Japanese tree frogs calling	Calling behaviour of Japanese tree frogs	2012	Hern'andez and Blum	[38]
Great salmon run	Salmon run phenomena	2012	Mozaffari	[39]

(continued)

Table 2 (continued)

Algorithm name	Inspiration	Year of development	Name of the developer	Refs.
Optbees	Process of collective decision making by bee colonies	2012	Maia et al.	[40]
Flower pollination algorithm	Pollination process of flowers	2012/2013	Yang	[41]
Atmosphere cloud model	The behaviour of cloud in the natural world	2013	Yan and Hao	[42]
Dolphin echolocation	Echolocation ability of dolphins	2013	Kaveh and Farhoudi	[43]
Egyptian vulture	Behaviour and key skills of the Egyptian vultures for acquiring food for leading their livelihood	2013	Sur et al.	[44]
Grey wolf optimization	The leadership hierarchy and hunting mechanism of grey wolves	2014	Mirjalili et al.	[45]
Lion optimization	Lifestyle of lions and their cooperation characteristics	2016	Yazdani and Jolai	[46]

domains are outlined in Sect. 3. Section 4 describes some critical analysis on the performance of the algorithm with some important untouched areas, where it may have broader scope of applications. Finally, Sect. 5 concludes the work with future directions (Table 3).

2 Chemical Reaction Optimization

CRO was an attempt to design an optimization inspired by a set of chemical reaction procedures. Although the algorithm is based on chemical processes, it does not strictly/fully simulate the reaction procedures. Rather only some methods have been adopted. Some typical chemical reaction terms such as endothermic, exothermic, kinetic energy, potential energy are used to simulate the behavioural aspects of algorithm. The procedure of converting from potential to kinetic energy has been simulated in the algorithm. Basically, four types of chemical reactions such as intermolecular collision, ineffective collision, synthesis, decomposition are considered for describing the whole process of CRO. Based on the chemical change in each of the above reactions and their outcomes, the algorithms may be simulated for any real-life applications. Energy conservation is considered in CRO as equivalent process of natural selection in genetic algorithm. After completion of all the four reactions, the generation of molecules is not the same, and for that, CRO is

Table 3 Various chemical-/physical-based optimization algorithms

Algorithm name	Inspiration	Year of development	Name of the developer	Refs.
Simulated annealing	Analogy with annealing in solids	1983	Kirkpatrick et al.	[47]
Stochastic diffusion search	Exhaustive search through the restaurant and meal combinations	1989	Bishop	[48]
Self-propelled particles	Self-ordered motion in systems of particles with biologically motivated interaction	1995	Vicsek	[49]
Harmony search	Works on the principle of music improvisation in a music player	2001	Geem et al.	[50]
Intelligent water drop	Actions and reactions that take place between water drops in the river	2007	Shah-Hosseini	[51]
River formation dynamics	River formation	2007	Rabanal et al.	[52]
Gravitational search	Works on the principle of gravitational force acting between the bodies	2009	Rashedi et al.	[53]
Central force optimization	Metaphor of gravitational kinematics	2007	Formato	[54]
Grenade explosion method	Principle of explosion of a grenade	2010	Ahrari and Atai	[55]
Galaxy-based search algorithm	Spiral arm of spiral galaxies to search its surrounding	2011	Shah-Hosseini	[56]
Spiral optimization	Analogy of spiral phenomena	2011	Tamura and Yasuda	[57]
Big bang–big crunch	Big bang and big crunch theory	2012	Zandi et al.	[58]
Black hole	Black hole phenomenon	2012	Hatamlou	[59]
Charged system search	The Coulomb law from electrostatics and the Newtonian laws of mechanics	2012	Kaveh and Talatahari	[60]
Water cycle algorithm	Water cycle process and how rivers and streams flow to the sea in the real world	2012	Eskandar et al.	[61]
Wind-driven optimization	Atmospheric motion	2010	Bayraktar et al.	[62]
Water wave optimization	Shallow water wave theory	2015	Zheng et al.	[63]

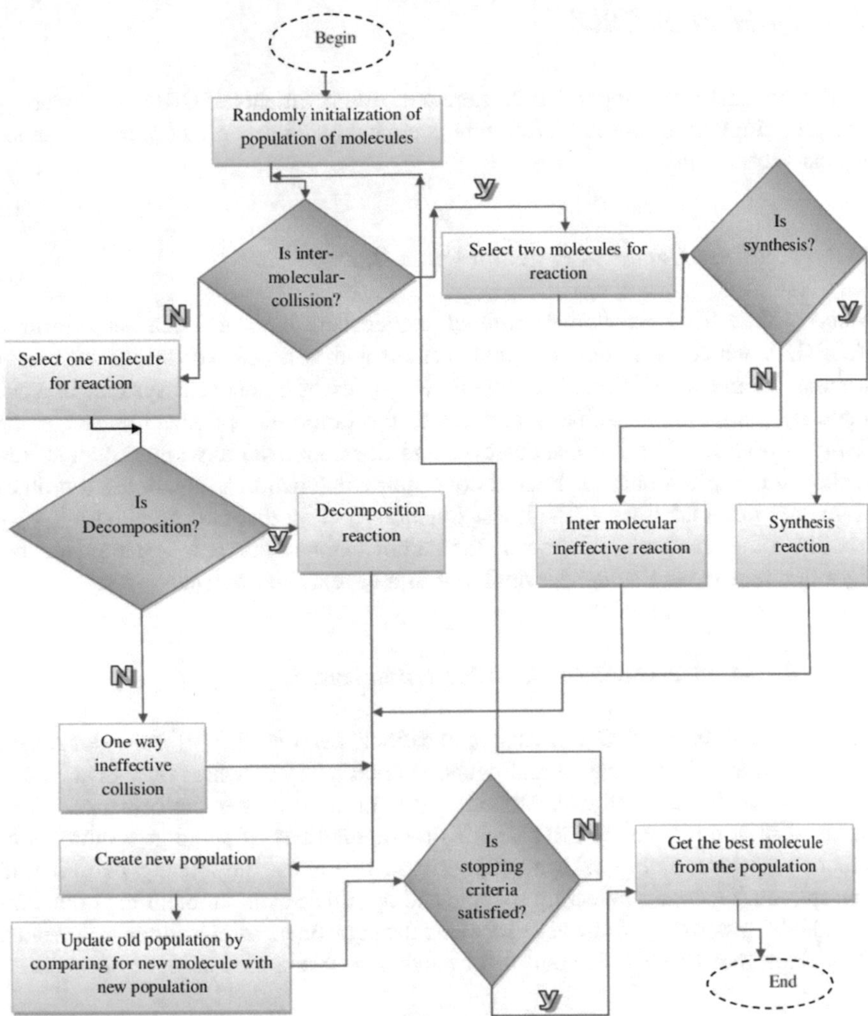

Fig. 1 Process of CRO

considered as a population-based metaheuristic with variability in nature. The detailed framework of CRO is illustrated in Fig. 1. Moreover, some other details such as processes of four elementary reactions, choosing the reactant molecules from each of the reactions can be found in [64].

2.1 Variants of CRO

Apart from firstly developed CRO, there are certain variants of CRO developed by using the simulating aspects of different chemical reactions and these are explained in next subsections.

2.1.1 Artificial Chemical Reaction Optimization

Alatas [65] in 2011 developed artificial chemical reaction optimization algorithm (ACROA), which is a computational method and is based on various factors of different chemical reactions. The factors are states of a chemical system, objects, processes, etc., and are suitably adhered to the proposed optimization. It has the ability at both local and global searches and does not need any other local search method for implementation. Except some normal function evaluations, the algorithm does not hold any special fitness functions and for that the complexity is very less than other methods. Moreover, the author claimed that it is a parameter-free algorithm and is well suited for applying in any real-life domains.

2.1.2 Real-Coded Chemical Reaction Optimization

After the success of CRO in many applications, Lam et al. [66] have developed another variant of CRO called real-coded version of CRO, where they made some modifications in originally developed CRO for more better performance. They modified in some areas of CRO such as representations of solution vectors, continuous searching ability with neighbourhood operators, introducing perturbation concept, implementing reflecting scheme and hybrid scheme for handling boundary constraints, proposing of adaptive schemes for adjustment of algorithm parameters. They claim that RCCRO is helpful for solving problems in continuous domain.

3 Applications of CRO

This section deals with various applications of CRO algorithm in diversified areas of research domains.

3.1 Bioinformatics

Alignment of protein sequence is considered to be important aspects of multiple sequence alignment problems, and the nature of such problem is NP-hard type.

Other than conventional approaches, nature-inspired techniques have shown their potentiality in solving such problems. Yadav and Banka [67] have proposed an improved version of CRO for solving the multiple alignment problems. They have emphasized on the development of initialization of the candidate solution through pairwise alignment method. So instead of using random population as like in traditional CRO, they used the pairwise alignment process which helps to obtain the global optimum solutions.

3.2 Developing and Testing of Software

Test case prioritization is a well-established area of research in software development, and choosing the correct test case is a challenging task. A regression-based test case method is developed by Mohapatra and Prasad [68] using CRO for object-oriented programming concepts. They measured the effectiveness of the considered test cases through average percentage fault detection method. The main focus of their method is to analyse on code coverage and execution time of the developed method. They claimed that with CRO-based technique, it is more easy and reliable to find the bugs at initial stage of execution of programs.

3.3 Graph Theory and Optimizations

The maximum flow problem in graph theory is a well-known research area and may be realized in many of the practical applications such as engineering, transport problems. The task in a maximum flow problem is to find out the optimality path/ solution in a weighted directed graph. A CRO-based strategy is developed by Barham et al. [69] to handle the maximum flow problem. With the experimentation of the proposed method, they found that CRO-based technique is more efficient than GA with the consideration of various factors like execution time, number of iterations. CRO is also successfully applied to solve capacitated ARP problem. The CARP problem can be defined as one of the NP-hard problems as it has to satisfy different demands from different clients situated at different places. Bensedira et al. [70] have proposed a new technique based on CRO to solve CARP problem. They have worked on the basic structure of molecules, and with the original CRO, they have tried to solve the problem efficiently. Another non-dominated sorting CRO is proposed by Bechikh et al. [71] for solving multi-objective optimization problems. They derived the method in terms of quasilinear average time complexity with a linear data structure. They tested their method with a number of other standard methods and found it more suitable than others. However, they also discussed some of the issues such as construction, internal and external validity.

3.4 Wireless Sensor Network

Wireless sensor network is one of the oldest research domains, and energy conservation is a sensitive area of this domain. Both clustering and routing are the frequently used techniques to conserve energy in WSN. Rao and Banka [72] have proposed some novel methods on clustering- and routing-based techniques with CRO to address the hotspot problems in WSN. For effective distribution of clusters, they have used cluster head algorithms with CRO mechanism for better performance in non-cluster heads. Also, CRO is used for routing purpose and with the consideration of a number of performance factors like number of live nodes, lifetime of the network, rate of convergence.

3.5 Task Scheduling

In any of the either homogeneous or heterogeneous networks, task scheduling plays a major role in reducing makespan or to decide about the further execution order. Xu et al. [73] have proposed an improved hybrid version of CRO to address the task scheduling problem in DAG-based system. To search the optimal local solutions, they have used a Gaussian random walk, and further to prove that their proposed algorithm is free from trapping at local optima, a left/right rotating procedure has been used. To maintain the diversity of molecules, a new selection strategy with normal distribution and a random shuffled-based method is used. The simulation results of their proposed method show the effectiveness of their method over some of the existing methods in solving the task scheduling problem.

3.6 Data Mining Problems

Solving various data mining problems such as classification, regression, clustering has always been a frequent interest among all researchers. Various techniques are adopted in solving all these problems. A novel higher-order pi–sigma neural network-based classification approach has been developed by Nayak et al. [64] to solve the data classification problem in data mining. They have used the basic procedures of CRO and used it to adjust the weights of the higher-order neural network with a quite reasonable error rate and high classification accuracy. They compared their proposed techniques with many of the existing standard methods like GA, PSO, and the experimental results seem to be superior to others. For dealing with the clustering problems, Nayak et al. [74] have implemented CRO with fuzzy c-means algorithm. Considering the limitations of FCM algorithm such as trapping at local optima, initial sensitiveness, they have implemented a hybrid approach with the integration of both CRO and FCM to address the clustering problem. After a rigorous experimentation with 13 real-world benchmark datasets, their proposed method is compared with many of the existing standard methods such as FCM, TLBO, k-means and hybrid algorithms of k-means to show the efficiency of the proposed method. Statistical methods and cluster validation index prove that the proposed method is quite well performed to address the clustering problem. Panigrahi [75] has proposed one hybrid approach with differential evolution and CRO-based method for training of neural network. Meeting with both diversification and intensification, his proposed method is suitably applied to solve parity and other classification problems.

3.7 Real-Life Problems

Some of the researchers have addressed few real-life problems with CRO techniques. Gan and Duan [76] have proposed a novel CRO-based method for intelligent detection of any particular visual object with the removal of noise. Their proposed method is quite similar to the capturing behaviour of human being. Due to fast training speed and convergence criteria of CRO, the method is well able to capture the visual objects without the more reduction of noise. Szeto et al. [77] have designed a new approach based on CRO to solve network design problem with considering some of the real factors like vehicle emission and noise. Their model is able to address the level of user's equilibrium as well as meeting the road capacity with less amount of emission. With the solution strategy to a number of

Table 4 Applications of CRO

Type of problem	Application area	Year	Refs.
Vehicle routing	Networking	2017	[78]
Flow shop scheduling	Operational research	2017	[79]
Community detection	Networking	2017	[80]
Optimizing refrigeration system	Electrical engineering	2017	[81]
Shop deteriorating scheduling	Mechanical engineering	2017	[82]
Vehicle repositioning	Transportation	2016	[83]
Complex surface coverage	WSN	2016	[84]
Fault diagnosis	Mechanical engineering	2015	[85]
Monopolist firm	Environmental engineering	2015	[86]
Economic emission	Electrical engineering	2014	[87]
Knapsack problem	Problem analysis	2013	[88]
Task mapping problem	Scheduling	2013	[89]
Motion problem	Robotics	2013	[90]
Job shop scheduling	Machine design	2012	[91]
Training of NN	Neural network	2012	[92]

environmental factors, their method outperforms the performance of GA and was found to be suitable for establishing the trade-off among the factors like cost and emission.

Apart from all the above-mentioned applications, CRO has also been used in some other important field of research domains and is summarised in Table 4.

4 Analytical Discussions

Although CRO is an optimization technique inspired through a series of chemical reactions, but like other optimization techniques (nature-inspired/swarm-based/ evolutionary-based), it is efficient in solving complex real-world problems. Due to the absence of a specific objective function with some complex computational parameters, its complexity is less than others. Many of the above-described literatures show that it is a well-performed optimization technique other than some competitive algorithms. On a keyword-based search in Google Scholar database, a total number of 2,770,00 records are found till this date (25 December 2017) which shows its applicability and popularity in many fields. CRO has been successfully used in the fields such as stock portfolio analysis, training neural networks, handling nonlinearity in clustering, wireless communication system, live streaming in peer-to-peer network, channel allocation problem, optimization of benchmark functions. Apart from all these application types, CRO has remained a frequent choice in the field of data mining and cloud computing-related areas. A clear depiction about the application areas of CRO has been illustrated in Fig. 2.

The main advantage of CRO is the maximum indication towards global minima with variable population size. With the use of stochastic processes, it is able to optimize the structure of any neural network and it gives the flexibility for not to change the network structure. Likewise, there are several other advantages that may be credited to it. However, few researches have also indicated some of the weak

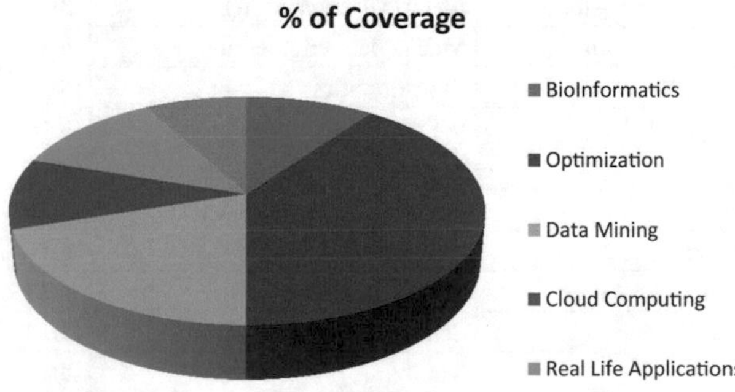

Fig. 2 Various applications of CRO

corners of this method, which are still in the limelight. Like other algorithms, CRO is not suitable for continuous domain and performs well in discrete domains. But later, the same authors developed another of its version called RCCRO, which is able to avoid that problem. In RCCRO with the help of perturbation, it is possible to change the parameters adaptively. Instead of using some standard complex mathematical models like the Gaussian method for adaptation purpose (as in RCCRO), some innovative ideas for self-adaption of all the parameters in all the reaction may be a challenging task.

5 Conclusion and Future Perspective

CRO is a chemical-inspired metaheuristic developed for solving real-life and optimization problems. Chemical concepts such as potential energy, kinetic energy, conservation of energy are used to govern the phases of chemical reactions in CRO. CRO does not use any additional function or parameter like fitness function of some other algorithms for determining the quality of reactants. Although many of the authors have indicated that it is a recently developed optimization algorithm, the literatures are showing much potential growth of this algorithm in its form of diversified application areas. Its population size is not fixed and always varies in the structure of population, for which it is able to maintain the same distance from diversification and intensification. As per 'no free lunch theorem', none of the algorithms is able to produce 100% of accuracy for all the applications, and depending on the type and nature of problems, the performance solution may vary. In concerned to that, CRO has remained quite successful than other evolutionary- and swarm-based algorithms in many of the major application areas. However, issues like large memory consumption, less error rate are still in embryonic position. Due to its adaptive nature, many of the real-life problems are solved with this and have remained successful. However, some of the other areas like load balancing and forecasting, economic despatching, public healthcare applications, pattern recognition are not yet being vastly considered with CRO. In future, these areas may be expanded with high-performance values and low computational cost.

References

1. Wolpert, D.H., Macready, W.G.: No free lunch theorems for optimization. IEEE Trans. Evol. Comput. **1**(1), 67–82 (1997)
2. Lam, A.Y.S., Li, V.O.K.: Chemical-reaction-inspired metaheuristic for optimization. IEEE Trans. Evol. Comput. **14**(3), 381–399 (2010)
3. Lucic, P., Teodorovic, D.: Bee system: modeling combinatorial optimization transportation engineering problems by swarm intelligence. In: Preprints of the TRISTAN IV Triennial Symposium on Transportation Analysis, pp. 441–445 (2001)

4. Kevin, M.: Passino Biomimicry of bacterial foraging for distributed optimization and control. Control Syst. IEEE **22**(3), 52–67 (2002)
5. Li, X.-L., Shao, Z.-J., Qian, J.-X.: Optimizing method based on autonomous animats: fish-swarm algorithm. Xitong Gongcheng Lilunyu Shijian/Syst. Eng. Theory Pract. **22**(11), 32 (2002)
6. Wedde, H.F., Farooq, M., Zhang, Y.: Beehive: an efficient fault-tolerant routing algorithm inspired by honey bee behavior. In: Lecture Notes in Computer Science (including subseries Lecture Notes in Artificial Intelligence and Lecture Notes in Bioinformatics). LNCS, vol. 3172, pp. 83–94 (2004)
7. Yang, X.-S.: Eng. Optim. Via Nat.-Inspired Virtual Bee Algoritm. **3562**, 317–323 (2005)
8. Krishnanand, K.N., Ghose, D.: Detection of multiple source locations using a glowworm metaphor with applications to collective robotics. In: Swarm Intelligence Symposium, 2005. SIS2005. Proceedings 2005 IEEE, pp. 84–91. IEEE (2005)
9. Pham, D.T., Ghanbarzadeh, A., Koc, E., Otri, S., Rahim, S., Zaidi, M.: The bees algorithm-a novel tool for complex optimisation problems. In: Proceedings of the 2nd Virtual International Conference on Intelligent Production Machines and Systems (IPROMS 2006), pp. 454–459 (2006)
10. Karaboga, D., Basturk, B.: A powerful and efficient algorithm for numerical function optimization: arti1ficialbee colony (abc) algorithm. J. Glob. Optim. **39**(3), 459–471 (2007)
11. Yang, X.-S., Lees, J.M., Morley, C.T.: Application of virtual ant algorithms in the optimization of cfrp shear strengthened precracked structures. In: Computational Science–ICCS 2006, pp. 834–837. Springer (2006)
12. Chu, S.-A., Tsai, P.-W., Pan, J.-S.: Cat swarm optimization. In: Lecture Notes in Computer Science (including subseries Lecture Notes in Artificial Intelligence and Lecture Notes in Bioinformatics), vol. 4099. LNAI:854–858 (2006). cited By (since 1996) 8
13. Su, S., Wang, J., Fan, W., Yin, X.: Goodlattice swarm algorithm for constrained engineering design optimization. In: International Conference on Wireless Communications, Networking and Mobile Computing, 2007. WiCom 2007, pp. 6421–6424. IEEE (2007)
14. Mucherino, A., Seref, O.: Monkey search: a novel metaheuristic search for global optimization. Data Min. Syst. Anal. Optim. Biomed. **953**, 162–173 (2007)
15. Chen, H., Zhu, Y., Hu, K., He, X.: Hierarchical swarm model: a new approach to optimization. Discret. Dyn. Nat. Soc. (2010)
16. de Comellas Padró, F.P., Mart´ınez Navarro, J., et al.: Bumblebees: a Multiagent Combinatorial Optimization Algorithm Inspired By Social Insect Behaviour (2011)
17. Yang, X.-S., Deb, S.: Cuckoo search via l´evy flights. In: World Congress on Nature & Biologically Inspired Computing, 2009. NaBIC 2009, pp. 210–214. IEEE (2009)
18. Tang, R., Fong, S., Yang, X.-S., Deb, S.: Wolf search algorithm with ephemeral memory. In: 2012 Seventh International Conference on Digital Information Management (ICDIM), pp. 165–172 (2012)
19. Gandomi, A.H., Alavi, A.H.: Krill herd: a new bio-inspired optimization algorithm. Commun. Nonlinear Sci. Numer. Simul. (2012)
20. Cuevas, E., Cienfuegos, M.: A new algorithm inspired in the behavior of the social-spider for constrained optimization. Expert Syst. Appl. **41**(2), 412–425 (2014)
21. Mirjalili, S.: The ant lion optimizer. Adv. Eng. Softw. **83**, 80–98 (2015)
22. Storn, R., Price, K.: Differential evolution–a simple and efficient heuristic for global optimization over continuous spaces. J. Glob. Optim. **11**(4), 341–359 (1997)
23. Abbass, H.A.: Mbo: Marriage in honey bees optimization-a haplometrosis polygynous swarming approach. In: Proceedings of the 2001 Congress on Evolutionary Computation, 2001, vol. 1, pp. 207–214. IEEE (2001)
24. Ferreira, C.: Gene Expression Programming: a New Adaptive Algorithm for Solving Problems (2001). c0102027
25. de Castro, L.N., Timmis, J.: Artificial Immune Systems: a New Computational Intelligence Approach. Springer Science & Business Media (2002)

26. Sung Hoon Jung: Queen-bee evolution for genetic algorithms. Electron. Lett. **39**(6), 575–576 (2003)
27. Eusuff, M.M., Lansey, K.E.: Optimization of water distribution network design using the shuffled frog leaping algorithm. J. Water Res. Plan. Manag. **129**(3), 210–225 (2003). cited By (since 1996) 297
28. Reza, A.: Mehrabian and C Lucas. A novel numerical optimization algorithm inspired from weed colonization. Ecol. Inf. **1**(4), 355–366 (2006)
29. Havens, T.C., Spain, C.J., Salmon, N.G., Keller, J.M.: Roach infestation optimization. In: Swarm Intelligence Symposium, 2008. SIS 2008, pp. 1–7. IEEE (2008)
30. de Lima Neto, F.B., Lins, A.J.C.C. Nascimento, A.I.S., Lima, M.P., et al.: A novel search algorithm based on fish school behavior. In: IEEE International Conference on Systems, Man and Cybernetics, 2008. SMC 2008, pp. 2646–2651. IEEE (2008)
31. Simon, D.: Biogeography-based optimization. IEEE Trans. Evol. Comput. **12**(6), 702–713 (2008)
32. He, S., Wu, Q.H., Saunders, J.R.: Group search optimizer: an optimization algorithm inspired by animal searching behavior. IEEE Trans. Evol. Comput. **13**(5), 973–990 (2009)
33. Zhang, L.M., Dahlmann, C., Zhang, Y.: Human-inspired algorithms for continuous function optimization. In: IEEE International Conference on Intelligent Computing and Intelligent Systems, 2009. ICIS2009, vol. 1, pp. 318–321. IEEE (2009)
34. Premaratne, U., Samarabandu, J., Sidhu, T.: A new biologically inspired optimization algorithm. In: 2009 International Conference on Industrial and Information Systems (ICIIS), pp. 279–284. IEEE (2009)
35. Hedayatzadeh, R., Salmassi, F.A., Keshtgari, M., Akbari, R., Ziarati, K.: Termite colony optimization: a novel approach for optimizing continuous problems. In: 2010 18th Iranian Conference on Electrical Engineering (ICEE), pp. 553–558. IEEE (2010)
36. Shi, Y.: An optimization algorithm based on brainstorming process. Int. J Swarm Intell. Res. (IJSIR) **2**(4), 35–62 (2011)
37. Parpinelli, R.S., Lopes, H.S.: An eco-inspired evolutionary algorithm applied to numerical optimization. In: 2011 Third World Congress on Nature and Biologically Inspired Computing (NaBIC), pp. 466–471. IEEE (2011)
38. Hern´andez, H., Blum, C.: Distributed graph coloring: an approach based on the calling behavior of Japanese tree frogs. Swarm Intell. **6**(2), 117–150 (2012)
39. Mozaffari, A., Fathi, A., Behzadipour, S.: The great salmon run: a novel bio–inspired algorithm for artificial system design and optimisation. Int. J. Bio-Inspired Comput. **4**(5), 286–301 (2012)
40. Maia, R.D., de Castro, L.N., Caminhas, W.M.: Bee colonies as model for multimodal continuous optimization: The optbees algorithm. In: 2012 IEEE Congress on Evolutionary Computation (CEC), pp. 1–8. IEEE (2012)
41. Yang, X.-S.: Flower pollination algorithm for global optimization. Unconv. Comput. Nat. Comput. 240–249 (2012)
42. Yan, G.-W., Hao, Z.-J.: A novel optimization algorithm based on atmosphere clouds model. Int. J. Comput. Intell. Appl. (2013)
43. Kaveh, A., Farhoudi, N.: A new optimization method: Dolphin echolocation. Adv. Eng. Softw. **59**, 53–70 (2013)
44. Sur, C., Sharma, S., Shukla, A.: Egyptian vulture optimization algorithm–a new nature inspired metaheuristics for knapsack problem. In: The 9th International Conference on Computing and Information Technology (IC2IT2013), pp. 227–237. Springer (2013)
45. Mirjalili, S., Mirjalili, S.M., Lewis, A.: Grey wolf optimizer. Adv. Eng. Softw. **69**, 46–61 (2014)
46. Yazdani, M., Jolai, F.: Lion optimization algorithm (LOA): a nature-inspired metaheuristic algorithm. J. Comput. Des. Eng. **3**(1), 24–36 (2016)
47. Kirkpatrick, S., Gelatt Jr., D., Vecchi, M.P.: Optimization by simulated annealing. Science **220**(4598), 671–680 (1983)

48. Bishop, J.M.: Stochastic searching networks. In: First IEE International Conference on (Conf. Publ. No. 313) Artificial Neural Networks, pp. 329–331. IET (1989)
49. Vicsek, T., Czir´ok, A., Ben-Jacob, E., Cohen, I., Shochet, O.: Novel type of phase transition in a system of self-driven particles. Phys. Rev. Lett. **75**(6), 1226–1229 (1995)
50. Geem, Z.W., Kim, J.H., Loganathan, G.V.: A new heuristic optimization algorithm: harmony search. Simulation **76**(2), 60–68 (2001)
51. Shah-Hosseini, H.: Problem solving by intelligent waterdrops. In: IEEE Congress on Evolutionary Computation, 2007. CEC 2007, pp. 3226–3231. IEEE (2007)
52. Rabanal, P., Rodr´ıguez, I., Rubio, F.: Using river formation dynamics to design heuristic algorithms. In: Unconventional Computation, pp. 163–177. Springer (2007)
53. Rashedi, E., Nezamabadi-Pour, H., Saryazdi, S.: Gsa: a gravitational search algorithm. Inf. Sci. **179**(13), 2232–2248 (2009)
54. Formato, R.A.: Central force optimization: a new metaheuristic with applications in applied electromagnetics. Prog. Electromagn. Res. **77**, 425–491 (2007)
55. Ahrari, A., Atai, A.A.: Grenade explosion method—a novel tool for optimization of multimodal functions. Appl. Soft Comput. **10**(4), 1132–1140 (2010)
56. Shah-Hosseini, H.: Principal components analysis by the galaxy-based search algorithm: a novel metaheuristic for continuous optimisation. Int. J. Comput. Sci. Eng. **6**(1), 132–140 (2011)
57. Tamura, K., Yasuda, K.: Spiral dynamics inspired optimization. J. Adv. Comput. Intell. Intell. Inf. **15**(8), 1116–1122 (2011)
58. Zandi, Z., Afjei, E., Sedighizadeh, M.: Reactive power dispatch using big bang-big crunch optimization algorithm for voltage stability enhancement. In: 2012 IEEE International Conference on Power and Energy (PECon), pp. 239–244. IEEE (2012)
59. Hatamlou, A.: Black hole: a new heuristic optimization approach for data clustering. Inf. Sci. (2012)
60. Kaveh, A., Talatahari, S.: A novel heuristic optimization method: charged system search. Acta Mech. **213**(3–4), 267–289 (2010)
61. Eskandar, H., Sadollah, A., Bahreininejad, A., Hamdi, M.: Water cycle algorithm–a novel metaheuristic optimization method for solving constrained engineering optimization problems. Comput. Struct. (2012)
62. Bayraktar, Z., Komurcu, M., Werner, D. H.: Wind driven optimization (WDO): a novel nature-inspired optimization algorithm and its application to electromagnetics. In: 2010 IEEE Antennas and Propagation Society International Symposium (APSURSI), pp. 1–4. IEEE (2010)
63. Zheng, Y.J.: Water wave optimization: a new nature-inspired metaheuristic. Comput. Oper. Res. **55**, 1–11 (2015)
64. Nayak, J., Naik, B., Behera, H.S.: A novel chemical reaction optimization based higher order neural network (CRO-HONN) for nonlinear classification. Ain Shams Eng. J. **6**(3), 1069–1091 (2015)
65. Alatas, B.: ACROA: artificial chemical reaction optimization algorithm for global optimization. Expert Syst. Appl. **38**(10), 13170–13180 (2011)
66. Lam, A.Y., Li, V.O., James, J.Q.: Real-coded chemical reaction optimization. IEEE Trans. Evol. Comput. **16**(3), 339–353 (2012)
67. Yadav, R.K., Banka, H.: An improved chemical reaction-based approach for multiple sequence alignment. Curr. Sci. **112**(3), 527 (2017)
68. Mohapatra, S.K., Prasad, S.: A chemical reaction optimization approach to prioritize the regression test cases of object-oriented programs. J. ICT Res. Appl. **11**(2), 113–130 (2017)
69. Barham, R., Sharieh, A., Sliet, A.: Chemical reaction optimization for max flow problem. Int. J. Adv. Comput. Sci. Appl. **7**(8), 189–196 (2016)
70. Bensedira, B., Layeb, A., Bouzoubia, S., Habbas, Z.: CRO-CARP: a chemical reaction optimization for capacitated arc routing problem. In: 2016 8th International Conference on Modelling, Identification and Control (ICMIC), pp. 757–762. IEEE (2016)

71. Bechikh, S., Chaabani, A., Said, L.B.: An efficient chemical reaction optimization algorithm for multiobjective optimization. IEEE Trans. Cybern. **45**(10), 2051–2064 (2015)
72. Rao, P.S., Banka, H.: Novel chemical reaction optimization based unequal clustering and routing algorithms for wireless sensor networks. Wirel. Netw. **23**(3), 759–778 (2017)
73. Xu, Y., Li, K., He, L., Zhang, L., Li, K.: A hybrid chemical reaction optimization scheme for task scheduling on heterogeneous computing systems. IEEE Trans. Parallel Distrib. Syst. **26** (12), 3208–3222 (2015)
74. Nayak, J., Naik, B., Behera, H.S., Abraham, A.: Hybrid chemical reaction based metaheuristic with fuzzy c-means algorithm for optimal cluster analysis. Expert Syst. Appl. **79**, 282–295 (2017)
75. Panigrahi, S.: A novel hybrid chemical reaction optimization algorithm with adaptive differential evolution mutation strategies for higher order neural network training. Int. Arab J. Inf. Technol (IAJIT) **14**(1) (2017)
76. Gan, L., Duan, H.: Chemical reaction optimization for feature combination in bio-inspired visual attention. Int. J. Comput. Intell. Syst. **8**(3), 530–538 (2015)
77. Szeto, W.Y., Wang, Y., Wong, S.C.: The chemical reaction optimization approach to solving the environmentally sustainable network design problem. Comput.-Aided Civ. Infrastruct. Eng. **29**(2), 140–158 (2014)
78. Li, H., Wang, L., Hei, X., Li, W., Jiang, Q.: A decomposition-based chemical reaction optimization for multi-objective vehicle routing problem for simultaneous delivery and pickup with time windows. Memet. Comput. 1–18 (2017)
79. Bargaoui, H., Driss, O.B., Ghédira, K.: A novel chemical reaction optimization for the distributed permutation flowshop scheduling problem with makespan criterion. Comput. Ind. Eng. **111**, 239–250 (2017)
80. Chang, H., Feng, Z., Ren, Z.: Community detection using dual-representation chemical reaction optimization. IEEE Trans. Cybern. (2017)
81. Hadidi, A.: A novel approach for optimization of electrically serial two-stage thermoelectric refrigeration systems using chemical reaction optimization (CRO) algorithm. Energy **140**, 170–184 (2017)
82. Fu, Y., Wang, Z., Zhang, J., Wang, Z.: A blocking flow shop deteriorating scheduling problem via a hybrid chemical reaction optimization. Adv. Mech. Eng. **9**(6), 1687814017701371 (2017)
83. Szeto, W.Y., Liu, Y., Ho, S.C.: Chemical reaction optimization for solving a static bike repositioning problem. Transportation research part D: transport and environment **47**, 104–135 (2016)
84. Yang, R., Fan, Y., Zhao, N., Cheng, Y.: Improved chemical reaction optimization algorithm for 3D complex surface coverage. In: 2016 IEEE 13th International Conference on Networking, Sensing, and Control (ICNSC), pp. 1–6. IEEE (2016)
85. Ao, H., Cheng, J., Yang, Y., Truong, T.K.: The support vector machine parameter optimization method based on artificial chemical reaction optimization algorithm and its application to roller bearing fault diagnosis. J. Vib. Control **21**(12), 2434–2445 (2015)
86. Choudhary, A., Suman, R., Dixit, V., Tiwari, M.K., Fernandes, K.J., Chang, P.C.: An optimization model for a monopolistic firm serving an environmentally conscious market: Use of chemical reaction optimization algorithm. Int. J. Prod. Econ. **164**, 409–420 (2015)
87. Roy, P.K.: Hybrid chemical reaction optimization approach for combined economic emission short-term hydrothermal scheduling. Electr. Power Compon. Syst. **42**(15), 1647–1660 (2014)
88. Truong, T.K., Li, K., Xu, Y.: Chemical reaction optimization with greedy strategy for the 0–1 knapsack problem. Appl. Soft Comput. **13**(4), 1774–1780 (2013)
89. Xu, C., Li, T.: Chemical reaction optimization for task mapping in heterogeneous embedded multiprocessor systems. In: Advanced Materials Research, vol. 712, pp. 2604–2610. Trans Tech Publications (2013)
90. Astudillo, L., Melin, P., Castillo, O.: Chemical optimization paradigm applied to a fuzzy tracking controller for an autonomous mobile robot. Int. J. Innov. Comput. Inf. Control **9**(5), 2007–2018 (2013)

91. Li, J.Q., Pan, Q.K.: Chemical-reaction optimization for flexible job-shop scheduling problems with maintenance activity. Appl. Soft Comput. **12**(9), 2896–2912 (2012)
92. James, J.Q., Lam, A.Y., Li, V.O.: Evolutionary artificial neural network based on chemical reaction optimization. In: 2011 IEEE Congress on Evolutionary Computation (CEC), pp. 2083–2090. IEEE (2011)

Shuffled Differential Evolution-Based Combined Heat and Power Economic Dispatch

S. Nagaraju, A. Srinivasa Reddy and K. Vaisakh

Abstract A novel metaheuristic algorithm SDE augments the features both shuffled frog-leaping algorithm and differential evolution algorithm by employing partitioning and shuffling. In order to verify the effectiveness of the shuffled-differential evolution (SDE) algorithm and also to identify the ideal solution of the CHPED problem, test systems having four units are considered. The outcomes attained from the projected technique are contrasted with the other optimization techniques and found that the projected technique shows remarkable performance in the resolution and the conjunction characteristics.

Keywords Combined heat and power economic dispatch (CHPED)
Shuffled frog-leaping algorithm • Differential evolution

1 Introduction

Combined heat and power economic dispatch (CHPED) comes across as a complex subject in electric power systems, carrying out a crucial job in the monetary operation of power system. CHPED happens to be one of the most complex issues because of its nonlinear and non-convex characteristics. The optimum exploitation of the combined heat and power units along with other units which produce either heat or electric power exclusively dictate the economics of the power system. Hence, the economic operation of the cogeneration units along with the other units

S. Nagaraju (✉)
Aditya Institute of Technology and Management, Tekkali 532201, India
e-mail: Subuddi_nagaraju@yahoo.com

A. Srinivasa Reddy
C.R. Reddy College of Engineering, Eluru, India
e-mail: srinivasareddyalla@yahoo.co.in

K. Vaisakh
Andhra University College of Engineering, Visakhapatnam, India
e-mail: vaisakh_k@yahoo.co.in

J. Nayak et al. (eds.), *Soft Computing in Data Analytics*,
Advances in Intelligent Systems and Computing 758,
https://doi.org/10.1007/978-981-13-0514-6_51

for minimizing the fuel cost becomes a critical hitch that requires potent procedures to resolve. The generation of electric power from fossil fuels is now seen as an incompetent procedure. Even the latest combined cycle plants have an efficacy in the range of 50–60% only.

The main objective of the CHPED is to determine the optimum operating point of power and heat generation values of all the operating units with the less fuel cost, such that the system electric power, heat demands, and other constraints are met, while the CHP units are running in the feasible region bounded by heat versus power of the corresponding unit. Hence, a robust optimization technique is necessary to obtain a quality solution for CHPED problem. Hence, the usage of gradient-based optimization techniques fails to ensure in obtaining the optimal solution.

With advancements and new developments in the stochastic search techniques, these stochastic search techniques are becoming more popular to address the complex engineering issues [1]. Therefore, various intelligent techniques came up on the horizon, to crack the CHPED problem. A revised IGA-MU tactic has been introduced [2], genetic algorithm-oriented penalty function method (GA_PF) [3], multi-objective particle swarm optimization (PSO) [4], an innovative selective particle swarm optimization (SPSO) [5], differential evaluation (DE) [6], differential evolution with Gaussian mutation [7], A bio-enthused GA (BIGA) [8], A fish school search algorithm (FSS) [9], modified group search optimization (MGSO) [10], evolutionary programming (EP) [11], integrating tabu search method and genetic algorithm (GT) [12], cuckoo search algorithm [13], etc., have all been projected as ways to handle the CHPED issues.

Shuffled differential evolution (SDE) is build by merging features of the shuffled frog-leaping algorithm (SFLA) [14] with differential evolution (DE) [15] and benchmarked by applying it to ED problem [16, 17]. SDE is actively used to address CHPED problem with valve-point loading, transmission losses, and along with other system operational constraints. To explain the achievement of the SDE, one standard test system out five are considered and the outcomes have been accompanied by other evolving methods in vogue.

The following is an outline of this paper:

Section 2 explains the mathematical formulation of the CHPED.

Section 3 implementation of SDE.

Section 4 provides the application of the projected algorithm process for 2 trial cases. Finally, Sect. 5 concludes this paper.

2 Problem Statement

The main desideratum of the combined heat and power economic dispatch (CHPED) problem is to evaluate optimal operating of unit's heat and power values, while the constraints of system are taken care. Figure 1 demonstrates the bounded

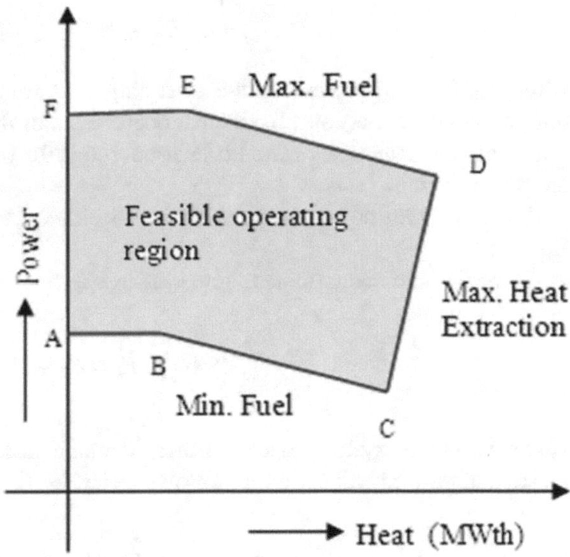

Fig. 1 Feasible area of operation for a CHP unit

viable area of a combined heat and power (CHP) unit. The ABCDEF curve is the operation region of CHPED unit.

The CHPED objective function is to

$$\text{Minimize}\, F = \sum_{j=1}^{N_p} C_j(P_j) + \sum_{k=1}^{N_c} C_k(P_k, T_k) + \sum_{l=1}^{N_o} C_l(T_l) \tag{1}$$

where C_j, C_k, and C_l are operating cost of the power-only, CHP, and heat-only units, respectively where N_p, N_c, and N_o are the number of the conventional thermal units, CHP, units, and heat-only units, respectively. j, k, and l are the indices used for the power-only, CHP, and heat-only units, respectively. F is the total cost; P_j and T_l are the power and heat output of the power-only and heat-only units, respectively; P_k and T_k indicate the power and heat output of the jth CHP unit.

The cost function for the conventional thermal unit in Eq. (1) is defined as follows:

$$C_j(P_j) = a_j(P_j)^2 + b_jP_j + c_j + \left|d_j \text{xsin}\left(e_j\text{x}\left(P_j^{min} - P_j\right)\right)\right| \quad (\$/\text{h}) \tag{2}$$

where $C_j(P_j)$ represents jth convention thermal unit's cost function for producing power $P_{j;}$ a_j, b_j, c_j, d_j, and e_j stands are the cost coefficients.

The kth CHP unit cost in Eq. (1) is defined as

$$C_k(P_k, T_k) = a_k(P_k)^2 + b_kP_k + c_k + d_k(T_k)^2 + e_kT_k + f_kT_kP_k \quad (\$/\text{h}) \tag{3}$$

And lth heat-only unit cost function in Eq. (1) is defined as in Eq. 4

$$C_l(T_l) = a_l(T_l)^2 + b_l T_l + c_l \quad (\$/\text{h}) \tag{4}$$

where $C_k(P_k, H_k)$ represents the operating cost function of the kth cogeneration unit; a_k, b_k, c_k, d_k, e_k, and f_k are cost coefficients of the kth CHP unit. $C_k(T_k)$ is the operating cost function of the lth heat-only unit for producing T_l heat; a_l, b_l, and c_l are its cost coefficients.

The following constraints should be considered for the problem of CHPED in Eq. (1):

Power balance constraint is given as in Eq. 5

$$\sum_{j=1}^{N_p} P_j + \sum_{k=1}^{N_c} P_k = P_d + P_{loss} \tag{5}$$

where P_d is the system electric power demand and P_{loss} is system transmission losses and can be calculated as follows as in Eq. 6:

$$P_{loss} = \sum_{i=1}^{N_p} \sum_{m=1}^{N_p} P_i B_{im} P_m + \sum_{i=1}^{N_p} \sum_{j=1}^{N_c} P_i B_{ij} P_j + \sum_{j=1}^{N_c} \sum_{l=1}^{N_c} P_j B_{jl} P_l \tag{6}$$

where B_{im}, B_{ij}, and B_{jl} are transmission loss coefficients.

Heat balance constraint is given in Eq. 7

$$\sum_{j=1}^{N_C} T_j + \sum_{k=1}^{N_o} T_k = T_d \tag{7}$$

where T_d is system heat demand.

Generations limit of the conventional thermal unit Eq. (2) is represented in Eq. 8

$$P_j^{min} \le P_j \le P_j^{max},\ \text{i} = 1, 2, 3, \ldots\ldots \text{N}_\text{p} \tag{8}$$

where P_j^{min} is lower power limit and P_j^{max} is upper limit of power for the jth power only unit.

Capacity limit of the CHP units Eq. (3) is represented in Eqs. 9 and 10

$$P_j^{min}(T_j) \le P_j \le P_j^{max}(T_j), \quad \text{j} = 1, 2, 3, \ldots\ldots \text{N}_\text{C} \tag{9}$$

$$T_j^{min}(P_j) \le T_j \le T_j^{max}(P_j), \quad \text{j} = 1, 2, 3, \ldots\ldots \text{N}_\text{C} \tag{10}$$

where $P_j^{min}(T_j)$ is the lower power limit and $P_j^{max}(T_j)$ is the upper power limits of the jth CHP unit and $T_j^{min}(P_j)$ is the lower heat limit and $T_j^{max}(P_j)$ is the maximum heat limits of the jth CHP unit.

Capacity limit of the heat-only unit in Eq. (4) is given in Eq. 11

$$T_k^{min} \leq T_k \leq T_k^{max}, \quad k = 1, 2, 3, \ldots\ldots N_o \tag{11}$$

where T_k^{min} is the lower heat frontier and T_k^{max} is the upper heat frontier of the *k*th heat-only unit.

3 Implementation of SDE Algorithm for Solving CHPED Problem

The economic load dispatch was successfully solved using SDE [14, 16]. The SDE is hybrid method which combines the features of both DE and SFLA. The inherent restrictions of SFLA and DE in solving large-scale complex economic load dispatch problems having non-differentiable and non-convex cost function is eliminated in the SDE. For detailed steps of this algorithm refer [14, 16].

4 Simulation Results and Analysis

SDE is exercised on four-unit test system problems for verifying its optimization capability. The computational results are compared to investigate the achievement of the SDE proposition with the other previous evolutionary methods. This SDE algorithm is implemented using MATLAB 10 and executed on Intel Core 2 Duo Processor, 1.66 GHz, and 2 GB personal computer.

4.1 Result Analysis

4.1.1 Test system 1

This SDE algorithm is applied first on a four-unit system containing 1 conventional thermal unit, two CHP units, and one heat-only unit for four load profiles to show the potency of this SDE algorithm in productive in accomplishing quality solutions. All the data for this test system are taken from [18].

As seen from Table 1, the proposed SDE could reach the global cost of \$9257.075 which is equal with the result of DE algorithm. From Fig. 2, it is clear that SDE is able to reach the final solution in less than 5 iterations whereas DE took around 15 iterations to reach the final solution.

Table 1 Optimum solutions for test system 1-load profile 1

	Power/heat								
	Unit 1	Unit 2		Unit 3		Unit 4			
Method	P_1	P_2	T_2	P_3	T_3	T_4	TP	TH	TC($)
DE	0.00	160.00	40.00	40.00	75.00	0.00	200.00	115.00	9257.0750
SDE	0.00	160.00	40.00	40.00	75.00	0.00	200.00	115.00	9257.0750

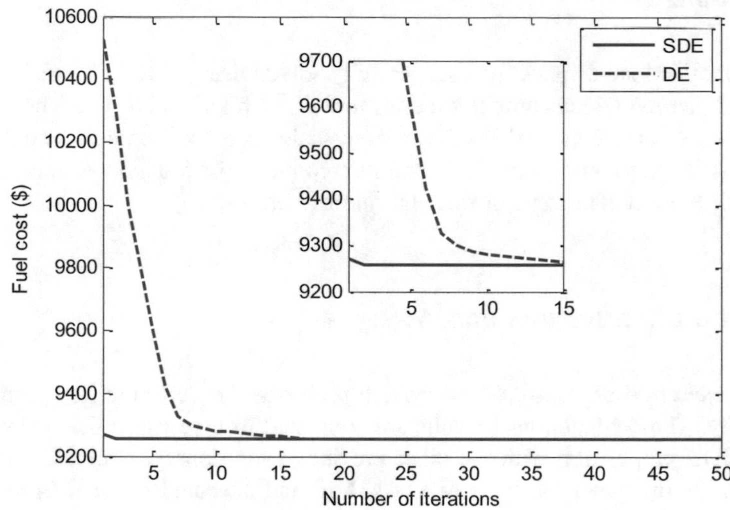

Fig. 2 Convergence characteristics of test system 1-load profile 1

4.1.2 Test System 1-Load Profile2 (LP2)

Table 2 contains the results obtained by different evolutionary algorithms for load profile 2 (P_{dx} = 175 MW and T_d = 110 Mwth) and are compared with the results obtained by proposed SDE algorithm. As seen from Table 2, this method could reach the global cost of $8555.9625 and is in accordance with the results of recent studies. SDE is able to converge in less iterations as compared with DE.

Table 2 Optimum solutions for test system 1-load profile 2

	Power/heat								
	Unit 1	Unit 2		Unit 3		Unit 4			
Method	P_1	P_2	T_2	P_3	T_3	T_4	TP	TH	TC($)
DE	0.00	135.00	35.00	40.00	75.00	0.00	175.00	110.00	8555.9625
SDE	0.00	135.00	35.00	40.00	75.00	0.00	175.00	110.00	8555.9625

5 Conclusion

For the problem of CHPED, a novel shuffled differential evolution optimization algorithm is adapted for 1 test system (with different demands) out of 5 test systems with fuel convex (non-convex) cost function are presented. The obtained simulation results of SDE are collated with the different intelligent algorithms published earlier in the literature. The comparison of results shows that this SDE method successfully attains the goal of reduced generation cost; and therefore, SDE can be used as a credible tool in cracking the CHPED problem. Considering the solutions of this paper, SDE algorithm can be applied efficiently to the numerous problems in the power system.

References

1. Spall, J.C.: Introduction to Stochastic Search and Optimization: Estimation, Simulation, and Control, vol. 65. Wiley (2005)
2. Su, C.T., Chiang, C.L.: An incorporated algorithm for combined heat and power economic dispatch. Electr. Power Syst. Res. **69**(2–3), 187–195 (2004)
3. Song, Y.H., Xuan, Q.Y.: Combined heat and power economic dispatch using genetic algorithm based penalty function method. Electr. Mach. Power Syst. **26**(4), 363–372 (1998)
4. Wang, L.F., Singh, S.: Stochastic combined heat and power dispatch based on multi-objective particle swarm optimization. Int. J. Electr. Power Energy Syst. **30**(3), 226–234 (2008)
5. Ramesh, V., Jayabarathi, T., Shrivastava, N., Baska, A.: A novel selective particle swarm optimization approach for combined heat and power economic dispatch. Electr. Power Compon. Syst. **37**(11), 1231–1240 (2009)
6. Basu, M.: Combined heat and power economic dispatch by using differential evolution. Electr. Power Compon. Syst. **38**, 996–1004 (2010)
7. Jena, C., Basu, M., Panigrahi, C.: Differential evolution with Gaussian mutation for combined heat and power economic dispatch. Soft Comput. 1–8 (2014)
8. Abdolmohammadi, H.R., Jafari, S., Rajati, M.R., Nazari, M.E.: A bio-inspired genetic algorithm applied to a constrained optimization problem in power systems. In: First Joint Congress on Fuzzy and Intelligent Systems, Mashhad, Iran, pp. 1–6 (2007)
9. dos Trigueiro, S.L., Costa e Silva, M.A., Santos, C.L.: combined heat and power economic dispatch by a fish school search algorithm. In: 13th Brazilian Congress of Thermal Sciences and Engineering, Uberlandia, MG, Brazil, No. ENC10-0571 (2010)
10. Hagh, M.T., Teimourzadeh, S, Alipour, M.: Combined heat and power dispatch using modified group search optimization method. In: International Power System Conference, Tehran, Iran, No. E-13-AAA-0000 (2013)
11. Wong, K.P., Algie, C.: Evolutionary programming approach for combined heat and power dispatch. Electr. Power Syst. Res. **61**(3), 227–232 (2002)
12. Sudhakaran, M., Slochanal, S.M.R.: Integrating genetic algorithms and tabu search for combined heat and power economic dispatch. In: TENCON 2003-Conference on Convergent Technologies for Asia-Pacific Region, Bangalore, India, pp. 67–71 (2003)
13. Nguyen, T.T., Dieu, N.V., Bach H.D.: Cuckoo search algorithm for combined heat and power economic dispatch. Electr. Power Energy Syst. **81**, 204–214 (2016)
14. Eusuff, M., Lansey, K., Pasha, F.: Shuffled frog-leaping algorithm: a memetic metaheuristic for discrete optimization. Eng. Optim. **38**(2), 129–154 (2006)

15. Storn, R., Price, K.: Differential evolution—a simple and efficient heuristic for global optimization over continuous spaces. J. Glob. Optim. **11**, 341–359 (1997)
16. Srinivasa Reddy, A., Vaisakh, K.: Shuffled differential evolution for large scale economic dispatch. Electr. Power Syst. Res. **96**, 237–245 (2013)
17. Srinivasa Reddy, A., Vaisakh, K.: Shuffled differential evolution for economic dispatch with valve point loading effects. Electr. Power Energy Syst. **46**, 342–352 (2013)
18. Mohammadi-Ivatloo, B., Moradi-Dalvand, M., Rabiee, A.: Combined heat and power economic dispatch problem solution using particle swarm optimization with time varying acceleration coefficients. Electr. Power Syst. Res. **95**, 9–18 (2013)

Fuzzy Granular Computing Controls Ill-Conditioned Matrix Definitions for Triple Inverted Pendulum

Srikanth Kavirayani and Umme Salma

Abstract Fuzzy logic control is applied for the control about the upright equilibrium position of a triple inverted pendulum by integrating time delay into system matrix definitions which makes the system well-conditioned from ill-conditioned. A novel fuzzy preview controller is proposed which uses the concept of fuzzy logic controllers which will define the controller for the plant. The output responses of the system are observed for efficient usage of energy with fuzzy controller. It is observed granular computing as speeded up the settling process and the fuzzy controller with granular computing has yielded improved response time.

Keywords Stability · Triple inverted pendulum · Time delay
Fuzzy granular computing · Preview control

1 Introduction

The inverted pendulum is a test bed in the control education which has been studied by researchers worldwide originating from the classical Indian rope trick. This test system has various models like the single inverted pendulum (SIP), double inverted pendulum (DIP), triple inverted pendulum (TIP) which are having linear motion in one plane, and also there are models with rotary aspects such as the rotary inverted pendulum (RIP) which have angular motion. The control studies of TIP can be applied for gymnast dynamic studies and aerobics as it closely resembles a gymnast

S. Kavirayani (✉) · U. Salma
Department of EEE, GITAM, Visakhapatnam 530045, Andhra Pradesh, India
e-mail: kavirayanisrikanth@gmail.com

U. Salma
e-mail: usalma123@gmail.com

S. Kavirayani
Department of EEE, Gayatri Vidya Parishad College of Engineering (A), Visakhapatnam 530048, Andhra Pradesh, India

J. Nayak et al. (eds.), *Soft Computing in Data Analytics*,
Advances in Intelligent Systems and Computing 758,
https://doi.org/10.1007/978-981-13-0514-6_52

in motion. The stabilization problem can be understood on similar lines as it is a classical model for studies on similar dynamic systems.

Eltohamy et al. [1] have discussed iterative control system design for a triple inverted pendulum where experimental studies were based on a DC motor driven by a microcomputer. Gupta et al. [2] and Sehgal et al. [3] have discussed the cart on pendulum model with three links, and an LQR controller was designed which is analyzed in MATLAB environment. Srikanth [4] has discussed the stabilization of the double inverted pendulum model using neural networks. Sivanandam et al. [5] have discussed ideology of comparing systems with and without time delays, and Amit [6] discussed the basic computational intelligence technique with emphasis on particle swarm optimization. The triple link pendulum being an extension is investigated in this work. Kamil et al. [7] have studied the robot gymnast stabilization about the upright equilibrium position using discrete linear quadratic regulator. The possible application of TIP systems has been studied in [7, 8] related to human postures and systems which learn from humans such as the humanoid robots. Huaguang [9] has discussed the general aspects of fuzzy logic and fuzzy control which could be applied to the system like the robot gymnasts or humanoid robots. Rajasekharan et al. [10] have discussed ideas of fuzzy logic, genetic algorithms, and neural network with applications specific to control systems which are partially extracted and applied in this work. Min et al. [11] have discussed the general aspects of time delay systems and the issues in dealing with time delay systems in terms of robustness and stability which are incorporated into the studies of the triple inverted pendulum system studies. Lai et al. [12] have discussed the granular computing technique where scaling down information into granules would help with the better study of a system in general and how it could be used in conjunction with fuzzy logic for improvising system studies. Drainkov et al. [13] have discussed the basic aspects of fuzzy control and how it could be extrapolated and tested for large-scale systems. Nagesh et al. [14] have studied aspects of rotary inverted pendulum control with integrated time delay which is combined with fuzzy control in this paper. Authors [15–17] have discussed various control methodologies for efficient design of controllers for nonlinear plant models. Yao et al. [18] have elaborated on the system identification aspects using fuzzy, and Tharun et al. [19] have applied fuzzy granular computing using PID controllers to the single inverted pendulum which is scaled up to higher-level modeling of the triple inverted pendulum in this paper.

In this paper, a new model of the TIP is proposed with integrated time delay into the system matrix which is not investigated so far in the literature. The proposed method not only considers the case of integrating the time delay into the system matrix but also investigates a method of using fuzzy lookup as a basis for the optimization methodology adopted for control of the TIP system. The control effort being an important guiding parameter in system design for intelligent control is expected that minimum control effort is taken by the system in order to stabilize the system when the system moves from the unstable equilibrium to the stable equilibrium. The TIP system being more complex than the double inverted pendulum (DIP) system is susceptible to unstable states which needed to be controlled with a single input. The system being underactuated is controlled by taking the studies using a pole placement-based full-state feedback controller which is then fed to the lookup table of the fuzzy controller. The parameters of settling time and overshoot for a particular state when the linearized model is taken for stabilization under the vicinity of the unstable equilibrium state are analyzed as the pole locations predominantly indicate the stability of the system. The studies are compared to classical techniques such as LQR with and without time delays.

2 Mathematical Modeling of the Triple Inverted Pendulum with Delay

The system model that would represent the triple inverted pendulum on a cart is shown below as in Sucheta et al. [3].

$$H1(\bar{z})\ddot{Z} = H_1(\bar{Z})\dot{Z} + H_3 Z + h_0 U \quad (1)$$

$$Z = (\mathrm{x}, \theta_1, \theta_2, \theta_3)^{\mathrm{T}} \quad (2)$$

$$\mathrm{h}_0 = (1 \quad 0 \quad 0 \quad 0)^{\mathrm{T}} \quad (3)$$

Equation 1 represents the basic dynamic model representation with the states defined for the three angle variations about the vertical upright equilibrium position where Z in Eq. 2 represents the four important parameters cart position, lower pendulum position, middle pendulum position, and upper pendulum position and h0 indicate the initialized values in Eq. (3). The state matrices defined are as follows with the defining matrices in Eq. 4 based on Euler equation defined for robotic systems.

$$H_1(\bar{z}) = \begin{bmatrix} a_0 & a_1\cos\theta_1 & a_2\cos\theta_2 & m_3 l_3\cos\theta_3 \\ a_1\cos\theta_1 & b_1 & a_2 L_1\cos(\theta_2-\theta_1) & m_3 L_1 l_3\cos(\theta_3-\theta_1) \\ a_2\cos\theta_2 & a_2 L_1\cos(\theta_2-\theta_1) & b_2 & m_3 L_2 l_3\cos(\theta_3-\theta_2) \\ m_3 l_3\cos\theta_3 & m_3 L_1 l_3\cos(\theta_3-\theta_1) & m_3 L_2 l_3\cos(\theta_3-\theta_2) & J_3+m_3 l_3^2 \end{bmatrix} \quad (4)$$

$$a_0 = m_0 + m_1 + m_2 + m_3 \quad (5)$$

$$a_1 = m_1 l_1 + m_2 L_1 + m_3 L_1 \quad (6)$$

$$a_2 = m_2 l_2 + m_3 L_2 \quad (7)$$

$$b_1 = J_1 + m_1 l_1^2 + m_2 l_1^2 + m_3 l_1^2 \quad (8)$$

$$b_2 = J_2 + m_2 l_2^2 + m_3 l_2^2 \quad (9)$$

In Eqs. 5–9, m_i, l_i, L_i, and J_i indicate the mass, length to center of mass, pendulum length, and inertia of the cart, lower, middle, and upper pendulum masses, respectively.

And

$$H_2(\overline{z},\dot{\overline{z}}) = \begin{bmatrix} -f_0 & a_1 \sin\theta_1 \cdot \dot{\theta}_1 & a_2 \sin\theta_2 \cdot \dot{\theta}_2 & m_3 l_3 \sin\theta_3 \cdot \dot{\theta}_3 \\ 0 & -f_1 - f_2 & a_2 L_1 \sin(\theta_2) \cdot \dot{\theta}_2 & m_3 L_1 l_3 \sin(\theta_3 - \theta_1) \cdot \dot{\theta}_3 \\ 0 & -a_2 L_1 \cos(\theta_2 - \theta_1) \cdot \dot{\theta}_1 + f_2 & -f_2 - f_3 & m_3 L_2 l_3 \sin(\theta_3 - \theta_2) \cdot \dot{\theta}_3 + f_3 \\ 0 & -m_3 L_1 l_3 \sin(\theta_3 - \theta_1) \cdot \dot{\theta}_1 & -m_3 L_2 l_3 \sin(\theta_3 - \theta_2) \cdot \dot{\theta}_2 + f_3 & -f_3 \end{bmatrix} \tag{10}$$

$$h_3(\overline{z}) = [0 \quad a_1 g \sin\theta_1 \quad a_2 g \sin\theta_2 \quad m_3 g l_3 \sin\theta_3]^T \tag{11}$$

which yields the state matrix as follows

$$\dot{X} = \begin{bmatrix} 0 & I_{4*4} \\ E^{-1}H & E^{-1}G \end{bmatrix} X + \begin{bmatrix} 0 \\ E^{-1}H \end{bmatrix} U \tag{12}$$

$$Y = CX + DU \tag{13}$$

The state vector governing Eqs. 12 and 13 is given below in Eq. 14

$$X = \begin{bmatrix} x & \theta_1 & \theta_2 & \theta_3 & \dot{x} & \dot{\theta}_1 & \dot{\theta}_2 & \dot{\theta}_3 \end{bmatrix}' \tag{14}$$

$$E = \begin{bmatrix} a_0 & a_1 & a_2 & m_3 l_3 \\ a_1 & b_1 & a_2 L_1 & m_3 L_1 l_3 \\ a_2 & a_2 L_1 & b_2 & m_3 L_2 l_3 \\ m_3 l_3 & m_3 L_1 l_3 & m_3 L_2 l_3 & J_3 + m_3 l_3^2 \end{bmatrix} \tag{15}$$

$$H = \begin{bmatrix} 0 & 0 & 0 & 0 \\ 0 & a_1 g & 0 & 0 \\ 0 & 0 & a_2 g & 0 \\ 0 & 0 & 0 & m_3 l_3 g \end{bmatrix} \tag{16}$$

$$G = \begin{bmatrix} -f_0 & 0 & 0 & 0 \\ 0 & -f_1 - f_2 & 0 f_2 & 0 \\ 0 & f_2 & -f_2 - f_3 & f_3 \\ 0 & 0 & f_3 & -f_3 \end{bmatrix} \tag{17}$$

The submatrices in the state equation can be defined as above in Eqs. 15, 16, and 17 which complete the formulation of the system definition before incorporation of the time delay.

3 Time Delay Integration by Reconstruction of the Dynamics

Let 'X' be the displacement of the triple pendulum cart which is the only parameter which is influenced by time delay defined as in Eq. 18.

$$G_D(s) = Gp(s)e^{-sT} \tag{18}$$

e^{-sT} can be approximated with Pade approximation as defined in Eq. 19

$$e^{-st} = (\frac{1 - s\tau/2)}{1 + s\tau/2)} \tag{19}$$

If Xm indicates the new measured state, the relationship between X and Xm with the time delay can be given as follows based on Eq. 20

$$X_m = X(\frac{1 - s\tau/2)}{1 + s\tau/2)} \tag{20}$$

Rewriting this equation, we would have the time delay incorporated as in Eq. 21

$$X_m(1 + s\tau/2) = X(1 - s\tau/2) \tag{21}$$

Rearranging the terms in this equation, we would have Eq. 22

$$X_m - X = \frac{s\tau}{2}(X + X_m) \tag{22}$$

The function in the frequency domain is converted into time domain by the Eq. 23

$$\dot{X}(s) = sX(s) - X(0) \tag{23}$$

Rewriting the equation given in (22), we have Eq. 24 which is given as

$$2X_m - (X_m + X) = \frac{s\tau}{2}(X + X_m) \tag{24}$$

Reorganizing the equation by defining λ such that we define a new state in Eq. 25

$$\lambda = (X + X_m) \tag{25}$$

The equation in (25) can be rewritten as follows:

$$2X_m - \lambda = \frac{s\tau}{2}\lambda \tag{26}$$

Rearranging the terms and using the identity defined in (23)

$$\frac{2}{\tau}(2X_m - \lambda) = s\lambda = \dot{\lambda} \tag{27}$$

which yields the new state equation for the delay state equation defined in Eq. 29

$$\dot{\lambda} = \frac{4}{\tau}X_m - \frac{2}{\tau}\lambda \tag{28}$$

Now if the typical three link inverted pendulum is represented by the following equations, then the equation represented would get reconstructed with row and column generation as follows: If the equation below represents the state space model

$$\dot{X} = \begin{bmatrix} 0 & I_{4*4} \\ E^{-1}H & E^{-1}G \end{bmatrix} X + \begin{bmatrix} 0 \\ E^{-1}H \end{bmatrix} U \tag{29}$$

The new set of state variables with the new state variable λ is given as follows:

$$X_{New} = \begin{bmatrix} X_m & \theta_1 & \theta_2 & \theta_3 & \dot{x} & \dot{\theta}_1 & \dot{\theta}_2 & \dot{\theta}_3 & \lambda \end{bmatrix}' \tag{30}$$

The state equation would be represented as follows

$$\dot{X} = \begin{bmatrix} \begin{bmatrix} 0 & 0 & 0 & 0 \\ 0 & 0 & 0 & 0 \\ 0 & 0 & 0 & 0 \\ 0 & 0 & 0 & 0 \end{bmatrix} & \begin{bmatrix} 1 & 0 & 0 & 0 \\ 0 & 1 & 0 & 0 \\ 0 & 0 & 1 & 0 \\ 0 & 0 & 0 & 1 \end{bmatrix} & 0 \\ E^{-1}H & E^{-1}G & 0 \\ \begin{bmatrix} \frac{4}{\tau} & 0 & 0 & 0 \end{bmatrix} & \begin{bmatrix} 0 & 0 & 0 & 0 \end{bmatrix} & \frac{2}{\tau} \end{bmatrix} \begin{bmatrix} x_m \\ \theta \\ \theta \\ \theta \\ x \\ \theta \\ \theta \\ \theta \\ \lambda \end{bmatrix} + \begin{bmatrix} 0 \\ E^{-1}h_0 \\ 0 \end{bmatrix} U \tag{31}$$

Comparing with the standard form $\underset{NEW}{\dot{X}} = AX_{NEW} + B_{NEW}U$, the values can be categorized as follows as defined in Eq. 31 and the matrix coefficients are given as in Eq. 32 with the time delay parameters included.

$$A = \begin{bmatrix} a_{11} & a_{12} & a_{13} & a_{14} & a_{15} & a_{16} & a_{17} & a_{18} & a_{19} \\ b_{21} & b_{22} & b_{23} & b_{24} & b_{25} & b_{26} & b_{27} & b_{28} & b_{29} \\ c_{31} & c_{32} & c_{33} & c_{34} & c_{35} & c_{36} & c_{37} & c_{38} & c_{39} \\ d_{41} & d_{42} & d_{43} & d_{44} & d_{45} & d_{46} & d_{47} & d_{48} & d_{49} \\ e_{51} & e_{52} & e_{53} & e_{54} & e_{55} & e_{56} & e_{57} & e_{58} & e_{59} \\ f_{61} & f_{62} & f_{63} & f_{64} & f_{65} & f_{66} & f_{67} & f_{68} & f_{69} \\ g_{71} & g_{72} & g_{73} & g_{74} & g_{75} & g_{76} & g_{77} & g_{78} & g_{79} \\ h_{81} & h_{82} & h_{83} & h_{84} & h_{85} & h_{86} & h_{87} & h_{88} & h_{89} \\ i_{91} & i_{92} & i_{93} & i_{94} & i_{95} & i_{96} & i_{97} & i_{98} & i_{99} \end{bmatrix} \tag{32}$$

With the values as defined as per appendix A. The ith column would have the integrated values of the time delay that is incorporated into the system.

$$i_{91} = 4/\tau\,;\, i_{92} = 2/\tau\,;\, i_{93} = 2/\tau\,;\; i_{94} = 2/\tau\,;\, i_{95} = 0;\, i_{96} = 0;\; i_{97} = 0;\; i_{98} = 0;\; i_{99} = 0;$$

3.1 Modification of Ill-Conditioned Matrices

Equations defined in Eqs. 30 and 31 change the system definition from ill-conditioned to well-conditioned taking the time delay parameter into the system definition. Systems are said to be ill-conditioned when small changes in the system definitions will aid to large change in solutions [20]. The variations in time delay signify this particular aspect when investigated from the perspective that we have added the column and row by generating the same for new matrix definition. The transmission matrix and the input matrices also have one element each added with a null value which changes the structure of these matrices; however, the performance aspect is not completely affected due to these changes; however, this change plays a crucial role in aspect definition of the system. The analysis for the definition of the augmentation and the improvement in the matrix definition is based on Eqs. 33 and 34 which yield the matrix definition for well-conditioned matrices. S_i is the estimate of the addition of elements in matrix where a_{ij} corresponds to the element within row i and column j, and S_i indicates the measured value. And, K_{meas} is the value that indicates whether the matrix definition has become well-conditioned or ill-conditioned based on the condition number. This definition thus makes the system well-conditioned mathematical definition because the condition number typically would be infinite without the row and column augmentation.

$$S_i = \left[a_{il}^2 + a_{i2}^2 + \cdots\right]^{1/2} \tag{33}$$

$$K_{meas} = \frac{|A|}{s_1 s_2 s_3 \ldots s_n} \tag{34}$$

4 Novel Fuzzy Preview Control Design

Preview-based logic is the study of methods and principles of reasoning in abstract manner with the knowledge of future, where logical reasoning means obtaining new propositions from existing propositions that are predefined [5]. The systematic framework for fuzzy modeling and fuzzy control of nonlinear systems with uncertainties is based on three types of models mostly which could be the Mamdani model or Takagi Sugeno model and fuzzy hyperbolic model. The Mamdani model constructs a bridge between the operator's knowledge and condition statements that are framed. The preview-based fuzzy controller is designed for the triple inverted pendulum taking into consideration the highly unstable triple pendulum on cart as shown in Fig. 1.

A rule base composed by rules where the rule antecedent contains the state variables $X = (x_1, x_2, x_3, x_4, x_5, x_6, x_7, x_8, x_9)$ represents the controller input variables. The consequent output contains a single control output U based on [5]. The process states variables X can take n_1 and n_2 linguistic values, respectively. Under these conditions, the maximum number of rule base is $n_1 * n_2$. The region of study for the analysis of the preview-based fuzzy controller in the state space is

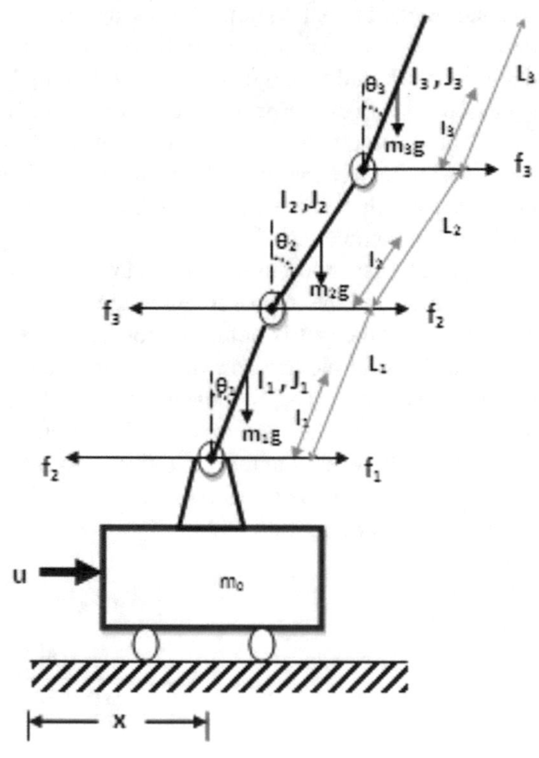

Fig. 1 Triple inverted pendulum on a cart model [3]

normally bounded by some finite values (min, max) for $X_{i=1 \text{ to } 9}$. A Type-1 fuzzy logic controller has been designed with two inputs—error and rate of error, and one output—F. The controller uses traditional lookup tables which are defined based on fuzzy members for the fuzzy membership values defined based on the possible inputs. The rate of error is additionally differentiated for further granular computing which results in faster lookup. The analysis is to be done by taking the time derivative of error and error itself and passing on to the fuzzy lookup table where the lookup table has data which is a set of possible fuzzy numbers defined for each of the adjectives of distortion of the pendulum from the unstable equilibrium position which decides the deviation. The output of the fuzzy lookup table is given as the input to the plant model as shown in Fig. 2, and then, a feedback gain matrix defined is used for controlling the plant for a given reference cart position. The model that is developed can be considered for the preview of stepping that happens during a staircase stepping case of the humanoid robot or for studies of robot gymnast models. The preview data is looked up using the fuzzy controller and is passed on to the plant model and is analyzed. The variation of the rate of error and the error itself are studied as different cases for analyzing the stability issues for the highly nonlinear dynamic system that is perturbed with variations. The initial conditions are assumed that the pendulums are in the vicinity of their unstable equilibrium position. It is assumed and the pendulums swing up from the downward equilibrium position to the upward position is already achieved. The studies on moving the three links from their stable equilibrium position to the upward equilibrium position are not considered here in this work. The work strictly concentrates on aspects of efficient energy usage of the pendulum systems about its unstable equilibrium position where the control energy is studied in terms of peak overshoots, settling time, and variation in parameters like the error and rate of errors. The block diagram clearly depicts the basic control loops that are involved in the design of the fuzzy preview controller using lookup tables.

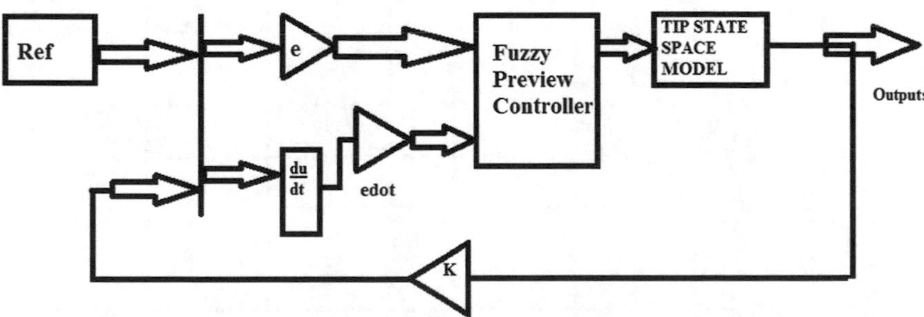

Fig. 2 Fuzzy preview control plant model

5 Results and Discussion

The output obtained with a basic linear quadratic controller with the integrated model is given in Fig. 3 wherein the integrated plant model is taken and peak overshoot is seen for the delay component of the plant dynamics. The maximum permissible variation in displacement is less than 10°. Table 1 indicates the various case studies performed on the triple inverted pendulum system. The analysis was performed to understand the significance of error influences on the dynamic response of the system. The analysis is done as per Table 1 indicated the case wherein the stability is difficult to achieve with fuzzy preview control. This indicates the practical case wherein it is virtually impossible to obtain zero error and also the case when the error is in the extremity of being at a very high value. In cases 2, 3, 4, it is clearly seen that incorporating the delay as a separate state has yielded a condition wherein maximum effort is required for stabilizing the state representing and corresponding to the time delay component. This categorically implies that the effort that is required in terms of the input for stabilization of other states is less. The other important observation is that the variations indicate that the system being highly unstable would oscillate continuously which indicates the difficulty involved in stabilization of a system such as the triple inverted pendulum plant.

The relative importance of error gain variation with error rate gain which act as the inputs for the fuzzy lookup tables is projected in Fig. 4. The variations are as per Table 1 which indicates that the variation of error gain with error rate gain should be relatively less in order to better stabilize a system smoothly without much variation.

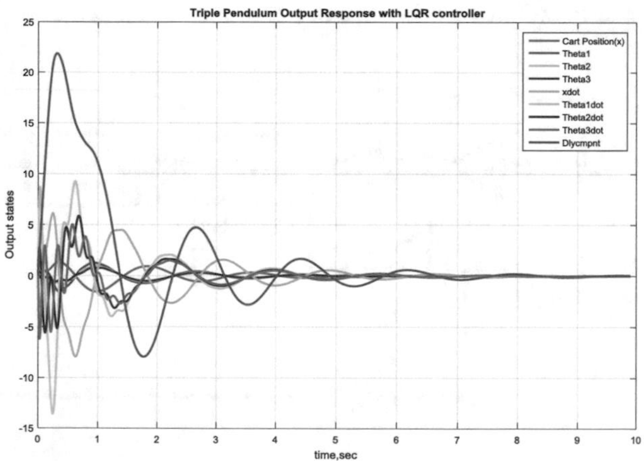

Fig. 3 Linear quadratic controller response for integrated plant

Table 1 Case studies for preview control

Case ID	Error gain	Error rate gain	Control status
1	20	0.02	Stable
2	2	0.002	Stable
3	0	0	Unstable
4	10	0.001	Stable
5	100	0	Unstable

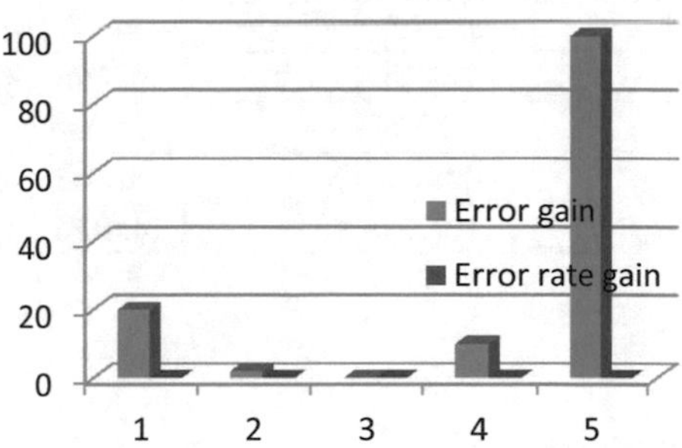

Fig. 4 Relative analysis for error gain and error rate

6 Conclusions

The major conclusions that can be drawn from the analysis are as follows. Time delays can be integrated into the system matrices for efficient studies of complex dynamic systems which here are analyzed by taking the particular control problem of the stabilization of the triple inverted pendulum system control. The integrated models change the system from ill-conditioned matrices to well-conditioned matrices, and granular computing makes lookup faster. Computational intelligence can be better utilized for complex dynamic systems to reduce the energy consumption and reduce the control efforts required for stabilization of system such as the inverted pendulums. Fuzzy preview control can be achieved by making use of lookup tables and then tweaking the parameters such as error gain and rate of error gain could be used to evolve systems such as triple inverted pendulum system to respond if the system is trained using a neural network to better learn the system.

Acknowledgements The authors would like to thank Gayatri Vidya Parishad College of Engineering (Autonomous) for providing the test facilities for conducting the research.

Appendix A

Matrix parameter					Value	Parameter	Value
a_{11}, a_{12}, a_{13}, a_{14}, b_{21}, b_{22}, b_{23}, b_{24}, c_{31}, c_{32}, c_{33}, c_{34}, d_{41}, d_{42}, d_{43}, d_{44}, e_{59},; f_{69}, g_{79}					0	f_{62}	67.1071
a_{17}, a_{18}, a_{19}, b_{26}, b_{28}, b_{29}, c_{36}, c_{37}, c_{38}, c_{39}, d_{46}, d_{47}, d_{48}, e_{51}, f_{61}, g_{71}, h_{81}, a_{16}, b_{27}, c_{38}, d_{49}, h_{89}					1	e_{52}	−12.4928
Parameter	Value	Parameter	Value	Parameter	Value	Parameter	Value
f_{66}	0.0039	e_{55}	−5.1127	g_{76}	−0.4334	f_{63}	65.2564
e_{53}	−2.0824	e_{56}	0.0075	g_{77}	1.1287	f_{64}	−71.9704
e_{54}	2.2956	e_{57}	0.0024	g_{78}	−0.7492	f_{68}	0.1659
h_{87}	−1.3621	e_{58}	−0.0053	h_{86}	−10.8077	f_{67}	−0.1948
h_{88}	0.826;	f_{65}	14.0176	h_{86}	0.6476	g_{75}	5.2021

References

1. Eltohamy, K.G., Kuo, C.Y.: Real time stabilisation of a triple link inverted pendulum using single control input. IEEE Proc. Control Theory Appl. **144**(5), 498–504 (1997)
2. Gupta, M.K., Bansal, K., Singh, A.K.: Stabilization of triple link inverted pendulum system based on LQR control technique. Recent Adv. Innov. Eng. (ICRAIE) Jaipur 2014, 1–5 (2014)
3. Sehgal, S., Tiwari, S.: LQR control for stabilizing triple link inverted pendulum system. In: 2012 2nd International Conference on Power, Control and Embedded Systems (ICPCES), Allahabad, pp. 1–5 (2012)
4. Srikanth, K.: Classical and Neural Net Control and Identification of Non-linear Systems with Application to the Two-joint Inverted Pendulum Control Problem. University of Missouri-Columbia, Diss (2005)
5. Sivanandam, S.N., Deepa, S.N.: Control Systems Engineering using MATLAB. Vikas Publishing house Pvt Ltd (2009)
6. Konar, A.: Computational Intelligence Techniques Principles, Techniques and Applications. Springer (2010)
7. Kamil, H.G., Eldukhri, E.E., Packianather, M.S.: Balancing control of robot gymnast based on discrete-time linear quadratic regulator technique, modelling and simulation (AIMS). In: 2014 2nd International Conference on Artificial Intelligence, Madrid, pp. 137–142 (2014)
8. von Bremiere, Y., Ribreau, C.: A double inverted pendulum model for studying the adaptability of postural control to frequency during human stepping in place. Biol. Cybern. **79**, 337–345(1998)
9. Zhang, H., Liu, D.: Fuzzy Modeling and Fuzzy control. Springer Science and Business Media, Inc. (2006)
10. Rajasekaran, S., Vijayalakshmi Pai, G.A.: Neural Networks, Fuzzy Logic, and Genetic Algorithms Synthesis and Applications. PHI Learning Pvt Ltd (2003)
11. Wu, M., He, Y., She, J.-H.: Stability analysis and robust control of time-delay systems. Springer, Science Press, Beijing
12. Lai, R., Chiang, D.: Constraint-based granular computing for fuzzy modeling, FUZZ-IEEE'02. In: Proceedings of the 2002 IEEE International Conference on Fuzzy Systems, vol. 1, no. pp. 584–589 (2002). https://doi.org/10.1109/fuzz.2002.1005057

13. Drainkov, D., Hellendoorn, H., Reinfrank, M.: An Introduction to Fuzzy Control. Narosa Publications
14. Kavirayani, S., Nagesh Kumar, G.V.: Rotary inverted pendulum control and the impact of time delay on switching between stable and unstable states with enhanced particle swarm optimization. Int. J. Comput. Commun. Syst. Eng. (IJCCSE) **02**(04) (2015)
15. Feifei, R., Lei, Z., Le, T., Yanhai, H., Pengpeng, Z.: Design and research of double closed-loop control strategy for inverted pendulum system. In: Third International Conference on Intelligent System Design and Engineering Applications (ISDEA), pp. 550–553, 16–18 Jan 2013
16. Mehrez, O., Ramadan, A.: A control strategy for designing an intelligent controller for highly dynamic/perturbed systems. In: 2012 12th International Conference on Control, Automation and Systems (ICCAS), pp. 924–929, 17–21 Oct 2012
17. Li, W., Ding, H., Cheng, K.: An investigation on the design and performance assessment of double-PID and LQR controllers for the inverted pendulum. In: 2012 UKACC International Conference on Control (CONTROL), pp. 190–196, 3–5 Sept 2012
18. Hongwei, Y., Xiaorong, M., Xianyi, Z.: Identification and design of multivariable fuzzy neural network system, 2000. In: Proceedings of the 3rd World Congress on Intelligent Control and Automation, vol. 3, pp. 2181–2185 (2000)
19. Tharun, P.S., Nigam, M.J.: Hierarchical fuzzy PID controller using granular computing for inverted pendulum. Int. J. Electr. Electron. Data Commun. Dir. Open Access J. (Sweden) (2013)
20. Sastry, S.S.: Introductory methods of numerical analysis. PHI Learning (2012)

Fuzzy Logic-Based Decision Making for Selection of Optimized Liquid Insulation Blend

S. Vedhanayaki, R. Madavan, Sujatha Balaraman, S. Saroja, S. Ramesh and K. Valarmathi

Abstract The introduction of ester oils replaces existing liquid insulation system (mineral oil) in transformers. Since ester oils are eco-friendly and biodegradable in nature. In this work, mineral oil and rapeseed oil are used as base fluids. Further, MO and RO are blended together at various ratios from 10 to 90%. The important characteristics like breakdown voltage, acidity, and dielectric loss of the oil samples are analyzed. Using fuzzy logic method, best optimistic liquid insulation sample is identified.

Keywords Transformer · Liquid insulation · Ester oil · Mixed insulating liquids · Optimistic sample · Fuzzy logic

---

S. Vedhanayaki · S. Balaraman
Department of Electrical and Electronics Engineering, Government College of Technology, Coimbatore 641013, India
e-mail: vedhanayakiselvaraj@gmail.com

S. Balaraman
e-mail: sujaengg@gmail.com

R. Madavan (✉) · S. Ramesh · K. Valarmathi
Department of Electrical and Electronics Engineering, P.S.R. Engineering College, Sivakasi 626140, India
e-mail: srmadavan@gmail.com

S. Ramesh
e-mail: s.ramesh@psr.edu.in

K. Valarmathi
e-mail: valarmathi@psr.edu.in

S. Saroja
Department of Information Technology, MEPCO Schlenk Engineering College, Sivakasi 626005, India
e-mail: activeroja@gmail.com

J. Nayak et al. (eds.), *Soft Computing in Data Analytics*, Advances in Intelligent Systems and Computing 758,
https://doi.org/10.1007/978-981-13-0514-6_53

1 Introduction

Most of the countries in the world suffered due to environmental issues. In order to have control over environmental issues, most of the countries follow strict norms in all sectors. Likewise, in power sector, in transformers, non-biodegradable mineral oil (MO) is replaced by biodegradable ester oil [1]. As of now, in most of the countries petroleum-based mineral oil is used as liquid insulation system in transformers, because of its good electrical, chemical and physical properties, availability, and low cost [2].

Ester oils are suitable alternative for replacing MO, because of its biodegradable and eco-friendly nature. Moreover, ester oils have good dielectric and physical properties. In particular, ester oils have excellent fire resistance properties [3]. Even though it has many advantages, it also has some disadvantages such as high cost, poor viscosity, and poor oxidation stability [1]. In order to overcome disadvantages of MO and ester oil, both the oils were mixed together at various ratios and some of the important properties were analyzed [4, 5]. Among this, selection of sample having better characteristic performance is a difficult task.

In today's competitive, challenging, and highly changing world, decision making is a complex process. It is mainly because of the involvement of fuzzy nature in the criteria and alternatives in the decision-making process. Fuzzy logic is essential to solve the decision-making problems. As of now, many researchers monitored the condition of the transformers using fuzzy logic-based decision-making system; accordingly, diagnostics required diagnostic process maybe followed [6]. In this work, an attempt has been made to choose the optimistic liquid insulation sample among all the samples by using fuzzy logic.

2 Samples Preparation Process

The required quantity of MO and rapeseed oil (RO) are purchased from reputed companies and these two oils are considered as base fluids. Then the oils are treated to remove the moisture and particles. Mixed fluids are prepared by mixing of two base fluids at various ratios with the help of magnetic stirrer for time span of 1 h. The prepared mixed insulating fluids are numbered as given in Table 1.

Table 1 Samples description

Sample no	Ratio
S1	MO 100%
S2	RO 100%
S3	MO 90% + RO 10%
S4	MO 80% + RO 20%
S5	MO 70% + RO 30%
S6	MO 60% + RO 40%
S7	MO 50% + RO 50%
S8	MO 40% + RO 60%
S9	MO 30% + RO 70%
S10	MO 20% + RO 80%
S11	MO 10% + RO 90%

3 Experimental Results

To analyze the characteristics performance of mixed liquid insulation, different important characteristics like breakdown voltage, acidity, and dielectric loss are observed in laboratory oriented tests. Moreover, experimental results of mixed insulating liquids are shown in Figs. 1, 2, and 3.

Breakdown voltage is the measure of electrical stress withstanding ability of insulating liquid. A fully automatic breakdown voltage measurement kit is used to measure the breakdown voltage of mixed insulating liquids as per IEC 60156. The breakdown voltage of the mixed insulating liquids is shown in Fig. 1, it is the average of six successive measurements with the equal time delay of 2 min with the sphere electrodes of 2.5 mm spacing [7]. It is clear from Fig. 1, when comparing breakdown voltage of both pure MO and RO oils, RO has higher breakdown than MO. Further, in the mixed insulating liquids, with the lower mixing ratios, breakdown voltage is somewhat higher than pure MO and for higher mixing ratios, breakdown voltage is also high.

Acidity is one of the most important parameters when assessing the condition of the liquid insulation. Generally, it is the measure of presence of organic and inorganic acids in the liquid insulation and, further, it is represented as milligram of potassium hydroxide required to neutralize the acids in 1 g of liquid insulation [8]. When looking over Fig. 2, compared with acidity of MO, RO have higher acidity. Even though it has higher acidity, it contains organic long-chain fatty acids which are less aggressive compared to organic short-chain fatty acids of MO. Moreover, with the increase in mixing ratio, acidity gets increased.

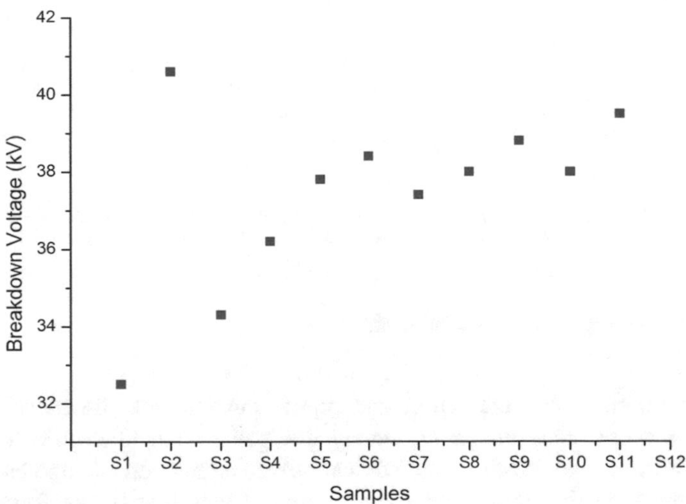

Fig. 1 Breakdown voltage of mixed insulating liquids

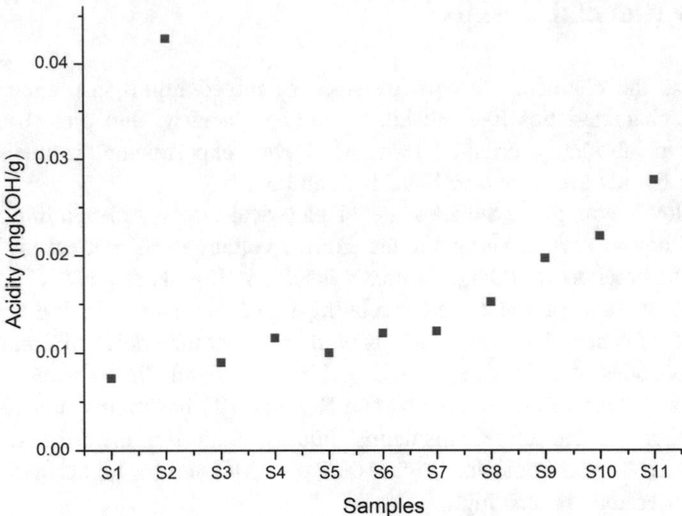

Fig. 2 Acidity of mixed insulating liquids

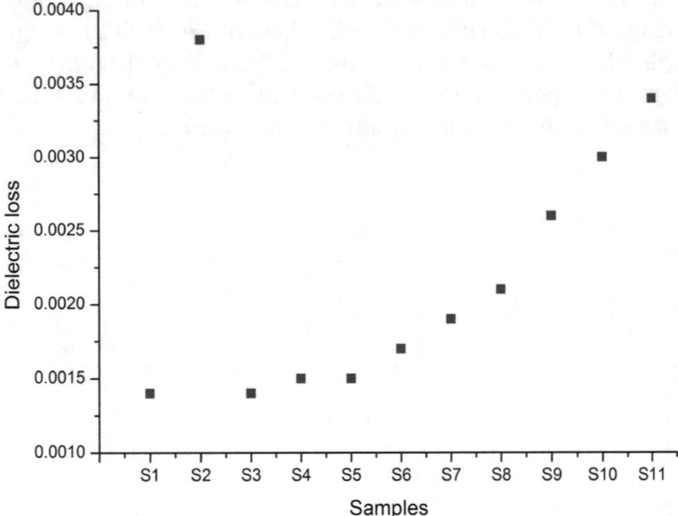

Fig. 3 Dielectric loss of mixed insulating liquids

In a material, ionic conduction and dipole moment are the most important mechanism cause electromagnetic energy dissipation, i.e., dielectric loss in the material. With the application of electrical stress, interaction of dipoles and ions generates heat on the entire material and energy gets lost in the form of heat. Dielectric loss of liquid insulations determined at 90 °C as per IEC 60247.

When comparing dielectric loss of MO and RO, dielectric loss of MO is less compared with RO as shown in Fig. 3. In the higher mixing ratios, the dielectric loss of the mixed insulating liquids gets increased with mixing ratio.

4 Fuzzy Logic

Fuzzy logic is an information system that imitates the decision-making ability of human knowledge into engineering systems. Fuzzy logic is made of three parts that are working memory, knowledge base, and inference engine. Working memory stores information from the user. In addition to the information about the problem, the final decision is made using inference engine [6].

In this section, fuzzy logic model is developed to take a decision to determine the optimal liquid insulation sample. Fuzz logic model is developed in accordance with fuzzy inference model as shown in Fig. 4.

Developed model is having three inputs with triangular membership function. Breakdown voltage, acidity, and dielectric loss are the three inputs with six ranges of linguistic variables like very good, good, moderate, bad, extremely bad, and disastrous. Then, the output corresponds to each input is also having triangular membership function at different linguistic variables like excellent, sufficicnt, barely sufficient, and critical. The output mainly depends upon breakdown voltage, acidity, and dielectric loss. Moreover, based on the results, the output is classified. If the output is critical then the oil contains more contaminated particles, and it is not suitable in transformers as insulating and cooling medium. The input and output membership functions are shown in Figs. 5, 6, 7, 8, and 9.

First input breakdown voltage is scaled from 0 to 90 and above, second input acidity is scaled from 0 to 2 and above, and third input is 0–0.6 and above. Output is from 0 to 1 point scale.

When the breakdown voltage is in good range, acidity is in moderate range and dielectric loss is in moderate range, then the output lies in the sufficient range. This shows that the state of the oil is acceptable, and thus the oil contains lesser amount of contaminated particles.

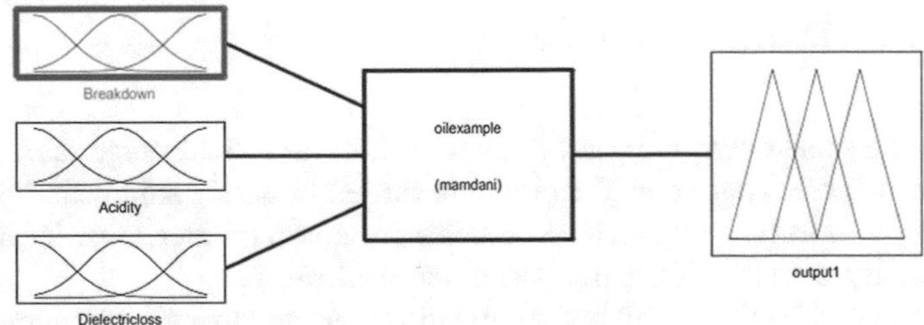

Fig. 4 Fuzzy inference model

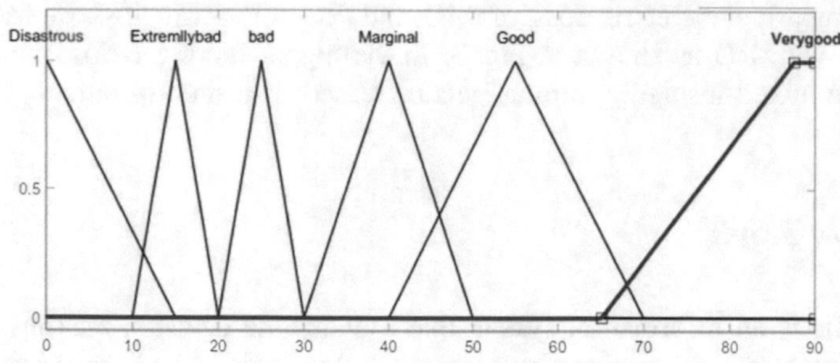

Fig. 5 Membership functions of breakdown voltage

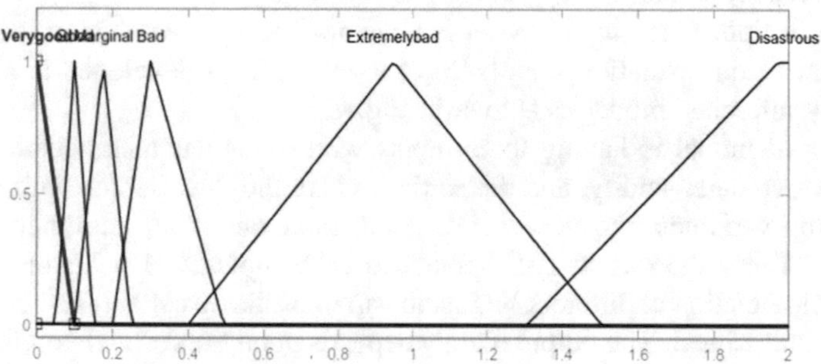

Fig. 6 Membership functions of acidity

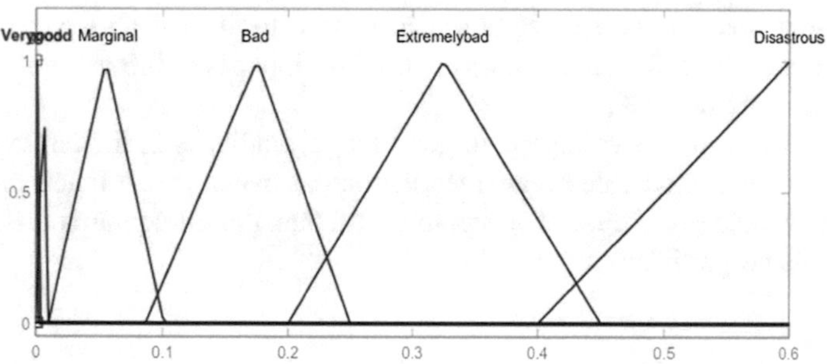

Fig. 7 Membership functions of dielectric loss

Again, the breakdown is in good range, acidity is in bad range, and dielectric loss is in moderate range, now the output is shifted to barely sufficient range. With the increase in acidity, contaminated particles present in oil increases. It can be seen that the acidity directly affects the performance of oil.

With the combination of above input and output membership functions, totally 216 rules are developed. With the help of centroid method, the output fuzzy set is

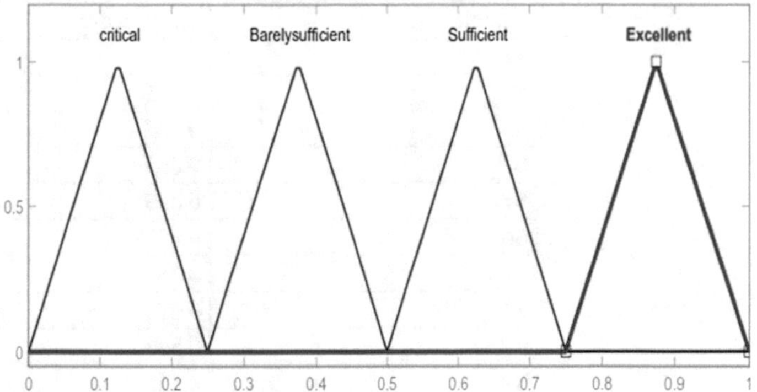

Fig. 8 Membership functions of output

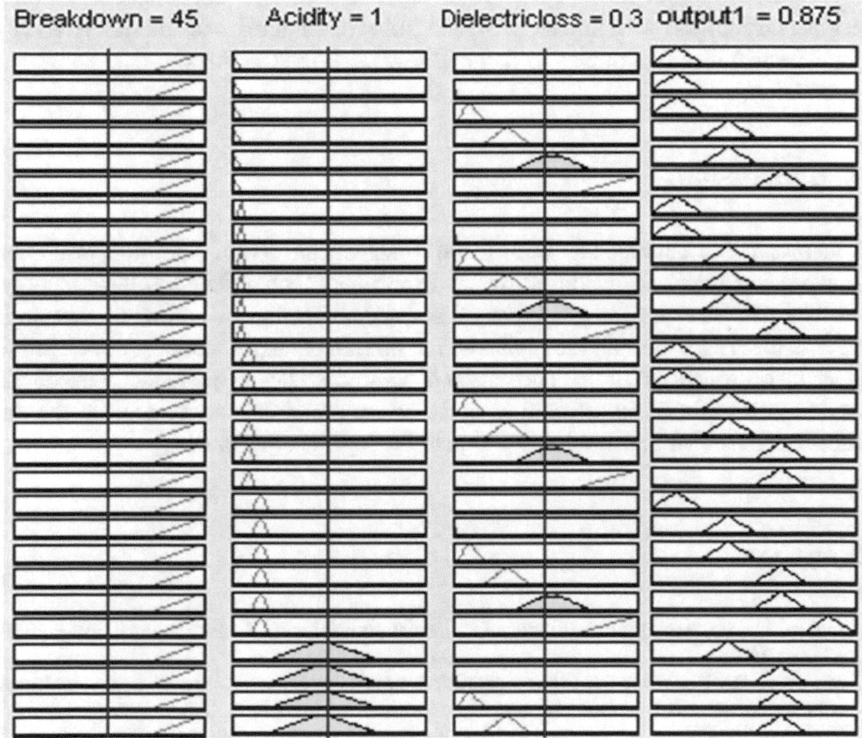

Fig. 9 All the three inputs with output

properly converted into a crisp set. From Table 2, it clearly indicates that initially the fuzzy score of the samples gets increased. After sample S5, fuzzy scores of the samples get reduced because of increment in acidity and dielectric loss. Since with

Table 2 Fuzzy scores of individual samples

Sample no	Fuzzy score
S1	0.766
S2	0.835
S3	0.782
S4	0.815
S5	**0.875**
S6	0.823
S7	0.786
S8	0.638
S9	0.537
S10	0.456
S11	0.394

the increase in sample ratios, acidity, and dielectric loss of liquid insulation samples gets increased, this ha an impact in membership function regions and moves acidity and dielectric loss of samples to bad and extremely bad regions.

5 Conclusion

In this work, an attempt has been made to select the optimistic liquid insulation sample using fuzzy logic. Based on the experiments, breakdown voltage, acidity, and dielectric loss are the parameters discussed. Selection of optimistic sample is not possible by looking all these values. For this, fuzzy logic model is developed to select the optimistic solution. The output of sample S5 lies in the range of excellent. Therefore, sample S5 is selected as optimistic sample and it may replace the traditional mineral oil and it maybe alternate for high-cost ester oils.

References

1. Fofana, I.: 50 years in the development of insulating liquids. IEEE Electr. Insul. Mag. **29**(5), 13–25 (2013)
2. Madavan, R., Balaraman, S.: Investigation on effects of different types of nanoparticles on critical parameters of nano-liquid insulation systems. J. Mol. Liq. **230**, 437–444 (2017)
3. Madavan, R., Balaraman, S.: Comparison of antioxidant influence on mineral oil and natural ester properties under accelerated aging conditions. IEEE Trans. Dielectr. Electr. Insul. **24**(5), 2800–2808 (2017)
4. Karthik, R., Raja, T.S.R., Shunmugam, S.S., Sudhakar, T.: Performance evaluation of ester oil and mixed insulating fluids. J. Inst. Eng. India Ser. B **93**(3), 173–178 (2012)
5. Fofana, I.: Challenge of mixed insulating liquids for use in high-voltage transformers, Part 2: investigations of mixed liquid impregnated paper insulation. IEEE Electr. Insul. Mag. **18**(4), 5–16 (2002)

6. Tanaka, K.: An Introduction to Fuzzy Logic for Practical Applications. Springer, New York (1996)
7. Madavan, R., Balaraman, S.: Failure analysis of transformer liquid—solid insulation system under selective environmental conditions using Weibull statistics method. Eng. Fail. Anal. **65**, 26–38 (2016)
8. Madavan, R., Balaraman, S., Saroja, S.: Multi criteria decision making methods for ranking the liquid insulation system based on its performance characteristics with anti-oxidants under accelerated aging conditions. IET Gener. Trans. Distrib. **11**(16), 4051–4058 (2017)

Adaptive Histogram Equalization and Opening Operation-Based Blood Vessel Extraction

Amiya Halder, Apurba Sarkar and Sneha Ghose

Abstract Retinal blood vessel detection is a fundamental procedure for automatic detection of retinal diseases and infections. This paper presents a method for blood vessel detection of retinal images using adaptive histogram equalization and morphological operations. After proper intensity adjustments, the image is subjected to morphological operations and then passed through a median filter. Threshold of the filtered image is then carried out to give a resultant image. The final image is obtained using vessel width dependent morphological filters to remove all the connected components that have fewer than the required pixel width; i.e., all the small components are removed. The performance of the proposed is very promising as compared to the existing techniques.

Keywords Retinal image · Opening operation · Median filter · Contrast enhancement

1 Introduction

Retinal blood vessels mainly imply the blood vessel of the retina. Retinal blood vessel detection plays an important role for retinal image analysis. Their different features such as width, length, and branching play a vital part in screening, diagnosis, detection, and early treatment of various cardiovascular [1] and ophthalmologic diseases and disorders such as diabetes [2], vessel occlusions, glaucoma, aneurysm, arteriosclerosis, hemorrhage, and stroke [3]. Analysis of retinal blood vessels and

A. Halder (✉) · S. Ghose
St. Thomas' College of Engineering and Technology, Kolkata 700023, India
e-mail: amiya.halder77@gmail.com

S. Ghose
e-mail: snhghose1313@gmail.com

A. Sarkar
IIEST, Shibpur, Howrah 711103, India
e-mail: as.besu@gmail.com

J. Nayak et al. (eds.), *Soft Computing in Data Analytics*,
Advances in Intelligent Systems and Computing 758,
https://doi.org/10.1007/978-981-13-0514-6_54

early detection of abnormalities is necessary so that the disease can be treated in time when it is still at an early stage [4, 5]. However, the manual analysis of retinal vessel images is tedious and requires a lot of concentration and observational skills [6]. Therefore, it is necessary to develop automatic detection techniques that can detect any abnormality in retinal blood vessel images. As a result, the necessity of vascular analysis system grows rapidly, with blood vessel detection as its foremost stage. The extraction of blood vessels using morphological operations and other techniques have been reported in [7–12]. In this paper, we propose a method for blood vessels extraction from the retinal images using adaptive histogram equalization and opening operation. In several previous works, it has been observed that mostly the green channel has been used for the further processing. In this paper, we have also used the red channel in certain cases as we have observed that working with the green channel or the red channel of the RGB images in case of blood vessel detection and analysis gives a more accurate result than that with the blue channel or with the image as a whole.

2 Proposed Work

The retinal blood vessel extraction mainly employs the contrast between the immediate background and the blood vessels. Since the method is automated, it makes it possible to analyze numerous images simultaneously; thus, it is more efficient than non-automated manual methods which are observation-driven. The suggested work in this paper is based on retinal blood vessel extraction using contrast-limited adaptive histogram equalization and morphological opening operations.

2.1 Algorithm

The proposed method is summarized as given below:

1. Extract green/red channel of the retinal image as required.
2. Perform complementing of the channel.
3. Apply contrast-limited adaptive histogram equalization.
4. Perform morphological opening operation on the image using a structuring element.
5. Subtract the morphologically opened image from the histogram equalized image to remove the optic disk.
6. Pass the result through a median filter.
7. Apply morphological opening on the resultant image, and this gives the background.
8. Remove the background from the filtered image by subtraction.
9. Enhance the contrast of the resulting image.
10. Perform threshold of the result to get a binary image.

11. Apply binary morphological opening of the image using a certain width to remove the small components.
12. Obtain resultant extracted image.

2.2 Data

The data required for retinal blood vessel extraction are colored eye fundus images taken from the DRIVE database.

2.3 Preprocessing

The preprocessing stage includes the following steps:

1. Green or red channel selection as preferred for the image,
2. Complementing of the preferred channel, and
3. To improve contrast enhancement of the blood vessel images using contrast-limited adaptive histogram equalization.

At first, the green plane or the red plane of the retinal image is extracted as required depending on the image to be processed. Usually, the green channel is preferred more during vessel detection as the blood vessels are dark in contrast in the green channel of the image and the blood vessels appear more clearly than in any other plane. It may so happen that in some cases the red plane may give a better result. In that case that particular plane is to be considered. Image $I_{org} = f(\alpha, \beta, \gamma)$ where $f : A \rightarrow B$, green channel $I_{green} = f(\alpha, \beta, 2)$ or red channel $I_{red} = f(\alpha, \beta, 1)$. The next step is to complement the channel. $g_{inv} = w|w \notin I_{green} = I^C_{green}$ or $r_{inv} = w|w \notin I_{red} = I^C_{red}$. The next step is to enhance the contrast of the resultant grayscale image by transforming the values using contrast-limited adaptive histogram equalization (CLAHE). CLAHE operates on small regions in the image, called tiles, rather than the entire image. The contrast of every single tile is enhanced, so that the histogram of the output region approximately matches the histogram specified by the 'Distribution' parameter. The neighboring tiles are then combined using bilinear interpolation to eliminate the boundaries that were formed artificially. The contrast, especially in homogeneous areas, can be limited to avoid amplifying any noise that might be present in the image. If a digital image has η pixels distributed in μ discrete intensity levels and ψ_k is the number of pixels with intensity level θ_k, then the probability distribution function (PDF) is given as (Eq. 1):

$$f_\theta(\theta_k) = \frac{\psi_k}{\eta} \tag{1}$$

And the cumulative distribution function (CDF) is given in Eq. 2.

$$F_k(\theta_k) = \sum_{j=0}^{k} f_\theta(\theta_j) \tag{2}$$

The uniform distribution is considered as the basis for creating the contrast transform function. Let θ_{c_max} and θ_{c_min} be the maximum and minimum permissible intensity levels. Let $F_k(\theta_{c\_in})$ be the cumulative distribution function for the relative input tile θ_{c_in}. Then, the resultant image using the uniform distribution function in Eq. 3.

$$\theta_{c\_out} = [\theta_{c\_max} - \theta_{c\_min}] * F_k(\theta_{c\_in}) + \theta_{c\_min} \tag{3}$$

2.4 Morphological Operations

The important requirement to get effective results in morphological image processing is to select the appropriate structuring elements. A suitable structuring element is chosen for the morphological opening operation of the image obtained after preprocessing. Equation 4 for morphological opening of image τ and structuring element σ is given as follows:

$$imgopen = \tau \circ \sigma = (\tau \ominus \sigma) \oplus \sigma \tag{4}$$

where $\tau \ominus \sigma(erosion) = z|(\sigma)_z \subseteq \tau, \tau \oplus \sigma(dilation) = z|(\dot{\sigma})_z \cap \tau \neq \phi$ and $\dot{\sigma}$ = reflection of $\sigma = w|w = -b, forb \in \sigma$. The optic disk is then removed from the image by subtracting the image obtained after preprocessing (I_{pre}), and the image on which morphological opening has been performed ($imgopen$) The resultant image is given by Eq. 5:

$$Godisk = I_{pre} - imgopen = \{w|w \subseteq I_{pre}, w \subseteq imgopen\} = I_{pre} \cap imgopen^C \tag{5}$$

The resultant image is then passed through a median filter [5, 6] to preserve the edges while removing any existing noise. The median filter is used for removing noise, i.e., smoothening of the resultant image in such a way that the contrast of the image is not reduced. The equations (Eq. 6) after normalization are given as:

$$I_{new\_normalized}(\alpha, \beta) = \frac{1}{\sum_{j=-1}^{1} \sum_{i=-1}^{1}} \sum_{j=-1}^{1} \sum_{i=-1}^{1} 1 * I_{old}(\alpha + i, \beta + j) \tag{6}$$

The background of the image is extracted and then removed from the filtered image, and the resulting image is then subject to contrast enhancement using the following equation (Eq. 7):

$$I_{out} = \frac{\acute{G}L_{max} - \acute{G}L_{min}}{GL_{max} - GL_{min}} [I_{in}(\alpha, \beta) - GL_{min}] + \acute{G}L_{min} \tag{7}$$

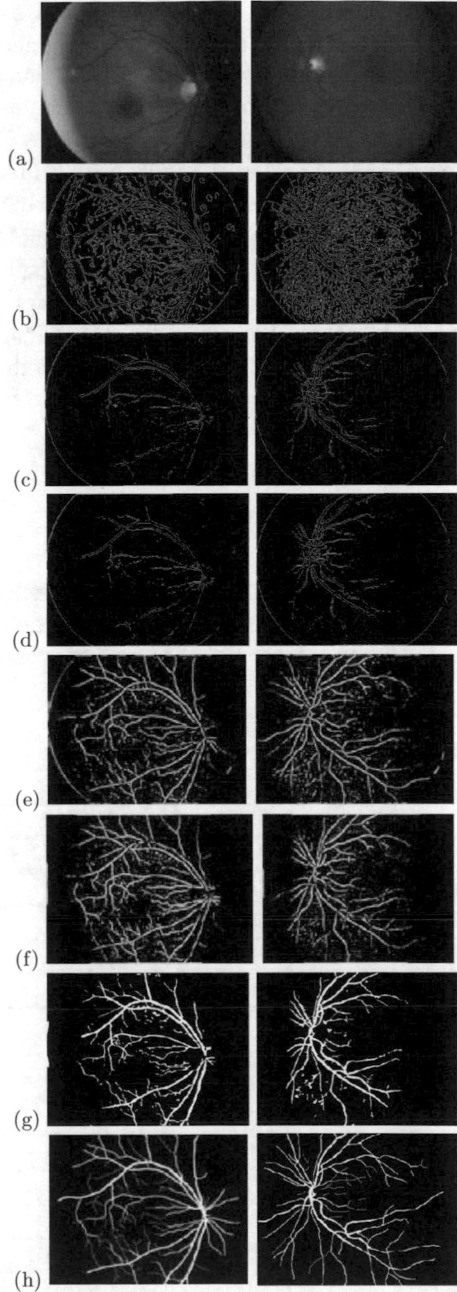

Fig. 1 Comparison of the detected images of retinal image1 and image2 using **b** Canny, **c** Prewit, **d** Sobel, **e** Paper1 approach, **f** Paper2 approach, **g** proposed method, and **h** ground truth image, **a** is the original retinal images

where GL_{max} and GL_{min} are the respective maximum and minimum gray level values of the original image I_{in}, and $\acute{G}L_{max}$ and $\acute{G}L_{min}$ are the respective maximum and minimum gray level values of the new image I_{out}. The next step is to threshold the image obtained to convert the grayscale image to a binary image for proper comparison. The method used for binarization is Otsu's threshold method to find the optimum threshold value. The gray level is divided into L values. Intra-class variance : $\sigma_{\lambda}^2(k) = \lambda_0(k)\sigma_0^2(k) + \lambda_1(k)\sigma_1^2(k)$, where the weights λ_0 and λ_1 are the probabilities of the two classes separated by a threshold k, and σ_0^2 and σ_1^2 are the variances of these two classes. The final image is obtained after the binary area opening of the binary image I using P pixels. It removes all connected components (objects) that have fewer than P pixels in the binary image I, producing another binary image I_b. It basically removes the small objects from the image; that is, it allows components and vessels above a certain width to exist in the final binary image.

2.5 *Post-processing*

The results after the blood vessel extraction are compared pixel by pixel with the results obtained by manual vessel detection of the retinal images. The performance of the proposed technique in terms of accuracy is measured and recorded.

Table 1 Accuracy value of the retinal image1

Methods	MA	FA	OE	Accuracy (%)
Prewit	3731	34044	37775	91.46
Canny	32491	30041	62532	85.86
Sobel	3801	33999	37800	91.45
Paper1	15643	14946	30589	93.08
Paper2	13104	17028	30132	93.18
PM	11891	13575	25466	94.24

Table 2 Accuracy value of the retinal image2

Methods	MA	FA	OE	Accuracy (%)
Prewit	5189	30509	35698	91.93
Canny	40378	27609	67987	84.63
Sobel	5193	30505	35698	91.93
Paper1	14620	14260	28880	93.47
Paper2	8358	18616	26974	93.90
PM	11587	13634	24822	94.38

3 Results

The performance of the proposed method is shown by comparing with different existing blood vessel extraction methods such as Prewit, Sobel, Canny edge detection methods, paper1 [13], and paper2 [14]. The extracted output of the retinal images for the above algorithms and ground truth images are shown in Fig. 1. Tables 1 and 2 show that the proposed algorithm gives more accuracy values than the existing methods. False alarm (FA), miss alarm (MA), and overall error (OE) are also calculated of the Prewit, Sobel, Canny edge detection methods, paper1, and paper2. It is noticed that the proposed method gives more encouraging results than the other existing methods.

4 Conclusion

The proposed method employs an easy technique for implementing blood vessel detection. This method gives very good experimental results as compared to the previous methods. The only drawback in this method is that it misses to detect the very thin blood vessels. The blood vessels with intensity values having good contrast with the background are detected easily, whereas those whose intensity values are very close to that of the background pixels are overlooked. The easy implementation and detection of blood vessels make this method good and convenient for blood vessel detection. The future work will be aimed at further improvement of the result such that even the very thin and fine retinal vessels can be detected and that the computational time required is improved.

References

1. Lowell, J., Hunter, A., Steel, D., Basu, A., Ryder, R., Kennedy, R.L.: Measurement of retinal vessel widths from fundus images based on 2-D modeling. IEEE Trans. Med. Imag. **23**(10), 1196–1204 (2004)
2. Staal, J., Abramoff, M.D., Niemeijer, M., Viergever, M.A., van Ginneken, B.: Ridge-based vessel segmentation in color images of the retina. IEEE Trans. Med. Imaging **23**, 501–509 (2004)
3. Wong T.Y., Shankar, A., Klein, R., Klein, B.E.K., Hubbard, L.D.: Prospective cohort study of retinal vessel diameters and risk of hypertension. BMJ, 1–5 (2004)
4. Udayakumar, R., Khanaa, V., Saravanan, T., Saritha, G.: Retinal image analysis using curvelet transform and multistructure elements morphology by reconstruction. Middle-East J. Sci. Res. **12**(12), 1668–1671 (2013)
5. Yang, Y., Huang, S., Rao, N.: An automatic hybrid method for retinal blood vessel extraction. Int. J. Appl. Math. Comput. Sci. **18**(3), 399–407 (2008)
6. Salazar-Gonzalez, A., Kaba, D., Li, Y., Liu, X.: Segmentation of the blood vessels and optic disc in retinal images. IEEE J. Biomed. Health Inform. 2168–2194 (2014)

7. Niemeijer, M., Staal, J., Ginneken, B., van Loog, M., Abrmoff, M.D.: Comparative study of retinal vessel segmentation methods on a new publicly available database. Proceedings of the SPIE Medical Imaging **5370**, 648–656 (2004)
8. Fritzsche, K.H., Can, A., Shen, H., Tsai, C.L., Turner, J.N., Tanenbaum, H.L., Stewart, C.V., Roysam, B.: Automated model-based segmentation, tracing and analysis of retinal vasculature from digital fundus images. In: Angiography and Plaque Imaging, pp. 225–297 (2003)
9. Mustafa W.A.B.W., Yazid, H., Bin Yaacob, S., Bin Basah, S.N.: Blood vessel extraction using morphological operation for diabetic retinopathy. In: 2014 IEEE Region 10 Symposium (2014)
10. Hassan, G., El-Bendary, N., Hassanienc, A.E., Fahmy, A., Shoeb, A.M., Snasel, V.: Retinal blood vessel segmentation approach based on mathematical morphology. Procedia Comput. Sci. **625**, 612–622 (2015)
11. De, I., Das, S., Ghosh, D.: Vessel extraction in retinal images using morphological filters. In: International Conference on Research in Computational Intelligence and Communication Networks (2015)
12. Walter, T., Klein, J.C.: Segmentation of color fundus images of the human retina: detection of the optic disc and the vascular tree using morphological techniques. Lecture Notes Computer Science **2199**, 282–287 (2001)
13. Wankhede, P.R., Khanchandani, K.B.: Noise removal and background extraction from retinal fundus images for segmentation of blood vessels. Int. J. Graph. Image Process. **3**(1) (2013)
14. Halder, A., Bhattacharya, P.: An application of bottom hat transformation to extract blood vessel from retinal images. In: 2015 International Conference on Communications and Signal Processing, pp. 1791–1795 (2015)

Scheduling Task to Heterogeneous Processors by Modified ACO Algorithm

M. Premkumar, V. Srikanth Babu and R. Somwya

Abstract Heterogeneous computing environment is having diverse computational requirements and utilizes a dispersed set of various high-performing machine with high-speed interlinks to perform varieties of computational applications. To meet out the demand of large and group of computational task, heterogeneous computing environment will be the best environment. The set of regular task assigned to heterogeneous processor has been a problem of determining the respective task with timing, and it is set to be NP-hard. In this research, a new optimization algorithm is proposed for scheduling and allocation. To improve the assigning and scheduling task to meet out the resource utilization and energy consumption, a local search algorithm can be used. In addition to feasible assignment solution, the algorithm can optimize the processor's energy consumption. The optimization algorithm is simulated with Extensive Java Agent Development Framework (JADE) simulator, and its results show that the approach offers accurate, efficient, and effective method for lower energy consumption.

Keywords Scheduling · Heterogeneous processors · Ant colony optimization Local search heuristic · Real-time task · Energy-aware scheduling

M. Premkumar (✉) · V. Srikanth Babu
Faculty of Electrical and Electronics Engineering, GMR Institute of Technology, Rajam, AP, India
e-mail: premkumar.m@gmrit.org

V. Srikanth Babu
e-mail: srikanthbabu.v@gmrit.org

R. Somwya
Electrical and Electronics Engineering, National Institute of Technology Tiruchirapalli, Tiruchirapalli, TN, India
e-mail: sowmyanitt@gmail.com

J. Nayak et al. (eds.), *Soft Computing in Data Analytics*, Advances in Intelligent Systems and Computing 758,
https://doi.org/10.1007/978-981-13-0514-6_55

1 Introduction

The set of heterogeneous processors are assigned to check the assigning of set of periodic tasks without deadline violations. Implementation of multiprocessors with real-time application is very difficult. The scheduling algorithm needs to execute the order of tasks along with determination of specific processor to be used for the respective task. All processors in homogeneous multiprocessor platform are identical in nature, and the algorithm needs resolving bin packing problem (BPP) while assigning the task to the processor [1]. When computing environment is built with heterogeneous multiprocessor, the assignment of task becomes more difficult. The part of the code requires various execution times and depends on different processors which will be the additional complexity for the scheduling algorithm. The periodic/regular tasks are allowed to use different utilization requirement on various processor, and this extension makes bin packing problem as invalid simulation model. Static allocation of task is performed well in heterogeneous processor because deadline of the assigned task and time for the task computation are known a priori and do not refer to the current system state [2].

One of the key challenges of heterogeneous systems is the scheduling. The primary objective of the scheduling algorithm is to minimize the execution time of the task. By reducing the energy consumption, the scheduling algorithm will optimize the solution effectively. The issues of reliability of such systems need to be addressed.

2 Problem Description

Heterogeneous multiprocessor platform (HMP) is denoted by HMP = {P1, P2, Pm} arbitrary with 'm' proactive processor based on CMOS processor technology [3]. The processors P_i from HMP are restricted to operate at one instruction per clock cycle and executed with various execution speed. $S_{i,j}$ in Eq. 1 represents the instruction clock frequency, i.e., the speed of each processor P_j for the respective task. The task execution time is represented by $e_{i,j}$ on individual processor. The Eq. 1 shows the correlation between $e_{i,j}$ and $S_{i,j}$.

$$e_{i,j} = \frac{c_i}{S_{i,j}} \tag{1}$$

where c_i represents the number of clock cycles to run the diverse task T_i. The consumption of energy $E_{i,j}$ of different assignment on different processor per period is given in Eq. 2.

$$E_{i,j} = Power_{i,j} * e_{i,j} \tag{2}$$

A periodic task set is represented by PTS = {T1, T2…, Tn}, and it consists of n real-time task. For given HMP and periodic task set, the tree augmentation problem (TAP) is proposed. TAP is nothing but the set optimization issue which consists of two main objectives. The first objective is called resource in which the tasks are assigned to a definite processor. The total utilization should not exceed and bound to this problem; the earliest deadline first (EDF) algorithm is proposed. EDF is also called as least time active scheduling algorithm, and it is used in real-time operating systems (RTOSs) to fit the task in priority queue.

The decision problem is represented by the resource objective, and it is confirmed to be NP-hard. The second objective is to be energy objective which minimizes the total energy consumption of all assigned tasks. In this work, the resource objective of TAP would be preferred over the second objective. The connection between TAP objectives might be valid to the critical systems and the assigned task could have hard limits.

3 Design Goals

Design of efficient scheduling for heterogeneous processor achieves the following goals: (i) to lower energy consumption of processor; (ii) dynamic allocation of task; (iii) to minimize make-span.

3.1 Metaheuristic Information

3.1.1 Local Search Algorithms

Local search algorithm (LSA) starts with initial solution and finds better and suitable solution in defined current solution of neighborhood, and it is shown in Fig. 1. This algorithm is an iterative algorithm and begins with initial assignment and finds the solution within the neighborhood task. If the algorithm finds suitable solution, then it replaces the existing solution and continues to search. The LSA stops its execution once it finds better solution, and the existing current solution may be called as local optimum. LSA algorithm, known as an iterative procedure and it keeps searching its neighborhood for getting better solution [4]. The steps will be repetitive if and only if there is no improvement in neighbor solution and stop if it finds in the neighborhood of the current solution. The procedure gives better neighbor solution if it is available; otherwise, the algorithm sends the current solution and stops its execution.

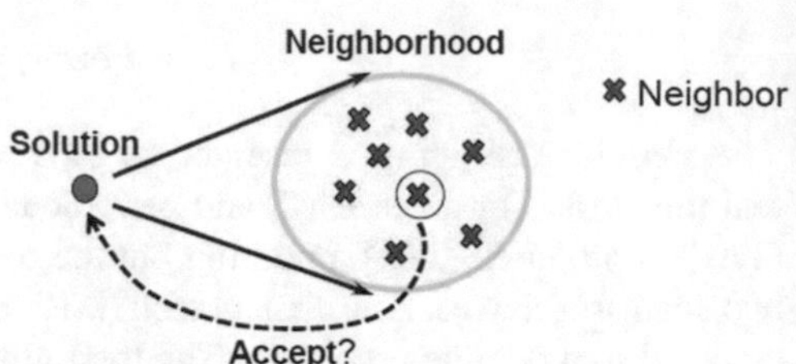

Fig. 1 Overview of local search algorithm

The performance of local search algorithm is choice of an appropriate neighborhood structure and it has been developed in a problem specific. The appropriate neighborhood structure comprises set of neighborhood solution, and it can be reached in single step. An example, local search with k-opt is the most widely used heuristic method for the traveling salesman problem (TSP) neighborhood in which neighbor solutions differ by at most [5].

The 2-opt algorithm is a heuristic algorithm and it tests the current tour, and by replacing the two edge, test can be improved. To specify the local search algorithm, a neighborhood examination is done that defines how it searches the neighborhood solution. In an iterative algorithm, the pivoting rule selects the neighbor solution by improving the objective function, and the improvement rule uses the first improved neighborhood solution and replaces the current solution. A problem with LSA is local minima, and results depend on the initial solution. Hence, to search for an improvised solution, the local search is designed and solution requires small fraction of the computing capacity.

3.2 *Ant Colony Optimization*

A metaheuristic algorithm is an adaptive algorithm, and it defines the heuristic methods which are applicable to different problems [6]. The most effective metaheuristic algorithm is Ant Colony Optimization (ACO), and ACO algorithm is based on the behavior of ants, which will find least path between food and nest. The ants are alive with lowly eye vision and some ants do not have eye vision but the real ants communicate among themselves using smell of chemicals, and its chemical is called pheromones. The ant will select a specific path based on the concentration of the chemical in that path. The intensity of the path is more when many ants are moving on the path and thus high deposition of pheromones, and it attracts more ants. If no ants select the path, then no pheromone or small deposition and thus pheromone evaporates when the time passes.

ACO is joint behavior of ants, and the behavior of ants is shown in Fig. 2. ACO allots agents called artificial ants to replicate the behavior of real ant. These agents start to communicate with each other with the help of environmental variables and

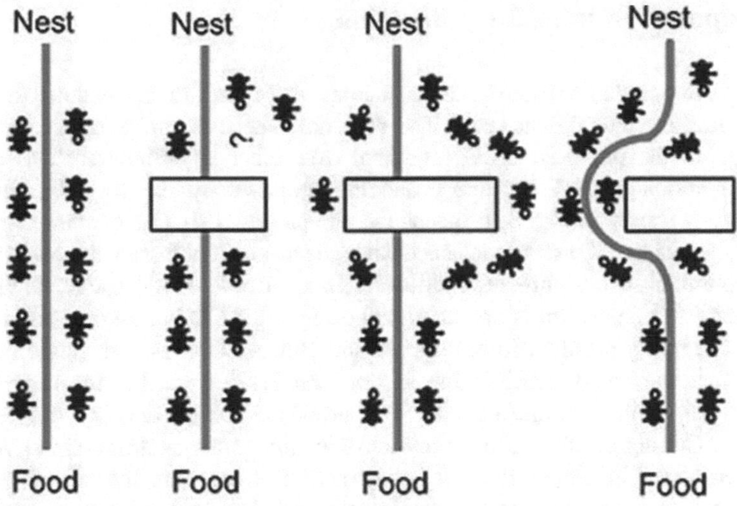

Fig. 2 Natural behavior of ant

apply this to solve set of optimization problems [4, 6, 7]. Starting with an ant system, a many algorithms were developed and applied to various optimization problems in real-world applications.

3.3 *Max–Min Ant Systems*

Many researches has been done on ACO, the researchers show that high performance is achieved by a strong manipulation of the obtained best solutions, and the analysis on search space gives details of the fact. Using a materialistic search potentially magnifies the problem of early inactivity of the search. To achieve the high performance on ACO algorithm, combine the obtained best solution during the search with an effective method to avoid early inactivity search.

Max–min ant system is a combined algorithm to meet the above-mentioned requirement, and it differs in various aspects from conventional ant system. The best solution is achieved during iteration search or during the execution of the algorithm, and after each search, the ant adds pheromone. The ant from the previous step might be the best solution in present iterative search will be called as an iterative best ant or the ant which finds the best solution from the global best ant. To avoid the inactivity of the search, the range of pheromone tracks is limited to a particular interval. Additionally, initializing the pheromone tracks to τ_{max} and solution at start of the algorithm is achieved in this way of higher exploration.

4 Proposed System for Allocation

To meet all the deadlines, heterogeneous multiprocessor platform is scheduled to a set of implicit deadline intermittent tasks. The dynamic change of problem instance is the advantage of the proposed ACO algorithm over other algorithms such as genetic algorithm and simulated annealing, and the proposed system for allocation and scheduling is shown in Fig. 3. It means that the proposed ACO algorithm can familiarize to the change of problem instance when the task is running continuously [7].

To optimize the assignment solution in terms of both TAP objectives, the new optimized ACO algorithm is proposed and designed. ACO is designed and used to reduce scheduling problem in multiprocessor. But, ACO considers single instance task so that the small modification is done on ACO to tackle the regular and periodic task with deadline checks. The proposed modified ACO differs from ordinary ACO algorithm because the main scheduler such as artificial ant will not use experimental information during construction of solution. Instead of heuristic information, the modified ACO uses local search heuristic because constructive heuristic is expensive to compute and ACO will not be useful for large problem instance with reasonable performance.

$$p(s,i,j) = \frac{[\tau(i,j)]\cdot[\eta(i,j,s)]^{\beta}}{\sum_{(i',j')N(s)}[\tau(i',j')]\cdot[\eta(i',j',s)]^{\beta}}\ if\ (i,j)\epsilon N(s) \quad (3)$$

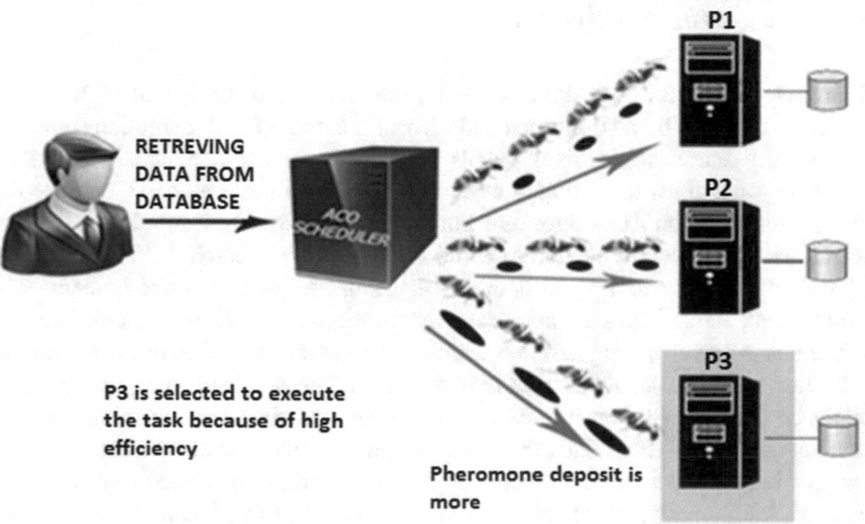

Fig. 3 Dynamic allocation of task to processor

where β is the heuristic weight, and it is expressed in Eq. 3 and will be another improbability of conventional ACO algorithm. The multiplication operation in ACO is computationally expensive compared with the additive operation. For large problem instance, the heuristic weight β will significantly slow down the execution with random value other than 0. The elimination of constructive heuristic is another advantage of modified ACO to find the optimal heuristic weight, but it requires a lot of experimental trials.

5 Architecture of the Proposed Scheduler

The architecture of the scheduler is shown in Fig. 4, and modules in the architecture will be discussed in the subsequent section.

5.1 *Module Description*

5.1.1 Main Scheduler (Artificial Ants)

The main scheduler is used to submit/schedule the tasks to the nodes based on local search algorithm. The artificial ant will allocate the task to the processor one by one till each task is allotted to processor or without exceeding the extra computing

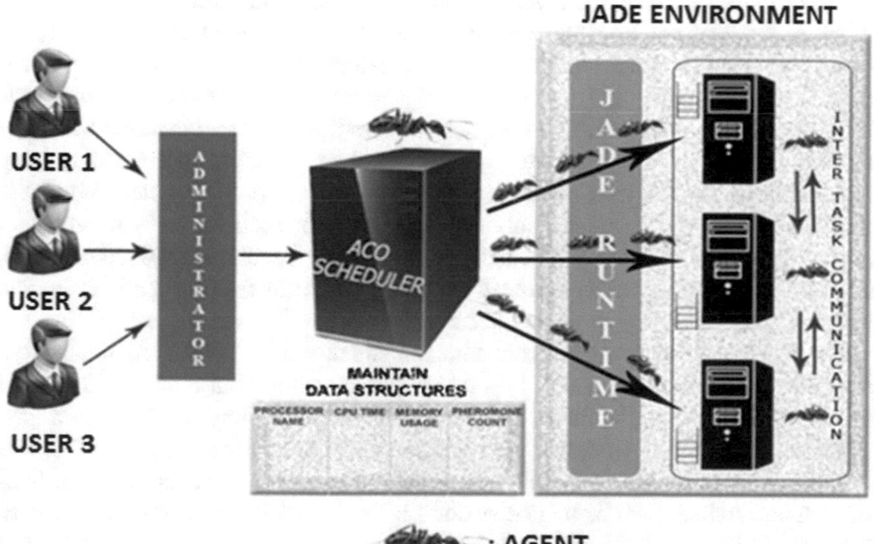

Fig. 4 Architecture of the scheduler

capacity; no remaining task is allotted to the processor. The scheduler starts with set of periodic tasks and set of processors from heterogeneous processor. If an artificial ant stops scheduling with minimum of one unassigned task, then the tour is called infeasible tour. At the same time, the tour may be called as feasible tour, if and only if an ant stops with the beginning of task being allocated. If an ant at no stop condition, the tour is called partial tour. An ant uses distribution of average of pheromone for selecting next task and processor over eligible pair (task, processor).

5.1.2 Node Agents

Node agents receive a task which has to be executed in the node and after execution of tasks send back information's like time taken to execute the task, CPU usage, memory usage to the main scheduler.

5.1.3 Local Search Algorithm Implementation

To improve the artificial ant solution, local search procedure needs to be followed after completion of construction procedure. Initial assignment solution will be the start in local search heuristic, and it will try to find the optimal solution by following neighborhood operations. (i) Opt: Take away the task from the processor and assign the same task to the different processor; (ii) Exchange: Take away the two tasks from two processors and assign the two tasks to same processor when the other task was assigned.

The proposed local search investigation is an iteration procedure and begins with initial assignment solution and continues the search within neighborhood. If the local search finds a better solution, then it substitutes the current solution and it is called local optimum, and it continues the search again. Hence, the search is designed and implemented to find the optimal solution and it requires fraction of the computing capacity in heterogeneous multiprocessor. To identify the task quickly by an artificial ant, the data structure is designed and the processor neighbor list is implemented. The list contains only the first processor such as the processor with less task utilization for a task. The neighbor list of the processor is a fixed structure, and it is reused by successive imitated ant to execute local search effectively so that the price of constructing can be paid back.

The concept of the don't-look bit is implemented, and it significantly reduces the execution/run period by permitting a minor rise in suboptimal tour. For the given task of T1, an artificial/imitated ant fails to determine an improved solution, and neighbors of T1's and processor remain same when an ant looks again at T1; it is doubtful to find an improved step. The proposed work implements vector of flags for each task, which may be named as don't look bit, and it is turned off at the start. If the search finds an improving move with T1 = T, the bit for the task T is on, and if T is one of the two exchanged tasks, the bit is turned off. An artificial ant ignores all the tasks whose bits are on when entrants for T1 are considered. The total number of T1's consideration is reduced by local search from O(x * n) to O (2x + n), where x is 2-Exchange moves from solution to the local optimum. The search time for the task by the local search is reduced from $O(x * n^2)$ to O(2x + n).

5.1.4 Min–Max Pheromone Update

Min–Max ant system plays a vital role in proposed system, and it is used to update the pheromone trails [8]. The selection of Best-so-far solution or iteration-best solution is based on the following way:

- The chance of using best solution is progressively increasing during the successive iterations that no improved solution was found.
- When an improved solution is found, re-initialization of pheromone traces will happen and the probability usage of best solution will be changed to a small value.

5.1.5 Communication Layer and GUI

JADE is used for the agent development and communication layer. We need to send the tasks' information to the nodes and get the information about how the tasks executed back to the main agent (main scheduler) and represent that data in Graphical User Interface [9].

5.2 Technical Architecture Using JADE

The proposed modified ACO proceeds with advantage of agent platform and technology, and an ant is implemented as a software agent because it can be simply created, modified and removed in the proposed system. The advantage of the modified ACO algorithm is that the solution construction is distributed to the processor which enhances the computational performance of the processor. If there is any failure of agents, the performance of the algorithm will not be affected because the algorithm is more reliable and strong for the scheduling system.

Java-based agent platform (JADE)-based modified ACO is presented in Fig. 5, and it is built on JADE. ACO scheduler agent in the architecture is accountable for initialization, termination stage and iteration control of the algorithm. The multiple JADE agents are allocated for iteration stage because the iteration stage is more computationally exhaustive. The JADE agents which are ants exist in various container will accomplish the iterative construction of the scheduling. The JADE agents are coordinated with ACO scheduler agent to update the iteration counter. Pheromone information is exchanged, and it will be achieved by predefined agent messages.

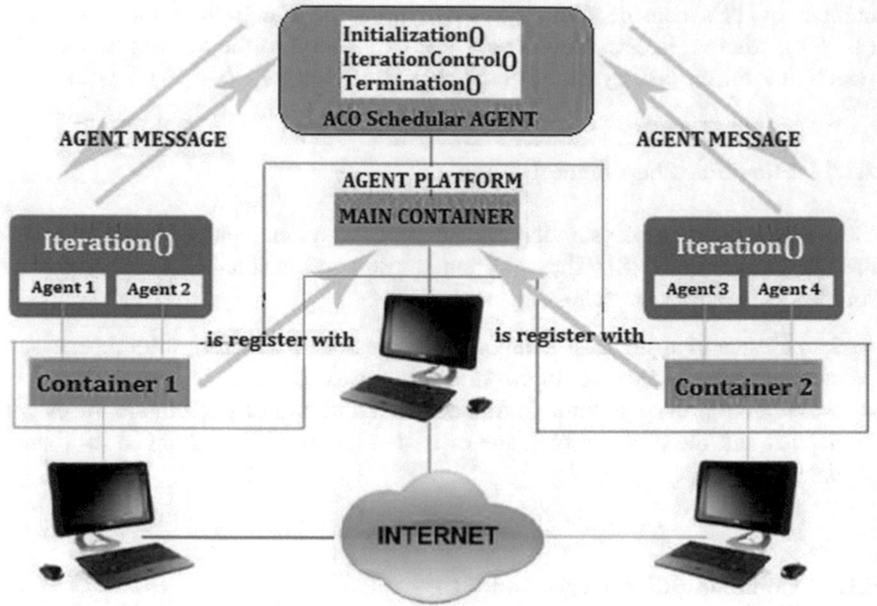

Fig. 5 Technical architecture

6 Effectiveness of Simulation

The proposed algorithm uses the simulation technique of Java agent development framework. JADE is one of the recent simulators, and it is used in efficient way to produce accurate result. The average running/execution time for the algorithms is smaller than Baruah's. The average running time and feasible solution are listed in Table 1.

The combined results are shown in Table 1, and it concluded that Genetic Algorithm (GA) has less running time for all problem sets. But GA is not able to find optimal solution and it stops execution. By comparing the average execution time of the proposed algorithm and GA, for the problem instances (37%), all three algorithms find a feasible solution but the average running time of proposed algorithm is smaller than GA.

Table 1 Average running time in ms

Problem set	Modified ACO		GA		Baruah's	
Size	Time	Feas.	Time	Feas.	Time	Feas.
u4 × 100	6739.63	70	9987.2	55	1389.1	20
u4 × 80	3751.23	93	5455.8	43	878.65	48
u5 × 150	71.67	147	4754.2	130	757.62	98
u8 × 60	895.5	90	8922.4	30	2334.8	0

Table 2 Average energy consumption by modified ACO

Problem set	Average standardized energy
Dimension	Modified ACO
u4 × 100	0.635
u4 × 80	0.647
u5 × 150	0.748
u8 × 60	0.843

The investigation of the performance of modified ACO algorithm from the energy consumption perceptive is shown in Table 2 and gives the average energy consumptions for all feasible solutions. From the optimization objective, the proposed algorithm has accomplished energy saving of 17.2% compared to other algorithms for the listed problem instances. Finally, it can be concluded that the energy consumption is less in modified ACO and it offers low computation cost.

7 Conclusion

A novel idea for task allocation and scheduling by modified ACO is proposed in this work. By keeping energy consumption low, the feasible assignment solution is searched through novel approach. The local search algorithm exploited insight into the problem so that ACO as well as other algorithms handles huge neighborhood efficiently. This paper concludes that, at any time, the processor executes single instruction per clock cycle and the speed of the assigned task, processor proves its heterogeneity. The general task of the model has low degree of instruction and equitable to stipulate a speed for the assigned task according to all tasks execution on the respective processor. The proposed approach may be practically tested in various computer clusters under variable load.

References

1. Aydin, H., Yang, Q.: Energy-aware partitioning for multiprocessor real-time systems. In: Proceedings of International Parallel and Distributed Processing Symposium, Nice, France (2003)
2. Braun, T., et al.: A comparison of eleven static heuristics for mapping a class of independent tasks onto heterogeneous distributed computing systems. J. Par. Dist. Comp. **61**, 810–837 (2001)
3. Baruah, S.: Task partitioning upon heterogeneous multiprocessor platforms. In: Proceedings of International Conference on Real-Time and Embedded Technology and Applications, Ontario, Canada, pp. 1–8 (2004)
4. Levine, J., Ducatelle, F.: Ant colony optimization and local search for bin packing and cutting stock problems. J. Oper. Rese. Soc. **55**, 705–716 (2004)
5. Bentley, J.L.: Fast algorithms for geometric traveling salesman problems. ORSA J. Comp. **4**, 387–411 (1992)

6. Ritchie, G.: Static multi-processor scheduling with ant colony optimization & local search. In: Thesis of School of Informatics, University of Edinburgh, UK (2003)
7. Jin, H., Wang, H., Wang, H., Dai, G.: An ACO-based approach for task assignment and scheduling of multiprocessor control. In: Lecture Notes in Computer Science, vol. 3959, pp. 587–598 (2006)
8. Bunde, D.P.: Power-aware scheduling for makespan and flow. J. Sch. **12**, 1245–1252 (2009)
9. Burd, T.D., Brodersen, R.W.: Energy efficient CMOS microprocessor design. In: Proceedings of Hawaii International Conference on System Science, Wailea, USA, pp. 288–297 (1995)

Process Framework for Modeling Multivariate Time Series Data

Jobiya John and Sreeja Ashok

Abstract In multivariate time series (MTS) data, the formation of high-quality, reliable, and statistically sound information by analyzing and interpreting large data set is becoming a challenging task due to its increased complexity and over-fitting problems. Preprocessing steps play an important role in overcoming the performance issues of MTS data analysis. Feature and data subset selections are important preprocessing steps before applying any data mining functionalities like clustering and classification to identify the efficient and valuable predictors and relevant instances that better represents the underlying process of the data. Here we introduced an optimized preprocessing step using process control charts to extract a subset of key instances that form the representative set of the core group and utilized two classification algorithms to analyze the performance. The results are also compared for different test scenarios by adding standard dimensionality reduction methods and numerosity reduction approaches.

Keywords Multivariate time series data · Dimensionality reduction
Numerosity reduction

1 Introduction

Multivariate time series (MTS) is a sequence of data points collected over multiple intervals of time. Each data point in MTS is a measurement captured at different time stamps [1]. Nowadays, the increased use of MTS data coming from different areas like weather forecasting, health care, pattern recognition, telecommunication, manufacturing arises the need of efficient feature selection processes for analyzing

J. John · S. Ashok (✉)
Department of Computer Science & IT, Amrita School of Arts & Sciences Amrita Vishwa Vidyapeetham, Kochi 682024, India
e-mail: sreeja.ashok@gmail.com

J. John
e-mail: jobiyajohn@gmail.com

J. Nayak et al. (eds.), *Soft Computing in Data Analytics*,
Advances in Intelligent Systems and Computing 758,
https://doi.org/10.1007/978-981-13-0514-6_56

and interpreting these huge data sets without losing the characteristics of the original data [2]. MTS data involves a tremendous number of data points because different sources measure and generate a huge amount of information even in one second time gap. The instances collected at multiple intervals of time may have some sort of internal relationship that should be identified and extracted. There exist a number of hidden and recurring patterns, and only a subset of these may be needed for building models. Techniques that are capable of extracting composite features from the huge data set are extremely useful for building models for data analysis [3].

Multivariate time series classification aims at categorizing the data sets to a previously known labeled class. Preprocessing steps help in reducing the computational complexity and improve the classification efficiency of MTS data through feature selection and data reduction methods. Feature selection technique helps in generating features that are low in dimension by removing unwanted features. This process retains the features useful to the classifier for improving the performance of the model and ensures that the transformed data does not alter the characteristics of the original data set [4]. Decision-making process can be significantly strengthened through the selection of efficient feature selection and data reduction methods. A data set of electroencephalogram (EEG) signals of epilepsy patients is used in this work for analysis [5].

2 Related Work

Peña and Poncela [6] compared and analyzed different dimensionality reduction methods both in stationary and non-stationary MTS data and found all the linear combinations of the features in time series with different approaches like principal component analysis (PCA), reduced rank models, canonical analysis (CA), scalar component models, and factor models [6]. With canonical correlation analysis, the dimension of the statistical correlation between the time series is tested. Spiegel [7] used a three-way approach for identifying patterns hidden in MTS signals [7]. In the first stage, suitable features were extracted to build models with PCA. Then, a singular value decomposition (SVD) model based on bottom-up segmentation was applied for identifying internally homogeneous time series data points. Through agglomerative hierarchical clustering those optimized segments are then categorized into multiple classes according to statistical characteristics of the segments.

Guerrero-Mosquera et al. [8] used different feature extraction methods in the preprocessing stage and created an electroencephalogram (EEG) feature matrix [8]. Then, its dimension is reduced with forward–backward procedure based on mutual information (MI) which measures the relationship between features (attributes) with the output variable (response variable). Among different classifiers, SVM classifier provides a good alternative for solving many classification problems with better accuracy. With tenfold cross-validation SVM model parameters were selected and the statistical significance of evaluation results has been verified by a Kruskal–Wallis test. Hallac et al. [9] proposed a different alternative for the classification of MTS data where the data is broken into multiple segments and each segment is modeled as independent samples based on a Gaussian distribution [9]. This top-down approach for segmentation helps in maximizing the covariance-regularized log likelihood in an efficient manner and can be easily scalable to vectors with high dimension and time series with different lengths. This is used to detect breakpoints in multivariate time series and used as a black box method which can automatically find an appropriate number of breakpoints, and the model parameters within each segment. Here we are focusing on defining a proper framework for classifying MTS data sets by introducing a segmentation process for improved accuracy.

3 Proposed Approach

An overview of the proposed approach is shown in Fig. 1. An optimized preprocessing step for segmentation is introduced to filter out statistically significant data points using process control charts. We have analyzed and compared fragmentation with feature selection processes like principle component analysis (PCA) and co-relation-based feature selection process (CFS) to check the accuracy of the model developed. A random selection method for identifying the key instances is also used for evaluating the performance of the model. The following sections detail the structure and representation of MTS data, different preprocessing steps including dimensionality and numerosity reduction and different classification algorithms used for comparison.

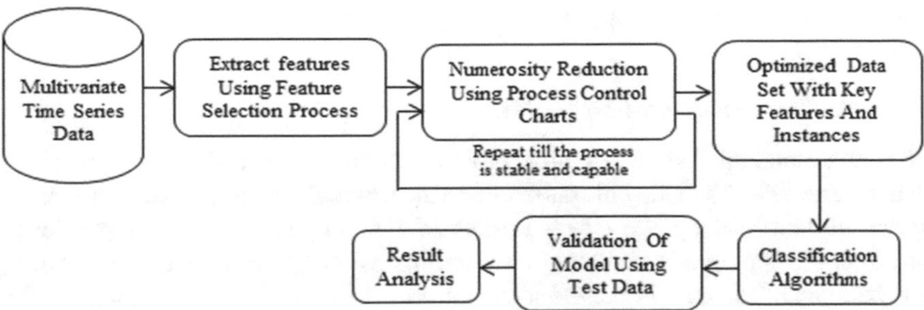

Fig. 1 Process flow for model construction and analysis of multivariate time series data

3.1 *Multivariate Time Series (MTS)*

Generally, a multivariate time series data can be denoted as $m \times n$ two-dimensional matrix in Eq. 1.

$$X = \begin{bmatrix} x11 & x12 & \cdots & x1n \\ x21 & x22 & \cdots & x2n \\ \vdots & \vdots & \vdots & \vdots \\ xm1 & xm2 & \cdots & xmn \end{bmatrix} \quad (1)$$

Here m is the count of given input variables, n is the total number of time stamps, and x_{ij} denotes the data point measured on input variable i at time stamp t_j ($1 \leq i \leq m$ and $1 \leq j \leq n$). This representation means that all m input variables are measured at same time sequence (t_1, t_2… t_n) and the time intervals between each nearest pair are equal [10]. Here time interval is a time unit measuring the sampling rate of a time series. There are situations where the time series collects various and irregular time intervals due to different data sampling rates which can be represented in Eq. 2, where m denotes the number of input variables in multivariate time series.

$$X = \begin{bmatrix} (x11, t11) & \cdots & (x1j, t1j) & \cdots & (x1n1, t1n1) \\ \vdots & \ddots & \vdots & \ddots & \vdots \\ (xi1, ti1) & \cdots & (xij, tij) & \cdots & (xini, tini) \\ \vdots & \ddots & \vdots & \ddots & \vdots \\ (xm1, tm1) & \cdots & (xmj, tmj) & \cdots & (xmnm, tmnm) \end{bmatrix} \quad (2)$$

In general, n_1, n_2… n_m denote data observation numbers of each variable in an m-variate time series where each (x_{ij}, t_{ij}) denotes the data point of the *i*th variable measured at *j*th time point.

3.2 *Preprocessing Steps*

3.2.1 Dimensionality Reduction Approaches

Principle Component Analysis (PCA)

PCA is a standard feature selection process commonly used in high dimensional data set due to its efficiency of transforming the original data to a lower-dimensional space without altering the characteristics of the original data set. It provides a one-to-one mapping of variables to a lower-dimensional platform while maximizing the variance of the data in the reduced platform [11]. The covariance matrix of the normalized data is first computed, and the eigenvectors and eigenvalues of this covariance matrix are extracted. The eigenvectors with highest eigenvalues are identified as the principal components of the given data set. In general, the original data set with high dimension has been reduced to low dimensional space by means of some transformed components called principle components.

Correlation-Based Feature Selection Process (CFS)

Feature selection process based on correlation is a statistical method for finding the stability between the input variables/attributes. The central idea of CFS is to extract features with high correlation to the labeled class, and uncorrelated with each other [12, 13]. Different methods are used to perform correlation analysis based on the density distribution between the variables: Mostly, Pearson parametric correlation test (r) is used to measure a linear relationship between the attributes under the assumption that all the variables are selected from a normally distributed data set. If the data is not normally distributed, other nonparametric rank-based correlation methods like Kendall tau and Spearman rho are used to find the stability between the input variables. The magnitude of the relationship between variables is specified as coefficient value of correlation.

3.2.2 Numerosity Reduction Methods

Recent research focuses on creating models that contain an optimized set of data points for classification since the evaluation can be extremely affected by unusual or outliers in the data. In order to avoid this, process control chart-based segmentation is introduced to improve the accuracy for the model and the same is compared with standard random selection process.

Process control charts

Control charts are excellent statistical tool for looking data points that can cause unusual performance and identifying whether they are sufficient for investigation [14]. It is used to monitor a process variation over time and to achieve process stability. The consistency of data points within the outer limits is based on three standard deviations (three sigma) of the average line (centerline). Figure 2 shows

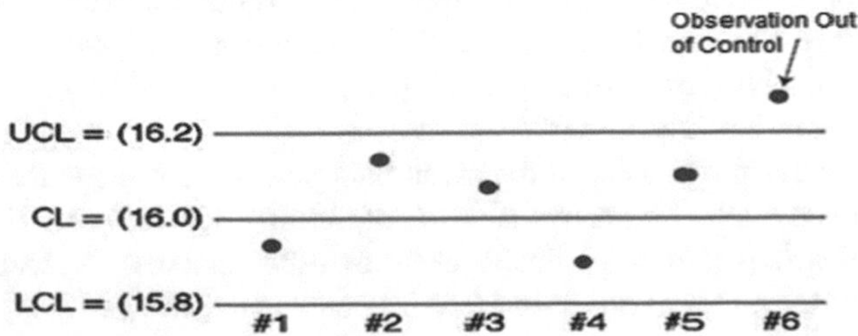

Fig. 2 Process control chart for a sample data set

how some sample data points are plotted with center line (CL), upper control limit (UCL), and lower control limit (LCL).

Commonly used charts are X-bar chart and R chart for finding variables in the data set. X-Bar chart measures the average of all the continuous data points collected over time intervals. If the observations are found out of control, that must be eliminated to reduce the effect of the out-of-control points by leaving them from the estimation of the average and control limits. We used X-bar charts to monitor the stability of the process for each class label and removed the outliers in an iterative manner till reaching the optimum set of data points which are significant in terms of both stability and consistency.

Random Selection Process

Random selection process is a basic and simple selection process in which we randomly select instances from each class under the assumption that the extracted data set will represent the original one without much losing the accuracy of the original data set [15]. In random selection processes, all the instances and attributes have the same probability of being selected as a sample for classification. The random selection process can be performed either manually or by using a function call with parameters containing the sample size and the total size of the data set.

3.3 Classification Algorithms

Classification algorithms are supervised learning procedures, which map data items presented in the input data set to a known label class [16]. Different classification algorithms use different methods for determining the relationships of attributes to the labeled class. Naïve Bayes' theorem represents a group of individual random variables and their correlation with predictor class. It works based on Bayes' theorem and centers on conditional probability. Moreover, Naïve Bayes generate models with an assumption that the presence of a particular variable/attribute in a class is independent of the presence of other variables. This model is easy to implement, simple to understand, and particularly useful for very large data sets and also used for highly sophisticated classification processes [17, 18].

Support vector machine (SVM) is another classification method in which many researches are taking place based on a kernel function. The word "kernel" is used to denote a weighting function for nonlinear SVM classification. Traditional SVMs provide a linear mapping of attributes to a labeled class, but with kernels we can make nonlinear mapping. The aim of SVM classifier is to find out an optimal break line which maximizes the margin of the given data points from the classes. Practically, SVM builds model with the available set of training data and each data points are mapped and grouped to a particular class or other classes. Typically, SVM classifiers plot the attributes as points in a multidimensional space and are grouped in a way that the attributes of different classes are divided by a clear break line called separating hyperplane that should be as wide as possible. The newly formed data items are then mapped to the previously created space and determined to which class it belong depending on which side of the break point they laid [19, 20].

4 Experiments and Result Analysis

Here an EEG data set (http://archive.ics.uci.edu/ml/datasets/Epileptic+Seizure+Recognition) of five categories is used for analysis; each contains 100 files and corresponds to a single person, and the file contains recording of brain functions for 23.6 s time interval. The corresponding time series is sampled into 2300 data points. Each data item is the measure of the EEG recording at different time intervals. Table 1 gives the details about the data set.

Description of each class label: class label 1—recording of seizure activity; class label 2—the EEG measurements from the part where the tumor grown; class label 3—the EEG measurements from the healthy part of the brain after identify the location of the tumor; class label 4—eyes closed, the patient closed his eyes at the time of recording EEG measurements; class label 5—eyes open, the patient opened his eyes at the time of recording EEG measurements.

For evaluation, the time series is analyzed with Naïve Bayes and SVM classifiers for nine different test cases: analyzed with/without using dimensionality reduction and numerosity reduction approaches. The combined approach of feature selection with PCA and the proposed segmentation process using process control chart is showing an improved performance when compared with all other test scenarios. Figure 3 shows the PCA components against the proportion of variance. The plot shows that ~50 components explain around 99% variance, and we have reduced 178 predictors to 50 without compromising the explained variance for further analysis.

With process control chart (PCC), we fragment each class of EEG signals with data points that seems to be sufficient for investigation by eliminating the instances that are out of control from each class. Following set of Fig. 4a–e shows the results of data sets before and after applying process control chart to each class label.

Tables 2 and 3 represent the accuracy measures for each test scenarios when modeled using Naïve Bayes and SVM classifiers. It is clearly observed that

Table 1 Data set description

Data set	Attributes	Label class y no. of categories	No. of instances in each class
EEG multivariate time series	178	5	2300

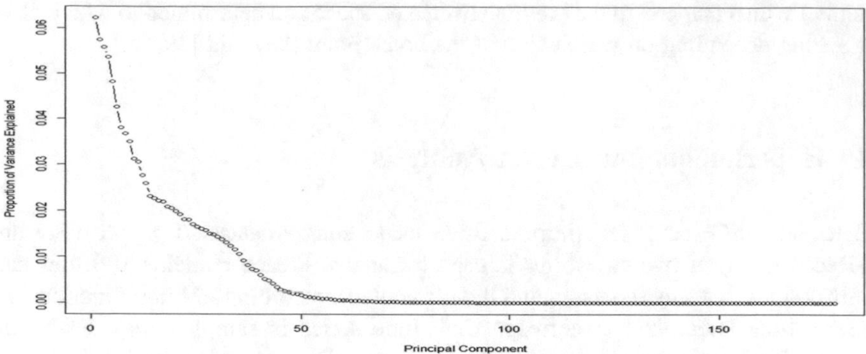

Fig. 3 PCA components against proportion of variance

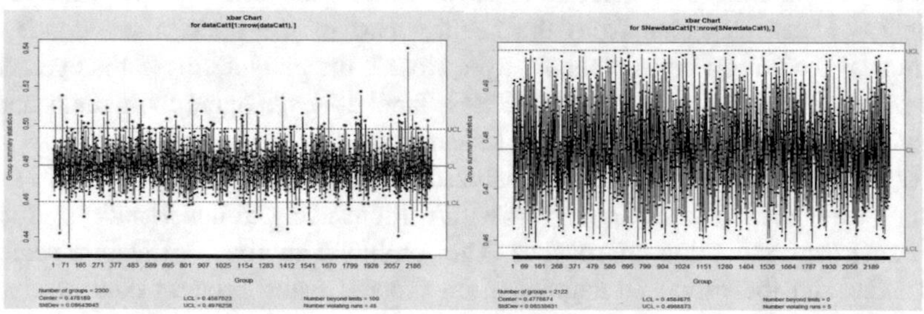

Fig. 4 **a** Before and after applying PCC for class label 1; **b** before and after applying PCC for class label 2; **c** before and after applying PCC for class label 3; **d** before and after applying PCC for class label 4; **e** before and after applying PCC for class label 5

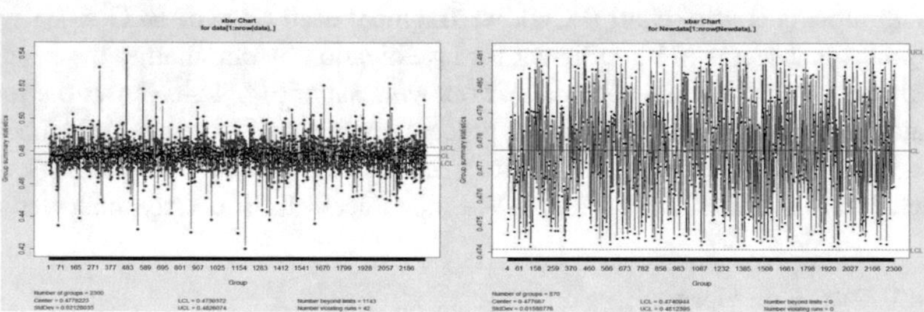

Fig. 4 (continued)

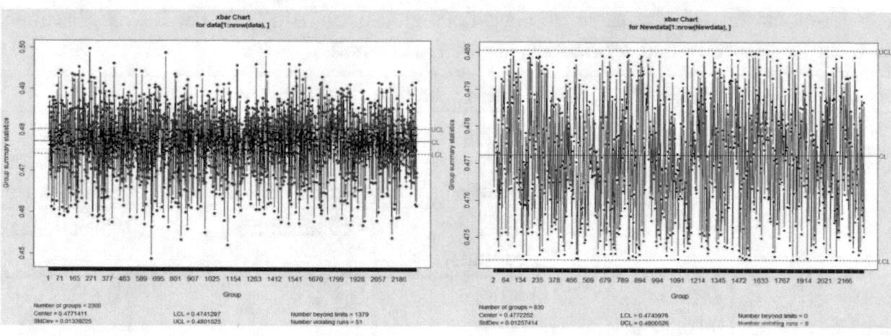

Fig. 4 (continued)

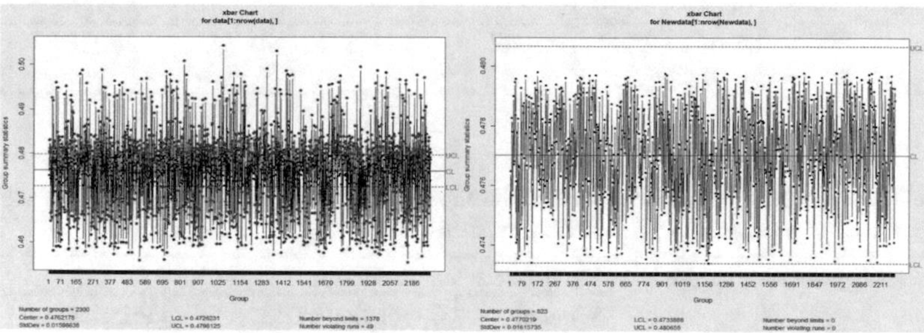

Fig. 4 (continued)

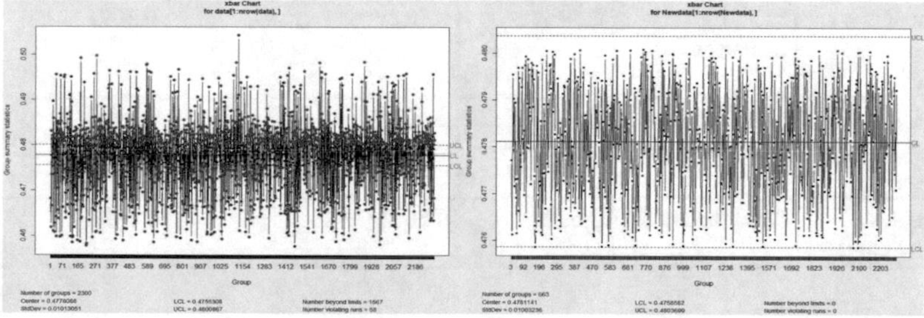

Fig. 4 (continued)

integrated model using PCA and PCC is showing a better accuracy when compared with other scenarios. This necessitates the importance of preprocessing steps in modeling multivariate time series data.

Table 2 Comparison of accuracy of classification model using Naive Bayes classifier with different feature selection and numerosity reduction methods

Numerosity reduction methods		Feature selection methods		
		Without feature selection	Correlation-based feature selection process (CFS)	Principal component analysis (PCA)
		(178 attributes)	(20 attributes)	(50 attributes)
Without numerosity reduction	11500 instances	0.4398	0.4121	0.6503
Random selection process	500 instances from each class	0.5081	0.4461	0.5962
Process control chart	5308 instances	0.5648	0.5326	**0.7387**

Table 3 Comparison of accuracy of classification model using SVM classifier with different feature selection and numerosity reduction methods

Numerosity reduction methods		Feature selection methods		
		Without feature selection	Correlation-based feature selection process (CFS)	Principal component analysis (PCA)
		(178 attributes)	(20 attributes)	(50 attributes)
Without numerosity reduction	11500 instances	0.5987	0.4821	0.5991
Random selection process	500 instances from each class	0.5321	0.4785	0.6442
Process control chart	5308 instances	0.6196	0.5631	**0.6938**

5 Conclusion

Multivariate time series analysis is becoming more dominant with the drastic increase of modern equipment and its corresponding chronologically collected enormous observations. In order to improve the performance of MTS analysis models, an optimized preprocessing framework is needed for extracting a set of discriminative features that traces the characteristics of the original data set. In this work, we introduced a standardized technique for preprocessing that adopts process control chart (PCC)-based segmentation process for identifying the significant data instances for further analysis. The results when analyzed using two different classification models demonstrated that the combined approach that includes both dimensionality reduction using principal component analysis (PCA) and numerosity reduction technique using the proposed approach enhances the performance of the system. This signifies the importance of efficient preprocessing steps for conceptualizing large data sets, reducing the computational complexity and improving the performance of analysis models.

References

1. Box, G.E., Jenkins, G.M., Reinsel, G.C.: Multivariate time series analysis. In: Time Series Analysis, 4th Edn., pp. 551–595 (2008)
2. De Gooijer, J.G., Hyndman, R.J.: 25 years of time series forecasting. Int. J. Forecast. **22**(3), 443–473 (2006)
3. Tiao, G.C., Box, G.E.: Modeling multiple time series with applications. J. Am. Stat. Assoc. **76** (376), 802–816 (1981)
4. Geurts, P.: Pattern extraction for time series classification. In: PKDD, vol. 1, pp. 115–127, Sept., 2001
5. Sanei, S., Chambers, J.A.: EEG Signal Processing. Wiley (2013)
6. Peña, D., Poncela, P.: Dimension reduction in multivariate time series. In: Advances in Distribution Theory, Order Statistics, and Inference, pp. 433–458 (2006)
7. Spiegel, S., Gaebler, J., Lommatzsch, A., De Luca, E., Albayrak, S.: Pattern recognition and classification for multivariate time series. In: Proceedings of the Fifth International Workshop on Knowledge Discovery from Sensor Data, pp. 34–42, Aug., 2011. ACM
8. Guerrero-Mosquera, C., Verleysen, M., Vazquez, A.N.: EEG feature selection using mutual information and support vector machine: a comparative analysis. In: 2010 Annual International Conference of the IEEE Engineering in Medicine and Biology Society (EMBC), pp. 4946–4949, Aug. 2010. IEEE
9. Hallac, D., Nystrup, P., Boyd, S.: Greedy Gaussian segmentation of multivariate time series (2016). arXiv:1610.07435
10. Hamilton, J.D.: Time series analysis, vol. 2. Princeton University Press, Princeton (1994)
11. Maćkiewicz, A., Ratajczak, W.: Principal components analysis (PCA). Comput. Geosci. **19** (3), 303–342 (1993)
12. Nanditha, J., Sruthi, K.N., Ashok, S., Judy, M.V.: Optimized defect prediction model using statistical process control and correlation-based feature selection method. In: Intelligent Systems Technologies and Applications, pp. 355–366. Springer, Cham (2016)
13. Hall, M.A.: Correlation-based feature selection of discrete and numeric class machine learning (2000)
14. Leavenworth, R.S., Grant, E.L.: Statistical Quality Control. Tata McGraw-Hill Education (2000)
15. Guyon, I., Elisseeff, A.: An introduction to variable and feature selection. J. Mach. Learn. Res. **3**(Mar), 1157–1182 (2003)
16. Suma, V.R., Renjith, S., Ashok, S., Judy, M.V.: Analytical study of selected classification algorithms for clinical dataset. Indian J. Sci. Technol. **9**(11) (2016)
17. Lewis, D.D.: Naive (Bayes) at forty: the independence assumption in information retrieval. In: European Conference on Machine Learning, pp. 4–15, Apr, 1998. Springer, Berlin, Heidelberg
18. John, G.H., Langley, P.: Estimating continuous distributions in Bayesian classifiers. In: Proceedings of the Eleventh Conference on Uncertainty in Artificial Intelligence, pp. 338–345, San Mateo. Morgan Kaufmann (1995)

19. Schölkopf, B., Smola, A.J.: Learning with Kernels: Support Vector Machines, Regularization, Optimization, and Beyond. MIT press (2002)
20. Muller, K.R., Mika, S., Ratsch, G., Tsuda, K., Scholkopf, B.: An introduction to kernel-based learning algorithms. IEEE Trans. Neural Netw. **12**(2), 181–201 (2001)

Traffic Data Classification for Security in IoT-Based Road Signaling System

Srijanee Mookherji and Suresh Sankaranarayanan

Abstract Traffic congestion is one of the major problems faced by people in large cities. The traffic controlling systems at present are time stamp oriented which is semiautomatic in nature. With the introduction of IoT in traffic signaling systems, researches are being done considering density as a parameter for automating the traffic signaling system and regulate traffic dynamically. Security is a concern when sensitive data of great volume is being transmitted wirelessly. To prevent the issues on security protocols, we here developed a secured IoT-based system with intelligence that will analyze the traffic data patterns for good or bad and accordingly block attacks to the system. In here, we are addressing man-in-the-middle attack (MITM) only on data for security analysis as a first step. So toward this, support vector machine learning algorithm deployed at the Edge that would classify the traffic data as good or bad based on training before being processed for regulating the traffic signal. The classification of data for applying SVM at the Edge is implemented by taking the raw traffic data set of three cities of Greater London region for 5 years ranging from 2011 to 2016 amounting to 3577. The implementation carried out using Raspberry Pi3 and Scikit.

Keywords IoT · MITM · SVM · Pi3 · Scikit

1 Introduction

Currently, our road traffic signaling systems are microcontroller based which are based on time stamp in regulating the traffic which is semiautomated. The drawback with this system is that it does not take traffic density into account in managing the

S. Mookherji (✉) · S. Sankaranarayanan
Department of Information Technology, SRM Institute of Science and Technology
Kattankulathur Campus, Chennai 603203, Tamil Nadu, India
e-mail: bobmookherji@gmail.com

S. Sankaranarayanan
e-mail: suresh.sa@ktr.srmuniv.ac.in

J. Nayak et al. (eds.), *Soft Computing in Data Analytics*,
Advances in Intelligent Systems and Computing 758,
https://doi.org/10.1007/978-981-13-0514-6_57

traffic as it is purely based on timer. So toward regulating the traffic signals dynamically based on traffic density, some amount of research been carried out by employing Internet of things technology.

Internet of things (IoT) is slowly and steadily becoming a part of almost every technology. IoT in transportation system can assist in integration of communications, control, and information processing. So based on IoT technology, there has been many research pertaining to traffic controlling where sensors employed toward collecting live data and accordingly transmit data wirelessly using backbone network for regulating the traffic [1–4].

Now, in regard to an IoT-based automated traffic signaling system, security is much of a concern in regard to data been communicated for regulating the traffic signaling system.

IoT architecture as such is widely distributed into three layers which are the perception layer, the network layer, and the application layer. These layers are susceptible to various security threats and attacks. The attacks that the systems are vulnerable to are the denial-of-service attack, the DY intruder attack and privacy attacks like eavesdropping, traffic analysis, and finally data mining. Along with these common attacks, there are many more vulnerabilities present in each layer that can lead to many more recent and common attacks along with various zero-day vulnerability attacks [5, 6].

Lot of IoT security protocols have been suggested, and many frameworks have also been researched upon that involves the use of IDPS in an IoT network to detect and block any incoming attacks from an end device [7–11]. There has been no research work performed in regard to securing the IoT-based automated traffic signaling system.

So based on the security issues and challenges in IoT-based automated traffic signaling system, we have developed an effective and intelligent system for detecting the anomalies against attack. Lot of research work has been done in the past in developing cryptosystem in securing the system against attacks. But none of these systems are intelligent or smart enough in detecting the attacks or anomalies based on data pattern which are heterogeneous in nature.

There has been research done with machine learning techniques being implemented in a normal network system in general. The work till date has been achieved to a level of automatically labeling data by studying its content and classifying into various security classes in a normal networked system and not an IoT-based system specific to traffic signaling system [11, 12].

So toward implementing an intelligent system for analyzing against security attacks in an IoT-based automated traffic signaling system, an efficient machine learning algorithm called support vector machine (SVM) be employed at the Edge that could study the traffic data pattern coming from IoT-based end devices called sensors and accordingly classify them as good or bad data. So toward classifying the data as good or bad by the Edge processor, the Edge processor be trained by applying SVM machine learning algorithm. The classification of data by applying SVM at the Edge is implemented by taking the raw traffic data set of three cities of Greater London region for 5 years ranging from 2011 to 2016 amounting to 3577 which is total vehicle count per hour. Raw traffic data is preprocessed and accordingly 70% be taken for training and remaining 30% for testing which is on an average 1464 preprocessed data per city.

The rest of paper is organized as follows. Section 2 gives an in-depth literature review pertaining to security and machine learning. Section 3 talks on intelligent IoT-based secured traffic signaling system. Section 4 shows the implementation results and analysis. Section 5 is the conclusion and future work.

2 Literature Review

Before going into the details of our IoT-based architecture for traffic signaling system toward intelligent security analytics, we will review some of the existing system in vogue.

Researchers in [1] proposed an architecture consisting of an IoT network integrated with machine learning technology which is used to create effective communication within the heterogeneous Web decentralized devices within the IoT. The architecture involves RFID sensors and Internet-based information systems with tagged traffic objects. The framework involves a large number of wireless devices, thus creating a great amount of sensitive data. Security of smart objects is a field of concern that has not been explored.

Researchers in [1–4] discussed two traffic light control strategies related to consideration of downstream traffic conditions. The suggested framework is compared to similar approaches where it outperformed congestion control as well as controlling air pollution and fuel consumption. The system implemented digital pheromones for data communication. The system lacks shortest path rerouting strategies and the presence of any machine learning system for uncertainty prediction.

In addition, researchers in [13] applied the Kerner's three-phase traffic theory to create a synchronized system for automatic traffic management and establishing communication between the traffic controllers. The research is based on the V2I0T Approach which identifies significant and recurring traffic. The WSN-based system calculates the density of the roads and transfers the data to each traffic controller the densest road is given the priority for immediate clearance.

Also, researchers in [12] have chalked out a survey paper focused on machine learning and data mining methods for cyber analytics in support of intrusion detection. The paper particularly describes how machine learning has been implemented to detect known attacks. The techniques involve clustering and support vector machine but the system lacks the ability to handle zero-day attack.

Researchers in [11] proposed a security classification system using machine learning that inspects data content and classifies information objects through labels. The project implemented clustering and analysis of "raw text" but was restricted to classify into only two security classes.

Researchers in [12] use machine learning for network intrusion detection. They work with logs generated from both network end and host end and thus analyzing the logs to find vulnerabilities. The machine learning supported system helps in tracing most types of attacks prevailing in the system.

Researchers in [10] proposed a bio-inspired system that uses machine learning techniques to differentiate between fraudulent nodes and good nodes in a wireless network system. The system employees a supervised learning machine along with an unsupervised k-mean-based learning as well as an anomaly detecting algorithm that can be used to disable the fraudulent nodes.

3 Intelligent Secured IoT-Based Traffic Signaling System

As Internet of things is slowly gearing up and being applied in every aspect of life, we see that there has been research carried out pertaining to traffic controlling where sensors employed toward collecting live data and accordingly transmit data wirelessly using backbone network for regulating the traffic.

Now, in terms of IoT-based automated traffic signaling system, security is much of a concern. Lot of Security protocols and framework have been proposed in IoT against security attacks. But the challenge in such security protocols is that it can only defend against known attacks, and also, they cannot handle heterogeneous data and purely based on cryptosystem only.

So there is need for an intelligent system which can defend the security attacks based on data pattern rather than relying on security protocols. So based on above drawbacks in analyzing the security attack in an IoT-based traffic signaling system, we here propose an IoT-based security architecture with intelligence that employ the machine learning in Edge toward classification of traffic data as good or bad and accordingly prevent attacks before data being processed for automating the traffic signals. Figure 1 shows the complete system architecture, and Fig. 2 shows the system block diagram.

In Fig. 1 which is the system architecture of our secured IoT-based traffic signaling system with intelligence, we have IP-based sensor cameras which gives the traffic image with density information. These IP-based sensor cameras as shown in Fig. 1 are deployed in junction roads alongside the traffic signals. These sensor cameras' data from the junction roads are aggregated and sent to an aggregator which are wireless enabled and accordingly communicated to the Edge router where real-time decision taken toward regulating the traffic. The decision signals are then sent to aggregator back in actuating the sensor device in regulating the traffic signals. The Edge router got the intelligence in analyzing the real-time traffic data for regulating the traffic signal. But these Edge router are vulnerable to security attack which possess the machine learning intelligence in analyzing the traffic data pattern for detecting anomalies and defending against attacks. The complete traffic control and decision center reside in the cloud server with traffic information and so forth.

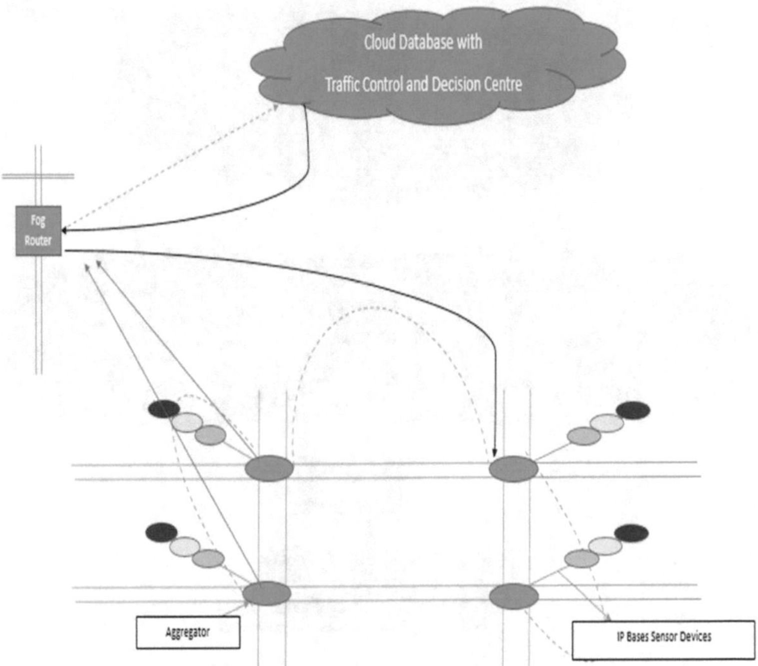

Fig. 1 System architecture of secured IoT traffic signaling system

Figure 2 gives the system design toward security analytics of the traffic data by employing machine learning intelligence against attacks. The IP-based sensors send their raw data to an aggregator which is then wirelessly communicated to the Edge router. Edge or Fog router would take the raw traffic data set and accordingly preprocess it. Preprocessed data is then trained by applying SVM machine learning algorithm for classifying the data pattern for good or bad so as to avoid any attacks to system. The metric pool consists of samples of untampered data and various samples of tampered data. In the last phase of machine learning, the data is sent to the security analysis centre for report and finally the report is sent to the cloud for storage. Raw traffic data set here refers to total vehicle count per hour for each city. Preprocessed data refers to hourly average vehicle count and also frequency of vehicle count/pattern minute-wise for each hour slot for every city.

Pseudocode algorithm of SVM machine learning algorithm for traffic data classification for good or bad data classification is given below:

Input: Sequence of 12 h Minute wise vehicle count for a city.

X = {(Time,Vehicle Count)....}

y = {Class(1,0)} where 1 is Good Data and 0 is Corrupt Data

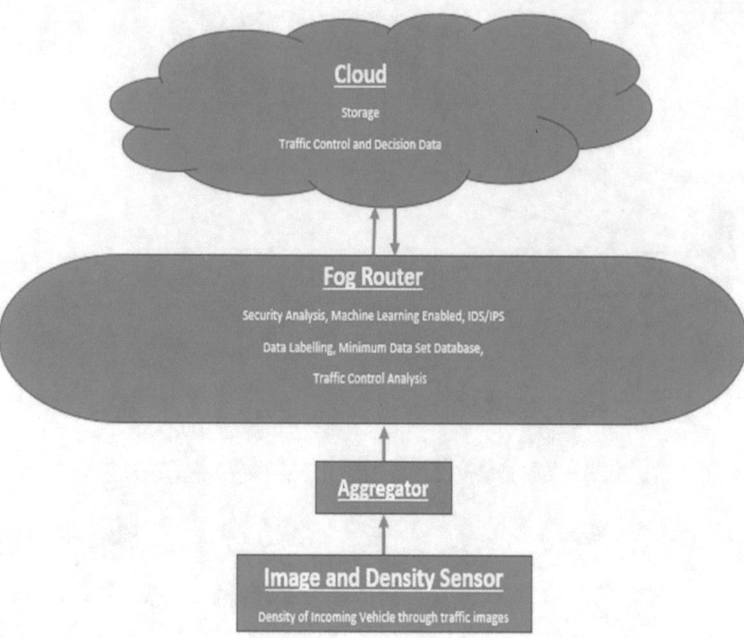

Fig. 2 System flow diagram

Output: Predicted outcome for Time vs Vehicle Count input, generated scatter points on a 2D plane and a separation SVM hyper plane,

1: Import required python libraries

numpy
matplotlib
sklearn

2: Load data sets for input using np.genfromtxt()
3: Randomly split data set into train and test data using train_test_split()
5: Implement SVM algorithm using predefined SVM features present in sklearn, svm.SVC(kernel"").fit()
6: Generate prediction using svm.predict()
7: To plot the test data points plt.scatter()
8: For each test data generates a score using svm.decision_function()

4 Implementation Results and Analysis

The implementation of machine level intelligence in classifying the traffic data at the Edge router as good or bad based on pattern would be helpful in regulating the signal properly. The bad data refers to tampered data, and good data refers to untampered data. In here, we have taken man-in-the-middle attack over wireless channel where data is tampered or modified as it reaches the Edge from sensor devices.

So toward this, support vector machine been employed which classifies the traffic data as good or bad based on training set. Also for Edge router, Raspberry Pi3 processor is deployed where SVM algorithm works in predicting the traffic data based on training set.

So for our implementation, we considered raw data set of vehicle count of Greater London region comprising 33 cities) over the year 2011–2016 which accounts to 45685 [14]. From that raw data set of Greater London region, we narrow down to three cities total vehicle count of about 3577 which are Merton, Harrow, and Hillingdon. The reason for narrowing down to three cities from 33 cities is that the main idea of our work is analyzing the traffic data sent by applying SVM based on training for predicting data as tampered or untampered. So it would be enough to do for three cities for now as the same SVM methodology is applied for other 33 cities of Greater London region.

So before training data set, the first step of machine learning is to preprocess raw data set of the three cities. Raw data cities give only total vehicle count per hour for each city and according preprocessing done toward computing the average vehicle count on hourly basis from morning 6 a.m to evening 6 p.m which would give pattern of traffic data as good or bad on hourly basis. In addition to average vehicle count per hour, we also computed the frequency of vehicles appearing minute-wise in an hourly basis for regulating the traffic which would give minute-wise traffic data pattern from morning 6 a.m to evening 6 p.m toward analyzing data as good or bad. This preprocessing of data from raw data set been done for three cities from 2011 to 2016. The total preprocessed data accounted to 1464 per city which is 732 good and 732 bad data on an average. The same holds good for other two cities too. The preprocessed data is nothing but the average vehicle count minute-wise in an hour slot against time. The sample screenshot of good and bad data pattern based on preprocessing for one city Merton is shown in Figs. 3 and 4.

The preprocessed is fed to Edge router which is Raspberry Pi3 where SVM machine learning algorithm applied for training data set for prediction good and bad data. From total preprocessed data, 70% of data taken for training which is 1024 and remaining 30% for testing which is 440 on an average for three cities. So before applying SVM, Edge router which is Raspberry Pi3 needs to be configured. We have opted for Raspberry Pi3 Model B with 1.2 GHZ quad-core ARM Cortex A53 processors with 1 GB memory and network support of 10/100 Mbps Ethernet, 802.11n Wireless LAN, and Bluetooth 4.0. Raspberry Pi3 been configured with Raspbian OS.

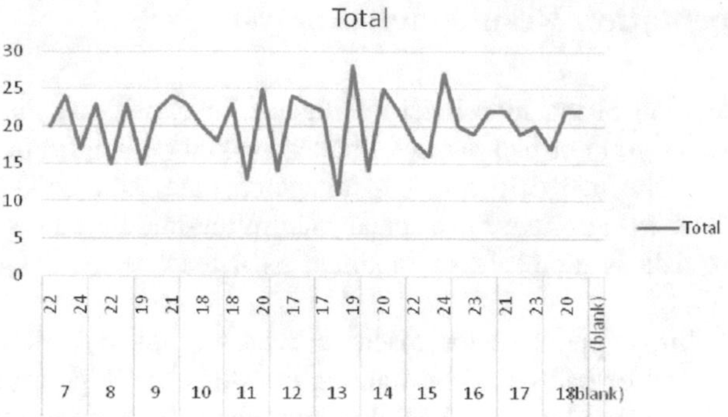

Fig. 3 Preprocessed good data

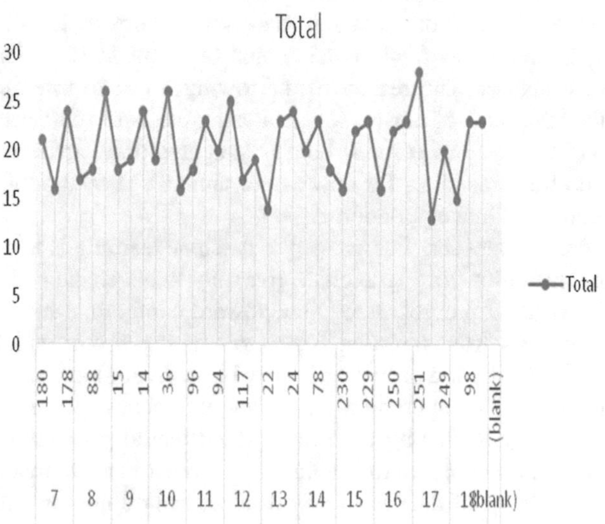

Fig. 4 Preprocessed bad data

Now for training the preprocessed data set in Pi3 processor, SVM machine learning algorithm deployed. So toward this Scikit-learn (formerly Scikits.learn), a free software machine learning library for the Python programming language has been employed. It features various classification, regression, and clustering algorithms and is designed to interoperate with the Python numerical and scientific libraries NumPy and SciPy.

The preprocessed data for three cities Merton, Harrow, and Hillingdon amounting to 1464 per city been fed to Edge router which is Raspberry Pi3 for training and prediction. SVM machine library takes 70% for training which 1024 is and remaining 30% which is 440 for testing for every city. The testing data set been

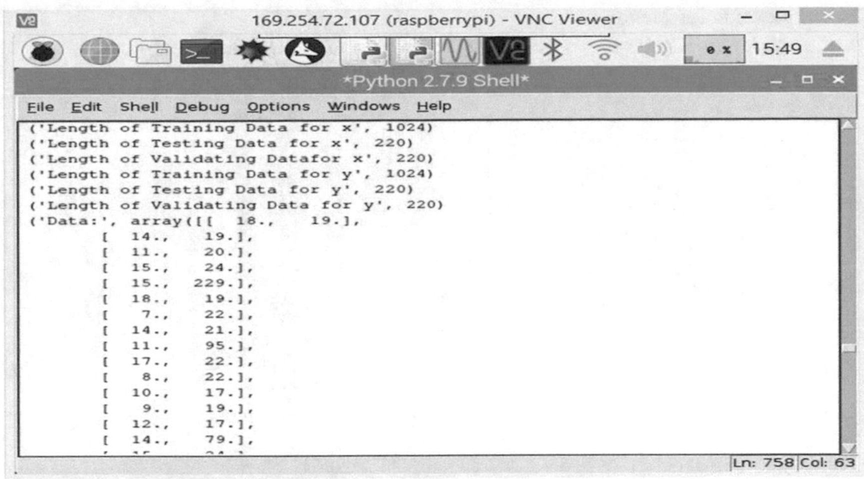

Fig. 5 Scikit training and testing

used for predicting good or bad traffic data at the Edge router. To differentiate good and bad data in prediction, we have used 1 for good and 0 for bad data. Figure 5 shows a sample screenshot in Pi3 for Merton city toward training and testing using Scikit with data separation. Figure 6 shows sample screenshot of prediction graph with testing data for Merton city. Figure 7 shows the predicted output for Merton.

Fig. 6 Prediction graph for Merton

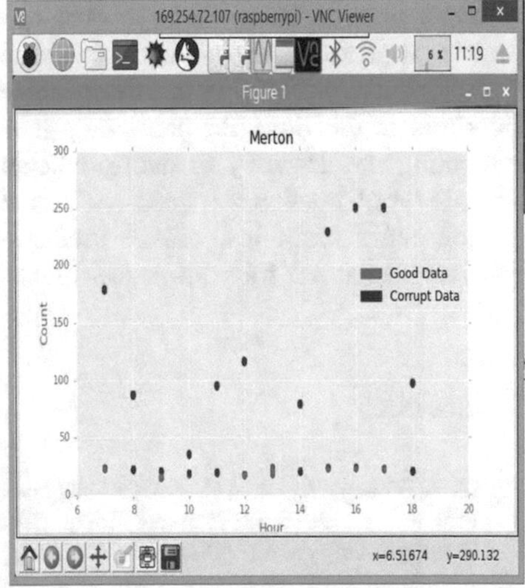

Fig. 7 Prediction for Merton

Merton Prediction
1=Good Data and 0=Corrupt Data

HOURS	TVC	PREDICTION
13	17	0
11	18	1
9	14	0
14	78	0
10	19	1
15	230	0
15	23	1
8	86	0
12	18	1
11	96	0
10	19	0
8	22	1
10	35	1
15	22	1
15	229	0
11	95	0
17	21	1
7	178	0
12	115	0
8	23	0
9	20	1
16	251	0
16	23	1
14	79	0
10	35	1
13	22	1
14	79	0
11	18	1

5 Conclusion and Future Work

So in conclusion, we here have developed an intelligent data classification in IoT by applying SVM machine learning algorithm which classifies data as good or bad data at the Edge router. So toward implementing SVM at the Edge router, real traffic data set of London area been taken from 2011 to 2016 and same preprocessed toward traffic data pattern pertaining to average vehicle count minute-wise in an hour slot from morning 6 a.m to evening 6 p.m. This is nothing but frequency of vehicles appearing minute-wise in an hour slot. Preprocessed data is then trained for each city and accordingly prediction carried out by applying SVM machine learning algorithm at the Edge for predicting good and bad data. In future, we would be computing the accuracy of data prediction of SVM machine learning algorithm. Also, level of intelligence needs to be more refined with multiple level of classification rather than just binary prediction seriousness of attack and proposed defense mechanism for man-in-the-middle attack.

References

1. Li, J., Zhang, Y., Chen, Y.: A self-adaptive traffic light control system based on speed of vehicles. In: 2016 IEEE International Conference on Software Quality, Reliability and Security Companion (QRS-C). IEEE (2016)
2. Al-Sakran, H.O.: Intelligent traffic information system based on integration of Internet of Things and Agent technology. Int. J. Adv. Comput. Sci. Appl. (IJACSA) **6**(2), 37–43 (2015)
3. Mone, S.P., et al.: An intelligent traffic light controlling system. Int. J. Recent Innov. Trends Comput. Commun. (ijrittc) **3**(3) (2015)

4. Cao, Z., et al.: A unified framework for vehicle rerouting and traffic light control to reduce traffic congestion. IEEE Trans. Intell. Trans. Syst. **18**(7), 1958–1973 (2017)
5. Abomhara, M., Køien, G.M.: Security and privacy in the Internet of Things: current status and open issues. In: 2014 International Conference on Privacy and Security in Mobile Systems (PRISMS). IEEE (2014)
6. Granjal, J., Monteiro, E., Sá Silva, J.: Security for the internet of things: a survey of existing protocols and open research issues. IEEE Commun. Surv. Tutor. **17**(3), 1294–1312 (2015)
7. Shanmugavadivu, R., Nagarajan, N.: Network intrusion detection system using fuzzy logic. Indian J. Comput. Sci. Eng. (IJCSE) **2**(1), 101–111 (2011)
8. Sinha, S., et al.: Automated traffic control network using secure wireless. Int. J. Sci. Res. Publ. **3**(9) (2013)
9. Hammcr, II., et al.: Automatic security classification by machine learning for cross-domain information exchange. In: Military Communications Conference MILCOM 2015–2015 IEEE. IEEE (2015)
10. Rathore, H., Jha, S.: Bio-inspired machine learning based wireless sensor network security. In: 2013 World Congress on Nature and Biologically Inspired Computing (NaBIC). IEEE (2013)
11. Singh, N., Chandra, N.: Integrating machine learning techniques to constitute a hybrid security system. In: 2014 Fourth International Conference on Communication Systems and Network Technologies (CSNT). IEEE (2014)
12. Buczak, A.L., Guven, E.: A survey of data mining and machine learning methods for cyber security intrusion detection. IEEE Commun. Surv. Tutor. **18**(2), 1153–1176 (2016)
13. Anass, R., Yassine, H., Mohammed, B.: IoT for ITS: a dynamic traffic lights control based on the Kerner three phase traffic theory. Int. J. Comput. Appl. **145**(1) (2016)
14. Traffic Data (n.d): https://data.gov.uk/dataset/gb-road-traffic-counts

Analysis of Various Fraud Behaviors Using Soft Computing Techniques

D. Priyanka and H. V. Bhagya Sri

Abstract These days, most associations, organizations, and government offices have embraced electronic trade to expand their profitability or effectiveness in exchanging items or administrations; in territories, for example, credit card, media transmission, healthcare insurance, computer intrusion, automobile insurance, online auction, and so forth are likewise utilized by genuine clients and fraudsters. There are some issues and difficulties that ruin the execution of FDS. This paper presents an extensive survey of soft computing methodologies used to protect against difficulties which are faced by the fraud detection system in three distinct fraud areas, i.e., credit card, media transmission, healthcare insurance.

Keywords Fraud · Fraud detection system (FDS) · Credit card
Media transmission · Healthcare insurance · Large dataset

1 Introduction

Fraud means a portrayal of a self-evident reality—regardless of whether by words or by lead, by false or deluding assertions, or by disguise of what ought to have been revealed—that misdirects and is proposed to deceive another with the goal that the individual will follow up on it to her or his legitimate damage. In the fast-changing technological environment, all mechanical structure that incorporates money and organizations can be exchanged off by counterfeit acts, such as credit card, media transmission, and healthcare insurance [1]. The most common areas for detecting fraud are shown in Fig. 1.

D. Priyanka (✉) · H. V. Bhagya Sri
Department of Computer Science and Engineering, Sri Sivani College of Engineering, Srikakulam 532410, Andhra Pradesh, India
e-mail: dabbeerupriyanka90@gmail.com

H. V. Bhagya Sri
e-mail: bhagya.hanumanthu@gmail.com

J. Nayak et al. (eds.), *Soft Computing in Data Analytics*,
Advances in Intelligent Systems and Computing 758,
https://doi.org/10.1007/978-981-13-0514-6_58

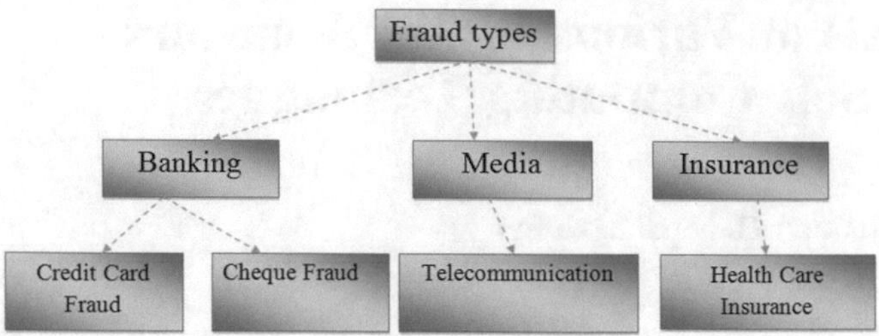

Fig. 1 Various types of fraud

An insight of distributed business is identified with the three regions from 2010 to 2017; it is unavoidable that the most investigated region is banking. Insurance is the third mainstream zone, which has been the principle subject of few investigations since it might incorporate and mixed with different territories, for example, medicinal services insurance extortion, automobile insurance misrepresentation, and home insurance extortion. Media transmission and Internet promoting are the minimum examined regions amid the predefined time frame.

The main focus of this paper will lead toward the use of various soft computing techniques in different diversified applications of fraud detections such as credit card, media transmission, and healthcare insurance framework. The rest of this paper is organized as follows: Sect. 2 describes background study; the main role of soft computing techniques for detecting fraud in three different areas is then expressed in Sect. 3. In Sect. 4, some critical analyses on different types of soft computing methods used for detecting the fraud behaviors are briefly analyzed. Lastly, the concluding remarks are summarized in Sect. 5 with some of the important future directions.

2 Background Study

All in all, the goal of misrepresentation location is to maximize remedy desires and keep up mistaken forecasts at an attractive level [2]. A high right analytic capacity can be suggested by controlling the undetected extortion and the fake cautions. Later some specific terms are depicted. The level of true blue exchanges that are incorrectly distinguished as fake is termed as false alert rate. Misrepresentation getting rate is the level of fake exchanges those are accurately distinguished as fake. False-negative rate means the level of fake exchanges those are mistakenly recognized as real. A few misrepresentation identification methods utilize measurements like the location rate, false alert rate, and average time of location. The general misrepresentation recognition techniques endeavor to augment precision rate and limit false alert rate [3].

The fraud cases must be distinguished from the accessible gigantic informational indexes, for example the logged information and client conduct. At introduce, fraud identification has been executed by various strategies, for example information mining, measurements, and artificial intelligence. Extortion is found from irregularities in information and examples. In this paper, the fraud areas incorporate credit card, media transmission, and healthcare insurance system.

2.1 Credit Card Fraud

Credit cards are exceptionally well known and assumed to be one of the critical part in electronic business and online cash exchange region, which is developing each year. Because of the developing use of this credit card, deceivers try to find probabilities to submit fake that can make gigantic misfortunes to cardholders and banks [1]. It is of two types; one is online fraud and the other is offline.

The fraud committed via Web or net is nothing but online fraud. Just the information of card is required, and a manual mark and agenda banner are not appropriate at the time of procurement. Offline is conferred by utilizing a stolen physical card at call center. As a rule, the establishment of issuing the card can bolt it before utilizing it as a part of a deceitful way [3].

2.2 Media Transmission System

Broadcast or media or telecommunication fraud has drawn the consideration of numerous analysts lately not just because of the immense financial burden on organizations' accountings yet in addition because of the fascinating aspect of client conduct portrayal [4]. Different types of frauds in media transmission system are contractual, hacking, technical and procedural frauds [5].

2.2.1 Contractual Fraud

Contractual frauds produce income through the ordinary utilization of an administration while having no expectation of paying for utilize. Cases of such fraud are subscription and premium rate misrepresentation.

2.2.2 Hacking Fraud

Hacking fraud produces income for the fraudster by breaking into insecure frameworks, and misusing or offering on any accessible usefulness. Cases of such fraud are PABX and network attack.

2.2.3 Technical Fraud

This type of frauds involves assaults against weakness in the innovation of the portable framework. Technical frauds normally require some underlying specialized information and capacity; although once a weakness has been found, this data is regularly and immediately conveyed in a frame that non-specialized individuals can utilize. Cloning and technical internal fraud are some of the examples of technical fraud.

2.2.4 Procedural Fraud

Procedural frauds include strikes against the frameworks completed to restrict introduction to blame and frequently assault the inadequacies in the business strategies used to offer access to the structure. A couple of cases of this kind of frauds is voucher ID duplication, roaming extortion, and faulty vouchers.

2.3 Healthcare Insurance

National Health Care Anti-Fraud Association characterized healthcare insurance deception as an intentional double dealing or misrepresentations made by a man, or an element, with the learning that the misdirection could bring about some unauthorized advantage to him or some other elements and social insurance can handle as the supplier practices that are conflicting with financial, business, or therapeutic practices, and result in an superfluous cost, or in repayment of administrations that are not medicinally fundamental or that neglect to meet professionally perceived norms for human services [5, 6].

3 Role of Soft Computing Methods in Detecting Fraud

Data mining is tied in with discovering experiences which are factually dependable, obscure previously, and significant from information [6]. It is difficult to be sure beyond a shadow of a doubt about the authenticity of what is more, expectation behind an application or exchange. Given the reality, the best financially viable option is to tease out conceivable confirmations of misrepresentation from the accessible information utilizing soft computing techniques [7]. Developed from various research groups, particularly those from created nations, the investigative motor inside these arrangements and programming is driven by various techniques like artificial neural networks, support vector machine, hidden Markov model, artificial immune system, genetic algorithm, fuzzy, and so on.

3.1 Fraud Detection in Credit Card System

Credit card is a standout among the most explored spaces of fraud or extortion detection [1]. Sherly et al. construct an effective fraud detection system which analyzes the approaching exchange against the exchange history to recognize the peculiarity utilizing BOAT method. In any case, these systems are insufficient to prevent credit card extortion. Thus, there is a need to utilize misrepresentation discovery approaches like artificial neural network which investigate information that can recognize and dispose of frauds in credit card [8]. Abnormality-based extortion discovery is generally utilized for charge card misrepresentation identification framework in which the cardholder's profile is made up by breaking down the cardholder spending conduct design. In doing as such, any approaching exchange that is conflicting with the cardholder's profile would be considered as suspicious [3, 9]. Sam Maes applied two methods such as Bayesian belief network and artificial neural networks for the issue to demonstrate their suggestive outcome on financial information of the credit card system [10]. This is because of its inefficiency to distinguish unusual fraud. Table 1 represents the research on various soft computing techniques used for detecting fraud in credit card system.

3.2 Fraud Detection in Media Transmission System

In telecommunication scheme, circumstances of fraud have a critical business effect. Because of the huge measures of information dealt with, extortion identification remains as an extremely troublesome and testing effort [10]. Detecting the fraud in media transmission or telecommunication system by mining outlier is based on the coefficient sum [11]. The fraud detection in media transmission was explored in the European project Advance Security for Personal Communications Technologies (ASPeCT) [12]. In the age of behavioral examples of certifiable subscribers and also fraudsters, a special technique for identifying deceitful calls by presenting Bayesian inferencing and Dempster–Shafer hypothesis is proposed by Panigrahi et al. [13]. Table 2 summarizes the research on various soft computing techniques used for detecting fraud in media transmission system.

3.3 Fraud Detection in Healthcare Insurance System

Healthcare insurance is the major issue whose prevalent development has profoundly influenced the US government [14]. The generally utilized crude information in healthcare misrepresentation discovery is insurance assert which includes the investment of a service provider and an insurance subscriber (Table 3) [15]. Moore enforced a fraud detection framework for analyzing the healthcare insurance

Table 1 Various soft computing techniques for detecting credit card fraud

Fraud area	Technique	Authors	Year	Performance parameters	References
Credit card	Artificial neural networks	Patidar and Sharma, Sahin and Duman	2011	Weight, network type, number of layers, number of nodes	[17, 18]
		Lei and Ghorbani	2012	Number of clusters, accuracy (%), recall (%), precision (%)	[19]
		Carneiro et al.	2015	Attribute (a), clusters, IG (T, a)%, IG (T, Ga)%	[20]
		Murli et al.	2015	Customer id, average balance, tenure, expense	[21]
		Gulati et al.	2017	Customer id, service area, transaction date, JV reference, date, and value	[22]
	Decision tree	Minegishi and Niimi	2011	Accuracy (%), tree size, run time	[23]
		Sahin et al.	2013	True-positive rate, saved loss rate	[24]
		Save et al.	2017	Address mismatch and degree of outlierness	[25]
	Artificial immune systems	Wong et al.	2012	Detection rate, FP rate, FP ratio	[26]
		Halvaiee and Akbari	2014	Number of nodes, map function, threshold computing time, memory cell generation time	[27]
	Hidden Markov model	Falaki et al.	2012	Card holder's profile, % of true- and false-positive rate, accuracy %	[28]
		Mule and Kulkarni, Kumari and Kannan	2014	Depth, flexibility, and quantitative nature	[29, 30]
	Support vector machine	Chen et al.	2004	Cost and accuracy difference	[31]
		Huang et al.	2007	Nominal and numeric features, hit rate	[32]
	K-nearest neighbor	Harris	2015	Training accuracy, testing accuracy, bac, auc, training time	[33]
	Genetic algorithm	Duman and Ozcelik	2011	Region points, threshold, daily amount, and count point	[34]
	Fuzzy logic	Bentley et al.	2000	FL with non-overlapping and overlapping MFs, MP-FL with non-overlapping and overlapping MFs	[35]
		Aburrous et al.	2010	e-banking phishing rating	[36]

Table 2 Various soft computing techniques used for media transmission

Fraud area	Technique	Authors	Year	Performance parameters	References
Media transmission	Artificial neural networks	Hilas and Sahalos	2005	Basic profiles, weekly aggregated behavior	[37]
		Hilas and Sahalos	2006	Distance measure, correct clustering, agglomerative coefficient, cophenetic coefficient	[38]
		Mohamed et al.	2009	–	[39]
		Farvaresh and Sepehri	2011	–	[40]
		Akhter and Ahamad	2012	Average fraud loss per case, detection time	[41]
	Decision tree	Verbeke et al.	2012	–	[42]
		Hilas et al.	2014	Monotone effect of a variable, degree of monotonicity	[43]
	Support vector machine	Subudhi and Panigrahi	2015	–	[44]
		Subudhi and Panigrahi	2016	Accuracy, TPR (true-positive rate), FPR(false-positive rate)	[45]
	Fuzzy neural network	Folasade	2011	Pattern recurrency	[46]
		Akhter and Ahamad	2012	Detection time and average fraud loss	[41]
		Moudani and Chakik	2013	Insolvent customers and error rate for solvent customers	[47]

Table 3 Fraud behaviors and their description for healthcare insurance system

Fraud behaviors	Description	References
Subscribers' fraud	Deceive records of work/eligibility for getting a lower premium rate	[15]
	Utilizing other people's card to unlawfully assert the protection benefits	
	Recording claims for medical administrations which are definitely not in reality	
Carriers' fraud	Misrepresenting advantage/benefit articulations	
	Adulterating repayments	

Table 4 Various soft computing techniques used for healthcare insurance

Fraud area	Technique	Authors	Performance parameters	Year	References
Healthcare insurance	Artificial neural networks	Ortega et al.	Fraud probability, number of cases (%)	2006	[48]
		Huang et al.	Patient id, total no. of physicians, total no. of top service codes	2008	[49]
		Pawar		2016	[50]
	Decision tree	Pawar		2016	[50]
		Ghuse et al.	Accuracy percentages, transaction analysis	2017	[51]
	Support vector machine	France		2011	[52]
		Pawar		2016	[51]
	Hidden Markov	Tang et al.	Group id, identification id, number of consumers, outlierness score	2011	[53]
	Rule based	Shan et al.	Rules violated, specialists with compliance record, average number of records	2008	[54]

based on fuzzy system which utilizes rules obtained from human specialists for recognizing abnormal conduct designs [16].

Table 4 represents the research on various soft computing techniques used for detecting the fraud in healthcare insurance system.

4 Analysis

In this paper, a number of fraud areas such as credit card detection, media transmission, healthcare insurance are considered with their applicability and developments. A brief analytical survey has been made on the above-mentioned techniques with the applications of soft computing techniques. The soft computing techniques are the most evolving and challenging tools for detecting any type of fraud either in real life or through online transactions. The methods of soft computing are quite uncertain, tolerance precessive, and adaptive to cope with any type of real-life complex problem like fraud detection. They are not only suitable for accurate detection of various frauds, but also they can handle those complex situations which are quite difficult in case of some traditional methods of early days computing. The methods of soft computing are so efficient that, on a keyword-based search for the above domain, it is found more than 3 lakhs of documents. But it is not possible to indicate all the documents in this paper, as we have filtered those applications into three areas such as credit card detection, media transmission, and healthcare insurance. Moreover, the soft computing methods that have been considered are only artificial neural network, fuzzy-based method, support vector machine, and decision trees, as it is found most of the problems are solved by using these methods only. In all the areas, most of the researchers have considered the factors like accuracy with the percentage of true-positive and false-positive rates, fraud loss factors, degree of correctness, etc. Although, the performance of all these methods is quite successful in dealing the fraud-related problems, but still there are a number of untouched issues which are under urgent attention. In the area of credit card detection, especially with meta-learning information, a deep attention is needed for adding the spatial information for better visibility. Due to the excess growth in volumes of data, the detection rate must be accurate so that the retrieval process will be easier for all level of users. Also, minimizing the false alarm rate is another productive risk, which has to be taken care. Some statistical approaches may be more helpful for minimizing the false alarm rate. Also, use of unsupervised learning for any of the above tasks is a difficult task, as those techniques require some prior information before proceeding for any other further identification. So, the complexity of such learning methods will be more as compared to those supervised methods. A brief analysis has been illustrated in Fig. 2 about the percentage of the application areas in these fraud areas, and it is found that the maximum ratio in fraud is in credit card detection. A more attention with some advance soft computing methods like deep learning methods is to be developed to handle this problem. Figures 3, 4, and 5 are the illustrations for a number of research articles published in recent years in all those considered areas. The maximum fraud detection areas are concerned with the decision tree method as those are adaptive to direct solutions rather the influence of algorithmic specific parameters. Apart from decision tree method, neural network, support vector machine, and fuzzy-based approaches also have been remained the hot favorite of all level of researchers. Also, everytime upgrading the information for current detection is another challenging task for all the methods.

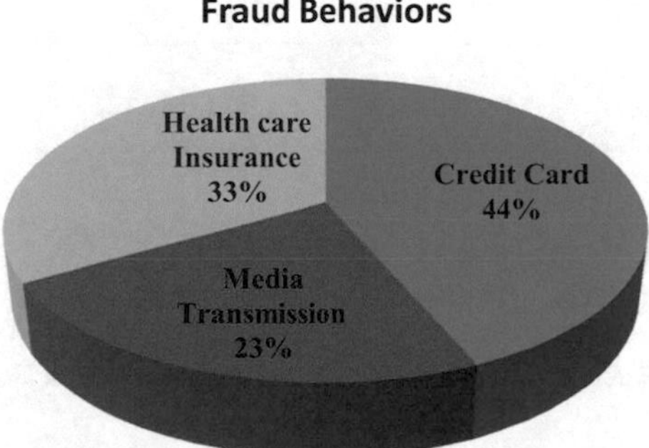

Fig. 2 Percentage of applications in detecting fraud behaviors

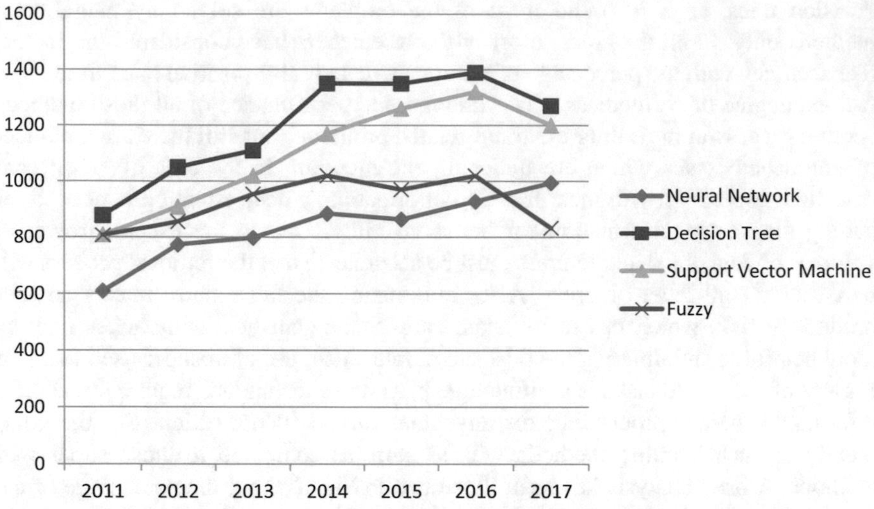

Fig. 3 Number of research articles published for detecting fraud in the area of credit card using soft computing techniques

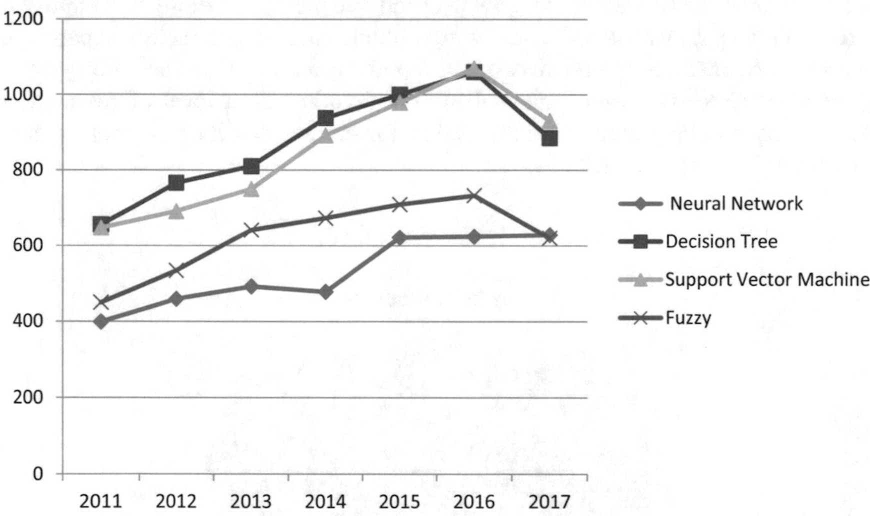

Fig. 4 Number of research articles published in media transmission fraud detection using soft computing techniques

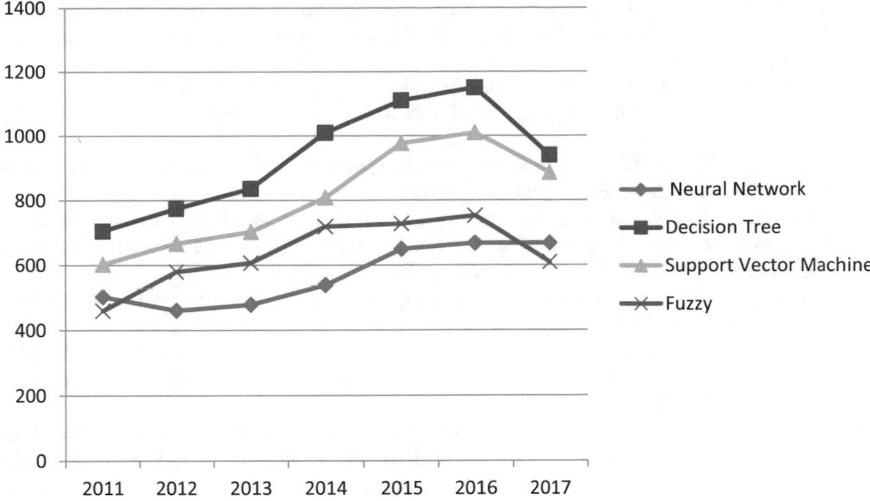

Fig. 5 Number of research articles published in healthcare insurance fraud detection using soft computing techniques

5 Conclusion

In this paper, the detection of fraud is examined in three different areas such as credit card, media transmission, and healthcare insurance. It shows the attributes of various types of fraud and techniques used for detecting the fraud and the probability of future work. Because of the security issues, just a couple of methodologies for credit card identification is publicly accessible. Among them, neural network is an extremely well-known approach. Despite, it is hard to execute due to the absence of accessible data set. Mostly, media transmission fraud detection methods investigate data collection of toll tickets and distinguish extortion from call designs. These structures are suitable against a couple of sorts of tricks, yet at the same time have some principle issues such that it cannot support extortion rates that do not take after the profiles and require exceptionally exact meanings of edges and parameters. In future, the research may likewise inspect contrasts in fake behavior among various sorts of fraud and investigate potential outcomes for making smart inferred credits more precisely.

References

1. Bolton, R.J., Hand, D.J.: Statistical fraud detection: a review. Stat. Sci. 235–249 (2002)
2. SAS Institute: Using Data Mining Techniques for Fraud Detection: A Best Practices Appmach to Government Technology Solutions. Whitepapers. hnp:ii (1996). www.sas.com

3. Kou, Y., Lu, C.T., Sirwongwattana, S., Huang, Y.P.: Survey of fraud detection techniques. In: 2004 IEEE International Conference on Networking, Sensing and Control, vol. 2, pp. 749–754. IEEE (2004)
4. Hilas, C.S., Mastorocostas, P.A.: An application of supervised and unsupervised learning approaches to telecommunications fraud detection. Knowl.-Based Syst. **21**(7), 721–726 (2008)
5. Yang, W.S., Hwang, S.Y.: A process-mining framework for the detection of healthcare fraud and abuse. Expert Syst. Appl. **31**(1), 56–68 (2006)
6. Elkan, C.: Magical thinking in data mining: lessons from CoIL challenge 2000. In Proceedings of the Seventh ACM SIGKDD International Conference on Knowledge Discovery and Data Mining, pp. 426–431. ACM Aug 2001
7. Phua, C., Lee, V., Smith, K., Gayler, R.: A comprehensive survey of data mining-based fraud detection research (2010). arXiv:1009.6119
8. Sherly, K.K., Nedunchezhian, R.: BOAT adaptive credit card fraud detection system. In: 2010 IEEE International Conference on Computational Intelligence and Computing Research (ICCIC), pp. 1–7. IEEE, Dec 2010
9. Malekian, D., Hashemi, M.R.: An adaptive profile based fraud detection framework for handling concept drift. In: 2013 10th International ISC Conference on Information Security and Cryptology (ISCISC), pp. 1–6. IEEE Aug 2013
10. Maes, S., Tuyls, K., Vanschoenwinkel, B., Manderick, B. (2002, January). Credit card fraud detection using Bayesian and neural networks. In *Proceedings of the 1st international naiso congress on neuro fuzzy technologies* (pp. 261–270)
11. Alves, R., Ferreira, P.M.S., Belo, O., Ribeiro, J.T.S.: Detecting telecommunications fraud based on signature clustering analysis (2007)
12. Burge, P., Shawe-Taylor, J.: An unsupervised neural network approach to profiling the behavior of mobile phone users for use in fraud detection. J. Parallel Distrib. Comput. **61**(7), 915–925 (2001)
13. Panigrahi, S., Kundu, A., Sural, S., Majumdar, A.K.: Use of dempster-shafer theory and Bayesian inferencing for fraud detection in mobile communication networks. In: Australasian Conference on Information Security and Privacy (pp. 446–460). Springer, Berlin, Heidelberg, July, 2007
14. Dua, P., Bais, S.: Supervised learning methods for fraud detection in healthcare insurance. In Machine Learning in Healthcare Informatics, pp. 261–285. Springer, Berlin, Heidelberg (2014)
15. Liu, Q., Vasarhelyi, M.: Healthcare fraud detection: a survey and a clustering model incorporating Geo-location information. In: 29th World Continuous Auditing and Reporting Symposium (29WCARS), Brisbane, Australia (2013)
16. Wong, W.K., Moore, A.W., Cooper, G.F., Wagner, M.M.: Bayesian network anomaly pattern detection for disease outbreaks. In: Proceedings of the 20th International Conference on Machine Learning (ICML-03), pp. 808–815
17. Patidar, R., Sharma, L.: Credit card fraud detection using neural network. Int. J. Soft Comput. Eng. (IJSCE) 1(32–38) (2011)
18. Sahin, Y., Duman, E.: Detecting credit card fraud by ANN and logistic regression. In: 2011 International Symposium on Innovations in Intelligent Systems and Applications (INISTA), pp. 315–319. IEEE, June 2011
19. Lei, J.Z., Ghorbani, A.A.: Improved competitive learning neural networks for network intrusion and fraud detection. Neurocomputing **75**(1), 135–145 (2012)
20. Carneiro, E.M., Dias, L.A.V., da Cunha, A.M., Mialaret, L.F.S. Cluster Analysis and Artificial Neural Networks: A Case Study in Credit Card Fraud Detection. In: 2015 12th International Conference on Information Technology-New Generations (ITNG), pp. 122–126. IEEE, Apr 2015
21. Murli, D., Jami, S., Jog, D., Nath, S.: Credit Card Fraud Detection Using Neural Networks. Int. J. Stud. Res. Technol. Manage. **2**(2), 84–88 (2015)

22. Gulati, A., Dubey, P., MdFuzail, C., Norman, J., Mangayarkarasi, R.: Credit card fraud detection using neural network and geolocation. In: IOP Conference Series: Materials Science and Engineering, vol. 263, no. 4, p. 042039. IOP Publishing Nov 2017
23. Minegishi, T., Niimi, A.: Detection of fraud use of credit card by extended VFDT. In: 2011 World Congress on Internet Security (WorldCIS), pp. 152–159. IEEE, Feb 2011
24. Sahin, Y., Bulkan, S., Duman, E.: A cost-sensitive decision tree approach for fraud detection. Expert Syst. Appl. **40**(15), 5916–5923 (2013)
25. Save, P., Tiwarekar, P., Jain, K. N., Mahyavanshi, N.: A novel idea for credit card fraud detection using decision tree. Int. J. Comput. Appl. **161**(13) (2017)
26. Wong, N., Ray, P., Stephens, G., Lewis, L.: Artificial immune systems for the detection of credit card fraud: an architecture, prototype and preliminary results. Inf. Syst. J. **22**(1), 53–76 (2012)
27. Halvaiee, N.S., Akbari, M.K.: A novel model for credit card fraud detection using Artificial Immune Systems. Appl. Soft Comput. **24**, 40–49 (2014)
28. Falaki, S.O., Alese, B.K., Adewale, O.S., Ayeni, J.O., Aderounmu, G.A., Ismaila, W.O.: Probabilistic credit card fraud detection system in online transactions. Int. J. Softw. Eng. Appl. **6**(4), 69–78 (2012)
29. Mule, K., Kulkarni, M.: Credit Card Fraud Detection Using Hidden Markov Model (HMM) (2014)
30. Kumari, N., Kannan, S., Muthukumaravel, A.: Credit card fraud detection using hidden markov model-a survey. Middle-East J. Sci. Res. **19**(6), 821–825 (2014)
31. Chen, R.C., Chiu, M.L., Huang, Y.L., Chen, L.T.: Detecting credit card fraud by using questionnaire-responded transaction model based on support vector machines. In: Intelligent Data Engineering and Automated Learning–IDEAL 2004, pp. 800–806 (2004)
32. Huang, C.L., Chen, M.C., Wang, C.J.: Credit scoring with a data mining approach based on support vector machines. Expert Syst. Appl. **33**(4), 847–856 (2007)
33. Harris, T.: Credit scoring using the clustered support vector machine. Expert Syst. Appl. **42** (2), 741–750 (2015)
34. Duman, E., Ozcelik, M.H.: Detecting credit card fraud by genetic algorithm and scatter search. Expert Syst. Appl. **38**(10), 13057–13063 (2011)
35. Bentley, P.J., Kim, J., Jung, G.H., Choi, J.U.: Fuzzy darwinian detection of credit card fraud. In: the 14th Annual Fall Symposium of the Korean Information Processing Society, vol. 14 Oct 2000
36. Aburrous, M., Hossain, M.A., Dahal, K., Thabtah, F.: Intelligent phishing detection system for e-banking using fuzzy data mining. Expert Syst. Appl. **37**(12), 7913–7921 (2010)
37. Hilas, C.S., Sahalos, J.N.: Testing the fraud detection ability of different user profiles by means of FF-NN classifiers. In: International Conference on Artificial Neural Networks, pp. 872–883. Springer, Berlin, Heidelberg, Sept 2006
38. Hilas, C.S., Sahalos, J.N.: User profiling for fraud detection in telecommunication networks. In: 5th International Conference on Technology and Automation, pp. 382–387, Oct 2005
39. Mohamed, A., Bandi, A.F.M., Tamrin, A.R., Jaafar, M.D., Hasan, S., Jusof, F.: Telecommunication fraud prediction using back propagation neural network. In: 2009 SOCPAR'09 International Conference of Soft Computing and Pattern Recognition, pp. 259–265. IEEE Dec 2009
40. Farvaresh, H., Sepehri, M.M.: A data mining framework for detecting subscription fraud in telecommunication. Eng. Appl. Artif. Intell. **24**(1), 182–194 (2011)
41. Akhter, M.I., Ahamad, M.G.: Detecting telecommunication fraud using neural networks through data mining. Int. J. Sci. Eng. Res. **3**(3), 601–606 (2012)
42. Verbeke, W., Dejaeger, K., Martens, D., Hur, J., Baesens, B.: New insights into churn prediction in the telecommunication sector: a profit driven data mining approach. Eur. J. Oper. Res. **218**(1), 211–229 (2012)

43. Hilas, C.S., Kazarlis, S.A., Rekanos, I.T., Mastorocostas, P.A.: A genetic programming approach to telecommunications fraud detection and classification. In: Proceedings of the 2014 International Conference on Circuits, System Signal Processing, Communications and Computing, pp. 77–83 (2014)
44. Subudhi, S., Panigrahi, S.: Quarter-Sphere support vector machine for fraud detection in mobile telecommunication networks. Proced. Comput. Sci. **48**, 353–359 (2015)
45. Subudhi, S., Panigrahi, S.: Use of fuzzy clustering and support vector machine for detecting fraud in mobile telecommunication networks. Int. J. Secure. Network. **11**(1–2), 3–11 (2016)
46. Folasade, I.O.: Computational intelligence in data mining and prospects in telecommunication industry. J. Emerg. Trends Eng. Appl. Sci. **2**(4), 601–6051 (2011)
47. Moudani, W., Chakik, F.: Fraud detection in mobile telecommunication. Lect. Notes Softw. Eng. **1**(1), 75 (2013)
48. Ortega, P.A., Figueroa, C.J., Ruz, G.A.: A medical claim fraud/abuse detection system based on data mining: a case study in chile. DMIN **6**, 26–29 (2006)
49. Li, J., Huang, K.Y., Jin, J., Shi, J.: A survey on statistical methods for health care fraud detection. Health Care Manag. Sci. **11**(3), 275–287 (2008)
50. Pawar, M.P.: Review on data mining techniques for fraud detection in health insurance. IJETT **3**(2) (2016)
51. Ghuse, N., Pawar, P., Potgantwar, A.: An improved approch for fraud detection in health insurance using data mining techniques. Int. J. Sci. Res. Netw. Secur. Commun. **5**(3), 27–33 (2017)
52. France, F.R.: eHealth in Belgium, a new "secure" federal network: role of patients, health professions and social security services. Int. J. Med. Inf. **80**(2), e12–e16 (2011)
53. Tang, M., Mendis, B.S.U., Murray, D. W., Hu, Y., Sutinen, A.: Unsupervised fraud detection in Medicare Australia. In: Proceedings of the Ninth Australasian Data Mining Conference, vol. 121, pp. 103–110. Australian Computer Society, Inc., Dec 2011
54. Shan, Y., Jeacocke, D., Murray, D. W., &Sutinen, A.: Mining medical specialist billing patterns for health service management. In: Proceedings of the 7th Australasian Data Mining Conference, vol. 87, pp. 105–110. Australian Computer Society, Inc., Nov 2008

Digital Report Grading Using NLP Feature Selection

R. Shiva Shankar and D. Ravibabu

Abstract The trend of online examination has been so common in today's world for the ones testing their English proficiency. As the number of persons increasing to take the tests such as GRE, TOEFL, GMAT that involves the essay writing. Due to the lack of efficient, grading of these essays by the graders led to the automatic evaluation of essays and textual summaries. It is a hectic for the graders to provide feedback with stable interface, mindset, and time bounds. Features like bag of words, sentence, and word count along with their average length, structure, and organization of an essay are used for achieving maximum accuracy in grading. The sequential forward feature selection algorithm is used to compare the accuracy and select the best subset in this way. An efficient subset is formed from a single empty set because of this algorithm and its operations made it easy to implement and perform well on small data sets.

Keywords Sequential forward feature selection · Bag of words
Sentence and word count

1 Introduction

Since many ages, there are huge volumes of text data and are developing day by day which provides with the extraction of implicit, previously unknown and potentially useful information lead to text mining where the collection of documents contains the automated extraction of information [1]. Different text mining techniques are introduced basing on the inherent nature of textual data. Those

R. Shiva Shankar (✉) · D. Ravibabu
Department of CSE, SRKR Engineering College, Bhimavaram 534204, India
e-mail: shiva.srkr@gmail.com

D. Ravibabu
e-mail: ravibabu.devareddi@gmail.com

J. Nayak et al. (eds.), *Soft Computing in Data Analytics*,
Advances in Intelligent Systems and Computing 758,
https://doi.org/10.1007/978-981-13-0514-6_59

techniques are imposed in two ways. One is applied, when the specific structure on the textual data is imposed and then applies the technique to the data. The other one is to develop a technique for mining that exploits the inherent characteristics of the textual data [2].

The competitive world gives much preference to an individual based on his/her the ability in both verbal communication and written communication. In order to train the students verbally and help them to improve their writing skills, essay writing has been the part of curriculum of high school education [3].

Each essay has to be evaluated and given a feedback by the teachers which is a very hectic task if there are number of essays. Moreover, the grading of an essay varies from a person to person according to their psychological condition, their depth of knowledge, and opinion on certain subjects and also on their capability. This consumes a lot of time and energy of graders making it difficult for them to give feedback on every single essay, thus lacking in effective grading system [4]. Thus, the process of evaluating the essays manually has many problems that are to be resolved which motivated the researchers to work on text mining using feature methods, natural-language processing (NLP) for developing software tools that would be able to grade essays.

In this specified model that is proposed here, feature extraction methods are used with the most common features being word frequency, sentence validation, spell checker, stop words, sentence separation, and so on. Thus, based on automated essay scoring and by the use of feature extraction methods, a general essay is evaluated first and then the final result is produced.

2 Literature Survey

Two persons named Jill Burstein and Daniel Marc in 2000 had researched on this automated easy scoring system. The e-rater combined with several NLP tools to identify modules such as syntactic, discourse, and vocabulary-based features for grading. The importance of applications modularity with regard to (a) experiments that evaluate the integration of new features and (b) the reuse of modules for evaluations that contributes to the adaption of the system toward the generation of feedback [5].

Beata Beigman Klebanov et al. [6] had cited their research on selection of information from external sources to improve automated scoring of essays. The evaluation will proceed as every essay E responds to a test that contains a lecture L and a reading R. It identifies the essay's overlap with the lecture and assigns a score by the content importance model.

The evaluation of an easy and allotment of scores as well as feedback are provided in fraction of seconds by automated easy scoring systems. The total score and feedback are provided by the AES that have four stages introduced and researched by Semire Dikli in 2006. Training stage and scoring stage are the two stages included in Project Essay Grader (PEG). It does not provide any feedback but trains the system. Intelligent Essay Assessor (IEA) analyses and scores an essay using a semantic text analysis method called latent semantic Analysis. It measures similarity among words and texts [7].

Siddhartha Ghosh and Sameen S Fatima in 2010 had researched in implementing the AES system which includes Project Essay Grade (PEG), Intelligent Essay Assessor (IEA), and Electronic Essay Rater (E-Rater). These are the basic models for using various methods in scoring. Using NLP techniques and linear regression, it gives the grade and a proper feedback when the essay is given as input [8].

In scoring, the essay involves the four stages by Randy Elliot Bennett and Anat Ben-simon in 2005 had cited their research for AES system [9]. Every essay is independently assigned with some initial weight on a 0–100 scale with the sum of dimensions equal to 100. Checking and analyzes of essay considering the stage one result in third stage by using the NLP techniques.

The focus of the research done by Danielle S. McNamara et al. which is to describe a new method of AES using hierarchical classification and report on its reliability as the existing system takes in the choices and preferences of the grader that gives unfair grading. First, target essays are clustered into training set and a test set. The text calculation is done by the computational algorithm contained in the training set. The essay score calculates by multiple linear regressions [10].

As all the AES systems do follow the similar requirements, and same pattern of working process needs to be trained through the creation of a knowledge base is proposed by John K. Lewis in his research [11]. Most of the systems grade essays based on style and content in accordance with the training materials have been inputted. For score submissions, the usage of natural-language processing (NLP) tools is prominent. Comparisons are not allowed if the AES system evaluation of essay is done by PEG and e-rater techniques.

3 Proposed System

This paper instigates with "model creation" using input data set with essays in it. Values for extracted features are computed, and rubrics are designed. Every essay is preprocessed with steps like stop word removal, tokenizing, and stemming in it. This also includes "case folding," i.e., converting into lower case as considered in this particular scenario. Later input test data is subjected to the same steps as the training data. Then computations are done for all features, and score is generated as output within a range of 1 to 5. These can be further discretized and graded if needed.

Word tokenization: ***Tokenization*** is a process of converting a hefty text into single pieces called tokens. Here, we have tokenized each essay before using them for feature extraction. In this step, essays are subjected to word tokenizing and sentence tokenizing. Here is output of a tokenized essay.

Removing Stop Words: ***Stop*** words are referred as a collection of words, that is mostly present in most of the texts and that are of least significant in research. Exclusion of such words reflects in the quality of the result. These lists contain words like a, an, the. And also the punctuations like brackets and other symbols.

Stemming: ***The*** phenomenon of trimming is the basic words to its base form or the stem form. For example, a set of words "am, is, are" are counter-down to their own individual base forms, i.e., "be." Similarly Car, cars, cars, etc. to the base form of car. Basing on the context, the words are taken in their own perspective manner/form for a grammatical and meaningful structure to the sentence. This algorithm is mostly used for reducing the dimensionality.

Methodology: ***Few*** of the existing grading systems are Project Essay Grader, Intelligent Essay Assessor; E-Rater, etc. They have been relying on different NLP approaches to a great extent to judge the contents of essay submitted. Training essays are referred to a dictionary for matching the content. Special implementations are made for the access of efficient features with improved score and performance of the system.

Hypothesis: We hypothesize that a good prediction for an essay score would involve a range of feature types such as language fluency and dexterity, diction and vocabulary, structure and organization, orthography and content. More efficient model is extracted from the above features which is subsequently tolerable at certain fields. The content test of the essay is not frequently possible because most of the essays will not have the source content at store.

Drawbacks in Existing System: ***Every*** feature considered either quantifiable or unquantifiable may not contribute to make the system more efficient. So it can be made better by including only a subset of extracted features that are best scoring the given essay. Besides all these to rate the performance of system, they considered "accuracy" as the only metric.

System Architecture: The proposed system as shown in Fig. 1 will ensure you the best possible score by using this feature selection algorithm as shown in Fig. 2, i.e. the "sequential forward feature selection" algorithm to select the efficient set of features and generate a better result with outmost consistency.

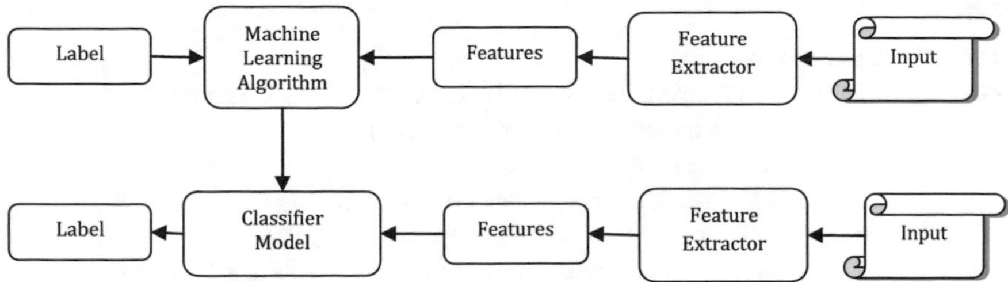

Fig. 1 System architecture for proposed system

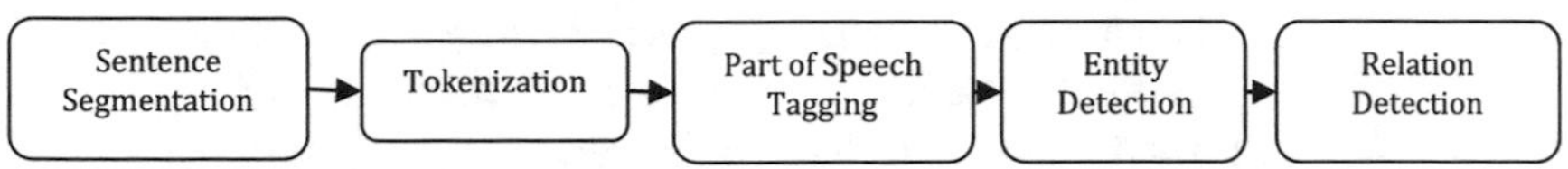

Fig. 2 Block diagram to feature extractor

4 Implementation

4.1 Scope and Features of the Proposed Method

1. The proposed algorithm is appropriate for identifying the validity and relevance of the essay to the topic given.
2. The method is capable of identifying the frequency of words occurred in the essay, compares the relevance of essay to the topic, and checks spellings of the words in the essay.
3. This method can give the validity of a sentence using a pos tagger. These sentence forms can be increased, and more efficiency could be obtained.
4. The algorithm is fashioned in a step-by-step manner considering the five features for scoring.

This method can be applied to a set of students belonging to the same class.

4.2 Proposed Algorithm

Algorithm: **1** for BagWords

Procedure: BagWords

Input: keywords in the essay are extracted by comparing all the words

Output: Bagwords are removed from the given sentences, and rest of the sentence is shown.

Initialisation: intersect, unionr, simratioo, first

Method:

```
Step1:topics=new
DataOperations().getBagwords(classname,topic)
            setkeywords(topics,db)
            unionr=getUnion()
            intersect=getIntersection()
simratio=((double)intersect/unionr)*100
Step2: if simratio is greater than or equal to 20
            return true ;    else  return false;
Step3: for each String sc in stopchars
            if(term.contains(sc)
            term=term.replace(sc," ");
            return term;     end for
Step4: for each String st in keywords
            temp=removeStopchars(st)
if(!stopwordsProvider.getINSTANCE().contains(temp))
            first.add(temp)
keywords=secondlist.toString().split(" ")
            end for
Step5: for each String st in keywords
            temp=removeStopchars(st)
if(!stopwordsProvider.getINSTANCE().contains(temp))
            first.add(st);   end for
Step6: return first
```

Algorithm: **2** for StopWordRemoval
Procedure: ArrayStopWord
Input: keywords in the essay will remove the word
Output: stopwords are removed from the given sentences, and rest of the sentence is shown.

```
Initialization: cnt=0, size=0, bt [] =null, dsen, words []
Method:
Step1: File fp=new File ("d:\inputs\stopwords.txt");
            size=length of FP; bt=new char[size];
Step2: FileReaderfis=new FileReader(fp);
Step3: for each integer i in stwdlength do
            ifstwd ignores the case then
            goto flag=false; else goto flag=true;
            end for
Step4: sz=counttokens from the given stwd
            while String hasMoreTokens
            cnt++
Step5: for each integer j in wordslength do
            dsen=dsen+words[j]+" ";
            end for
Step6: return dsen
```

Algorithm: **3** for POSTaggerClass
Procedure: POSTagger
Input: keywords in the essay are extracted for the POSTaggerClass
Output: white spaces are removed from the given sentences, and rest of the sentence is shown.

Initilisation:`speech,sentences[],tags[],wst(WhitespaceTokenizer)`
Method:

```
Step1: filelocation="inputs"+posmodelfile;
            File inputfile=new File(filelocation);
             speech=new ArrayList<String>
Step2: If model is not equal to NULL then
POSTaggerMe tagger=new POSTaggerME(model);
            If tagger is not equal to NULL then
sentences=contents.toString().split("\n");
Step3: for sentence: sentences
tags[]=tagger.tag(wst.INSTANCE.tokenize(sentence));
for each integer i in length of the wstLine
word=whitespaceTokenizerLine[i].trim();
            tag=tags[i].trim();
            speech.add(word+":"+tag);
            end forend for
Step4: tagger=null;return speech
```

5 Experimental Analysis and Results

In this paper, we collected the data from 15 students that will be considered as a test data set. From the set of test data, we'd score computations are done for every essay and stored. Later to measure the variations from the expected score and actual score visually, Database table is dumped into Excel files as shown in Table 1 and represented as Fig. 3.

Accuracy

It can be defined as the amount of uncertainty in a measurement with respect to an absolute standard. Accuracy specifications usually contain the effect of error due to gain and offset parameters.

$$\text{Accuracy\%} = 100 - \text{Error\%}$$

Error: All measurements are subject to error, which contributes to the uncertainty of the results.

$$\text{Error\%} = ((|\text{Experimental} - \text{Manual}|)/\text{Manual}) * 100$$

Figure 4 shows the performance of the valuation of 15 students for system and manual generated scores. The X-axis and Y-axis represent the student register

Table 1 Scores of each student generated manually and by the proposed system

Student regd no.	Manual scores	System scores	Student regd no.	Manual scores	System scores
1	45	43	9	49	48
2	39	38	10	47	46
3	37	36	11	34	33
4	48	45	12	43	42
5	39	38	13	35	34
6	43	42	14	38	37
7	46	45	15	40	39
8	36	34			

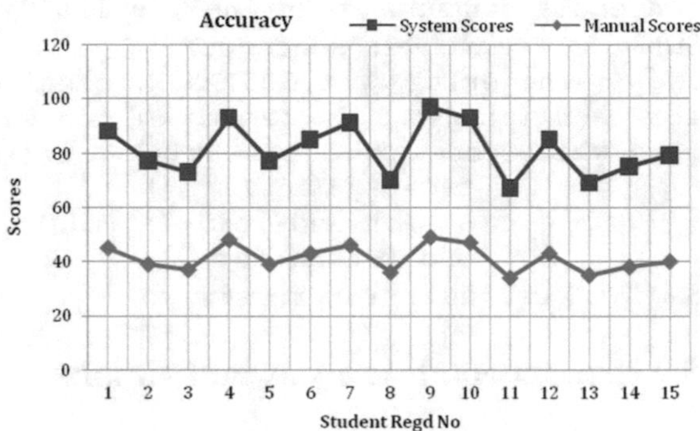

Fig. 3 Scores of each student generated manually and by the proposed system

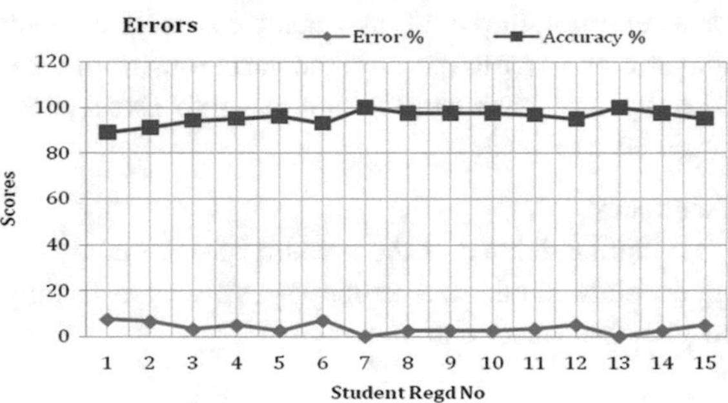

Fig. 4 Performance comparison for system and manual generated scores

Table 2 Error % and accuracy % of scores

Student regd no.	Error %	Accuracy %	Student regd no.	Error %	Accuracy %
1	7.48	89.05	9	2.63	97.37
2	6.54	91.23	10	2.5	97.5
3	3.2	94.3	11	3.22	96.77
4	4.89	95.1	12	5.12	94.87
5	2.56	96.2	13	0	100
6	6.97	93.02	14	2.5	97.5
7	0	100	15	4.87	95.13
8	2.56	97.44			

number and the scores of each student, respectively. It can be interpreted from the graph that there is very less difference between both the scores. Hence, our proposed system is valid. Table 2 shows the error % and accuracy % of each student. The graph is drawn to show the efficiency of the proposed system.

6 Conclusions

This research shows that automated essay grading will help writers to know their own level of competency. However, there are many areas of improvement especially with the complexity for languages around the world even in this system some of the critical grammatical errors are quite impossible to register and they have gone unnoticed. Nonetheless, it is quite evident what machine learning can do if the agent can be trained properly with large set of relevant data and reasonably good algorithm will even make the learning easier. This research suggests that, although statistically significant differences existed between instructor grading and AES grading, it would be quite preferable to use this interface. Consequently, depending on the intent of individual instructors for their chosen assignment, these systems may be more acceptable as a formative, as opposed to summative, assessment of student learning.

References

1. Burstein, J., Marc, D.: Benefits of modularity in an automated essay scoring in creative. Commons attribution teaching and research, vol. 5, pp. 22–29 (2000)
2. Kumar, L.: Contrasting automated and human scoring of essays, research paper, vol. 2, pp. 1–5 (2001)
3. Tang, D.: A joint segmentation and classification in conference on empirical analysis in conference on empirical methods. Nat. Lang. Process. **5**, 477–487 (2004)
4. Ruden, L.M.: An Evaluation intelli-metric essay scoring system using responses to GMAT. J. Manag. Adm. Counc. **2**, 1–12 (2005)
5. Herzberg: Automated essay scoring in the classroom, Research paper, vol. 5, pp. 45–57 (2011)
6. Klebanov, B.B., Madanani, N., Burstein, J., Somasundaran, S.: Content importance models for scoring writing from sources. In: Association For Computational Linguistics, vol. 4, pp. 247–252, June 2014
7. Dikli, S: Automated essay scoring. Turk Online J. Distance Educ. **7**, 49–62 (2006)
8. Gosh, S., Fatima, S.S.: Design of an automated essay grading system in Indian context. Int. J. Comput. Appl. **1**, 60–65 (2010)
9. Bennett, R.E., Simon, A.B.: Toward Theoretically Meaningful Automated Essay Scoring, vol. 1, pp. 60–77. National Center for Education Statistics (2005)
10. Mcnamara, D.S., Crossley, S.A., Roscoe, R.D., Allen, L.K., Dai, J.: A hierarchical classification approach to automated essay scoring, vol. 23, pp. 35–59, Jan 2015
11. Lewis, J.K.: Ethical implementation of an automated essay scoring system in faculty and staff-articles & papers, vol. 7, pp. 47–61, Jan 2013

Brain Subject Segmentation in MR Image for Classifying Alzheimer's Disease Using AdaBoost with Information Fuzzy Network Classifier

P. Rajesh Kumar, T. Arun Prasath, M. Pallikonda Rajasekaran and G. Vishnuvarthanan

Abstract Alzheimer's disease (AD) is a neurodegenerative brain disorder which develops gradually over several years of time period. It is not always understandable at initially because which starts with few symptoms can partly cover with some other disorders. In some case, it could be very difficult to differentiate Alzheimer's from mild cognitive impairment which can be understood in normal aging. In clinical visualization of brain disorders, magnetic resonance imaging takes place an important role in diagnosis and period before a surgical operation planning. In order to extract the features from the MRI brain image, segmentation process has been performed to segment different brain matters. The segmentation task is executed with the help of patch-based clustering principle, and three different labels were assigned for grouping gray matter (GM), white matter (WM), and skull region. Various features which are chosen from the segmented brain GM and WM to feed as input to the AdaBoost with information fuzzy network classifier for classifying the stages of the AD/MCI.

Keywords Alzheimer's disease · Patch-based clustering · AdaBoost
Information fuzzy network

P. Rajesh Kumar (✉) · T. Arun Prasath · M. Pallikonda Rajasekaran · G. Vishnuvarthanan
School of Electrical and Electronics Engineering, Kalasalingam Academy of Research and Education, Virudhunagar 626126, Tamil Nadu, India
e-mail: rkrrajesh74@gmail.com

T. Arun Prasath
e-mail: arun.aklu@gmail.com

M. Pallikonda Rajasekaran
e-mail: m.p.raja@klu.ac.in

G. Vishnuvarthanan
e-mail: gvvarthanan@gmail.com

J. Nayak et al. (eds.), *Soft Computing in Data Analytics*,
Advances in Intelligent Systems and Computing 758,
https://doi.org/10.1007/978-981-13-0514-6_60

1 Introduction

Alzheimer's disease is discriminated by progressive death of brain cells and memory function, and over a course of time period, it leads to the neurodegenerative disorder in the form of dementia. The brain function changes either increasing or decreasing while the variation in both glucose metabolism and the blood flow [1]. Vast literatures have been studying about the classification of AD and MCI one or multiple imaging techniques, which are magnetic resonance imaging (MRI), positron emission tomography (PET), and computed tomography (CT) but MRI is preferable because of its better visualization among various. Most of the AD/MCI studies have been focusing on classification analysis of AD and MCI subject from normal subjects. However, an explicit diagnosis of AD and MCI is very crucial for appropriate treatment and feasible delay of the disease [2]. In this work, we analyzed the segmentation of brain tissues in conjunction with the classification of Alzheimer's disease, in which segmentation seems to be the preliminary process. Various types of brain tissues have segmented with the help of patch-based differential clustering principle [3]. In this clustering task, higher intensity pixels are labeled as skull region, another cluster has created for segment the gray matter (GM), and final cluster has made for segment the white matter (WM) region. After making the segmentation process, volume of the brain matters has estimated for classifying the severity of the disease [4]. Generally, Alzheimer's disease initially affects the hippocampus region. According to the changes in the volume of hippocampus region and gray matter, the physician could take the preclinical decisions. Beheshti et al. [5] have proposed CAD system with feature ranking using genetic algorithm. Voxel-based morphometry process has executed to investigate atrophy and gray matter. Liu et al. [6] developed morphometric based on pattern analysis with multi-template for the process of feature ranking for classifying AD/MCI. Platero and Tobar [7] have proposed patch-based labeling principle for segmenting the various brain subjects with multi-atlas-based wrapping non-grid registration. Rajesh Kumar et al. [8] worked on PSO K-means clustering principle to segment various brain matters to estimate the volume for classifying the abnormality of the disease. Zhan et al. [9] were used a network-based statistic (NBS) principle for clustering the brain associated region for selecting the features to feed classifier input to classify AD/MCI.

2 Materials and Methods

In this research work, brain images of T1 MRI scans images have preferred, because of it is having white matter and gray matter with better contrast when compared to T2 weighted images. For this work, brain MRI scan images have taken from Open Access Service of Imaging Studies (OASIS) to examine the data of brain image. This work focuses on classification analyze of AD and MCI subjects. 150 MRI scan images have accessed from the OASIS database. Various brain subjects have segmented with the help of patch-based clustering principle. After the segmentation task, various features have taken and which are given to the AdaBoost with info fuzzy network classifier. The classification results state 22 subjects were at MCI stage out of 50 subjects.

2.1 Preprocessing

Intensity normalization is the process of changing the intensity of the input image. The intensity of the all taken input images has set in the range of [0–100], and then the normalization task has performed with [N] stated method with Eq. (1). Intensity normalization process can be written as

$$I_N = (I - Min)\frac{newMax - newMin}{Max - Min} + newMin \tag{1}$$

Laplacian windowing filter is used to remove additive noise from the input image and edge preserving also possible by implementing the filtering process. Generally laplacian is applied for smoothening an image but continuous filtering may cause an image to become blurred, so this process should be confined to limited iterations.

2.2 Patch-Based Differential Clustering

Patch preselection process has performed initially for the purpose of reducing the computational time. Preselection of patches can be achieved with the help of simple static mean and variance. In order to discard the patches which are having most dissimilarity, patch preselection can be achieved with contrast luminance properties of an image [10]. The structure similarity can be represented in Eq. (2).

$$ss = \frac{2\mu_i\mu_{s,j}}{\mu_i^2\,\mu_{s,j}^2} \times \frac{2\sigma_i 2\sigma_{s,j}}{\sigma_i^2\sigma_{s,j}^2} \tag{2}$$

μ mentions mean and σ mentions standard deviation of cantered patch on voxel x_i, and voxel $x_{s,j}$ at the particular location j in subject s. The threshold value has fixed 0.95 for the entire experiment.

2.3 Label Propagation Based on Pairwise Similarity

Pairwise similarity process is initiating with input image L and T which has mentioned anatomical image for the corresponding labeled image [11]. Patch image labeling principle has represented with the help of Eq. (3) for grouping the GM, WM of the brain input image.

$$\forall x \epsilon \Omega, L(x) = \frac{\sum_{y \epsilon N(x)} w(x,y) l(y)}{\sum_{y \epsilon N(x)} w(x,y)} \tag{3}$$

L mentions proportion of each label and $l(y)$ is total number of labels.

Label propagation based on groupwise similarity.

Groupwise similarity has designed according to fuzzy-based labeling method. This is the similarity measure among the pixel in the group which labeled with arbitrary numbers. Groupwise similarity measured using Eq. (4).

$$\forall x \epsilon \Omega, L(x) = \frac{\sum_{i=1}^{n} \sum_{y \epsilon N(x)} w_i(x,y) l_i(y)}{\sum_{i=1}^{n} \sum_{y \epsilon N(x)} w_i(x,y)} \tag{4}$$

Final segmentation task has set as L. If we consider any one final labeling, it will not be like pairwise similarity concept. Patches are labeled using Eq. (5).

$$\forall x \epsilon \Omega, P_L(x) = \sum_{i=1}^{n} \sum_{y \epsilon N(x)} w_i(x,y) p_{l_i}(y) \tag{5}$$

The classification analysis of AD/MCI can be predicted with the help of volume changes in WM, GM [12]. Existing works show that the major reason for AD starts from the size shrinkage of hippocampus and gray matter volume reduction. Volume of GM, WM is calculated using Eqs. (6) and (7).

$$Volume_{WM} = \sum_{slice=1}^{n} \sum_{i=1}^{x} \sum_{j=1}^{y} f(i,j) > thres \tag{6}$$

$$Volume_{GM} = \sum_{slice=1}^{n} \sum_{i=1}^{x} \sum_{j=1}^{y} f(i,j) == thres \tag{7}$$

2.4 Classification Based on Image Segmentation Features

As denoted above, feature selection is an essential task to design a classifier. For the feature extraction process of an image, we have chosen cortical and non-cortical thickness of segmented image [13]. Various thicknesses of 50 different cortical regions as well as the volume of 22 different brain structures are measured initially.

After estimating the volume and thickness, next go for shape feature which has included 12 lateral shape variations and 30 hippocampus shape difference. All the segmentation features are fed as input to the AdaBoost with info fuzzy classifier for classifying the AD/MCI subjects.

2.5 AdaBoost with Info-Fuzzy Network Classifier

AdaBoost is a process which has been used to improve the accuracy of taken learning algorithm. This principle is simple in construction and quick in process. In this concept, it selects the weaker learner at initially with lowest error in order to produce efficient classifier. And it will neglect remaining. It does not need any preceding knowledge particularly about the weak leaner [4]. In binary classification purpose, we have chosen it and the dataset s of N samples has taken for the input I.

$$F_T(x) = \sum_{t=1}^{T} f_t(x) \tag{8}$$

$S = (x_{i,} y_i), i = 1, \ldots .N$ $x_i \in X$ feature vector of taken dataset and $y_i \in Y = \{0, 1\}$ respective class label. The labels 1 and 0 represent abnormality and normal correspondingly from Eq. (8). Number of rounds it is repeating in entire round, every training example declares particular weight, and it represents the probability of the particular example which is selected to the training set for a classifier. In beginning, the particular weight will be allocated for all examples of the training set. Then incorrect classifies the samples which are increasing in each round. At last, resulting hypothesis is generated with linear combination of weak hypothesis in every round.

$$E_t = \sum_i E[F_{t-1}(x_i) + \alpha_t h(x_i)] \tag{9}$$

Info-fuzzy network (IFN) method has developed for the purpose of knowledge discover. The mutual cooperation between input and objective attributes of various types is mentioned with information theoretic connected network [14]. This model is having three different layers. Input layer is represented as $L1$ and in $L2$ layer fuzzy membership function can be executed and same compared units also present with membership function is represented Eq. (9)

$$net_i = \sum_{j}^{n} out_j w_{ij} \tag{10}$$

out_j Denotes the output for the membership function j that respect to the input x_i. Layer $L2$ and $L3$ mention double layer feed forward network. Layer $L3$ mentions output layer in Eq. (10)

$$out_j = \frac{1}{1 + exp\{net_i + \varnothing\}} \tag{11}$$

Height between $L2$ and $L3$ are selected random manner and learning updated subsequently in Eq. (11). Based on minimum and maximum values, the membership function can be calculated.

3 Results and Discussion

Table 1 shows various features for feeding the inputs to the AdaBoost with info fuzzy network classifier, and the classification accuracy has validated with the help of above values. Four various features have selected, and those are given as input to the classifier. The efficiency of the classifier has validated, and it has proved that it has given better performance than support vector machine (SVM) as well as nearest particle interconnect classifier. From Fig. 1, 1.1(b) is the output of enhanced image, intensity normalization process has implemented for better visualization with the help of Eq. (1) for the respective input sagittal view of image 1.1(a). Laplacian windowing filter has applied with the help of best matching kernel mask to remove the additive noise, and also the smoothening process can be achieved by filtering process for the enhanced image 1.1(b). Labeling operation is executed to the filtered image output 1.1(c). Three various strategies have applied for achieving the labeling task which has been executed with the help of patch-based clustering principle, and the structure similarity has measured using Eq. (2).

After the structure similarity task, label propagation has executed with pairwise and groupwise similarity measures using Eqs. (5)–(7) to the filtered image 1.1(c). The above-mentioned similarity measures have applied to make three various clusters for grouping GM, WM, and skull region.

Table 1 Feature selection for AD/MCI classification

Features	Image 1	Image 2	Image 3	Image 4	Image 5	Image 6
Cluster prominence	2.6e+03	2.4e+03	2.4e+03	2.3e+03	2.5e+03	2.4e+03
Correlation	0.8674	0.8624	0.8841	0.8765	0.8854	0.8735
Entropy	0.5386	0.4818	0.4612	0.5762	0.4723	0.5312
Energy	0.7252	0.7677	0.7711	0.7354	0.7522	0.7465
Maximum probability	0.8420	0.8669	0.8527	0.8634	0.8554	0.8726
Sensitivity	95.2381	95.2521	95.2318	95.1154	95.8564	95.2318
Specificity	97.9167	97.5462	97.8624	97.9167	97.6524	97.9166
Recall	95.2381	95.6452	95.2381	95.8564	95.2384	95.3341
Accuracy	96.3964	95.5462	96.5562	96.2312	95.6423	96.7319
Kappa index	0.9269	0.4756	0.7568	0.8563	0.4853	0.6549

From Fig. 2, 2.1–2.3(b) show the output of segmented region of the gray matter with skull and 2.1–2.3(c) show the segmentation output of white matter with skull region. Exact segmentation output of Gray matter has shown in 2.1–2.3(d) as well

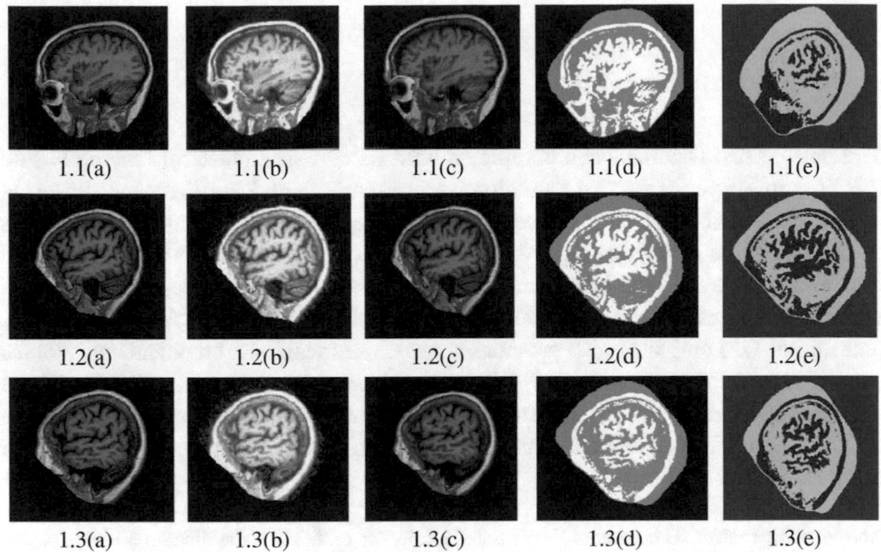

Fig. 1 1.1–1.3 **a**—Input images, 1.1–1.3 **b**—Enhanced images, 1.1–1.3 **c**—Filtered images, 1.1–1.3 **d**—Labeled images, 1.1–3 **e**—Clustered images

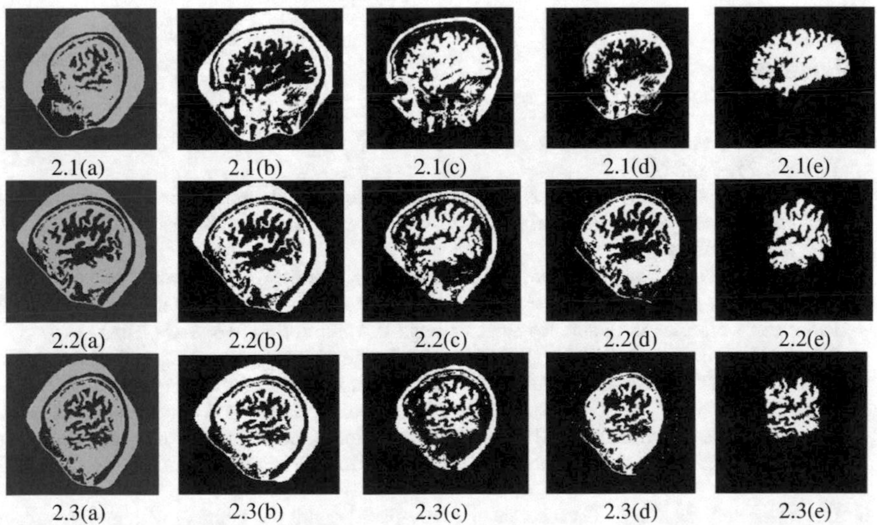

Fig. 2 2.1–2.3 **a**—Labeled images, 2.1–2.3 **b**—Gray matter with skull, 2.1–2.3 **c**—White matter with skull, 2.1–2.3 **d**—Segmented gray matter, 2.1–2.3**e**—Segmented white matter

as the accurate segmentation of white matter has shown in 2.1–2.3(e). The volume of gray matter and white matter volume has calculated using Eqs. (6) and (7) for classification analysis of AD. Classifier has given 2.1–2.2(e) which are on the stage of MCI, and 2.3(e) is on AD and the result which has validated by the radiologist.

4 Conclusion

The present research has been completed with patch-based clustering methods, and this principle's performance evaluation has compared with PSO K-means clustering and SOM-based clustering concept [8]. The segmentation accuracy as well as executional time of present study has compared with the state of the art and proved that it has achieved better performance. Patch preselection process has done with various strategies for providing exact segmentation of GM and WM. Volumetric measure of GM and WM also provides some more meaningful medical analysis for staging the severity of AD, and this has compared with normal subjects. These above results will give better information's to the medical society for classifying AD/MCI, and further work is going to deal with game theory-based classifier for exact staging of AD.

Acknowledgements The authors thank the Department of ECE, Kalasalingam University, for permitting to use the computational facilities available in Center for Research in Signal Processing and VLSI Design which was set up with the support of the Department of Science and Technology (DST), New Delhi, under FIST Program in 2013 (Reference No: SR/FST/ETI-336/2013 dated November 2013).

References

1. Chatterje, P., Milanfar, P.: Patch-based near-optimal image de-noising. IEEE Trans. Image Process. **21**, 4 (2012)
2. Khanal, P., Lorenzi, M., Ayache, N., Pennec, N.: A biophysical model of brain deformation to simulate and analyze longitudinal MRIs of patients with Alzheimer's disease. NeuroImage. **134**, 35–52 (2016)
3. Tokuchi, R., Hishikawa, N., Sato, K., Hatanaka, N., Fukui, Y., Takemotoa, M., Ohta, Y., Yamashita, T., Abe, K.: Age-dependent cognitive and affective differences in Alzheimer's and Parkinson's diseases in relation to MRI findings. J. Neurol. Sci. **365**, 3–8 (2016)
4. Nayak, D.R., Dash, R., Majhi, B.: Brain MR image classification using two-dimensional discrete wavelet transform and AdaBoost with random forests. Neurocomputing (2012)
5. Beheshti, I., Demirel, H., Matsuda, H.: Classification of Alzheimer's disease and prediction of mild cognitive impairment-to-Alzheimer's conversion from structural magnetic resource imaging using feature ranking and a genetic algorithm. Comput. Biol. Med. **83**, 109–119 (2017)
6. Liu, M., Adeli, E., Zhang, D.: Inherent structure-based multiview learning with multitemplate feature representation for alzheimer's disease diagnosis. IEEE Trans. Biomed. Eng. **63**, 7 (2016)

7. Platero, C.M., Tobar, C.: A fast approach for hippocampal segmentation from T1-MRI for predicting progression in Alzheimer's disease from elderly controls. J. Neurosci. Methods **270**, 61–75 (2016).
8. Rajesh Kumar, P., Arun Prasath, T., Pallikonda Rajasekaran, M., Vishnuvarthanan, G.: Brain subject estimation using PSO K-means clustering—an automated aid for the assessment of clinical dementia. In: Satapathy, S., Joshi, A. (eds.) Information and Communication Technology for Intelligent Systems (ICTIS 2017)—Volume 1. ICTIS 2017. Smart Innovation, Systems and Technologies, vol 83. Springer, Cham (2018)
9. Zhan, Y., Yao, H., Wang, P., Zhou, B., Zhang, Z., Guo, Y., Ningyu An, Jianhua Ma, Xi Zhang, and Yong Liu.: Network-based statistic show aberrant functional connectivity in alzheimer's disease. IEEE J. Sel. Topics Signal Process. 10, 7 (2016)
10. Coupé, P., José, V., Manjón, Fonov,V., Pruessner, J., Robles, M., Collins, L.D.: Patch-based segmentation using expert priors: application to hippocampus and ventricle segmentation. NeuroImage **54**, 940–954 (2011)
11. Roussea, U., Habas, A.P., Studholme, C.: A supervised patch-based approach for human brain labeling. IEEE Trans. Med. Imaging **30**, 10 (2011)
12. Shanthi, K.J., Ravish, D.K.: Image segmentation an early detection to alzheimer's disease (2013). 978-1-4799-2275-8/13/$31.00 2013 IEEE
13. Mirzaei, G., Adeli, A., Adeli, A.: Imaging and machine learning techniques for diagnosis of Alzheimer's disease (2016). 10.1515/revneuro-2016-0029
14. Kulkarni, A.D., Cavanaugh,C.D.: Fuzzy neural network models for classification. Appl. Intell. **12**, 207–215 (2000)

Soft Computing-Based Intrusion Detection Approaches: An Analytical Study

D. Neelima, J. Karthik, K. Aravind John, S. Gowthami and Janmenjoy Nayak

Abstract In the recent days, intrusion detection system using soft computing methods is one of the most interesting and attractive areas of research. Soft computing majorly undergoes with uncertainty, partial truth to gain robustness and low-cost solution. Intrusion detection system is a device or software application that monitors a network or system from unwanted activities. If the system detects any malware, it reports to the administrator or collects using security information. To develop efficient security methods, many researchers used fuzzy logic, neural network, machine learning, support vector machines, evolutionary computation, and probabilistic reasoning techniques. In this paper, a brief analysis has been conducted on various soft computing techniques for the successful detection of intrusive behaviors. Further, some future challenges with advantages and limitations are highlighted in a successive manner.

Keywords Intrusion detection · Soft computing · Neural network
FCM · Swarm intelligence

D. Neelima · J. Karthik · K. Aravind John · S. Gowthami · J. Nayak (✉)
Department of Computer Science & Engineering, Sri Sivani College of Engineering, Chilakapalem, Srikakulam 532410, AP, India
e-mail: mailforjnayak@gmail.com

D. Neelima
e-mail: neelima947@gmail.com

J. Karthik
e-mail: karthik.jallu@gmail.com

K. Aravind John
e-mail: aravind.mahima@gmail.com

S. Gowthami
e-mail: simhadrigowthami@gmail.com

J. Nayak et al. (eds.), *Soft Computing in Data Analytics*, Advances in Intelligent Systems and Computing 758,
https://doi.org/10.1007/978-981-13-0514-6_61

1 Introduction

Now-a-days, Internet plays a very important role in our society. The benefits provided by the Internet are communication, data transfers, electronic mail, social media, business purpose, etc. So, protection has become an important issue because if there is no protection then sensible data of any organization may be violated. So, security of data is considered as a challenging threat. For providing the security, intrusion prevention/detection system is introduced by some researchers. Intrusion prevention/detection system is a software application or a device which encounters an illegal activity. If it is identified for sensible data, then it gives an alert to the computer system or to system administrator. It avoids active and passive attacks by using firewalls [1]. Active attack is easy to detect because the data may alter. But in passive attacks, it is difficult to locate due to eavesdropping the data by the intruders. IDS can be of misuse-based (signature) and behavior (anomaly)-based. Misuse-based uses one database for storing and gives a clear description of malicious patterns and also for comparisons of sounds of an alarm. Behavior (anomaly)-based intrusion prevention system detects both network and computer intrusions.

The process of intrusion prevention system is monitoring, analyzing, detecting, and recovery/resposing. For this process, IDS contains detection engine which includes log and action: Log is for maintaining data, and action is to take necessary operations. There is variety of intrusion prevention systems such as host-based, network-based, application-based. Host-based approach is for an individual network, and network intrusion detection system works on large scale. It monitors network traffic and examines the traffic, and based on the observation it categorizes the traffic into normal or suspicious. Traffic monitoring is done at firewall, hub, switch, etc. Final one is the application-based approach which works on one specific protocol. Although it is implemented periodically, it does not meet some factors such as completeness, timeliness, performance, fault tolerance, accuracy. So day-by-day, number of researchers are adding the features for IDS to get a fruitful solution. For evaluating the performance, some tools such as Snort, OOSEC, fragroute, Suricata, Bro IDS, OpenWIPS-ng, Security Onion have been used.

Only for labeled data, some proposed algorithms were worked. But for unlabeled data, the current technologies are not detecting new intrusions. So for those, some authors proposed a new variety of algorithm named clustering-based intrusion detection algorithm, unsupervised anomaly detection. Their main intention is to classify the trained data which are real in nature. Also, variety of number of data mining and soft computing-based techniques such as fuzzy logic, evolutionary algorithms, neural networks, swarm-based algorithms, is incurred to take better implementation of IDS.

The goal of this paper is to provide an efficient study of research which cooperates an analysis of various soft computing techniques that involves neural networks, evolutionary algorithms, fuzzy logic, swarm-based algorithms. Also, an effective way of analysis for advantages and limitations of different soft computing techniques in IDS is done. The rest of the paper has been described as follows.

Section 2 demonstrates some of the preliminaries such as intrusion detection system and background structure of IDS along with the description about various soft computing techniques. Section 3 outlines the applications of soft computing techniques in IDS. Some major analytical study on the performance of important methods of soft computing is critically analyzed in Sect. 4. Section 5 concludes the work with some necessary future directives.

2 Background Study

2.1 Intrusion Detection

Intrusion detection system helps to find active attacks and policy of violations. It is a software application which notifies to the system administrator when any threat is there for sensible data. The input to an IDS is the data which describes the activities, network/system [2]. The data are to be sent to data preprocessor for getting activity records which are important for security. There is a demand to analyze those activity records by detection engine and use detection model which is already described for IDS and stored in IDS. If the system detects any suspicious access, then it raises an alert to decision engine which uses decision table for appropriate action. Then finally, it gives the response with recovery option. Figure 1 depicts about the abstract working model of IDS.

IDS used two most common techniques for identifying suspicious intrusions. Those are pattern matching, statistical anomaly. Pattern matching is for identifying known attacks by their signatures. And each suspicious activity is compared with their signatures which are already defined. Normally, this approach helps to make the baseline as normal. Finally, anomaly-based approach is better than the signature-based approach because it may find new intrusions by comparing the statistical anomaly-based approach consisting of one baseline with normal behavior. If it observes as abnormal and abnormal behaviors [3].

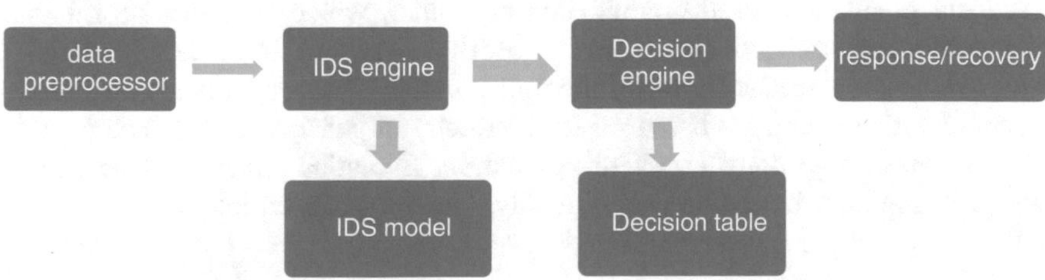

Fig. 1 Working model of IDS

2.2 *Soft Computing*

Soft computing sometimes refers as computational intelligence, but still now it does not have any obeyed definition and it deals with complex tasks like NP-complete problems (unsolvable problems for a given polynomial time), unclear, unpredictable, partial truth, and uncertainty solutions [4]. There are various techniques for soft computing such as fuzzy c-means clustering, neural networks, evolutionary algorithms, and swarm-based algorithms.

2.2.1 Fuzzy C-Means Clustering

Fuzzy c-means clustering is one of the popular clustering algorithms which is most frequently used in diversified applications and is very similar to k-means algorithm with some differences in selection of cluster centers [5]. The process of FCM is to select number of clusters, then allocate coefficients to each data point for their presence, and repeat this step until the algorithm is converged. Finally, compute the centroid for each cluster using the fuzzy-based objective function [6].

2.2.2 Neural Network

Neural networks are the connection-oriented models based on the computation process which are equivalent to processing unit of human brain. This technology is inspired by biological neural networks (series of interconnected networks). It is a powerful approach to solve real-world problems using artificial intelligence, and it has the capability to deal with incomplete information [7]. The training procedure of neural networks can be performed by various learning methods such as supervised and unsupervised types.

2.2.3 Evolutionary Algorithm

These are population-based optimization algorithm, which combines all the candidate solutions collectively. These algorithms do not have any underlying assumptions about the landscape of the fitness values. These are mainly inspired by genetic population and are based on the hierarchy of generations [8]. The types of EA are genetic algorithms are Neuro evolution, Evolution strategy, Gene expression programming, Evolution programming, Genetic programming etc.

2.2.4 Swarm-Based Algorithm

Swarm-based algorithms are developed by the behavioral properties of various swarm elements that exist in nature. It is the group of decentralized, self-organized systems, artificial or natural. Swarm intelligence system consists of agents, mingling one another with their environment. These algorithms are taken from biological systems. Every agent has some rules and regulations for their individual or group behavior [9]. There are various swarm-based algorithms such as particle swarm optimization, cat swarm, water wave, back-tracking search, whale optimization, hybrid whale optimization, ant bee colony.

3 Soft Computing Methods for Intrusion Detection System (IDS)

Soft computing is the mixture of methodologies that were drafted to model and find solutions to problems, which are not too difficult to model, mathematically. Soft computing techniques are the flexible and efficient way of handling the complex problems [10]. Soft computing consists of a group of methods which shares business interest of methodologies that works together and provides, in one form or another, flexible information altering competence for handling obscure situations. Its aim is to exploit uncertainty, partial truth to gain robustness and low-cost solution. Soft computing varies from conventional (hard) computing in a sense of variability in computing, and its process is similar to that of processing nature of human mind [4].

3.1 Fuzzy Logic for IDS

Fuzzy logic can be used in computing where the focus is not only on true or false results, but also the level of the truthiness in the solution. Fuzzifiers are used to generate fuzzy values from scalar values. Sometimes, the fuzzifiers are also called as membership functions. Different methods can be used to generate the fuzzy rules [11]. The fuzzy rules can be defined by the security administrator with fuzzy input sets. Therefore using the test data, the percentage of intrusion can be calculated against the fuzzy rules.

Longzhi yang et al. [12] proposed variety of data-driven approaches for intrusion detection system which are helpful for avoiding complexity of systems. In their paper, they introduce fuzzy interpolation to reduce complexity of systems. They proved that their research meets online performance of intrusion detection system as well as high detection rate. Also, the experimental results proved that their approach gives a better result compared to existing one.

Security is needed not only for data but also for resources (servers, etc.). So if there is any threat or eavesdropping in resources, then sensible data may be violated. It is referred as active attack in that denial of service. The authors Lalithasaini et al. [13] proposed two approaches MCA and fuzzy logic based on intrusion detection system. Their approach experimentally proved that it gives high speed of detection rate and better performance, while any occurrence of network attacks.

IDS specialty is to detect some hard intrusions which cannot be identified by firewall system. So, many of the IDS are developed by using machine learning techniques, especially with neural networks. Lin et al. [14] have proposed a feature representation method which is a pattern classifier named as cluster center and nearest neighbor approach. The empirical progress shows as it gives high computational accuracy of IDS.

Now-a-days, it is a very challenging task to implement intrusion detection system with better feature for protecting the important data. Shamshirband et al. [15] introduced bio-inspired method, and they named as cooperative fuzzy artificial immune system which is extracted by danger theory. The proposed approach notifies a better accuracy and detection rate.

Anomaly-based intrusion detection system specialty is to find new intrusion attacks. Although its duty is to find new attacks, it contains some limitations and Geramiraz et al. [16] have proposed a novel anomaly intrusion detection system with fuzzy controller for decreasing the no. of false alarms. The simulated results evaluated by the proposed method state that it is able to increase the accuracy of IDS. Apart from all these works, some other notable methods (Table 1) are being proposed and suitably applied for solving the IDS problem.

3.2 Neural Networks and Other Clustering Methods for IDS

Due to the efficiency in solving the real-life problems in diverse domains, several researchers have applied these to solve IDS problem. At first, the information is to be gathered from the source, which is then transformed into a suitable form for analysis. Hence, the analysis may be done to find the degree of uncertainty where the uncertainty is an essential feature for intrusion detection system [7].

Fast increasing growth of network technologies and Internet may lead to data violations. So, there is a need to identify and rectify those types of intrusions. Manzoor et al. [17] have developed an intelligent intrusion detection system which does feature ranking using correlation. Their approach separates rest of the data into nonattack and attack classes. Their proposed system tested on variety of datasets, and their results have been reported with supportive results.

Sunitha et al. [18] described the problem of networks and the limitations of rapid technology due to attackers. They implemented MLP-based IDS. The experimental results show that the present approach also detects the type of attack whether it is normal or abnormal attack. Even though the growth of security protection of IDS increases to reduce malicious activities, but there are some complaints about new

Table 1 Various fuzzy-based methods for IDS

Year	Author(s)	Algorithm	Performance metrics	Dataset used	Advantages	Limitations	References
2012	Alsubhi et al.	Alert rescoring	Accuracy	DARPA 2000	IDS alert management system is used to provide optimal results	It contains configuration and specification issues	[34]
2013	Selman	–	Achieved 99.3% performance	KDD Cup 99	Data mining and machine learning techniques are used to reduce the dependency in the literature	Various security mechanisms were designed, but there is no reliable to protect the system from attacks	[35]
2014	Sharma et al.	Machine learning algorithm	–	DARPA98 + KDD Cup 99	Adaptive system rules for IDS are used for detecting new attacks	Many current IDSs are unable to find unknown attacks	[36]
2016	Kudłacik et al.	Improved cross-semi-global alignment	Computational complexity	KDD Cup 99	Flexibility in terms of features, low, linear computational used according to input data	The problem of detecting masqueraders in computer system is major issue	[37]
2016	Desai et al.	Identification algorithm + fuzzy genetic algorithm + signature matching	Internal attack detection, external attack detection	KDD Cup 99	Robust intrusion detection system is used to avoid internal and external attackers	Exploration of different types of vulnerabilities threats is increasing	[38]

type of attacks. So there is a need to implement one effective IDS, and Haidar and Boustany proposed anomaly-based network IDS. This helps to find new attacks. The progress notified is that this approach is feasible as compared to the existing method [19]. With the consideration of basic issue of implementing an IDS, Kumar and Yadav [20] have developed an artificial neural network-based approach with extended features for complete KDD'99 dataset. As compared to existing approaches, it occupies best detection rate.

Security is a very crucial component for sensible data of any industry or any organization. Realizing the role of IDS in protecting data from unauthorized activity, Devaraju and Ramakrishnan [21] compared different IDSs and performances. They used five classifiers such as generalized regression, forward neural network, Elman neural network, probabilistic neural network, and radial base neural network. Their comparison on featured and reduced KDD Cup 99 datasets shows that identified reduced data performance is better than featured dataset. Some other neural network-based approaches developed for IDS problem are illustrated in Table 2.

3.3 Evolutionary Algorithms for IDS

Evolutionary algorithm is an effective approach which is entirely based on Darwinian's evolution principle and survival of fittest for making efficient a candidate solution's population against predefined fitness. In real-time environment, a set of classification rules are generated using network audit. These rules are used for categorizing network connections which is useful for efficient intrusion detection [22].

Fouad and Hameed [23] proposed a genetic algorithm-based clustering method which distinguishes the network traffic packets for checking the normal or abnormal attacks. The proposed approach consists of two techniques called as GA#1 and GA#2, which are helpful for increasing accuracy, rate of detection, and true negative rate.

An intrusion detection system can be capable for detecting malicious attacks. But sometimes, it suffers from conventional firewall so the authors proposed two techniques to avoid this problem those are particle swarm optimization and graphical search. These are designed for neural networks to train. These results check some other methods like decision tree, artificial neural network (ANN)-based genetic algorithm. And it gives an encouraging result compared to other ANN-based genetic algorithm [24].

Data mining techniques play a vital role to improve detection/prevention system accuracy and performance of IDS. Das [25] proposed an evolutionary-based neural network approach with interselection criteria to solve IDS problem. The proposed method is able to produce high-accuracy value with less no. of features compared to earlier studies.

Table 2 Neural network-based methods for IDS

Year	Author (s)	Method	Performance matrices	Dataset used	Advantages	Limitations	References
2013	Singh et al.	Neural network	Accuracy + error rate	NSL-KDD	Multilayer perception neural network algorithm is providing more accurate results	Complexity	[39]
2014	Ranjan et al.	K-means + K-medoids	Detection rate + false alarm rate + false positive rate + false negative rate	KDD Cup 99	Clustering techniques of data mining are used for detecting possible intrusion and attacks	K-means include dependence on initial centroids, dependence on number of clusters and degeneracy	[40]
2015	Esmaily et al.	Decision tree algorithm	False alarm rate + detection rate + MLP	–	Machine learning and data mining techniques, IDS are able to diagnose attacks and system anomalies more effectively. ANN is possible to identify the attacks, even when the dataset is nonlinear, limited, or incomplete	Abnormal traffics attackers and intruders are penetrating systems and causing damage to users and network	[41]
2016	Hodo et al.	–	Accuracy	–	ANN is used as an off-line IDS to get information from IOT network and identify DOS attack	Attackers can take advantage of flaws in firmware running on the devices and run their software to disrupt the IOT architecture	[42]
2017	Yin et al.	–	Accuracy + true positive rate + false positive rate	NSL-KDD	RNN-IDS model of the intrusion detection improves the accuracy and provides a new research method	Machine learning cannot effectively solve the intrusion data classification problem that comes in real network application	[43]

An efficient IDS by alternating the existed genetic algorithm to detect network intrusion is being proposed by Pal and Parashar [26]. For achieving this, they used generated rule which reduces training time and complexity of system. The result clarifies their method shows higher rate of detection and lesser false positive rate.

IDS is a problem-solving approach for network attacks or any unauthorized access. Hassan claimed that the existed IDS is not efficient and not capable to tackle the problem of false detections. So, they proposed a new intrusion detection system by providing genetic algorithm and fuzzy logic for efficient and fast detection of intrusions. The empirical progress reported the proposed approach got feasible accuracy [27]. Some other notable research works with evolutionary algorithms are listed in Table 3.

3.4 Swarm-Based Algorithm for IDS

Swarm-based algorithms are the most frequently used soft computing techniques which were introduced by inspiring from the behaviors of some of the nature's creatures. For solving IDS problem, various swarm-based algorithms such as ant colony optimization, bee colony optimization, firefly algorithm, particle swarm optimization [28] are being frequently used. Intrusion detection/prevention system is introduced for network security purpose, which gives alert when any unauthorized access observed. But IDS has one technical challenging problem such as high dimensionality. To deal with this problem, Sedghi and Mirnia [29] introduced the algorithm named bat algorithm with k-means method. The experimental results noted as proposed IDS are able to detect intrusion activities with high accuracy.

Aghdam and Kabiri [30] proposed an intelligent ID, providing facilities efficiency and effectiveness for reducing nonlinear behavior of the intrusion activities. For achieving this, they introduced a method based on ant colony optimization algorithm with some added features for solving IDS problem.

Varma et al. [31] proposed the fuzzy entropy-based approach with ant colony optimization to search global smallest set of network traffic which facilities and extracts the real-time intrusion detection system. The experimental results proved that the proposed IDS is efficient for real-time detection of intrusive activities.

Anomaly-based IDS is an emerging technology for these days because it identifies new intrusion activities as well as facilitates the speed of Internet. But a good IDS is to be considered as variety of intrusions with high accuracy and high detection rate. With the drawback of earlier system such as high false positive rates and high negative rates, Aldwairi's [32] idea is to increase IDS performance by developing an artificial bee colony algorithm-based anomaly intrusion detection system.

Neshat et al. [33] have developed a novel artificial fish swarm algorithm which is inspired by movements of fishes and their different natures to handle IDS problem. In their research, they have realized and stated the importance of artificial fish swarm algorithm to achieve high speed of convergence and flexibility, high

Table 3 Evolutionary algorithm-based methods for IDS

Years	Author(s)	Algorithm	Performance matrices	Dataset used	Advantages	Limitations	References
2013	Zaman et al.	Genetic algorithm + particle swarm optimization + differential evolution	They achieved training time + classification accuracy + false positive rate	KDD Cup 99	IDS undergoes with large amount of data containing redundant features, which leads to heavy computational resources, low detection accuracy	Elimination of useless data enhances the accuracy of the classification while speeding up the computation	[44]
2014	Benaicha et al.	Genetic algorithm	False positive rate + detection rate	NSL-KDD 99	By combining IDS with GA increases the performance of network intrusion detection model and the false positive rate is reduced	–	[45]
2015	Hadded et al.	Cluster head election algorithm, NSGA-II algorithm	Efficiency	–	By using spacing, spread, generation distance approaches, it can achieve effectiveness	–	[46]
2016	Kumar et al.	–	Accuracy + detection rate	–	GA is capable to obtain answers always and gets better with time	It is not capable to store domain knowledge	[47]
2017	Mahmood et al.	Evolutionary algorithm	Accuracy + detection rate + false alarm rate	NSL-KDD	The performance is evaluated against two baseline models feature vitality-based reduction method and NB	High-dimensionality models based on multiobjective optimization using MOEA/D	[48]

Table 4 Swarm intelligence-based methods for IDS

Years	Author(s)	Algorithm	Performance matrices	Dataset used	Advantages	Limitations	References
2013	Bansal et al.	Artificial bee colony algorithm	Efficiency	–	ABC is not sensitive, to initial parameter values, and not affected by increasing dimension of the problem	Premature convergence or stagnation that leads to loss of exploration and exploitation capability of ABC	[49]
2014	Enache et al.	Swarm intelligence algorithm	Accuracy + attack detection rate false alarm rate	NSL-KDD dataset	Attack detection rate, false alarm rate, and accuracy	–	[50]
2015	Mahamod et al.	Artificial bee colony algorithm	Accuracy + error rate	NSL-KDD dataset	IDS encryption, authentication, and other hardware or software solutions. No duplicate records	–	[51]
2016	Azad et al.	Novel + particle swarm optimization algorithm	Classification accuracy + classification error rate	KDD Cup 99 + DARPA	Enhanced heuristic algorithm is used for sensitivity and specificity factors for increasing the success rate in the intrusion detection	The traditional network takes the multiple passes in learning process	[52]
2017	Kanakavardhini et al.	Ant colony optimization + swarm intelligence algorithm	Accuracy + detection rate	KDD dataset	Integrity, availability, reliability of the data	Due to heavy traffic in the network, IDS had many limitations	[53]

accuracy, fault tolerance. Table 4 states some other important applications of an IDS using swarm-based algorithms.

4 Critical Analysis

The problems of intrusion detection and to solve those issues through various advance computational techniques such as computational intelligence, soft computing, artificial intelligence have always been a challenging task for all level of researchers. In this paper, types of soft computing approaches applied to solve IDS problem have been considered with their detailed analytical study and issues related to those techniques. Most of the soft computing techniques are quite adaptive to the solution of the problem, fault tolerant, and suitable to apply in various types of problems in IDS. In this paper, four major techniques such as fuzzy logic, neural network, evolutionary algorithms, and swarm-based algorithms are considered, which have been most frequently used in IDS. Apart from hundred numbers of swarm- and evolutionary-based algorithms, mostly ACO, PSO algorithms have been simulated for analyzing the detection of intrusive behaviors. Also, neural network and fuzzy logics are used in much research of IDS with a higher capacity to solve the problem with high accuracy and low false rate. Most of the researchers have considered both the true and false positive rates along with the rate of detection for solving the problem. Although they successfully proved that their methods are quite successful in solving the issues-related IDS, still there are certain areas to be covered in which more analytical and experimental studies are required. Some of the machine learning and soft computing techniques are able to produce high initial rate of success, but with the increased number of types of attacks they are considered to be failed in many directions. Some other IDSs are developed through signature-based authentication system and are having good performance throughput. But these methods suffer from high memory consumption along with more complexity. Some of the methods based on statistical approach are being suffered due to the adaptivity procedures, and for that their output in generation of high false alarm rate is more. However, the current trend of fuzzy logic in various IDS applications is quite good and especially using the automatic rule generation procedure many of the research works are in breeding stage. In case of swarm intelligence algorithms, artificial immune system-based and some of the other nature-inspired algorithms are in hot seat of development as they have proved substantial amount of success rate in detecting the imperfect activities, distinguishing unique features and adaptability in dynamic environments. On the other hand, the self-arrangement/adjustment with distributive properties of both ACO and PSO makes those more proficient for the relative use in IDS problem.

So, after a careful observation of all the methods of soft computing, it may be realized that almost many methods are efficient in solving IDS problem with their own advantages. The performance of every individual differs from all others due to their variety in nature of problem solving. For example, neural networks are always being developed to minimize the error in the output and swarm-based algorithms are able to handle the nonlinearity situations due to their suitable design of objective function. A global scenario of publication of research articles using those four methods in solving IDS problem is depicted in Fig. 2. Figure 3 depicts the overall research of all the soft computing methods considered in this paper and their percentage of applications in solving IDS problem. It is found that, compared to other techniques such as evolutionary algorithm, swarm intelligence, and fuzzy logic, the neural network techniques have been mostly used in the IDS research areas. The analysis indicates that day by day the applicability of all these soft computing techniques is increasing with a number of more suitability in the use and practice of such methods.

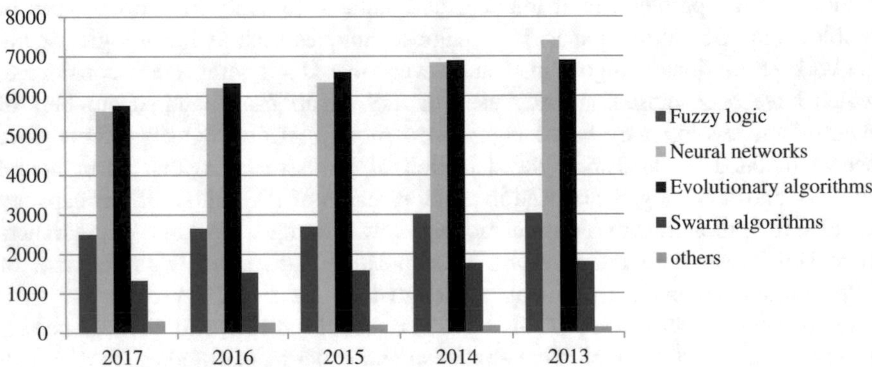

Fig. 2 No. of research articles published in IDS areas using soft computing techniques

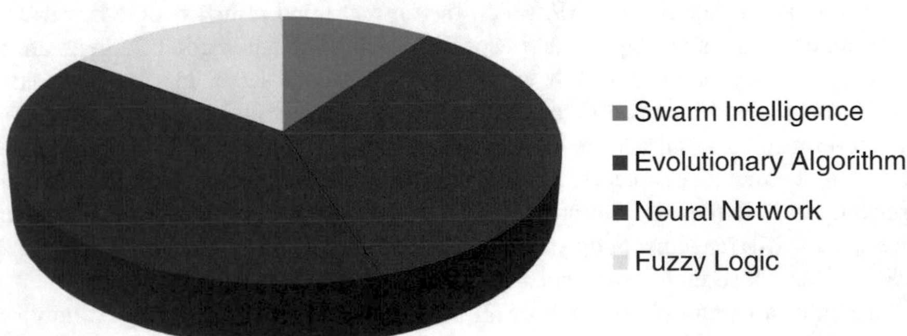

Fig. 3 Various applications of soft computing techniques in solving IDS problem

5 Conclusion

Now-a-days, security has become a major issue to be taken care of in all the applications, as there are many threats towards it. Though IDS is one of the promising solutions for such threats, but due to some more advanced security mechanisms, more research is required in upcoming scenario of advance computing. This paper deals with a rigorous survey on soft computing techniques such as fuzzy logic, neural networks, evolutionary algorithms, swarm intelligence in solving IDS problem. This survey provides a detailed applicability with mcthods and procedures with pros and cons of the above-mentioned technologies in intrusion detection systems. Soft computing has the potency to combine the powers of those technologies which helps to solve those disadvantages. The properties such as fault tolerance, adaptability to environment, fault tolerance make them more effective to use in IDS problems. The applications, pros, cons, and their future perspectives are summarized in this research report. Further, it may be assumed that this research work will be very much helpful in guiding to do effective research work in IDS problem.

References

1. Wu, S.X., Banzhaf, W.: The use of computational intelligence in intrusion detection systems: a review. Appl. Soft Comput. **10**(1), 1–35 (2010)
2. Vaidya, H., Mirza, S., Mali, N.: Intrusion detection system. Int. J. Adv. Res. Eng. Sci. Technol. **3** (2016)
3. Catania, C.A., Garino, C.G.: Automatic network intrusion detection: current techniques and open issues. Comput. Electr. Eng. **38**(5), 1062–1072 (2012)
4. Nassif, A.B., Capretz, L.F., Ho, D.: Enhancing use case points estimation method using soft computing techniques (2016). arXiv:1612.01078
5. Nayak, J., Naik, B., Behera, H.S.: Fuzzy C-means (FCM) clustering algorithm: a decade review from 2000 to 2014. In Computational Intelligence in Data Mining, vol. 2, pp. 133–149. Springer, New Delhi (2015)
6. Zadeh, Lotfi A.: Fuzzy logic, neural networks, and soft computing. Commun. ACM **37**(3), 77–84 (1994)
7. Zhang, C., Jiang, J., Kamel, M.: Intrusion detection using hierarchical neural networks. Pattern Recogn. Lett. **26**(6), 779–791 (2005)
8. Dhopte, S., Chaudhari, M.: Genetic Algorithm for Intrusion Detection System. IJRIT Int. J. Res. Inf. Technol. **2**(3), 503–509 (2014)
9. Aljarah, I., Ludwig, S.A.: Mapreduce intrusion detection system based on a particle swarm optimization clustering algorithm. In: 2013 IEEE Congress on Evolutionary Computation (CEC), pp. 955–962. IEEE, June 2013
10. Chandrasekaran, M., Muralidhar, M., Krishna, C.M., Dixit, U.S.: Application of soft computing techniques in machining performance prediction and optimization: a literature review. Int. J. Adv. Manuf. Technol. **46**(5–8), 445–464 (2010)
11. Dickerson, J.E., Dickerson, J.A.: Fuzzy network profiling for intrusion detection. In: 2000 19th International Conference of the North American Fuzzy Information Processing Society, NAFIPS, pp. 301–306. IEEE (2000)

12. Yang, L., Li, J., Fehringer, G., Barraclough, P., Sexton, G., Cao, Y.: Intrusion detection system by fuzzy interpolation. In: 2017 IEEE International Conference on Fuzzy Systems (FUZZ-IEEE), pp. 1–6. IEEE, July 2017
13. Saini, L., Suryawanshi, N.Y.: A System for Denial-of-service attack detection using MCA and IDS-based on fuzzy logic. Int. J. Comput. Appl. **141**(2) (2016)
14. Lin, W.C., Ke, S.W., Tsai, C.F.: CANN: An intrusion detection system based on combining cluster centers and nearest neighbors. Knowl.-Based Syst. **78**, 13–21 (2015)
15. Shamshirband, S., Anuar, N.B., Kiah, M.L.M., Rohani, V.A., Petković, D., Misra, S., Khan, A.N.: Co-FAIS: cooperative fuzzy artificial immune system for detecting intrusion in wireless sensor networks. J. Netw. Comput. Appl. **42**, 102–117 (2014)
16. Geramiraz, F., Memaripour, A.S., Abbaspour, M.: Adaptive anomaly-based intrusion detection system using fuzzy controller. IJ Netw. Secur. **14**(6), 352–361 (2012)
17. Manzoor, I., Kumar, N.: A feature reduced intrusion detection system using ANN classifier. Expert Syst. Appl. **88**, 249–257 (2017)
18. Sunita, S., Chandrakanta, B.J., Chinmayee, R.: A hybrid approach of intrusion detection using ANN and FCM. Eur. J. Adv. Eng. Technol. **3**(2), 6–14 (2016)
19. Haidar, G.A., Boustany, C.: High perception intrusion detection system using neural networks. In: 2015 Ninth International Conference on Complex, Intelligent, and Software Intensive Systems (CISIS), pp. 497–501. IEEE, July 2015
20. Kumar, S., Yadav, A.: Increasing performance of intrusion detection system using neural network. In: 2014 International Conference on Advanced Communication Control and Computing Technologies (ICACCCT), pp. 546–550. IEEE, May 2014
21. Devaraju, S., Ramakrishnan, S.: Detection of accuracy for intrusion detection system using neural network classifier. Int. J. Emerg. Technol. Adv. Eng. **3**(1), 338–345 (2013)
22. Li, W.: Using genetic algorithm for network intrusion detection. In: Proceedings of the United States Department of Energy Cyber Security Group, vol. 1, pp. 1–8 (2004)
23. Fouad, N., Hameed, S.M.: Genetic Algorithm based Clustering for Intrusion Detection
24. Dash, T.: A study on intrusion detection using neural networks trained with evolutionary algorithms. Soft Comput. **21**(10), 2687–2700 (2017)
25. Desale, K.S., Ade, R.: Genetic algorithm based feature selection approach for effective intrusion detection system. In: 2015 International Conference on Computer Communication and Informatics (ICCCI), pp. 1–6. IEEE, Jan 2015
26. Pal, D., Parashar, A.: Improved genetic algorithm for intrusion detection system. In: 2014 International Conference on Computational Intelligence and Communication Networks (CICN), pp. 835–839. IEEE, Nov 2014
27. Hassan, M.M.M.: Network intrusion detection system using genetic algorithm and fuzzy logic. Int. J. Innov. Res. Comput. Commun. Eng. **1**(7) (2013)
28. Bonabeau, E., Dorigo, M., Theraulaz, G.: Swarm Intelligence: from Natural to Artificial Systems, No. 1. Oxford University Press (1999)
29. Sedghi, S., Mirnia, M.: Integration bat algorithm with k-means for intrusion detection system. Int. J. Comput. Sci. Netw. Secur. **17**(7), 315–319 (2017)
30. Aghdam, M.H., Kabiri, P.: Feature selection for intrusion detection system using ant colony optimization. IJ Netw. Secur. **18**(3), 420–432 (2016)
31. Varma, P.R.K., Kumari, V.V., Kumar, S.S.: Feature selection using relative fuzzy entropy and ant colony optimization applied to real-time intrusion detection system. Proced. Comput. Sci. **85**, 503–510 (2016)
32. Aldwairi, M., Khamayseh, Y., Al-Masri, M.: Application of artificial bee colony for intrusion detection systems. Secur. Commun. Netw. **8**(16), 2730–2740 (2015)
33. Neshat, M., Sepidnam, G., Sargolzaei, M., Toosi, A.N.: Artificial fish swarm algorithm: a survey of the state-of-the-art, hybridization, combinatorial and indicative applications. Artif. Intell. Rev. pp. 1–33 (2014)
34. Alsubhi, K., Aib, I., Boutaba, R.: FuzMet: A fuzzy-logic based alert prioritization engine for intrusion detection systems. Int. J. Netw. Manage. **22**(4), 263–284 (2012)

35. Selman, A.H.: Intrusion detection system using fuzzy logic. Southeast Eur. J. Soft Comput. **2** (1) (2013)
36. Sharma, S., Kumar, S., Kaur, M. (2014). Recent trend in intrusion detection using fuzzy-genetic algorithm. Int. J. Adv. Res. Comput. Commun. Eng. 3(5)
37. Kudłacik, P., Porwik, P., Wesołowski, T.: Fuzzy approach for intrusion detection based on user's commands. Soft Comput. **20**(7), 2705–2719 (2016)
38. Desai, A.S., Gaikwad, D.P.: Real time hybrid intrusion detection system using signature matching algorithm and fuzzy-GA. In: 2016 IEEE International Conference on Advances in Electronics, Communication and Computer Technology (ICAECCT), pp. 291–294. IEEE, Dec 2016
39. Singh, S., Bansal, M.: Improvement of intrusion detection system in data mining using neural network. Int. J. Adv. Res. Comput. Sci. Softw. Eng. **3**(9) (2013)
40. Ranjan, R., Sahoo, G.: A New Clustering Approach for Anomaly Intrusion Detection (2014). arXiv:1404.2772
41. Esmaily, J., Moradinezhad, R., Ghasemi, J.: Intrusion detection system based on Multi-Layer Perceptron Neural Networks and Decision Tree. In: 2015 7th Conference on Information and Knowledge Technology (IKT), pp. 1–5. IEEE, May 2015
42. Hodo, E., Bellekens, X., Hamilton, A., Dubouilh, P.L., Iorkyase, E., Tachtatzis, C., Atkinson, R.: Threat analysis of iot networks using artificial neural network intrusion detection system. In: 2016 International Symposium on Networks, Computers and Communications (ISNCC), pp. 1–6. IEEE, May 2016
43. Yin, C., Zhu, Y., Fei, J., He, X.: A deep learning approach for intrusion detection using recurrent neural networks. IEEE Access **5**, 21954–21961 (2017)
44. Zaman, S., El-Abed, M., Karray, F.: Features selection approaches for intrusion detection systems based on evolution algorithms. In: Proceedings of the 7th International Conference on Ubiquitous Information Management and Communication, p. 10. ACM, Jan 2013
45. Benaicha, S.E., Saoudi, L., Guermeche, S.E.B., Lounis, O.: Intrusion detection system using genetic algorithm. In: 2014 Science and Information Conference (SAI), pp. 564–568. IEEE, Aug 2014
46. Hadded, M., Zagrouba, R., Laouiti, A., Muhlethaler, P., Saidane, L.A.: A multi-objective genetic algorithm-based adaptive weighted clustering protocol in vanet. In: 2015 IEEE Congress on Evolutionary Computation (CEC), pp. 994–1002. IEEE, May 2015
47. Kumar, K., Singh, S.: Intrusion Detection Using Soft Computing Techniques (2016)
48. Mahmood, D.I., Hameed, S.M.: A Multi-objective Evolutionary Algorithm based Feature Selection for Intrusion Detection
49. Bansal, J.C., Sharma, H., Jadon, S.S.: Artificial bee colony algorithm: a survey. Int. J. Adv. Intell. Paradig. **5**(1–2), 123–159 (2013)
50. Enache, A.C., Patriciu, V.V.: Intrusions detection based on support vector machine optimized with swarm intelligence. In: 2014 IEEE 9th International Symposium on Applied Computational Intelligence and Informatics (SACI), pp. 153–158. IEEE, May 2014
51. Mahmod, M.S., Alnaish, Z.A.H., Al-Hadi, I.A.A.: Hybrid intrusion detection system using artificial bee colony algorithm and multi-layer perceptron. Int. J. Comput. Sci. Inf. Secur. **13** (2), 1 (2015)
52. Azad, C., Jha, V.K.: Fuzzy min–max neural network and particle swarm optimization based intrusion detection system. Microsyst. Technol. **4**(23), 907–918 (2016)
53. Vardhini, K.K., Sitamahalakshmi, T.: Enhanced intrusion detection system using data reduction: an ant colony optimization approach. Int. J. Appl. Eng. Res. **12**(9), 1844–1847 (2017)

Handling Small Size Files in Hadoop: Challenges, Opportunities, and Review

Mohd Abdul Ahad and Ranjit Biswas

Abstract Recent technological advancements in the field of computing have been the cause of voluminous generation of data which cannot be handled effectively by traditionally available tools, processes, and systems. To effectively handle this big data, new techniques and frameworks have emerged in recent times. Hadoop is a prominent framework for managing huge amount of data. It provides efficient means for the storage, retrieval, processing, and analytics of big data. Although Hadoop works very well with large files, its performance tends to degrade when it is required to process hundreds or thousands of small size files. This paper puts forward the challenges and opportunities that may arise while handling large number of small size files. It also presents a comprehensive review of the various techniques available for efficiently handling small size files in Hadoop on the basis of certain performance parameters like access time, read/write complexity, scalability, and processing speed.

Keywords Hadoop · HDFS · Big data · HAR · SequenceFile Map file · ADS

1 Introduction

With Hadoop, we can process huge amount of data in a much effective and efficient manner which was not possible with the conventional data handling frameworks [1–4]. Hadoop "small size file problem" may be defined as the inability to efficiently process the files which are less than 64 MBs in size. "Small files" is an ambiguous term and has several definition attached to it. However in the context of

M. A. Ahad (✉) · R. Biswas
Department of Computer Science and Engineering, School of Engineering Sciences and Technology, Jamia Hamdard University, New Delhi 110062, India
e-mail: itsmeahad@gmail.com

R. Biswas
e-mail: ranjitbiswas@yahoo.com

J. Nayak et al. (eds.), *Soft Computing in Data Analytics*, Advances in Intelligent Systems and Computing 758,
https://doi.org/10.1007/978-981-13-0514-6_62

the current paper, a file may be called as small if it is less than 64 MBs in size. Mathematically, we can say categorize a "file" as small or large on the basis of its size. That is,

If File_size $\geq$ 64 MB, it is a "Large file," else it is a said to be a "Small File."

There are several tools and techniques for handling the small size file problem in Hadoop. Some of the most prominent ones are Hadoop Archive Files (HAR), SequenceFiles, Map Files, Atrain Distributed System (ADS), getmerge, FileCrush, Hmfs, NHAR, Federated Namenode, and CombineFileInputFormat, etc.

1.1 Overview of Hadoop and HDFS

Hadoop has emerged as a default option for handling big data. It works like the "client–server" and "master–slave" architecture and proceeds by dividing the data files into fixed size data packets or segments (blocks) of 64 MBs. Hadoop uses MapReduce model for executing and processing jobs in parallel. The task of data management and storage is carried out by Hadoop distributed file system (HDFS) [1–4]. HDFS allows only one user to write the data at any given instance of time but two or more users are allowed for simultaneous reading of data [1–4]. Figure 1 presents the architecture of HDFS [4]. The two inherent parts of HDFS are known as "Namenode" and "Datanode." The brief details about these components are given below.

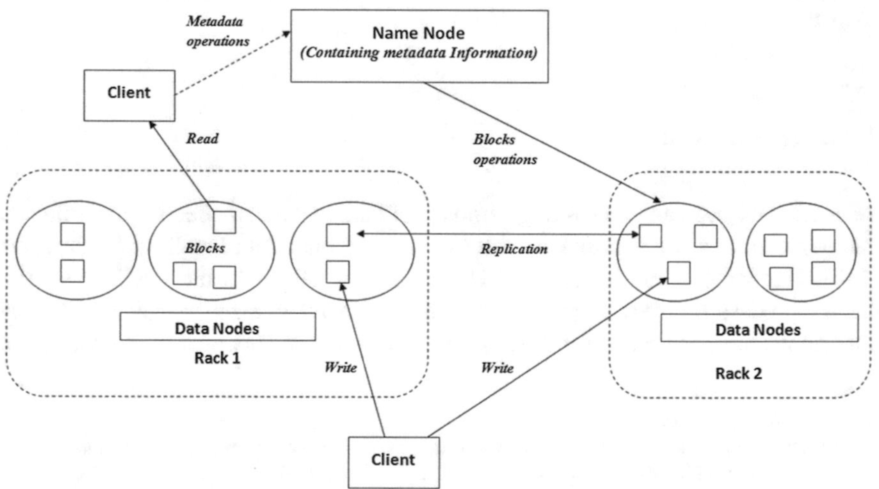

Fig. 1 HDFS architecture

1.1.1 Namenode

Namenode is considered as the brain of HDFS. It consists of the metadata information about the data stored in HDFS. Namenode stores the information like "Which Datanode contains what data" [4]. It also resolves important issues like: "when to replicate blocks," "how to balance the load among different Datanodes" [4]. Further details can be found in [1–4].

1.1.2 Datanode

They are the slave in the Hadoop master–slave architecture. The actual big data is stored in Datanodes. Apart from this, they perform the operations like "creation and deletion of blocks," "block replications" [4]. Further details can be found in [1–4].

1.2 Challenges with Small Size File Handling in Hadoop

Hadoop is primarily designed for processing large files. The large files are split into smaller chunks of prefixed size and are processed in parallel. Hadoop works very well for large size files; however, its performance degrades when we have to frequently process several small size files. This is because of the following reasons [5–10]:

- Namenode memory issues: When we have large number of small size files, each of these files will have separate metadata information which will be stored in the Namenode. Thus, it will consume more memory of Namenode [5, 10].
- MapReduce performance issues: Large number of small size files will invite large number of Disk I/O, and thus, more time will be wasted in performing the I/O instead of actual processing of data. Secondly, each file in a MapReduce framework is handled by a separate mapper; thus, if we have large number of small size files, we must use equal number of mappers which causes severe overhead and in turn reduces the performance of the system as a whole [5, 10].
- High access latency: In case of large number of small size files, for performing any operation the metadata of each file is required by HDFS client for each file access; thus, the latencies are bound to happen. Furthermore, the storage mechanism of the classical Hadoop also causes higher access latencies [5, 10].

1.3 Opportunities with Small Size File Handling in Hadoop

Since the inception of IoT technologies, the generation of real-time data has taken an exponential leap. The capabilities of providing real-time data analytics are proving to be a vital parameter in measuring the success of any enterprise. In an IoT ecosystem, everyday objects are embedded with multiple chip-based sensors having the capabilities of sensing the environment and its surroundings. This information is then transmitted to the base station for further processing. The information captured by the sensors is very small in size but the pace at which it is generated makes it massive. Real-time IoT systems involve processing the massive information stored in large number of small size files. Since IoT is rapidly finding its applications in almost every domain of human life like healthcare, education, transportation, etc., it becomes a necessity to develop new tools and techniques for effective processing of the gigantic information stored in these large number of small size files [11–13].

If Hadoop is made capable of effectively handling the small size files of such IoT ecosystems, it will surely revolutionize the computing era and takes it to new horizons.

In order to explore these opportunities and overcome the limitations of the conventional Hadoop system, a variety of solutions have been proposed in recent years. This paper provides a comprehensive evaluation of some of the most prominent techniques used for solving small size file handling problem in Hadoop.

2 Related Works

The authors of [5] discussed about an optimized approach for handling small size file on cloud. They classified the files into three categories on the basis of file correlation and termed them as logically related files, structurally related files, and unrelated files. They had used file merging, pre-fetching, and grouping techniques in their proposal. The authors of [6] used the pre-fetching and caching mechanism in their approach. The researchers in [7] proposed an efficient middleware model by the name "Hmfs" for processing small size files. It makes use of three components: file operation interface, file management, and file buffer for carrying out different related tasks. The researcher in [9] proposed an improved version of HDFS for storing small size files using a three layer approaches by segregating the user interface, data storage, and data processing layer. The researchers in [14] surveyed different file handling techniques in HDFS like HAR, SequenceFile, and TLB-MapFile. Wei Qu et al. in [15] proposed an approach resolve the problem of small files in Hadoop using the concept of merging and indexing. The authors in [16] proposed merging, indexing and hashing-based technique for handling small size files. The researchers of [17] talked about improving the performance of Hadoop by proposing File Manager and MapCombine Reduce techniques. The work of File Manager is to manage the metadata information of the Namenode while "MapCombineReduce (CMR)" is used to processes the actual data. The researchers in [18] introduced a dynamic merging and indexing technique for handling the small size files in Hadoop. In [19], the authors talked about De-Duplication techniques for merging uniquely identifiable small files into a large file.

3 Techniques for Handling Small Size Files in Hadoop

There are several techniques for solving the small size file problem of Hadoop. Some of the most prevalent techniques are explained in this section.

3.1 Hadoop Archive (HAR)

HAR stands for Hadoop archive files. In this, several small size files are wrapped within a single file and are termed as HAR file [20–22]. In other words, we can say that a HAR is a collection of small size files which are referred by a common name. To identify the individual files in HAR, the concept of indexing is used. This techniques solve the small size file problem in Hadoop to some extent; however, its limitation is that it requires two index reads and one data read to actually access the data, which is fairly time-consuming and has a cascading effect on the performance of the system in a negative way. The master index consists of the indexes of the base index file. Figures 2 and 3 present the layout of the HAR file and the process of creating HAR files, respectively [20–22].

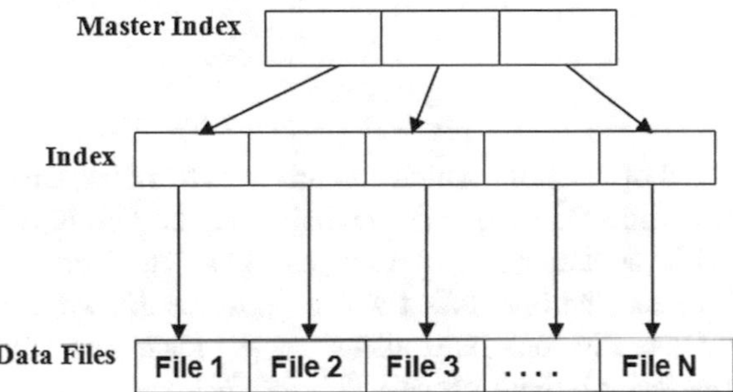

Fig. 2 HAR file layout

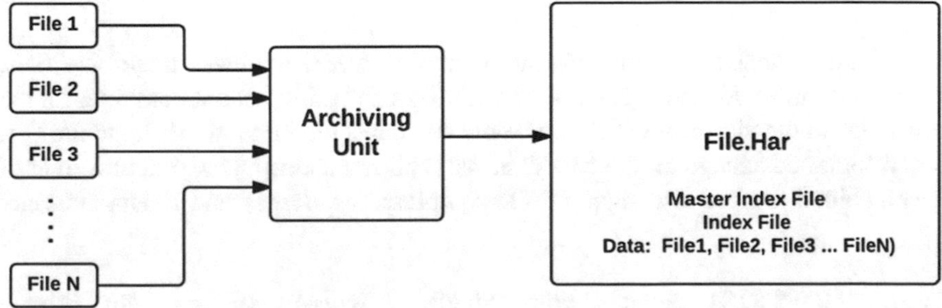

Fig. 3 Process of creating HAR files

Key (FileName)	Value (Contents of File)	Key (FileName)	Value (Contents of File)	Key (FileName)	Value (Contents of File)

Fig. 4 Layout of the SequenceFile

3.2 SequenceFile

SequenceFile is a kind of flat file which consists of "binary key/value" pairs. The small size files are stored in a SequenceFile as "key/value pair" wherein the file name is taken as a key and its contents serves as the value [23]. Figure 4 presents the layout of the SequenceFile. These files can be processed in a streamed fashion. The main advantage of these types of files is the capability to split itself at block level or record level thereby making them to be used in parallel for faster processing. Furthermore, they are extensively used in MapReduce framework and supports only append operations. There are three types of classes supported by SequenceFile, namely Writer class, Reader class, and Sorter class which are used for writing, reading, and sorting, respectively. Furthermore, the SequenceFile are available in three different formats, viz. "uncompressed key/value records," "record compressed key/value records," and "block compressed key/value records" [21–23].

3.3 Map Files

A Map File is a kind of directory which comprises of two files. One file contains the index information while the second file contains the data. Both of these files are a type of SequenceFile containing a key/value pair [24]. The index file consists of the actual index of file and the byte offset which performs the work of lookup on the basis of key. The data file consists of filenames as indices and their contents as its value. They allow faster lookup process [21, 22, 24].

3.4 Hadoop File Crush Utility

It is a utility which converts several small size files into fewer large size files. A prefixed threshold size is given to the "File Crush" utility, and it combines all the small files in the directory which are below this threshold into a single large file [25, 26]. Another advantage of "filecrush" is that it allows choosing the output format of the combined file and also supports file compressions. Syntax and Example usage:

Syntax

Crush [OPTION] <input dir> [Single Space] <output dir> [Single Space] <timestamp>

where input dir: input folder where the small files are stored, output dir: output folder where you want to store the merged files, time stamp: a 14 digit number to uniquely name the files. For example:

Crush/mypc/example/input/mypc/example/output 20110211185655.

3.5 HDFS Getmerge Command

The HDFS "getmergc" command is used to copy the multiple files present in folder in HDFS to a single concatenated file in the local file system [27]. It has the following syntax:

Hadoop fs -getmerge -nl <source file path> <local system destination path>

Example:

hadoop fs -getmerge/user/hadoop/myfiles single_merged.txt

3.6 CombinedFileInputFormat

It is a kind of abstract class that can be explicitly used to combine several "small files" into a single "large file." It is an inbuilt class in Hadoop. It works by combining multiple small files in each split. Here we can process multiple small files in a single map task [10, 28].

3.7 Atrain Distributed System (ADS)

"ADS" is a fairly new architecture capable of handling structured and unstructured big data in a relatively simple and efficient manner. It makes use of "r-train" and "r-atrain" data structures for storing homogeneous and heterogeneous big data, respectively [4, 29]. Here the whole memory is represented in terms of coaches, and each coach is linked with its previous and next coach via backward and forward links, respectively (analogous to Doubly Linked List). Furthermore, the coaches can be of variable size. The various components or nodes in the ADS follow multi-horse cart topology as depicted in Fig. 5 (making it highly flexible and scalable) [4, 29]. New nodes can be added either in width or height or both. It follows master–slave architecture (analogous to Hadoop) wherein the Pilot Computer (PC) works like the master and coaches (DCs) works like slaves. Since the memory is not divided into fixed size blocks, ADS does not suffer from small size file handling like in conventional Hadoop [4, 29]. Therefore, ADS can be used as an alternative for Hadoop for processing large number of small size files. Further, details about ADS can be found in [4, 29].

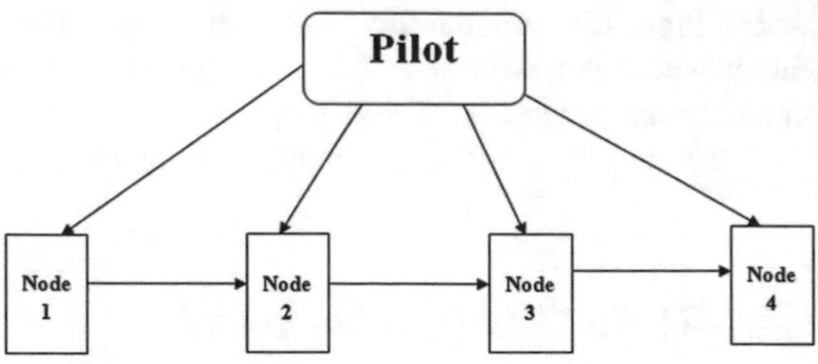

Fig. 5 Multi-horse cart topology

3.8 New Hadoop Archive (NHAR)

This technique has been inspired from the conventional HAR. It overcomes the limitations of the conventional HAR by allowing insertion of additional files in the existing Hadoop archive, thus making it extensible. The authors of NHAR claim to improve the performance by 85% as compare to the conventional HAR [8].

4 Comparative Analysis

This section presents the comparative analysis of the various techniques for handling small size file problem in Hadoop [5–10, 14, 17, 18, 20, 23–30].

5 Conclusion

The paper discusses the various challenges and opportunities that arise with the effective handling of small size files in Hadoop. Several techniques have already been proposed for tackling this issue. Some of the techniques are inbuilt while others are custom made explicitly for effective handling of small size files in terms of improving the read, write, and processing capabilities of Hadoop. From the comparative analysis given in Table 1, it can be concluded that ADS is looking a promising distributed system having the capabilities of handling any size of historical or real-time data. Furthermore, the FMP-SSF proposal and FGP-LSF proposal take the lead with their proven experimental analysis and theoretical computations. HAR is best suited for archival purposes while NHAR and Hmfs can be useful in storing data requiring real-time analysis.

The use of any of these techniques is dependent on the type of data to be handled. It can be concluded that the structure of files and their correlation play a vital role while selecting any of these techniques. Few of the techniques discussed here have changed the philosophy of the classical Hadoop system while others

Table 1 Comparative analysis

Parameters of comparison	HAR	SequenceFile	Map file	File crush	NHAR	Combine file input format	ADS	Hmfs
Support of small size files	Yes	Yes	Yes	Yes	Yes	Yes	Yes	Yes
Scalability	No, once created no additional files are allowed	Yes, we can add key/value pair for each additional file	Yes, we can add key/value pair for each additional file	Yes, we can add additional files in the directory and perform crush operation	Yes	Yes	Yes, additional files can be added anytime	Yes
Support for compression	No	Yes	Yes	Yes	Yes	No	No	Yes
Updation of files	No	Yes	Yes	Yes	Yes	Yes	Yes	Yes
Memory usage	Moderate	Moderate	Moderate	Moderate	Low	Moderate	Low	Low
File splitting capabilities	Yes	Yes	Yes	No	Yes	No	No	Yes
Access rate	Slow, as it requires two index reads	Moderate	Better than SequenceFile	High	High	High	High	High
Conversion into multiple formats	No	Yes	Yes	No	Yes	No	No	Yes
Read/write capabilities	Medium	Medium	Medium	High	High	High	High	High
Method for storage	Archive based	File based	File based	File based	Archive based	File based	Linked based	File based

proposed the enhancement in the existing Hadoop. Largely, we have conclude that these techniques are primarily dependent on the type of data (structured or unstructured) to be handled.

References

1. Shvachko, K., Kuang, H., Radia, S., Chansler, R.: The Hadoop Distributed File System. IEEE (2010)
2. Borthakur, D.: HDFS Architecture Guide, Apache Foundation (2008). https://hadoop.apache.org/docs/r1.2.1/hdfs_design.pdf
3. Gupta, L.: HDFS—Hadoop Distributed File System Architecture Tutorial (2015). http://howtodoinjava.com/big-data/hadoop/hdfs-hadoop-distributed-file-system-architecture-tutorial/
4. Ahad, M.A., Biswas, R.: Comparing and analyzing the characteristics of hadoop, cassandra and quantcast file systems for handling big data. Indian J. Sci. Technol. [S.l.] (2017). ISSN 0974-5645. http://www.indjst.org/index.php/indjst/article/view/105400, https://doi.org/10.17485/ijst/2017/v10i8/105400
5. Dong, B., Zheng, Q., Tian, F., Chao, K.-M., Ma, R., Anane, R.: An optimized approach for storing and accessing small files on cloud storage. J. Netw. Comput. Appli. **35**, 1847–1862 (2012)
6. Bao, Z., Xu, S., Zhang, W., Chen, J., Liu, J.: A strategy for small files processing in HDFS. In: Che W. et al. (eds) Social Computing. ICYCSEE 2016. Communications in Computer and Information Science, vol. 623. Springer, Singapore (2016)
7. Yan, C., Li, T., Huang, Y., Gan, Y.: Hmfs: efficient support of small files processing over HDFS. In: Sun X. et al. (eds) Algorithms and Architectures for Parallel Processing. ICA3PP 2014. Lecture Notes in Computer Science, vol. 8631. Springer, Cham (2014)
8. Vorapongkitipun, C., Nupairoj, N.: Improving performance of small-file accessing in hadoop. In: 2014 11th International Joint Conference on Computer Science and Software Engineering (JCSSE), pp. 200–205. IEEE (2014)
9. Liu C.: An improved HDFS for small file. In: 2016 18th International Conference on Advanced Communication Technology (ICACT) Advanced Communication Technology (ICACT), pp. 478–481, IEEE
10. White, T.: The small files problem (2009). http://www.cloudera.com/blog/2009/02/02/the-small-files-problem/
11. Ashton, K.: That "Internet of Things" thing. RFiD J. (2009)
12. Gubbi, J., Buyya, R., Marusic, S., Palaniswami, M.: Internet of things (IoT): a vision, architectural elements, and future directions. Future Gener. Comput. Syst. **29**, 1645–1660 (2013)
13. Kim, T.-H., Ramos, C., Mohammed, S.: Smart city and IoT, future. Gener. Comput. Syst. **76**, 159–162 (2017)
14. Revathi, S., Kauser, S., Sushmitha, S., Vinodini, G.: A survey on different file handling mechanisms in HDFS. Int. Res. J. Eng. Technol. (IRJET), **04**(06), 1994–1996 (2017)
15. Qu W., Cheng S., Wang H.: Efficient file accessing techniques on hadoop distributed file systems. In: Che W. et al. (eds.) Social Computing. ICYCSEE 2016. Communications in Computer and Information Science, vol. 623. Springer, Singapore (2016)
16. Mir, M.A., Ahmed, J.: An Optimal Solution for small file problem in Hadoop. Int. J. Adv. Res. Comput. Sci. **8**(5) (2017)
17. Prasad, G., Nagesh, M.S., Swathi Prabhu, H.R.: An efficient approach to optimize the performance of massive small files in hadoop MapReduce framework. Int. J. Comput. Sci. Eng. IJCSE **5**(6), 112–120 (2017)

18. Bhandari, S., Chougale, S.: An approach to solve a small file problem in hadoop by using dynamic merging and indexing scheme. Int. J. Recent Innov. Trends Comput. Commun. **4** (11), 227–230 (2016)
19. Vaidya, I., Nath, R.: An improved de-duplication technique for small files in hadoop. Int. Res. J. Eng. Technol. (IRJET) **04**(07), 2040–2045 (2017)
20. Hadoop Archives: Hadoop archives guide (2011). http://hadoop.apache.org/common/docs/current/hadoop_archives.htmlS
21. Solving Small size problem in Hadoop. http://blog.cloudera.com/blog/2009/02/the-small-files-problem/
22. Solving Small size problem in Hadoop. https://pastiaro.wordpress.com/2013/06/05/solving-the-small-files-problem-in-apache-hadoop-appending-and-merging-in-hdfs/
23. SequenceFile: Sequence file wiki (2011). http://wiki.apache.org/hadoop/SequenceFileS
24. MapFile: Mapfile api (2011). http://hadoop.apache.org/common/docs/current/api/org/apache/hadoop/io/MapFile.htmlS
25. White, S., Hadoop, T.: The Definitive Guide. Yahoo Press (2010)
26. Hadoop File Crush Utility. https://github.com/edwardcapriolo/filecrush
27. Hadoop getmerge command. https://hadoop.apache.org/docs/current/hadoop-project-dist/hadoop-common/FileSystemShell.html
28. Combined File Input Format. https://gist.github.com/airawat/6647007
29. Biswas, R.: Atrain distributed system (ADS): an infinitely scalable architecture for processing big data of any 4Vs. In: Computational Intelligence for Big Data Analysis, Volume 19 of the series Adaptation, Learning, and Optimization, pp. 3–54. Springer International Publishing, Switzerland, (2015). online ISBN: 978-3-319-16598-1
30. Bendea, S., Shedgeb, R.: Dealing with small files problem in hadoop distributed file system. In: 7th International Conference on Communication, Computing and Virtualization 2016, Procedia Computer Science, vol. 79, pp. 1001–1012, Elsevier (2016)

A TOPSIS Approach of Ranking Classifiers for Stock Index Price Movement Prediction

Rajashree Dash, Sidharth Samal, Rasmita Rautray and Rasmita Dash

Abstract Predicting future stock index price movement is equivalent to a binary classification problem with one class label for increasing movement and other for decreasing movement. In the literature, a wide range of classifiers are tested for this application, but the decision regarding a better technique varies with the choice of performance measures. Hence, assessing classifiers can be considered as a multi-criteria decision-making (MCDM) problem. In this study, a TOPSIS-based MCDM framework is suggested for ranking five classifier models such as radial basis function, Naïve Bayes, decision tree, support vector machine, and k-nearest neighbor with respect to four criteria in application to prediction of future stock index price movements. Historical stock index prices of two benchmark stock indices such as BSE SENSEX and S&P500 are taken for the empirical validation of the model. The results reveal that ranking a classifier with respect to multiple evaluation measures is better compared to selecting one considering single criterion.

Keywords Financial timeseries analysis · Stock index movement prediction
MCDM · TOPSIS

R. Dash · S. Samal (✉) · R. Rautray
Computer Science & Engineering Department, Siksha O Anusandhan
(Deemed to be University), Bhubaneswar 751030, India
e-mail: sidharthsamal95@gmail.com

R. Dash
e-mail: rajashree_dash@yahoo.co.in

R. Rautray
e-mail: rasmitarautray@yahoo.co.in

R. Dash
Computer Science & Information Technology Department,
Siksha O Anusandhan (Deemed to be University),
Bhubaneswar 751030, India
e-mail: rasmita02@yahoo.co.in

J. Nayak et al. (eds.), *Soft Computing in Data Analytics*,
Advances in Intelligent Systems and Computing 758,
https://doi.org/10.1007/978-981-13-0514-6_63

1 Introduction

Being a popular channel of financial investment, analysis of stock market is receiving increasing attention day to day. But several factors such as inflation, deflation, interest rates and monetary policies of national banks, changes in economic policies of a country, political shock highly influence the stock market. The price of stock can rise or fall overnight. Hence, predicting the future upswings or downswings on the stock index price movement has been appeared as a challenging task both for investors and researchers involved in financial analysis.

Prediction of stock index movement problem can be explained as a binary classification problem with two class labels, one representing the increasing movement and other representing the decreasing movement. In the literature, a wide range classifiers such as neural networks [1–5], support vector machines (SVM) [6–9], Naïve Bayes model (NB) [9, 10], decision tree (DT) [11, 12], K-nearest neighbor (KNN) model [13, 14] are proposed to predict the future movements of stock market index or its return and to devise financial trading strategies to translate the forecasts into profits. In [1, 2], authors have revealed the better prediction accuracy of a simple multilayer perceptron (MLP) compared to various traditional statistical techniques. In [3], a probabilistic neural network (PNN) has shown better result in predicting the direction of index returns and also provides higher profit than the other investment strategies utilizing the market direction generated by the forecasting methods. In [4], a radial basis function network (RBFN) trained by a ridge extreme learning machine (RELM) is proposed to capture the relationship between the future trend and the past values of the technical indices. Due to the use of kernel functions and the capacity control of the decision function, support vector machine (SVM) has also been appeared as promising trend prediction model [6, 7]. A hybrid genetic algorithm optimized decision tree–SVM model has also shown better trend prediction accuracy compared to an artificial neural network (ANN)-based system and a Naïve Bayes (NV)-based system [9]. The design of a decision tree-rough set-based stock market trend prediction system for predicting the future trend of the SENSEX is presented in [11, 12].

Evaluation and selection of classification algorithm is an important issue in many disciplines. The selection may depend on the choice of performance measures included in the study. Assessing the performance of an algorithm involves inclusion of multiple criteria and weighting one criterion against the gains of other complementary criteria. Hence, the task of efficient classifier selection can be considered as a multicriteria decision-making (MCDM) problem [15–18]. In this study, stock index price movement prediction problem is combined with MCDM problem to rank classification models. One of the popular MCDM approaches termed as Technique for Order of Preference by Similarity to Ideal Solution (TOPSIS) is suggested for ranking five popular classifier models such as RBF, NB, DT, SVM, and KNN with respect to four criteria such as accuracy, F-measures, precision, and recall. Historical stock index prices of two benchmark stock indices such as

BSE SENSEX and S&P500 collected from January 01, 2015 to November 13, 2017 are taken for the empirical validation of the model.

The remainder of this paper is organized as follows: Section 2 specifies the detail of the proposed model for stock index movement prediction using classifiers and ranking of those classifiers using TOPSIS. Section 3 provides experimental result acquired from classification and TOPSIS. In the end, Sect. 4 contains the concluding remarks.

2 Proposed Model

This section outlines the details of stock index price movement prediction using classifiers and assessment of those classifiers using TOPSIS approach. Figure 1 depicts the architecture of the proposed model. Initially, historical stock index prices comprising open, high, low, and closing prices are collected within a particular period. Then, the raw timeseries data are passed through the data preprocessing step to reduce redundancy, duplication, invalid, and empty data fields and to produce an acceptable input–output form for the classifiers. In input preparation step, the acceptable normalized input samples for the classifiers are generated using technical indicator calculation and data normalization steps. As surveyed in the literature, researchers have selected various technical indicators for their experiment in stock index movement prediction. In this study, six commonly used technical indicators such as simple moving average (SMA), stochastic K%, D%, Larry William's R%, moving average convergence and divergence (MACD), and relative strength index (RSI) are chosen as input to the classifiers. The detailed mathematical expressions for calculating these technical indicators are represented in [4, 5]. The six technical indicator values are scaled between 0 and 1 using the max–min normalization as mentioned below.

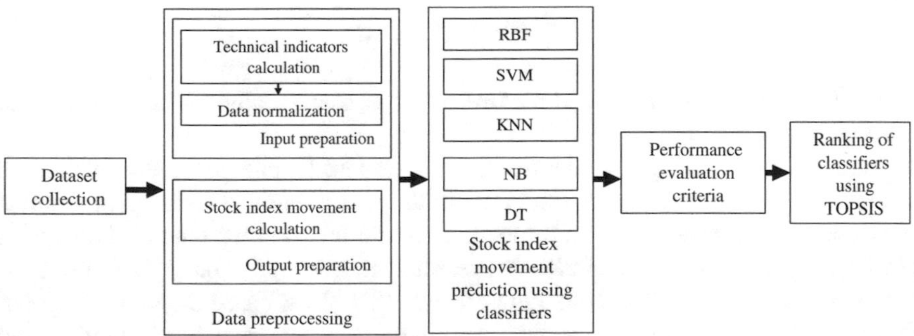

Fig. 1 Proposed model for stock index price movement prediction

$$Nv = \frac{Cv - V_{\min}}{V_{\max} - V_{\min}} \tag{1}$$

where Nv = Normalized value, Cv = Current value, V_{min} = Minimum value among all data, V_{max} = Maximum value among all data.

After normalization, the input patterns are prepared by arranging the normalized technical indicator values corresponding to each trading day as a row vector. The output pattern comprising the 1 and 0 values representing the up and down movements, respectively, is calculated from historical closing prices using the following rule:

$$\begin{aligned} &\text{If } Cprice(i) > Cprice(i-1) \text{ then IM(i)} = 1 \text{ (increase)} \\ &\text{If } Cprice(i) < Cprice(i-1) \text{ then IM(i)} = 0 \text{ (decrease)} \end{aligned} \tag{2}$$

where Cprice(i) = closing price of *i*th row and IM(i) = index movement of *i*th row.

After the preparation of input and output samples, two-thirds of these samples are used for training and validation of the classifiers and the remaining are used for testing. In this study, the performance of five popular classifiers such as RBF, SVM, KNN, NB, and DT is evaluated for stock index movement prediction as up or down movements. Then, the performance of individual classifier is measured in terms of four criteria such as accuracy, F-measure, precision, and recall by using following formulas:

$$Accuracy = \frac{TN + TP}{TN + FN + TP + FP} \tag{3}$$

$$F - measures = \frac{2 \times Precision \times Recall}{Precision + Recall} \tag{4}$$

$$Precision = \frac{TP}{TP + FP} \tag{5}$$

$$Recall = \frac{TP}{TP + FN} \tag{6}$$

where TP, TN, FP, and FN stand for true positive, true negative, false positive, and false negative, respectively.

Instead of evaluating the classifiers with respect to one particular evaluation criterion, they are ranked by considering all the four criteria. Hence, the problem is treated as a MCDM problem with five models and four criteria. One of the popular MCDM approaches, i.e., TOPSIS, is presented in this study for ranking the classifiers considering multiple evaluation criteria. TOPSIS is based on the principle of determining the best alternative with minimum distance from ideal solution and maximum distance from negative ideal solution [18]. The ideal solution is derived from the combination of best performance values exhibited by any alternative for each criterion, whereas the negative ideal solution is derived from the combination of the worst performance values. Assuming a problem with M alternatives (classifiers) and N criteria, the procedure of TOPSIS can be summarized as follows:

- Generate a decision matrix in which the rows represent alternatives, i.e., the classifiers, and columns represent the performance evaluation criteria.
- Derive a standardized decision matrix by using the following formula:

$$SDM_{ij} = \frac{DM_{ij}}{\sqrt{\sum_{i=1}^{m} DM_{ij}^2}}, j = 1, \ldots, J \quad (7)$$

  DM_{ij} = Decision matrix, SDM_{ij} = Standardized decision matrix
  M = Number of alternatives, j = Number of criterion
- Assign weights to each criterion and generate a weighted standardized decision matrix as follows:

$$WN_{ij} = w_i \times SDM_{ij}, j = 1, \ldots, J \quad (8)$$

  where w_i is the weight of the *i*th criteria and $\sum_{i=1}^{n} w_i = 1$
- Find the ideal solutions A^+ from the combination of best performance values exhibited by any alternative for each criterion and the negative ideal solution A^- from the combination of the worst performance values using the following formula:

$$A^+ = \{WN_1^+, \ldots, WN_m^+\} = \{(\max_j WN_{ij} | i \in h), (\min_j WN_{ij} | i \in l)\} \quad (9)$$

$$A^- = \{WN_1^-, \ldots, WN_m^-\} = \{(\max_j WN_{ij} | i \in h), (\min_j WN_{ij} | i \in l)\} \quad (10)$$

  where h is the set of benefit criteria having the property of more is better and l is the set of negative criteria having the property of less is better.
- Calculate separation measures S^+ from ideal solution and S^- from negative ideal solution by using n-dimensional Euclidean distance as follows:

$$S_j^+ = \sqrt{\sum_{i=1}^{m} (WN_{ij} - WN_i^+)^2}, j = 1, \ldots, J \quad (11)$$

$$S_j^- = \sqrt{\sum_{i=1}^{m} (WN_{ij} - WN_i^-)^2}, j = 1, \ldots, J \quad (12)$$

- Calculate the relative closeness to the ideal solution using the following equation:

$$C_j^+ = \frac{S_j^-}{(S_j^+ + S_j^-)}, j = 1, \ldots, J \tag{13}$$

- Finally ranking of alternatives is done according to the decreasing order of the relative closeness values.

3 Experimental Setup and Result Analysis

The aim of this experiment is to examine the relative performance of different classifiers for stock index movement prediction by means of a TOPSIS-based MCDM framework. Experiments have been carried out on two benchmark stock indices such as BSE SENSEX and S&P 500 considered within the period from January 01, 2015 to November 13, 2017. After preprocessing, total 685 samples are generated for both the datasets. Then, the samples are divided into training and testing sets. For training a total of 485 samples consisting of six normalized technical indicator values and their respective class level, i.e., 0 or 1 are set as input and output, respectively, to the classifiers. After the training phase, testing is done by using remaining 200 samples as the input to the classifiers. The classifiers used in the experiment are DT, NB, KNN with three different K values, SVM with five different kernel functions, and RBF with Gaussian basis function. The results from classifying the test samples have been evaluated with four performance evaluation criteria such as accuracy (CR1), precision (CR2), recall (CR3), and F-measure (CR4) and analyzed with TOPSIS. Table 1 summarizes the value of each performance evaluation measure for all the 11 classifiers for BSE SENSEX and S&P 500 dataset.

For each criterion, the best performing classifier is underlined. Analyzing Table 1, it is observed that there is no classifier that achieves the best results across all measures and ranking of the best performing model varies based on the chosen performance evaluation measure. Therefore, it leads to a realistic situation of making a more reliable decision considering multiple criteria. Hence, further analysis of classifiers is done by a TOPSIS-based MCDM framework. Table 1 is also considered as two decision matrices of size 11 $\times$ 4, one for BSE SENSEX and other for S&P 500 dataset. The standardized decision matrix derived by using Eq. (7) for both the datasets is given in Table 2. Then, weights are assigned to each criterion according to its relative importance, and then, they are normalized in the interval [0, 1] such that the sum of all weights is equal to 1. The weighted standardized decision matrix obtained by using Eq. (8) is represented in Table 3 for both the datasets. Then, the ideal solution and negative ideal solution obtained by using Eqs. (9) and (10) are also represented in Table 3. Finally Table 4 represents the separation measures from ideal solution (S^+), separation measures from negative ideal solution (S^-), corresponding relative closeness (C^+), and ranking of classifiers. Rank 1 is considered as the best classifier, and rank 11 is considered as

Table 1 Classification results

Classifiers notation	BSE SENSEX				S&P 500			
	CR1	CR2	CR3	CR4	CR1	CR2	CR3	CR4
CM1	0.6500	0.5750	0.5610	0.5679	0.8100	0.7763	0.7375	0.7564
CM2	0.7300	0.7414	0.5244	0.6143	0.7850	0.6989	0.8125	0.7514
CM3	0.7350	0.6835	0.6585	0.6708	0.7700	0.6518	0.9125	0.7604
CM4	0.7200	0.7407	0.4878	0.5882	0.8250	0.7711	0.8000	0.7853
CM5	0.7400	0.7143	0.6098	0.6579	0.7950	0.8095	0.6375	0.7133
CM6	0.7850	0.7826	0.6585	0.7152	0.8150	0.8116	0.7000	0.7517
CM7	0.7900	0.7273	0.7805	0.7529	0.7900	0.7273	0.7805	0.7529
CM8	0.8200	0.7347	0.8780	0.8000	0.7450	0.6261	0.9000	0.7385
CM9	0.8100	0.7444	0.8171	0.7791	0.8100	0.7917	0.7125	0.7500
CM10	0.8200	0.8194	0.7195	0.7662	0.8100	0.7692	0.7500	0.7595
CM11	0.6350	0.6154	0.2927	0.3967	0.6450	0.6452	0.2500	0.3604

Decision Tree (CM1), KNN with K = 3 (CM2), KNN with K = 4 (CM3), KNN with K = 5 (CM4), SVM with linear kernel (CM5), SVM with Gaussian kernel (CM6), SVM with polynomial kernel—order 2 (CM7), SVM with polynomial kernel—order 3 (CM8), SVM with polynomial kernel—order 4 (CM9), RBF (CM10), NB (CM11)

Table 2 Standardized decision matrix

Classifier	BSE SENSEX				S&P 500			
	CR1	CR2	CR3	CR4	CR1	CR2	CR3	CR4
CM1	0.2609	0.2410	0.2583	0.2540	0.3118	0.3175	0.2980	0.3144
CM2	0.2930	0.3107	0.2414	0.2748	0.3022	0.2858	0.3283	0.3124
CM3	0.2950	0.2865	0.3032	0.3000	0.2964	0.2665	0.3687	0.3161
CM4	0.2890	0.3105	0.2246	0.2631	0.3176	0.3153	0.3233	0.3264
CM5	0.2970	0.2994	0.2807	0.2943	0.3060	0.3310	0.2576	0.2965
CM6	0.3151	0.3280	0.3032	0.3199	0.3137	0.3319	0.2828	0.3125
CM7	0.3171	0.3048	0.3593	0.3368	0.3041	0.2974	0.3154	0.3130
CM8	0.3292	0.3079	0.4042	0.3578	0.2868	0.2560	0.3637	0.3070
CM9	0.3251	0.3120	0.3762	0.3485	0.3118	0.3237	0.2879	0.3118
CM10	0.3292	0.3435	0.3313	0.3427	0.3118	0.3146	0.3030	0.3157
CM11	0.2549	0.2579	0.1347	0.1774	0.2483	0.2638	0.1010	0.1498

the worst classifier. By considering these ranked classifiers, it is found that CM8, i.e., SVM with 'polynomial' kernel function having third-order polynomial is the best suited classifier for BSE SENSEX dataset. Similarly, CM4, i.e., KNN with K = 5 is the best suited classifier for S&P dataset. Above experiment shows that classifiers are data-dependent. If the dataset changes, the performance of the classifier also changes. So, it is not effective to choose a single classifier for multiple datasets.

Table 3 Weighted standardized decision matrix

Classifier	BSE SENSEX				S&P 500			
	CR1	CR2	CR3	CR4	CR1	CR2	CR3	CR4
CM1	0.1044	0.0241	0.0517	0.0762	0.1247	0.0317	0.0596	0.0943
CM2	0.1172	0.0311	0.0483	0.0824	0.1209	0.0286	0.0657	0.0937
CM3	0.1180	0.0287	0.0606	0.0900	0.1186	0.0267	0.0737	0.0948
CM4	0.1156	0.0310	0.0449	0.0789	0.1270	0.0315	0.0647	0.0979
CM5	0.1188	0.0299	0.0561	0.0883	0.1224	0.0331	0.0515	0.0890
CM6	0.1260	0.0328	0.0606	0.0960	0.1255	0.0332	0.0566	0.0937
CM7	0.1268	0.0305	0.0719	0.1010	0.1216	0.0297	0.0631	0.0939
CM8	0.1317	0.0308	0.0808	0.1073	0.1147	0.0256	0.0727	0.0921
CM9	0.1301	0.0312	0.0752	0.1045	0.1247	0.0324	0.0576	0.0935
CM10	0.1317	0.0343	0.0663	0.1028	0.1247	0.0315	0.0606	0.0947
CM11	0.1020	0.0258	0.0269	0.0532	0.0993	0.0264	0.0202	0.0449
A^+	**0.1317**	**0.0343**	**0.0808**	**0.1073**	**0.1270**	**0.0332**	**0.0737**	**0.0979**
A^-	**0.1020**	**0.0241**	**0.0269**	**0.0532**	**0.0993**	**0.0256**	**0.0202**	**0.0449**

Table 4 Ranking of classifiers

Classifier	BSE SENSEX				S&P 500			
	S^+	S^-	C^+	Rank	S^+	S^-	C^+	Rank
CM1	0.0517	0.0338	0.3955	**10**	0.0148	0.0684	0.8216	**7**
CM2	0.0436	0.0399	0.4776	**8**	0.0119	0.0701	0.8546	**3**
CM3	0.0305	0.0526	0.6333	**6**	0.0111	0.0757	0.8717	**2**
CM4	0.0487	0.0349	0.4177	**9**	0.0092	0.0747	0.8900	**1**
CM5	0.0340	0.0490	0.5900	**7**	0.0244	0.0592	0.7081	**10**
CM6	0.0239	0.0601	0.7155	**5**	0.0177	0.0667	0.7898	**9**
CM7	0.0126	0.0704	0.8483	**3**	0.0131	0.0689	0.8405	**4**
CM8	0.0036	0.0822	0.9586	**1**	0.0156	0.0722	0.8221	**6**
CM9	0.0072	0.0762	0.9136	**2**	0.0169	0.0667	0.7976	**8**
CM10	0.0153	0.0706	0.8221	**4**	0.0138	0.0692	0.8335	**5**
CM11	0.0824	0.0017	0.0201	**11**	0.0806	0.0008	0.0096	**11**

4 Conclusion

In this paper, a TOPSIS-based MCDM framework is discussed to evaluate the performance of different classifiers considering four criteria such as accuracy, F-measures, precision, and recall for prediction of future stock index price movements. An experimental study including 11 classifiers and four criteria is conducted over two benchmark stock indices such as BSE SENSEX and S&P500 to validate the proposed model. For BSE SENSEX dataset, SVM classifier performs better than other classifiers, and for S&P 500 dataset, KNN classifier performs better than other models included in the study. The results reveal that ranking a classifier using single performance criteria may lead to unreliable conclusions about the best performing model, whereas ranking the models including multiple criteria will generate more robust rankings.

References

1. Guresen, E., Kayakutlu, G., Daim, T.U.: Using artificial neural network models in stock market index prediction. Expert Syst. Appl. **38**(8), 10389–10397 (2011)
2. Mostafa, M.M.: Forecasting stock exchange movements using neural networks: empirical evidence from Kuwait. Expert Syst. Appl. **37**(9), 6302–6309 (2010)
3. Chen, A.S., Leung, M.T., Daouk, H.: Application of neural networks to an emerging financial market: forecasting and trading the Taiwan stock index. Comput. Oper. Res. **30**(6), 901–923 (2003)
4. Dash, R., Dash, P.K.: A comparative study of radial basis function network with different basis functions for stock trend prediction. In: IEEE Power, Communication and Information Technology Conference (PCITC), pp. 430–435 (2015)
5. Dash, R., Dash, P.K.: Stock price index movement classification using a CEFLANN with extreme learning machine. In: IEEE Power, Communication and Information Technology Conference (PCITC), pp. 22–28 (2015)
6. Huang, W., Nakamori, Y., Wang, S.Y.: Forecasting stock market movement direction with support vector machine. Comput. Oper. Res. **32**(10), 2513–2522 (2005)
7. Kara, Y., Boyacioglu, M.A., Baykan, Ö.K.: Predicting direction of stock price index movement using artificial neural networks and support vector machines: the sample of the Istanbul Stock Exchange. Expert Syst. Appl. **38**(5), 5311–5319 (2011)
8. Shen, S., Jiang, H., Zhang, T.: Stock Market Forecasting Using Machine Learning Algorithms. Department of Electrical Engineering, Stanford University, Stanford, CA, pp. 1–5 (2012)
9. Nair, B.B., Mohandas, V.P., Sakthivel, N.R.: A genetic algorithm optimized decision tree-SVM based stock market trend prediction system. Int. J. Comput. Sci. Eng. **02**(9), 2981–2988 (2010)
10. Mahajan Shubhrata, D., Deshmukh Kaveri, V., Thite Pranit, R., Samel Bhavana, Y., Chate, P. J.: Stock market prediction and analysis using Naïve Bayes. Int. J. Recent Innov. Trends Comput. Commun. (IJRITCC) **4**(11), 121–124 (2016)
11. Tiwari, S., Pandit, R., Richhariya, V.: Predicting future trends in stock market by decision tree rough-set based hybrid system with HHMM. Int. J. Electron. Comput. Sci. Eng. **1**, 1578–1587 (2012)
12. Nair, B.B., Mohandas, V.P., Sakthivel, N.R.: A decision tree-rough set hybrid system for stock market trend prediction. Int. J. Comput. Appl. **6**(9), 1–6 (2010)
13. Imandoust, S.B., Bolandraftar, M.: Application of K-Nearest Neighbor (KNN) approach for predicting economic events: theoretical background. Int. J. Eng. Res. Appl. **3**(5), 605–610 (2013)
14. Ban, T., Zhang, R., Pang, S., Sarrafzadeh, A., Inoue, D.: Referential kNN regression for financial time series forecasting. Neural information processing. In: International Conference on Neural Information Processing (ICONIP), pp. 601–608 (2013)
15. Kou, G., Lu, Y., Shi, Y.: Evaluation of classification algorithms using MCDM and rank correlation. Int. J. Inf. Technol. Decis. Making **11**, 197–225 (2012)
16. Mehdiyev, N., Enke, D., Fettke, P., Loos, P.: Evaluating forecasting methods by considering different accuracy measures. Procedia Comput. Sci. **95**, 264–271 (2016)

17. Pacheco, A.G.C., Krohling, R.A.: Ranking of classification algorithms in terms of mean-standard deviation using A-TOPSIS. arXiv preprint arXiv:1610.06998 (2016)
18. Sánchez, J.S., García, V., Marqués, A.I.: Assessment of Financial risk prediction models with multi-criteria decision making methods. Neural information processing. In: International Conference on Neural Information Processing (ICONIP), pp. 60–67 (2012)

IoT in Education: An Integration of Educator Community to Promote Holistic Teaching and Learning

Gautami Tripathi and Mohd Abdul Ahad

Abstract The technological advancements in the field of computing are swiftly bringing a paradigm shift in education system across the globe. The term 'Education' today is replaced by 'Smart Education,' wherein teaching–learning resources are connected with each other. With the emergence of IoT, interconnecting objects of everyday life has become an easy task. These connected objects work in synchronization and can be controlled using handheld devices. Due to its ubiquitous and scalable nature, IoT is finding its application in almost every area of human life. The prospective applications of IoT are invading today's world, initiating a pedagogical advancement and bridging the gap between the real and the virtual world by bringing together all the objects of everyday life under a single umbrella of IoT. This paper proposes an effective IoT model tailored for an educational environment to make the task of teaching and learning more simple and efficient.

Keywords Internet of Things (IoT) · Education · Software-defined networking (SDN) · Sensor

1 Introduction

The term 'Internet of Things' was coined by Kevin Ashton in a presentation to Proctor & Gamble in 1999 [1]. The adaptation of a variety of wireless technologies like RFID tags, actuators, sensors, etc., has fueled the development of Internet of Things. The concept of IoT has brought a technological revolution making every day devices and communication smarter, intelligent, and informative. As per Gartner [2], twenty-five billion devices will be connected to the Internet by 2020, enabling the user

G. Tripathi (✉) · M. A. Ahad
Department of Computer Science and Engineering, School of Engineering Sciences and Technology, Jamia Hamdard University, New Delhi 110062, India
e-mail: gautami1489@gmail.com

M. A. Ahad
e-mail: itsmeahad@gmail.com

J. Nayak et al. (eds.), *Soft Computing in Data Analytics*, Advances in Intelligent Systems and Computing 758,
https://doi.org/10.1007/978-981-13-0514-6_64

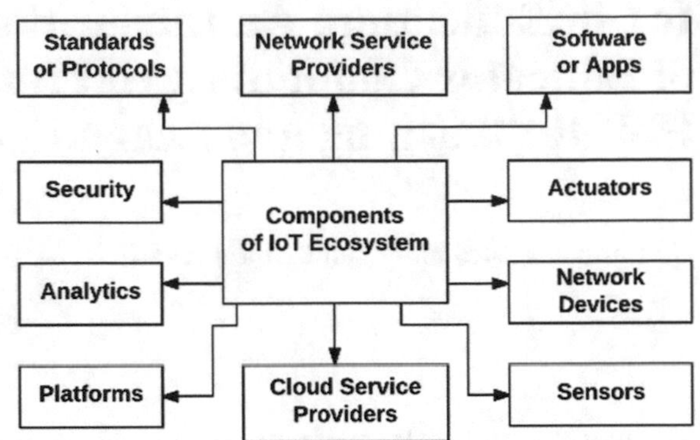

Fig. 1 Components of IoT ecosystem

data to be pre-planned, analyzed, managed to make intelligent decisions. Today, the agglomeration of several technologies is leading us to an era where objects of everyday life are becoming smart. In the past few decades, the most profound technological evolution has come into picture by combining the virtual and physical worlds using embedded sensors, actuators, processors, and transceivers leading to the concept of Internet of Things. Figure 1 presents the different components and elements of IoT ecosystem.

IoT has added a new dimension to the existing Internet by facilitating communication between everyday-life objects and humans. With the evolution of varied applications of IoT in recent years, a number of contextual definitions have emerged. The 'cluster of European research projects on the Internet of Things' defines the 'Things' as the active entity capable of interacting and communicating with the environment and among themselves [3]. In context of the proposed model, IoT can be defined as the interconnected network of all the elements of modern-day education uniquely addressable using standard protocols and facilitating the sharing of information across all the participating entities. With the technological advancement, IoT has shown profound impact on several aspects of everyday life and its users. The convergence of several technologies under the umbrella of IoT has allowed the development of a number of intelligent applications. Although all these applications are not fully functional as yet, study shows that IoT has a great potential in improving the quality of life by facilitating home automation, fitness tracking, environment protection, developing smart cities, modifying health care, smart transportation, smart domotics, e-governance, assisted living, smart education, retail, logistics, agriculture, automation, industrial manufacturing, and business management [4, 5]. This paper presents a model for IoT-based education system and discusses its implementation in the real-world scenario.

2 Related Work

In recent years, Internet of Things has attracted the attention of the researchers from across the globe. The ever-increasing application areas of IoT have led numerous researches in the field. IoT has been studied in light of several other technologies accomplishing the goal of a smarter world. Sethi and Sarangi [4] have comprehensively covered the major technologies related to IoT and have proposed a novel taxonomy for IoT technologies, highlighting some of the most important technologies, and profiles some applications that have the potential to make a striking difference in human life, especially for the differently abled and the elderly. Shin and Jin Park [6] performed a study offering a contextualized socio-technical analysis of IoT, providing insight into its challenges and opportunities. This insight helps to conceptualize how IoT can be designed and situated within human-centered contexts. Ray [5] has summarized the current Internet of Things architectures in various domains and discussed how they can improve the understanding about the various tools, technologies, and methodologies to aid developers' requirements. The author also investigated the research challenges in the area. Weber and Weber [7] have pointed out the legal barriers that may stand in the way of adopting the Internet of Things. The authors in [8] proposed a model for integrating objects to Virtual Academic Communities (VAC). The model was tested using a case study, and the results show that the application of IoT leads to a more engaging learning environment for the students and provides more data to the teachers for analysis. The investigators in [9] discussed the current status of smart campus and indicate the difference between digital campus and smart campus. They also introduced the application of the Internet of Things and the cloud computing in education. Atzori et al. [10] give a detailed vision of the IoT paradigm from the viewpoint of several scientific communities and focus on the major research issues that need to be faced. The authors of [11] discussed the ways IoT can transform education, and they also presented the history of the IoT and the examples of IoT usage in education systems. The author in [12] presented the growth, potential, and applications of IoT in educational institutes. In [13], the researchers categorized the application of IoT into four groups like energy management, health care, classroom control, and improving teaching–learning. The authors in [14] talked about leveraging the asset intelligence from the IoT ecosystem for adding value to the teaching–learning processes. The authors of [15] discussed the potential of IoE and the various challenges it possesses to the formal education including the integration of big data with IoE. In [16], the researchers talked about IoT and its related technologies.

3 IoT in Education: The Proposed Model

IoT is gaining popularity for its potential to converge several technologies into a single and effective servicing unit. This paper presents a proposal which aims to automate every aspect of education system and the entities involved in it. To take the advantage of IoT technology, the entities of the educational institute are embedded with sensors having sensing, computing, and wireless communication capabilities. Figure 2 shows the different elements of proposed model for smart education system.

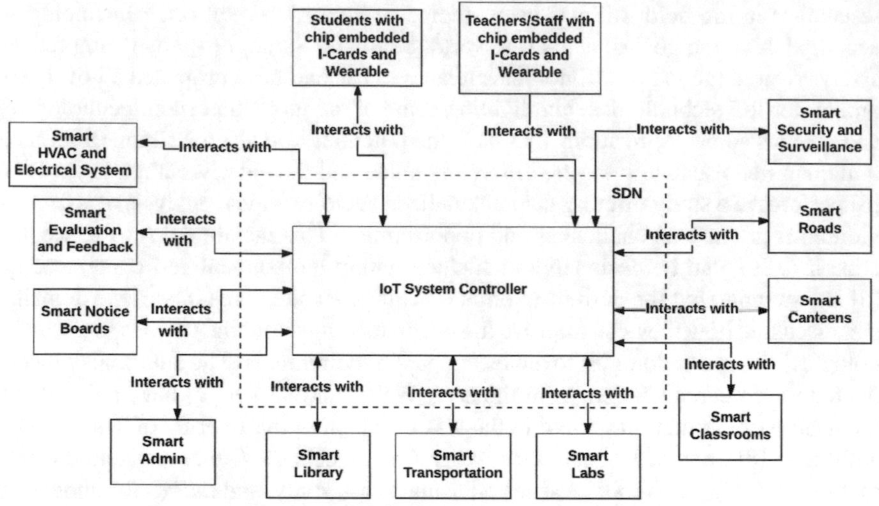

Fig. 2 Elements of proposed smart education system

In the proposed IoT-based education system model, every entity of educational system is implanted with an IoT chip (sensor) for sensing and storing the information about itself and its surroundings. Teacher/staff and student are the primary entities in the proposed system. They (students and teachers/staff) have IoT-based smart identity cards and smart wearable which constantly stores, transmits, and receives information about them and their surroundings to and from the IoT system controller. For better understanding of the proposed approach, we have divided it into the following core components.

3.1 Teaching and Learning

The classrooms are equipped with smart cameras, microphone, air purifiers, HVAC systems, and smart boards. The lectures of the class are captured in the form of videos, audios, texts, and images. These lectures are simultaneously stored in real time on the cloud for future references. IoT assists and supports the students to connect and interact with the teachers using varied modalities and helps in creating a fruitful teaching–learning environment. For conducting examinations, the students are provided a smart tab, which can be used to write text or draw figures. These contents are automatically transferred to the teachers' account on the cloud for evaluations. After evaluation, the teacher can directly send feedback to the respective students. This technology ensures timely evaluation and comprehensive feedback mechanism helping student in identifying the areas that need improvements and adopt corrective measures. The laboratories consist of virtual setups with open-source software suites that can be accessed using basic Internet connectivity from the computers, laptops, or other handheld devices. These laboratories are equipped with multiple simulations on the related topics. The students can get hands-on experience of the laboratory setup using their respective devices. The laboratory work conducted by the students is also stored in real time on the cloud for evaluation and reference purposes.

3.2 Security

The educational institutions have local IoT system controllers installed at multiple sites within the campus which works in synchronization with each other. The security measures in the campus include door access control systems, wherein the exit and entry doors of the premises, classrooms, washrooms, library, admin, laboratories and canteens have chip-enabled locks which can be opened or closed only by legitimate personnel by means of an embedded chip in their identity cards or using remote devices, thereby avoiding any security breaches or trespassing. The attendance of students and staff is automatically tracked by reading the information from the chip embedded in their identity cards. The roads within the campus are equipped with speed sensors which constantly monitor the speed of the vehicles inside the campus. If the speed of any vehicle exceeds the predefined speed limit, a notification is sent to the driver in real time, which is also reflected in the digital log book of the particular driver (student or staff). In this way, the admin can control the speed limits of vehicles in the campus. The campus premises have smart security alarms and cameras which are capable of intelligently monitoring the surroundings, thus allowing any abnormal activity to be tracked and reported in real time. Furthermore, since every component is embedded with IoT chips, they can be tracked and monitored in real time and thus prevent possible chances of thefts. Live announcements, automatic water sprinklers, smart fire alarms, specific emergency tones, automated notification to the disaster management units, pre-recorded instructional messages can be used in case of any emergency situations within the campus premises.

3.3 Management

The smart notice boards are capable of broadcasting any message that is displayed on it to all the connected display devices in real time making the users aware of any important notice(s). The teacher can directly broadcast any information to their students in real time using smart apps. They can also keep track of their academics and other related work submitted to the admin using the tracking applications software on any handheld device. Since all information is stored on the cloud and is securely accessed by the participating entities, it will reduce the manual paperwork, ensure better tracking, monitoring, and management, and in turn help in saving the time and environment. The admin, on the other hand, can effectively manage all the activities like tracking the issued library books, fees dues, academic and personal details of the staff and students using dedicated software and apps.

3.4 Environment

The IoT-based smart HVAC and electrical systems (fans, tube lights, projectors, monitors, air purifiers, etc.) are responsible for automatically adjusting the atmospheric conditions within the premises of the university. The temperature, air quality, and ambience of the classrooms, laboratories, and other facilities can be dynamically and automatically controlled and adjusted according to the preference of the users to provide conducive teaching–learning environment. The electrical systems are equipped with sensors, which sense the presence of students or staff inside a classroom or other facilities and can take intelligent decision about when to switch on/off or adjust the controls of the systems and equipments. This technology ensures a favorable environment (which is neither too cold nor too hot) and thus helps in increasing the productivity of students and staffs as they do not feel exhausted because of the controlled temperature and enhanced air quality. Moreover as the systems and equipments are automatically controlled using the sensing technology, it helps in reducing the electricity bills and in turn helps in preserving the atmosphere to a certain extent by reducing the carbon emission from the systems and equipments.

3.5 Health and Hygiene

The canteens in the proposed model are equipped with IoT-based sensors. Every food item of the canteen has a bar code mentioning the calorie count and expiry date. The wearable of the student/staff keeps track of the total calorie count in a day. If the calorie count exceeds the predefined limit, an instant notification is sent to them. In this way, the staff and the students can keep track of the calorie intake by them and stay healthy. By tracking the expiry dates of the food items, the wearable ensures that the user is always taking hygienic food as it notifies the user in case any expired food item is picked by the user. The wearable is also capable of automatically monitoring the vitals of the users and instantly notifies the concerned departments in case of any abnormal deviations. Enabling the proposed model will ensure better teaching–learning environment, improved interaction and attentiveness among students and staff, increased responsibility and productivity, better tracking of progress and development of students, ubiquitous availability of resources (laboratories and lectures), and above all, the proposed model will help in saving the environment as a whole.

4 Working of the Proposed Approach

The various participating entities in an educational system are embedded with IoT-based microchips (sensors) which store the information about the entity and its surroundings. The IoT system controller is also responsible for mining the relevant information using various data preprocessing techniques like data cleaning, pruning, stemming. These entities are connected to an IoT control system which guides them about the needed actions in case of any user request. Moreover, the system controller has the ability to adapt and improvise according to the situation by making use of highly powerful artificial intelligence algorithms. Figure 3 presents the overview of the technical working of the proposed model.

In order to dynamically control the network and make an agile, adaptable, and scalable networking environment, the concept of software-defined networking (SDN) [17–19] is incorporated in the proposed approach. Using SDN, the IoT system controller gets the flexibility to dynamically monitor and control the network traffic, routing algorithms, packet forwarding protocols, and congestion control using simple programming constructs [17–19]. It makes use of OpenFlow [19] protocol standards for interacting with the underlying IoT devices. The various sensors and actuators work in synchronization with each other to convert energy into electrical data and vice-versa. All the data from the participating devices and the sensors gets stored in the cloud.

5 Discussions

This paper focuses on the prospective impact of IoT technology on the education sector. Although this concept is in its embryonic stage, the underlying technologies and protocols are exhibiting signs of maturity. Moreover, as per the 'Moore law,' the cost of implementing the model will be inversely proportional to the advancement in technology in the coming years. Today, the 'Internet of Things' is bringing a pedagogical advancement in the education sector and promoting collaborative, interactive, safer, faster, and retained learning environment. It also helps in showcasing higher level of personalized and adaptive learning environment by incorporating real-life problems into classroom teaching and involving students to get hands-on experience about the challenges of the society as a whole. The future of this continued transformation is showing signs of its exponential growth since IoT has a preeminent potential to eradicate the obstacles in current education system such as demographics, gender biases, languages, and economic status [20]. Hence, IoT-based education can effectively satisfy the demands and needs of student and educator community in a much larger context.

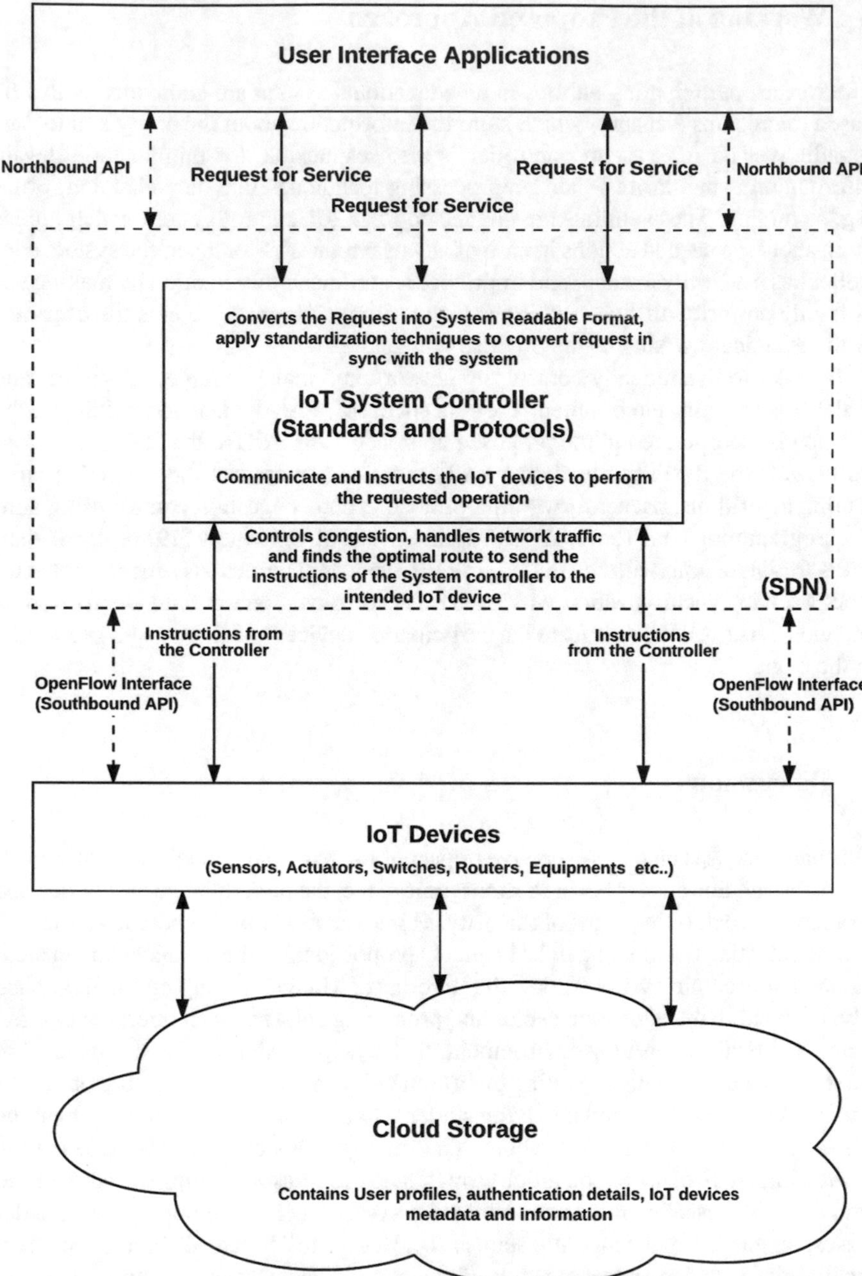

Fig. 3 Proposed model for IoT-based education system

References

1. Ashton, K.: That Internet of Things thing. RFiD J. (2009)
2. http://www.gartner.com/newsroom/id/3598917
3. Sundmaeker, H., Guillemin, P., Friess, P., Woelffl, S.: Vision and challenges for realising the Internet of Things. Cluster of European Research Projects on the Internet of Things-CERP IoT (2010)
4. Sethi, P., Sarangi, S.R.: Internet of Things: architectures, protocols, and applications. J. Electr. Comput. Eng. 1–25 (2017)
5. Ray, P.P.: A survey on Internet of Things architectures. J. King Saud Univ. Comput. Inf. Sci. (2016)
6. Shin, D.H., Jin Park, Y.: Understanding the Internet of Things ecosystem: multi-level analysis of users, society, and ecology. Digit. Policy, Regul. Gov. **19**(1) (2017)
7. Weber, R.H., Weber, R.: Internet of Things: Legal Perspectives, vol. 49. Springer Science & Business Media (2010)
8. Marquez, J., Villanueva, J., Solarte, Z., Garcia, A.: IoT in education: integration of objects with virtual academic communities. In: WorldCIST (1), pp. 201–212, Mar 2016
9. Nie, X.: Constructing smart campus based on the cloud computing platform and the internet of things. In: Proceedings of the 2nd International Conference on Computer Science and Electronics Engineering (ICCSEE 2013), pp. 1576–1578. Atlantis Press, Paris, France, Mar 2013
10. Atzori, L., Iera, A., Morabito, G.: The internet of things: a survey. Comput. Netw. **54**(15), 2787–2805 (2010)
11. Clarity Innovations, Internet of Things (2016). https://www.clarity-innovations.com/sites/default/files/publications/clarity-iot-in-education.pdf
12. Benson, C.: The Internet of Things, IoT systems, and higher education. EDUCAUSE Rev. **51**(4), 6 (2016)
13. Bagheri, M., Movahed, S.H.: The effect of the Internet of Things (IoT) on education business model. In: 2016 12th International Conference on Signal Image Technology & Internet Based Systems (SITIS), pp. 435–441. IEEE, Nov 2016
14. Zebra Technologies: How the Internet of Things Is Transforming Education. http://www.zatar.com/sites/default/files/content/resources/Zebra_Education-Profile.pdf. Accessed 20 July 2017
15. Selinger, M., Sepulveda, A., Buchan, J.: Education and the Internet of Everything: How ubiquitous connectedness can help transform pedagogy. Cisco Consulting Services and Cisco EMEAR Education Team (2013)
16. Gaggl, L.: Education and the internet of things. https://www.brightcookie.com/wp-content/uploads/2011/11/IoT_in_Education-3.pdf
17. Software-defined networking: the new norm for networks, Open Networking Foundation (ONF), White Paper, 1–12 (2012)
18. Braun, W., Menth, M.: Software-defined networking using Openflow: protocols, applications and architectural design choices. Future Internet **6**, 302–336 (2014)
19. Kreutz, D., Ramos, F.M.V., Uhlig, S.: Software-defined networking: a comprehensive survey. Proc. IEEE **103**(1), 14–76 (2015)
20. Mishra, S.: 12 Modern Learning Practices with Internet of Things i.e. IoT in Education. http://www.clickonf5.org/internet/modern-learning-practices-iot-education/126745. Accessed 12 July 2017

Visual Encryption by Pixel Property Permutation

Ayush Surana, Tazeen Ajmal and K. Annapurani

Abstract Data security is a challenging problem in the sphere of storage and transmission of data. With the development of heavy computational resources, various encryption algorithms are getting more prone to subsequent attacks or threats either via brute force, cipher attacks or statistical attacks, and side-channel attacks. Thus, for better security there is a need for more robust encryption of data especially unstructured data. This paper has tried to propose a novel image encryption technique known as Pixel Property Permutation (PPP) on the basis of Visual Cryptography. In this algorithm, any image RGB or greyscale is first converted to Black and White and compressed of (200 * 200) pixels. Although this is not mandatory, one can even use the RGB image however this guideline is assumed only for simplicity purposes. The pixel values of the image are extracted and stored. These pixel values are then permuted with one another using random permutations. Thus, the encrypted image is obtained. Now, using the encrypted image and the original image a key is generated (superimposing one on the other). The key and the encrypted image are both subjected to the 128-bit AES Encryption (Roeder, Symmetric-Key Cryptography, [1]). The decryption process is also very similar. The two AES encrypted images are decrypted using the same key obtained earlier. In the next step we use the decrypted AES image and the cipher key image to recover the original image by superimposing one on top of another. This process of PPP uses AES encryption algorithm as an added layer of defence which is not essentially necessary but encouraged, this technique is cost effective and can be used to secure highly sensitive data.

A. Surana · T. Ajmal
Department of CSE, SRM University, Chennai 603203, India
e-mail: suranaayush97@gmail.com

T. Ajmal
e-mail: ajmaltazeen@gmail.com

K. Annapurani (✉)
Department of CSE, SRM Institute of Science and Technology, Chennai 603203, India
e-mail: annapoorani.k@ktr.srmuniv.ac.in

J. Nayak et al. (eds.), *Soft Computing in Data Analytics*,
Advances in Intelligent Systems and Computing 758,
https://doi.org/10.1007/978-981-13-0514-6_65

1 Introduction

In this data extensive world, there are huge amount of data available for various purposes. Accessing these data has become increasingly easy and important. Every corporation wants to secure their data in the most prolific manner possible. In this modern era, with the development and the vast influence of the Internet makes it highly difficult for enterprises to maintain their information as confidential. The secrecy of the data has to be maintained both at the senders and the receivers end. One such method to secure data is cryptography. It is the mechanism in which one can study the various mathematical techniques involved in the sphere of securing information and data. Visual Cryptography is the type of cryptography to be applied for visual authentication and identification of images both normal and digital. Visual Cryptography is the new technique that uses simple mathematical models for its purposes unlike heavy highly complex models used by the traditional cryptography. Here, the fundamental principle is to encrypt the data in such a way that the decryption can be performed by the human visual systems. This simple technique of Visual Cryptography Systems (VCS) encrypts an image in different shares and uses these images to give back the original image by superimposing them.

There are many traditional and highly complex encryption algorithms available to us like Blowfish, Twofish, Triple DES, IDEA, AES. This paper has tried to use the advanced encryption standard (AES), recognized by the US National Institute of Standards and Technology (NIST) in 2001, and is among the most secure algorithm available to us using the Rijndael Cipher to encrypt the different shares (Key and the PPP Image) obtained after performing the PPP encryption process. However, the AES algorithm is no longer the most secure. There are several methods like cache timing attacks [2] and biclique cryptanalysis [3].

This paper has been dissected as described ahead. Section 2 describes the related works on the topic of Visual Cryptography. Section 3 gives the overview of the proposed algorithm Pixel Property Permutation. Section 4 deals with the in-depth execution of the algorithm and all of its major steps. Section 5 deals with the various different cryptanalytic methods available to show the robustness of the proposed algorithm. Section 6 gives the streamline of the simulation result when the proposed algorithm was executed. The last section gives a brief summary and future prospects and improvements that can be made on this project which if implemented would immensely be useful for keeping our precious data secure.

2 Related Works

Here, we will discuss some of the works/ideas we studied from the various research papers and illustrations. These ideas have been classified under suitable sub-headings as illustrated below.

2.1 Chaos-Based Image Encryption

S. El Assad, M. Farajallah, C. Vladeanu [4] proposed a chaos-based block cipher technique for visual encryption In their proposed method, they have built a chaos-based generator in which a plain image is sent to a substitution/diffusion chamber and through that chamber it is sent to the 2D cat map which is highly efficient in generating a secret key from the aforesaid chaotic generator. This in turn gives us the cipher image.

2.2 DES and AES Encryption Algorithm

Zarko S. Stanisavljevic of Bulgaria, Serbia, in his paper on data encryption standard [5] visual representation has illustrated a method for visual representation using the DES Method. The prescribed method was implemented in the COALA System (CryptOgraphic ALgorithms visual simulAtion). It is designed for the better understanding of the cryptographic algorithms. The specifications about these kinds of systems are thus highly useful in understanding the efforts and effects of the COALA System on the young minds.

2.3 Visual Transformation-Based Algorithm

Xiaowei Xu, Scott Dexter and Ahmet M. Eskicioglu [6] present a hybrid image protection scheme to establish a relation between the data encryption key and the watermark. This scheme completes in three steps, 1. Generate shares of the image, and then embed watermark to this image share. Finally, this embedded image has to be encrypted with any cipher scheme to get final encrypted image.

2.4 Value Transformation-Based Permutation

Aloka Sinha and Kehar Singh [7] have proposed a new technique. In this technique, the digital signature of the original image is added to the encoded version of the original image. An error code that is best suited is followed to do the encoding of the image, e.g. Bose–Chaudhuri–Hocquenghem (BCH) code. After decryption of that image at the receiver's end, the digital signature verifies the authenticity of the image.

3 Proposed System and its Merits—Pixel Property Permutation (PPP)

The system proposed in this paper deals with Visual Cryptography on the basis of Pixel Property Permutation (PPP). This paper deals with the idea of extracting the pixel information of an Image; converting it into black and white or keeping as RGB depending on the computational complexity and compressing it to 200 * 200 pixels. The new image obtained is stored, and its pixel values are randomly permuted into various transparent shares. Each share combined constitutes an image. This image along with the original image is used to find the key for decryption process. This key along with the encrypted image is processed through the infamous AES encryption technique. The keys of the images encrypted are stored.

These keys are used to decrypt the cipher key and the encrypted image. These two images are superimposed to each other to give the resultant image. The resultant image passes through some denoising algorithms to give the original image.

This seemingly complex process has many merits as over the existing systems as has been illustrated below:

- Brute forcing this algorithm would be futile as the possible number of combinations available are : - (3.8 * 10 to the power 38) * 2 along with (200 * 200)! permutations. It would take billions of years for the fastest supercomputer to break through it.
- It is very easy to implement. If one uses only PPP for encryption even then it would be highly secure and would consume very little time and space complexity.
- It is highly portable and can be implemented on any platform irrespective of the system it is being implemented on.
- The space it consumes is also not very large as after compressing the image the size of the image is considerably reduced and this image is used for further illustrations.
- The PPP can be implemented using JAVA/C/C++/Python/OpenCV/MATLAB and various other languages thus it gives us the platform independency much essential for any algorithms.

Thus, this algorithm successfully removes most of the demerits of the existing systems and produces a highly cost effective and fast method for encrypting images for safe and secure transmission and storage.

4 Architecture of Pixel Property Permutation

The architecture of the proposed Pixel Property Permutation Algorithm (PPP) has been illustrated via the block diagrams Figs. 1 and 2. Figure 1 describes the encryption process, and Fig. 2 describes the decryption process.

The step by step process for securing the visual data through PPP has been illustrated in the modules given below which describes the detailed analysis of the proposed algorithm.

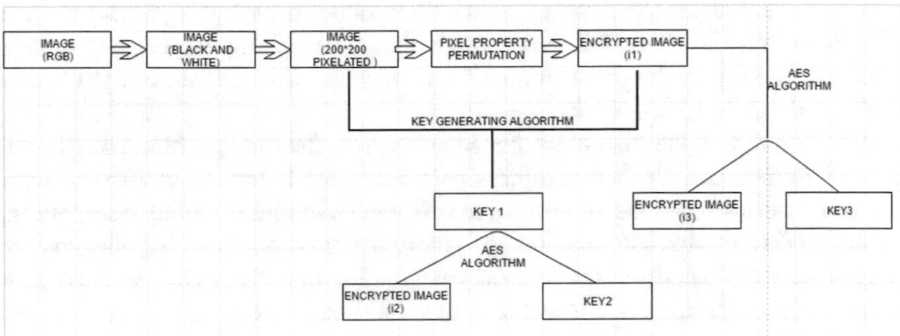

Fig. 1 Block diagram of the encryption process by converting an RGB image to Black and White

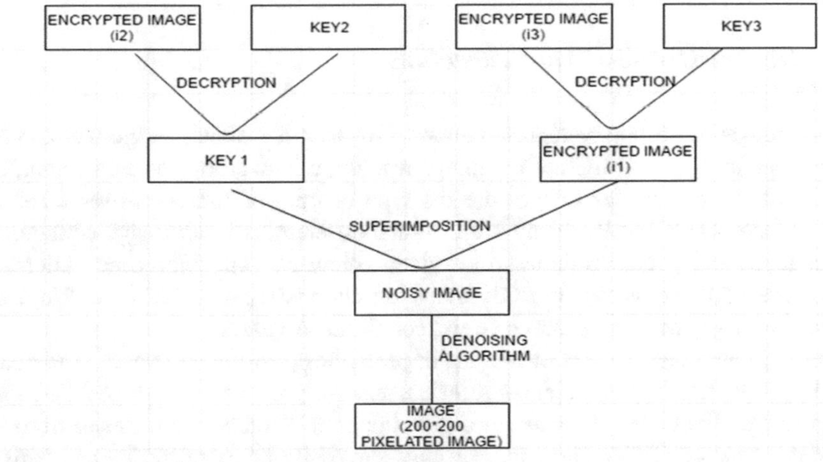

Fig. 2 Block diagram for the decryption process

4.1 Compression and Conversion of Image

Unstructured data comprises of various different types of images like RGB colour image which stores three different set of pixel values red, green and blue and greyscale reflect the amount of light for each pixel, little light generates dark pixels, and the large light generates the bright pixels. The first step of this encryption technique is that the image either in RGB or in greyscale needs to be converted to Black and White Binary Image. One approach towards conversion of an RGB colour image into the black and white image is to take each of the RGB values and converts it into one value that reflects the brightness of that pixel. This can be accomplished by taking the average of the three values (R + G + B)/3. But this will be highly erroneous. So, what we can do is that we can allow the brightness to be perceived majorly by the green colour and thereby take a weighted average of it, for example, 0.6G + 0.2B + 0.2R.

After converting the image, we have to compress the image without any loss to most of its features. Such a compression technology is termed as lossless compression. There are various algorithms to accomplish lossless compression differing on the quality of the compression to its varying complexity. There are various image formats that perform the lossless compression like PNG, GIF, whereas there are some that uses some lossy methods like TIFF and MNG.

In our procedure, the image is converted and compressed because it would save a lot of complexity while performing random permutations. However, if necessary then this algorithm could also work without converting and compressing the image thereby rendering this step null and void.

4.2 Digital Image Pixel Extraction

Digital images are a binary representation of a two-dimensional image that can be stored depending upon whether the image resolution is fixed or not, in a vector, or else a raster form. Raster images are the type of images that comprises a set of digital values called pixels. They are the smallest element in an image containing values representing the brightness of a given colour at a specified time. After we have successfully converted an RGB image to its corresponding Black and White or greyscale image, we now proceed to extract all those values.

Pixels in an image consist of two major properties position and value. The pixel position is defined by the (x, y) coordinates, and the pixel values indicate the RGB or in our case the greyscale value. A digital image essentially comprises of m rows and n columns of pixels. Thus, these values can easily be extracted by converting the image into a textual image. If the image is an RGB image then each pixel value would have to store three different values red, green and blue. All these values are stored in an n * m matrix which will then be used to perform permutation of over these values.

4.3 Random Permutation of the Pixel Property

This is one of the major steps proposed in this project. Here, we will try to permutate the value of the pixels of an image. This pixel matrix will be exposed to various permutating algorithms which can later be reversed to give us the decrypted image. This encryption scheme uses a secret key sharing algorithm which requires two Keys. This encryption process is completely based on changing the pixel position and its values. To increase the efficiency of the encryption process, the values are permuted using a special random permutations. It is the random permutations that can be useful to derive the desired results. For easier understanding, we are going to assume that the image is in Black and White. The same can be accomplished in an RGB image also but with higher complexity. In the previous step of this process, we had extracted the pixel values in an m * n matrix. Now all these pixel values need to be encoded. For each cell, we can have multiple values. If an image is strictly Black and White, then we can simply encode the values nearing Black to 0 and the values nearing White to 1. If not then we can keep the values in their 8-bit binary form. Once we have accomplished encoding the matrix, we will now perform permutations. The following enlists the proper process to accomplish it:

- The random permutations used here is accomplished by using the Fisher–Yates Shuffle Algorithm also known as the Knuth Shuffle algorithm to shuffle the finite sequence of Numbers.
- Shuffle the elements present in the *i*th row with each other by selecting randomly a number and swapping it with the first index.
- The number that was selected is thus stroked out and is not taken into consideration.
- The same process is repeated for all the elements in a single row and in all the rows in the matrix.
- The above can also be accomplished using columns instead of rows, but whichever convention is chosen needs to be followed.

When the plain text image is a greyscale image, the pixel values in the image is constructed depending upon the amount or intensity of light. It carries only intensity information. The pixel values in this matrix comprise an 8-bit binary value which indicates different shades of grey. These values are stored in the W * H matrix, where W = width and H = height. All the bits of the 8-bit binary value of each and every pixel in the image are permuted with each other using the method described above.

The process explained above takes about $O(n^2)$ complexity which will exponentially increase if more colours are to be encoded and is taken into consideration.

4.4 Key-Generating Algorithm

The image obtained above is used to generate the key required to be used for the purpose of Encryption. The image obtained above is superimposed with the original image to give us another image. This image can be then be superimposed with the PPP image to give us the Original image.

Superimposing can be described as the process of combining multiple images and overlaying them on top of each other. Superimposition of two-dimensional images can cause the production of moiré patterns. Superposition of parallel and correlated layers gives rise to line moiré patterns. When Superimposition takes place over two identical patterns, there may arise randomly selected patterns at a small scaling difference or available at a steep angle. All the above factors may cause a distortion in the image, and hence needs to be dealt with. The glass moiré and shape moiré patterns may also arise due to the complex shapes that are sometimes created. Once we get this image, then we can use this image as our key for the decryption of the encrypted image obtained after performing the PPP process.

The encrypted PPP image obtained from the above step, and the plaintext image obtained is used to generate the key used for the decryption process. The key is generated using each of the 8-bit binary value of the image complementing them with each of the 8-bit binary pixel value of the PPP image. This key obtained can thus be used for the purpose of decrypting the greyscale image.

4.5 AES Encryption of the Key and the PPP Image

Another step of security is added by encrypting the key and the PPP image with AES encryption technique, the most widely known encryption technique which has been found to be much faster than the earlier known standard DES. The AES encryption algorithm is used here as an added layer of defence which is not essentially necessary but encouraged for much greater security essential for highly sensitive data. The AES encryption algorithm is best defined as a symmetric block cipher encryption developed by using a Rijndael Cipher comprising of 128, 192 and 256-bit keys. It is among the most secure software available in the industry today. We use AES algorithm to encrypt the key and the Image obtained after performing encryption via the PPP method. This would enable us to secure the data already secured in a more prolific manner.

4.6 *Decryption Process*

The keys and the image encrypted through the PPP technique needs to be decrypted back to obtain the original image. The decryption process is very similar to the encryption process.

The encryption of the visually encrypted image through AES algorithm gives rise to a key. This unique key varies from time to time, and this key is used to decrypt the image back to the visually encrypted format. The process of decryption of an AES ciphertext is similar to the encryption process in the reverse order. Since sub-processes in each round are in reverse manner, unlike for a Feistel Cipher, the encryption and decryption algorithms need to be separately implemented, although they are very closely related. The key is also simultaneously decrypted back to give us the key image. This key along with the visually encrypted image via PPP is superimposed on each other to give us the original image. However, while superimposing these images we tend to get some noise in our image in the form of moiré patterns. Thus, we use a sophisticated denoising algorithm to remove that unwanted noise and give us the resultant image.

If our plaintext image is greyscaled, then superimposition will not work. The image can be converted to a Black and White Image which can later be converted greyscale but there is a better approach. During the random permutations instead of permuting the 8-bit binary value with each other, we can permute the 8-bit binary values of an index with each other bit value. This random permutation can be recovered by superimposing the values of the key image obtained in the step number 4.4.

5 Cryptanalysis

Cryptanalysis is the process of analysing the information systems to study the hidden aspects of the systems. They are used to breach the cryptographic security systems and breach the hidden encrypted material even without knowing the key. Mathematical analysis on the other hand analyses the cryptographic algorithm for the study of side-channel attacks that take into consideration not the algorithm as its own but also its implementation. Some of these techniques are discussed below.

5.1 *Brute Force Attack*

In a brute force attack, the algorithm is implemented as a black box and we try all the possible combinations of the key until the correct key is developed. The complexity of such an attack depends on the key-generating algorithm and the size of the key.

Key Size: the maximum size of the key obtained depends on the encoding performed as described in the module 4.2. Let us take the best-case scenario in which the image is converted into Black and White. This encoding is performed on a 200 * 200 pixelated image. Now that means there are 40,000 pixels in our image. Depending on the random permutation of these pixels, we get that each pixel could have either a value of 0 or a value of 1 and this thing would continue for 40,000 indices. Such a huge complexity would make it impossible to brute force with the current computing ability. Now, during decryption of an RGB image the encoding of the coloured values gives rise to vectors in place of indices. Now trying to identify all these values in each of the 40000 vectors would cause the encryption process to be increased to a much higher complexity.

5.2 Ciphertext-Only Attack

This is the method which uses various different stages for the decryption process and determines the candidates for each stage. A perfect set of candidates can be used to successfully obtain the decrypted process. There can be various methods using which one can establish a ciphertext attack. Reverse Transposition [8] is the process of analysing the transposed vector of the given matrix and using this transformation one can calculate the number of iterations required to calculate the value of the different possible candidates. The iterations will be the factorial of its indices. As reverse engineering of the random permutation cannot be accomplished. Thus, the average number of decryptions needed to get the correct solution would still be half of the total number of combinations.

5.3 Chosen Plaintext Attack

The chosen plaintext attack [9] is the type of attack in which the cryptanalyst has the idea and the knowledge of the encryption algorithm which uses the random permutations (RP). Even after the knowledge of these random permutations is not useful for the process of analysis as the analyst cannot significantly determine the correct value of a particular index value by reverse engineering the algorithm. Hence, this strategy will not be an efficient way of decrypting the encrypted image.

The basic method at disposal of the cryptanalyst is the encryption algorithm. Thus, he takes various plaintexts and subjugates them to the encryption algorithm to try and understand the algorithm in an effort to find and establish the key. He takes into account various different scenarios and different plaintexts until he is in a position to make a calculated guess on the Key obtained. However, as the encryption process is established using the method of random permutation using the Fisher–Yates Shuffle Algorithm the process is much more time dependent and using the same plaintext different keys are obtained thereby restricting the analyst to analyse and reverse engineer the algorithm. Thus, even after knowing the encryption algorithm it becomes effectively difficult for the person to establish a common link between the number plaintext and the key obtained and how to use the two to get the decrypted image.

5.4 Known Plain Text Analysis

In this analysis, we assume that the cryptanalyst has the knowledge of the plaintext and the ciphertext. He knows the width W and the height H of the image matrix. He even knows the encoding criteria, i.e., whether Black and White or RGB and also knows the way the image was encoded. Thus, because of knowing these things performing reverse transposition took about half the factorial of the total number of iterations. However, here the number of candidate keys obtained is significantly decreased. However, due to the large values of the width W and the height H the value of the iterations are still quite large for easy extraction and decryption.

6 Simulation Results

This section provides the basic simulation results obtained in trying to accomplish the task of encryption. A basic image was taken in our simulation and converted to a Black and White image. This Black and White 'water droplet' image is compressed to a size of 200 * 200 pixels. This image was now used to perform PPP on it. The PPP Encrypted Image along with the key is shown below. This experiment was constructed on a 2.25 GHz Intel Core i5 Second Generation with 5 GB Memory and has been depicted in Fig. 3. Figure 3a depicts the image, and Fig. 3b

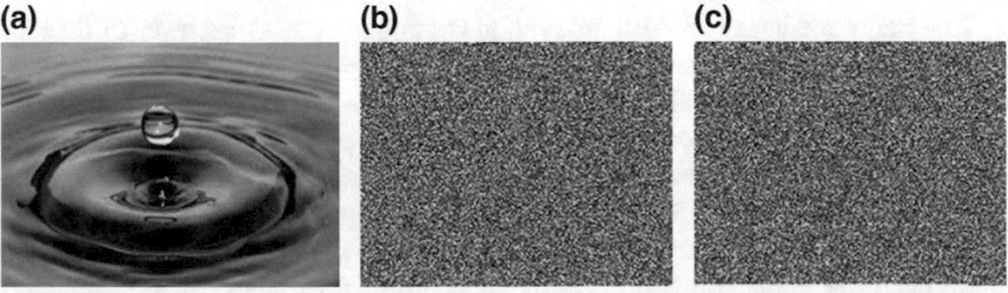

Fig. 3 'Water Droplet' plain image (**a**), the PPP Encrypted Image (**b**), the key generated (**c**) during a simulation

Table 1 Decryption and encryption time as the size increases

Image	Compressed (Black and White) image size	PPP image size	Encryption time (in s)	Decryption time (s)
1	200 * 200	200 * 200	5.1	2.4
2	300 * 300	300 * 300	6.5	3.7
3	400 * 400	400 * 400	42.3	16

depicts the PPP image obtained. Figure 3c depicts the key obtained by following the key-generating algorithm. The algorithm was carried out on a Java Platform with NetBeans as the IDE.

The simulation time for all these experiments with varying size of the image has been described in Table 1. It has the dimensions of the plaintext image and the ciphertext image along with the encryption time (ET), decryption time (DT) taken for performing each of those random permutations.

7 Conclusion and Future Work

This paper streamlines on the basis of privacy and data security. It embodies itself as privacy and data security being a fundamental right of the citizens. Thus, in this project a model image encryption technique has been illustrated that has been at the foundation of this project all along. The image encryption practiced under this project has been accomplished using Pixel Property Permutation (PPP).

The basic idea behind the PPP is that any digital image is made up of pixels small atomic units having a value and showcasing a colour. These pixel values are then permuted with each other so as to divide the entire report into different shares which in turn would give us nothing but a hazy image impossible to decipher. Thus, this is a major advantage and boost in the era of visual encryption. These techniques when used for the proper data encryption procedure enabled us to save the data in a more secure form than many of the technologies available to man. The proposed algorithm is secured against brute force, ciphertext-only attacks, plaintext attacks and chosen plaintext attacks.

This technique has used AES encryption technique as an outer layer of defence, and it uses this defensive measure to make it impenetrable via brute force or cipher attacks. Thus, these methods made the encryption more secure.

This algorithm works perfectly well when the image is converted to a Black and White image. However, once the image is taken and is to be stored as an RGB image it assumes it to be a 24-bit image. This results in losing some of the features of the image and might not give us the lossless image decryption. However, in future once the implementation becomes more robust theoretically we should be able to perform easily taking much less time the algorithm of Pixel Property Permutation.

References

1. Roeder, T.: Symmetric-Key Cryptography
2. Bernstein, D.J.: Cache-timing attacks on AES (2005)
3. Bogdanoy, A., Khovratovich, D., Rechberger, C.: Biclique Cryptanalysis of the Full AES
4. El Assad, S., Farajallah, M., Vladeanu, C.: Chaos-Based Block Ciphers: An Overview (2011)
5. Stanisavljevic, Z.S.: Data Encryption Standard Visual representation (1997)
6. Lee, S.-S.: Phase Masking Visual Cryptography using Interferometer (2008)
7. Schmeh, K.: Cryptography and Public Key Infrastructure on the Internet, p. 45 (2001)
8. Iyer, K.C., Subramanya, A.: Image Encryption by Pixel Property Separation
9. Koh, M.-S., Rodriguez-Marek, E., Talarico, C.: A Novel Data Dependent Multimedia Encryption Algorithm Secure Against Chosen Plaintext Attacks. In: ICME 2007 (2007)

Social Context Based Naive Bayes Filtering of Spam Messages from Online Social Networks

Cinu C. Kiliroor and C. Valliyammai

Abstract Nowadays, online social networking (OSN) sites are an inevitable way of communication. Almost all the OSNs have attained an explosive growth in the last few years. Spammers or hackers found that OSN is an important and easy way to spread spam messages over the network because of its popularity. Spammers use different strategies to spread spam. So, spam detection must be strong enough to detect spam effectively. Though several spam detection techniques are available, it is necessary to increase the accuracy of spam detection techniques. In this paper, a spam detection technique is proposed to detect and prevent spam messages. In addition to the usage of basic classifiers, social context features such as shares, likes, comments (SLC) are also done. Experimental evaluations and comparisons prove that the proposed system with SLC factors provides a higher accuracy than accuracy of basic classifier.

Keywords Naive Bayes · Social networks · Social context features
Spam filter

1 Introduction

Rapid growth happens in the field of OSNs such as Facebook, LinkedIn, Twitter. The large number of users and the exposure to the outside world are the two reasons for the unpredicted growth of the OSNs. Several meritorious things happen in the field of social networks, but neglecting the demerits is not a good solution toward the disadvantages in the social networks. The most irritating thing happening in the

C. C. Kiliroor · C. Valliyammai (✉)
Department of Computer Technology, MIT, Anna University,
Chennai 600044, India
e-mail: cva@annauniv.edu

J. Nayak et al. (eds.), *Soft Computing in Data Analytics*,
Advances in Intelligent Systems and Computing 758,
https://doi.org/10.1007/978-981-13-0514-6_66

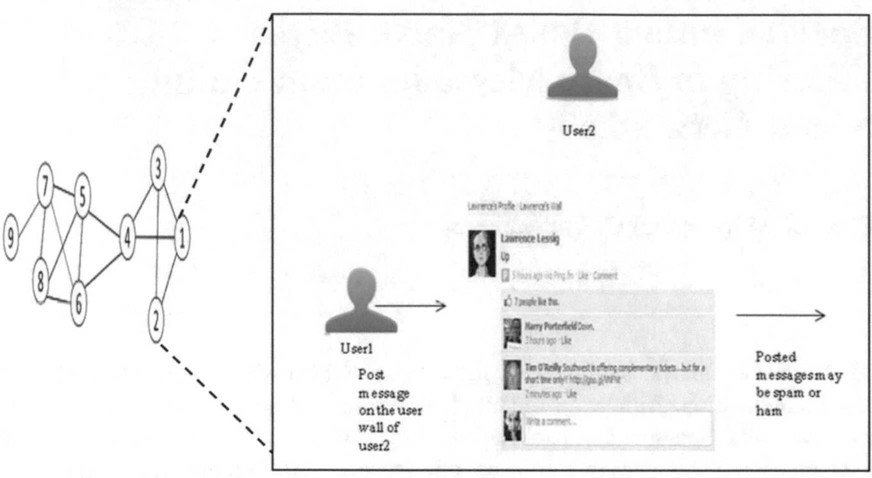

Fig. 1 Message sending from user 1 to user 2 in OSN

social networks is sending and receiving a lot of unsolicited messages over the networks. A lot of fake or unwanted messages are roaming over the networks, and this will lead to several social issues. Several methods exist to filter the spam messages. But there is no effective reduction in the number of spam messages in the social networks. An OSN is nothing but a connected graph with nodes and edges. Figure 1 shows the detailed communication scenario between two users from the entire social network graph. The user 1 in the network sends message to user 2, and the posted message in the user wall of the user 2 may be spam or not spam. Several methods such as analysis of the URLs, short text classifiers to classify short messages, classification based on the analysis of the relations in the social networks are currently available. None of the methods are very effective in the spam filtering in social networks. The proposed model is used to filter the spam messages from the OSN user walls by considering social context features in addition to the basic spam classifier. The remainder of this paper is arranged as follows. Section 2 contains the related work, Sect. 3 contains the proposed methodology, Sect. 4 contains the experimental results and evaluation, and the final section summarizes and concludes the results.

2 Related Work

Different types of unsolicited messages spread by spammers and several spam detection techniques are also famous. Mapping URLs from starting to end was described in [1]. The URL analysis was taken place in which filtering of unwanted

URLs takes place. Certain patterns with URLs are identified, and filtering happens in [2]. Posts or comment spam in social networks [3] was detected by analyzing features such as the similarity between post and comments, length of comments, interval between the posts and comments. Spammers were identified using various techniques like spam comments detection through knowledge model comparison and spam user detection using single attribute [4]. Different techniques were there to detect anomalies, namely behavior-based techniques, structure-based techniques (using graphs), and spectral-based techniques [5]. In behavior-based techniques, content-based filtering was done where anomalous behavior was detected by analyzing the internal content of sent and received messages. Some conventional approaches were there to detect spam in OSNs [6]. Some of them include co-classification framework and social tagging systems. Classifiers were built to detect comment spam in social networks [7]. Tangram was a spam filtering system that performs an online inspection on a stream of user-generated messages [8]. Spam can be detected in two ways, one by analyzing social network and the alternative is by extracting the user information from social attributes and textual contents [9]. Content-based filtering was done to filter out unsolicited texts. Posts were filtered after determining the social contexts and relationships. Filtering was based on the spam templates. Web pages might contain spam data [10]. So those data should be identified and filtered out efficiently. Contents of Web pages were analyzed, and then spam data were filtered out.

Researchers proposed a technique called personalized fine-grained filtering (PIF) with privacy preservation. PIF consists of: social-assisted filtering distribution, coarse-grained filtering, fine-grained filtering, filter authentication, and update scheme. Spam messages can be detected using some patterns [11]. The new spam is added to the dictionary if the new words are identified as spam. A self-extensible spam dictionary was used for this purpose. The words dictionary might cause negative effects to users and aim at promoting the purchase of luxury overseas. Super sequences were used to create templates for spam messages [12]. The problems of finding shortest common supersequences (SCSs) and longest common subsequences (LCSs) are two well-known NP-hard problems. Short text classification differs from traditional text classification. Many techniques have been proposed by researchers to efficiently and effectively classify short texts. Short text classifier might be a hierarchical one [13]. It contains two levels, namely hard classifier level and short classifier level.

Feature sparseness is there in short texts. Some researchers proposed a technique that exploits feature sparseness of short texts [14]. The above classification might increase the complexity. So, it is necessary to classify short texts without adding additional information [15]. Neural probabilistic language model and word distributed representation for short text classification solved this problem. User-to-user relationship plays an important role in OSN structure [16]. The proposed model uses social context features in addition to the basic classifier in order to improve the effectiveness in filtering.

3 Proposed Methodology

The proposed model is used to detect the unwanted or unsolicited messages from the OSN user walls. In addition to a normal Bayesian classifier, the proposed system takes SLC features to calculate the relationship between the sender and the receiver.

The workflow of the proposed system is shown in Fig. 2. The proposed filtering system gets active whenever the user tries to post message in another user's wall. Most of the spam detecting systems do not consider the relationships between the people involved, it just analyzes the messages and checks whether the message is spam or not by using some classifiers. The proposed system controls the overhead of running the spam classifier on obvious spam messages by considering the relationships among the users who are involved. SLC factors are calculated by using the probabilities of likes, comments, shares between the users and the type of relationship between them which is shown in the Algorithm 1: SLC calculation. The calculated SLC values are also included into Bayesian classifier to check whether the incoming messages are spam or not.

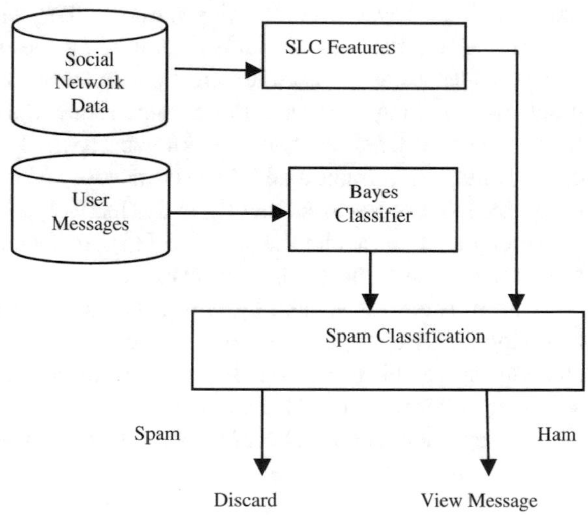

Fig. 2 Workflow of the proposed spam filtering system

Algorithm 1: SLC Calculation

```
Function: slc_calculate (a, b, k)
Variables: P_s(y, z) – probability of shares between y and z
           P_l(y, z) – probability of likes between y and z
           P_c(y, z) – probability of comments between y and z
Input: two users a and b with relationship k
Output: SLC between user a and b

  Begin

    Relationship p[]={1,0.5,0.1}
    SLC = p[k] *(P_s(y, z) + P_c(y, z)+ P_l(y, z));
  End
```

Spam classification involves reading the message and checking it for spam. The message gets filtered before the user receives it, if the message contains spam. Otherwise, it is left as it is. Classification is the task of choosing the correct class label for a given input. This helps to assign a label to each type or class of objects to differentiate their attributes. A text classifier is used, because the messages that are forwarded, shared, or sent in social media are often shortened and not in the natural language form. The classification of spam or ham is given in the Algorithm 2: Test spam/ham.

Algorithm 2: Test spam/ham

```
Function: test()
Variables: us_p – User who posts the message
           us   – User that receives the message
Input: Test Set
Output: Reports messages as spam or ham
Begin
    Get us_p, us and message
    SLC= slc_calculate(us, us_p, relation)
        P=( w*SLC+z*message_classify)/(2)
    if (P<threshold) then
        report spam
    else
        report ham
    end if
End
```

4 Experimental Results and Evaluation

The proposed system is used to detect and reduce spam messages in an OSN user's wall. Though many spam detection techniques are already available, it is necessary to increase the accuracy of spam detection techniques in an effective way. Existing spam filtering techniques are based on normal classifiers. Hence, the objective of the proposed system is to create an effective filtering system by adding several parameters into the basic classifiers. The dataset used is a collection of messages which contains friends of a user, likes, comments, and shares for the post of a particular user which are compiled from different sources. The spam dataset is taken from the Web site csmining.org, and this is a repository which consists of various spam datasets. The relationship dataset consists of friends of a particular user, relationship with that user, and likes, comments, and shares for the posts of that user. The dataset is collected from different sources. Likes, comments, shares, and relationships are the attributes taken for calculation of SLC, and it is taken from UCI repository. SLC is calculated using the formula which is given in Eq. 1:

$$\text{SLC} = \text{p}[\text{k}] * (\text{Rp}[0] * \text{P s}(\text{x}, \text{y}) + \text{Rp}[1] * \text{P c}(\text{x}, \text{y}) + \text{Rp}[2] * \text{P l}(\text{x}, \text{y})) \quad (1)$$

where Rp is the reward point, and p is the relationship weightage array. To classify the messages, Naïve Bayes is used. Using these factors, one can predict whether the user is a spammer or not. In spam analysis, the dataset is used to train the data and get the accuracy. The accuracy can be calculated using the formula which is given in Eq. 2:

$$\text{accuracy} = \frac{\text{number of messages correctly classified}}{\text{total number of messages}} \quad (2)$$

The number of correctly found spam can be identified by checking the result of test set against the already known results. NB with SLC provides a high accuracy of 91.18% when compared with the basic classifier, and it is clearly mentioned in Fig. 3. It clearly states that NB with SLC provides a higher accuracy than NB. The true positive and true negative rate of the proposed system is shown in Fig. 4 and Fig. 5, respectively.

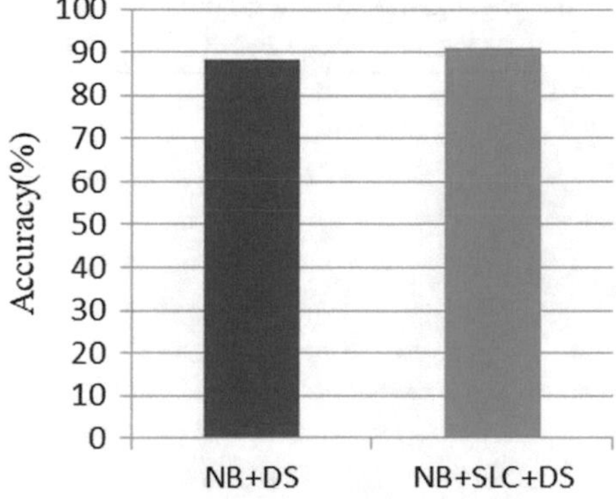

Fig. 3 Accuracy of spam classification

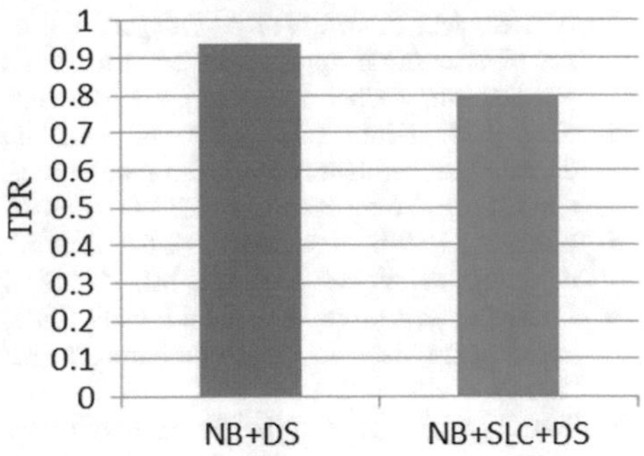

Fig. 4 True positive rate

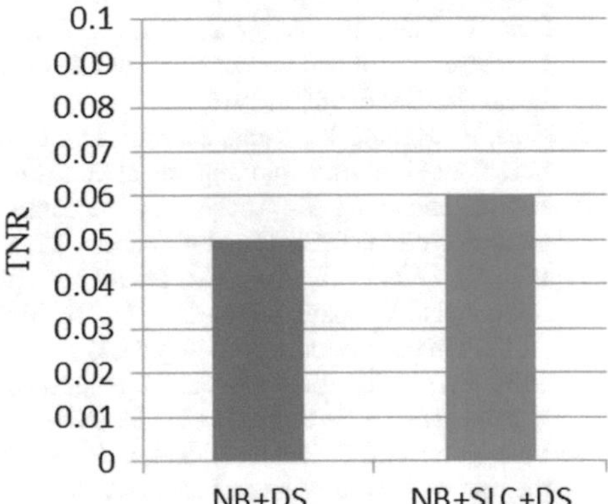

Fig. 5 True negative rate

5 Conclusion

The proposed spam filtering mechanism considers social context features in addition to the spam classification with basic NB classifier. The experiment results prove that NB with SLC factors provides a higher accuracy than compared with basic NB. Future work envisions taking additional social context features from the OSNs with spam template matching in order to increase the accuracy.

Acknowledgements This work is financially supported under grants provided by the Visvesvaraya Ph.D. Scheme for Electronics and IT.

References

1. Wang, D., Pu, C.: BEAN: a behaviour analysis approach of URL spam filtering in Twitter. In: International Conference on Information Reuse and Integration, San Francisco, CA, pp. 403–410 (2015)
2. Chen, C., Zhang, J., Xie, Y., Xiang, Y., Zhou, W., Hassan, M.M., AlElaiwi, A., Alrubaian, M.: A performance evaluation of machine learning based streaming spam tweets detection. IEEE Trans. Comput. Soc. Syst. **2**, 65–76 (2015)

3. Alsaleh, M., Alarifi, A., Al-Quayed, F., Al-Salman, A.: Combating comment spam with machine learning approaches. In: 14th International Conference on Machine Learning and Applications (ICMLA), Miami, pp. 295–300 (2015)
4. Kim, J.M. Kim, Z.M., Kim, K.: An approach to spam comment detection through domain-independent features. In: International Conference on Big Data and Smart Computing (BigComp), Hong Kong, pp. 273–276 (2016)
5. Kaur, R., Singh, S.: A survey of data mining and social network analysis based anomaly detection techniques. J. Egypt. Inf. **17**, 199–216 (2016)
6. Chakraborty, M., Pal, S., Ravindranath Chowdary, C., Pramanik, R.: Recent developments in social spam detection and combating techniques. J. Inf. Process. Manag. **52**, 1053–1073 (2016)
7. Yin, R., Wang, H., Liu, L.: Research of integrated algorithm: establishment of spam detection system. In: 4th International Conference on Computer Science and Network Technology (ICCSNT), Harbin, pp. 584–589 (2015)
8. Zhu, T., Gao, H., Yang, Y., Bu, K., Chen, Y., Downey, D., Lee, K., Choudhary, A.N.: Beating the artificial chaos: fighting OSN spam using its own templates. IEEE/ACM Trans. Netw. **24**, 3856–3869 (2016)
9. Wua, F., Huang, Y., Yuan, Z., Shu, J.: Co-detecting social spammers and spam messages in microblogging via exploiting social contexts. J. Elsevier Neuro Comput. **201**, 51–65 (2016)
10. Hua, J., Huaxiang, Z.: Analysis on the content features and their correlation of web pages for spam detection. IEEE China Commun. **12**, 84–94 (2015)
11. Liu, C., Wang, J., Lei, K.: Detecting spam comments posted in micro—blogs using self-extensible spam dictionary. In: IEEE International Conference on Communications (ICC), Kuala Lumpur, pp. 1–7 (2016)
12. Song, C., Ge, T.: Window-chained longest common subsequence: common event matching in sequences. In: 31st International Conference on Data Engineering, Seoul, pp. 759–770 (2015)
13. Vairagade, A.S., Fadnavis, R.A.: Automated content based short text classification for filtering undesired posts on Facebook. In: World Conference on Futuristic Trends in Research and Innovation for Social Welfare (Startup Conclave), Coimbatore, pp. 1–5 (2016)
14. Gao, L., zhou, S., Guan, J.: Effectively classifying short texts by sparse representation with dictionary filtering. Inf. Sci. J. Elsevier **323**, 130–142 (2015)
15. Yao, D., Bi, J., Huang, J., Zhu, J.: A word distributed representation based framework for large scale short text classification. In: International Joint Conference on Neural Networks (IJCNN), Killarney, pp. 1–7 (2015)
16. Cheng, Y., Park, J., Sandhu, R.: An access control model for OSN using user-to-user relationships. IEEE Trans. Dependable Secure Comput. **13**, 424–436 (2016)

Data De-duplication Using Cuckoo Hashing in Cloud Storage

J. Sridharan, C. Valliyammai, R. N. Karthika and L. Nihil Kulasekaran

Abstract Cloud computing facilitates on-demand and ubiquitous access to a centralized pool of resources such as applications, networks, and storage services. Redundant copies of the same data are stored in multiple places, thus occupying more space in servers. Recent increase in computing leads to enormous volume of data, which are backed up and stored in cloud and made available to address consent, real-time insights and in regulating the data. In order to address the above problem, an enhanced cuckoo hashing algorithm is proposed for identifying duplicate data. Cuckoo hashing performs insertion, deletion, and retrieval in constant time. The metadata of files such as the basic file attributes, user-defined attributes, and user principal owner attributes is preserved after de-duplication. The experimental results show the efficiency of cuckoo hashing in the de-duplication of data chunks in the cloud environment.

Keywords Data de-duplication · Cuckoo hashing · Compression
Metadata

J. Sridharan · C. Valliyammai (✉) · R. N. Karthika · L. Nihil Kulasekaran
Department of Computer Technology, MIT Campus, Anna University, Chennai 600044, Tamil Nadu, India
e-mail: cva@annauniv.edu

J. Sridharan
e-mail: jjsridharan@gmail.com

R. N. Karthika
e-mail: karthirn@mitindia.edu

L. Nihil Kulasekaran
e-mail: nihilkulasekaran@gmail.com

J. Nayak et al. (eds.), *Soft Computing in Data Analytics*,
Advances in Intelligent Systems and Computing 758,
https://doi.org/10.1007/978-981-13-0514-6_67

1 Introduction

Today, cloud computing is growing popular in terms of computing capability and storage [1]. Cloud data centers are awash in digital data, easily possessing petabytes and exabytes of information, and the complexity of data management escalates in big data. Cloud storage is a popular service model which maintains, manages, backs up data remotely, and makes it available to users over a network connection. Millions of users are now storing their data in a cloud environment. There are lots of issues to be focused on the modern cloud storage which include security to data, performance measures, and providing a user-friendly environment. The data integrity of the files has to be ensured along with the preservation of data attributes of the file.

Data de-duplication [2] has become a key component in modern backup systems due to its demonstrated ability in improving storage efficiency. Data de-duplication [3] is a form of compression in which repeated chunks of information are identified and stored once, to save memory space in a cloud environment. Cuckoo hashing [4] is a technique, which is used in computer programming for resolving collisions of keys, with worst-case constant lookup time. A cuckoo hash-based de-duplication method is proposed in this paper, and performance is compared against standard Java hash table. This paper is organized as follows: Section 2 explores the literature survey which briefs the existing work in de-duplication and cloud storage. Section 3 describes proposed method with upload and download files algorithm. In Sect. 4, implementation and results are explored. Section 5 provides conclusion of the work carried out in this paper and future enhancements.

2 Literature Survey

Cuckoo hashing usually is implemented with O(1) retrieval time and constant time for insertion [5]. But, cuckoo hashing suffers from infinite loop problem during insertion. To overcome infinite loops during data insertion, an efficient approach was proposed in which the data is selected from un-busy kicking-out routes [4]. Un-busy kicking-out data units are determined by the frequency it is accessed with a counter for each value in the hash table. The fine-grained locks were implemented to improve the concurrency. Space overhead is incurred as every value uses a counter variable [4]. This method is called as Min Counter method as they have selected data unit based on the frequency of elements which is minimum. Cuckoo hashing is now being used in modern switches [6] to find the path to route in constant time. Cuckoo hashing is used in applications where more retrieval information is present. De-duplication-assisted cloud-of-clouds (DAC) [6] supports the following features in client side, data de-duplication, data distribution, performance evaluation, and cost evaluation. Data de-duplication takes care of identifying duplicate data using SHA1 or MD5 hash value.

Fixed length block-level de-duplication reduces the probability of finding the duplicates. Data de-duplication is special form of compression [7]. For identification of identical chunks, a single pointer is sufficient whereas for chunks that are similar different approaches were followed. For identifying similar chunks, a cubic time complexity method with quadratic space was proposed for storing and retrieving the information. Based on the information that is already present, new data and its pointer length were decided which ensures dynamic block-level de-duplication.

Attribute-based sharing is widely used in environments where data providers outsource their data to cloud service providers in encrypted format [8]. Dedupe storage nodes are responsible for storing raw data. Directory provides the logical separation among data. Clients data is separated using directory and stored in dedupe storage node. In content-defined chunking, files were split into fixed and variable length blocks. For variable length blocks, cut points were decided based on features of file. Computational complexity of deciding cut points is high. Rapid Asymmetric Maximum (RAM) [9] based on asymmetric chunking algorithm was discussed. Instead of using hashes, RAM uses bytes value to declare the cut points. In post-processing de-duplication algorithm, users' uploaded file will be saved in server and then subjected to de-duplication algorithm. History-aware inline de-duplication algorithm (HIDC) [10] described about encoding and decoding can be applied over compressed data.

3 Proposed Work

Data duplication is a problem in cloud storages where same or different users upload the same data pattern to their storage. Duplication of the same data again and again just in different names of storage space or different file names is simply waste of valuable storage space. The proposed system provides a suitable method to de-duplicate cloud storage such that the performance is enhanced and storage space is saved, and it is shown in Fig. 1.

Since all the users may provide duplicated data, there arises a need to store all the data together. A separate folder is visualized as cloud environment where each user is considered as a subdirectory. Let the name of the folder be "Cloud" and "$User_1$", "$User_2$" and up to "$User_n$" be the user's subdirectories under the "Cloud" directory. The user stores all the files under the user subdirectory in the Cloud folder, and during the de-duplication process, the data from all the users irrespective of the file type should be stored centrally so that the data can be de-duplicated. The de-duplication will be applied on entirely on "Cloud" folder. That is once the user uploads all the files to the folder, de-duplication process is run once for all the files in that folder and duplicated data is removed. The process supports all types of files and is entirely done on chunk level. Since all type of files can be treated as binary data, all files can be de-duplicated using this process. De-duplication can be done both in file-level and on a chunk-level. Here, we do it on a chunk-level basis as it

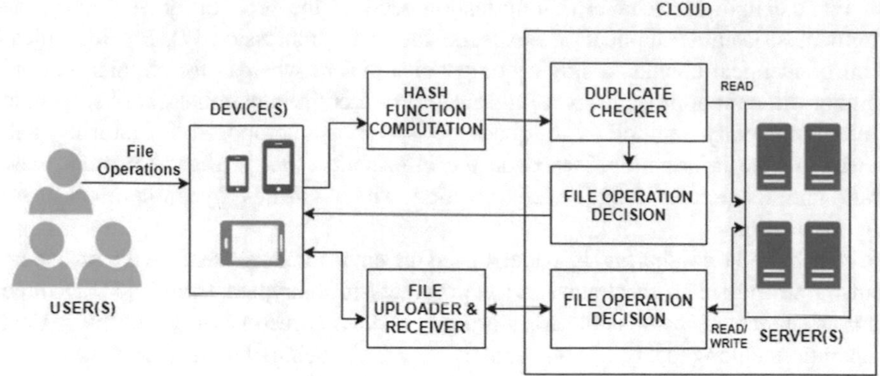

Fig. 1 Architecture diagram for proposed system

provides better de-duplication as the level becomes finer and finer. Fixed-sized chunks of data are used as data de-duplication units rather than going for the entire file. As a user uploads a file of size 24 MB, if the chunk size is about 4 MB, then file is divided into six parts, and rest of the de-duplication process is done. The entire chunks are stored in a centralized file as it is present in the user file. This file is a large file as it stores all the data of all the user files. The data in this file should not be duplicated. That is this file store should only contain unique data so that the process is achieved. In order to achieve that, hashing-based scheme is adopted. The duplicated chunks are to be identified among all the chunks. The chunks are to be matched with the fore coming chunks so that we may know which part is duplicated. But comparing large chunks of saying size 4 MB every time is a tedious process and is not encouraged given the performance should be higher and not much time can be placed in the de-duplication process as user satisfaction is more important.

So, the chunks are hashed and a 48-bit hash value is generated for each chunk of the specified size. A good key generation algorithm is to be chosen so that the keys do not collide much. As the hash is generated, a hash map needs to be maintained so that the hashes generated by the hash function are stored separately. The hash table must be so efficient and use optimal space. Cuckoo hashing is preferred in this case as it provides constant time complexity. As each hash is generated by the function, it is to be compared with the previous value right away. The granularity of the chunk decides how efficient this algorithm is going to be. If the chunk size is chosen to be very small, then the de-duplication efficiency will be higher as for each smaller chunk size definitely a large part of the original data may be found duplicated. But taking large chunk size may or may not give better results as compared to finer chunks. The problem of varying chunk sizes should also be dealt with as it cannot be said sure every time that the duplicate chunk exists from a starting point and an ending point in the file. The process of checking the hash with the previous values of hash map is only for storing non-duplicated data in the central store, whereas each corresponding file must store all the hash values of the chunk in that file.

Algorithm 1: Upload Files Algorithm

```
Input  :Pathname of the list of files to be uploaded
Output :True if files uploaded successfully, else false

  Procedure UPLOADFILES(filelist)

  hl ← Ø
  N  ← filelist
  cs ← 4MB
  si ← 0
  M←Create Hash Map(fn,si,li)
  begin
    for each fd in N
        fw open(fd.src)
         while  fd!=EOF
        Chf ← read(fd,cs)
        si  ← si + cs
        li  ←length(Chf)
        hf  ← Hf (Chf )
        append(fw , hf)
        add hf to hl
        if ∃hf ∈ M
          write(fw, hf)
        else
          M(hf)  ← (fd,si,li)
          write(fw, hf)
        end if
      end while
        Storeattributes(fd,fw)
        Upload(fw)
    end for
    rl ← Upload(hl)
    ft ← open("data.txt")
    for each  hk in rl
        fr,si,li ← M(hk)
        fd ← open(fr)
        seek(fd,si)
        Chf ← read(fd,li)
        append(ft, Chf)
    end for
        Upload(ft)
  end
```

The file details corresponding to each user should be preserved using basic file attributes, user-defined attributes, and user principal owner attributes as though data may get de-duplicated, each file is different in its permissions, ownership, creation time, etc. The time complexity of the Algorithm 1: UPLOADFILES is linear as the content of the files is read at most twice if there is duplication. Each operation within two for loop takes constant time, which results in linear time complexity. Space complexity of the algorithm is constant, as there is no extra space needed for finding duplicated data. Similarly, for the Algorithm 2: DOWNLOADFILE, the time complexity is linear. Each file takes linear amount of time, and thus for downloading "n" files, it will take at most "n" times the size of file as time complexity. Space complexity for DOWNLOADFILE can be given as the space required for reconstructing "n" files, as these files are formed along with contents central file store. Table 1 shows the list of all notations used in the algorithm. Space used by hash table will be very less when compared with actual data size. For each chunk of size 4 MB, there will be only a 48-byte hash value.

Algorithm 2: Download Files Algorithm

```
Input : Name of list of files to be downloaded.
Output: List of files downloaded.

  Procedure DOWNLOADFILES(filelist)

    N  ← filelist
    cs ← 48 B
    d_e ← "dedupe.txt"
    begin
      f_r ←  read(d_e)
      M ← convert (f_r)
      for each f_d in N
            f_o  ← create(f_d)
            f_w   ← open(src(f_d))
            while f_w != EOF
                  h_f ← read(f_w,cs)
                  d_r,s_i,l_i  ← M(h_f)
                  f_d ←  open(dedupe(d_r))
                  seek(f_d,s_i)
                  Ch_f  ←  read(f_d,l_i)
                  write(f_o,Ch_f)
            end while
            Storeattributes(f_w,f_o)
            FTPtransfer(f_o)
      end for
    end
```

4 Implementation Details and Result

The testbed is created using the dataset provided in [11]. When unzipped, the given dataset has four text files, each ranging around 20 GB of size. We duplicated 30% content of the unzipped text files. Files duplicated are copied into a "Test" folder. Before de-duplication "Test" folder disk space size is 81.4 GB. After applying

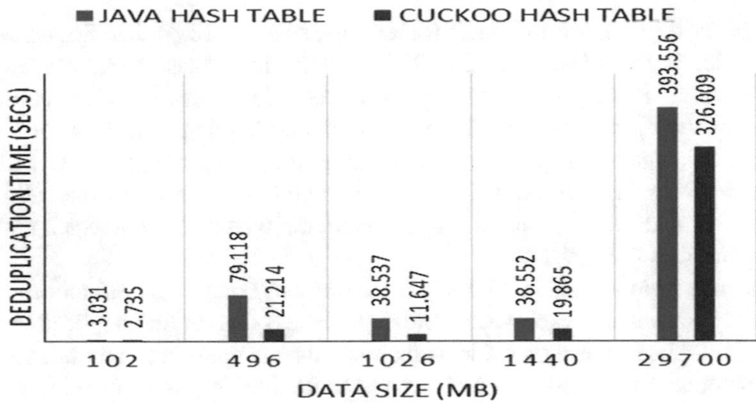

Fig. 2 Comparison chart for different datasets

Table 1 List of notations used in the algorithm

Symbols	Description	Symbols	Description
f_x	File pointer $\forall x\ x \in (a, b, c, \ldots z)$	*Read* (f_x, *len*)	Read *len* bytes from f_x file pointer
M	Cuckoo hash map object	*Write* (f_x,Ch_f)	Writes content (Ch_f) into f_x file pointer
s_i	Start index of *i*th data chunk	*Convert* (f_x)	Converts file pointed by f_x into map object
l_i	Offset of *i*th data chunk	*Open* (*file*)	Opens mentioned file and returns file pointer
N	Input file list	*Copy* (*target*, *source*)	Copy attributes from source to target
cs	Chunk size	*Upload* (f_x)	Uploads the content in f_x to server
h_f	Hash value of the content	*Append* (f_x,Ch_f)	Appends content (Ch_f) to the file pointed by f_x
H_f	Hash function used (MD5 hash)	*Seek* (f_x,s_i)	Moves file pointer f_x to s_i position
Ch_f	Content of chunk	*Storeattributes* (f_d,f_w)	Stores all attributes in f_d to f_w
x_l	Hash list $\forall x\ x \in (a, b, c, \ldots z)$	*FTPtransfer* (f_x)	Transfers the file f_x to the requested client

Table 2 Transmission dataset size after de-duplication

Dataset size (GB)	Amount of data transmitted after de-duplication (bytes)
0.49	18,893
0.80	26,421
1.02	37,043
1.51	45,298
19.0	7,024,768

de-duplication procedure to "Test" folder, disk space is 18.84 GB. So, percentage of space that is consumed is given as 23.14%. The given dataset saves space around 76.86% which is very much higher than other de-duplication algorithms used previously. Also, the proposed system uses cuckoo hashing to retrieve data which makes it efficient in terms of time required to perform de-duplication. Metadata information like last accessed time, last modified time, created time, read/write permissions, and owner of the file is preserved and included in file which makes to retrieve the files correctly.

Java hash table is a Standard Template Library (STL) used as default hash table. Figure 2 shows that cuckoo hash table requires less time to find duplication. Once data is de-duplicated at server side, if the same data is uploaded again from another client, amount of bandwidth required to upload the files is massively reduced which is shown in Table 2.

5 Conclusion and Future Work

The proposed cuckoo hashing-based de-duplication technique identifies the duplicate data with less time and saves bandwidth. Thus, the proposed system saves data storage space to a great extent. Further, the proposed system can be extended to client–server architecture implemented as a simulation of cloud. In the simulated model, de-duplication across users can be applied in real time to determine the amount of space saved. Data integrity can also be ensured using Redundant Array of Inexpensive Disks (RAID)-level storage using Reed–Solomon code.

References

1. Patidar, S., Rane, D, Jain. P.: A survey paper on cloud computing. In: Second International Conference on Advanced Computing & Communication Technologies, Rohtak, Haryana, pp. 394–398 (2012)
2. Aldar, B.D., Devmane, V.: A survey on secure deduplication of data in cloud storage. Int. J. Innov. Eng. Technol. **6** (2015)
3. Vijay Kakde, M., Kadu, N.B.: Survey paper on deduplication data and secure auditing in Cloud. Int. J. Compt. Sci. Inf. Technol. (IJCSIT) **7**, 94–95 (2016)

4. Sun, Y., Hua, Y., Feng, D., Yang, L., Zuo, P., Cao, S., Guo, Y.: A Collision-mitigation cuckoo hashing scheme for large-scale storage systems. IEEE Trans. Parallel Distrib. Syst. **28**, 619–632 (2017)
5. Pagh, R., Rodler, F.: Cuckoo Hashing. Springer, Berlin, Heidelberg (2001)
6. Wu, S., Li, K., Mao, B., Liao, M.: DAC: Improving storage availability with deduplication-assisted cloud-OF-clouds. Future Gener. Comput. Syst. **74**, 190–198 (2017)
7. Hirsch, M., Ish-Shalom, A., Klein, S.T.: Optimal partitioning of data chunks in deduplication systems. Discrete Appl. Math. **212**, 104–114 (2016)
8. Cui, H., Deng, R.H., Li, Y., Wu, G.: Attribute-based storage supporting secure deduplication of encrypted data in Cloud. IEEE Trans. Big Data vol. 1–1 (2017)
9. Widodo, Ryan N.S., Hyotaek, Li, Mohammed, A.: A new content-defined chunking algorithm for data deduplication in cloud storage. Future Gener. Comput. Syst. **71**, 145–156 (2017)
10. Fegade, A., Bharathi, R.D.: cloud idedup: history aware in-line deduplication for cloud storage to reduce fragmentation by utilizing cache knowledge. In: International Conference on Computing, Analytics and Security Trends (CAST), Pune, pp. 244–249 (2017)
11. Carnegie Mellon University: ClueWeb12 Remove Duplicate Records. https://lemurproject.org/clueweb12/ClueWeb12-RemoveduplicateRecords.php. Accessed Jan 2013

A Secure Chaotic Image Encryption Based on Bit-Plane Operation

Dasari Sravanthi, K. Abhimanyu Kumar Patro, Bibhudendra Acharya and Saikat Majumder

Abstract Today security is an issue in the transmission of digital information, basically digital images through the internet. In this paper, we have proposed an image encryption technique based on bit-plane operation by using Piece-wise Linear Chaotic Map (PWLCM) and 2-D Logistic-adjusted-Sine map. At first, bit-plane diffusion operation is performed by using the PWLCM system and secondly, the row-shuffling and column-shuffling operations are performed by using the 2-D Logistic-adjusted-Sine map. Apart from that, the Secure Hash Algorithm SHA-256 is used to update the secret keys of the proposed cryptosystem to resist known-plaintext attack and chosen-plaintext attack. The main significance of this algorithm is bit-plane operation. The bit-plane operation not only confuses the pixels but also diffuses the pixels simultaneously. The simulation results show the better encryption results of the proposed cryptosystem while the security analysis shows the stronger resistivity against the most known common attacks.

Keywords Image encryption · Bit-plane operation · Piece-wise linear chaotic map · 2-D Logistic-adjusted-Sine map · Secure hash algorithm SHA-256

D. Sravanthi (✉)
Department of Information Technology, National Institute of Technology Raipur, Raipur 492010, Chhattisgarh, India
e-mail: sravanthi.dasari1994@gmail.com

K. Abhimanyu Kumar Patro · B. Acharya · S. Majumder
Department of Electronics and Telecommunication Engineering, National Institute of Technology Raipur, Raipur 492010, Chhattisgarh, India
e-mail: abhimanyu.patro@gmail.com

B. Acharya
e-mail: bacharya.etc@nitrr.ac.in

S. Majumder
e-mail: smajumder.etc@nitrr.ac.in

J. Nayak et al. (eds.), *Soft Computing in Data Analytics*,
Advances in Intelligent Systems and Computing 758,
https://doi.org/10.1007/978-981-13-0514-6_68

1 Introduction

In the present world, encryption is an essential criterion for secure communication of digital images. So, efficient encryption algorithm is necessary for secure communication of images. Chaotic systems give efficient encryption algorithm because of its features like sensitivity to system parameters as well as initial conditions, pseudo-randomness, and ergodicity. The chaotic sequence helps to permute and diffuse image pixels efficiently to generate secure encrypted images.

Many chaotic systems based encryption algorithms are developed in recent years. Among them, Wang et al. [1] proposed an encryption algorithm in which simple row and column switching is done using Logistic map. Xu et al. [2] proposed bit-plane based image encryption using PWLCM system. Li et al. [3] proposed a simple encryption algorithm in which image pixels are diffused with tent map chaotic sequence. Chai [4] proposed an encryption algorithm in which bit-planes of image undergo permutation and diffusion using Logistic-Sine system and Logistic-Tent system. Pak and Huang [5] developed a color image encryption algorithm by using combinations of 1-D chaotic maps. Zhou et al. [6] proposed an image encryption algorithm based on new 1-D chaotic maps such as Logistic-Sine map, Logistic-Tent map, and Tent-Sine map. The use of new 1-D chaotic maps provides better security in encryption cryptosystem. Mollaeefar et al. [7] developed a color image encryption scheme in which high-level chaotic maps are used to perform image encryption. This paper also uses chaotic maps to provide higher security in image encryption.

The main contributions includes in this paper are as follows:

- Bit-planes of plain images are utilized to perform image encryption. The use of bit-plane operations not only permutes the pixels but also diffuses the pixels simultaneously. So there is no need of pixel operations separately.
- A large sized pseudo-random key image (same size of the plain image) is utilized to confuse and diffuse pixels in plain images. The large sized key image does more confusion and diffusion of the pixels as compared to the small block based key image.
- SHA-256 hash algorithm is used to update the secret keys of the proposed cryptosystem. This process resists the cryptosystem against known-plaintext attack and chosen-plaintext attack.
- To ensure high confusion among image pixels, a 2-D Logistic-adjusted-Sine map is utilized to scramble the rows and columns of images.

This paper is organized in the following way: Sect. 2 defines the chaotic maps used in this paper. Section 3 presents the key generation and image encryption algorithms. Section 4 presents the simulation results and the security analysis of the proposed algorithm. Finally, Sect. 5 concludes the paper.

2 Chaotic Maps

2.1 PWLCM System

The Piece-wise Linear Chaotic Map (PWLCM) [8] is expressed as follows in Eq. 1:

$$x_{n+1} = \begin{cases} \frac{x_n}{\mu} & if\ 0 \leq x_n < \mu \\ \frac{x_n - \mu}{0.5 - \mu} & if\ \mu \leq x_n < 0.5 \\ 1 - x_n & if\ 0.5 \leq x_n < 1 \end{cases} \tag{1}$$

where μ is the system parameter. The PWLCM has uniform chaotic behavior over the range and so it gives better performance than Logistic map.

2.2 2D Logistic-Adjusted-Sine Map

The 2D Logistic-adjusted-Sine map [9] is expressed as follows in Eq. 2:

$$\begin{cases} x_{n+1} = \sin(\pi \times r \times (y_n + 3) \times x_n \times (1 - x_n)) \\ y_{n+1} = \sin(\pi \times r \times (x_{n+1} + 3) \times y_n \times (1 - y_n)) \end{cases} \tag{2}$$

where r is the system parameter. It has high range of chaotic behavior as compared to Logistic map.

3 Proposed Algorithm

3.1 Generation of Secret Keys

The initial values and system parameters of the chaotic maps are the secret keys in the proposed algorithm. The process involved for updating the secret keys of the proposed algorithm is as follows:

Step-1: Choose a gray scale image *img*.
Step-2: Apply SHA-256 hash function on the input plain image to generate 256-bits of hash values which are then expressed into 64-hex values. The expression of 64-hex values is as follows in Eq. 3:

$$\text{hash values} = imhx_1,\ imhx_2,\ \ldots,\ imhx_{63},\ imhx_{64} \tag{3}$$

Step-3: Then, the 64-hex values are converted into 32-decimal values. They are expressed as follows in Eq. 4:

$$\text{decimal values} = s_1,\ s_2,\ \ldots,\ s_{31},\ s_{32} \tag{4}$$

Step-4: By using these 32-decimal values, the initial values and system parameters of PWLCM system are updated as in Eq. 5,

$$\begin{cases} xp = xp(1) + \left(\left(\frac{mod((s7:s12),256)}{2^9}\right) \times 0.1\right) \\ \mu p = \mu p1 + \left(\left(\frac{mod((s1:s6),256)}{2^9}\right) \times 0.1\right) \end{cases} \tag{5}$$

The initial values and system parameters of 2D Logistic-adjusted-Sine map are updated as in Eq. 6,

$$\begin{cases} xlas = xlas(1) + \left(\left(\frac{mod((s19:s25),256)}{2^9}\right) \times 0.1\right) \\ ylas = ylas(1) + \left(\left(\frac{mod((s26:s32),256)}{2^9}\right) \times 0.1\right) \\ rlas = rlas1 + \left(\left(\frac{mod((s13:s18),256)}{2^9}\right) \times 0.1\right) \end{cases} \tag{6}$$

3.2 Encryption Algorithm

This section presents the process for image encryption. Figure 1 shows the corresponding block diagram to show the encryption process. The steps for encryption process are as follows:

Step-1: We choose a gray scale image *img* of size $m \times n$.
Step-2: The 8 bit-planes of the image are extracted and denoted as $bbp1, bbp2, \ldots, bbp8$
Step-3: The initial values and system parameters of PWLCM system are updated according to the method described in Sect. 3.1. The chaotic sequence of PWLCM system of size m × n is generated by using Eq. (1) and represented in Eq. 7.

$$xp = (xp(1),\ xp(2),\ xp(3),\ \ldots,\ xp(m \times n)) \tag{7}$$

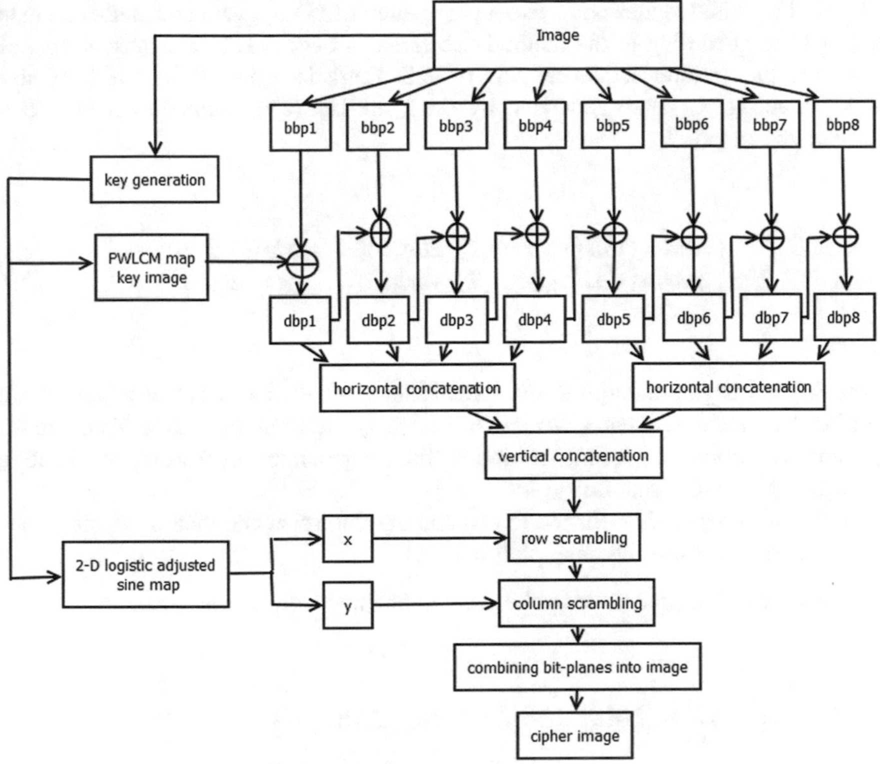

Fig. 1 Encryption block diagram of the proposed cryptosystem

Step-4: The obtained chaotic sequence is converted into key image K of size $m \times n$.
Step-5: The bit-planes of the input plain image are XORed with the key image K which is as follows in Eq. 8:

$$\begin{aligned} dbp1 = xor(bbp1, K),\ dbp2 = xor(bbp2, dbp1),\ dbp3 = xor(bbp3, dbp2), \\ dbp4 = xor(bbp4, dbp3),\ dbp5 = xor(bbp5, dbp4),\ dbp6 = xor(bbp6, dbp5), \\ dbp7 = xor(bbp7, dp6),\ dbp8 = xor(bbp8, dbp7) \end{aligned} \tag{8}$$

Step-6: The obtained first four XORed bit-planes $dbp1$, $dbp2$, $dbp3$, $dbp4$ are horizontally concatenated and the last four XORed bit-planes $dbp5$, $dbp6$, $dbp7$, $dbp8$ are horizontally concatenated and finally both are vertically concatenated as image I.

Step-7: The initial values and system parameters of 2D Logistic-adjusted-Sine map are updated according to the method described in Sect. 3.1. The chaotic sequence *xlas* and the chaotic sequence *ylas* of 2-D Logistic-adjusted-Sine map of size $2 \times m$ and $4 \times n$ are generated by using the Eq. (2) presented in Sect. 2 and represented in Eq. 9.

$$\begin{cases} xlas = (xlas(1), xlas(2), xlas(3), \ldots, xlas(2 \times m)) \\ ylas = (ylas(1), ylas(2), ylas(3), \ldots, ylas(2 \times m)) \end{cases} \quad (9)$$

Step-8: The image *I* undergoes row scrambling using the index values generated by sorting the chaotic sequence *xlas* and forms image *Ir*. Then, the image *Ir* undergoes column scrambling by using the index values generated by sorting the chaotic sequence *ylas* and forms image *Ic*.
Step-9: The image *Ic* is reshaped to original bit-plane size and then these bit-planes are combined to form the encrypted image *C*.

The decryption process is in the reverse of the encryption process.

4 Simulation Results and Security Analysis

This section presents the security analysis and the simulation results of “Lena” image of size 512×512 and “Cameraman” image of size 256×256 by using the simulating software MATLAB R2012a. The initial values and system parameters of PWLCM system and 2D Logistic-adjusted-Sine map are taken as $xp =$ 0.205432670203452, $\mu p =$ 0.302364574565767, $xlas =$ 0.608464853256365, $ylas =$ 0.044504744652543, $rlas =$ 0.857729492676456. Figure 2 shows the results of simulation by using the proposed cryptosystem. As observed from Fig. 2, the plain images are properly encrypted and the encrypted images are properly decrypted by using the proposed cryptosystem.

4.1 Key Space Analysis

The secret keys used in this proposed cryptosystem are:

- The initial value and system parameter of PWLCM system and 2-D Logistic-adjusted-Sine map,
- The 256-bit hash value.

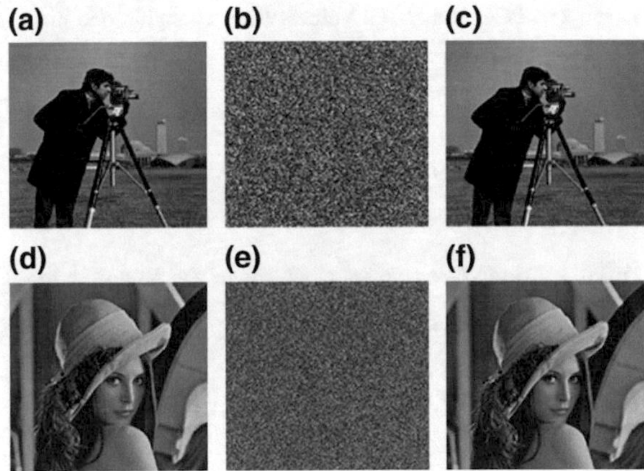

Fig. 2 **a**, **d** Plain "Cameraman" and "Lena" images; (**b**), **e** Corresponding encrypted images of (**a**), (**d**) respectively; (**c**), **f** Corresponding decrypted images of (**b**), (**e**) respectively

The key space of an encryption algorithm should be larger than 2^{128} to resist brute-force attack [10]. Total key space of the proposed cryptosystem is $\left(10^{15} \times 10^{15}\right) \times \left(10^{15} \times 10^{15} \times 10^{15}\right) \times 2^{128} = 1.1038 \times 2^{377}$ which is large enough to evade brute-force attack.

4.2 *Key Sensitivity Analysis in the Encryption and Decryption Process*

Table 1 shows the NPCR and UACI results of "Cameraman" and "Lena" images by changing key *rlas* to to $rlas + 10^{-15}$ in encryption and decryption processes by using the same proposed algorithm. By observing Table 1, we can see that 99% of pixels are changed by changing only one key. This proves the higher sensitivity of keys in the proposed cryptosystem.

4.3 *Plaintext Sensitivity Analysis*

Table 2 shows the NPCR and UACI results when one pixel changes in the plain images by using the proposed cryptosystem. By observing the results in Table 3, we can see that 99% of pixels are changed by changing only one pixel value in the

Table 1 NPCR and UACI values when encryption and decryption key is changed

Images	Key	Encryption process		Decryption process
		NPCR	UACI	NPCR
Cameraman (256 × 256)	$rlas + 10^{-15}$	99.6124	33.4088	99.6597
Lena (512 × 512)	$rlas + 10^{-15}$	99.6098	33.5235	99.6006

Table 2 NPCR and UACI values when changing one pixel value in the plain image

Images	NPCR			UACI		
	Min.	Max.	Avg.	Min.	Max.	Avg.
Cameraman	99.5544	99.6780	99.6098	33.1379	33.7670	33.4164
Lena	99.5747	99.6475	99.6098	33.3841	33.5843	33.4707

Table 3 Correlation coefficient results of the proposed cryptosystem

Algorithms	Images	Plain images			Encrypted images		
		Horizontal	Vertical	Diagonal	Horizontal	Vertical	Diagonal
Ours	Cameraman	0.9734	0.9856	0.9615	–0.0079	–0.0154	–0.0028
	Lena	0.9263	0.9587	0.9074	0.0125	–0.0174	–0.0065
Ref. [2]	Lena	–	–	–	–0.0230	0.0019	–0.0034

plain image. This proves the higher sensitivity of plaintext in the proposed cryptosystem.

4.4 Histogram Analysis

Figure 3 shows the histogram analysis results of both "Cameraman" and "Lena" images. In encrypted images, the histograms should have uniform range of gray scale values and should be different than the histograms of plain images [11, 12]. In Fig. 3, we can see the uniformity of gray scale values in histograms of encrypted images and also the large differences in the histograms of encrypted images and plain images. This proves the strong resistivity of the statistical attack by using the proposed cryptosystem.

4.5 Correlation Coefficient Analysis

In this paper, random selections of 3000 pairs of adjacent pixels are used to analyze the correlation coefficient. Table 3 presents the results for correlation coefficient. By

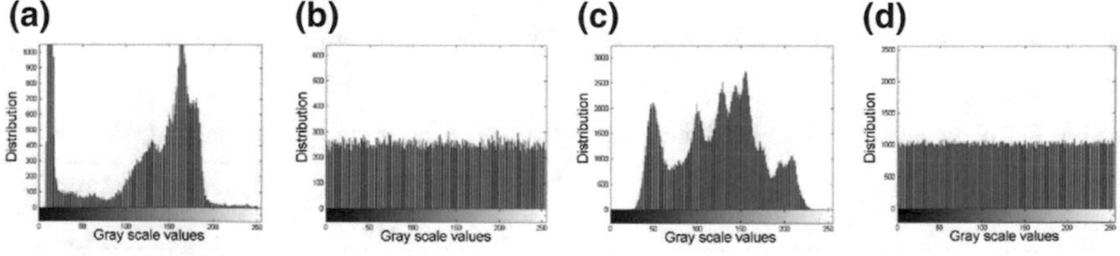

Fig. 3 **a**, **c** Histograms of plain "Cameraman" and "Lena" images, respectively; **b**, **d** Histograms of corresponding encrypted "Cameraman" and "Lena" images, respectively

Table 4 Information entropy results of the proposed cryptosystem

Algorithms	Images	Ours	
		Plain images	Encrypted images **(Ideal value = 8)**
Ours	Cameraman	7.0097	7.9973
	Lena	7.4451	7.9993
Reference [2]	Lena	–	7.9974
Reference [13] (7 rounds)	Lena	–	7.9970

observing Table 3, we can see that the correlation coefficient values of encrypted images for all the three directions are very nearer to 0 and also better than [2]. That means weaker correlation occurs in the adjacent pixels of encrypted images. This proves the higher resistivity of the statistical attack in the proposed cryptosystem.

4.6 *Information Entropy Analysis*

Table 4 shows the information entropy results of "Cameraman" and "Lena" image by using the proposed cryptosystem and also shows the comparison of information entropy of "Lena" image with other reported articles [2, 13]. By observing Table 4, we can see that the encrypted images have information entropy value nearer to 8 and also better than those in [2, 13]. This proves the stronger resistivity of the information entropy attack of the proposed cryptosystem.

4.7 *Known—Plaintext Attack and Chosen—Plaintext Attack Analysis*

In the proposed cryptosystem, all the secret keys, are not only dependent on the given initial values and system parameters of the chaotic maps but also on the plain image. So, for every input of plain image, the secret keys are changed in an encryption process. Hence attackers cannot extract useful information by encrypting some predesigned special images because the encrypted output is related to those chosen images only. This concludes that the proposed algorithm is highly image-dependent and capable of effectively repelling known-plaintext attack and chosen-plaintext attack.

5 Conclusion

This paper proposed an image encryption algorithm based on bit-plane operations by using the chaotic maps. In this paper, firstly, bit-plane diffusion operations were performed, secondly, row-shuffling then column-shuffling operations are performed by using the chaotic maps. The proposed algorithm produced better encryption results and provided strong sensitivity to the secret keys as well as plaintext. The proposed algorithm can withstand known-plaintext attack and chosen-plaintext attack as the secret keys depend on the input plain image. The entropy values, histogram plots and correlation coefficient results show that the proposed algorithm can evade statistical attack. So, the proposed algorithm is encryption efficient and secure.

Acknowledgements This research work is supported by Information Security Education Awareness (ISEA) project phase-II, Department of Electronics and Information Technology (DeitY), Govt. of India.

References

1. Wang, X., Wang, Q., Zhang, Y.: A fast image algorithm based on rows and columns switch. Nonlinear Dyn. **79**, 1141–1149 (2015)
2. Xu, L., Li, Z., Li, J., Hua, W.: A novel bit-level image encryption algorithm based on chaotic maps. Opt. Laser Eng. **78**, 17–25 (2016)
3. Li, C., Luo, G., Qin, K., Li, C.: An image encryption scheme based on chaotic tent map. Nonlinear Dyn. **87**, 127–133 (2017)
4. Chai, X.: An image encryption algorithm based on bit level Brownian motion and new chaotic systems. Multimed. Tools Appl. **76**, 1159–1175 (2017)
5. Pak, C., Huang, L.: A new color image encryption using combination of the 1D chaotic map. Signal Process. **138**, 129–137 (2017)
6. Zhou, Y., Bao, L., Chen, C.P.: A new 1D chaotic system for image encryption. Signal Process. **97**, 172–182 (2014)
7. Mollaeefar, M., Sharif, A., Nazari, M.: A novel encryption scheme for colored image based on high level chaotic maps. Multimed. Tools Appl. **76**, 607–629 (2017)
8. Wang, X.Y., Yang, L.: Design of pseudo-random bit generator based on chaotic maps. Int. J. Mod. Phys. B **26**, 1250208 (2012)
9. Hua, Z., Zhou, Y.: Image encryption using 2D Logistic-adjusted-Sine map. Inf. Sci. (Ny) **339**, 237–253 (2016)
10. Kulsoom, A., Xiao, D., Abbas, S.A.: An efficient and noise resistive selective image encryption scheme for gray images based on chaotic maps and DNA complementary rules. Multimed. Tools Appl. **75**, 1–23 (2016)
11. Zhu, Z.L., Zhang, W., Wong, K.W., Yu, H.: A chaos-based symmetric image encryption scheme using a bit-level permutation. Inf. Sci. (Ny) **181**, 1171–1186 (2011)
12. Zhang, Y.Q., Wang, X.Y.: A symmetric image encryption algorithm based on mixed linear–nonlinear coupled map lattice. Inf. Sci. (Ny) **273**, 329–351 (2014)
13. Wang, X., Liu, L., Zhang, Y.: A novel chaotic block image encryption algorithm based on dynamic random growth technique. Opt Laser Eng. **66**, 10–18 (2015)

Edge Detection Method Using Richardson's Extrapolation Formula

Amiya Halder, Pritam Bhattacharya and Aritra Kundu

Abstract Edge detection is a technique used in image processing for locating boundaries of objects within images. This paper proposes a novel approach to edge detection based on Richardson's extrapolation technique. The measure of edge strength is computed using this technique. Several experiments are being carried out to evaluate and compare the proposed technique with preexisting techniques, and the results are encouraging.

Keywords Edge detection · Sobel operator · Richardson extrapolation

1 Introduction

Intensity changes in images occur from reflectance or illumination or surface discontinuities. The points where this intensity abruptly changes that consists of edges [1]. Edge detection includes a variety of mathematical techniques to identify these points of discontinuity or abrupt change. It is a fundamental tool for image segmentation and areas such as feature detection, feature extraction, image processing, machine vision, and computer vision. In image processing, there are different types of techniques that are used in performing edge detection such as gradient-based and derivative-based [2]. Sobel, Prewitt, and Robert operators are the classical gradient edge detectors [3]. The gradient-based techniques try to compute the rate at which a pixel intensity value is changing in a particular direction. If we see that the gradient is large in a particular direction, then it is a candidate point for being an edge pixel. We use the same process in different directions to try and determine the

A. Halder · P. Bhattacharya · A. Kundu (✉)
STCET, Kolkata 700023, India
e-mail: aritrakundu99@gmail.com

A. Halder
e-mail: amiya.halder77@gmail.com

P. Bhattacharya
e-mail: prb2794@gmail.com

J. Nayak et al. (eds.), *Soft Computing in Data Analytics*,
Advances in Intelligent Systems and Computing 758,
https://doi.org/10.1007/978-981-13-0514-6_69

edges and therefore try and determine the points of rapid intensity change in different direction. The second-order derivative methods correctly locate edges over a large area; however, they produce false edges due to high intensity changes at corners and curves. Joshi [4] has proposed edge detection based on Haar wavelet transform. They concluded Canny detector [5] works best even under blurred conditions. A hybrid approach has been proposed in Kalra et al. [6] by combining Sobel and Canny operators and have reduced false edges. However, none of these techniques give an ideal result in a noisy environment. Mahmood et al. [7] concluded that Canny edge detector works best under these circumstances. Yi [8] has proposed using wavelet transform to localize edges. Perona and Malik [9] have proposed modifying the linear scale-space paradigm and using anisotropic diffusion for edge detection. Piotr and Zitnik [10] have proposed edge detection based on a structured learning approach using random forests. In the paper, an efficient edge detection algorithm is proposed using Richardson's extrapolation formula for calculating approximate derivative at a point in both x and y directions.

2 Proposed Algorithm

2.1 *Richardson's Extrapolation*

Richardson's extrapolation is actually used in numerical analysis to improve the rate of convergence of a sequence. This technique is invented by Lewis Fry Richardson [11]. The technique tries to approximate a series $\lambda(\omega)$ that depends on a small step size ω by approximating the series for two different values of ω, ω and $\frac{\omega}{\beta}$ for some value of β. A general recurrence relation can be defined as (in Eq. 1):

$$\Delta(i+1) = (\beta^k)\lambda(\omega/\beta) - \lambda(\omega))/(\beta^k - 1) \tag{1}$$

where i represents the ith step. In our technique, we use this concept and apply it in the Taylor series and try to approximate the second-order derivative with a higher rate of convergence. In Taylor's expansion (Eq. 2)

$$\psi(x+\omega) = \psi(x) + \omega\psi'(x) + \omega^2\psi''(x) + ... \tag{2}$$

The derivative of $\psi(x)$ is given by (Eq. 3):

$$\psi'(x) = (\psi(x+\omega) - (\psi(x))/\omega) - \omega\psi''(x)/2 + ... \tag{3}$$

The initial approximations are chosen as in Eq. 4:

$$\kappa(\omega) = \frac{\psi(x+\omega) - \psi(x)}{\omega} \tag{4}$$

The first formulae extrapolated for $\kappa 1$ would be in Eq. 5:

$$\kappa 1 = 2\kappa(\omega/2) - \kappa(\omega) \tag{5}$$

We can extrapolate again to obtain

$$\kappa = \frac{(4\kappa 1(\omega/2) - \kappa 1(\omega))}{3} \tag{6}$$

The concepts of our proposed method depend on this expression. We try to formulate the edge detection technique using Eq. 6.

2.2 *Algorithm*

In this method, we consider 5×5 window size with center pixel $\mu(\eta, \lambda)$. In proposed method, we calculate the strength of $\mu(\eta, \lambda)$ based on the neighboring pixel values according to the formula obtained from Eq. 6 and replace the pixel with this value. Finally, a threshold is applied. The approach consists of two main steps. At step 1, replace each pixel by its approximate derivative value, and in step 2, apply a threshold to quantify how well a pixel point responds to the technique. The proposed algorithm is shown in Algorithm1.

Algorithm 1: PROPOSED ALGORITHM

Input: An image $\Omega(\chi, \psi)$
Output: Edge detected image
for *each pixel* $\mu(\eta, \lambda)$ *of the image* $\Omega(\chi, \psi)$ **do**
  find the 5×5 mask centering $\mu(\eta, \lambda)$
  $\alpha = \frac{(\mu(\eta+1)(\lambda-1)+\mu(\eta+1)(\lambda)+\mu(\eta+1)(\lambda+1)-\mu(\eta-1)(\lambda-1)-\mu(\eta-1)(\lambda)-\mu(\eta-1)(\lambda+1))}{2}$
  $\beta = \frac{(-\mu(\eta-1)(\lambda-1)+\mu(\eta-1)(\lambda+1)-\mu(\eta)(\lambda-1)+\mu(\eta)(\lambda+1)-\mu(\eta+1)(\lambda-1)+\mu(\eta+1)(\lambda+1))}{2}$
  $\alpha 1 = \frac{(-\mu(\eta-2)(\lambda-1)-\mu(\eta-2)(\lambda)-\mu(\eta-2)(\lambda+1)+\mu(\eta+2)(\lambda-1)+\mu(\eta+2)(\lambda)+\mu(\eta+2)(\lambda+1))}{4}$
  $\beta 1 = \frac{(-\mu(\eta-1)(\lambda-2)+\mu(\eta-1)(\lambda+2)-\mu(\eta)(\lambda-2)+\mu(\eta)(\lambda+2)-\mu(\eta+1)(\lambda-2)+\mu(\eta+1)(\lambda+2))}{4}$
  $\gamma = \sqrt{\alpha^2 + \beta^2}$
  $\delta = \sqrt{\alpha 1^2 + \beta 1^2}$
  $\mu(\eta, \lambda) = \frac{(4 * \gamma - \delta)}{3}$
for *each pixel* $\mu(\eta, \lambda)$ *of the image* $\Omega(\chi, \psi)$ **do**
  if $\mu(\eta, \lambda) \leq \theta$ **then**
    $\mu(\eta, \lambda) = 0$
  else
    $\mu(\eta, \lambda) = 255$

---

3 Experimental Results

This proposed algorithm is implemented using Dev C++ and MATLAB of a large number of 8-bit grayscale images. Comparison of edge maps is obtained from different existing algorithm such as statistical [12] method and novel edge detection (NED) method [13] and Sobel, Prewitt, and Robert operators with proposed algorithm using the following statistical indices [14] $\zeta(C) = \frac{TP}{max(F1,F2)}$, $\zeta(NC) = \frac{FN}{max(F1,F2)}$ and $\zeta(I) = \frac{FP}{max(F1,F2)}$, where $\zeta(C)$, $\zeta(NC)$, $\zeta(I)$ are correctly identified, not detected, and incorrectly detected pixels, respectivcly. *TP*, *FP*, and *FN* represent the true positive, false positive, and false negative values, respectively. $F1$ and $F2$ are the number of edge pixel of the output and ground truth images, respectively. Also, calculate the similarity between two contours: $M_P = \frac{1}{max(F1,F2)} \sum \frac{1}{1+\eta * \rho_i^2}$ and the distance: $(\chi_f^4) = \sqrt{(\zeta(C)-1)^2 + (M_P - 1)^2 + (\zeta(NC))^2 + (\zeta(I))^2}$, where ρ_i is the distance between a edge pixel and the nearest edge pixel of the ground truth and η is a constant and value of the $\eta = \frac{1}{9}$, optimal value established by Pratt [15]. The results of the different images (`Lena`, `Wheel`, `Parrot`, `Tool`, `Zelda`, `Pepper` images) are tabulated in Tables 1, 2, 3, 4, 5, and 6, and also output of the images is shown in Fig. 1. From the above results, it is noticed that the proposed algorithm gives the better results than the existing statistical [12] method and NED method [13] and Sobel, Prewit, and Robert operators. In proposed method, threshold is chosen by trial and error method.

Table 1 Pratt evaluation using Robert, Prewitt, Sobel, NED, statistical, and proposed method for `Wheel` image

Method	Statistical	Robert	Prewitt	Sobel	NED	Proposed
M_P	0.7945	0.2889	0.2325	0.2329	0.6130	0.8971
$\zeta(NC)$	0.8042	0.1146	0.0489	0.0485	0.8568	0.7057
$\zeta(C)$	0.1958	0.1862	0.1884	0.1893	0.1432	0.2943
$\zeta(I)$	0.2636	0.8138	0.8116	0.8107	0.1304	0.5071
χ_f^4	1.1854	1.3576	1.3816	1.3803	1.2787	1.1241

Table 2 Pratt evaluation using Robert, Prewit, Sobel, NED, statistical, and proposed method for `Lena` image

Method	Statistical	Robert	Prewitt	Sobel	NED	Proposed
M_P	0.8176	0.3328	0.3297	0.3307	0.6622	0.8999
$\zeta(NC)$	0.7935	0.1986	0.1111	0.1150	0.8310	0.6717
$\zeta(C)$	0.2065	0.1583	0.2328	0.2304	0.1690	0.3283
$\zeta(I)$	0.2305	0.8417	0.7672	0.7696	0.1919	0.4133
χ_f^4	1.1600	1.3789	1.2802	1.2829	1.2378	1.0407

Table 3 Pratt evaluation using Robert, Prewitt, Sobel, NED, statistical, and proposed method for `Pepper` image

Method	Statistical	Robert	Prewitt	Sobel	NED	Proposed
M_P	0.8484	0.3271	0.3383	0.3406	0.6556	0.8888
$\zeta(NC)$	0.7802	0.1804	0.0979	0.1015	0.8556	0.6986
$\zeta(C)$	0.2198	0.1662	0.2510	0.2502	0.1444	0.3014
$\zeta(I)$	0.2237	0.8338	0.7490	0.7498	0.1372	0.2031
χ_f^4	1.1360	1.3696	1.2528	1.2529	1.2656	1.0148

Table 4 Pratt evaluation using Robert, Prewit, Sobel, NED, statistical, and proposed method for `Parrot` image

Method	Statistical	Robert	Prewitt	Sobel	NED	Proposed
M_P	0.8278	0.3312	0.4023	0.4073	0.7390	0.8970
$\zeta(NC)$	0.7887	0.2184	0.1673	0.1769	0.7842	0.6506
$\zeta(C)$	0.2113	0.1408	0.2584	0.2554	0.2158	0.3494
$\zeta(I)$	0.2244	0.8592	0.7416	0.7446	0.1257	0.3322
χ_f^4	1.1508	1.4041	1.2186	1.2213	1.1463	0.9837

Table 5 Pratt evaluation using Robert, Prewitt, Sobel, NED, statistical, and proposed method for `Zelda` image

Method	Statistical	Robert	Prewitt	Sobel	NED	Proposed
M_P	0.8212	0.2752	0.3494	0.3499	0.7122	0.9009
$\zeta(NC)$	0.7889	0.1713	0.1560	0.1566	0.8035	0.6550
$\zeta(C)$	0.2111	0.1276	0.2150	0.2150	0.1965	0.3450
$\zeta(I)$	0.4792	0.8724	0.7850	0.7850	0.2287	0.4999
χ_f^4	1.2273	1.4411	1.2962	1.2960	1.1943	1.0572

Table 6 Pratt evaluation using Robert, Prewitt, Sobel, NED, statistical, and proposed method for `Tool` image

Method	Statistical	Robert	Prewitt	Sobel	NED	Proposed
M_P	0.7755	0.2087	0.2185	0.2183	0.6897	0.8934
$\zeta(NC)$	0.7780	0.1085	0.0425	0.0428	0.8221	0.6677
$\zeta(C)$	0.2220	0.1113	0.1802	0.1799	0.1779	0.3202
$\zeta(I)$	0.4951	0.8887	0.8198	0.8201	0.2174	0.6798
χ_f^4	1.2272	1.4891	1.3988	1.3993	1.2229	1.1753

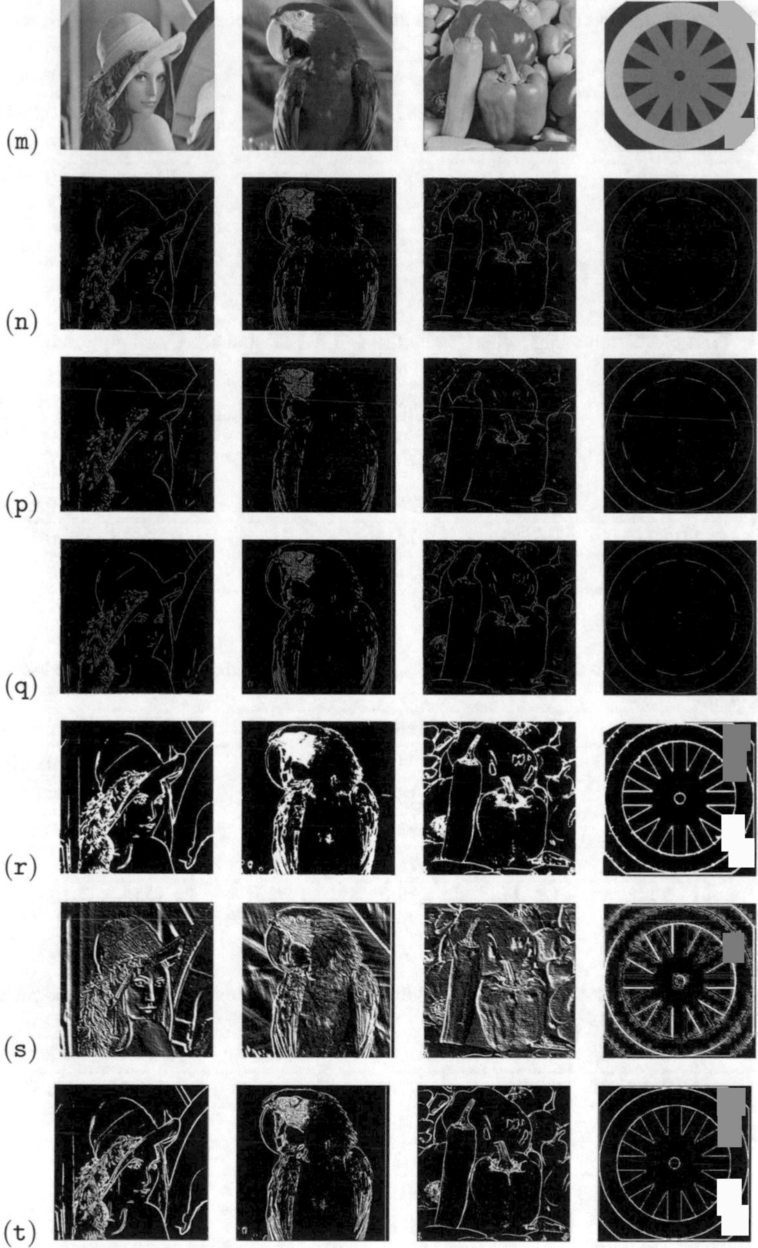

Fig. 1 Comparison of the output images of Lena, Parrot, Pepper, and Wheel using **n** Prewit, **p** Robert, **q** Sobel, **r** Statistcal approach, **s** NED method and **t** proposed method and **m** original images

4 Conclusions

A novel approach to edge detection using Richardson's extrapolation technique for calculating strength of gradient in different directions has been proposed. The approach approximates the second-order derivative by using Richardson's extrapolation technique which is used to approximate the Taylor series. The formulae are applied to different grayscale images and a threshold is applied so that the final image is a binary one i.e. edge or no edge. The final output image is being compared with preexisting techniques for edge detection based on various factors as described, and the results obtained are promising.

References

1. Gonzalez, R.C., Woods, R.E.: Digital Image Processing. Pearson Education (2002)
2. Yasri I., Hamid, N.H., Yap V.V.: An FPGA implementation of gradient based edge detection algorithm design. In: International Conference on Computer Technology and Development (2009)
3. Marr, D., Hildreth, E.: Theory of edge detection. In: Proceedings Of The Royal Society B, pp. 187–217 (1980)
4. Joshi, S.R., Koju, R.: Study and comparision of edge detection algorithms, In: 3rd Asian Himalayas International Conference on Internet, pp. 1–5 (2012)
5. Canny, J.: A computational approach to edge detection. IEEE Trans. Pattern Anal. Mach. Intell. **8**(6), 679–698 (1986)
6. Kalra, A., Chhokar, R.L.: A Hybrid approach using sobel and canny operator for digital image edge detection. In: International Conference on Micro-Electronics and Telecommunication Engineering, pp. 305–310 (2016)
7. Mahmood, A.M., Maras, H.H., Elbasi, E.: Measurement of edge detection algorithms in clean and noisy environment. In: 8th International Conference on Application of Information and Communication Technologies, pp. 1–6 (2014)
8. Yi, S., Labate, D., Easley, G.R., Krim, H.: A shearlet approach to edge analysis and detection. IEEE Trans. Image Process. **18**(5), 929–941 (2009)
9. Perona, P., Malik, J.: Scale-space and edge detection using anisotropic diffusion. IEEE Trans. Pattern Anal. Mach. Intell. **12**(7), 629–639 (1990)
10. Doll, P., Zitnick, C.L.: Fast edge detection using structured forests. IEEE Trans. Pattern Anal. Mach. Intell. **37**(8), 1558–1570 (2015)
11. https://en.wikipedia.org/wiki/Richardson_extrapolation
12. Halder, A., Chatterjee, N., Kar, A., Pal, S., Pramanik, S.: Edge detection: a statistical approach. In: 3rd International Conference on Electronics Computer Technology, pp. 306–309 (2011)
13. Prathusha, P., Jyothi, S.: A Novel edge detection algorithm for fast and efficient image segmentation. In: Data Engineering and Intelligent, Computing, pp. 283–291 (2017)
14. Gonzaga, A.: Method to evaluate the performance of edge detector. In: International Conference on Intelligent Systems Design and Applications, pp. 341–346 (2007)
15. Abdou, I.A., Pratt, W.: Quantitative design and evaluation of enhancement/thresholding edge detectors. Proc. IEEE **67**(5), 753–766 (1979)

An Efficient Framework for Document Retrieval in Relationship Strengthened Personalized Ontologies

S. Vigneshwari, A. Mary Posonia and S. Gowri

Abstract Personalization is crucial in the prevailing Internet scenario. An efficient information retrieval not only relies on personalization but also upon security. There is need of a semantic approach of document retrieval to boost the search quality. The incorporation of knowledge extracted from multiple ontologies is the significant research direction to get the most appropriate documents from the knowledge base. In this paper, local and global ontologies are used for mining the semantic associations. Multiple ontologies can share a variety of concepts, and automatic merging of them is feasible. Any input query is often matched with the semantic association for finding the relevant words. This is achieved by analyzing the intensified concept relationships in the local repository or local ontology and finding out the possibilities of automation by similarity calculation. A user profiling model is constructed which is trained for personalization. A novel framework has been proposed which provides document retrieval. The results are analyzed using benchmark SWETO ontology. Due to the query property enhancements, the error rate is reduced and the local ontology gets optimized for synchronization with the global Web ontology, thereby enhancing user personalization.

Keywords Local ontology · Global Web ontology · Query enhancement Automated ontologies · Document rctrieval · Security · Cloud

S. Vigneshwari (✉) · A. Mary Posonia · S. Gowri
School of Computing, Sathyabama Institute of Science & Technology,
Chennai 600119, Tamil Nadu, India
e-mail: vikiraju@gmail.com

A. Mary Posonia
e-mail: soniadelicate@gmail.com

S. Gowri
e-mail: gowriamritha2003@gmail.com

J. Nayak et al. (eds.), *Soft Computing in Data Analytics*,
Advances in Intelligent Systems and Computing 758,
https://doi.org/10.1007/978-981-13-0514-6_70

1 Introduction

Recently, the number of Web-based data accessible has been intensely inflated. Gathering useful information from the Web has become an ambitious issue for users. The Web information gathering systems [1] prevailing are endeavoring to satisfy user needs by capturing their data needs. In personalized Web information gathering system, ontology a knowledge description and formalization model is accomplished to simulate user concept models, and these ontologies are named as ontological user profiles [2]. Recently, researchers have tried to explore user background knowledge through global or local analysis [1, 3].

The proposed system focuses on two different ontologies or otherwise RDF-OWL structures, which are local repository and global web ontology. The size of the local data repository is approximately 5 MB, since the size depends upon the number of concepts used for searching, whereas the size of the global ontology is very huge. Due to the mass volume, global Web ontology is stored in the cloud. Private cloud is used in the analysis of the current paper. This paper focuses on concept strengthening for local ontology optimization, query enhancement for unambiguous query search which is also a part of ontology optimization and secured storage and retrieval of documents from the cloud.

2 Literature Review

Chin et al. [3] and Stumme et al. [4] set up a framework for multidimensional association mining. Here data are fetched from heterogeneous resources. Hence, the ontology development model is mandatory for personalization.

Francesca et al. [5] proposed architecture for managing RDF data in a distributed manner. Amazon Web Services (AWS) was manipulated for the study. The key value of AWS identified RDF datasets efficiently. This paper initiated a cloud-based approach for ontology management in the proposed system.

Jayasri and Manimegalai [6] suggested ontology-based approach upon multi-ontologies. Distance-based similarity measures using cross-ontology mining techniques were discussed. The proposed system is based on Jayasri et al's idea of multi-ontologies. Lin et al. [7] discussed calculation of similarity measures. The similarity calculation between two documents was based on the occurrence or non-occurrence of certain features, contained in the documents. Utilizing Lin's idea the fuzzy mutual similarity measure approach was followed in the proposed system.

Parameswaran et al. [8] proposed a two-level access control for mapping files and retrieving data in the cloud. This paper speaks about cloud security and initiates a cloud-based approach for the proposed system.

Literature on different research areas like cloud, information security, and Semantic Web provided a great instinct on the development of the proposed document retrieval framework.

3 Proposed Framework and Methodologies

The proposed system comprises two important modules, namely local query search model with query enhancement and secured cloud search module as shown in Fig. 1. The user given query will be searched for pattern match in local repository. If match is found, then the results will be retrieved from the cached offline local storage. If the query does not match with the local repository, then query enhancement takes place. This action requires strengthening of relationships of the contexts or otherwise concepts which are represented in form of RDF-OWL structure. The offline search is always synchronized with the online search. The online repository is also in the RDF-OWL format but comprises more concepts which are mostly linked to online Web links. The global ontology thus formed is a dynamic one and will be updated based on user's query search. The entire framework of global Web ontology is stored in encrypted form in the cloud infrastructure using Blowfish algorithm. This forms a secured semantic database for effective data extraction. The objective is to train or strengthen the local repository to cope up with the dynamicity of the global ontology.

3.1 Strengthening the Local Query Search with Query Enhancement

The ontology is constructed based on the topic 'database.' It is possible to assign relations dynamically and automatically by prioritizing the values based on fuzzy mutual similarity (FMS). In [10], the retrieval time of the Web documents is reduced upon establishing relationships in user profiling ontologies and is analyzed. FMS is calculated using the following fuzzy entropy equation:

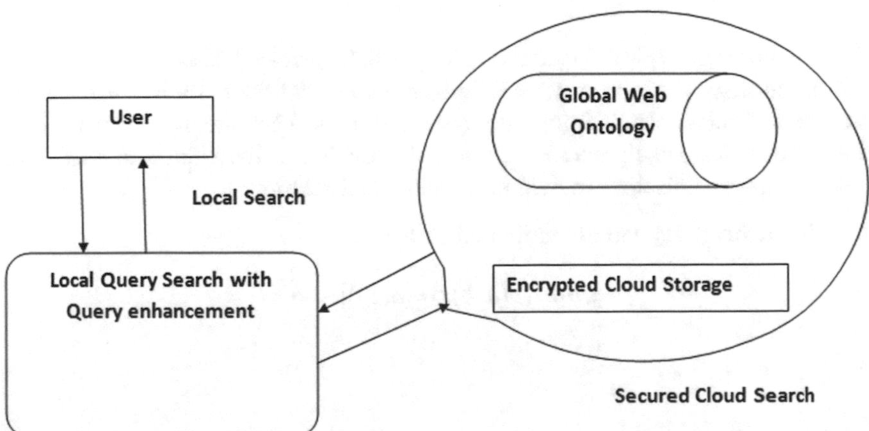

Fig. 1 Overall framework of the proposed approach

$$fH(X,Y) = -\frac{1}{n*m}\sum_{i=1}^{n}\sum_{j=1}^{m}\log\frac{\mu(co_{Xi}, co_{Yj})}{n*m} \quad (1)$$

In Eq. 1, fH() represents fuzzy entropy function, X and Y are the different ontologies (RDF-OWL formats) for different Web pages available in the Web for ‘n’ number of concepts in X and ‘m’ number of concepts in Y.

3.2 Algorithm for Strengthening Relationships

Input: Generated local ontology based on the past query search experience
Output: Semantic Relatedness between manual and automated relationships
Step 1: Perform concept similarity of local and global ontologies using FMS.
Step 2: Apply relations to the concepts based on the values of FMS
If FMS range is *High* then
Relation= *is_a*
Else if FMS range is *Medium* then
Relation = *has*
Else if *low*<FMS range<*Med*ium then
Relation=*union_of*
Else
Relation=*part_of*
End if
Step 3: Select the concepts, where t_r is the threshold value
$FMS > t_r$
Step 4: Apply relation count
Step 5: Select top ranked concepts (Schenkel et al., [10]).
Step 6: Extract web information and personalize local repository
$C_{local} = [C^1_{global}, C^2_{global}, C^3_{global}, \ldots, C^n_{global}]$
Step 7: Compare the similarity of the concepts by establishing manual relations with FMS value
Step 8: Return the similarity index
End

3.3 Analysis of the Strengthened Local Repository Using SWETO Dataset

SWETO ontology [9] is a standard ontology which is created based on the Semantic Web technology data. Two types of properties are considered while establishing strengthened relationships from ontological concepts. They are data properties and object properties. Data properties disclose the relation of individuals to data types. Object properties disclose individual to individual relation

Functional property rule is indicated in Eq. 2:

$$\forall a, b, c\, [(p(a,b) \cap p(a,c))] \rightarrow b = c \quad (2)$$

Sample relation1: Tej has_father Gana
Sample relation2: Tej has_father Raj
Inferred property: Gana = Raj (Gana and Raj refer to the same person)

Inverse functional property rule is given in Eq. 3:

$$\forall a, (a = b) \rightarrow (b = a) \tag{3}$$

For example, 'Anna University is_university of India' and 'India has_university Anna University' represents the same semantics. Use of inverse functional property is that since both relations share the same context, one relation can be ignored without any loss of information.

3.4 *Query Enhancement Algorithm*

Input: Reduced Document Vector (RDV), weight of each term t in document D ∈ RDV

Output: $\overrightarrow{DI}$, Indexed documents

Step 1: Initialize w as the average weight of all the terms in the RDV

Step 2: For each term t_i

Step 2.1: Train the data set by assigning initial value of w

Step 2.2: Calculate the error: e=Expected term weight - Actual term weight

Step 2.3: Calculate the normalized coefficient in terms of e

Step 2.4: Calculate the Reformulated weight of the query

// is the query coeffient and the value is assigned to 0.5 after training

$W'_{t',q'} = (1-)W_{t,q} + .tf_t$

Step 2.5: Set threshold for measure value

Step 2.6: If $W'_{t',q'}$ > threshold,

Then select query term t'

Step 3: Index the documents by initializing the document index DI=

Step 3.1: Calculate Total Query Frequency, TQF(i) from the selected query terms, using the formula:

$\mathrm{TQF(i)} = \sum_{j=1}^{m} \sum_{j=1}^{m} \mathrm{qf_{ij}} \ \forall \mathrm{j} > 0 \ and\ i > 0$

Step 3.2: For each Document Di

//Comparing TQF with Mean Query Frequency(MQF)

IF TQF(i)>MQF

Then $\overrightarrow{DI} = \overrightarrow{DI}$ U Di

End If

Step 4: Return $\overrightarrow{DI}$

4 Results and Discussion

The properties of the instances in the generated local RDF structure are tested with 50, 100, 150, 200, 250 documents for different values of the training coefficient α. The size of the database is 5 MB with initially assigning α as 0.1 through 0.8 to train the data. We fix the value of α by checking with different values, and we obtained good performance and a saturated state when α = 0.5.

In Fig. 2, the relevant information obtained based on the current study over the ontologies SWETO small and SWETO medium is presented. Table 1 depicts the information extracted for the input query concept. The average execution time used by the present approach for extracting the relevant information is 843778 ms, and the memory utilized is given as 1.1 MB.

Table 2 shows the comparison of relevancy score obtained using FMS and relationship ranking using SWETO dataset. The error rate is calculated using root mean square error (RMS Error) technique. RMS is used to calculate the difference between the automated FMS value obtained and the relevancy score obtained upon establishing manual relationships. For a concept pair $\{co_i, co_j\}$, the FMS-based relevancy score is denoted as 'f' and the relationship ranking-based relevancy score is denoted by 'r'. RMS error is calculated using the formula in Eq. 4:

Fig. 2 Relevant information

```
[Journels, Has_financial_organization_of, IJAMM]
[Journels, Has_financial_organization_of, IJPA]
[Journels, Has_financial_organization_of, IJSE]
[Journels, Has_financial_organization_of, IJAP]
[Journels, Has_financial_organization_of, JCIB]
[Journels, Has_financial_organization_of, IJWCS]
[Journels, Has_financial_organization_of, IJCIR]
[Journels, Has_financial_organization_of, MMAC]
```

Table 1 Searching a concept in SWETO

Query concept	Information extracted based on the query	Time (ms)	Memory (MB)
Victoria	[England, Has_airport, Victoria], [Victoria, Is_airport, England]	81114	0.85
Italy	[Italy, Has_airport, Cagliari_Elmas], [Italy, Has_university_of, Politecnico_Di_Milana], [Italy, Has_university_of, Politecnico_Di_Bari], [Italy, Has_city, Venice], [Italy, Has_city, Rome]	86547	1.18
Anna University	[Anna_University, Is_university_of, India] [India, Has_university, Anna_University], [Anna_university, part_of,national_university]	84789	1.1
Srilanka	[Srilanka, Has_university_of, Open_University_Of_Srilanka], [Eastern_University_Of_Srillanka, Is_university_of Srilanka], [Srilanka, Has_university_of, Eastern_University_Of_Srillanka], [Nuwara_Eliya, Is_city, Srilanka], [Srilanka, Has_city, Nuwara_Eliya], [Srilanka, Has_city, Galle], [Colombu,part_of, Srilanka]	85123	1.4

Table 2 Calculation of error rate between automated and manual relevancy score using SWETO dataset

SWETO instance pair	Relevancy score		Difference	RMS error
	FMS	Relevancy based on relations		
{country, airport}	0.71	0.86	0.16	0.11
{country, city}	0.88	0.92	0.04	0.03
{financial_organization, book}	0.21	0.19	0.01	0.01
{university, country}	0.53	0.63	0.11	0.07
{city, university}	0.61	0.71	0.11	0.08
{book, city}	0.11	0.05	0.05	0.04
{company, university}	0.16	0.09	0.07	0.05
{country, journal}	0.46	0.58	0.12	0.08
{country, book}	0.48	0.71	0.23	0.16
{Airport, financial_organization}	0.19	0.33	0.14	0.10
{financial_organization, city}	0.68	0.81	0.13	0.09
Average value	**0.45**	**0.53**	**0.11**	**0.07**

$$RMS_{Error} = \sqrt{\frac{(f-r)^2}{2}} \quad (4)$$

From Table 2, it is inferred that the average relevancy measure obtained using FMS measure is 0.45 and the average score obtained using semiautomated ontology generation by establishing relations like 'is_a', 'has', 'part_of', 'union_of' is 0.53. If the values are directly subtracted, the average difference or error obtained is 0.11. In order to obtain precise difference, RMS error is calculated using Eq. 4 and the average value obtained is 0.07 which is negligible. The evaluation shows that FMS value could be utilized for relevancy score calculation where human interpretation of relationships becomes infeasible. This evaluation helps in personalizing the local repository automatically with the help of FMS.

5 Conclusion

An improved cross-ontology mining method is used for above-mentioned Web information extraction. In order to make the ontologies more effective, manual input is needed. Accordingly, relations like 'is_a', 'has_a', 'union_of', 'part_of' are introduced in the ontology, so that the generated ontology is effective. Ontology-based mining procedure is devised with the consideration of all the properties and relation included in it. The effectiveness of information mined from those ontologies is analyzed. The results show that the mean error rate of automated and manual relevancy score is negligible.

For semantic information retrieval, a better understanding of the input query is necessary, which can be achieved through query enhancement of the functional properties of the ontology. To ensure secure information retrieval, encrypted RSA with Blowfish is utilized to overcome brute force attacks. By using better association rule, mining algorithms for cross-ontology mining can be utilized for the betterment of the proposed system and this in turn is considered as a future work of the current research.

References

1. Berendt, B., Andreas, H., Dunja, M.: A roadmap for web mining from web to semantic web. In: Institute of Technical and Business Information Systems, Ottovon–Guericke–University Magdeburg, vol. 18, pp. 1–21 (2008)
2. Toledo, C.M., Ale, M.A., Chiotti, O., Galli, M.R.: An ontology-driven document retrieval strategy for organizational knowledge management systems. Electron Notes Theor. Comput. Sci. **281**, 21–34 (2011)
3. Chin, A.W., Wen, Y.L., Chuan, C.W.: An active multidimensional association mining framework with user preference ontology. Int. J. Fuzzy Syst. **12**(2), 125–135 (2010)
4. Gerd, S., Alexander, M.: Ontology Merging for Federated Ontologies on the Semantic Web, Institute for Applied Computer Science and Formal Description Methods (AIFB) University of Karlsruhe, vol. 3(12), pp. 1–9 (2005)
5. Francesca, B., Francois, G., Zoi, K., Ioana, M.: RDF data management in the amazon cloud. In: Proceedings of the 2012 Joint EDBT/ICDT Workshops. ACM, New York, pp. 61–72 (2012)
6. Jayasri, D., Manimegalai, D.: An efficient cross ontology based similarity measure for Bio-Document retrieval system. J. Theor. Appl. Inf. Technol. **54**(2), 245–254 (2013)
7. Yung-Shen, L., Jiang, J.-Y., Lee, S.-J.: A similarity measure for text classification and clustering. IEEE Trans. Knowl. Data Eng. **26**(7), 1575–1590 (2014)
8. Parameswaran, T., Vanitha, S., Arvind, K.S.: An efficient sharing of personal health records using DABE in secure cloud environment. Int. J. Adv. Res. Comput. Eng. Technol. 2278–1323 (2013)
9. SWETO: archive.knoesis.org/library/ontologies/sweto. Last Accessed Mar 2017
10. Vigneshwari, S., Aramudhan, M: An approach for ontology integration for personalization with the support of XML. Int. J. Eng. Technol. **5**(6), 4556–4571 (2011)

Student Performance Analysis with Using Statistical and Cluster Studies

T. PanduRanga Vital, B. G. Lakshmi, H. Swapna Rekha and M. DhanaLakshmi

Abstract Education is the backbone and significant factor in development of a country. The research on education system and performance of student's learning are very important for educational institutions and government to make decisions on quality education. This study analyzes the student's performance by using statistical and unsupervised machine learning (hierarchical and k-means) algorithms. These statistical reports are useful for student's educational strategies and their performance. As per statistical reports, one student education is mainly dependent on the family background, his personal profile, and his activities. Interestingly, some of factors like alcohol consumption, outing (going outside with friends), and romance are also impacted on his education and his result. The unsupervised machine learning algorithms like k-means and hierarchical cluster studies give the good results for predicting performance (pass or fail) of the student. The hierarchal cluster study projects the cause of pass and failure of students by different factors like family size, alcoholic consumption on working days and weekends, address (rural/urban), sex (male/female), and student regularity.

Keywords Education · Student's performance · Data mining
Clustering · Projections

T. PanduRanga Vital (✉) · B. G. Lakshmi · H. Swapna Rekha · M. DhanaLakshmi
Department of CSE, Sri Sivani College of Engineering, Srikakulam 532402,
Andhra Pradesh, India
e-mail: vital2927@gmail.com

B. G. Lakshmi
e-mail: bglaks.05@gmail.com

H. Swapna Rekha
e-mail: swapnarekha23@gmail.com

M. DhanaLakshmi
e-mail: dhana5831@gmail.com

J. Nayak et al. (eds.), *Soft Computing in Data Analytics*,
Advances in Intelligent Systems and Computing 758,
https://doi.org/10.1007/978-981-13-0514-6_71

1 Introduction

Presently, educational organizations like schools, colleges, and universities accumulate and store huge dimensions of data, such as student enrollment, attendance or participation records, reading and learning performance benchmarks, and additionally their examination outcomes or results about [1, 2]. Data mining (DM) is very important for educational data analysis where it affords numerous methods for this [3, 4]. The substantial measure of data at present in databases of student beats the human capacity to analyze and dig out the large amount of useful information without help from automated analysis techniques [5]. Knowledge discovery (KD) is the procedure of nontrivial extraction of understood, obscure, and conceivably helpful data from an expansive database [6, 7].

Van et al. are analyzed performance of the secondary education online indicators. The main study is about student overall learning actions and its progress. They focused in this examination is to decide such markers or indicators by accepting an extensive variety of factors into thought concerning general learning movement and substance handling. A genetic algorithm is utilized for the determination procedure and factors are assessed in light of their perceptive power in an arrangement responsibility. Activities intended to well-trained students in comprehension and applying objects are observed to be particularly useful. Kim et al. are analyzed day by day data of student's learning on consolidating different controls including detecting innovations, sensing technologies, disappointment material science, machine learning, modern statistics, and reliability engineering [8].

Eagles, Michael et al. are analyzed on student modeling and intelligent tutors. They have studied on earlier period examinations that have demonstrated that Bayesian Knowledge (BKT) Tracing. It can predict performance of the students and their skill set and implementations effectively. Standard BKT individualizes parameter gauges for aptitudes, likewise submitted to as knowledge components (KCs) [9]. This student's schoolwork investigates whether individual student limitations, in particularly, individual difference (IDWs) weights can be gotten from students actions preceding mentor utilize [10]. They find that performance of student's dealings in appraisal instructional content and in conceptual knowledge and they find learning efficiency with best corresponding IDWs.

Rao, HS Sushma et al. studied on student admissions by education data mining. Each school year, the organization receives its students from various areas and gives its important assets to each student to accomplish their graduation [11]. Computational data mining strategy finds designs in huge datasets utilizing artificial intelligence (AI), machine learning, and database frameworks. Educational DM inclined to these sensitive issues utilizing a huge procedure of data mining for admission. In this paper, the assessment of admissions is based on area wise and correlation is done in as per admissions in each year. The reports of evaluated data are informed and visualized for the decision making authorities.

Fernandes, Eduardo, et al. analyzed student's performance with using data mining techniques. In this, they presented reports on pedantic presentation of students at public high school [12, 13]. Utilizing CRISP-DM data mining approach, they could accomplish more noteworthy information disclosure than thinks about utilizing conventional distinct measurable investigation. This investigation presents data demonstrating other often announced ascribes significant to potential scholarly disappointment in this specific circumstance, and a point by point clarification of the procedure, and the means taken to get this data.

2 Methodology

The student dataset is collected from UCI machine learning repository [14, 15] and analyzed by DM unsupervised machine learning algorithms statistical methods. There are 33 attributes in the dataset (Student's-School, Student's-Sex, Student's-Age, Student's-Home-Address-Type, Parent's-Cohabitation-Status, Mother's-Education, Mother's-Job, Father's-Education, Father's-Job, Student's-Guardian, Family-Size, Quality-Of-Family-Relationships, Reason-To-Choose-School, Home-To-School-Travel-Time, Weekly-Study-Time, Number-Of-Past-Class-Failures, Extra-Education-School-Support, Family-Educational-Support, Extra-Curricular-Activities,Extra--Paid-Classes,Internet-Access-At-Home, Attended-Nursery-School, Wants-To-TakeHigher-Education, With-Romantic-Relationship, Free-Time-After-School, Going-Out-With-Friends,Weekend-Alcohol-Consumption, Workday-Alcohol-Consumption, Current-Health-Status, Number-Of-School-Absences, Grades-In-First-Period, Grades-In-Second-Period, Final-Grading, Grading, Result).

2.1 K-Means (KM) Algorithm

The KM algorithm is dividing the set of observation points into k clusters [9]. Let R is the real number set and Rd said to be the d dimensional object space. Given a finite set X, that is determined by Eq. (1).

$$X = \{x_1, x_2 \ldots x_n\} \tag{1}$$

where 'n' indicates the objects number.

The KM algorithm partitions the set into subset S, whose subsets mentioned in Eq. (2)

$$S = \{S_1, S_2, \ldots\ldots\ldots., S_k\} \tag{2}$$

where k is a predefined number.

Each cluster C is represented by an object. It is mentioned in Eq. (3)

$$C = \{c_1, \ldots, c_k\} \quad (3)$$

where C is the center set in the object area.

The Euclidean distance estimation measures with using the distance among items and cluster centers. Eq. (4) represents the objective function that it should be minimized.

$$f = \sum_{i=1}^{k} \sum_{i=1}^{n} \left(\left\|x_i(j) - C_j\right\|\right)^2 \quad (4)$$

where

$x_i(j)$	is a particular ith data point at jth cluster
C_j	is centroids or cluster center
n	is number of instances or data points
$\left\|x_i(j) - C_j\right\|$	is Euclidean distance(ED) between $x_i(j)$ and C_j The cluster centers are derived by Eq. (5)

$$C_j = \left(\frac{1}{n}\right) \sum_{x \in s}^{n} X_j \quad (5)$$

where C_j is the count of data objects.

Each cluster is characterized by its center point called the centroid. Generally, the distance measures used in clusters do not represent the spatial distance values. In general, the solution to find the global minimum which is the exhaustive choice of starting points. But usage of many replicates with random initial point leads to a solution called global solution. A centroid is a point whose coordinates are gotten by calculating the mean of each coordinate of the points of samples classified into the clusters. The KM algorithm steps are following below.

2.2 *KM Algorithm Steps*

Input: number of clusters k and n objects

- Step 1: Select k objects as initial cluster centers
- Step 2: Repeat
  - Step 2.1: Do
    - Step 2.1.1: Compute the distances between the cluster centers and the data objects with using Eq. (4)
    - Step 2.1.2: Allot each object or item to corresponding cluster group according to the minimum distance.

Step 2.1.3: When all the items or objects were allocated, recalculate the perspectives of k-mean centroids using Eq. (5)

Step 2.1.4: Until converging has been accomplished (cluster centers do not change).

Step 2.2: Done

Output: Set of K clusters which minimize the squared error function.

3 Results and Discussions

Fig. 1 shows the student personal details like age, sex, address (urban or rural), studied nursery school (attended or not), and present studying school.

In this analysis, the student's age (Fig. 1a) is grouped in 8 distant values, the minimum value is 15 and the maximum is 22. The mean age is 16.70 ± 1.27 and median is 17 years. The studied female students (208 female instances) are more

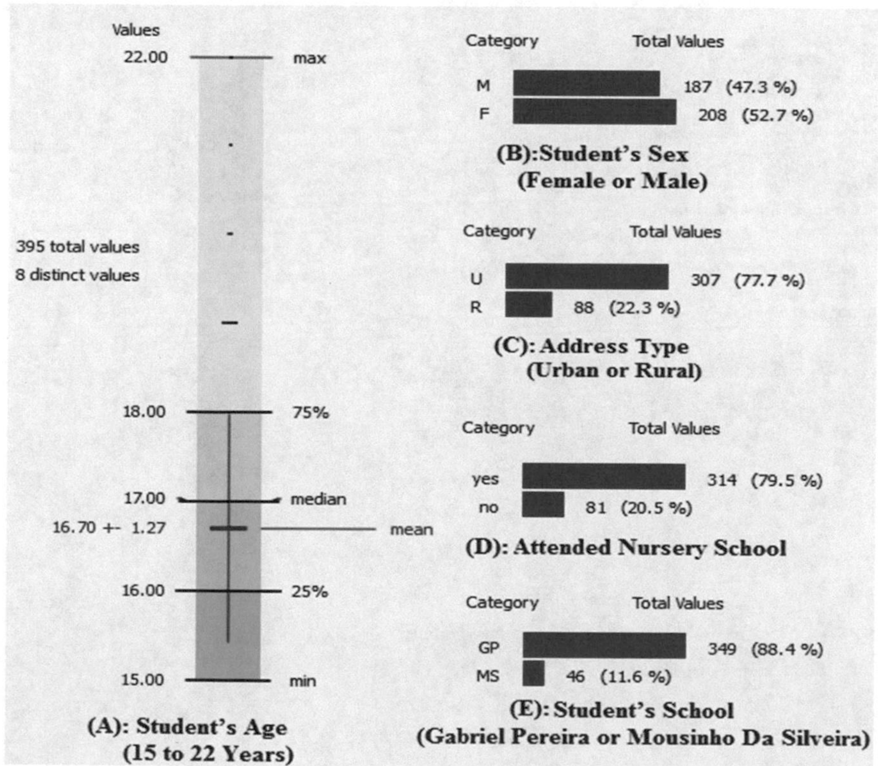

Fig. 1 Student personal details

than male students (187). Most of the students come from urban area (77.7%) and only 20.5% of students come from rural. Most of the students (349 out of 395) attended nursery school. Many students choose the Gabriel Pereira School for the mathematics subject.

Figure 2 shows the analysis of student's family background. The analysis shows that it is the very important factor for student's behavior and his education where it is directly impacted on student's mind and his studies. Comparatively, most of the student's parents education above secondary education (shown in Fig. 2a, b) and higher education instances are indicated that the student's mothers (131(33.2%)) more than the father's higher education (0.96(24.3%)).

Most of the student's parents placed in good jobs and higher positions that they give services (higher positions that mother's job 103 and father's 111 in services shown in Fig. 2c, d). Some student's mothers (14.9%) are staying at home (no job) compare to student's fathers (5.1%). Students' mothers are very responsible than students' fathers where guardian of student is mother (273(69.1%) shown in Fig. 2e). 301(76.2) instances are indicated that the students family relationship is

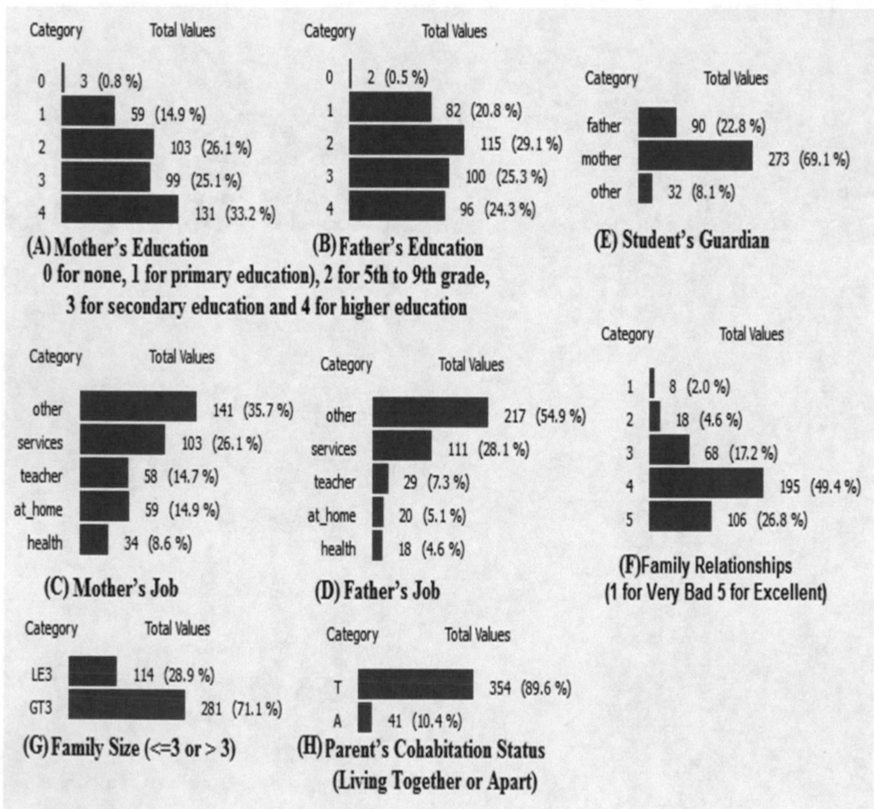

Fig. 2 Student's family background

very good and excellent (shown in Fig. 2f). 26 instances indicate that 6.6% of family relationships are bad and very bad. As per analysis, this feature impacts on the attribute parent's cohabitation status (Fig. 2h) where 41 instances are indicated that the status is apart. 71.1% of the students are coming from family size is greater than 3 and 28.9% less than or equivalent 3 (Fig. 2g).

Other important factors for the student education are external education supports and extracurricular activities. Fig. 3 shows student educational activities and supports. In this analysis, the schools give less extra educational support (12.9% only—shown in Fig. 3a) but the students get big support from family (61.3% shown in Fig. 3b). This is also one of the causes that impacts student result on subject mathematics.

As per Fig. 3c, some of students get extra educational support with paid classes (181(45.8%)). It is one of good choices for good results for subject maths. Internet access at home is somewhat useful resource to improve the knowledge of students. Most of students (83.3%) use Internet at home that it describes in Fig. 3d. Above 50% of students involve in extracurricular activities (Fig. 3e).

Fig. 3f shows that 94.9% (375 out of 395) students want to take higher education. 290(73.4%) students studied more than 2 h in a week (Fig. 3g). Majority of

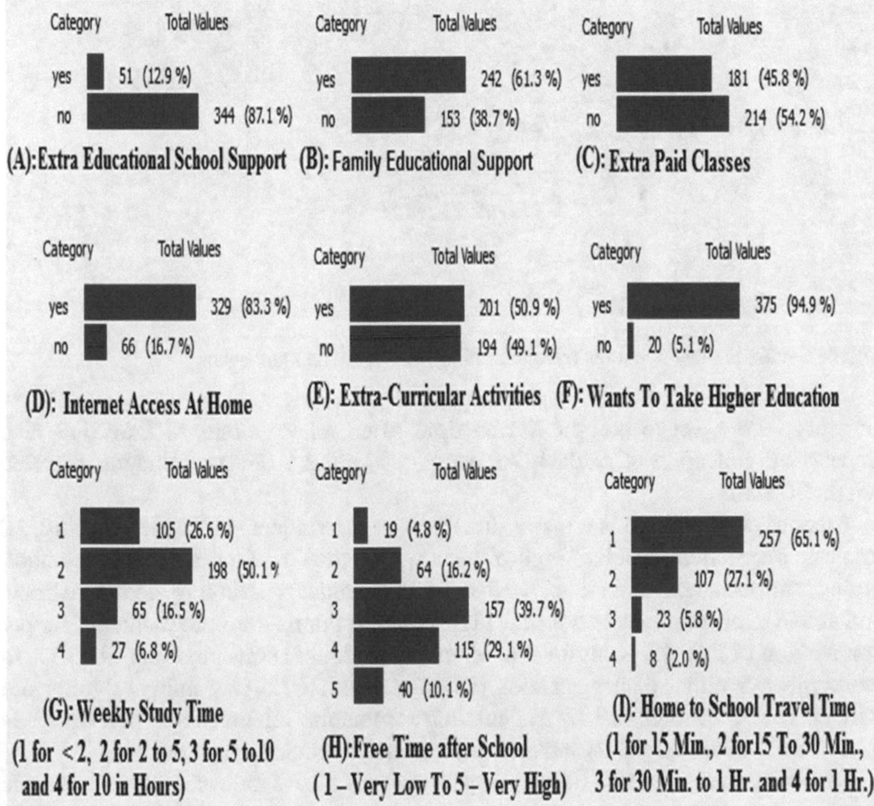

Fig. 3 Student educational activities and supports

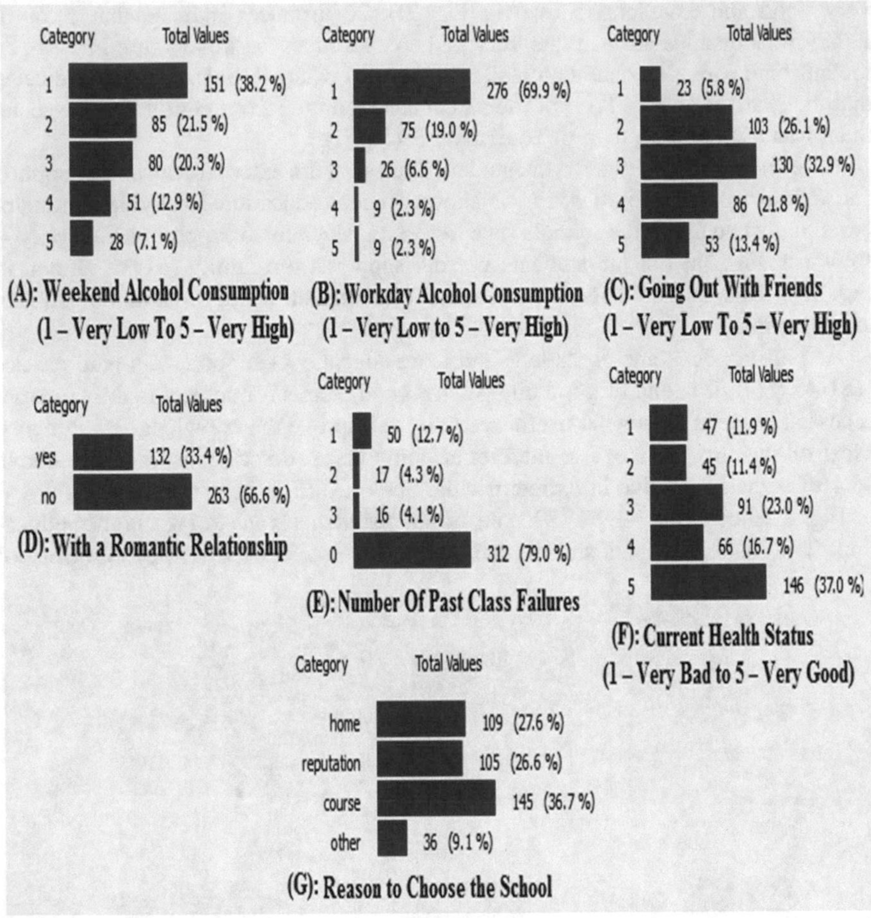

Fig. 4 Student's other activities (alcohol, outing, and romance) and status

students take average and high free time after school timings. Traveling also impacts on education of student. As per Fig. 3i, 92.2% (364) of students traveled within 30 min.

Alcohol consumption is a major drawback of the student studies. In this study, it impacts on student results. Figure 4 shows student's other activities (alcohol, outing, and romance) and status. 40.3% (159) of students' drinking alcohol average and above average in weekends and 11.2% (34) of pupils drunk in workday. As per comparison of Fig. 4a, b, high value of very low alcohol consumption (69.9%) is in working days rather than weekends (38.2%). Above 67% of pupils are going out with their friends Fig. 4c.132 students have romantic relationships, percentage is 33.4. 66.6% (263) of pupils haven't any romantic relationships (Fig. 4d).

79% of students have no failures in past classes. 12.7% (50) of students have one failure. 8.4% of pupils have more than one failure. These statistics are shown in

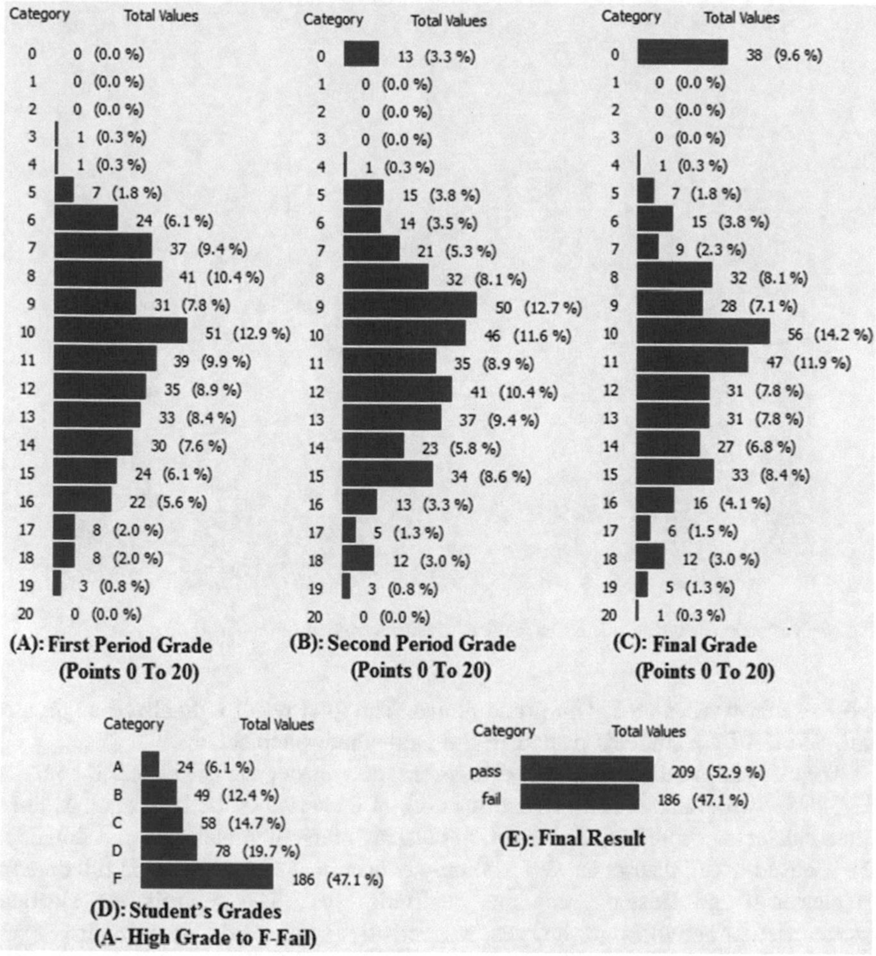

Fig. 5 Performance of the students

Fig. 4e Health factor is also impacts on education of student. As per statistics of Fig. 4f, 37% of students have very good health, 23% average health, 16.7% good health and 23.3% below average health condition (bad (11.4%) and very bad (11.9%). Students choose the school for some reasons. Course and reputation of the school are main factors to choosing the school. Fig. 4g shows that course and reputation occupy the 63.3% (250). Some of students choose the nearest school at their home (27.6%).

Fig. 5 shows the performance of the students on subject mathematics. The final results are assessing with using student's grades. The student grade assess with using first-period grade, second-period grade, and final grade. Level of grades A, B, C, D, and F show the main performance of the student. A is higher grade and F is

Fig. 6 K-means clustering related to result and cluster values

the Fail that it calculated with grade points. The total result is finalized as pass or fail. 52.9% of the students passed in the mathematics subject.

The clusters are divided into two (cluster 0 and cluster 1). The cluster 0 contains 52% of instances and cluster 1 contains 48% of instances in the total data. Mainly, these clusters specified as sex—male in cluster 0, female in cluster 1 and female is the centroid in full dataset as well as family educational support 'yes' in full data no in cluster 0 and cluster 1 contains centroids 'yes'. The centroids of attribute extracurricular activities no, no, yes, respectively. Extra-paid classes centroids are yes, yes, and no. The result attribute gives pass in total in full data pass in cluster 0 and fail instances are clustered in cluster 1.

Figure 6 shows k-means clustering related to result and cluster values. Cluster values (cluster 0 and cluster 1) located on y-axis and result values are put in x-axis. The red color attributes specify the failing instances in result attribute and blue colored passing attributes in result.

Fig. 7 complete linkage hierarchical cluster with respect to all attribute of students performance data set. Fig. shows the detailed analysis about hierarchical structure. Some linkages of hierarchical cluster related to impact of student results. Mainly the attributes reasons, activities, failures, romantic, and absences are related to the grades (G1, G2, G3, and grading (A, B, C, D, F (Fail)). Another cluster shows the family background and supporting for education that attributes like father education and job, mother education and job, and family support and some other features like paid education support and providing Internet at home. As per the analysis, these features are most important for the student result (pass or fail). Some

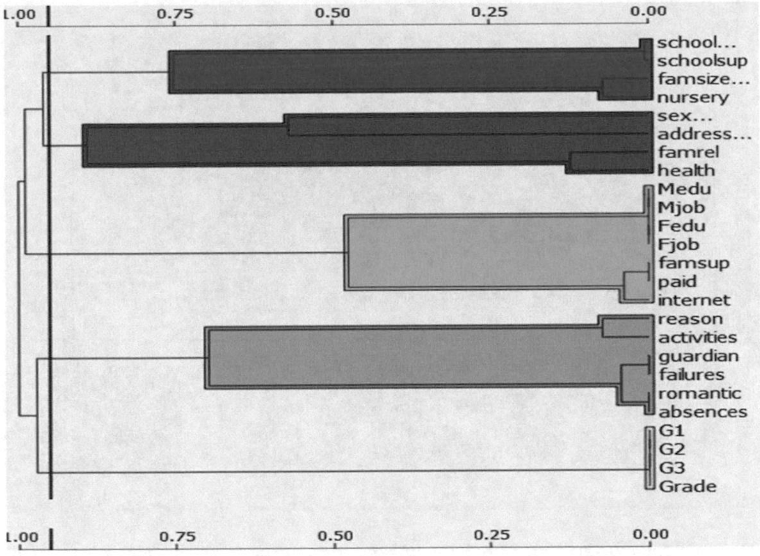

Fig. 7 Hierarchical cluster with respect all attributes of students' performance dataset

other clusters pink and blue colored have their own attribute features affected on student result on subject mathematics shown in Fig. 7. The height of dendrogram value is 0.94788.

Figure 8 shows the hierarchical cluster structure with complete linkage respect to attribute result on subject mathematics of student's performance dataset. It shows the result pass or fail related to other relative attributes. The height of dendrogram value is 4.9888.

Figure 9 shows the cluster projections family size, working day alcoholic consumption, weekend alcoholic consumption, address (rural or urban), sex (male or female), and number of absences. These attributes are shown high-rank projections that affect on the result on subject mathematics. The blue elements show the result pass and the red elements show the failed in result of maths. The most of passing instance indicates the size of family size is greater than 3, working day and weekend daily alcohol consumption is very less value, and very less absent.

Figure 10 shows the cluster projection attributes with respect to parent status HC-class, and grade and final grade G3. The blue elements show the pass instances and the red elements show the failed instances. Most of the failed instances have not Internet. These elements under the hierarchical cluster 0 and 1. Some of these elements have the parent status is apart (A) and some are under together (T). Maximum of failed elements got the below 7 final grade points and grade is F.

Figure 11 shows that quality of projections in peaks. In this, the X-axis contains the projection score and the Y-axis contains number of projections. Clearly, Fig. 12 shows the relation between 'number of projections' related to 'projection scores'

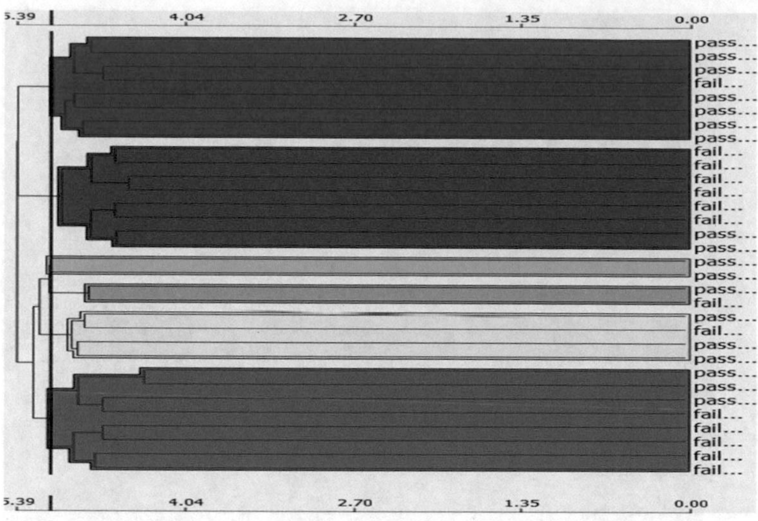

Fig. 8 Hierarchical cluster with complete linkage with respect to attribute result of student's performance dataset

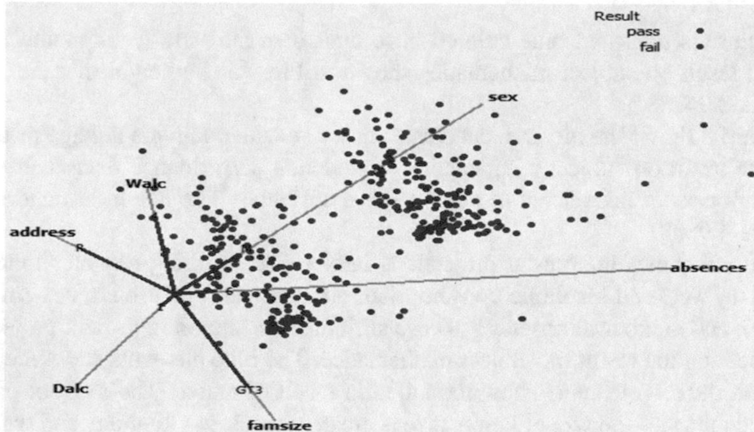

Fig. 9 Cluster projections family size, working day alcoholic consumption, weekend alcoholic consumption, address (rural or urban), sex (male or female), and number of absences

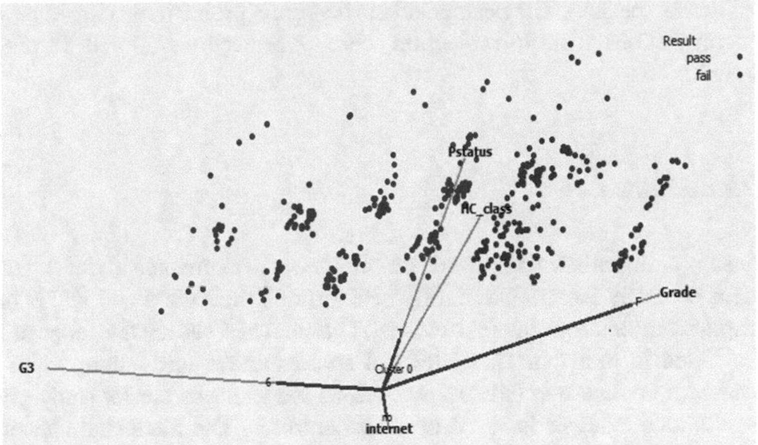

Fig. 10 Projections for Internet, parent status, HC-class, and grade and final grade G3

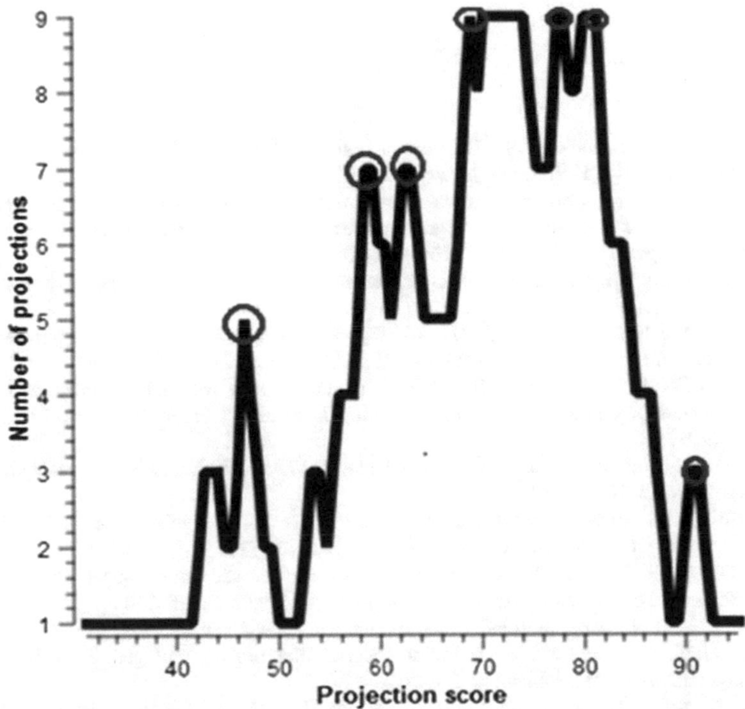

Fig. 11 Quality of projections

with red marks (peaks.). The peak positions that are 3 projections with 42 projection score, 5 projections with 46 projection score, 7 projections 60 and 65 projection scores, and so on.

4 Conclusion

The cluster and statistical analysis on students' performance (result) that it is dependent on many factors like family background, their good and bad habits and supporting from external internal events. The analysis shows the most of failure students failed in past that 12.7% (50) of students have one failure and 8.4% of pupils have more than one failures. As well as the k-means cluster study gives the good results and analysis is based on their centroids. The hierarchal cluster study projections clearly describe the cause of pass and failures of students that the main combination factors are family size, working day alcoholic consumption, weekend alcoholic consumption, address (rural or urban), sex (male or female), and number of absences. The study also shows the quality and performance of the projections.

References

1. Baker, R.S., Inventado, P.S.: Educational data mining and learning analytics. Learning Analytics, pp. 61–75. Springer, New York (2014)
2. Xing, W.: Participation-based student final performance prediction model through interpretable genetic programming: integrating learning analytics, educational data mining and theory. Comput. Hum. Behav. **47**, 168–181 (2015)
3. Xu, J., et al.: Progressive Prediction of Student Performance in College Programs. AAAI (2017)
4. Ahmed, A.B.E.D., Elaraby, I.S.: Data mining: a prediction for student's performance using classification method. World J. Comput. Appl. Technol. **2**(2), 43–47 (2014)
5. Villegas-Ch, W., Luján-Mora, S.: Analysis of data mining techniques applied to LMS for personalized education. In: World Engineering Education Conference (EDUNINE), IEEE. IEEE (2017)
6. Romero, C., Ventura, S.: Data mining in education. Wiley Interdiscip. Rev. Data Min. Knowl. Discov. **3**(1), 12–27 (2013)
7. Shahiri, A.M., Husain, Wahidah: A review on predicting student's performance using data mining techniques. Procedia Comput. Sci. **72**, 414–422 (2015)
8. van Diepen, P., Bert, B.: Performance indicators for online secondary education: a case study. In: Benelux Conference on Artificial Intelligence. Springer, Cham (2016)
9. Eagle, M., et al.: Estimating individual differences for student modeling in intelligent tutors from reading and pretest data. In: International Conference on Intelligent Tutoring Systems. Springer International Publishing (2016)
10. Kim, Nam-Ho, An, Dawn, Choi, Joo-Ho: Introduction. Prognostics and Health Management of Engineering Systems, pp. 1–24. Springer, Cham (2017)
11. Rao, H.S.S., Aishwarya S., Vinayak, H.: Academic dashboard—descriptive Analyti-cal Approach to analyze student admission using education data mining. In: Information and

Communication Technology for Sustainable Development, pp. 423–432. Springer, Singapore (2018)
12. Fernandes, E., et al.: Educational data mining: discovery standards of academic performance by students in public high schools in the federal district of Brazil. In: World Conference on Information Systems and Technologies. Springer, Cham (2017)
13. Daud, A., et al.: Predicting student performance using advanced learning analytics. In: Proceedings of the 26th International Conference on World Wide Web Companion. International World Wide Web Conferences Steering Committee (2017)
14. Cortez, P., Silva, A.: Using data mining to predict secondary school student performance. In: Brito, A., Teixeira, J. (eds.), Proceedings of 5th Future Business Technology Conference (FUBUTEC 2008), pp. 5–12, Porto, Portugal, Apr, 2008, EUROSIS (2008). ISBN 978-9077381-39-7
15. https://archive.ics.uci.edu/ml/datasets/student+performance

Early Prediction of Five Major Complications Ascends in Diabetes Mellitus Using Fuzzy Logic

Aruna Pavate, Pranav Nerurkar, Nazneen Ansari and Rajesh Bansode

Abstract A diabetes proliferation gives growth to many health issues. Diabetes complications can be avoided and or prevented with the recommended lifestyle and treatments. Analysis and administration of diabetes mellitus are the inspiring problems for the present-day exploration of risk classification. Diabetes mellitus is a chronic condition noticeable by higher levels of blood glucose. The objective of this work is to expect the major five complications which arise because of diabetes mellitus. In the proposed model, fuzzy logic is used to classify the risk level of major five complications like Vision Loss, Kidney failure, Neuropathy, Stroke diseases, Heart Problem. The correct classification of a rate of FL measured 92.5% for prediction of disease.

Keywords Diabetes mellitus · Fuzzy logic · Neuropathy · Heart problem Vision Loss · Kidney failure · Stroke disease

A. Pavate (✉)
Athrava College of Engineering, Mumbai University, Mumbai 400095, Maharashtra, India
e-mail: arunapavate@atharvacoe.ac.in

P. Nerurkar
Veermata Jijabai Technological Institute, Mumbai, Maharashtra, India
e-mail: pranavn91@gmail.com

N. Ansari
St. Francis Institute of Technology, Mumbai University, Mumbai 400130, Maharashtra, India
e-mail: naz2ensfit@gmail.com

R. Bansode
Thakur College of Engineering, Mumbai University, Mumbai 400101, Maharashtra, India
e-mail: rajesh.bansode1977@gmail.com

J. Nayak et al. (eds.), *Soft Computing in Data Analytics*,
Advances in Intelligent Systems and Computing 758,
https://doi.org/10.1007/978-981-13-0514-6_72

1 Introduction

The prevalence of diabetes is growing at a fast speed due to lots of factors. In type 2 diabetes patients, a kind of dissimilar organs are below stress and are prejudiced by the different metabolic situation. Hence, the disease has to be properly coped within specified boundaries to reduce other further health complications affected by diabetes mellitus and to promise the prolonged existence of life. This work deals with designing and building a system for diabetes mellitus patients to predict the further complications in diagnosis which are common health disorders in many people. This system is designed in simple English language where user is expected to give input and based on the input system ascends the output. The system uses a fuzzy logic to decide the further five complications like Heart Disease with the risk level that is caused because of diabetes in an individual. The three altered levels are categorized as low risk, medium risk, and high risk. The disease is any situation that harms the normal function of the body and is therefore associated with paralysis of usual steadiness [1]. Chronic disease is a long-lasting situation that cannot be healed but can be controlled by proper care. Chronic disease affects the population worldwide. Diabetes is one of the chronic diseases that sneak up on millions of Indian people. It is gradually renowned as one of the major cause of ill health and death; on the other hand, many people who are suffering from diabetes mellitus are naïve of unbearable side effects of the disease and further complications that rises for the reason that of disease like Stroke problem, Nerve Damage, Heart Disease, Neuropathy, blindness, Nephropathy, High blood Pressure [1].

The main reason for this research was to bring together the knowledge and improvement of the system which is able to completely diagnose the chronic disease and the complications that arise from diabetes disease. The system "Disease Analysis for Diabetes Diagnosis and Complications" is a medical system for diagnosis of the disease using fuzzy rule-based approach and has shown better knowledge representation to improve the decision-making process, and it supports diagnosis of complications. The system provides an interactive way to the user and asks questions about personal details and symptoms, to answer whether the diabetes symptom is experienced by the individual or not. This system is not meant to replace the human endocrinologist. This system is firstly assessed with the available test. The result of the evaluation is precise.

1.1 *Objectives of the Research*

Diabetes is a most appropriate disease that can be predicted using soft computing techniques for a number of reasons [2]. The major reason is that the diabetic patient data are easily available. Another reason, early treatment helps the people to save the money and prohibit further complications in the early stage as diabetes can yield awful diseases such as Kidney failure, blindness, and heart failure. Lastly, general practitioners need to identify how to quickly he/she can diagnose the cases and avoid further complications and how to quickly identify and diagnose potential cases. In the present investigation, we apply soft computing techniques such as fuzzy logic-based new test cases and the diabetes data that is available [1]. The objective of this work is to use soft computing technique to ascend the risk level that recognizes the best mode of treatment for diabetes in the early stage across different age groups in India. The ideas of this research are:

- To recognize the various complications that are rising in Indian people because of diabetes using soft computing technique like fuzzy logic.
- To develop a system which helps the general practitioners and individuals to identify the risk level such as low risk, medium risk, and high risk.

2 Literature Review

The literature review consists of critical appraisal of previous works done in this field. There are lots of experiment done in the past to predict the diabetes disease. A few of them are explained below:

To classify the cardiovascular disease, there are numerous techniques used in the previous work such as artificial neural network (ANN) [3–5], linear discriminant analysis (LDA), decision trees [6], support vector machine (SVM), and fuzzy logic system [7, 8]. Primarily, many researchers applied support vector machine and fuzzy logic in their research work [9–11].

P. K. Anooj in his work presented a system based on weighted fuzzy rule base for diagnosing the Heart Disease. The proposed system generates the fuzzy rules automatically. The weighted fuzzy rule system is an advantage for operative learning of the system. The proposed system gives better result as compared with the network-based system in terms of accuracy, sensitivity, and specificity [12].

Anu Sebastian, Surekha Mariam Varghese present a fuzzy-based system to classify and predict the life probability of patient's life [13].

Randa El-Bialy, in this study, aims to create a fuzzy logic-based handy sensory technology such as smartwatch which monitors the risk factors. This work presented here introduced the concept of home-based falls prevention and management using vital signals for instantaneous risk estimation, employing predictive modeling and fuzzy logic-based decision-making [14].

3 Working

Diabetes can cause severe complications in both changeable and long-standing. The changeable complications are hyperglycemia, diabetic ketoacidosis, and hypoglycemia [15]. Long-standing complications include Vision Loss, Kidney failure, Neuropathy, Stroke diseases, Heart Problem.

3.1 Fuzzy System

The concept of fuzzy logic was introduced by Lotfi A. Zadeh in the 1960s to make available rules of mathematics and the functions which help in solving queries generated by natural language. Fuzzy logic is used to predict the values which are not precise or vague [16].

A fuzzy logic methodology has been developed in a number of tactics for predicting the disease. The features entered by the user are considered as input to the fuzzification module and their membership functions are outlined as Low, Medium, and High as explained in Table 1. The system structure can be précised in the following steps and carried out in order as shown in Fig. 1:

1. **Outline linguistic variable for the system**:
   Linguistic variables are the entered and displayed variables defined using simple terms like word or sentences. For this system, Low, Medium, High, Present, Absent, Maybe are linguistic variables represented in Table 1.
2. **Building Membership functions for the linguistic variables**:
   Membership function can be described as a method to explain the real-world problems by understanding in place of knowledge. Membership functions are signified by graphical diagrams like trapezoidal, triangular, and rectangular [16]. Figure 2 represents the graphical representation of membership function for the heart rate. We have used triangular and trapezoidal membership function for the fuzzy set such as for heart rate the membership function defined as follows:

   $\mu low(y) = 1$ when $y \in [1, 70]$ $\mu low(y) = (70 - y)/(70 - 90)$ when $y \in [70, 90]$.
   $\mu low(y) = 0$ otherwise.
   $\mu medium(y) = (y - 90)/(110 - 90)$ when $y \in [90, 110]$ $\mu medium(y) = (111 - y)/(130 - 111)$ when $y \in [111, 130]$.
   $\mu medium(y) = 0$ otherwise.
   $\mu high(y) = 0$ when $y \in [1, 130]$ $\mu high(y) = (y - 135)/(216 - 135)$ when $y \in [135, 216]$ $\mu high(y) = 1$ when $y \in [135, 250]$.

3. **Assembling knowledge base rules**:
   In this module, a matrix is created for relevant features versus symptom values that diagnosis system is projected to offer. A set of if–then rules and the related membership functions have been constructed. The obtained result is to be assigned to output variable.

Table 1 Diabetes complication symptoms used in our experimentation

Disease name	Attribute description with linguistic variables
Kidney disease	Fatigue: indicates person suffering from fatigue values: present = high = 1, absent = low = 0
	Nocturia indicates person suffering from Nocturia values: present = high = 1, absent = low = 0
	Poor appetite: indicates person suffering from poor appetite values: present = high = 1, absent = low = 0
	Legs swelling: indicates person suffering from fatigue values: present = high = 1, absent = low = 0
	Itchy and dry skin: indicates person suffering from itchy and dry skin values: present = high = 1, absent = low = 0
	Blood pressure: indicates person suffering from blood pressure values: severe = high = 1, normal = low = 0
Heart disease	Cholesterol level (mg/dl): indicates person cholesterol values: (1) less than 175: low (2) 176–251: medium (3) more than 251: high
	Blood sugar: indicates person cholesterol values: (1) less than 125: low (2) 126–200: medium (3) more than 200: high
	Heart rate: indicates person cholesterol values: (1) less than 90: low (2) 70–130: medium (3) more than 110: high
Neuropathy	Neuropathy: indicates person suffering from numbness values: present = high = 1, absent = low = 0
	Tingling: indicates person suffering from tingling values: present = high = 1, absent = low = 0
	Pricking: indicates person suffering from pricking values: present = high = 1, absent = low = 0
	Weakness in hands, arms, and feet: indicates person suffering from weakness in hands values: present = high = 1, absent = low = 0
	Loss of sense: indicates person suffering from loss of sense values: present = high = 1, absent = low = 0
Stroke	Sudden weakness: indicates person suffering from numbness values: present = high = 1, absent = low = 0
	Difficulty in speaking: indicates person suffering from numbness values: present = high = 1, absent = low = 0
	Difficulty in swallowing: indicates person suffering from numbness values: present = high = 1, absent = low = 0
	Sudden intense headache: indicates person suffering from numbness values: present = high = 1, absent = low = 0
	Loss of consciousness: indicates person suffering from numbness values: present = present = high = 1, absent = low = 0
Vision Loss	Blurred vision: indicates person suffering from blurred vision values: present = present = high = 1, absent = low = 0
	Fluctuating vision: indicates person suffering from fluctuating vision values: present = high = 1, absent = low = 0
	Dark or empty areas: indicates person suffering from dark or empty areas values: present = high = 1, absent = low = 0
	Color perception difficulty: indicates person suffering from color perception difficulty values: present = high = 1, absent = low = 0

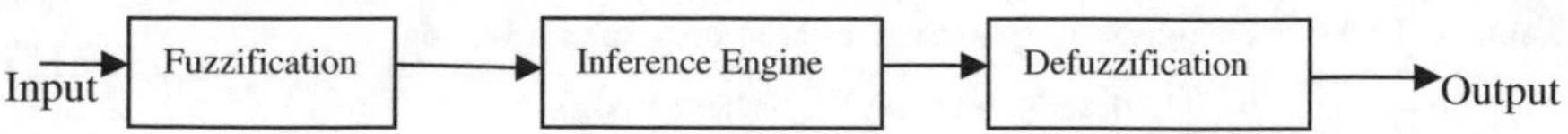

Fig. 1 General structure of fuzzy logic

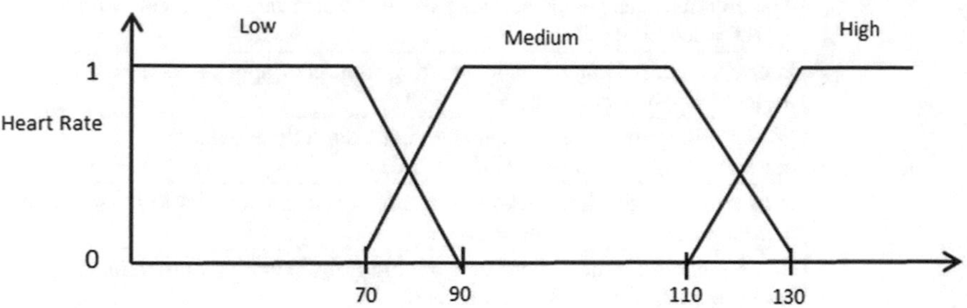

Fig. 2 Membership function representation for heart rate

Sample rules created using fuzzy rule base for disease complications are listed as follows:

- If ((numbness = Absent) and (tingling = Absent) and (prickling = Absent) and (weakness_in_hands & arms = Absent) and (loss_of_sense = Absent) Then (Neuropathy_Risk = Low)).
- If ((Sudden_Weakness = Absent) and (Difficulty_in_speaking = Absent) and (Difficulty_in_swallowing = Absent) and (Sudden_intense_in headache = Absent) and (Loss_of_consciousness = Absent) Then (Stroke_Disease_Risk = Low)).
- If ((blurred_vision = Present) and (Fluctuating_vision = Absent) and (Dark_empty_areas = Absent) and (color_perception = Absent) Then (Vision Loss_Risk = Low)).
- If ((cholesterol < 175) and (blood_pressure = Normal) and (blood_sugar < 125) and (heart_rate < 90) Then (Heart_Disease_Risk = Low)).

4. **Attaining Fuzzy values**:
   Set of operations are performed on generated rules. The different operations performed like max, min, respectively. The evaluation result is combined together to get the final fuzzy result.
   $max_val = max($per_very_high,$per_high,$per_medium,$per_low,
   $per_very_low);

5. **Perform Defuzzification**: It converts the fuzzy output into a crisp output value. Converting fuzzy output values into crisp values is an optional step which is used when it is useful. Many defuzzification methods are presented [17]. The maximum and centroid methods are the most commonly used techniques. The established fuzzy expert system prototype would query the user for the applicable patient symptoms. The strength of every single symptom is identified by a fuzzy value such as low, moderate, and high for those symptoms that cannot be measured quantitatively. Other measurable symptoms such as the cholesterol level or blood sugar, for instance, would be input directly as numeric values that would be properly fuzzified.

4 Result

The performance testing has been performed on both training and testing datasets, and we merely used diabetes disease datasets to evaluate our system [1]. Figures 3, 4, and 5 show the diabetes complications, Vision Loss Module, and result of Vision Loss, respectively. Accuracy can be obtained by sum of True classified cases (TP) and True No classified cases (TN) divided by the total number of instances which is given in Eq. 1.

$$\text{Accuracy} = (\text{TP} + \text{TN}) \,/\, \text{Total number of Instances} \tag{1}$$

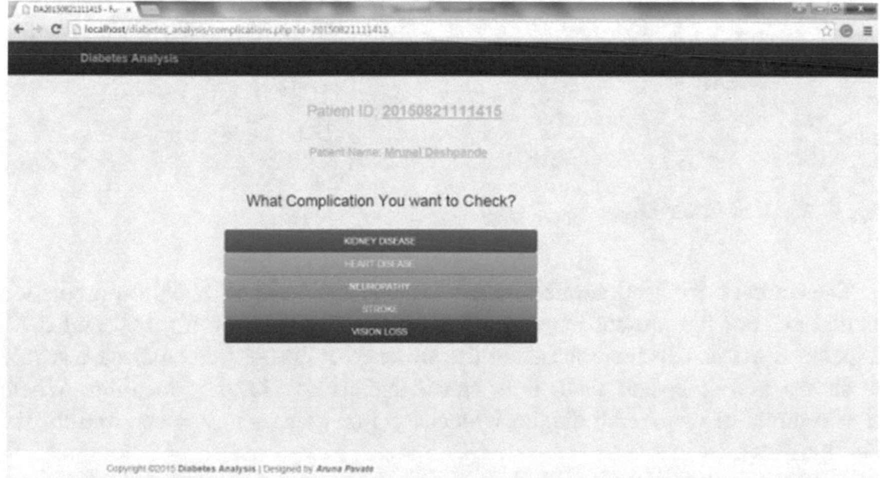

Fig. 3 Diabetes complications

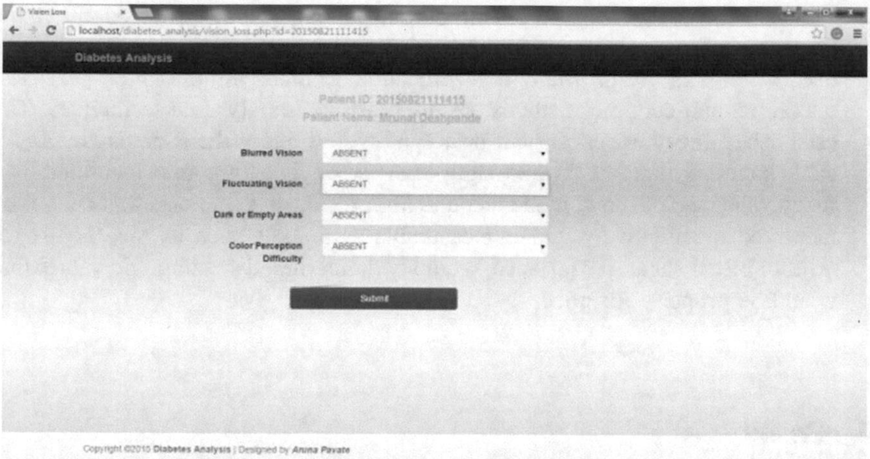

Fig. 4 Vision Loss module

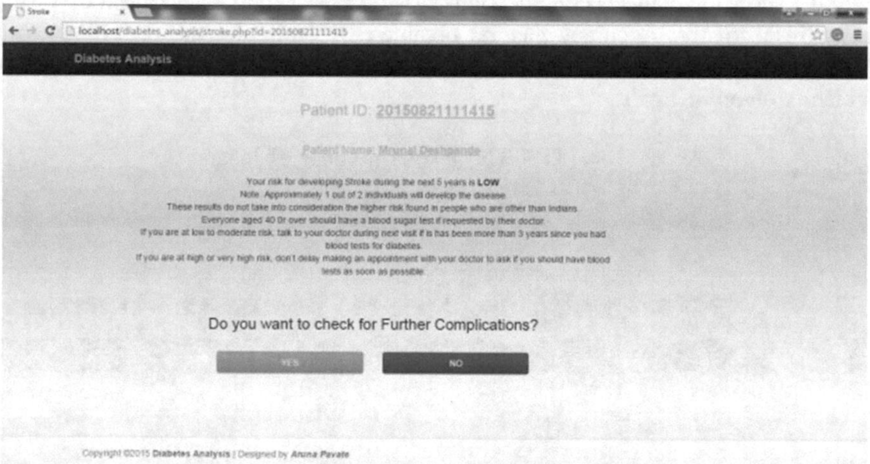

Fig. 5 Result of Vision Loss

The correct classification rate of fuzzy logic is measured as 92.5% for prediction of disease. For the present example, the values of xi and xy are 0.69 and 0.77, respectively. This tells that the risk of the presence of the considered disease is 79% as shown in Fig. 6, and then, it is considered as risk level is medium. Where xi = centroid of the overall diagnosis decision fuzzy set and xy = centroid for the Yes fuzzy set.

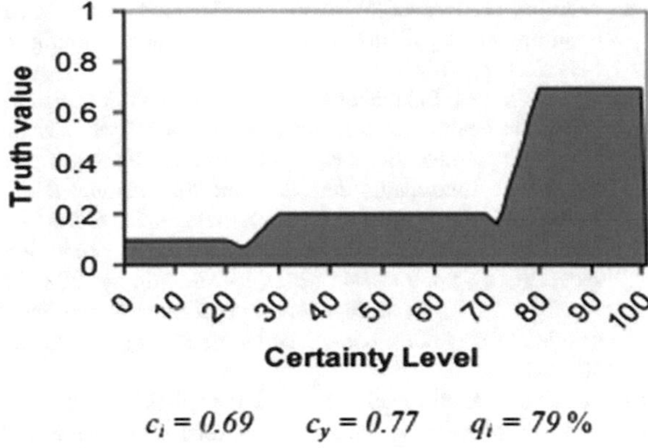

Fig. 6 Defuzzification result of Vision Loss

5 Conclusion

The proposed system is not a substitute for the physician; this system will provide a basic conclusion based on user input. The system is designed as a dynamic Web application that offers an interface to the user to enter the symptoms, and system gives predication of five major complications using fuzzy logic. The accuracy of the system is intended, and it gives 92.5% correctly classified cases.

References

1. Pavate, A., Ansari, N.: Risk prediction of disease complications in type 2 diabetes patients using soft computing techniques. In: 2015 Fifth International Conference on Advances in Computing and Communications. IEEE, pp. 371–375 (2015)
2. Zainuddin, Z., Pauline, O., Ardil, C.: A neural network approach in predicting the blood glucose level for diabetic patients. World Academy of Science, Engineering and Technology, 2009, vol. 6, no. 50, pp. 981–988 (2009)
3. Silipo, R., Marchesi, C.: Neural techniques for ST-T change detection. In: Proceedings of Computers in Cardiology 1996, pp. 677–680 (1996)
4. Ng, B.A.: An intelligent approach based on principal component analysis and adaptive neuro fuzzy inference system for predicting the risk of cardiovascular diseases. In: 2013 Fifth International Conference on Advanced Computing (ICoAC). IEEE, pp. 241–245 (2013). 978-1-4799-3448-5/13
5. Geman, O., Chiuchisan, I., Toderean, R.: Application of adaptive neuro-fuzzy inference system for diabetes classification and prediction. In: The 6th IEEE International Conference on E-Health and Bioengineering—EHB 2017 Grigore T. Popa University of Medicine and Pharmacy, Sinaia, Romania, 22–24 June 2017
6. El-Bialy, R., Salamay, M.A., Karam, O.H., Khalifa, M.E.: Feature analysis of coronary artery heart disease data sets. 1877–0509. The Authors. Published by Elsevier B.V. (2015)
7. Tsipouras, M.G., Voglis, C., Fotiadis, D.I.: A framework for fuzzy expert system creation application to cardiovascular diseases. IEEE Trans. Biomed. Eng. **54**, 2089–2105 (2007)

8. Akbulut, H., Barisci, N., Arinc, H., Topal, T., Luy, M.: Prediction of coronary angiography requirement of patients with fuzzy logic and learning vector quantization. In: IEEE 2013, ICECCO, pp. 1–4
9. Pasupathi, P., Bakthavathsalam, G., Rao, Y.Y., Farook, J.: Cigarette smoking–effect of metabolic health risk: a review. Diabet. Metab. Syndr. Clin. Res. Rev. **3**(2), 120–127 (2009)
10. Ghosh, D., Midya, B.L., Koley, C., Purkait, P.: Wavelet aided SVM analysis of ECG signals for cardiac abnormality detection. In: 2005 Annual IEEE INDICON, pp. 9–13 (2005)
11. Abbasi, H., Unsworth, C.P., McKenzie, A.C., Gunn, A.J., Bennet, L.: Using type-2 fuzzy logic systems for spike detection in the hypoxic ischemic EEG of the preterm fetal sheep. IEEE, pp. 938–941 (2014). 978-1-4244-7929-0/14
12. Anooj, P.K.: Clinical decision support system: risk level prediction of heart disease using weighted fuzzy rules. King Saud University. Production and hosting by Elsevier B.V., pp. 27–40 (2011)
13. Sebastian, A, Varghese, S.M.: Fuzzy logic for child-pugh classification of patients with cirrhosis of liver. In: 2016 International Conference on Information Science (ICIS). IEEE, pp. 168–171 (2016)
14. Iakovakis, D.E., Papadopoulou, F.A., Hadjileontiadis, L.J.: Fuzzy logic-based risk of fall estimation using smart watch data as a means to form an assistive feedback mechanism in everyday living activities. Healthc. Technol. Lett. 1–6 (2016)
15. What is diabetes, causes. https://www.medicalnewstoday.com/info/diabetes. Accessed 25 Nov
16. Zadeh, L.A.: Fuzzy sets. Inf. Control **8**(3), 338–353 (1965)
17. Klir, J.K., Yuan, B.: Fuzzy Sets and Fuzzy Logic: Theory and Applications. Prentice- Hall (1995)

An Enhanced Deep Learning Model for Duplicate Question Pairs Recognition

K. Abishek, Basuthkar Rajaram Hariharan and C. Valliyammai

Abstract Quora is a question-and-answer site, where questions are asked, answered, edited, and organized by its community of users. It is a platform to ask questions and connect with people who contribute unique insights and quality answers. But, huge number of questions ends up being duplicates of other questions asked by different user with various words or expressions. Preferably, these duplicate inquiries would be consolidated into a solitary accepted query, as doing as such would give various advantages. Quora uses a random forest model for duplicate question pair's identification. This method is not efficient for a large number of questions. The proposed model uses GloVe embedding which consists of 100d Wikipedia vectors and improves the prediction accuracy.

Keywords Deep learning · Neural networks · CNN · LSTM

1 Introduction

Identifying duplicate questions spares the enquirer time, if their query has just been addressed already on the site. Rather than sitting tight for minutes or hours for an answer, they can find their solution immediately. This lessens the time it takes users to locate the best answers and enables analysts to better comprehend the connection among questions and their answers. We say that two questions are copies, if the inquiry askers communicated a similar aim when composing the inquiry. That is, any legitimate response to one inquiry is likewise a substantial response to the next. For instance, the inquiries "What's the easiest way to become a millionaire?" and

K. Abishek · B. R. Hariharan · C. Valliyammai (✉)
Department of Computer Technology, Anna University, Chennai 600025, India
e-mail: cva@annauniv.edu

K. Abishek
e-mail: abishekkumar2@gmail.com

B. R. Hariharan
e-mail: basuthkarhariharan@gmail.com

J. Nayak et al. (eds.), *Soft Computing in Data Analytics*,
Advances in Intelligent Systems and Computing 758,
https://doi.org/10.1007/978-981-13-0514-6_73

"How do I get to become a millionaire in the simplest way possible?" are said to be copies. It is significant that a few questions have characteristic uncertainty in view of their writings, and we cannot state for sure that they express a similar expectation. For instance, "How would I make $100 k USD?" and "How would I get a hundred thousand?" could be copies, in the event that we expect that the "hundred thousand" wanted is in US dollars, however, that may not be valid. Frequently, any human naming procedure will mirror this uncertainty and acquaint some measure of such dataset noise. Our research goal in this paper is to develop learning models that can automatically induce representations of human language, in particular its structure and meaning in order to solve multiple higher-level language tasks.

There has been great progress in delivering technologies in natural language processing (NLP) such as extracting information from big unstructured data on the Web, sentiment analysis in social networks, or grammatical analysis for essay grading. In NLP and machine learning, we often develop an algorithm and then force the data into a format that is compatible with this algorithm. For instance, a common first step in text classification or clustering is to ignore word order and grammatical structure and represent texts in terms of unordered lists of words, so-called bag of words. This leads to obvious problems when trying to understand a sentence. For example, the two sentences: "Unlike the surreal Leon, this movie is weird but likeable." and "Unlike the surreal but likeable Leon, this movie is weird." The overall sentiment expressed in the first sentence is positive. Our model learns that while the words compose a positive sentiment about Leon in the second sentence, the overall sentiment is negative, despite both sentences having exactly the same words. This is in contrast to the above-mentioned bag-of-words approaches that cannot distinguish between the two sentences. Another common simplification for labeling words with, for example, their part-of-speech tag is to consider only the previous word's tag or a fixed-sized neighborhood around each word. Our models do not make these simplifying assumptions and are still tractable. The most learning systems depend crucially on the feature representations of the input. Instead of relying only on word counts to classify a text, state-of-the-art systems use part-of-speech tags, special labels for each location, person, or organization (so-called named entities), parse tree features or the relationship of words in a large taxonomy such as WordNet.

In this paper, we propose a Siamese neural network architecture for recognizing identical question pair that matches and applies three distinctive encoding procedures to produce feature vectors for each inquiry. These encoded vectors are connected into a solitary input vector that is given through a multilayer perceptron to induce the output label likelihood. The encoding techniques investigated incorporate a convolution neural network (CNN). The proposed models overcome the above-mentioned shortcomings.

2 Related Work

Many models have been proposed for the process of text classification over the years. CNNs, LSTMs, and a hybrid model have used the Siamese network architecture [1]. However, the accuracy is limited because of using pre-trained word embeddings not related to the dataset. There is a possibility to move away from Siamese architecture toward multi-perspective matching. Basic Bi-CNN based on Siamese architecture with two weights sharing CNNs [2], each sharing one of the two sentences and a final layer that solves sentence pair tasks. Also, three different ABCNN techniques were used to compute attention weights. But, no significant improvements were observed for two layers of convolution over one layer. Deep ABCNNs show the improved results over large datasets. Multiple models to obtain distributed representation of words in vector space were also proposed [3]. The models include feedforward neural net language model (NNLM), recurrent neural net language model (RNNLM), continuous bag-of-words and skip-gram model. However, this cannot be applied to automatic correction of words in knowledge bases. It can be extended for verification of correctness of existing facts. Another method involves encoding the sentences using bidirectional long short-term memory network (BiLSTM) [4]. Then, the sentences are matched in two directions, P -> Q and P <– Q. Then, another BiLSTM layer is utilized to aggregate the results into fixed length matching vector. Finally, based on the matching vector, a decision is made through fully connected layer. However, the accuracy of bilateral multi-perspective matching (BiMPM) is not impressive for answer selection tasks and can be improved further. This kind of method can work well on a specific task or dataset, but it is hard to generalize well to other tasks. In another model, LSTM-RNN model sequentially takes each word in a sentence, extracts its information, and embeds it into a semantic vector [5]. Due to its ability to capture long short-term memory, the LSTM-RNN accumulates increasingly richer information as it goes through the sentence, and when it reaches the end, the hidden layer provides information about whole sentence. However, the model can be further extended for important NLP takes like questions/answers. Also, attention mechanisms can be included with this model to improve performance and find out which words in the query are aligned to which words in the document. Topic-aware CNNs and topic-aware LSTMs for disambiguation in the topic models were used, but choosing right parameters for TopCNN and TopLSTM is necessary for improving the accuracy [6].

A novel neural net structure, feedforward sequential memory networks (FSMNs) [7] have also been proposed to model long-term dependence in time series without using recurrent feedback. However, the model fails to consider the semantic or contextual similarity among words and simply considers word embeddings from a trained corpus. An interactive representation to model the relationship between two sentences not only on word level, but also on phrase and sentence level by employing CNNs to conduct paraphrase identification by using semantic and syntactic features [8]. However, this model does not give attention to each particular word, and hence, attention-based models outperform this. There is no importance given to the context where the word occurs in the sentence. Experimental results in [9] show that the CNN with in-domain word embeddings achieves high performance even with limited training data. The word embeddings pre-trained on domain-specific data achieve very high performance and higher accuracy. However, some of the failed cases include questions where essential information is presented as an image. A question similarity computation model adopting deep learning method has also been proposed [10]. It models questions using BiLSTM and uses fully connected neural networks as similarity measurement. However, the questions' similarity is computed without considering the answer text. It does not include tasks related to question answering, paraphrase detection, and generation.

A simple CNN with little hyperparameter tuning and static vectors achieves excellent results on multiple benchmarks. However, trains a simple CNN with one layer of convolution on top of word vectors obtained from an unsupervised neural language model. A general educational domain ontology construction method has also been proposed with deep analysis on the relationships between subject knowledge in university majors [11]. However, there is no reuse ontology for the fields, so need to develop a new ontology for certain fields during the evaluation stage. The ontology is not automatic and extracts a set of keywords from each training document and automatically generates a domain ontology using the theorem of formal concept analysis and the keyword dictionary [12]. As a result, it classifies documents into proper nodes (classes) in the domain ontology. As a result, a new global log-bilinear regression model that combines the advantages of the two major model families in the literature: global matrix factorization and local context window methods [13]. But, it does not achieve the functions of Word2vec vectorizer. A method to reengineer the thesauri to ontology and especially the method to how to obtain the correct semantic relations has been proposed [14]. It also proposes a kind of way to get and adjust the semantic relations automatically. However, this does not have the versatility for other thesauri resources. The thesauri give only rough relations without clear identifications, and the errors exist in many relations of concepts, and these are seriously short of the requirements of ontological semantics.

In conventional natural language processing (NLP), w-shingling has been effectively used to evaluate the comparability between two text records. In any case, matching inquiries can be reworded from numerous points of view; thus, such methods that depend on word coincidence will not work. CNNs have indicated noteworthy guarantee over conventional NLP systems in the domains of sentence classification and sentiment analysis. Taking sentence inputs truncated to a most extreme length, the expressions of each sentence are changed over into a matrix of pre-trained word embeddings inferred utilizing Word2vec. This model accomplishes great execution over a diverse range of sentence classification tasks including sentiment analysis. With Quora's first open dataset3, latest scholarly concentration for matching question sets recognize has been watched.

3 Proposed Work

The questions pairs from the dataset are passed through a Siamese neural networks architecture which is identical for both questions. The first encoding is a CNN which is shown in Fig. 1. The questions from Quora are concatenated, and a vocabulary preprocessor is created. This vocabulary preprocessor is used to get word mappings and pre-trained words from GloVe model which is used to create word embeddings. These word embeddings are passed to CNN which three filters of sizes (3 × d), (4 × d), (5 × d). The CNN layer uses a stride size of [1, 1, 1, 1] and has valid padding which reduces the dimensions of the output. Rectified linear unit (ReLU) activation is used for nonlinearity, and this output is passed through a maxpool layer. The maxpool layer also has stride size [1, 1, 1, 1] and has valid padding. Following this, a dropout layer is added to reduce overfitting. The output from this layer is passed through a fully connected layer which has 1536 hidden units. The final output layer consists of a one-hot vector with the two elements of the two classes such as non-duplicate and duplicate.

Algorithm1: Word Embeddings

```
Input: Quora Dataset with 400K Questions
Output: Word embeddings
begin
    n ← number of words in word corpus
    m ← number of training examples
    glove ← pre- trained glove embeddings
    A ← [ ]
    questions ← [questions in dataset]
    for question in the questions
            X ← glove[question]
            A ← A + X
    end for
end
```

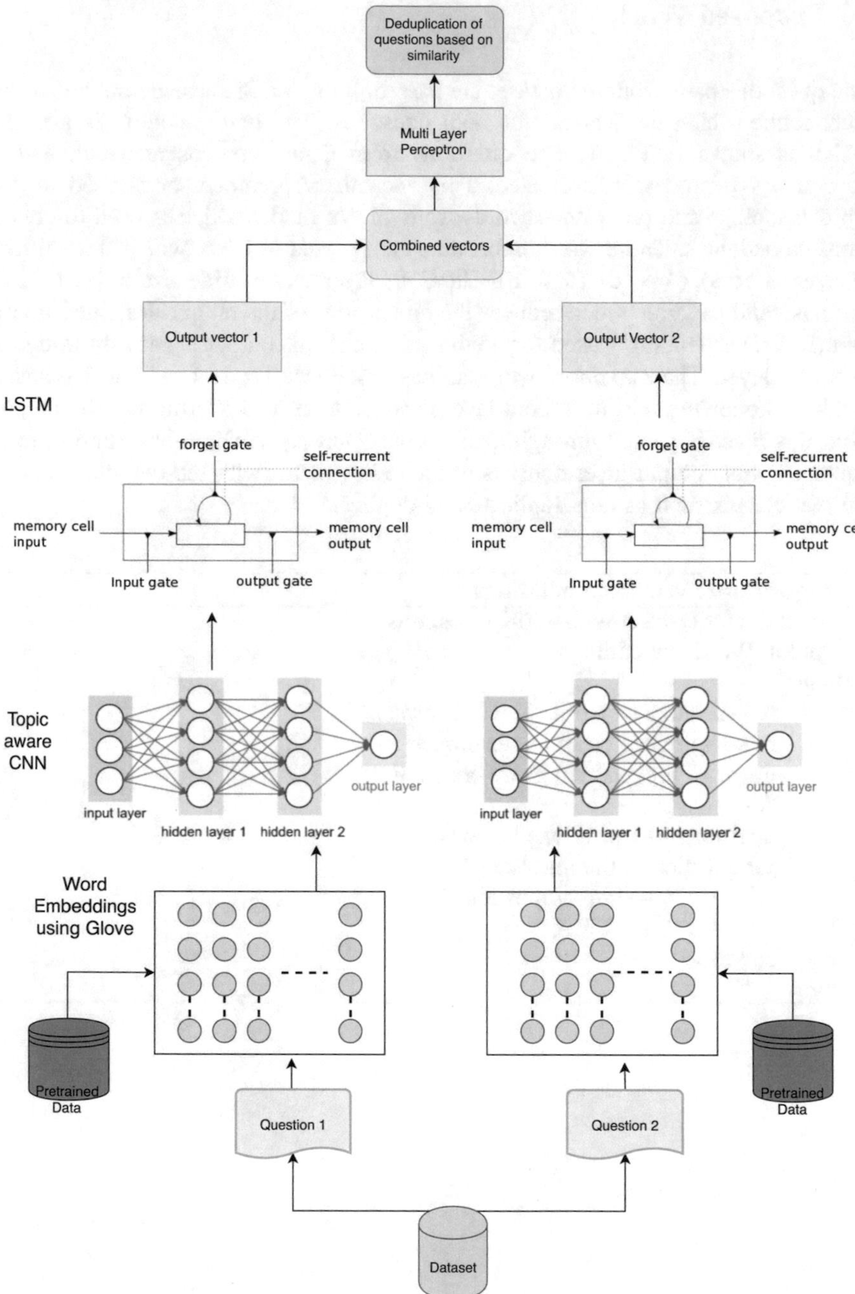

Fig. 1 Proposed system

Algorithm2: Convolution Neural Network

```
Input : Word embeddings, filter size, number of filters, stride size
Output : Prediction file
begin
        b ← Bias term
        padding ← VALID or SAME
        for filter in number_of_filters
                W ← Random initialized weight vectors based on filter size
                conv ← tf.conv2D(embeddings, W, strides, padding, name)
                non-linearity(conv)
                pooling ← tf.max_pool(size, strides, padding, name)
                combined_outputs ← combine all filter outputs.
                predictions ← softmax(combined_outputs)
                loss ← compute_loss(predictions)
                accuracy ← compute_accuracy(predictions)
        end for
end
```

The Quora dataset is used for analysis, and it consists of nearly 400 K question pairs which are labeled as duplicates or non-duplicates. During the preprocessing step, the questions are cleaned of punctuations and non-ASCII characters using regular expressions from Python. Also, any questions which had less than five characters long were or any questions which had more than 59 words are removed, as these questions were less likely to be duplicated. After preprocessing, GloVe-based word embeddings are used for converting the words to vector space.

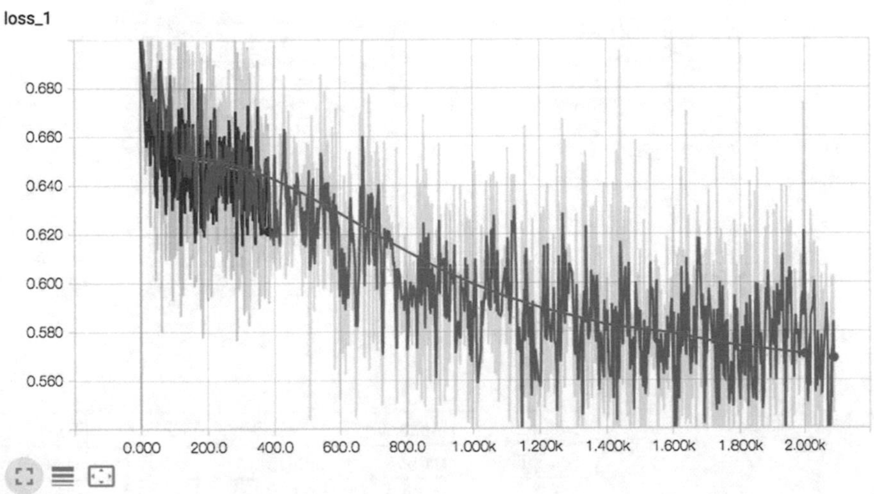

Fig. 2 Loss of model over multiple training steps

In order to achieve this, vocabulary preprocessor from Tensorflow is used. Tensorflow embedding lookup function is used for getting word embeddings. The vocabulary had around 85,000 words which were mapped to unique ids, and 100d GloVe embeddings are used for the conversion to vector space.

4 Results and Discussion

The experiment is carried out using Quora dataset for 30 epochs and used a batch size of 16. For introducing nonlinearity, ReLU activation function is used and loss is calculated using Adam optimizer. These functions are implemented using Tensorflow—GPU, with processes distributed over multiple cores. The training set has 361,745 question pairs, and it was split into 90–10% for training and testing. Hence, the training set had 325,571 question pairs and test set had 36,174 question pairs. The loss of the model decreased rapidly in the initial epochs but flattened out into a plateau as the number of training steps increased which is shown in Fig. 2. The model accuracy is increased when the numbers of training steps are increasing which is shown in Fig. 3.

The 7% performance gain is achieved over simpler multilayer perceptron architecture, by the Siamese network architecture. And a 14% gain over the shingling technique. 80% of accuracy has been accomplished by the Siamese CNN architecture, as shown in Table 1, and the loss is less than 1, which is superior to the past works.

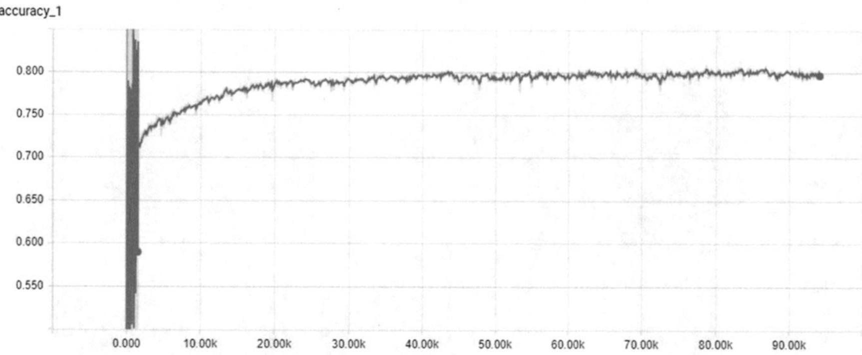

Fig. 3 Accuracy of the model over multiple training steps

Table 1 Performance comparison

Methods	Accuracy	Precision	Recall	F1
Shingling	0.6657	0.5151	0.7297	0.6039
MLP	0.7263	0.5878	0.7245	0.6490
CNN	0.8044	0.7003	0.7688	0.7330

5 Conclusion and Future Work

Deep learning strategies plainly beat the baselines for the task of duplicate question pair recognition. The proposed Siamese CNN architecture significantly improves the accuracy of detecting duplicated question pairs. The training on Quora dataset using word embeddings improves the performance. This can be further improved by using LSTM model following the CNN model. It is fascinating to figure out to accomplish multi-perspective matching by combining question features within a CNN layer.

References

1. Addair, T.: Duplicate question pair detection with deep learning. Stanf. Univ. J. (2017)
2. Yin, W., Schütze, H., Xiang, B., Zhou, B.: ABCNN: attention-based convolutional neural network for modeling sentence pairs. Arxiv J. (2016). arXiv:1512.05193v3
3. Mikolov, T., Chen, K., Corrado, G., Dean, J.: Efficient estimation of word representations in vector space. Arxiv J. (2013). arXiv:1301.3781v3
4. Wang, Z., Hamza, W., Florian, R..: Bilateral multi-perspective matching for natural language sentences. Arxiv J. (2017). arXiv:1702.03814
5. Palangi, H., Deng, L., Shen, Y., Gao, J., He, X., Chen, J., Song, X., Ward, R..: Deep sentence embedding using long short-term memory networks: analysis and application to information retrieval. Arxiv J. (2016). arXiv:1502.06922v3
6. Zhao, R., Mao, K.: Topic-aware deep compositional models for sentence classification. IEEE/ACM Trans. Audio Speech Lang. Process. **25**(2), 248–260 (2016)
7. Zhang, S., Jiang, L.C., Wei, S., Da, L., Hu, Y.: Non-recurrent neural structure for long-term dependence. IEEE/ACM Trans. Audio Speech Lang. Process. **25**(4), 871–884 (2017)
8. Zhang, X., Rong, W., Liu, J., Tian, C., Xiong, Z.: Convolution neural network based syntactic and semantic aware paraphrase identification. In: International Joint Conference on Neural Networks, Anchorage, Alaska, USA, pp. 2158–2163 (2017)
9. Bogdanova, D., Santos, C., Barbosa, L.: Detecting semantically equivalent questions in online user forums. In: Conference on Computational Language Learning, pp. 123–131 (2015)
10. Kim, Y.: Convolutional neural networks for sentence classification. In: Proceedings of Conference on Empirical Methods in Natural Language Processing, pp. 1746–1751 (2014)
11. Luo, Z., Deng, M., Yongjian, L., Jingling, Y.: An ontology construction method for educational domain. In: Fourth International Conference on Intelligent Systems Design and Engineering Applications, pp. 99–102 (2013)
12. Chang, Y.: Automatically constructing domain ontology for document classification. In: Proceedings of the Sixth International Conference on Machine Learning and Cybernetics, vol. 4, pp. 1942–1947 (2007)
13. Pennington, J., Socher, R., Manning, C.D.: GloVe: global vectors for word representation. In: Proceedings of Conference on Empirical Methods in Natural Language Processing, pp. 1532–1543 (2014)
14. Sun, W., Jia, M., Zheng, D., Cao, H., Yan, B., Yu, H.: Automatic domain ontology construction based on thesauri. In: Sixth International Conference on Fuzzy Systems and Knowledge Discovery, pp. 415–418 (2009)

Differentiation of Speech and Song Using Occurrence Pattern of Delta Energy

Arijit Ghosal, Ghazaala Yasmin and Debanjan Banerjee

Abstract Differentiation of speech and song from acoustic signal is a challenging issue. It is a significant part of automatic classification of audio. Most of the previous works have been done for classifying speech and non-speech, but comparatively less work has been done for differentiating speech and song. Mostly, frequency and perceptual domain features were common in those works. In this work, a small dimensional acoustic feature has been proposed. Speech differs from song due to the absence of instrumental part within it which is present in song and causes increase of energy for song signal compared to speech signal. Short-time energy (STE), an acoustic feature, can reflect this observation. For precise study of energy variation, features based on very small change of energy, Delta Energy, and co-occurrence matrix of it are considered. For classification purpose, some well-known classifiers have been employed. Experimental result has been compared with existing methodologies to reflect the efficiency of the proposed system.

Keywords Speech–song classification · Acoustic features · Delta Energy
Co-occurrence matrix

A. Ghosal (✉)
Department of Information Technology, St. Thomas' College of Engineering and Technology, Kolkata 700023, West Bengal, India
e-mail: ghosal.arijit@yahoo.com

G. Yasmin
Department of Computer Science and Engineering, St. Thomas' College of Engineering and Technology, Kolkata 700023, West Bengal, India
e-mail: me.ghazaalayasmin@gmail.com

D. Banerjee
Department of Management Information Systems, Sarva Siksha Mission Kolkata, Kolkata 700042, West Bengal, India
e-mail: debanjanbanerjee2009@gmail.com

J. Nayak et al. (eds.), *Soft Computing in Data Analytics*,
Advances in Intelligent Systems and Computing 758,
https://doi.org/10.1007/978-981-13-0514-6_74

1 Introduction

Volume of multimedia resources especially audio data is increasing day by day. They need to be properly categorized for proper retrieval in future. Audio data or signal is not similar to other signals. Lots of meaningful information can be extorted from audio data. If such knowledge is extracted from audio data, then those can be accumulated as content depiction in an orderly fashion. These contentment descriptions will help in applications related to proper storage and retrieval of audio. This will also help in other applications like audio indexing, audio retrieval, music, and instrument classification. Speech and song are the two basic categories of audio data. So, classification of audio data can be started by differentiating speech and song. Discrimination of speech and song is very interesting and popular research domain because of its applicability toward search engines, media services, and intelligent human–computer systems. Speech and song separation can also be considered as a basic step toward making of automatic speech recognition system. Advancement in the domain of signal processing and data mining technologies also caused enhancement of research work in the area of audio retrieval. Being a promising research area, very less work has been performed in this domain so far. Previous approaches were mostly based on perceptual and frequency domain acoustic features like pitch, rhythm, spectral flux, and spectral roll-off. Pitch-based features are overused in the field of speech signal processing. Inspiring from these, this work aims to suggest a simple as well as low-dimensional acoustic feature set which is neither based on frequency domain features nor based on perceptual domain features for differentiating speech and song.

The work is organized as follows: Description of previous work done is described in Sect. 2. The approach of the proposed scheme is explained in Sect. 3. Section 4 describes the experimental result with comparative study followed by conclusion.

2 Related Works

Haralick and Shapiro [1] have discussed method of extraction of statistical features. This concept is helpful in other fields also. Gerhard [2] has used pitch-based features for classifying speech and song. Bugatti et al. [3] have compared statistical and neural approach for classifying speech and music. Gerhard [4] has applied perceptual features in fuzzy speech–song discrimination. He has used pitch. Some research works have been performed so far related to the domain of this work. But their numbers are not much. Several acoustic features have been used in those works in search of good feature set: voicedness, rhythm, rhyme in his work. Gerhard [5] has studied influence of silence and rhythm for analysis of speech and song. Tzanetakis [6] has studied the characteristics of song-specific singing voice. In every 20 ms, a classification decision is taken in his work. Lin and Chen [7] have proposed a real-time classification scheme to categorize audio signals into several basic categories. Umbaugh [8] has discussed method to generate co-occurrence matrix of a certain feature. This is very much helpful in different fields of pattern recognition. Zhang and Zhang [9] have used harmonic structure modeling for separation of music signals. Ruinskiy and Lavner [10] have proposed a method for detection of sounds of breath in song and speech signals. For classification and segmentation of speech and music, Lavner and Ruinskiy [11] have proposed a decision tree-based algorithm. Gallardo-Antolín and Montero [12] have categorized speech, music as well song using histogram equalization-based features. Salselas and Herrera [13] have studied melodic and rhythmic pattern in speech and song. Sonnleitner et al. [14] have used spectral features for identification of speech in mixed audio signal. A review on applicability of support vector machine (SVM) for categorization of data has been carried out by Bhavsar and Panchal [15]. Velayatipour and Mosleh [16] have reviewed different methods aiming to differentiation of speech and music. A wavelet-based feature has been proposed by Ramalingam and Dhanalakshmi [17] for discrimination of speech and music.

3 Proposed Methodology

From the past research works, it is quite understandable that perceptual and frequency domain acoustic features are used while dealing with speeches. Among them, pitch-based features are the mostly used features in speech signal processing causing overuse of this feature. Excessive use of pitch and perceptual domain features has made a scope to delve into this field to propose an alternative acoustic feature in this domain. In this work, the goal is paid to differentiate speech and song by using a good acoustic feature.

By hearing through bare ear, it is quite clear that energy distribution of speech signal and song signal is not same. When human beings talk something, there is lots of silence zone in that signal. But in case of song, such type of silence zone does not exist, and rather most of the times instrumentals are being played in the background. This observation has motivated to consider energy-based features which are time domain acoustic features. Zero crossing rate (ZCR) could have been considered, but it has not been considered because of some reason. If human beings speak something with solo voice, then it is being considered as speech while the same speech accompanied with instrumental in the background might be considered as song. In that case, there will not be significant variation in the number of zero crossings measuring how many times the audio signal crosses the zero axes for speech and song. In the next subsections, computation of acoustic features and classification schemes used have been illustrated, respectively. Figure 1 reflects the outline of the proposed scheme.

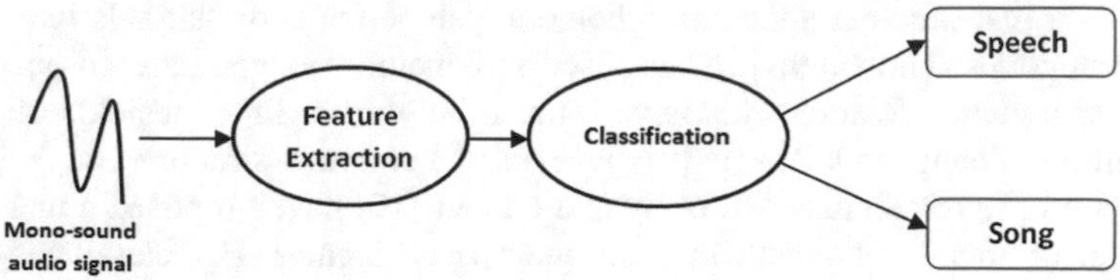

Fig. 1 Outline of proposed scheme

3.1 *Feature Computation*

It is the nature of a good feature set that it will help classifier to produce high accuracy of classification while keeping its dimension small. If the features are chosen judiciously, then only it will be able to generate a good feature set. It is observed that for speech signal, frequent occasion of silence produces a characteristic mark while distribution of energy for song signal spreads over a wide range frequency.

This observation has inspired to propose a feature set which is able to reflect the characteristics of energy distribution for speech and song well to achieve the aim maintaining high accuracy and keeping the length of feature vector small.

3.1.1 Delta Energy-Based Features

To know how energy is varying with time for speech and song, specifically how energy is associated with short-term region of speech and song, short-time energy (STE) needs to be measured. This can be measured for audio data considered as discrete signals and divided into P frames $\{x_i(m): 1 \leq i \leq P\}$ by Eq. (1) as:

$$En_n = \frac{1}{len} \sum_{m=1}^{r} [x_n(s)]^2 \tag{1}$$

where En_n indicates nth frame energy, len is the length of frame, and $x_n[s]$ denotes sth sample in nth frame. These frames are overlapped in nature to avoid the loss of any boundary characteristics for a certain frame. The values of the Delta Energy give discriminative nature for homogenous set of values. This is why this feature has been adopted for the proposed feature set.

As energy variation for speech signal is quite different from that of song signal, changes of energy variation for speech signal between two successive frames will also differ from song signal. This change of energy variation between two consecutive frames is denoted as 'Delta Energy' (written as ΔEnergy).

Let, for kth frame, energy is computed as E_k, and for k + 1th frame energy is computed as E_{k+1}. Then, Delta Energy is computed by Eq. (2) as:

$$\Delta\text{Energy} = \text{E}_{\text{k}+1} - \text{E}_{\text{k}} \quad (2)$$

After computing Delta Energy, mean and standard deviation of it have been considered.

Delta Energy plots for speech and song are depicted in Fig. 2a and Fig. 2b, respectively. From the plots, it is evident that distribution of Delta Energy changes noticeably for speech and song. In case of speech signal, some peaks are only visible in the plot of Delta Energy when there is human utterance; otherwise, it comes down near to zero value because of the frequent presence of silence in speech. But this observation is not visible for song signal because of the presence of instrumentals in background.

It is known that mean and standard deviation of a feature give an overall idea about its distribution only. It fails to depict the precise information related to it. For precise study of characteristics of Delta Energy, concept of co-occurrence matrix has been utilized [8]. In case of image, occasion of different intensity values contained by a neighborhood produces a pattern which is utilized to parameterize the texture or appearance of an image. Same concept has been implemented here. Delta Energy is calculated using Eq. (2). Thus, a series of Delta Energy, $\{\Delta EN_i\}$, is attained for input audio signal. Occurrence of different Delta Energy values surrounded by a neighborhood echoes the pattern which characterizes the nature of the audio signal. So, a matrix C_m of dimension $B \times B$ (where $B = max\{\Delta EN_i\} + 1$) is generated. In this work, an element in the matrix Cm(r, s) indicates the number of occurrences of Delta Energy r and s in successive time instances. Finally, statistical measures, based on co-occurrence matrix [1] like entropy, energy, contrast, homogeneity, and correlation, are computed.

Co-occurrence matrix plots of Delta Energy for speech and song are depicted in Fig. 3a and Fig. 3b, respectively. From the plots, it is evident that occurrence pattern of Delta Energy is not same for speech and song.

To achieve overall as well as precise characteristics of Delta Energy, a seven-dimensional feature vector comprising of mean and standard deviation of

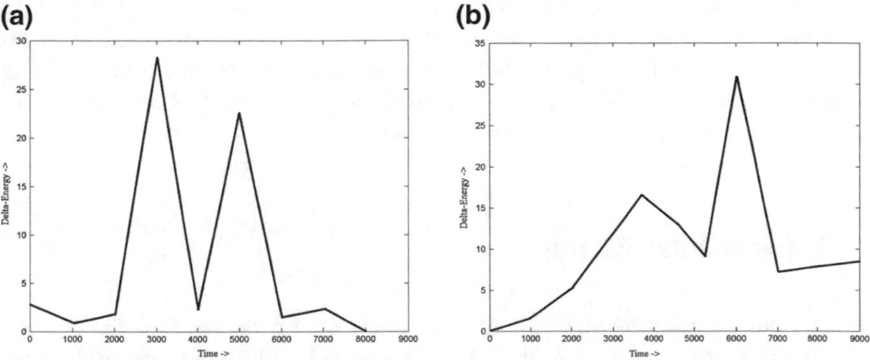

Fig. 2 Delta Energy plot for **a** speech signal and **b** song signal

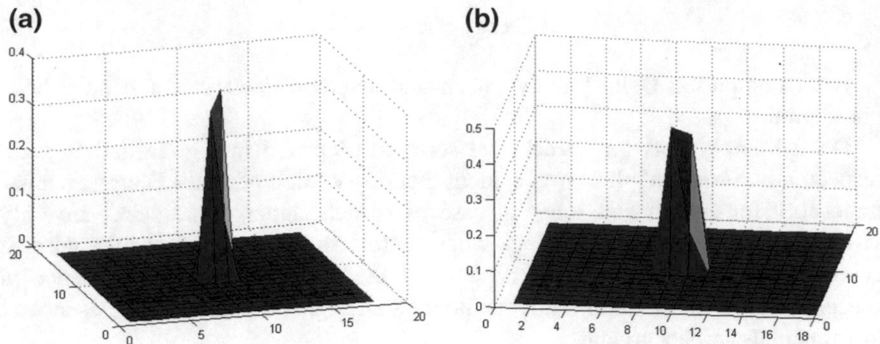

Fig. 3 Co-occurrence matrix plot of Delta Energy for **a** speech signal and **b** song signal

Delta Energy along with five co-occurrence-based features—entropy, energy, contrast, homogeneity, and correlation—has been generated.

3.2 *Classification*

As the main objective of this work is to evaluate the discriminating power of the proposed feature set for differentiating speech and song, simple and popular supervised classifiers like support vector machine (SVM), neural network, and k-nearest neighbors (k-NN) have been used. Multi-layer perceptron (MLP) has been applied here as neural network. During classification, the whole dataset used in the system is expected to be split into two parts—each part symbolizes either testing or training data. A total of 200 audio files have been used here where each category of audio has equal division. Testing and training dataset has been formed by splitting the total dataset into two equal halves that is each category of audio files has 50 files each for training and testing resulting in a total of 100 audio files for testing and training dataset. SVM has been applied with quadratic kernel type. Neural network (MLP) has been applied by considering seven neurons to be present in input layer reflecting seven feature values, two neurons in output layer corresponding to two categories of audio files—speech and song—and four neurons in the one and only hidden layer. k-NN has been employed considering $k = 3$, 'City Block' as distance metric and Nearest Neighbor rule for tie-break.

4 Experimental Results

To carry out the experiment, a custom audio dataset has been prepared consisting of 100 speech files and 100 song files. All of these audio files are of monotypes and have duration of 90 s. Different languages and different age groups have been

Table 1 Classification accuracy for classification of speech and song

Classification scheme	Classification accuracy (in %) for proposed work	
	Speech	Song
SVM	93.5	92.0
Neural network	91.0	90.5
k-NN	89.5	89

considered for both speech and song. These files are collected from recording of different live programs, CD recording, as well as downloaded from Internet. A few audio files in the custom audio dataset are noisy too to make the dataset reliable for real outline.

All the audio files are split into frames having 50% overlapped with previous frame to avoid missing of any boundary characteristics of a frame. For training purpose, 50% of the whole dataset has been used and remaining 50% has been left for testing purpose. Again, same classification work is repeated by reversing training and testing dataset. Average of these two steps is given in Table 1.

4.1 Comparative Analysis

Classification performance of proposed feature set has been judged against some past works related to differentiation of speech and song. Same custom audio dataset used in this work has been used to implement the systems proposed by David Gerhard [2, 4] in his two works. Gerhard has used pitch-based features for achieving the goal in his first work, and he has used perceptual features like pitch, rhythm in his second work which deals with fuzzy classification of speech and song. The comparative performance has been reflected in Table 2. From the comparative performance, it is evident that Delta Energy-based proposed feature set is able to differentiate speech and song in a much better way.

Table 2 Comparative performance of proposed work with other works in % accuracy

Precedent method	Speech	Song
David Gerhard [2]	88.5	88.0
David Gerhard [4]	89.0	88.5

5 Conclusion

For differentiating speech and song, a time domain acoustic feature-based simple feature set has been proposed in this work. The proposed feature set relies on Delta Energy-based features. Experimental results indicate that proposed feature set performs better than some other past works based on perceptual features. Proposed feature set is computationally simple, and at the same time it is of low dimension too. In future, it is aiming to incorporate advance techniques like deep learning instead of using these classifiers.

References

1. Haralick, R.M., Shapiro, L.G.: In: Computer and Robot Vision, vol. 1 (1992)
2. Gerhard, D.: Pitch-based acoustic feature analysis for the discrimination of speech and monophonic singing. Can. Acoust. **30**(3), 152–153 (2002)
3. Bugatti, A., Flammini, A., Migliorati, P.: Audio classification in speech and music: a comparison between a statistical and a neural approach. In: EURASIP J. Adv. Signal Process. **4** (2002)
4. Gerhard, D.: Perceptual features for a fuzzy speech-song classification. In: IEEE International Conference on Acoustics Speech and Signal Processing, vol. 4, pp. 4160–4160 (2002)
5. Gerhard, D.: Silence as a cue to rhythm in the analysis of speech and song. Can. Acoust. **31** (3), 22–23 (2003)
6. Tzanetakis, G.: Song-specific bootstrapping of singing voice structure. In: ICME'04 2004 IEEE International Conference on Multimedia and Expo, 2004, vol. 3, pp. 2027–2030. IEEE (2004)
7. Lin, R.S., Chen, L.H.: A new approach for classification of generic audio data. Int. J. Pattern Recognit. Artif. Intell. **19**(01), 63–78 (2005)
8. Umbaugh, S.E.: Computer Imaging: Digital Image Analysis and Processing. CRC press (2005)
9. Zhang, Y.G., Zhang, C.S.: Separation of music signals by harmonic structure modeling. In: Advances in Neural Information Processing Systems, pp. 1617–1624 (2006)
10. Ruinskiy, D., Lavner, Y.: An effective algorithm for automatic detection and exact demarcation of breath sounds in speech and song signals. IEEE Trans. Audio Speech Lang. Process. **15**(3), 838–850 (2007)
11. Lavner, Y., Ruinskiy, D.: A decision-tree-based algorithm for speech/music classification and segmentation. EURASIP J. Audio Speech Music Process. **2009**(1) (2009)
12. Gallardo-Antolín, A., Montero, J.M.: Histogram equalization-based features for speech, music, and song discrimination. IEEE Signal Process. Lett. **17**(7), 659–662 (2010)
13. Salselas, I., Herrera, P.: Music and speech in early development: automatic analysis and classification of prosodic features from two Portuguese variants. J. Portuguese Linguist. **10**(1) (2011)
14. Sonnleitner, R., Niedermayer, B., Widmer, G., Schlüter, J.: A simple and effective spectral feature for speech detection in mixed audio signals. In: Proceedings of the 15th International Conference on Digital Audio Effects (2012)

15. Bhavsar, H., Panchal, M.H.: A review on support vector machine for data classification. Int. J. Adv. Res. Comput. Eng. Technol. (IJARCET), **1**(10), 185 (2012)
16. Velayatipour, M., Mosleh, M.: A review on speech-music discrimination methods. Int. J. Comput. Sci. Netw. Solut. **2**(2), 67–78 (2014)
17. Ramalingam, T., Dhanalakshmi, P.: Speech/music classification using wavelet based feature extraction techniques. J. Comput. Sci. **10**(1), 34 (2014)

Empirical Analysis on Cancer Dataset with Machine Learning Algorithms

T. PanduRanga Vital, M. Murali Krishna, G. V. L. Narayana, P. Suneel and P. Ramarao

Abstract For the medical database, the usage of data mining methods in particular for cancer and related diseases has been recently escalating. The patient's life is dependent upon the outcome of these methodologies—and hence utmost care is taken to predict or suggest the correct medicines for the concerned disease. In this, we collected cancer data from zone 1 districts of Andhra Pradesh (AP) and conduct the statistical analysis and analyzed performance of the dataset for predicting cancer with supervised machine learning algorithms. The study shows good statistical results for preventing the cancer in zone 1 of AP as well the classification algorithms like decision tree, ADT, Naïve Bayes, BayesNet, K-Star, and Random Forest algorithms give the good results for prediction of cancer. All applied algorithms show above 96% accuracy. In comparison, k-star model is performed 100% with less model construction time (0.01).

Keywords Cancer · Machine learning · Data mining · Classification

T. PanduRanga Vital (✉) · M. Murali Krishna · G. V. L. Narayana · P. Suneel · P. Ramarao
Department of CSE, Sri Sivani College of Engineering, Srikakulam 532402, Andhra Pradesh, India
e-mail: vital2927@gmail.com

M. Murali Krishna
e-mail: maadugula@gmail.com

G. V. L. Narayana
e-mail: gvln60.gmrit@gmail.com

P. Suneel
e-mail: suneelpappala@gmail.com

P. Ramarao
e-mail: paraselli@yahoo.com

J. Nayak et al. (eds.), *Soft Computing in Data Analytics*,
Advances in Intelligent Systems and Computing 758,
https://doi.org/10.1007/978-981-13-0514-6_75

1 Introduction

Data mining (DM) is an effective approach to extract knowledge from large raw data. DM primarily formulated from the areas of machine learning (ML) and statistics as interdisciplinary areas, DM has progressed from this early development to include database design, pattern recognition, artificial intelligence (AI), visualization, and so forth. DM can reveal early biomedical and primary health care ability for clinical and legitimate basic leadership and additionally produce scientific theories from colossal exploratory data, biomedical lit and clinical databases [1, 2]. Medical DM is an interdisciplinary area with contribution of medicine and data mining. Machine learning is the technique of AI that admits the machines to learn various patterns in the complex datasets to unearth the hidden concepts. ML algorithms provide perfect measures and mechanisms to predict various kinds of cancers [3].

Predicting the results of a cancer disease are one of the most challenging and fascinating assignment in the present decades where the advancement is seen with DM patterns [4]. Medical information diseased data analysis is very difficult that it contains numerous variables with many hidden values [2]. DM methods are well known and powerful tool for medical analysts to recognize and accomplishment patterns, relationships among substantial number of variables and made them ready to diseased predicting results utilizing the recorded historical datasets [5, 6].

Oncology is one of the branches in medical field concerned with the treatment and examine of tumors that assures the prevention and diagnosis of cancer [7, 8]. The major intention of modern oncology is prediction of specific cancer by study of the tumors and to provide appropriate therapy.

Breast cancer is probable to persist in patients that have had their cancers removed [9, 10]. These two main problems are in the span of the classification problems. In this study, various evaluations and sensible editorials on breast cancer diagnosis and prediction are summarized. Gupta et al. presented an overview of the modern research using the DM methods to improve the breast cancer recognition and forecasting [11]. The research on colon cancer tissue statistics discovered with their novel global k-means algorithm. It can efficiently deal with high dimensions [12]. The global k-means may be able to avoid the pressure of noisy data.

In health care organization especially in cancer detection, the DM activities much important and influenced role for new trends [13]. It is very supportive for all related to this area. This survey shows many different services of DM techniques like clustering, classification, association, regression in this province. Tomar and Agarwal present techniques and their highlights. This study describes challenges and future issues of DM in health care. The appropriate choices of existing DM technique are recommended by the authors [14].

2 Methodology

The dataset has been analyzed by DM techniques. The frequency of patients has been analyzed for the collected data from Zone 1 (Srikakulam, Vizianagaram, and Visakhapatnam) Districts of Andhra Pradesh, India.

The data was collected with 1008 instances (504 cancer instances and 504 non-cancer) and 46 attributes and one class (Place, Age, Cancer Type, Family History, Drinking, Smoking, Tea, Coffee, Milk, Job, Morning-Eat, Lunch-Eat, Dinner-Eat, Travel, Living Status, Fruits, Vegetables, Sleeping hours, Tensions, Cool-Drinks and Ice-cream, Study, Height, Weight, Bp, Pains, Hair-Loss, Gutkha (chewing tobacco), Marital Status, Blood-Group, Treatment Type, Bathing, Oils, Fast-food, Other-disease, Morning-walk, Use-mobile, Drugs or Using Tablets, Treatment-Mode, Diagnosis Mode, Meditation, Mosquito-repellents, Injuries, Speak-level, Watching-TV, Think Levels) and class status of cancer (0 (no) for non-cancer and 1 (yes) for cancer).

2.1 *Naive Bayes Classification*

Naive Bayes applies the probabilistic Naive Bayes classifier. Naive Bayes informs additive version that processes one request at a time. It can use a kernel estimator but it is not meet for discretization. This process is founded on conditional probabilities. Naive Bayes uses Bayes theorem, and it was formulated with probability by considering the frequency of values and the historical data combination values. Bayes theorem determines the probability shown in equations.

$$P(S/X) = \frac{P(X/S)P(S)}{P(S)} \tag{1}$$

$$P(C/X) = P(X_1|C) \times P(X_2|C) \times \cdots \times P(X_n|C) \times P(C) \tag{2}$$

2.2 *K-Star (K*) Classification*

K* is an instance-based classifier that it is very simple and similar to k-nearest neighbors (KNN) algorithm. New data or information instances X_i where i = 1, 2, 3, …, k are allotted to the class that happens most often along with the k-nearest information or data instances or items Y_j where j = 1, 2, 3, …, k.

The most related items from the dataset are retrieved from entropic distance. By the method for entropic separation as a metric have various advantages, such as handling of missing values and real-esteemed features or attributes. The K* function can be analyzed as Eq. 3.

$$K*(I,x) = -\ln P*(y_i,x) \tag{3}$$

where P is the transformational path probability from data point or instance x to instance y. It is very helpful to understand the probability that instance x will reach the destination y through a random walk.

2.3 C4.5 Classification Algorithm

Decision tree (DT) algorithm is handled in the way that the element vector acts for a count of instances by training data samples. The classes for the newly generated instances are being found in this algorithm. This algorithm generates the rules for the prediction of the target variable. With the assistance of tree classification, the critical distribution of the data is easily understandable C4.5 is an expansion of ID3.

2.4 Alternating Decision Tree (AD Tree)

AD Trees have decision as well prediction nodes. In this, the decision node assigns a proclaim condition and the prediction node comprises an individual number. In AD Trees, prediction nodes have mutually leaves and root. An instance is arranged by AD Tree by tailing all ways in which all decision or choice nodes are valid and summing any expectation or prediction nodes that are traversed. This is dissimilar as of binary classification trees, for example, C4.5 and regression tree instance follows one and only path through the tree.

3 Results and Discussions

The analysis provided the relationship of the collected datasets. There are more number of female patients (64.3%) compared to male patients (35.7%). The family history that 19.4% cancer patient's family members had cancer. Hence, cancer is related to both family history and metabolic. Drinking and smoking habits are the factors of occurrence of cancer as per analysis 26.2% of cancer patients have the drinking alcohol habit and 33.3% of cancer patients have smoking habit.

The study shows more number of breast cancer (20.2%) patients followed by cervix (16.3%), stomach (7.3%) and blood (7.3%). Hence, there is a need to control these cancers in the zone 1 regions of AP. 44.2% patients are housewives which are observed to be more when compared to working women. Most of the cancer patients have tension (54.8%). Most of the patients have no education (47.2%). Most of them are having low blood pressure compared to high BP and normal. The patients preferred allopathic (89.5%) as primary treatment. Chemotherapy is a crucial treatment for the cancer patients. Most of the diagnosis was conducted by scanning and biopsy.

3.1 Alternative Decision Trees (ADT) Classification Analysis

The analysis of AD tree classification algorithm, the legend of negative (−ve) value specifies 0 (status non-cancer instances) and positive (+ve) specifies 1 (status cancer instances). Total tree size is 31 constructed with 21 leave nodes and time taken for the model 0.27 s and correctly classified instances are 917 (90.9722%). It describes the confusion matrix along with precision and recall values. The authors get the results as correctly classified instances are 917 (90.9722%), incorrectly classified instances are 91 (9.0278%), mean absolute error (MAE) is 0.1992, root mean squared error (Root MSE) is 0.277, relative absolute error (RAE) is 39.8423%, and root relative squared error (Root RSE) is 55.4081%.

Table 1 describes the accuracy of AD tree classification algorithm. The result, true positive rate weighted average is 0.91 as well precision average value is 0.91 and average of ROC value is 0.969. So ADT algorithm classifies the cancer dataset accurately. The confusion matrix of AD trees 0 denotes the non-cancer instances and 1 denotes the cancer instances. In this 462 status '0' and 455 status '1' instances are correctly classified, and 42 (status 0) and 49 (status 1) are incorrectly classified.

Figure 1 shows the receiver operating characteristic (ROC) curve of AD Tree. The value of ROC is 0.9689, so the dataset is correctly classified as AD tree algorithm.

Figure 2 shows the AD trees visualization classifies the cancer (1) and non-cancer instances (0). It classifies with using cancer status attribute (0 for non-cancer and 1 for cancer instance). The negative (−ve) value indicates 0 and positive (+ve) value indicates the 1. As per alternative decision tree (ADT) most of the cancer patients used mosquito replants (+0.273) than healthy people.

This analysis shows that Gutkha (chewing tobacco) is causes of cancer where it is used by cancer patients (values is 1.005). This analysis also gives about cancer instances symptoms that are hair loss (0.347) and BP (blood pressure) is not normal (positive value 1.042) than non-cancer instances. The tree indicates the most of the

Table 1 AD tree accuracy by class status (Cancer and non-Cancer)

Class	TP rate	FP rate	Precision	Recall	F-measure	ROC area
0	0.917	0.097	0.904	0.917	0.91	0.969
1	0.903	0.083	0.915	0.903	0.909	0.969
Average	0.91	0.09	0.91	0.91	0.91	0.969

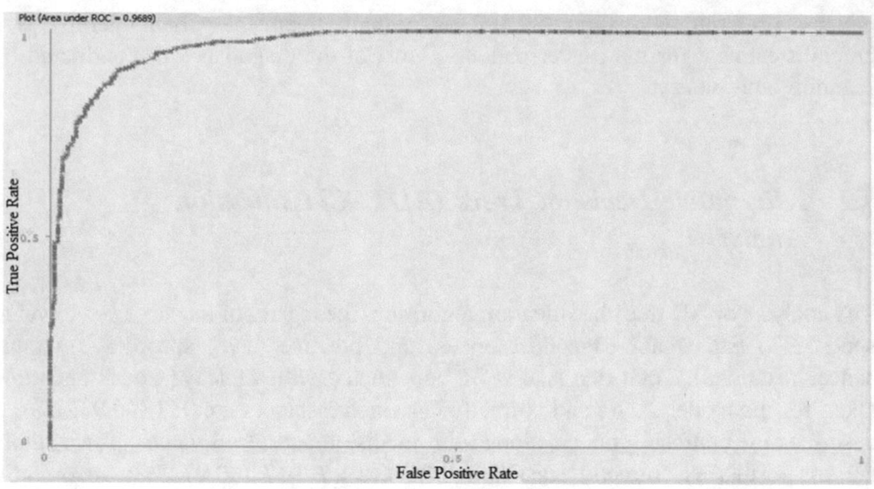

Fig. 1 ROC curve of AD tree

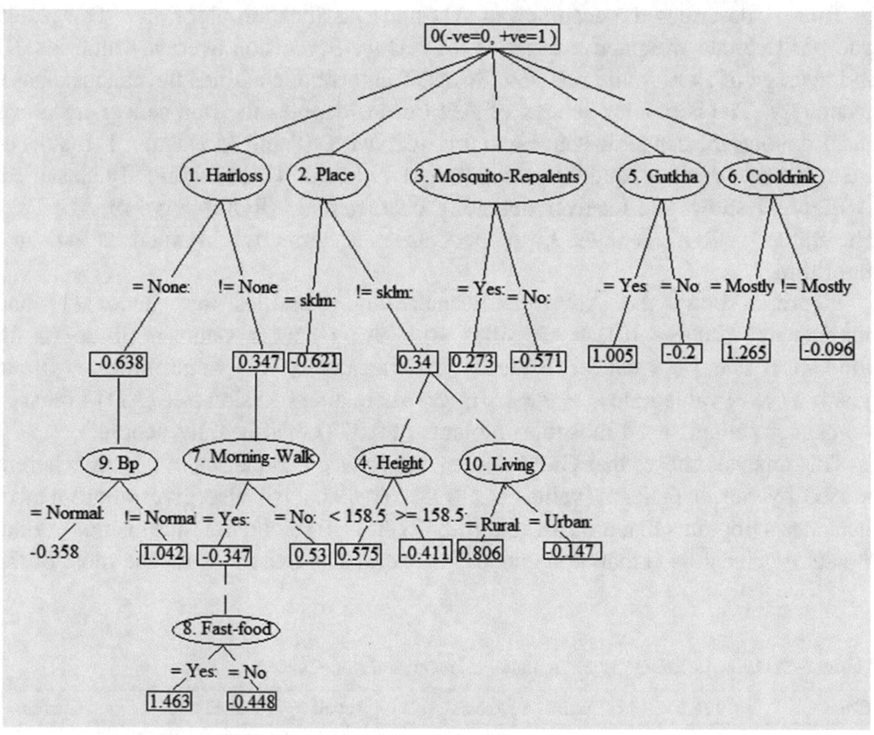

Fig. 2 AD trees visualization

cancer instances from rural area (value 0.806) and eats fast foods than healthy (non-cancer) instances. It also shows that mosquito replants one of the causes of cancer (value 0.273).

3.2 C4.5 Decision Tree Classification Analysis

In the analysis of C4.5 decision tree, the total tree size is 63 constructed with 92 leave nodes and time taken for the model 0.05 s. Correctly, classified instances are 985 (97.7183%). It shows the confusion matrix with cancer status (0 or 1) as well precision and recall values. The authors get the results with classification class cancer status attribute. The correctly classified instances are 985 (97.72%), incorrectly classified instances are 23 (2.28%), MSE value is 0.0406 Root MSE is 0.14, RAE is 8.12% Root RSE is 28.48%.

Table 2 shows the accuracy of C4.5 decision tree classification algorithm. ‘0’ (Zero) denotes the non-cancer (healthy) instances and ‘1’ represents the cancer instances. The results, true positive (TP) rate weighted average is 0.977, precision (P) weighted average is 0.978, and weighted average of ROC is 0.993. So C4.5 decision tree algorithm classification accuracy is good for collected dataset. The confusion matrix of C4.5 decision tree, ‘0’ denotes the non-cancer instances, and ‘1’ denotes the cancer instances. In this 486 cancer status ‘0’ and 499 status ‘1’ instances are correctly classified, and 18 (status 0) and 5 (status 1) are incorrectly classified.

Figure 3 shows the receiver operating characteristic (ROC) curve of C4.5 decision tree. The value of ROC is 0.9929, so the performance of the classification algorithm is good for collected cancer dataset.

Figure 4 shows the C4.5 decision tree visualization. It classifies the cancer and non-cancer with using attribute status. Most of the instances related to Gutka (chewing tobacco), place, marital status, BP, mosquito replants, and so on. Most of the cancer instances related to Gutkha (value is 167.0/25.0) and mosquito replants (192.0/16.0). Some cancer instances are related to low BP or high BP than non-cancer instances (value 0 (241.0/15.0)). The C4.5 decision tree gives the deep description of cancer and non-cancer instances.

Table 2 C4.5 decision tree accuracy by class status of cancer

Class	TP rate	FP rate	Precision	Recall	F-measure	ROC area
0	0.964	0.01	0.99	0.964	0.977	0.993
1	0.99	0.036	0.965	0.99	0.977	0.993
Average	0.977	0.023	0.978	0.977	0.977	0.993

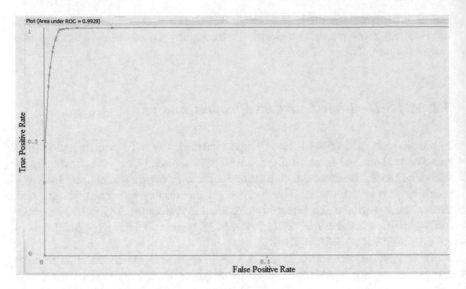

Fig. 3 ROC curve of C4.5 decision tree

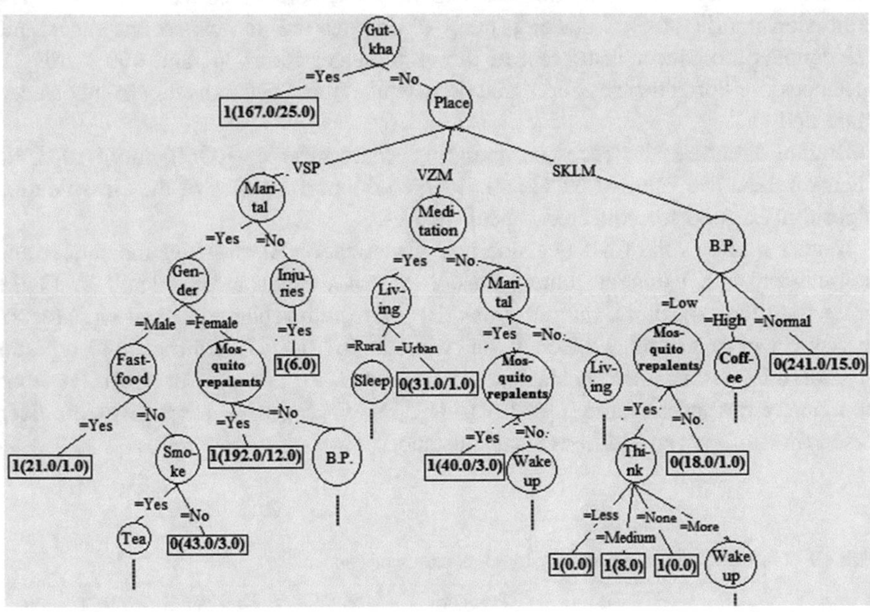

Fig. 4 C4.5 classification algorithm visualization

Table 3 Naive Bayes algorithm accuracy for the cancer and healthy dataset

Class	TP rate	FP rate	Precision	Recall	F-measure	ROC area
0	0.974	0.036	0.965	0.974	0.969	0.993
1	0.964	0.026	0.974	0.964	0.969	0.993
Average	0.969	0.031	0.969	0.969	0.969	0.993

3.3 Naive Bayes Classification

Naive Bayes classification algorithm model takes 0.02 s time for the construction. In this analysis, the algorithm classifies correctly 977 (96.9246%) instances out of 1008 instances. The confusion matrix describes accuracy of the algorithm. The kappa statistic, mean squared, precision, and recalls values give status of the algorithm. The authors get the below results with classification class cancer status attribute. The correctly classified instances are 977 (96.9246%), incorrectly classified instances are 31 (3.0754%), MAE is 0.0389, Root MSE is 0.1674, RAE is 7.7898% and Root RSE is 33.4867%.

Table 3 shows the accuracy of Naive Bayes classification algorithm. The results are true positive (TP) rate weighted average is 0.969 as well precision weighted average values is 0.969 and average of ROC value is 0.993. So Naive Bayes classification algorithm is accurate for collected dataset where the ROC value is nearer to one. The confusion matrix of Naive Bayes classification describes '0' for non-cancer and '1' for cancer instances. In this 491 non-cancer and 486 cancer instances are correctly classified. There are 13 (non-cancer) and 18 (cancer) instances incorrectly classified.

Figure 5 shows the receiver operating characteristic (ROC) curve of Naïve Bayes Classification. The value of ROC is 0.9932, so the performance of the classification algorithm is accurate.

3.4 Bayes Network Classification

Bayes network classification algorithm model takes 0.1 s time for the construction. In this analysis the algorithm classifies correctly 981 (97.3214%) instances. The confusion matrix describes accuracy of the algorithm. The kappa statistic, mean Ssquared, precision, and recalls values give status of the algorithm for class cancer status. The authors get the results of the Bayes network classification with class cancer status attribute. The correctly classified instances are 981 (97.33%) incorrectly classified instances are 27 (2.67%), MAE is 0.0322, Root MSE is 0.1499 RAE is 6.43% and Root RSE is 29.98%.

Table 4 shows the accuracy of Bayes network classification algorithm. The results are true positive (TP) rate weighted average is 0.973, precision average value

Fig. 5 ROC curve of Naive Bayes classification

Table 4 Bayes network accuracy by class cancer status

Class	TP rate	FP rate	Precision	Recall	F-measure	ROC area
0	0.974	0.028	0.972	0.974	0.973	0.996
1	0.972	0.026	0.974	0.972	0.973	0.996
Average	0.973	0.027	0.973	0.973	0.973	0.996

is 0.973, and weighted average of ROC value is 0.996. So Bayes network algorithm is a good classifier for the cancer dataset as per ROC value.

The confusion matrix of BayesNet classification describes with '0' for non-cancer and '1' for cancer instances. In this 491 non-cancer and 490 cancer instances are correctly classified. There are 13 (non-cancer) and 14 (cancer) incorrectly classified instances.

Figure 6 shows the receiver operating characteristic of BayesNet classification. The value of ROC is 0.9957, so the performance of the BayesNet classification algorithm is accurate.

3.5 *K-Star Classification Algorithm*

K-Star classification algorithm model takes 0.01 s time for the construction. In this analysis the algorithm classifies 100%. The authors get the below results with class attribute cancer status. The correctly classified instances are 1008 (100%). The accuracy of K-Star classification algorithm. The values of true positive (TP),

Fig. 6 BayesNet classification ROC curve

precision value and average of ROC value are one (1). The FP rate is zero (0). So, the K-Star classification algorithm is very accurate for classification of this dataset. The confusion matrix of K-Star classification algorithm for cancer dataset and it also shows '0' for non-cancer '1' for cancer.

3.6 Random Forest Classification Algorithm

Random forest of 100 trees, each constructed while considering six random features. Random forest classification algorithm model takes 0.87 s time for the construction. Out of bag error value is 0.0327. In this analysis, the algorithm classifies 100%. The authors get the results with class cancer status (0-no 1-yes) attribute. The correctly classified instances 1008 (100%), incorrectly classified instances are 0, kappa statistic value is 1, MAE is 0.0674, Root MSE is 0.0833, RAE is 13.4767%, and Root RSE is 16.6636%.

The accuracy of random forest classification algorithm with the values of true positive (TP), precision value and average of ROC value are one (1). The FP rate is zero (0). So the random forest classification algorithm is very accurate for classification of this dataset. The confusion matrix of random forest classification algorithm for cancer dataset and it also shows '0' for non-cancer '1' for cancer instances.

Figure 7 was shown classification algorithms with classified instances in percentage. In this, the blue colored bars indicate correctly classified instances and brown colored bars represent the incorrectly classified instances. In the top of the bar mentioned a value in percentage that indicates the correctly or incorrectly

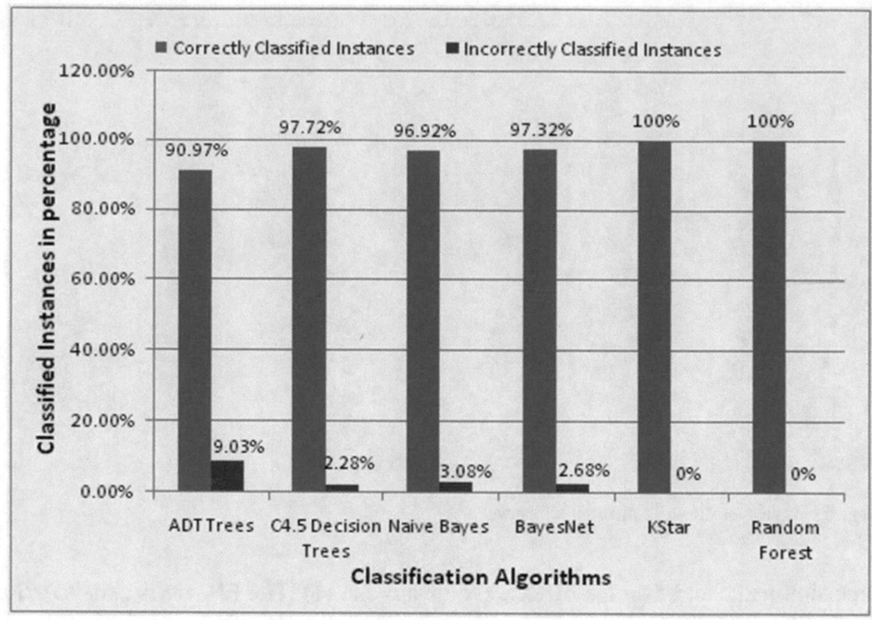

Fig. 7 Comparisons of classification algorithms with classified instances

classified instances. The ADT classification algorithm classified 90.97% correctly and 9.03% incorrectly. In the comparison, K-Star and RF algorithms are classified 100% correctly for collected dataset.

4 Conclusion

The data provides more number of breast cancer instances in the regions from zone 1 districts of AP, India. In this, the most of classification algorithms like decision tree, ADT, Naïve Bayes, BayesNet, K-Star, and random forest algorithms give the good results for prediction of cancer. The dataset is very accurate for predicting the cancer with results of ROC. The K-star algorithm takes very less time (0.01 s) and classifies the 100% instances of the dataset.

References

1. Varlamis, I., et al.: Application of data mining techniques and data analysis methods to measure cancer morbidity and mortality data in a regional cancer registry: the case of the island of Crete, Greece. Comput. Methods Programs Biomed. **145**, 73–83 (2017)

2. Dubey, A.K., Umesh, G., Sonal, J.: Breast cancer statistics and prediction methodology: a systematic review and analysis. Asian Pac. J. Cancer Prev. **16**(10), 4237–4245 (2015)
3. Kourou, K., et al.: Machine learning applications in cancer prognosis and prediction. Comput. Struct. Biotechnol. J. **13**, 8–17 (2015)
4. Lynch, C., et al.: Prediction of lung cancer patient survival via supervised machine learning classification techniques. Int. J. Med. Inform. **108**, 1–8 (2017)
5. Dora, L., et al.: Optimal breast cancer classification using Gauss-Newton representation based algorithm. Expert Syst. Appl. **85**, 134–145 (2017)
6. Cheriguene, S., et al.: Optimized tumor breast cancer classification using combining random subspace and static classifiers selection paradigms. Applications of Intelligent Optimization in Biology and Medicine. Springer International Publishing, pp. 289–307 (2016)
7. Tseng, C.-J., et al.: Integration of data mining classification techniques and ensemble learning to identify risk factors and diagnose ovarian cancer recurrence. Artif. Intell. Med. (2017)
8. Chauhan, D., Varun J.: An efficient data mining classification approach for detecting lung cancer disease. In: International Conference on Communication and Electronics Systems (ICCES). IEEE (2016)
9. Mandal, S.K.: Performance analysis of data mining algorithms for breast cancer cell detection using Naïve Bayes, logistic regression and decision tree. Int. J. Eng. Comput. Sci. **6**(2), 20388–20391 (2017)
10. Ferrari, D.G., De Castro, L.N.: Clustering algorithm selection by meta-learning systems: a new distance-based problem characterization and ranking combination methods. Inf. Sci. **301**, 181–194 (2015)
11. Gupta, S., Kumar, D., Sharma, A.: Data mining classification techniques applied for breast cancer diagnosis and prognosis. Indian J. Comput. Sci. Eng. (IJCSE) **2**(2), 188–195 (2011)
12. Xie, J., Jiang, S., Xie, W., Gao, X.: An efficient global K-means clustering algorithm. J. Comput. **6**(2), 271–279 (2011)
13. Panduranga Vital, T., et al.: Data collection, statistical analysis and machine learning studies of cancer dataset from north costal districts of AP, India. Procedia Comput. Sci. **48**, 706–714 (2015)
14. Tomar, D., Agarwal, S.: A survey on data mining approaches for healthcare. Int. J. Bio-Sci. Bio-Technol. **5**(5), 241–266 (2013)

Efficient High-Utility Itemset Mining Over Variety of Databases: A Survey

U. Suvarna and Y. Srinivas

Abstract High-utility itemset mining (HUIM) is creating new heights of challenge in the areas of research in big data analytics blending data structures, AI, and machine learning techniques to find efficient utility patterns. In the recent past, the merchandise market researchers were pretty much interested in analyzing the purchasing patterns and forecast new profitable area of business. HUIM is that one kind where we get business intelligence and has become the most basic method of finding the knowledge in decision making and optimizing the decisions of market analysis, streaming analysis, biomedicine, mobile computing, stock exchanges, etc. HUIM was initially applied widely in the transactional databases, later applied recursively in incremental databases, further moved to dynamic databases like temporal, spatial, and data stream databases. This chapter highlights the insights of various existing algorithms present in the literature based on several applications and tries to update the various availabilities of different approaches for coining latest research in this domain.

Keywords Associate rule · Itemset/pattern mining · High-utility itemset mining (HUIM) · Frequent · Periodic · Incremental

1 Introduction

Initially before data mining came into existence, huge amount of data was collected on a daily basis and are stored in the databases for years together and of no business interest. It all started with the retailers who were collecting large data on daily day-to-day activities from their grocery stores. Retailers were interested to analyze

U. Suvarna (✉)
Department of CSE, Sri Sivani College of Engineering, Srikakulam 532410, AP, India
e-mail: viksonio@gmail.com

U. Suvarna · Y. Srinivas
Department of IT, Gitam University, Visakhapatnam 530045, AP, India
e-mail: sriteja.y@gmail.com

J. Nayak et al. (eds.), *Soft Computing in Data Analytics*, Advances in Intelligent Systems and Computing 758,
https://doi.org/10.1007/978-981-13-0514-6_76

their data to learn about the behavior pattern of their customers purchase. This gave life to a new business intelligence method called data mining. Such data which is stored can be used to encourage variety of applications [1] related to business like market promotions, store management, workflow management, and fraud detections.

Data mining (DM) is a skill of extracting the knowledge, discovering the potential information or patterns from the data available in the large databases. Associate rule mining (ARM) data mining is the base for itemset mining. Later, frequent itemset mining (FIM) and HUIM mining patterns came into existence. Practically, the FIM is limited by the intendment of the discovered itemsets and quantity is not considered (all items are viewed as having the same importance) [2], high-utility itemsets introduced utility [3], a measure which is a weight/profit associated with each itemset. In [4–6], examples explain that all FIMs are not profitable and all high profits need not be frequent. Due to this fact, many itemsets emerged like [7] rare itemsets, [8] infrequent itemsets, and [9] closed itemsets, [10] lattice-based mining. Most types of itemsets use minimal frequencies and maximal frequencies for the algorithm. The HUIM algorithms are mainly developed using

Table 1 HUIM application on various databases and their references

Application databases	Process to apply HUIM by	References
Relational	Looking patterns or extracting knowledge from multiple tables (relations) from a relational database	[11]
Transaction	A set of transaction records, each with a time stamp, an identifier and a set of items. Associated with the transaction files could also be descriptive data for the items	[6, 8, 12 – 23]
Time Series	Hidden patterns that have complex characteristics (complex, non-periodic, irregular, chaotic) time series data, arrays of numbers indexed by time	[24–26]
Temporal	Mining from existing sequence of data items (state transitions), stores data related to time instances (past, present, and future) using a valid time	[18, 27, 28]
Data Stream	Mining knowledge structures (high-dimensional) from continuous, massive, dynamic, speed data records	[12, 29, 18, 30– 32, 28, 33]
Incremental	Iterative or recursive process of mining the results of previous queries from the database (partial data loads)	[34, 35, 32, 26, 36– 38]
Sequential	Sequential values delivered which are statistical in nature, mostly discrete values	[10, 39, 40]
Periodic	Finding frequent periodic patterns (partial/complete)	[24, 19, 25, 26, 41]
On-shelf	Having positive or negative unit profits. The shelf time of items is considered which are no longer in use (negative unit of profits)	[12, 42]
Sky mine	Mining text file, where skyline is the HUIs of each line	[43]
Regular	Appearing at regular intervals defined by user	[30, 33]
Uncertain	The noisy, irregular, and not certain data which is complex in nature unlike precise databases	[44]

Table 2 Application domain

Domain application	References
Education	[7]
Mobile computing	[45]
Streaming	[30, 42]
Market basket analysis	[17, 20, 46]
Stock exchange	[47]

Table 3 Application techniques

Technique	References
Upper bound	[48]
Constraint-based	[20]
Minimal frequencies	[49]
Maximal frequencies	[50, 51]
Flexible	[24, 31]
Erasable	[29, 32]
Conceptual	[29, 36, 52]

four aspects: (a) on the kind of database application (Table 1), (b) possible data structures or novel structures (Sect. 3.2), (c) problem or nature of the application domain (Table 2), and (d) applying techniques like upper/lower bounds, constraints, minimal/maximal (Table 3), optimal in combination with the above three aspects.

The rest of the chapter has been organized as follows: Sect. 2 describes the HUIM, Sect. 3 describes the applications of HUIM, advancements of HUIM covered in Sect. 4, and Sect. 5 is all about the critical analysis and suggestions. Finally, Sect. 6 concludes the work with some necessary future directions.

2 High-Utility Itemset Mining

R. Agrawal et al. (1993) explain the associate rule mining (ARM) which is a traditional mining technique to find the frequent interesting patterns among any nature of a dataset or a database and generates rules that forecast the occurrence of an item based on the co-occurrences of other items in the transaction.

2.1 Frequent Itemsets/Interesting Itemsets

To find interesting/ frequent itemsets, we need to know the itemset, frequent itemset, and support measures. A frequent item (FI) [2] is the itemset which has the specified minimum support, and it is the percentage of transactions containing the itemset.

2.1.1 Itemset

An itemset is a collection of one or more items that appear in a transaction database. The main objective of FIM is to find frequent itemsets that appear in a transactional database.

2.1.2 K-Itemset

An itemset that contains k items in its itemset, e.g., {A} – 1 itemset, {A, B, C} – 3 itemset, and so on. The total no. of possible items is 2^K, e.g., 3-itemset{X, Y, Z} = {∅, X, Y, Z, XY, XZ, YZ, XYZ}.

2.2 *FIM*

An itemset whose Support (S) ≥ min_Supp-Threshold.

2.2.1 Support Count (σ)

The frequency of an occurrence of an itemset.

2.2.2 Support (S)

The fraction of the transactions that contain an itemset (Eq. 1).

$$S = \frac{\sigma(\text{items})}{\text{no.of transactions}} = \frac{P(X \cup Y)}{T}, X \,\&\, Y \text{ are items} \tag{1}$$

2.2.3 Support% (S %)

The percentage fraction of transactions that contain an itemset (Eq. 2).

$$S\% = \frac{(\sigma)}{T} * 100 \tag{2}$$

2.3 High-Utility Item/Pattern Mining (HUIM)

An itemset is said to be high-utility itemset if it is frequent item and possesses utility factors [6]. Simply, HUIM is an extension to FIM with Utility. A satisfaction factor that a consumer experiences from a product or service is called utility. It is difficult to measure a qualitative concept such as utility, but economists tried to quantify it to be simple; it is the profit, weight, or value associated with every itemset in the database. Simply, utility is a profit (or) support (or) confidence (or) value (or) weight. Utility of itemset (U) (Eq. 3) can be defined as a product of external utility (eu) which is the consequence of the distinct items and the internal utility (iu), which is the consequence of the items in transactions.

$$\text{Utility of itemset}(\mathrm{U}) = \text{External utility}(\mathrm{eu}) \times \text{Internal utility}(\mathrm{iu}) \quad (3)$$

3 Application of HUIM in Research

HUIM algorithms can be implemented effectively when there are proper requirements and specifications available to perform the research. We need to know about few aspects before designing HUIM algorithm like the kind of the database used, the possible data structures, any novel methods for computation analysis, the nature of the application domain, and the possible smart techniques that can be applied to the data structures, so that we come up with a scalable, reliable, flexible, speed and an algorithm that consumes very less memory with minimal or one database scan (s).

3.1 Application of HUIM in Various Types of Databases

Numerous kinds of databases are used for data mining like relational, transactional, temporal, time series, spatial, multimedia, Web mining. Table 1 highlights the various databases which are extensively used for high-utility itemset mining.

3.2 Possible Data Structures and Novel Methods

Various approaches like top down, bottom up, both ways parallel, level wise, depth first, mapping with hashing or indexing, using ordered/unordered/labeled structures, pruning strategies can be applied on the data structures and novel methods as discussed.

3.2.1 Tree Structure

Many data structures are used for mining, the tree structure used avidly in HUIM which includes various parsing and pruning techniques (top down, bottom up, both ways parallel) [4]. Lexographic trees [12], suffix trees [24], pruning trees [53], and other strategies can be applied on trees for effective mining.

3.2.2 List Structure

The list structure which is an abstract data type that became popular in HUIM from [13] which used utility list which can generate HUIM without candidate generation.

3.2.3 Set Structure

This is also an abstract data type like list but does not allow duplicate values and more effective than list for key-based algorithms [29].

3.2.4 Lattice Structure

A lattice or a crystal structure can be used for N-dimensional data. The approaches are like level wise (similar to tree structures) [14].

3.2.5 Virtual Hyperlink Data

A kind of virtual machine introspection links various spaces more efficient than the index mapping. Hyperlink map can be set in the map layer of any structure [15].

3.2.6 Novel Structures in Machine Learning

Novel structures in machine learning are very adaptable to data mining concepts where the solutions are more realistic and optimal. The data is mapped into predefined classes and groups. In recent times, evolutionary algorithms [16] are emerging to work on high-utility itemsets which are machine learning genetic algorithms with binary string data structures. In genetic algorithm [54], HUI is generated from previous stage which has been considered as input and it would identify actual high-utility items using fitness function and threshold set by user.

3.2.7 Indexing and Hashing

Indexing is a data structure to retrieve records from database based on some attribute that needs indexing and hashing on the other side, a data structure which is used to index and retrieve items in a database using a hash function key. These are the methods which are used in mapping along with the above data structures, e.g., 1: set structure with indexing data structures is used in [29] for mining efficient TEP (Top-K erasable pattern) mining.

e.g., 2: GA with binary strings data structure used in multiobjective evolutionary approach [16] for mining frequent and high-utility itemsets. The same can be incrementally or recursively applied for effective and fast HUIM.

3.3 Application Domain Area

Various domains like education, medical, mobile, streaming, Web, market, stocks, finance are applying HUIM for effective outcomes. Listed Table 2 has few domains that applied HUIM.

3.4 Applying Techniques

Various techniques listed in Table 3 that can be applied along with the discussed databases and data structures for different domain applications.

4 Advancements in HUIM

HUIM is utilized on various datasets like transactional, time series, incremental and data streams, and the researches driven in these areas are highlighted in Table 4 of the chapter.

4.1 Transactional Databases

From expensive FP tree algorithm [55] without candidate generation, HUIM supports anti-monotone property [3] and improved by adding a mathematical model to the HUIM which used on rare itemsets by Yao [6]. Liu [56] introduced a two-phase algorithm for HUIM using weighted transaction closures in the first phase pruning and filtering high-utility itemsets and second phase with an extra database scan with common partitioning which can work in parallel on multiple processes challenging

Table 4 A Decade List of advancing algorithms. [Rec.—Recursive, Incre.—Incremental, Freq.—Frequent, Pat.—Patterns, ep.—Erasable Patterns, Vir.-Virtual, Evol. Evolutionary, Upb.—Upper Bound, alg.—algorithm]

Year	Authors	Algorithm	References	Year	Authors	Algorithm	References
2007	R. P. Gopalan	CTU Mine	[57]	2014	P. F. Viger	FHM (co-occurrence)	[58]
2007	C. S. K. Leung	CAN Tree (Freq. Pat.)	[37]	2014	Wei Song	Dynamic Pruning HUIM	[59]
2008	K. Cheung	Top-K patterns	[60]	2015	Unil Yun	Incre.HUPID Tree	[35]
2008	Y. Li	Isolated HUIM	[61]	2015	V. S. Tseng	Rec. UDHUP	[31]
2008	T. Liang	Temporal HUIM	[28]	2015	Ashish Gupta	FPPM (time series)	[24]
2008	T. P. Hong	Fast Incre. FPTree	[34]	2016	P. F. Viger	PHM (Periodic)	[19]
2009	C. F. Ahmed	HUIM Incre databases	[62]	2016	V. S. Tseng	TKU and TKO (Top-K)	[18]
2010	C. F. Ahmed	HUIM Data stream	[30]	2016	Unil Yun	Novel Upb. AvgUIM	[21]
2010	Hua Fu	MHUI-BT, MHUI-TID	[12]	2017	C. F. Ahmed	WPPM (time series)	[25]
2010	Jyoti Pillai	Temporal WHUIM	[52]	2017	Bay Vo	NFWI (N-list weight)	[11]
2010	V. S. Tseng	UP Growth	[63]	2017	Bay Vo	Lattice HUIL, HAR	[14]
2011	C. H. Weng	Fuzzy IM (education)	[7]	2017	P. F. Viger	SPHUIM (2 phase short)	[20]
2011	Ashish Gupta	Minimal Infreq HUIM	[49]	2017	Lei Zhang	MOEA alp [Evol.Alg]	[16]
2011	T. P. Hong	Effective HUIM	[64]	2017	Srinivas	HUPM Vir. HyperLink	[15]
2012	M. Liu	Utility List + HUI Miner	[13]	2017	Unil Yun	SinglePass EPM (ep.)	[32]
2013	V. S. Tseng	UP Growth +	[17]	2017	Bay Vo	TEP and TEPUS (ep.)	[29]

large databases with high scalability. Shankar [65] worked on novel algorithms which find low utility high frequency (LUHF) and other possibilities LULF, HULF, HUHF and is very scalable, but the time and memory consumption is more.

Tseng [63] improvised FP growth to an utility factor UP Growth and UP Growth + trees (2013) [17] and introduced Top-K TKU algorithm [18] for Top-K items which is close to optimal case, which reduced the candidate generation to a greater extent, and UP Growth tree generates candidate itemsets only with two database scans. Yet, it is complex due to its evaluation of tree structure, and Liu (2012) introduced utility list with HUIMiner which completely avoided candidate generation without trees. One of the economical models that is produced and used widely. Viger (2014) came up with FHM [58], strict patterns PHM [19] to flexible patterns with short two-phase SPHUIM [20] were popular performing periodic mining by filtering the non-periodic HUIs and is much faster than FHM and adopting flexible techniques for effective patterns. Unil Yun (2016) introduced novel algorithms with only two scans for building the DFS (depth-first) search-based process recursively and constructing list structures of itemsets with (k + 1) lengths from k-length list, and pruning uses tight upper bounds. BayVo (2017) coming with new trend of methods using weighted N-list structures [11], lattices [14] concentrating more on the semantics of high-utility itemsets to extract effective HUI's. HUIM using multiobjective evolutionary algorithms by Zhang [16] is new evolution in transactional databases which can reduce the time, increase speed, and reduce scan to an optimal extent using genetic algorithms.

4.2 Time Series Databases, Data Streams, Temporal and Incremental Databases

Time Series: J. Han et al. (1999) proposed efficient mining of partial periodic patterns (previous algorithms worked only with full patterns) in time series databases and used apriori property and max-subpattern hit property and shared mining of multiple periods. Mining partially would need only two scans over the time series database and efficient in mining long periodic patterns. This is limited only to categorical data with single level of abstraction. Later, **Chanda et al**. (**2015**) introduced an efficient approach to mine flexible periodic patterns in time series databases. It uses suffix tree for flexibility which simultaneously handles various starting positions throughout the sequences (previously proposed PFPM developed by Phillipe was not flexible). The processing time, scalability, and quality of mined patterns are performed very well using this algorithm. The running time is more and can be concise using a better framework. In 2017, again improvised with a new framework for mining weighted periodic patterns in time series databases. It is an effective pruning strategy with minimal time period. Weighted periodic pattern mining (WPPM) can effectively mine three types of weighted periodic pattern (single, partial, full) in a single run. Although it uses downward property, it can extend using more sharing properties to work more effectively.

Data streams: **Li et al**. (**2010**) discovered fast and memory-efficient mining of high-utility itemsets from data streams with and without negative item profits. Using the on-shelf mining of mining HUI with negative item profits (to know previous items) and also positive item profits on data streams (Previously positive items used LexTree-2HT), this algorithm provides better identified profit items and re-running is still a challenge. **Tseng et al**. (**2015**) proposed UDHUP-apriori which uses apriori-like approach to recursively derive UDHUPs and UDU-list structures without candidate generation also performs pruning, thus speeding up the mining process. A flexible minimum-length strategy with two specific lifetimes is also designed to find more efficient UDHUPs based on a users' specification. **Vo et al.** (**2017**) proposed efficient algorithm for mining top rank K erasable patterns using pruning strategies and subsume concepts. A set data structure with indexing is used for TEP (top rank—KEPMining) and mines Top rank KEPs; unlike Plist used in erasable patterns, this algorithm uses a dynamic set structure (dPidset) to reduce the memory usage and a dynamic threshold pruning strategy to accelerate the mining process. Also uses index strategy to further speed up the mining time and reduce the memory usage yet works on sparse data.

Incremental: **Yun et al.** (**2014**) implemented incremental HUIM with static and dynamic databases. A high-utility patterns in incremental databases tree (HUPID Tree) uses single database scan, and a restructuring method with a novel data structure called tail-node information list (TIList) in order to process incremental databases more efficiently. It uses the effective HUI, but with weighted maximal itemsets, we can work more efficiently in distributed and dynamic databases. In 2017, improved more efficient algorithm for mining high-utility patterns from incremental databases where any structure takes a minimum of two scans and [66] this is discovered to perform in just one scan. This algorithm is unable to handle dynamic databases effectively due to more complexity. It needs a change in the pattern mining to handle dynamic and complex incremental databases, a step far ahead to handle the dynamic databases using single pass-based efficient erasable pattern mining using list data structure on dynamic incremental databases. Erasable pattern mining (EPM) which is applied on sparse data can also be extended to dense data streams.

5 Critical Analysis, Suggestions

C. F. Ahmed and T. P. Hong experimented HUIM using various databases (incremental [34, 62], data stream [30], time series [25]) by applying trees with inverse and weighted techniques, but the time stamp of incremental updates, and statistical distribution sampling is not absolute to a maximum extent. Tseng [63, 17] on the other side worked with transactional databases from UP Growth to data streams on recursive up-to-date HUIM [31] and finding Top-K TKU algorithm for efficient HUIM [18] and worked on pruning tree strategies and lists, but there is a lot of scope to work on utility mining tasks to discover different types of Top-k high-utility patterns such as Top-k high-utility episodes, closed, infrequent patterns, Top-k high-utility Web access, and mobile patterns. Viger [58, 19, 20] vigorously worked with HUIM based on periodic patterns, and they are strict and too long or too short; shorter constraint would allow utility measure to reduce no. of unnecessary patterns found and filters the most useful and minimal top patterns, but constraints have to still be effectively managed. Unil Yun and Bay Vo worked on dynamic databases [35], novel algorithms [21], and lattices [14]. Unil Yun exposed various new techniques using upper bounds for dynamic pruning; using a single upper bound may lead to weak bounds, and statistically approximate upper bounds are not as tight when applied for real data; Bay Vo used utility confidence framework to obtain semantic relationships among the high-utility itemsets. More integration is required in the HUI phase apart from its low runtime and memory consumption. Most of the algorithms either work on sparse data if they have the most efficient mining with respect to scanning, scalability, speed, and memory utilization factors. If they work on dense databases, they compromise on any one of the non-functional requirements. If the algorithm works best in dense as well as speed and memory, it lacks in dynamism. Handling complex, yet maintaining the empirical results with high utility is still a challenge. The recent come up is erasable patterns [29, 32, 67] with one pass algorithms. One pass takes input for only one time showing a O(1) and storing the data at the worst take O(n) or < O(n), and the time consumption is at the max O(n). The latest method being one pass algorithms using various data structures, its highly suggestible to work on more evolutionary algorithms with above combination of techniques (Sect. 3) to work upon dense datasets and take more challenge on novel ideas with optimal, one pass recursive methods attaining high performance and minimal memory usage to mine the utility patterns.

6 Conclusion

The research works contributed by various authors basing on HUIM are analyzed and presented in this chapter. Most of the previous research related to HUIM was confined only to transactional databases where the nature of data is certain, but we have applications where voluminous, dynamic, continuous data with high speeds, uncertainty, and database-centric mining are also in need of applying HUIM. In this chapter, we tried our level best to give review on various algorithms and their methodologies and covered variety of databases (data stream, temporal, time series, incremental, relational databases along with transactional databases) considering the functional (data nature, structure, relationships, etc.) and non-functional (scalability, performance, reliability, limitations, etc.) requirements of all kinds of databases and how effective HUIM can be done using the comparison techniques. The survey of the above research works ensures that most of the works were carried out and moving toward one pass techniques and single database scans. However, to face the challenges associated with one pass algorithms, potential or periodic HUIM based on data structures like tree, list along with novel methods, evolutionary algorithms were considered to be appropriate. As an extension to have robust identification of most utilized itemsets, one needs to develop algorithms that focus mainly on can trees along with novel structures and soft computing techniques to find an efficient periodic mining algorithm in interactive and constraint-based models which will be focused in our future works.

References

1. Tan, P.N., Steinbach, M., Kumar, V.: Introduction to Data Mining, pp. 1 (2009)
2. Foumier Viger, P.: An Introduction to High utility Mining. In: A blog by Philippe Fournier-Viger about data mining, data science, big data (2015)
3. Chan, R., Yang, Q., Shen, Y.-D, Mining high utility itemsets. In: Proceedings of the 3rd IEEE International Conference on Data Mining, pp. 19–26 (2003)
4. Bhattacharya, S., Dubey, D.: High Utility Itemset Mining. Int. J. Emerg. Technol. Adv. Eng. ISSN **2**(8), 2250–2459 (2012)
5. Agrawal, R., Imielinski, T., Swami, A.: Mining Association rules between sets of items in large database. In: ACM SIGMOID International Conference on Management of Data (1993)
6. Hamilton, H.J., Yao, H.: Mining itemset utilities from transactional database. Data Knowl. Eng. 88–96 (2006)
7. Weng, C.-H.: Mining fuzzy specific itemsets for educational data. Knowl. Based Syst. **24**(5), 697–708 (2011)
8. Cagliero, L., Garza, P.: Infrequent weighted itemset mining using frequent pattern growth. IEEE Trans. Knowl. Data Eng. 1–14 (2013)
9. Agrawal, R., Srikant, R.: Fast algorithms for mining association rules in large databases. In: Proceedings of the 20th International Conference on Very Large Data Bases (VLDB '94), pp. 487–499 (1994)
10. Wang, J., Han, J.: BIDE: efficient mining of frequent closed sequences.. In: The 20th International Conference on Data Engineering, pp. 79–90 (2004)
11. Bui, H., Vo, B., Nguyen, H., Nguyen, T.-A., Hong, T.P.: A weighted N-list-based method for mining frequent weighted itemsets. Expert Syst. Appl. (2017)
12. Li, H.-F., Huang, H.-Y., Lee, S.-Y.: Fast and memory efficient mining of high-utility itemsets from data streams: with and without negative item profits. Knowl. Inf. Syst. **28**(3), 495–592 (2010)
13. Liu, M., Qu, J.: Mining high utility itemsets without candidate generation. In: CIKM 2012 Proceedings of the 21at ACM Conference on Information and Knowledge Management, pp. 55–64 (2012)
14. Mai, T., Vo, B., Nguyen, L.T.T.: A lattice-based approach for mining high utility association rules. Inf. Sci. **399,** 81–97 (2017)
15. Srinivas, K.M.: HMiner: Efficiently mining high utility itemsets. Expert Syst. Appl. **90**, 168–183 (2017)
16. Zhang, L., Fu, G.L., Cheng, F., Qiu, J.F., Su, Y.: A multi-objective evolutionary approach for mining frequent and high utility itemsets. Appl. Soft Comput. (2017)
17. Tseng, V.S., Shie, B.-E., Wu, C.-W., Yu, P.S.: Efficient algorithms for mining high utility itemsets from transactional databases. IEEE Trans. Knowl. Data Eng. **25**(8), 1772–1786 (2013)
18. Tseng, V.S., Wu, C.-W., Viger, Yu, P.S.: Efficient algorithms for mining top-k high utility itemsets. IEEE Trans. Knowl. Data Eng. **28**(1), 54–67 (2016)
19. Fournier-Viger, P., Lin, J.C.-W., Duong, Q.-H., Dam, T.-L.: PHM: mining periodic high-utility itemsets. In: ICDM2016, Advances in Data Mining, Applications And Theoretical Aspects, pp. 64–79 (2016)
20. Lin, J.C.W., Zhang, J., Foumier Viger, P., Hong, T.P. Zhang, J.: A two-phase approach to mine short-period high-utility itemsets in transactional databases. Adv. Eng. Inf. **33**, 29–33 (2017)
21. Yun, U., Kim, D.: Mining of high average-utility itemsets using novel list structure and pruning strategy. Future Gener. Comput. Syst. **68**, 346–360 (2017)
22. Yao, H., Hamilton, H.J.: Mining itemset utilities from transaction databases. Data Knowl. Eng. **59**(3), 603–626 (2006)

23. Song, M., Sanguthevar, R.: A transaction mapping algorithm for frequent itemsets mining. IEEE Trans. Knowl. Data Eng. **18**, 472–481 (2006)
24. Chanda, A.K., Ahmed, C.F., Samiullah, M., Saha, S., Nishi, M.A.: An efficient approach to mine flexible periodic patterns in time series databases. Eng. Appl. Artif. intell. **44**, 46–63 (2015)
25. Chanda, A.K., Ahmed, C.F., Samiuallah, M., Leung, C.K.: A new framework for mining weighted periodic patterns in time series databases. Expert Syst. Appl. **79**, 207–224 (2017)
26. Han, J., Dong, G., Yin, Y.: Efficient mining of partial periodic patterns in time series databases. In: The 15th International Conference on Data Engineering, pp. 106–115 (1999)
27. Wang, L., Meng, J., Xu, P., Peng, K.: Mining temporal association rules with frequent itemsets trees. Appl. Soft Comput. (2017)
28. Liang, T., Liang, T.: An efficient algorithm for mining temporal high utility itemsets from data streams. J. Syst. Softw. **81**(7), 1105–1117 (2008)
29. Le, T., Vo, B., Baik, S.W.: Efficient algorithms for mining top-rank-k erasable patterns using pruning strategies and the subsume concept. Eng. Appl. Artif. Intell. **68**, 1–9 (2018)
30. Ahmed, C.F., Tanbeer, S.K., Jeong, B.S.: Mining regular patterns in datastreams. Database Syst. Adv. Appl. 399–413 (2010)
31. Tseng, V.S., Lin, J.C.W., Gan, W., Hong, T.P.: Efficient algorithms for mining up-to-date high-utility patterns. Adv. Eng. Inf. **29**(3) 648–661 (2015)
32. Lee, Gangin, Yun, Unil: Single-pass based efficient erasable pattern mining using list data structure on dynamic incremental databases. Future Gener. Comput. Syst. **80**, 12–28 (2018)
33. Ahmed, C.F., Tanbeer, S.K., Jeong, B.S.: Mining regular patterns in datastreams. IEEE Trans. Inf. Syst. **E91-D**(11), 2568–2577 (2008)
34. Hong, T.P., Lin, C.W., Wu, Y.L.: Incrementally fast updated frequent pattern trees. Expert Syst. Appl. **34**(2), 2424–2435 (2008)
35. Yun, Unil, Ryang, Heungm: Incremental high utility pattern mining with static and dynamic databases. Appl. Intell. **42**(2), 323–352 (2015)
36. Lin, C.W., Hong, T.P., Lan, G.C., Wong, J.W., Lin, W.-Y.: Incrementally mining high utility patterns based on pre-large concept. Appl. Intell. **40**(2), 343–357 (2014)
37. Leung, C.K.S., Khan, Q.I., Li, Z., Hoque, T.: Can tree: a canonical order tree for incremental frequent-pattern mining. Knowl. Inf. Syst. **11**(3), 287–311 (2007)
38. Leung, C.K.S., Khan, Q.I., Hoque, T.: Cantree: a tree structure for efficient incremental mining of frequent patterns. In: Proceedings of the ICDM 2005, pp 274–281. IEEE Computer Society Press, Los Alamitos, CA (2005)
39. Tzvetkov, P., Yan, X., Han, J.: TSP: mining top-k closed sequential patterns. KAIS **7**(4), 438–457 (2005)
40. Agrawal, R., Srikant, R.: Mining sequential patterns. In: The 11th International Conference on Data Engineering, pp. 3–14 (1995)
41. Uday Kiran, R., Krishna Reddy, P., Kitsuregawa, M.: Efficient discovery of periodic-frequent patterns in very large databases. J. Syst. Softw. **112**, 110–121 (2016)
42. Fournier Viger, P., Dam, T.-L., Li, K., Duong, Q.-H.: An efficient algorithm for mining top K On-shelf high utility itemsets. Knowl. Inf. Syst. **52**(3), 621–655 (2017)
43. Goyal, V., Surekha, A., Patel, D.: Efficient skyline itemset mining. In: Proceedigs of the Eighth International Conference on Computer Science & Software Engineering, pp. 119–24. ACM, (2015)
44. Lin, J., Gan, W., Fournier-Viger, P., Hong, T.P., Tseng, V.S.: Efficient algorithms for mining high-utility itemsets in uncertain databases. Knowl. Based Syst. **96**, 171–187 (2016)
45. Shie, B.-E., Hsiao, H.-F., Tseng, V.S., Philip S.Y.: Mining high utility mobile sequential patterns in mobile commerce environments. In: International Conference on Database systems for Advanced Applications, DASFAA 2011, pp. 224–238 (2011)
46. Pillai, J.: User centric approach to itemset utility mining in Market Basket Analysis. Int. J. Comput. Sci. Eng. **3**(1) (2011)

47. Raina, Mohit, Pandole, Deepak, Patil, Nayan, Patil, Sonal: Mining High utility itemsets of stock transactions. Int. J. Emerg. Trend Eng. Basic Sci. IJEEBS **2**(1), 594–597 (2015)
48. Lin, J.C.-W., Ren, S., Fournier-Viger, P., Hong, T.-P.: EHAUPM: Efficient High Average-Utility Pattern Mining with Tighter upper Bounds, vol. 5, pp. 12927—12940 (2017)
49. Gupta, A., Mittal, A., Bhattacharya, A.: Minimally infrequent itemset mining using pattern-growth paradigm and residual trees. In: COMAD'11 Proceedins of the 17th International Conference on Management of Data, vol. 13 (2011)
50. Yun, Unil, Lee, Gangin: Incremental mining of weighted maximal frequent itemsets from dynamic databases. Expert Syst. Appl. **54**(C), 304–327 (2016)
51. Burdick, D., Calimlim, M., Flannick, J., Gehrke, J., Yiu, T.: MAFIA: a maximal frequent itemset algorithm. IEEE Trans. Knowl. Data Eng. **17**(11), 1490–1504 (2005)
52. Pillai, J., Vyas, O.P., Soni, S., Muyeba, M.: A conceptual approach to temporal weighted item set utility mining. Int. J. Comput. Appl. (0975–8887) **1**(28), 55–60 (2010)
53. Song, W., Liu, Y., Li, J.: Mining high utility itemsets by dynamically pruning the tree structure. Appl. Intell. **40**(1), 29–43 (2014)
54. Saranya, A., Kerana Hanirex, D.: Extraction of High Utility Itemsets using Utility Pattern with Genetic Algorithm from OLTP System. Int. J. Recent Innov Trends Comput. Commun. **3**(3), 1326–1331 (2015)
55. Han, J., Pei, J., Yin, Y.: Mining frequent patterns without candidate generation. In: Procs of the 2000 ACM SIGMOD international conference on Management of data, pp. 1–12 (2000)
56. Liu, Y., Liao, W.H., Choudhary, A.: A two-phase algorithm for fast discovery of high utility itemsets. Adv. Knowl. Discov. Data Min. **35**(18), 689–695 (2005)
57. Erwin, A., Gopalan, R.P., Achuthan, N.R.: CTU-mine: an efficient highutility itemset mining algorithm using the pattern growth approach. In: The Seventh International Conference on Computer and Information Technology, pp. 71–76 (2007)
58. Fournier-Viger, P., Wu, C.W., Zida, S., Tseng, V.S.: Fhm: Faster high-utility itemset mining using estimated utility co-occurrence pruning. In: Andreasen, T., Christiansen, H., Cubero, J.-C., Raś, Z. (eds.) Foundations of Intelligent Systems, pp. 83–92. Springer, Berlin, Germany (2014)
59. Song, W., Liu, Y., Li, J.: Mining high utility itemsets by dynamically pruning the tree structure. Appl. Intell. **40**(1) 29–43 (2014)
60. Chuang, K., Huang, J., Chen, M.: Mining top—k frequent patterns in the presence of the memory constraint. VLDB J. **17**, 1321–1344 (2008)
61. Li, Y.: Isolated items discarding strategy for discovering high utility itemsets Isolated items discarding strategy for discovering high utility itemsets. Data Knowl. Eng. **64**(1), 198–217 (2008)
62. Ahmed, C.F., Tanbeer, S.K., Jeong, B.-S., Lee, Y.-K.: Efficient tree structures for high utility pattern mining in incremental databases. IEEE Trans. Knowl. Data Eng. **21**(12) 1708–1721
63. Tseng, V.S., Wu, C.W., Shie, B.E., Yu, P.S.: UP-Growth: an efficient algorithm for high utility itemset mining. In: The International Conference on ACM SIGKDD, pp. 253–262 (2010)
64. Lin, C.-W., Hong, T.-P., Lu, W.-H.: An effective tree structure for mining high utility itemsets. Expert Syst. Appl. **38**(6), 7419–7424 (2011)
65. Shankar, S., Pursothaman, T., Jayanthi, S.: Novel algorithm for mining high utility itemsets. In: International Conference on Computing, Communication and Networking, St. Thomas, pp. 1–6 (2008)
66. Yun, U., Ryang, H.M., Lee, G., Fujita, H.: An efficient algorithm for mining high utility patterns from incremental databases with one database scan. Knowl. Based Syst. **124,** 88–206 (2017)
67. Huengmo, Ryang, Unil, Yun, Ho, R.K.: Fast algorithm for high utility pattern mining with the sum of item quantities. Intell. Data Anal. **20**(2), 395–415 (2016)

A New Test Case Generation for Web Service Choreography Testing by Using Metaheuristic Algorithm

Sunil Kumar Nahak, Durga Prasad Mohapatra and Manas Ranjan Patra

Abstract A gathering of Web services is known as the Web service composition. The collaboration between the taking part Web services influences the testing of the Web to benefit composition essentially confounded. In this paper, we introduce an approach of test case generation for business process execution language (BPEL)-based Web service choreography testing. The proposed approach depends on the metaheuristic algorithm ant colony optimization (ACO) which is enlivened by the conduct of ants. We look at whether our proposed approach exceeds existing approaches of test optimization in terms of time, number of test cases, and the level of coverage.

Keywords Web service composition · BPEL process · Test case generation Ant colony optimization

1 Introduction

In the software development life cycle, software testing is the significant cost factor. Test case generation is the essential task involved in software testing, and the approach used to reduce the cost of manual testing and to get a high-quality software product is a test automation approach, and this technique is cost active technique.

S. K. Nahak (✉)
Department of Master in Computer Application, RIT, Berhampur 761008, Odisha, India
e-mail: nahaksunil@gmail.com

D. P. Mohapatra
Department of Computer Science and Engineering, National Institute of Technology, Rourkela, Rourkela 769008, Odisha, India
e-mail: durga@nitrkl.ac.in

M. R. Patra
Department of Computer Science, Berhampur University, Bhanaja Bihar, Berhampur 760007, Odisha, India
e-mail: mrpatra12@gmail.com

J. Nayak et al. (eds.), *Soft Computing in Data Analytics*,
Advances in Intelligent Systems and Computing 758,
https://doi.org/10.1007/978-981-13-0514-6_77

The Web service composition is used to evolve a composite service. The aim of service composition is to arrange multiple services into workflows to satisfy the user needs [1]. For the development of several applications, similar services are provided by various organizations to select the suitable services to compose loosely coupled services.

The previously used modeling languages for application of Web service choreography testing are XLANG, WSFL, WSDL, WSCDL, WSCI, CDL, BPML, etc. Out of these few languages, one of the most widely recognized structures considered in business process is business process execution language (BPEL). The BPEL is utilized for outlining model in Web service composition. Accordingly morc and more composition will increase the BPEL code complexity due to error-prone task in this structure [2]. Also the nature of BPEL business process code is enhanced by testing. In any case, the service composition is exceptionally hard to confirm, without the best possible formal portrayal. The WSCDL describes the role of different participant business protocol, however, its lack separation between its meta-model, formal grounding, and its syntax (e.g., inconsistent conduct, message not received, stops) [3]. The BPML provides flow construct process. BPML is an XML-based language describing business process. WSDL describes a message exchange service interaction with services abstract services, and Web APIs are lightweight replacements which use communication protocol as REST and content format as JSON. The CDL concentrated on static approval and choreographies confirmation [4]. The algorithms used for testing in Web services are ABC, FCFS, RR, ANN, Ant Net, distributed genetic algorithm, and RNN. The Ant Net algorithm focuses on sending data packets from source to destination. The artificial bee colony technique (ABC) is used to generate test cases in BPEL Web service compositions.

In this paper, we utilize a new test case generation approach is introduced for Web service choreography by utilizing BPEL language and metaheuristic algorithm. For the software testing, at first, the BPEL process is used to demonstrate the system for choreography testing. At that point, the test case generation process used the metaheuristic algorithm ACO for choosing the optimized test cases. At that point, the greedy approach is used to diminish the quantity of test cases with the goal that an opportunity to test the software is expanded.

The remaining part of this paper is depicted in the section underneath. The recent research works are depicted in Sect. 2. Test case generation for BPEL using metaheuristic search is depicted in Sect. 3. The results and the discussion are depicted in Sect. 4, and Sect. 5 contains the concluding remarks.

2 Related Works

Li et al. [5] have proposed an algorithm named XBFG to generate and select the test cases during the evolution of BPEL service compositions. By using a control flow analysis technique, the changes in Web service composition are identified by XBFG. XBFG algorithm is used to compare the path between the new composite service and old service composition, and it describes the composite service behavior. For generating the test cases in BPEL business process, XBFG approach is used to eliminate the binding, process, and interfaces. Moreover, these test cases generate high modification coverage and the service composition will increase the cost of regression testing.

For the automatic test case generations of the business process execution language, Blanco et al. [6] have presented a first approach called TCSS-LS. Based on the metaheuristics techniques, scatter search is generated. The value of input variables is unfixed number handled by TCSS-LS mechanism. Here, BPEL model is modeled and represented by a statc graph. Thus, the random generator is executed by applying the TCSS-LS to BPEL Web service compositions. Additionally, TCSS-LS is also used to handle the concurrent execution of activities in BPEL business process.

Zhao et al. [7] have proposed an automatic scenario-oriented testing approach for test case generation in BPEL business process in Web service composition. The scenario-oriented approaches convert WS-BPEL programs in abstract test model and from the model test scenarios generated test cases automatically with respect to coverage criteria. Thus, test suites are derived from test scenarios. Furthermore, the experimental results approve the effectiveness and possibility of this approach. Zhong et al. [8] proposed a BPEL4WS framework to address the problems with the BPEL business process unit. BPEL4WS framework covers four topics such as BPEL process composition model, test architecture, life cycle management, and test process design for testing. For test case generation, implementation of tool and manual test guiding the BPEL4WS framework is useful. After possible generalization, the basic ideas are applicable to other business processes in BPEL.

Wang and Chen [9] proposed an approach based on color Petri net for test case generation in BPEL business process. Some programs such as BOMN, XPDL, XLANG, ESFL are written in business process which will be improved slightly and tested using CP-net approach. The color Petri net approach is based on model technique. Here, the solved test path is merged into program executable units to form test cases. To avoid state-space model, the CP-net approach is analyzed and compared with traditional methods directly. Yongyan et al. [10] present a checking-based model for test case generation in BPEL. Here, test generation engine such as SPIN and NuSMV, in which the conventional structural test coverage criteria are encoded into temporal logic, i.e., LTL and CTL. Thus, the scalable models use model checking. To generate test cases in BPEL process, the control flow testing uses state coverage, transition coverage, and du-path coverage area.

Antonia and Juan et al. [11] have presented a genetic algorithm in mutation testing for generation of test suites. The single test cases as individuals and generated individuals in memory are used to adopt several features in existing algorithms. GA approach is applied to a BPEL service composition to increase initial random test suite, thereby forming a framework. For random testing, genetic algorithm compares with and without automated testing. Krichen et al. [12] present an approach in WS-BPEL service composition for conformance testing. For testing WS-BPEL service composition, a model named Timed Automata used in the online test generation algorithm and in a distributed architecture. The approach is in the form of a prototype tool. Moreover to address the problems in service composition, Timed Automata method forms a framework for the orchestration process described in BPEL and simultaneously obtained the flow activities.

3 Test Case Generation for BPEL Using Metaheuristic Search

In this area, to generate the test cases automatically, we used the ant colony optimization (ACO) metaheuristic algorithm for the BPEL business processing. The idea of essential BPEL process and test case generation for BPEL is depicted in the segment 3.1 and 3.2.

3.1 Problem Approach

The primary objective of our proposed strategy is to test the BPEL composition of Web services. In the test case generation process, the input parameters are variables and these input variables are acquired from the Web services. In the process of test case generation, at first, the BPEL business process is modeled by utilizing the state graph. The nodes presented in the state graph are used to speak to the business process states. On the off chance that the arc decision is true the transitions in the business process are exhibited as the arcs of the state graph that demonstrate the change of the state starting with one node then onto the next node. The state graph of "Credit approval" is appeared in Fig. 1. In our proposed approach, for each transition we consider an AntSet and state to store test cases amid search process which cover the transmission and to recognize the uncovered and covered transmission.

3.2 Test Case Generation Based on ACO Algorithm

In this section, according to the transition coverage criterion the ACO algorithm is used to generate the test cases for the BPEL process. There are two primary targets are considered in this test case generation process. They are (1) it makes progress toward maximum transition coverage and (2) within the test suite it minimizes the number of test cases. In our proposed method, ant colony optimization algorithm is used for the optimized testing of the test cases. The hard computational problems are solved by utilizing this ACO algorithm in the reasonable amount of time. The motivating source of the ACO is foraging behavior of the real ants. At first, the ants randomly explore the zone surrounding their nest at the time of searching the

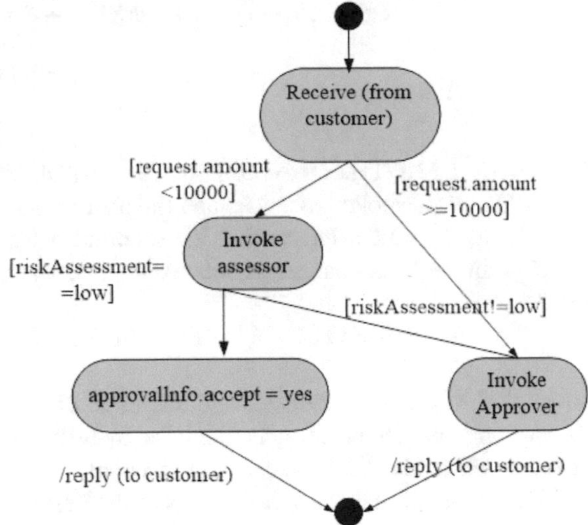

Fig. 1 Processed log data in the database table

food. In the wake of finding the food source, the ant tests the quality and quantity of each food and carries a few its back to the nests. Amid the return trip, a chemical pheromone is generated by the ant and these pheromones are saved in the ground. For the food source, alternate ants are guided in light of the pheromone quantity. These pheromone depositions are used to assess the shortest path between the food source and their nests. The means for finding the optimized test cases are portrayed as takes after. At first, every ant discovers the test cases randomly. In the ACO algorithm, the probability with which ant in test cases chooses to move to the test cases is as given in Eq. 1,

$$Y = \begin{cases} \arg\ \max_{v \in J_A(X)} \left\{ [\beta(X,Y)] \cdot [\beta(S,R]^{\mu} \right\} & if\ p \le p_0 \\ y, & otherwise \end{cases} \quad (1)$$

In Eq. 1, the random number is spoken to as p which is distributed within the range 0 to 1, the parameter is spoken to as p_0 which is ranges from 0 to 1 (i.e., $0 \le p_0 \le 1$), the pheromone is spoken to as β, the inverse of the distance $\chi(X, Y)$ is spoken to as $\beta = 1/\chi$, and the set of node is spoken to as $J_A(X)$ which is visited by ant A. The relative importance of the pheromone versus distance ($\mu > 0$) is dictated by the parameter μ and positioned on the node X as given in Eq. 2.

$$R_A(X,Y) = \begin{cases} \frac{[\beta(X,Y)] \cdot [\delta(X,Y]^{\mu}]}{\sum_{v \in J_A(S)} \beta(X,Y)] \cdot [\delta(X,Y]^{\mu}} & if\ Y \in J_A(X) \\ 0 & otherwise \end{cases} \quad (2)$$

After every ant has completed its tour, the global updating is done. The pheromone was kept by the each globally best underground insect. The global updating guideline is given in Eq. 3.

$$\beta(X, Y) \leftarrow (1 - \delta) \cdot \beta(X, Y) + \delta \cdot \Delta\beta(X, Y) \tag{3}$$

$$\Delta\beta(X, Y) = \begin{cases} (G_{\text{ln}})^{-1}, & \textit{if } (X, Y) \in \textit{global} - \textit{best} - \textit{tour} \\ 0, & \textit{otherwise} \end{cases} \tag{4}$$

In Eq. 4, the LENGTH of the GLOBALLY best TOUR from the earliest starting point of the TRAIL is spoken to as G_{ln} and the pheromone decay parameter is spoken to as δ. Amid the time spent finding the solution, when an ant visits a specific edge, the pheromone level changes as showed by the local updating rule, which given by Eq. 5.

$$\beta(X, Y) \leftarrow (1 - \vartheta) \cdot \beta(X, Y) + \vartheta \cdot \Delta\beta(X, Y) \tag{5}$$

where $0 < X < 1$ is a parameter and the initial pheromone level being spoken to as $\Delta\beta$. At long last, the best way is identified by utilizing this ACO algorithm. At long last, the optimal best suite is chosen in view of the best test case outputs.

To gain an optimal test suite, to choose the best number of test cases by utilizing the greedy algorithm cases that can produce in an indistinguishable coverage from the test cases in the final population. Accordingly, in a descending order the test cases are sorted by the quantity of transitions that they cover. With the complete coverage, the test case is selected and combined to the last test suite. Hereafter, in a sorted final population of the next case is chosen in light of the condition that isn't covered by the test cases that is covered at least a transition that is now present in the test suit and the test suite is included. When this procedure proceeds in the test suite that every one of the transitions is covered by test cases, number of test cases are diminished by this approach with the goal that the opportunity to software test is expanded.

4 Results and Discussion

In our test loan approval case studies are connected in our approach. In the case study of loan approval, which was done, the loan request is accepted and afterward decided denial or approval of the request in this process. Based on the loan amount and risk level, the decisions are made. In our approach, we compared the result with other existing TCSS-LS method, ABC method, and random test case generator technique. In our method, stopping criteria are achieving 100% transition coverage. On a dual-core processor 2.80 GHz all runs were done using a main memory of size 4 GB. To get the average values for every composition, we ran the technique 100 times.

Number of test cases, percentage of transmission coverage, and time are compared with our proposed ACO method and some existing methods like ABC, TCSS-LS, and random method that are shown in Fig. 2. In this case, the proposed method of number of generating test cases is lower when compared with other existing methods. The method of TCSS-LS and random have low percentage of transition coverage, when compared with other two ACO and ABC techniques. The proposed ACO got 100% transition coverage. In time, the proposed method got low time when compared

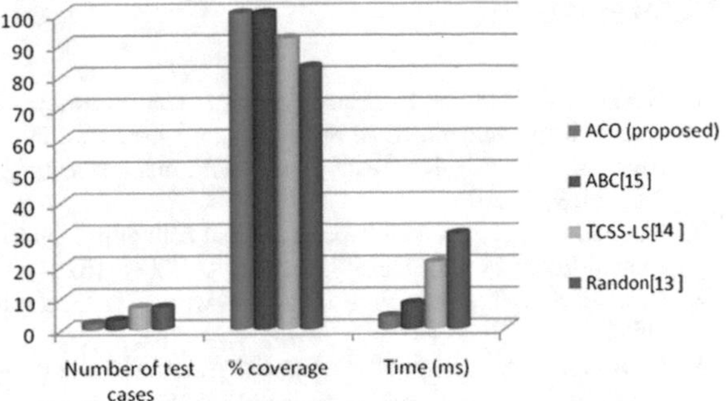

Fig. 2 Processed log data in the database table

Table 1 Test case for loan approval case study

Number	Input parameters	
	Amount	Risk assessment
1	99	Low
2	80	High
3	121	Low

with other three existing techniques. Table 1 shows the case studies of test case that are offered by the proposed method for loan approval.

Our experimental results shown in Fig. 2 illustrate that our method outdoes random generator in terms of number of test cases, percentage of transition coverage, and time. In addition, our method generates a lower number of test cases and better performance.

5 Conclusion

The proposed work introduces a metaheuristic algorithm-based test case generation approach for BPEL Web service composition. Here, the loan approval process is considered as the inspiration illustration. The BPEL process is used for building the motivation model. Here, the test case generation process used the ACO algorithm for the optimal test case choice for the BPEL business process. At long last, the greedy algorithm is used to choose the best test suite. In the future work, to make the system completely programmed, we intend to broaden this approach by using the propelled optimization algorithms and the propelled business arranged process languages.

References

1. Garousi, V., Felderer, M., Hacaloglu, T.: What we know about software test maturity and test process improvement. IEEE Softw. **35**(1), 84–92 (2017)
2. Afrin, A., Mohsin, K.: Testing approach: manual testing vs automation testing. Glob. Sci.-Technol. **9**, 1 (2017)
3. Lin, M., Lin, Z.: A cost-effective critical path approach for service priority selections in grid computing economy. Decis. Support Syst. **42**(3), 1628–1640 (2006)
4. Dustdar, S., Schreiner, W.: A survey on web services composition. Int. J. Web Grid Serv. **1**, 1 (2005)
5. Li, B., Qiu, D., Leung, H., Wang, D.: Automatic test case selection for regression testing of composite service based on extensible BPEL flow graph. J. Syst. Softw. **85**(6), 1300–1324 (2012)
6. Blanco, R., Tuya, J., Adenso-Díaz, B.: Automated test data generation using a scatter search approach. Inf. Softw. Technol. **51**(4), 708–720 (2009)
7. Sun, C., Zhao, Y., Pan, L., Liu, H., Chen, T.: Automated testing of WS-BPEL service compositions: a scenario-oriented approach. IEEE Trans. Serv. Comput. 1–1 (2015)
8. Li, Z., Sun, W., Du, B.: BPEL4WS unit testing: framework and implementation. Int. J. Bus. Process Integr. Manag. **3**, 2 (2008)
9. Wang, Y., Chen, P.: Test case generation of web service composition: an approach based on the color petri net. Appl. Mech. Mater. **336**, 2063–2070 (2013)
10. Zheng, Y., Zhou, J., Krause, P.: An automatic test case generation framework for web services. J. Softw. **2**, 3 (2007)
11. Palomo-Duarte, M., Garca-Domnguez, A., Medina-Bulo, I.: Automatic dynamic generation of likely invariants for WS-BPEL compositions. Expert Syst. Appl. **41**(11), 5041–5055 (2014)
12. Maalej, A., Krichen, M.: Study on the limitations of WS-BPEL compositions under load conditions. Comput. J. **58**(3), 385–402 (2014)
13. Bueno, P., Wong, W.E., Jino, M.: Improving random test sets using the diversity oriented test data generation. In: Proceedings of the 2nd international workshop on Random testing: co-located with the 22nd IEEE/ACM International Conference on Automated Software Engineering (ASE 2007), ACM (2007), pp. 10–17. (2007)
14. Blanco, R., Garćia-Fanjul, J., Tuya, J.: A first approach to test case generation for bpel compositions of web services using scatter search. In: Software Testing, Verification and Validation Workshops, 2009. ICSTW'09. International Conference On, IEEE (2009), pp. 131–140. (2009)
15. Daghaghzadeh, M., Babamir, M.: An abc based approach to test case generation for bpel processes. In: Computer and Knowledge Engineering (ICCKE), 2013 3th International Conference on, IEEE (2013), pp. 272–277. (2013)

Extractive Summarization of a Document Using Lexical Chains

Chirantana Mallick, Madhurima Dutta, Ajit Kumar Das, Apurba Sarkar and Asit Kumar Das

Abstract Nowadays, efficient access of information from the text documents with high-degree of semantic information has become more difficult due to diversity of vocabulary and rapid growth of the Internet. Traditional text clustering algorithms are widely used to organize a large text document into smaller manageable groups of sentences, but it does not consider the semantic relationship among the words present in the document. Lexical chains try to identify cohesion links between words by identifying their semantic relationship. They try to link words in a document that are thought to be describing the same concept to gather information. This method of text summarization helps to process the linguistic features of the document which is otherwise ignored in statistical summarization approaches. In this paper, we have proposed a text summarization technique by constructing lexical chains and defining a coherence metric to select the summary sentences.

Keywords Summarization · Extractive summary · Lexical chains · Hyponym Hypernym

C. Mallick · M. Dutta (✉) · A. K. Das (✉) · A. Sarkar · A. K. Das
Department of Computer Science and Technology, Indian Institute of Engineering Science and Technology, Shibpur, India
e-mail: madhurima.pg2016@cs.iiests.ac.in

A. K. Das
e-mail: writetoajit@yahoo.com

C. Mallick
e-mail: chirantana9@gmail.com

A. Sarkar
e-mail: as.besu@gmail.com

A. K. Das
e-mail: akdas@cs.iiests.ac.in

J. Nayak et al. (eds.), *Soft Computing in Data Analytics*, Advances in Intelligent Systems and Computing 758,
https://doi.org/10.1007/978-981-13-0514-6_78

1 Introduction

In the Internet era, enormous amount of online information is available to readers in the form of e-newspapers, content from various news feeds, journal articles, transcription reports, social media streams, emails, etc. It is a tedious job for any individual to extract meaningful information from a large amount of text letting alone multiple documents in a stipulated time. Hence, there is a need for automatic summarization system to digitally retrieve information from a vast number of online and offline sources in the form of a summary. A summary is a piece of text, containing a major segment of information compiled from one or more documents. Automatic text summarization has garnered a lot of interest in the recent years. It is the ability to encapsulate information automatically and present results to the end user in a compressed, yet complete form.

Summarization systems provide two canonical strategies, namely extractive summarization [1] and abstractive summarization [2] for constructing summaries from single source or multisource documents. The main challenge of "extractive summarization" is to select important sentences from the original document into the summary without any modification. It consists of concatenating source sentences into a summary. "Abstractive summarization" techniques tend to generate novel sentences based on information gathered from the document. These techniques are much harder to implement than extractive summarization techniques in general. Apart from these, several systems try to merge these concepts for providing summaries which are both extractive and abstractive in nature.

Modern extractive summarization systems work in three steps. The first step involves the construction of an intermediate representation of the original text. The second step is to analyze the intermediate representation and define a metric as a means of scoring the sentences. The score is commonly related to how well a sentence expresses some of the most important topics in the document or to what extent it combines information about different topics. Hence, in this step, the important sentences get classified. The third step involves selection of coherent sentences for construction of the summary. This step follows a greedy approach and generally involves selection of the top predefined number of weighted sentences.

A plethora of methods [3–5] have been employed to create an intermediate representation of the original document. There are "statistical approaches" which are employed to find word frequency, theme words, etc. Statistical approaches include identification of "topic words" or "topic signatures" [6] which are words directly related to the theme of a particular document. Metrics like term frequency and inverse document frequency, TF-IDF [7], also contribute to summarization systems.

Centroid summarization [8] used by open-source multidocument multilingual text summarizer and popularly known as MEAD system works by computing the centroid of the document. Other than these, entropy, mutual information, and other statistical models are used to find important terms or words in sentences. Emphasis is also given to dialogues, capitalized words, cue phrases, and so on. In query-focused summarization techniques [9], these metrics have been proved to produce meaningful results.

There are "semantic approaches" like lexical chaining [10], word sense disambiguation [11] which try to establish relationships between words or sentences and hence lead to partial understanding of the document. These are also known as "linguistic approaches." Several graph-based models like Pagerank [12], Lexrank [13] are also thought of as semantic approaches for text summarization. There are several similarity metrics like cosine similarity, Levenshtein distance [14], which are used to compute similarity between pairs of sentences. Hence, in case of graph models, sentences are taken as vertices and the edge between two vertices bears the similarity measure as weight.

Other approaches include designing probabilistic models [15] to assign weightage to terms and sentences in general to determine the suitable sentences to go into the summary. Several systems also use neural networks [16], Bayesian topic models [17], Markov chains [18] to generate summaries where documents are to be extracted from multiple sources. Over the years, research in this field has picked up pace and a lot of methods have been defined. In this paper, a summarization method has been defined using lexical chains to make a summary of a single document.

The rest of the paper is organized in the following manner. The concept of WordNet and lexical chains is thoroughly discussed in Sect. 2. The proposed method is discussed in Sect. 3. The results obtained are depicted in Sect. 4, and finally, future work and conclusion are drawn in Sect. 5.

2 Background Study

2.1 WordNet

WordNet [19] organizes lexical information in terms of word meanings, rather than word forms. Nouns, verbs, adjectives, and adverbs are grouped into sets of cognitive synonyms (synsets), each expressing a distinct concept. Synsets are linked by semantic relations [20], i.e., synonyms of each word, and also hyponym/hypernym (i.e., Is-A), and meronym/holonym (i.e., Part-Of) relationships, and word forms are linked by lexical relations. The most frequently encoded relation among synsets is the super–subordinate relation (also called hypernymy, hyponymy, or IS-A relation) [21].

The synsets provide a hierarchical tree-like structure for each term. WordNet in document clustering is used to improve the clustering performance for word sense disambiguation by Hotho et al. [22], and this work is extended by Sedding and Kazakov [23] using their part of speech tags. But the main bottleneck of both these approaches is the increase in dimensionality of the data. WordNet is a large lexical database of English organized by semantic relations. Suppose there is a semantic relation X between meaning $\{a_1, a_2, \ldots\}$ and meaning $\{b_1, b_2, \ldots\}$, then there is a relation X between $\{b_1, b_2, \ldots\}$ and $\{a_1, a_2, \ldots\}$. Lexical chain [24] is constructed by calculating the semantic distance between the words using WordNet. Lexical relationship exists between words, while semantic relationship occurs between the concepts synsets represent. The relationship between canine and dog is an example of a semantic relationship. WordNet precisely expresses this as hyponym from the synset containing the words canine, canid to the synset with the words dog, domestic dog, Canis Familiaris. Antonymy is an example of a lexical relationship. Dark is an antonym of light; however, darkness and lightness, which belong to the same synsets as dark and light, respectively, are not antonyms. WordNet provides semantic and lexical relationship between noun synsets.

2.2 *Lexical Chaining*

The various chaining algorithms used in text summarization where their implementations are different. In general, they are characterized by certain rules based on how they perform chaining as well as discover chains. The idea of lexical chaining was first implemented by Morris and Hirst [25]. It mainly deals with the problem of word sense disambiguation (WSD). Lexical chains are created based on the same topic words of the document. In this approach, identities, synonyms, and hypernyms or hyponyms are the relations among words which cause them to be grouped into the same lexical chain, and for this purpose, WordNet is widely used in this method. Lexical chain is constructed by determining the semantic distance between the words using WordNet. There are four common steps to be performed to construct desired summary using lexical chain.

1. **Segmentation of the original source text**: At first, each sentence is identified by tokenization from the input text document.
2. **Construction of lexical chains**: Lexical chain is constructed in three different steps. First, we select a set of candidate words and then for each candidate word find an appropriate chain relying on a relatedness criterion (like identities, synonyms, and hypernyms or hyponyms) among the other members of the chains. If any similar candidate word with certain relatedness criterion is found, insert it into the same chain and update the score of the chain accordingly.

3. **Identification of strong chains**: To identify the strong chains, first the overall score of each chain is determined by adding the weights of each individual element in the chain. The chains with score more than the mean score for every chain in the document are selected, and for each chain, identify the word whose contribution to the chain is maximum as the representative of the chain.
4. **Extraction of significant sentences**: Finally, pick the sentence that contains the representative word in the chain.

Overall, lexical chains provide a better indication of discourse topic than does word frequency simply because different words may refer to the same topic. Even without sense disambiguation, these approaches are able to derive concepts.

3 Proposed Methodology

3.1 *Preprocessing*

We have used news articles from BBC News feed. The BeautifulSoup library [26] of Python [27] is used as the Web scraping tool of our work. This library helps in extracting only the text portion of the HTML or XML document. The articles are successfully read into the program, in the form of a single piece of text or string. The texts of the document D are tokenized into individual sentences $\{S_1, S_2, \ldots S_n\}$ and each of them is Part of Speech tagged (i.e., CC, DT, JJ, NN, NNP, NNS, VB, VBN, VBG, RB, where CC is coordinating conjunction, DT is determiner, JJ is adjective, NN is singular noun, NNP is proper noun, NNS is noun plural, VB is verb, VBN is verb past participle, VBG verb past tense, and RB is adverb) using the NLTK (Natural Language Processing Toolkit) library [28] of Python. After Part of Speech tagging, the stop words are removed from the sentences $S_i \in D$ $\forall i = 1$ to n. Thus, after removing stop words, the processed sentences will be of the form, i.e., $S_i = \{w_i(CC) \cup w_i(DT) \cup w_i(JJ) \cup w_i(NNP) \cup w_i(NNS) \cup w_i(VB) \cup w_i(VBN) \cup w_i(VBG) \cup w_i(RB)\} \cap \neg STW$, where S_i is the ith sentence of the given document D, $w_i(X)$ is the word present in the sentence S_i with X tagging, and STW is the set of all stop words of the document obtained using NLTK library. The NLTK library tags the words using the same tags as of Penn Treebank [29].

3.2 *Lexical Chain-Based Summarization*

In this proposed method, we have considered semantic relationship between the words within a sentence as well as among the sentences of the given document. After preprocessing of the sentences, with the help of regular expressions, we chunk

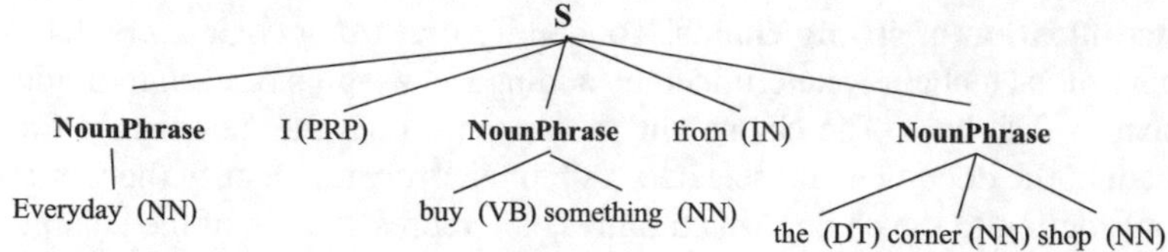

Fig. 1 Chunking a sentence into its noun phrases

out the noun phrases (w(NP)) from the document and subsequently the document is represented only by means of the noun phrases. We define noun phrases by using Expression (1), where the proper nouns are accumulated together.

$$\text{``} < DT > * < JJ.? > * < CC > * < NN.? > * < VB.? > * < NN.? > \text{''} \quad (1)$$

where DT refers to determiners, JJ refers to adjectives, CC refers to conjunctions, NN refers to nouns, and VB refers to verbs.

For the sentence, "Everyday I buy something from the corner shop.", Fig. 1 shows the chunked grammars.

A hypernym/hyponym set (H) is constructed with the words having noun (NN) and verb (VB) tag of the sentence in the document as follows:

$$H = \{hr(w(NN)) \cup hp(w(NN)) \cup hr(w(VB)) \cup hp(w(VB)) \; : \; w(NN), w(VB) \in S_i \forall i\}$$

where $hr(w(NN)$ is the set of hypernyms of the words having noun tag and $hp(w(NN)$ is the set of hyponyms of the words having noun tag and so for the verbs. Next, we have assigned weights for the nouns and the verbs of every sentences. For weight assignment, we have considered the following conditions:

i. If the word $w_i \in H$, then set the weight ($wt(w_i)$) as 0.5.
ii. If the word w_i is unique (i.e., term frequency ($tf(w_i)$) is 1), then set the weight ($wt(w_i)$) as 1.
iii. If the term frequency ($tf(w_i)$) > 1, set the weight ($wt(w_i)$) as $1/tf(w_i)$).

Then, a set of lexical chains is formed with nouns and verbs of the processed sentences along with their weights as shown below:

$$chain_i = \{w_i(VB), wt(w_i)(VB)\} \cup \{w_i(NN), wt(w_i)(NN)\} \forall i = 1, 2, \ldots, n.$$

After this, the relationships between the words within the sentences are examined. This is done for every noun and verb from the noun phrases for their hypernyms and

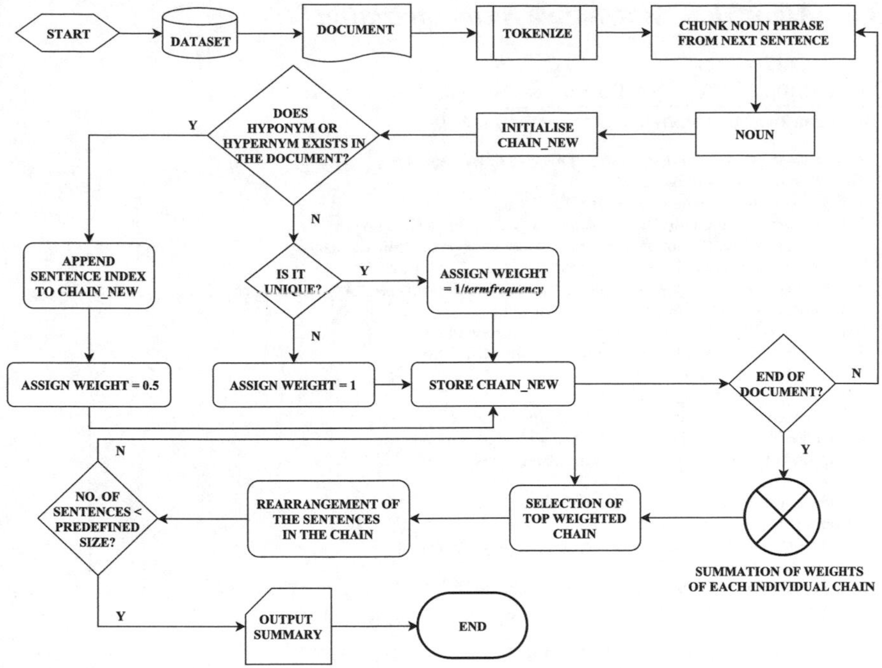

Fig. 2 Flowchart of the proposed summarization method

hyponyms if they are present in the passage. For every possible pair of sentences (S_i, S_j), we have examined the noun and verb from the noun phrases $\in (S_i, S_j)$ for matching of their hypernyms and hyponyms, and if any matching is found, construct a chain by appending their corresponding sentence indices. For an example, if matching is found between sentences S_i and S_j of the document D, then the chain contains i and j. In this way, all the noun phrases of all the sentences are examined and some of them are chained until and unless no pair of sentences is left for processing. In the next step, the weights of the individual chains are summed up and sorted in descending order. At the end, the highest weighted chains are picked iteratively until the length of the summary is reached. The final summary is obtained after arranging the selected sentences in order of their occurrence in the original document.

The flowchart of the proposed summarization technique is shown in Fig. 2.

3.3 Algorithm: Extractive Summarization

The outline of the algorithm is as follows:

Algorithm 3.1: MULTISENTENCE-LEXICAL-CHAINING(D, n)

Input: D= the target text document and n= size of the summary .
Output: Summary of the document D.
1. Encode D into D' using UTF-8 format;
2. D' is tokenized into individual sentences (S_i) using the NLTK library;
3. The sentences are Part-of-speech tagged using the tags of Penn Treebank;
4. Chunk each sentences by means of grammar using Expression (1);
5. Extract Hypernyms, Hyponyms for each noun of each processed tokenized sentence and construct a set of Hypernyms, Hyponyms ;
6. Initialise *index_value* value at zero for the sentence index;
7. **for each** $S_i \in D'$ in order of appearance in encoded D **do**
   7.1. Remove the stop words from the sentence;
   7.2. Increase the *index_value* by 1;
   7.3. Assign the *index_value* to the sentence;
8. If similarity between the nouns of other sentences is found:
   Append those sentence indices to the same chain;
9. Else
   Construct a chain with that sentence index;
10. **for each** word w in D' **do**
   10.1. If the word is present in the Hypernyms, Hyponyms set:
      Assign weight of w as 0.5;
   10.2. If the word is unique in the given document:
      Assign weight of w as 1;
   10.3. Else
      10.3.1. Calculate TF for that word w in D';
      10.3.2. Assign weight of w as $1/TF(w)$;
11. Calculate overall weight for each chain;
12. Arrange the weighted chain in descending order in a list;
13. **for each** chain in the list **do**
   13.1. If number of sentences $\leq n$:
      13.1.1. Store all sentences from that chain into the summary;
      13.1.2. remove the chain from the list;
      13.1.3. $n = n$ - number of sentences in the chain;
   13.2. Else
      13.2.1. Select top n weighted sentences from that chain and store the sentences in summary;
      13.2.2. Go to step 14;
14. Rearrange the sentences in order of their occurrence in D in summary;
return (*summary*)

4 Results

The method has been thoroughly tested on the news articles generated by the BBC News feed. We have used the BeautifulSoup library [26] and the NLTK library [28] of Python 2.7 [27] for extracting and preprocessing of the document.

Evaluation using Rouge: Since the early 2000s Recall-Oriented Understudy for Gisting Evaluation (ROUGE) [30] is widely used for automatic summaries and also for the performance evaluation. For the first time, it is used in DUC 2004. Some of the ROUGE methods, i.e., ROUGE_N, ROUGE_L, ROUGE_W and ROUGES_U are commonly used to measure the performance. ROUGE_N is the N-gram ($N \geq 1$)

Table 1 Rouge values of the proposed method and the traditional TF-IDF method with respect to ground truth

Method	Rouge values of the proposed method						Rouge values of traditional TF-IDF method					
Length	5%			10%			5%			10%		
Metric	Recall	Precision	f-score	Recall	Precision	f-score	Recall	Precision	f-score	Recall	Precision	f-score
Rouge-1	0.85	0.14358	0.24267	0.70968	0.20336	0.31549	0.15284	0.92105	0.26217	0.22008	0.96610	0.35849
Rouge-2	0.57759	0.10696	0.17863	0.36475	0.10839	0.1669	0.12719	0.63700	0.23730	0.18992	0.84483	0.31012
Rouge-L	0.71667	0.12712	0.21338	0.54033	0.15782	0.24385	0.12227	0.70830	0.20974	0.20077	0.88136	0.22704
Length	20%			25%			20%			25%		
Metric	Recall	Precision	f-score	Recall	Precision	f-score	Recall	Precision	f-score	Recall	Precision	f-score
Rouge-1	0.69731	0.61338	0.55534	0.6132	0.53305	0.5382	0.56371	0.62602	0.60206	0.58687	0.50915	0.58801
Rouge-2	0.35798	0.47314	0.35929	0.37022	0.37020	0.36085	0.32822	0.40222	0.30338	0.36822	0.36965	0.36193
Rouge-L	0.46729	0.59415	0.47193	0.50258	0.46665	0.46335	0.40560	0.52121	0.47660	0.45560	0.40736	0.41649

recall between a system summary and a set of sample summaries. It is used to estimate the fluency of summaries. Here, the value of N is considered as 1 and 2. ROUGE_1 and ROUGE_2 denote the overlap of 1-gram and bi-grams between the system and sample summaries, respectively.

Rouge-L is the longest subsequence-based statistics which is used to identify the longest co-occurring in sequence n-grams automatically. Suppose we assume that A is the set of sentences of the reference summary and B is the set of sentences of the candidate summary represented by the sequence of words and LCS-based f-score indicates the similarity between A (of length m) and B (of length n).

The unbiased ground truth or the reference summaries of the news articles were obtained from our fellow peers and research scholars having different field expertise. The evaluation of our proposed summarization technique is done by comparing it with the ground truth summary, as listed in Table 1. The size of the summary is set as 5, 10, 20, and 25% of the total sentences in the document and compared with the ground truth summary in terms of recall, precision, and f-score. For all three metrics, Rouge-1, Rouge-2, and Rouge-L, it is observed that summary with size 20% of the text document yields better result than the others. It is noticed that summary size contains 20% sentences of the document and ground truth summary sizes are almost same. This proves that the proposed technique provides the summary which is almost close to that obtained by the experts. Also the result is compared with the summary obtained by the work presented in [7] which shows the superiority of the method.

5 Conclusion and Future Work

This paper is based on the semantic relationship among the sentences of the given document using WordNet and chaining multiple sentences lexically. We have also identified cohesion links between words by identifying their semantic relationship. In this paper, the word sense disambiguation of the given document is handled efficiently and provided with more suitable summary preserving the central idea of the original document. The results of applying the proposed approach on extractive summarization are quite promising.

But the proposed algorithm does not take care of anaphora resolution problem. We can include this for better improvement. The proposed method may select more than one sentence imparting the same meaning into the summary. To eliminate this kind of duplication, we may consider any kind of clustering algorithm. As an extension of the paper, we will compare our method with existing state-of-the-art text summarization methods evaluating different performance measurement metrics.

References

1. Nenkova, A., McKeown, K.: A Survey of Text Summarization Techniques. Springer Science+Business Media (2012)
2. Dalal, V., Malik, L.: A survey of extractive and abstractive text summarization techniques. In: 2013 6th International Conference on Emerging Trends in Engineering and Technology, pp. 109–110, Dec 2013
3. Agrawal, N., Sharma, S., Sinha, P., Bagai, S.: A graph based ranking strategy for automated text summarization. DU J. Undergrad. Res. Innov. **1**(1) (2015)
4. Landauer, T.K., Foltz, W., Laham, D.: An introduction to latent semantic analysis. Discourse Process. **25**(2–3), 259–284 (1998)
5. Saggion, H., Lapalme, G.: Generating indicative-informative summaries with sumUM. Comput. Linguist. **28**(4), 497–526 (2002)
6. Lin, C.-Y., Hovy, E.: The automated acquisition of topic signatures for text summarization. In: COLING '00 Proceedings of the 18th Conference on Computational Linguistics, pp. 495–501. Association for Computational Linguistics, Stroudsburg, PA, USA 2000 (2000)
7. Seki, Y.: Sentence extraction by TF/IDF and position weighting from newspaper articles (2002)
8. Radev, D.R., Jing, H., Stys, M., Tam, D.: Centroid-based summarization of multiple documents. Inf. Process. Manag. **40**, 919–938 (2003)
9. Zhao, L., Lide, W., Huang, X.: Using query expansion in graph-based approach for query-focused multi-document summarization. Inf. Process. Manag. **45**(1), 35–41 (2009)
10. Wei, T., Yonghe, L., Chang, H., Zhou, Q., Bao, X.: A semantic approach for text clustering using wordnet and lexical chains. Expert Syst. Appl. **42**(4), 2264–2275 (2015)
11. Ghose, A.: Supervised lexical chaining. Master's thesis, Indian Institute Of Technology, Madras (2011)
12. Mihalcea, R.: Graph-based ranking algorithms for sentence extraction, applied to text summarization. In: Proceedings of the ACL 2004 on Interactive Poster and Demonstration Sessions, p. 20. Association for Computational Linguistics (2004)
13. Erkan, G., Radev, D.R.: Lexrank: graph-based lexical centrality as salience in text summarization. J. Artif. Intell. Res. 457–479 (2004)
14. Dutta, S., Ghatak, S., Roy, M., Ghosh, S., Das, A.K.: A graph based clustering technique for tweet summarization. In 2015 4th International Conference on Reliability, Infocom Technologies and Optimization (ICRITO) (Trends and Future Directions), pp. 1–6. IEEE (2015)
15. Barzilay, R., Lee, L.: Catching the drift: probabilistic content models, with applications to generation and summarization, vol. 34, pp. 1–34, Mar 2008
16. Kaikhah, K.: Automatic text summarization with neural networks. In: 2004 2nd International IEEE Conference on Intelligent Systems, 2004. Proceedings, vol. 1, pp. 40–44, June 2004
17. Daumé III, H.: Bayesian query-focused summarization. CoRR arxiv:abs/0907.1814 (2009)
18. Nenkova, A., Maskey, S., Liu, Y.: Automatic summarization. In: Proceedings of the 49th Annual Meeting of the Association for Computational Linguistics: Tutorial Abstracts of ACL 2011, HLT '11, pp. 3:1–3:86. Association for Computational Linguistics, Stroudsburg, PA, USA (2011)
19. Wordnet. http://wordnet.princeton.edu/. Accessed 30 Dec 2017
20. Miller, G.A.: Wordnet: a lexical database for English. Commun. ACM **38**(11), 39–41 (1995)
21. Fellbaum, C.: WordNet. Wiley Online Library (1998)
22. Hotho, A., Staab, S., Stumme, G.: Ontologies improve text document clustering. In: Third IEEE International Conference on Data Mining, 2003. ICDM 2003, pp. 541–544. IEEE (2003)
23. Sedding, J., Kazakov, D.: Wordnet-based text document clustering. In: Proceedings of the 3rd Workshop on Robust Methods in Analysis of Natural Language Data, pp. 104–113. Association for Computational Linguistics (2004)
24. Jain, A., Gaur, A.: Summarizing long historical documents using significance and utility calculation using wordnet. Imp. J. Interdiscip. Res. **3**(3) (2017)
25. Morris, J., Hirst, G.: Lexical cohesion computed by thesaural relations as an indicator of the structure of text. Comput. Linguist. **17**(1), 21–48 (1991)

26. Beautifulsoup documentation. https://www.crummy.com/software/BeautifulSoup/bs4/doc/. Accessed 29 Nov 2017
27. Python 2.7.14 documentation. https://docs.python.org/2/index.html. Accessed 29 Nov 2017
28. Bird, S., Klein, E., Loper, E.: Natural Language Processing with Python. O'Reilly (2009)
29. Penn treebank pos tags. https://www.ling.upenn.edu/courses/Fall_2003/ling001/penn_treebank_pos.html. Accessed 30 Dec 2017
30. Lin, C.-Y.: Rouge: a package for automatic evaluation of summaries. In: Proceedings of the ACL Workshop: Text Summarization Braches Out 2004, p. 10, 01 (2004)

Detecting Intrusive Behaviors using Swarm-based Fuzzy Clustering Approach

Debasmita Mishra and Bighnaraj Naik

Abstract Modern communication network relies on intrusion detection system which is an important phenomenon to monitor the suspicious activity and avoid network traffic. Due to security need and challenges, this has been focused by many researchers in last decade. In this paper, fuzzy clustering (FCM) and particle swarm optimization (PSO)-based unsupervised machine learning intrusion identification system have been proposed and tested with benchmark IDS datasets. In this proposed IDS, fuzzy clustering has been used to separate intrusion activities from normal activities and particle swarm optimization algorithm has been used to get optimal separations among these activities. Here, the PSO has been used to help FCM to start with better cluster centers (rather randomly initialized) which affects accuracy of the IDS system with an affordable computational burden. The proposed work has been found to be better as compared to other related existing models.

Keywords Fuzzy C-means (FCM) · Particle swarm optimization (PSO) Intrusion detection system (IDS)

1 Introduction

The safeguarding of security becomes more difficult nowadays as all the files and records are easily accessed through Web by different attackers. It is an easy task for the attackers to exploit vulnerabilities quickly once they enter the network.

D. Mishra · B. Naik (✉)
Department of Computer Application, Veer Surendra Sai University of Technology (VSSUT), Burla Sambalpur 768018, Odisha, India
e-mail: mailtobnaik@gmail.com

D. Mishra
e-mail: debasmitamishra21@gmail.com

J. Nayak et al. (eds.), *Soft Computing in Data Analytics*,
Advances in Intelligent Systems and Computing 758,
https://doi.org/10.1007/978-981-13-0514-6_79

Intrusion detection system (IDS) is being developed in response to the increasing number of attacks on major sites and networks. If some malicious activities or security policy violations is carried out by someone, then the IDS which is a type of security software, which alerts administrators automatically. The network security technology IDS originally detects vulnerability exploits against a computer or any other target application. It also monitors the Internet automatically for searching of any latest threats which may cause a future attack. There are several ways of detection performed by IDS. Generally, IDS follows a two-step process. The first one is host-based and includes different tasks like inspection of system configuration to find risky settings; inspection of password files to find imprudent passwords; and inspection of some other system areas for detecting policy violations. The second one is network-based that is an active component. Nowadays, computers are more connected to public accessible networks for which network intrusions occurred quickly. To minimize the risk and maximize the efficiency, different anomaly detection techniques based on clustering methods are used. Clustering is a process of organizing a set of objects into different clusters according to their similarities. Cluster analysis is a common unsupervised machine learning method, which can be used to detect unknown attacks of IDS. Intrusion detection systems are found to be one of the key areas of application of data mining techniques and

Table 1 Related works on IDS

Reference	Methods used
Liu et al. [1]	IDBGC
Jiang et al. [2]	CUBID
Wei and Wu [3]	KFDA + SVM
Wang et al. [4]	ANN + fuzzy clustering
Liu et al. [5]	CWSN + IIDS
Chen et al. [6]	BIRCH + SVM
Rajaram et al. [7]	K-means clustering + C4.5 decision tree
Zheng-wang [8]	Clustering algorithm
Fahmy et al. [9]	Weighted K-means + random forest
Furnell et al. [10]	D-FICCA + ICA
Muthukumar et al. [11]	NOF
Ravale et al. [12]	Hybrid IDS (K-means clustering +SVM)
Tsai et al. [13]	CANN
Chen M et al. [14]	FCM
Yaseen et al. [15]	Modified K-means + SVM + ELM
Qutayri et al. [16]	SMLC
Kumar et al. [17]	Information gain and correlation
Zhuo et al. [18]	OA-LS-SVM
Wang et al. [19]	SVM

principles. There are many applications of IDS used in the past few years. Many researchers have proposed various methods (Table 1) which are very effective for finding unknown intrusions present in a system. A new detection algorithm [1] called intrusion detection based on genetic clustering (IDBGC) algorithm which can create clusters and find intruders by labeling usual and unusual groups. Here, in the first stage the nearest neighbor method is used for grouping of network data and a genetic optimization technique used in the second stage for obtaining optimal detection result. The proposed algorithm was found to be feasible and effective for intrusion detection. A clustering-based unsupervised intrusion detection (CBUID) method has been proposed by Wei and Wu [2], and the results were found to be effective in terms of accuracy and detecting unknown intrusions. A new intrusion detection technique has been proposed by Wei et al. [3] where kernel Fisher discriminant analysis (KFDA) is used to detect the data points having the minimal within-class scatter and maximal between-class scatter. It was found to be easy to separate into different classes by using multiclass SVM. The authors found the method to be suitable for feature extraction that improves the classification accuracy with reduced training time. An another new intrusion detection method known as FC-ANN proposed by Hao et al. [4], where ANN and fuzzy clustering were used. Here, IDS problem was solved to obtain higher detection rate and better stability. The authors found the result of FC-ANN to be outperformed BPNN and some well-known methods such as decision tree, naïve Bayes in terms of detection stability and detection precision. Later Liu et al. (2011) proposed a mechanism of IDS known as integrated intrusion detection system (IIDS) [5] which creates a cluster-based wireless sensor network (CWSN) that provides the system to resist intrusions. This proposed method is focused on the goal to increase the detection rate and decrease the false positive rate. Another SVM-based intrusion detection system [6] has been developed by combing hierarchical clustering algorithm and a feature selection procedure. The balanced iterative reducing and clustering using hierarchies (BIRCH) that is a hierarchical clustering method, has been used to provide reduced and abstracted datasets, instead of original large dataset, which is helpful for training phase and testing phase as well. For detection of unusual and usual activities of the network, Rajaram et al. [7] have proposed a method based on cascade k-means clustering, where the training instances are divided into 'k' clusters by applying Euclidean distance similarity. Further, C4.5 decision tree method has been adopted to build decision tree on each cluster which is used to find the various anomalies. In these experimental results, the proposed algorithm has obtained impressive detection accuracy. An intrusion detection system based on clustering algorithm has been proposed by Zheng-wang [8], where both information entropy and relative entropy of different clusters are calculated by the algorithm. This system was able to find intrusions by integrating two kinds of improved detecting methods. Another hybrid framework by employing weighted K-means and random forest has been proposed by Fahmy et al. [9] for data mining-based network intrusion detection systems. This projected system is found to be good in terms of better false positive rates and detection rates, but it was unable to detect anomalous and normal clusters properly. A new hybrid clustering method is introduced by Furnell et al. [10] named as density-based fuzzy imperialist competitive clustering algorithm (D-FICCA) with imperialist competitive algorithm (ICA) is used in WSNs to get optimum clustering result. The proposed method enhances the accuracy of malicious detection. The results were found to be good in terms of higher detection accuracy and clustering quality. A novel method neighborhood outlicr factor (NOF) has been proposed by Muthukumar et al. [11] that measures the anomaly dataset. Here the outlier detection is used by the IDS to detect the anomalies present in the system. A hybrid learning approach [12] has been proposed by combining both K-means clustering and a SVM classifier which is used for classification purpose for large heterogeneous dataset. Only selected attributes are used for cluster formation as well as for classification purpose. As a result, complexities get reduced and consequently performances get increased. Cluster analysis and nearest neighbor-based classifier (CANN) [13] is used for intrusion detection, where each data sample represents intrusions. Here CANN is found to be outperformed the k-NN and SVM classifiers. Another novel FCM technique proposed by Chen et al. [14] for an efficient intrusion detection on MANET system and is implemented in a cloud storage environment. The proposed method gives better performance as compared to other models. A study proposed a new multi-level hybrid SVM and ELM model-based network IDS [15], which is marked by a significantly better performance than multi-level SVM and multi-level ELM models. Here the modified K-means is designated to preprocess the training dataset and provide new high-quality training datasets that may improve the training time and overall performance of SVM and ELM. A semi-supervised multilayered clustering model (SMLC) for intrusion detection is used by Al-Jarrah et al. [16] that is trained from partially labeled data to achieve a better detection performance. Experimental results shown that SMLC is superior and providing comparable detection accuracy for labeled instances of the training data. However, the SMLC has a relatively high testing time. The method uses the feature extraction concepts of information gain and correlation [17] with preprocessing technique to eliminate redundant and irrelevant data from the dataset, which improves resource utilization and time complexity. In order to justify the significance of proposed method, an experiment was performed in which feature reduced system was compared with a system without feature reduction mechanism. Here authors have suggested to extend work to improve the shortcomings in number of ways such as finding the optimal number of features to further increase the accuracy of intrusion detection system, optimization techniques such as genetic algorithms, big bang big crunch optimization can be used and fuzzy system-based preprocessing work can also be used. Zhuo et al. [18] have proposed a novel approach based on sampling with least square support vector machine (LS-SVM) for intrusion detection system which gives better performance in terms of accuracy and efficiency. Here IDS uses a sampling method known as the optimum allocation-based least square support vector machine (OA-LS-SVM). The result shows that the method is effective to detect intrusions for static as well as for incremental datasets. Another improved framework for effective intrusion detection based on support vector machine (SVM) is used by Wang et al. [19] for improving the capability of the model in a new and better quality. To evaluate the method, NSL-KDD dataset is used and the result was found to be good in terms of high accuracy, high detection rate, low false alarm rate, and training speed. As here only binary case intrusion problems are defined, In future, the authors have planned to include other attack types.

In this study, swarm intelligence-based fuzzy clustering approach has been presented for the detection of intrusive behaviors in the network.

2 Proposed Work

2.1 *Fuzzy Clustering*

Fuzzy C-means (FCM) algorithm [20] can be used by intrusion detection system to find out the differentiation between intrusion and normal activity of a network. The inseparability between one piece of data to several clusters can be achieved by FCM. It is a partitioned algorithm where the objective function can be minimized w.r.t. partition matrix (Eq. (1)) (Asyali et al. 2006) [21].

$$J(U,V) = \sum_{i=1}^{C} \sum_{j=1}^{N} u_{ij}^{m} \, ||x_j - v_i||^2 \tag{1}$$

In Eq. 1, ‘xj’ and ‘vi’ are the jth cluster point and ith cluster center, respectively. $u_{i,j}$ indicates the membership value of jth cluster point w.r.t. cluster ‘i’. ‘m’ is fuzzy controlling factor. It may result hard partition for setting the value as ‘1’ and complete fuzziness for setting the value as ‘∞.’ |||| is the norm function. Mainly fuzzy membership function, partition matrix and objective function are the three factors used by FCM (Lin et al. 2014) [22]. u_{ij}^{m} is calculated as described in Eq. (2) and cluster center as in Eq. (3).

$$U_{ij} = \left[\sum_{k=1}^{c} \left(\frac{||x_j - v_i||}{||x_j - v_k||} \right)^{\frac{2}{m-1}} \right]^{-1} \tag{2}$$

$$v_i = \sum_{j=1}^{N} u_{ij}^{m} x_j / \sum_{j=1}^{N} u_{ij}^{m} \tag{3}$$

In Eq. 3, $i \geq 1, i \leq c$.

This method is almost equivalent to K-means algorithm except the factor u_{ij}^{m} (fuzziness factor). The fuzziness level for different clusters can be determined by this fuzzy factor. The disadvantage is that here the results of FCM can be determined only by the initial values of clusters. The steps of the algorithm can be found in (Nayak et al. 2015) [23].

2.2 Particle Swarm Optimization

Particle swarm optimization (PSO) [24] is a widely used stochastic-based search algorithm, and it is able to search global optimized solution. Like other population-based optimization methods, the particle swarm optimization starts with randomly initialized population for individuals, and it works on the social behavior of particle to find out the global best solution by adjusting each individual's positions with respect to global best position of particle of the entire population (society). Each individual is adjusting by altering the velocity according to its own experience and by observing the experience of the other particles in search space by use of Eqs. 4 and 5. Equation 4 is responsible for social and cognition behavior of particles, where c1 and c2 are the constants in between 0 and 2 and rand(1) is random function which generates random number in between 0 and 1.

$$V_i(t+1) = V_i(t+1) + c_1 * rand(1) * (lbest_i - X_i) + c_2 * rand(1) * (gbest_i - X_i) \quad (4)$$

$$X_i(t+1) = X_i(t) + V_i(t+1) \quad (5)$$

2.3 Proposed Algorithm

Here FCM has been used to separate intrusion activities from normal activities, and PSO algorithm has been used to get optimal separations among these activities. In this proposed work, the PSO has been used to help FCM to start with better cluster centers (rather randomly initialized) which affects accuracy of the IDS system with an affordable computational burden. Algorithm 1 describes the proposed algorithm in flowchart form. The details implementations of FCM and PSO for clustering may be found from the previous related literature [25–34].

$$P = \{w_1, w_2 \ldots w_n\} \quad (6)$$

$$w_i = \{c_1^i, c_2^i \ldots c_m^i\}, i \in \{1, 2 \ldots n\} \quad (7)$$

$$c_j = \{v_1^j, v_2^j \ldots v_d^j\}, j \in \{1, 2 \ldots m\} \quad (8)$$

Algorithm -1 :Proposed PSO-FCM based Intrusion Identification

1. Start
2. Iter = 1
3. Randomly initialize population P of cluster centers as in Eq.6. In P each candidate solution is presented as in Eq. 7, which represents i[th] set of cluster centers, where c_j represents j[th] cluster center (Eq.8). Here, 'n', 'm' and 'd' denotes size of population, no. of cluster centers and attributes of the cluster centers respectively. Randomly initialize velocity 'V' for P.
4. if Iter = = 1
5. Then Local best Lbest = P
6. Else, Find out fitness value of all W_i in P by using Eq.9 and form a fitness vector.
7. Compare the fitness of all W_i in P with Lbest
8. Update Lbest by comparing fitness of all W_i in P with Lbest.
9. Compute the new velocity by using velocity equation Eq. 4.
10. Update the population by using Eq. 5.
11. If stopping criteria is meet
12. Goto Step - 14
13. Else Iter = Iter + 1, Goto Step - 4
14. Find out the best w_i in P. Set it to Eq.1 and evaluate partition matrix.
15. Find out U_{ij} by using Eq. 2
16. Update the new cluster centers by using Eq.3.
17. If Stopping Criteria
18. Stop
19. Else Goto 14.

$$F(w_i) = \frac{k}{\left(\sum_{j=1}^{m}\sum_{r=1}^{n}(o_r - c_{i,j})^2\right) + d} \tag{9}$$

In Eq. 9, 'w_i' is the 'ith' candidate solution, '$F(w_i)$' is the fitness of 'w_i', 'n' is the number of instance, 'm' is the number of cluster center and 'o_r' is the rth instance, '$c_{i,j}$' denotes jth cluster center of w_i candidate solution. 'k' and 'd' are the positive constants. The function 'F' is used to compute the generalized solutions (fitness functions). The '$o_r - c_{i,j}$' denotes Euclidean distance between the object o_r to the cluster center $c_{i,j}$.

3 Result and Discussion

This study includes six IDS detection model such as density-based clusterer [35], K-means [36], filtered cluster [37], FCM [20], GA-FCM [31], and PSO-FCM. The models FCM, GA-FCM, and PSO-FCM have been simulated by using MATLAB, and other models density-based clusterer, K-means, filtered cluster have been studied by using WEKA tool on KDD cup 99 dataset. Performances of these studied models are presented in Table 2 along with total number of activities,

Table 2 Performance comparison on KDD cup 99 dataset

Method	Identification accuracy (%)	Distribution	
		Intrusive activities	Normal activities
Density-based clusterer	78.492	213422 (43.201%)	280598 (56.798%)
K-mean	78.535	211477 (42.807%)	282543 (57.192%)
Filtered cluster	78.535	211477 (42.807%)	282543 (57.192%)
FCM	82	109236 (22.112%)	384784 (77.888%)
GA-FCM	85.213	80196 (16.233%)	413824 (83.766%)
PSO-FCM (Proposed)	87.703	77833 (15.755%)	416187 (84.245%)

number of incorrectly identified activities and their percentage. It is observed that the proposed PSO-FCM IDS system is found to be better as compared to other models.

4 Conclusion

In this paper, FCM and PSO-based unsupervised machine learning intrusion identification system has been proposed and tested with benchmark IDS datasets. It is observed that the FCM is found efficient to separate intrusion activities from normal activities. On the other hand, PSO algorithm is found to be helpful for FCM clustering process to start with better cluster centers that affect on the accuracy of the IDS system significantly. However, this proposed model is having some extra computation for PSO to find out optimal initial cluster centers. Along with an affordable computational burden, the proposed work has been found to be better as compared to other related existing models. This projected work may be helpful to solve other complex problem in IDS, in future.

References

1. Liu, Y., Chen, K., Liao, X., Zhang, W.: A genetic clustering method for intrusion detection. Pattern Recogn. **37**(5), 927–942 (2004)
2. Jiang, S., Song, X., Wang, H., Han, J.J., Li, Q.H.: A clustering-based method for unsupervised intrusion detections. Pattern Recogn. Lett. **27**(7), 802–810 (2006)
3. Wei, Y.X., Wu, M.Q.: KFDA and clustering based multiclass SVM for intrusion detection. J. China Univ. Posts Telecommun. **15**(1), 123–128 (2008)
4. Wang, G., Hao, J., Ma, J., Huang, L.: A new approach to intrusion detection using artificial neural networks and fuzzy clustering. Expert Syst. Appl. **37**(9), 6225–6232 (2010)
5. Wang, S.S., Yan, K.Q., Wang, S.C., Liu, C.W.: an integrated intrusion detection system for cluster-based wireless sensor networks. Expert Syst. Appl. **38**(12), 15234–15243 (2011)
6. Horng, S.J., Su, M.Y., Chen, Y.H., Kao, T.W., Chen, R.J., Lai, J.L., Perkasa, C.D.: A novel intrusion detection system based on hierarchical clustering and support vector machines. Expert Syst. Appl. **38**(1), 306–313 (2011)

7. Muniyandi, A.P., Rajeswari, R., Rajaram, R.: Network anomaly detection by cascading k-Means clustering and C4. 5 decision tree algorithm. Proced. Eng. **30**, 174–182 (2012)
8. Zheng-wang, Y.: The Research of intrusion detection algorithms based on the clustering of information entropy. Proced. Environ. Sci. **12**, 1329–1334 (2012)
9. Elbasiony, R.M., Sallam, E.A., Eltobely, T.E., Fahmy, M.M.: A hybrid network intrusion detection framework based on random forests and weighted k-means. Ain Shams Eng. J. **4**(4), 753–762 (2013)
10. Shamshirband, S., Amini, A., Anuar, N.B., Kiah, M.L.M., Teh, Y.W., Furnell, S.: D-FICCA: a density-based fuzzy imperialist competitive clustering algorithm for intrusion detection in wireless sensor networks. Measurement **55**, 212–226 (2014)
11. Jabez, J., Muthukumar, B.: Intrusion detection system (IDS): anomaly detection using outlier detection approach. Proced. Comput. Sci. **48**, 338–346 (2015)
12. Ravale, U., Marathe, N., Padiya, P.: Feature selection based hybrid anomaly intrusion detection system using K means and RBF kernel function. Proced. Comput. Sci. **45**, 428–435 (2015)
13. Lin, W.C., Ke, S.W., Tsai, C.F.: CANN: An intrusion detection system based on combining cluster centers and nearest neighbors. Knowl.-Based Syst. **78**, 13–21 (2015)
14. Chen, M., Wang, N., Zhou, H., Chen, Y.: FCM technique for efficient intrusion detection system for wireless networks in cloud environment. Comput. Electr. Eng. (2017)
15. Al-Yaseen, W.L., Othman, Z.A., Nazri, M.Z.A.: Multi-level hybrid support vector machine and extreme learning machine based on modified K-means for intrusion detection system. Expert Syst. Appl. **67**, 296–303 (2017)
16. Al-Jarrah, O.Y., Al-Hammdi, Y., Yoo, P.D., Muhaidat, S., Al-Qutayri, M.: Semi-supervised multi-layered clustering model for intrusion detection. Digit. Commun. Netw. (2017)
17. Manzoor, I., Kumar, N.: A feature reduced intrusion detection system using ANN classifier. Expert Syst. Appl. **88**, 249–257 (2017)
18. Kabir, E., Hu, J., Wang, H., Zhuo, G.: A novel statistical technique for intrusion detection systems. Future Gener. Comput. Syst. (2017)
19. Wang, H., Gu, J., Wang, S.: An effective intrusion detection framework based on SVM with feature augmentation. Knowl.-Based Syst. **136**, 130–139 (2017)
20. Bezdek, J.C., Ehrlich, R., Full, W.: FCM: the fuzzy c-means clustering algorithm. Comput. Geosci. **10**(2-3), 191–203 (1984)
21. Asyali, M.H., Colak, D., Demirkaya, O., Inan, M.S.: Gene expression profile classification: a review. Curr. Bioinform. **1**, 55–73 (2006)
22. Phen-Lan, L., Po-Whei, H.: Kuo CH, Lai YH. A sizeinsensitive integrity-based fuzzy c-means method for data clustering. Pattern Recogn. **47**, 2042–2056 (2014)
23. Nayak, J., Naik, B., Kanungo, D. P., Behera, H. S.: A hybrid elicit teaching learning based optimization with fuzzy c-means (ETLBO-FCM) algorithm for data clustering. Ain Shams Eng. J. (2016)
24. Kennedy, J., Eberhart, R.C.: Particle swarm optimization. In: IEEE International Conference on Neural Networks, Perth, Australia, pp. 1942–1948 (1995)
25. Nayak, J., Naik, B., Behera, H.S.: Fuzzy C-means (FCM) clustering algorithm: a decade review from 2000 to 2014. In: Computational Intelligence in Data Mining, vol. 2. Springer, New Delhi, pp. 133–149 (2015)
26. Naik, B. et al.: Cooperative swarm based clustering algorithm based on PSO and k-means to find optimal cluster centroids. In: 2012 National Conference on Computing and Communication Systems (NCCCS). IEEE (2012)
27. Kanungo, D.P., et al.: Hybrid clustering using elitist teaching learning-based optimization: an improved hybrid approach of TLBO. Int. J. Rough Sets Data Anal. (IJRSDA) **3**(1), 1–19 (2016)
28. Nayak, J. et al.: An improved swarm based hybrid k-means clustering for optimal cluster centers. Information Systems Design and Intelligent Applications. Springer, New Delhi, pp. 545–553 (2015)

29. Nayak, J. et al.: An improved firefly fuzzy c-means (FAFCM) algorithm for clustering real world data sets. In: Advanced Computing, Networking and Informatics, vol. 1. Springer, Cham, pp. 339–348 (2014)
30. Nayak, J. et al.: Evolutionary improved swarm-based hybrid K-means algorithm for cluster analysis. In: Proceedings of the Second International Conference on Computer and Communication Technologies. Springer, New Delhi (2016)
31. Nayak, J. et al.: Hybrid chemical reaction based metaheuristic with fuzzy c-means algorithm for optimal cluster analysis. Expert Syst. Appl. **79**, 282–295 (2017)
32. Nayak, J., Naik, B., Behera, H.S.: Cluster analysis using firefly-based k-means algorithm: a combined approach. In: Computational Intelligence in Data Mining. Springer, Singapore, pp. 55–64 (2017)
33. Naik, B. et al.: Fuzzy Clustering with improved swarm optimization and genetic algorithm: hybrid approach. In: Computational Intelligence in Data Mining. Springer, Singapore, pp. 237–247 (2017)
34. Nayak, J. et al.: A hybrid elicit teaching learning based optimization with fuzzy c-means (ETLBO-FCM) algorithm for data clustering. Ain Shams Eng. J. (2016)
35. Ester, M. et al.: A density-based algorithm for discovering clusters in large spatial databases with noise. Kdd **96**(34) (1996)
36. Hartigan, J.A., Wong, M.A.: Algorithm AS 136: a k-means clustering algorithm. J. Roy. Stat. Soc.: Ser. C (Appl. Stat.) **28**(1), 100–108 (1979)
37. Waikato Environment for Knowledge Analysis: Version 3.8.1, (c) 1999–2016, The University of Waikato Hamilton, New Zealand

TLBO Algorithm Optimized Fractional-Order PID Controller for AGC of Interconnected Power System

Tulasichandra Sekhar Gorripotu, Halini Samalla, Ch. Jagan Mohana Rao, Ahmad Taher Azar and Danilo Pelusi

Abstract The present study focuses on the problem of automatic generation control (AGC) by employing the design of fractional-order proportional–integral–derivative controller (FOPID). A hydrothermal power system with governor dead band (GDB) and generation rate constraint (GRC) is considered for investigation. FOPID controller optimal values are obtained by using teacher learning-based optimization (TLBO) technique, and the employed objective function is integral time-multiplied absolute error (ITAE). The supremacy of the proposed controller is also shown by comparing with PID controller. Further, the performance of the system is analyzed by varying parameter values of the system. From the simulation results, it reveals that TLBO optimized FOPID controller minimizes the errors in frequency of the control areas and tie-line power effectively.

Keywords AGC · FOPID · ITAE · TLBO

T. S. Gorripotu (✉) · H. Samalla · Ch.Jagan Mohana Rao
Department of Electrical & Electronics Engineering, Sri Sivani College of Engineering, Srikakulam 532410, Andhra Pradesh, India
e-mail: gtchsekhar@gmail.com

H. Samalla
e-mail: halini.samalla@gmail.com

Ch.Jagan Mohana Rao
e-mail: ch.jagan211@gmail.com

A. T. Azar
Faculty of Computers and Information, Benha University, Banha 13511, Egypt
e-mail: ahmad_t_azar@ieee.org

D. Pelusi
Communication Sciences, University of Teramo, Coste Sant'Agostino Campus, Teramo, Italy
e-mail: dpelusi@unite.it

A. T. Azar
School of Engineering and Applied Sciences, Nile University, Sheikh Zayed District—Juhayna Square, 6th of October City, Giza 12588, Egypt

J. Nayak et al. (eds.), *Soft Computing in Data Analytics*,
Advances in Intelligent Systems and Computing 758,
https://doi.org/10.1007/978-981-13-0514-6_80

1 Introduction

Nowadays, dealing with a power system is not an easy task as it is complicated and complex. The power system consists of so many number of generators and various loads which are interconnected. If any disturbance occurs in any area of the system, then the frequency in the control areas and tie-line powers are altered. That means, in an interconnected power system the interchange in tie-line power and system frequency has fluctuated in conjunction with the fluctuations in connected load. The primary controller with regulating system, i.e., with rotating mass of the generators can withstand the system for small load changes. Therefore, secondary controller should be included which minimizes frequency changes in the power system and change in tie-line power in an interconnected power system for large changes in the load. This mechanism is called as automatic generation control (AGC), and name itself signifies that it controls the power by maintaining constant frequency [1–3].

During last few decades, several techniques and optimization techniques have been acknowledged by the researchers in the area of AGC. From the literature, it has been observed that different control and optimization techniques have been suggested such as Ziegler Nicholos, classical controllers, fuzzy logic, genetic algorithm, differential evolution, artificial neural networks, firefly algorithm, particle swarm optimization, and many more advanced controllers [4–19].

2 System Under Investigation

In this study, a two-area hydrothermal power system has been considered. The thermal power system consists of a reheat turbine, and the power systems are incorporated with appropriate GDBs and GRCs. The considered system has been modeled in the SIMULINK (MATLAB) environment, and this model of the system is shown in Fig. 1. The appropriate parameters that are considered in Fig. 1 have been furnished in references [20].

In the present study, FOPID controllers are used in each of the area and the design of FOPID controller is shown in Fig. 2. For acceptable operation of any industry, it is obligatory to incorporate secondary controller in the system. In the literature survey, it is mentioned that PID controllers widely are used in the industries due to its simplicity in structure used, but to progress the system control performance, FOPID controllers are being used as it provides extra two degrees of freedom [21–23].

The main advantage of FOPID over PID controller is that it gives the control variables through that plane whereas PID controller offers controller parameters only on the coordinate axis. The area control errors (ACE) which are the errors to the controllers are specified by Eqs. (1) and (2):

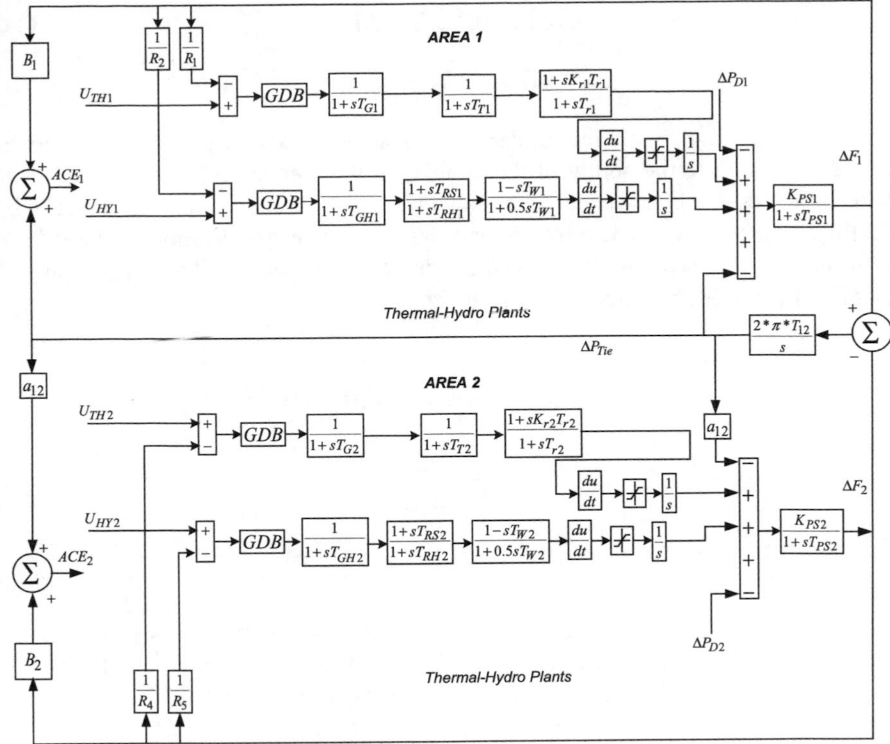

Fig. 1 SIMULINK model of hydrothermal power system

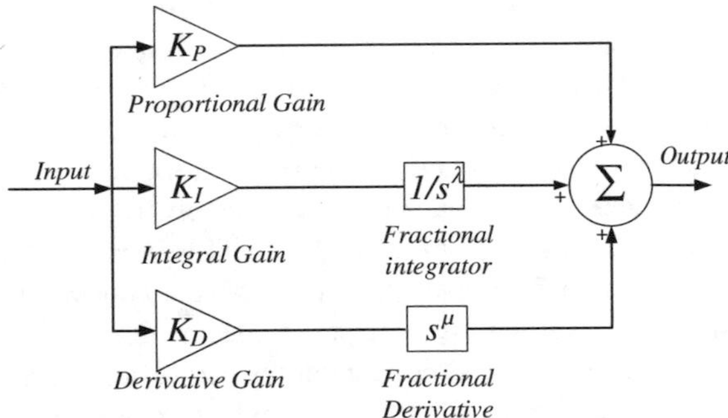

Fig. 2 Structure of FOPID controller

$$ACE_1 = B_1 \Delta F_1 + \Delta P_{Tie} \tag{1}$$

$$ACE_2 = B_2 \Delta F_2 - \Delta P_{Tie} \tag{2}$$

With the design of the controller, it is also necessary to define an objective function to analyze the ability of the system. In the literature, it is mentioned that there are various objective functions but out of which it is concluded that ITAE performs better [24]. ITAE provides better performance specifications such as less settling time, less peak overshoot, and better transient response. The expression for ITAE for the AGC problem is given in Eq. (3).

$$ITAE = \int_0^{t_{sim}} (|\Delta F_1| + |\Delta F_2| + |\Delta P_{Tie}|) \cdot t \cdot dt \tag{3}$$

3 Result Analysis

Primarily, the transfer function model of two-area hydrothermal power system is developed in MATLAB/SIMULINK environment as shown in Fig. 1. Four different FOPID controllers are considered for each generating unit to control the change in frequencies and tie-line power. The controller gains of FOPID controller are optimized by using TLBO technique [25–29] and the process of tuning the controller gain parameters and regarding TLBO elaborately discussed by the author in [20]. To show the preeminence of FOPID controller, PID controller is considered which is widely used in industrial process. The optimal control parameters of FOPID and PID controllers are given in Table 1. The ITAE provided in Eq. (3) has been evaluated by simulating the model considered for 1% increase of the step load in area-1. The simulated transient performance of system with both the controllers is shown in Figs. 3 and 4. The resultant performance index values such as settling time, peak overshoot, and ITAE are given in Table 2. From Figs. 3 and 4 and Table 2, it is apparent that proposed TLBO-based FOPID controller performs better compared to PID controller. Further, the capability of the proposed controller is shown by varying system parameters in the range from +25% to −25% of the nominal values. The corresponding simulated responses are illustrated in Figs. 5 and 6. From, Figs. 5 and 6, it is evident that FOPID controller can withstand any sudden changes in system parameters without retuning controller parameters.

Table 1 Optimal gain values

Parameters	PID	FOPID
K_{P1}	0.4312	0.6236
K_{P2}	0.6755	0.6725
K_{I1}	0.2110	0.4419
K_{I2}	0.0001	0.6034
K_{D1}	0.3779	0.3731
K_{D2}	0.0864	0.2827
u_1	–	0.1775
u_2	–	0.5944
λ_1	–	0.8173
λ_2	–	0.8823
K_{P3}	0.4118	0.1741
K_{P4}	0.1389	0.2634
K_{I3}	0.5179	0.7449
K_{I4}	0.1769	0.4156
K_{D3}	0.4026	0.1704
K_{D4}	0.2440	0.9909
u_3	–	0.2455
u_4	–	0.5585
λ_3	–	0.2531
λ_4	–	0.6156

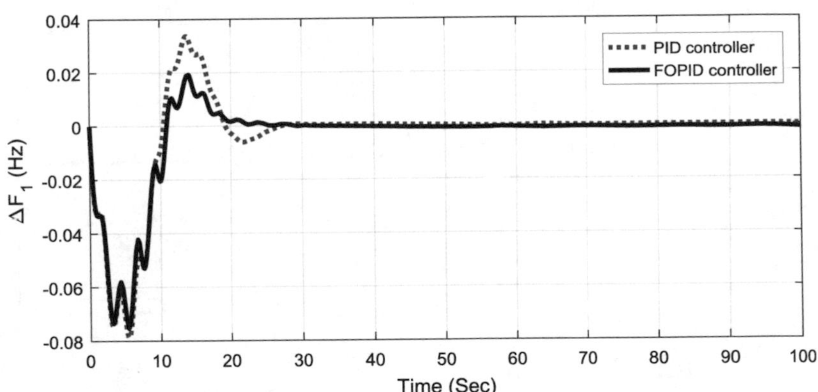

Fig. 3 Deviation in frequency of area-1

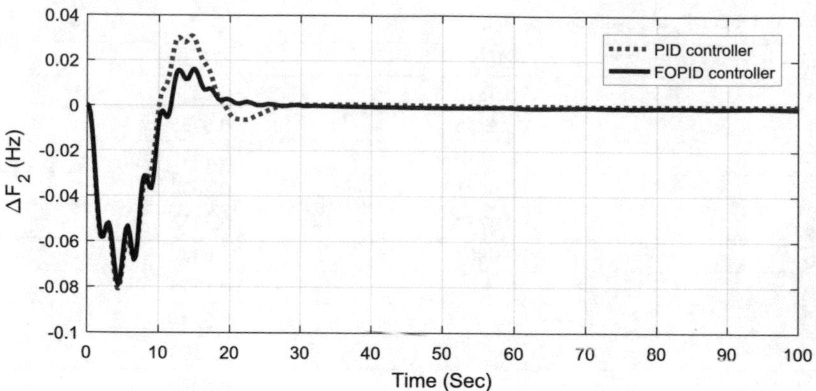

Fig. 4 Deviation in frequency of area-2

Table 2 Performance index values

Parameters		PID controller	FOPID controller
ITAE		12.8454	8.7134
T_S (sec)	ΔF_1	60.90	25.57
	ΔF_2	72.30	35.28
	ΔP_{Tie}	67.25	26.79
Peak overshoot	ΔF_1	0.0337	0.0193
	ΔF_2	0.0307	0.0160
	ΔP_{Tie}	0.0008	0.0011

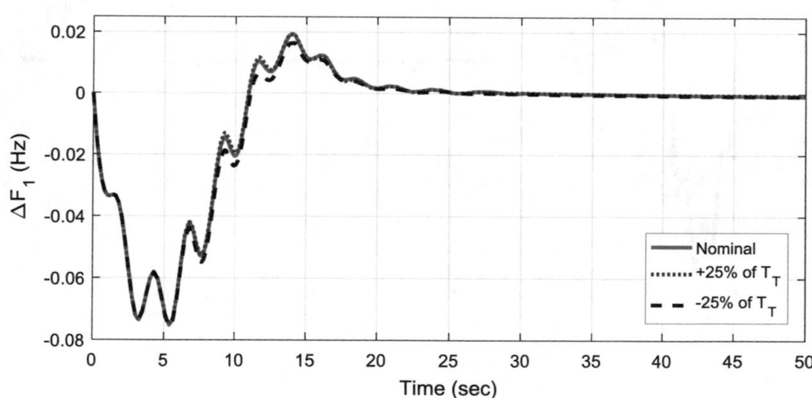

Fig. 5 Deviation in frequency of area-1 for turbine time constant

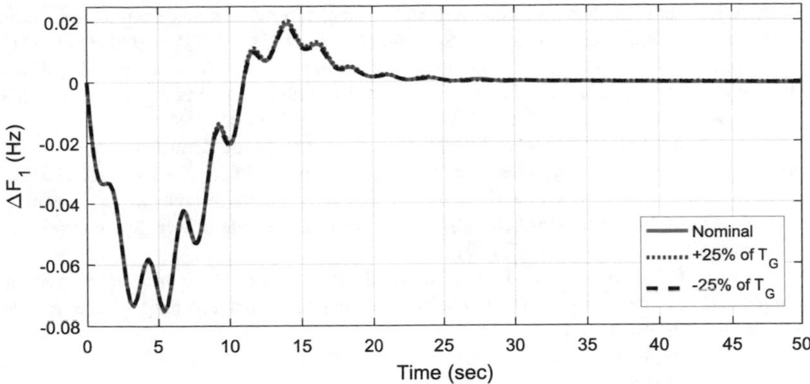

Fig. 6 Deviation in frequency of area-1 for the governor time constant

4 Conclusion

In this present work, an effort has been made to optimize FOPID controller by employing TLBO technique for a two-area hydrothermal power system. A practical power system has been examined in the present study by considering nonlinearities such as GDB and GRCs. A comparative analysis was made with PID controller to elucidate the supremacy of the proposed FOPID controller. Further, an analysis was made by varying system parameters to illustrate the ability of the proposed controller. From the responses of the system, it is evident that proposed controller provides better results.

References

1. Elgerd, O.I.: Electric Energy Systems Theory–An Introduction. Tata McGraw Hill, New Delhi (2000)
2. Bevrani, H.: Robust Power System Frequency Control. Springer, Berlin (2009)
3. Bevrani, H., Hiyama, T.: Intelligent Automatic Generation control. CRC Press, Boca Raton, Florida (2011)
4. Azar, A.T., Vaidyanathan, S.: Advances in Chaos Theory and Intelligent Control. In: Studies in Fuzziness and Soft Computing, vol. 337, Springer, Germany (2016). ISBN 9783319303383
5. Saikia, L.C., Mishra, S., Sinha, N., Nanda, J.: Automatic generation control of a multi area hydrothermal system using reinforced learning neural network controller. Int. J. Electr. Power Energy Syst. **33**(4), 1101–1108 (2011)
6. Ali, E.S., Abd-Elazim, S.M.: BFOA based design of PID controller for two area load frequency control with nonlinearities. Int. J. Electr. Power Energy Syst. **51**, 224–231 (2013)
7. Parmar, K.P.S., Majhi, S., Kothari, D.P.: Improvement of dynamic performance of LFC of the two area power system: an analysis using MATLAB. Int. J. Comput. Appl. **40**(10), 28–32 (2012)

8. Azar, A.T., Vaidyanathan, S.: Chaos modeling and control systems design. In: Studies in Computational Intelligence, vol. 581, Springer, Germany (2015). ISBN 9783319131313
9. Azar, A.T., Vaidyanathan, S.: Handbook of research on advanced intelligent control engineering and automation. In: Advances in Computational Intelligence and Robotics (ACIR) Book Series, IGI Global, USA (2015). ISBN 9781466672482
10. Azar, A.T., Zhu, Q.: Advances and applications in sliding mode control systems. In: Studies in computational intelligence, vol. 576. Springer, Germany (2015). ISBN: 9783319111728
11. Chandra Sekhar, G.T., Sahu, R.K., Panda, S.: AGC of a multi-area power system under deregulated environment using redox flow batteries and interline power flow controller. Eng. Sci. Technol. Int. J. **18**, 555–578 (2015)
12. Gorripotu, T.C.S., Sahu, R.K., Panda, S.: Application of firefly algorithm for AGC under deregulated power system. In: Computational Intelligence in Data Mining—Volume 1, Smart Innovation, Systems and Technologies, vol. 31, pp. 677–687 (2015)
13. Chandra Sekhar, G.T., Sahu, R.K., Baliarsingh, A.K., Panda, S.: Load frequency control of power system under deregulated environment using optimal firefly algorithm. Int. J. Electr. Power Energy Syst. **74**, 195–211 (2016)
14. Sahu, R.K., Gorripotu, T.C.S., Panda, S.: A hybrid DE–PS algorithm for load frequency control under deregulated power system with UPFC and RFB. Ain Shams Eng. J. **6**, 893–911 (2015)
15. Azar, A.T., Vaidyanathan, S.: Computational intelligence applications in modeling and control. In: Studies in Computational Intelligence, vol. 575, Springer, Germany (2015). ISBN 9783319110165
16. Zhu, Q., Azar, A.T.: Complex system modelling and control through intelligent soft computations. In: Studies in Fuzziness and Soft Computing, vol. 319. Springer, Germany (2015). ISBN: 9783319128825
17. Azar, A.T., Serrano, F.E.: Robust IMCPID tuning for cascade control systems with gain and phase margin specifications. Neural Comput. Appl. **25**(5), 983995. https://doi.org/10.1007/s005210141560x (2014)
18. Sahu, R.K., Chandra Sekhar, G.T., Panda, S.: DE optimized fuzzy PID controller with derivative filter for LFC of multi source power system in deregulated environment. Ain Shams Eng. J. **6**, 511–530 (2015)
19. Sahu, R.K., Panda, S., Chandra Sekhar, G.T.: A novel hybrid PSO-PS optimized fuzzy PI controller for AGC in multi area interconnected power systems. Int. J. Electr. Power Energy Syst. **64**, 880–893 (2015)
20. Rabindra Kumar, S., Tulasichandra Sekhar, G., Sidhartha, P.: Automatic generation control of multi-area power systems with diverse energy sources using teaching learning based optimization algorithm. Eng. Sci. Technol. Int. J. **19**(1), 113–134 (2016)
21. Taher, S.A., Fini, M.H., Aliabadi, S.F.: Fractional order PID controller design for LFC in electric power systems using imperialist competitive algorithm. Ain Shams Eng. J. **5**(1), 121–135 (2014)
22. HongSheng, L., Ying, L., YangQuan, C.: A fractional order proportional and derivative (FOPD) motion controller: tuning rule and experiments. IEEE Trans. Control Syst. Technol. **18**(2), 516–520 (2010)
23. Tahmine, V., Moghaddam, N., Bigdeli, Afshar, K.: Tuning a fractional order PID controller with lead compensator in frequency domain. Int. J. Electr. Electron. Eng. **5**(3), 139–144 (2011)
24. Shabani, H., Vahidi, B., Ebrahimpour, M.: A robust PID controller based on imperialist competitive algorithm for load-frequency control of power systems. ISA Trans. **52**(1), 88–95 (2013)
25. Nayak, J. et al.: A hybrid elicit teaching learning based optimization with fuzzy c-means (ETLBO-FCM) algorithm for data clustering. Ain Shams Eng. J. (2016)
26. Naik, B., Nayak, J., Behera, H.S.: A TLBO based gradient descent learning-functional link higher order ANN: An efficient model for learning from non-linear data. J. King Saud Univ.-Comput. Inf. Sci. (2016)

27. Kanungo, D.P., et al.: Hybrid clustering using elitist teaching learning-based optimization: an improved hybrid approach of TLBO. Int. J. Rough Sets Data Anal. (IJRSDA) **3**(1), 1–19 (2016)
28. Rao, R.V., Savsani, V.J., Vakharia, D.P.: Teaching–learning-based optimization: an optimization method for continuous non-linear large scale problems. Inf. Sci. **183**(1), 1–15 (2012)
29. Rao, R.V., Kalyankar, V.D.: Parameter optimization of modern machining processes using teaching–learning based optimization algorithm. Eng. Appl. Artif. Intell. **26**(1), 524–531 (2013)

Author Index

J. Nayak et al. (eds.), *Soft Computing in Data Analytics*,
Advances in Intelligent Systems and Computing 758,
https://doi.org/10.1007/978-981-13-0514-6

Printed in the United States
By Bookmasters